Osswald · Baur · Brinkmann
Oberbach · Schmachtenberg

International Plastics Handbook

The Resource for Plastics Engineers

HANSER

Hanser Publishers, Munich • Hanser Gardner Publications, Cincinnati

The Authors:
Prof. Dr. Tim A. Osswald, Department of Mechanical Engineering University of Wisconsin-Madison, Madison, WI, USA
Dr. Erwin Baur, M-Base GmbH, Aachen, Germany
Sigrid Brinkmann, Brinkmann Kunststofftechnologie und Redaktion Bad Aibling, Germany
Karl Oberbach, Leuchter Gemark 3, 51467 Bergisch Gladbach, Germany
Prof. Dr-Ing. Ernst Schmachtenberg, Institute of Polymer Technology, University of Erlangen-Nuremberg, Germany

Distributed in the USA and in Canada by
Hanser Gardner Publications, Inc.
6915 Valley Avenue, Cincinnati, Ohio 45244-3029, USA
Fax: (513) 527-8801
Phone: (513) 527-8977 or 1-800-950-8977
www.hansergardner.com

Distributed in all other countries by
Carl Hanser Verlag
Postfach 86 04 20, 81631 München, Germany
Fax: +49 (89) 98 48 09
www.hanser.de

The use of general descriptive names, trademarks, etc., in this publication, even if the former are not especially identified, is not to be taken as a sign that such names, as understood by the Trade Marks and Merchandise Marks Act, may accordingly be used freely by anyone.
While the advice and information in this book are believed to be true and accurate at the date of going to press, neither the authors nor the editors nor the publisher can accept any legal responsibility for any errors or omissions that may be made. The publisher makes no warranty, express or implied, with respect to the material contained herein.

Library of Congress Cataloging-in-Publication Data
International plastics handbook : the resource for plastics engineers / Tim A. Osswald ... [et al.]. -- 4th ed.
   p. cm.
 ISBN-13: 978-1-56990-399-5 (hardcover)
 ISBN-10: 1-56990-399-9 (hardcover)
 1. Plastics--Handbooks, manuals, etc. I. Osswald, Tim A.
 TP1130.I58 2006
 668.4--dc22
                      2006012879

Bibliografische Information Der Deutschen Bibliothek
Die Deutsche Bibliothek verzeichnet diese Publikation in der Deutschen Nationalbibliografie; detaillierte bibliografische Daten sind im Internet über <http://dnb.ddb.de> abrufbar.

ISBN-10: 3-446-22905-1
ISBN-13: 978-3-446-22905-1

All rights reserved. No part of this book may be reproduced or transmitted in any form or by any means, electronic or mechanical, including photocopying or by any information storage and retrieval system, without permission in writing from the publisher.

© Carl Hanser Verlag, Munich 2006
Production Management: Oswald Immel
Typeset by Sylvana García and Alejandro Roldán, USA
Coverconcept: Marc Müller-Bremer, Rebranding, München, Germany
Coverdesign: MCP • Susanne Kraus GbR, Holzkirchen, Germany
Printed and bound by Kösel, Krugzell, Germany

*In gratitude to our mentor,
Professor Dr.-Ing. Georg Menges,
a pioneer in the field of
plastics technology and engineering*

PREFACE

As the title suggests, this handbook was written as a source of reliable information for the practicing engineer, in a world which is increasingly driven by globalization. While it was written for the American market, the book also includes international standards and information that is now often necessary for any practicing engineer. Plastics remains the material group with the largest growth rate worldwide. This is due to their unique characteristics; the ease of production of complex parts with relatively economical processes, and the ability to tailor the properties of the material to suit a specific application. Due to these attributes, today we find that plastics are not only replacing traditional materials, but are the driving force in innovation in the fields of electronic products, medicine, automobiles, household goods, and construction. Thus, we find uncountable examples of how plastics have improved our standard of living in the last decade alone; from lighter automobiles to thinner cellular telephones, and from more fuel efficient composite aircraft to improved human heart valves, all made possible by advances in plastics and plastics technology. The plastics industry will continue to grow and develop, as the properties of polymers achieve higher limits, continuously replacing other materials and making new applications and products possible. The image of plastics has seen a positive turn in the last decade. Many studies have shown that plastics are the driving force in the solution of ecological problems; from making vehicles more fuel efficient, to allowing the manufacture of complex products with a fraction of the energy required when using metals. Furthermore, today we find ourselves revisiting the past, where plastics were made of renewable resources. At the end, it is really only a question of economics, whether we will make plastics from petroleum or from bio-materials. From this point of view, we can really say that the age of plastics is still in its infancy. *International Plastics Handbook* is based on a similar publication, which has been available to the German plastics engineer for the past 70 years. A new edition of the German book *Saechtling Kunststoff Taschenbuch*, now in its 29th edition, appears every three years at the K-Show, in Duesseldorf, Germany. In 1936, when the first edition of the German book was published, Dr. F. Pabst, the original author of the German handbook, wrote *"This book is intended to answer questions in the applied field of plastics."* For this handbook, that is our intent as well. This first edition of *International Plastics Handbook* will appear at the NPE show in Chicago, in June of 2006. Due to the rapid changes in the field of plastics we are planning to present new editions of the handbook at NPE shows, every three years. In addition, this book will be accessible through the World-WideWeb where the authors will maintain updated versions of the handbook, including its tables and list of trade names. Each handbook contains an individual code on the inside front cover which provides access for the download of the electronic version of the book at www.hanser.de/plasticshandbook. The authors cannot possibly acknowledge everyone who in oneway or another helped in the preparation of this handbook. First of all wewould like to thank Drs. Wolfgang Glenz and Christine Strohm of Hanser publishers for being the catalysts for this project. Additionally, Dr. Strohm gave us her input and support during the years it took to prepare this handbook. Chapter 6 comes from the German Saechtling Kunststoff Taschenbuch materials chapter, which was translated by Dr. Strohm. We are indebted to her for all of this. Special thanks are due to Luz Mayed D. Nouguez for the superb job of drawing the figures. We are grateful to Sylvana García and Alejandro Roldán for the long hours they put into generating tables for the book and preparing the camera-ready manuscript. We thank Juan Pablo Hernández-Ortiz for developing the typesetting template. We also thank Oswald Immel for his support throughout the development of the manuscript. Above all, we thank our families for their patience, encouragement and support.

TIM A. OSSWALD

Madison, Wisconsin
Spring, 2006

CONTENTS

Preface   vii

1 Introduction   1

  1.1 Statistical data   1
  1.2 Polymer and plastics categories   5
  1.3 Plastics Acronyms   8

2 Materials Science of Polymers   17

  2.1 Polymer Structure   17
      2.1.1 Chemistry   17
      2.1.2 Morphological Structure   35
      2.1.3 Thermal Transitions   39
  2.2 Material Modification of Plastics   53
      2.2.1 Polymer Blends   53

	2.2.2	Filled Polymers and Reinforced Composites	54
	2.2.3	Other Modifications	57
2.3	Plastics Recycling		58

3 Properties and Testing 63

3.1	Comparability of Material Properties		63
3.2	Thermal Properties		67
	3.2.1	Thermal Conductivity	67
	3.2.2	Specific Heat and Specific Enthalpy	74
	3.2.3	Density	76
	3.2.4	Thermal Diffusivity	82
	3.2.5	Linear Coefficient of Thermal Expansion	82
	3.2.6	Thermal Penetration	91
	3.2.7	Thermal Data Measuring Devices	92
3.3	Curing Behavior		99
3.4	Rheological Properties		103
	3.4.1	Flow Phenomena	103
	3.4.2	Viscous Flow Models	109
	3.4.3	Rheometry	115
	3.4.4	Surface Tension	122
3.5	Mechanical Properties		126
	3.5.1	The Short-Term Tensile Test	126
	3.5.2	Impact Strength	141
	3.5.3	Creep Behavior	160
	3.5.4	Dynamic Mechanical Tests	172
	3.5.5	Fatigue Tests	178
	3.5.6	Strength Stability Under Heat	185
3.6	Permeability properties		194
	3.6.1	Sorption	195
	3.6.2	Diffusion and Permeation	195
	3.6.3	Measuring S, D, and P	201
	3.6.4	Diffusion of Polymer Molecules and Self-Diffusion	203
3.7	Friction and Wear		205
3.8	Environmental effects		207
	3.8.1	Water Absorption	208

3.8.2 Weathering		211
3.8.3 Chemical Degradation		214
3.8.4 Thermal Degradation of Polymers		216
3.9 Electrical Properties		223
3.9.1 Dielectric Behavior		223
3.9.2 Electric Conductivity		229
3.9.3 Application Problems		235
3.9.4 Magnetic Properties		247
3.10 Optical Properties		249
3.10.1 Index of Refraction		250
3.10.2 Photoelasticity and Birefringence		251
3.10.3 Transparency, Reflection, Absorption and Transmittance		257
3.10.4 Gloss		259
3.10.5 Color		262
3.10.6 Infrared Spectroscopy		263
3.11 Acoustic Properties		266
3.11.1 Speed of Sound		266
3.11.2 Sound Reflection		266
3.11.3 Sound Absorption		268

4 Plastics Processes — 271

4.1 Raw Material Preparation		272
4.1.1 Mixing Processes		272
4.2 Mixing Devices		275
4.2.1 Mixing of Particulate Solids		275
4.2.2 Screw-Type Mixers		275
4.2.3 Granulators and Pelletizers		288
4.2.4 Dryers		290
4.3 Extrusion		294
4.3.1 The Plasticating Extruder		296
4.3.2 Troubleshooting Extrusion		304
4.3.3 Extrusion Dies		308
4.4 Injection Molding		314
4.4.1 The Injection Molding Cycle		314
4.4.2 The Injection Molding Machine		318

4.4.3 Special Injection Molding Processes	325
4.4.4 Troubleshooting Injection Molding	368
4.5 Compression Molding	369
4.5.1 Compression Molding of SMC and BMC	371
4.5.2 Compression Molding of GMT and LFT	373
4.5.3 Cold Press Forming	376
4.5.4 Troubleshooting Compression Molding	377
4.6 Composites Processing	381
4.6.1 Resin Transfer Molding and Structural RIM	382
4.6.2 Filament Winding	384
4.6.3 Pultrusion	385
4.7 Secondary Shaping	386
4.7.1 Fiber Spinning	387
4.7.2 Film Production	388
4.7.3 Thermoforming	397
4.8 Calendering	400
4.9 Coating	402
4.10 Foaming	407
4.11 Rotational Molding	410
4.12 Welding	412
4.12.1 Hot Tool Butt Welding	413
4.12.2 Ultrasonic Welding	416
4.12.3 Vibration Welding	424
4.12.4 Spin Welding	427
4.12.5 IR and Laser Welding	430
4.12.6 RF/Dielectric Welding	433
4.12.7 Hot Gas Welding	435
4.12.8 Extrusion Welding	437
4.12.9 Implant Induction Welding	439
4.12.10 Implant Resistance Welding	442
4.12.11 Microvawe Welding	444
4.13 Rapid Prototyping	444
4.13.1 Stereo-Lithography (STL)	445
4.13.2 Solid Ground Curing (SGC)	445
4.13.3 Selective Laser Sintering (SLS)	446

		4.13.4 3D Printing or Selective Binding	447
		4.13.5 Fused Deposition Modeling (FDM)	448
		4.13.6 Laminated Object Manufacturing (LOM)	449

5 Engineering Design — 451

5.1 Design Philosophy — 451
    5.1.1 Defining Product Requirements — 453
    5.1.2 Preliminary CAD Model — 454
    5.1.3 Material Selection — 455
    5.1.4 Process Selection — 458
5.2 Process Influences on Product Performance — 458
    5.2.1 Orientation in the Final Part — 459
    5.2.2 Fiber Damage — 465
    5.2.3 Cooling and Solidification — 467
    5.2.4 Shrinkage, Residual Stresses and Warpage — 470
    5.2.5 Process Simulation as Integral Part of the Design Process — 476
5.3 Strength of Materials Considerations — 485
    5.3.1 Basic Concepts of Stress and Strain — 485
    5.3.2 Anisotropic Strain-Stress Relation — 494
5.4 Functional Elements — 498
    5.4.1 Press Fit Assemblies — 498
    5.4.2 Living Hinges — 499
    5.4.3 Snap Fit Assemblies — 499
    5.4.4 Mechanical Fasteners — 504
5.5 Software — 505

6 Materials — 507

6.1 Polyolefins (PO), Polyolefin Derivates, and Copolymers — 508
    6.1.1 Standard Polyethylene Homo- and Copolymers (PE-LD, PE-HD, PE-HD-HMW, PE-HD-UHMW, PE-LLD) — 513
    6.1.2 Polyethylene Derivates (PE-X, PE + PSAC) — 523
    6.1.3 Chlorinated and Chloro-Sulfonated PE (PE-C, CSM) — 524
    6.1.4 Ethylene Copolymers (ULDPE, EVAC, EVAL, EEAK, EB, EBA, EMA, EAA, E/P, EIM, COC, ECB, ETFE — 525
    6.1.5 Polypropylene Homopolymers (PP, H-PP) — 533

        6.1.6 Polypropylene Copolymers and -Derivates, Blends (PP-C,
              PP-B, EPDM, PP + EPDM) 539
        6.1.7 Polypropylene, Special Grades 541
        6.1.8 Polybutene (PB, PIB) 542
        6.1.9 Higher Poly-α-Olefins (PMP, PDCPD) 545
   6.2 Styrene Polymers 546
        6.2.1 Polystyrene, Homopolymers (PS, PMS) 546
        6.2.2 Polystyrene, Copolymers, Blends 547
        6.2.3 Polystyrene Foams (PS-E, XPS) 553
   6.3 Vinyl Polymers 554
        6.3.1 Rigid Polyvinylchloride Homopolymers (PVC-U) 554
        6.3.2 Plasticized (soft) Polyvinylchloride (PVC-P) 560
        6.3.3 Polyvinylchloride: Copolymers and Blends 565
        6.3.4 Polyvinylchloride: Pastes, Plastisols, Organosols 566
        6.3.5 Vinyl Polymers, other Homo- and Copolymers (PVDC,
              PVAC, PVAL, PVME, PVB, PVK, PVP) 567
   6.4 Fluoropolymers 569
        6.4.1 Fluoro Homopolymers (PTFE, PVDF, PVF, PCTFE) 569
        6.4.2 Fluoro Copolymers and Elastomers (ECTFE, ETFE, FEP,
              TFEP, PFA, PTFEAF, TFEHFPVDF (THV), [FKM, FPM,
              FFKM]) 574
   6.5 Polyacryl- and Methacryl Copolymers 576
        6.5.1 Polyacrylate, Homo- and Copolymers (PAA, PAN, PMA,
              ANBA, ANMA) 576
        6.5.2 Polymethacrylates, Homo- and Copolymers (PMMA,
              AMMA, MABS, MBS) 577
        6.5.3 Polymethacrylate, Modifications and Blends (PMMI,
              PMMA-HI, MMA-EML Copolymers, PMMA + ABS blends 581
   6.6 Polyoxymethylene, Polyacetal Resins, Polyformaldehyde (POM) 584
        6.6.1 Polyoxymethylene Homo- and Copolymers (POM-H,
              POM-Cop.) 584
        6.6.2 Polyoxymethylene, Modifications and Blends (POM + PUR) 585
   6.7 Polyamides (PA) 586
        6.7.1 Polyamide Homopolymers (AB and AA/BB Polymers) (PA6,
              11, 12, 46, 66, 69, 610, 612, PA 7, 8, 9, 1313, 613) 586

6.7.2 Modifications	593
6.7.3 Polyamide Copolymers, PA 66/6, PA 6/12, PA 66/6/610 Blends (PA +: ABS, EPDM, EVA, PPS, PPE, Rubber)	594
6.7.4 Polyamides, Special Polymers (PA NDT/INDT [PA 6-3-t], PAPACM 12, PA 6-I, PA MXD6 [PARA], PA 6-T, PA PDA-T, PA 6-6-T, PA 6-G, PA 12-G, TPA-EE)	597
6.7.5 Cast Polyamides (PA 6-C, PA 12-C).	598
6.7.6 Polyamide for Reaction Injection Molding (PA-RIM)	598
6.7.7 Aromatic Polyamides, Aramides (PMPI, PPTA)	599
6.8 Aromatic (Saturated) Polyesters	599
6.8.1 Polycarbonate (PC)	600
6.8.2 Polyesters of Therephthalic Acids, Blends, Block Copolymers	605
6.8.3 Polyesters of Aromatic Diols and Carboxylic Acids (PAR, PBN, PEN)	610
6.9 Aromatic Polysulfides and Polysulfones (PPS, PSU, PES, PPSU, PSU + ABS)	613
6.9.1 Polyphenylene Sulfide (PPS)	613
6.9.2 Polyarylsulfone (PSU, PSU + ABS, PES, PPSU)	616
6.10 Aromatic Polyether, Polyphenylene Ether, and Blends (PPE)	617
6.10.1 Polyphenylene Ether (PPE)	617
6.10.2 Polyphenylene Ether Blends	618
6.11 Aliphatic Polyester (Polyglycols) (PEOX, PPOX, PTHF)	620
6.12 Aromatic Polyimide (PI)	622
6.12.1 Thermosetting Polyimide (PI, PBMI, PBI, PBO, and Others)	623
6.12.2 Thermoplastic Polyimides (PAI, PEI, PISO, PMI, PMMI, PESI, PARI)	627
6.13 Liquid Crystalline Polymers (LCP)	629
6.14 Ladder Polymers: Two-Dimensional Polyaromates and -Heterocyclenes	632
6.15 Polyurethane (PUR)	635
6.15.1 Fundamentals	635
6.15.2 Raw Materials and Additives	639
6.15.3 PUR Polymers	642
6.16 Biopolymers, Naturally Occurring Polymers and Derivates	650

6.16.1 Cellulose- and Starch Derivates (CA, CTA, CAP, CAB, CN, EC, MC, CMC, CH, VF, PSAC) — 650
6.16.2 Casein Polymers, Casein Formaldehyde, Artificial Horn (CS, CSF) — 656
6.16.3 Polylactide, Polylactic Acid (PLA) — 656
6.16.4 Polytriglyceride Resins (PTP®) — 656
6.16.5 Natural Resins — 657
6.17 Other Polymers — 657
6.17.1 Photodegradable, Biodegradable, and Water Soluble Polymers — 657
6.17.2 Conductive/Luminescent Polymers — 659
6.17.3 Aliphatic Polyketones (PK) — 662
6.17.4 Polymer Ceramics, Polysilicooxoaluminate (PSIOA) — 664
6.18 Thermoplastic Elastomers (TPE) — 664
6.18.1 Physical Constitution — 666
6.18.2 Chemical Constitution, Properties, and Applications — 666
6.19 Thermosets — 670
6.19.1 Chemical Constitution — 670
6.19.2 Processing, Forms of Delivery — 676
6.19.3 Properties — 678
6.19.4 Applications — 683
6.20 Rubbers — 686
6.20.1 General Description — 686
6.20.2 General Properties — 687
6.20.3 R-Rubbers (NR, IR, BR, CR, SBR, NBR, NCR, IIR, PNR, SIR, TOR, HNBR) — 689
6.20.4 M-Rubbers (EPM, EPDM, AECM, EAM, CSM, CM, ACM, ABM, ANM, FKM, FPM, FFKM) — 693
6.20.5 O-Rubbers (CO, ECO, ETER, PO) — 695
6.20.6 Q-(Silicone) Rubber (MQ, MPQ, MVQ, PVMQ, MFQ, MVFQ) — 696
6.20.7 T-Rubber (TM, ET, TCF) — 697
6.20.8 U-Rubbers (AFMU, EU, AU) — 698
6.20.9 Polyphosphazenes (PNF, FZ, PZ) — 699
6.20.10 Other Rubbers — 699

7 Polymer Additives — 701

7.1 Antiblocking Agents	701
7.2 Slip Additives	702
7.3 Plasticizers	702
7.4 Stabilizers	702
7.4.1 Antioxidants	703
7.4.2 Flame Retardants	704
7.4.3 UV Stabilizers	705
7.4.4 PVC Stabilizers	706
7.5 Antistatic Agents	706
7.6 Antimicrobial Agents	707
7.7 Antifogging Agents	708
7.8 Blowing Agents	709
7.9 Colorants	710
7.10 Fluorescent Whitening Agents	712
7.11 Fillers	712

Appendix A: Material Property Tables — 717

Appendix B: Literature — 775

B.1 Books in the Plastics Technology Field	776
B.2 Journals and Trade Magazines	788
B.2.1 Trade Magazines	788
B.2.2 Archival Journals	789

Appendix C: Polymer Research Institutes — 791

Appendix D: Tradenames — 805

D.1 Introduction	806
D.2 Tradenames Table	807

Topic Index — 877

CHAPTER 1

INTRODUCTION

The word plastics has been deeply ingrained into our society and culture, to the point that many consider this the age of plastics. The word itself applies to materials that can be shaped and formed, however, today we use it to describe a polymer which contains additives, such as pigments, fillers, antioxidants, and UV-protectors, to name a few. Polymers are materials composed of molecules of high molecular weight. The unique material properties of plastics and versatile processing methods are attributed to their molecular structure. The ease with which polymers are processed and with which one can consolidate several parts into a single part, as well as their high strength - to - weight ratio, make them the most sought after materials today.

1.1 STATISTICAL DATA

In the last century, plastics have gained significant importance in the technological as well as economic arena. Fig. 1.1 compares the production of plastics' resins in the past 55 years to steel and aluminum. Before 1990, the figure depicts the production in the Western World and after that year, when the iron curtain came down, the worldwide production. Table 1.1 presents the per capita polymer resin use, by region, for 1980 and 2002 as well as the projected 2010 yearly production.

In general, the plastics industry can be broken down into three distinct sub-categories:

- Plastics resin manufacturers and suppliers
- Plastics product manufacturing (Original Equipment Manufacturers (OEM))
- Plastics machinery (Machinery supplier)

The over 18,000 different grades of resins, available today in the U.S., can be divided into two general categories - thermosetting and thermoplastic polymers. Of the over 31 million tons of polymers produced in the United States in 1993, 90% were thermoplastics. Figures 1.2 and 1.3 show a percentage break down of U.S. polymer production of thermoplastics and thermosets, respectively. Each is

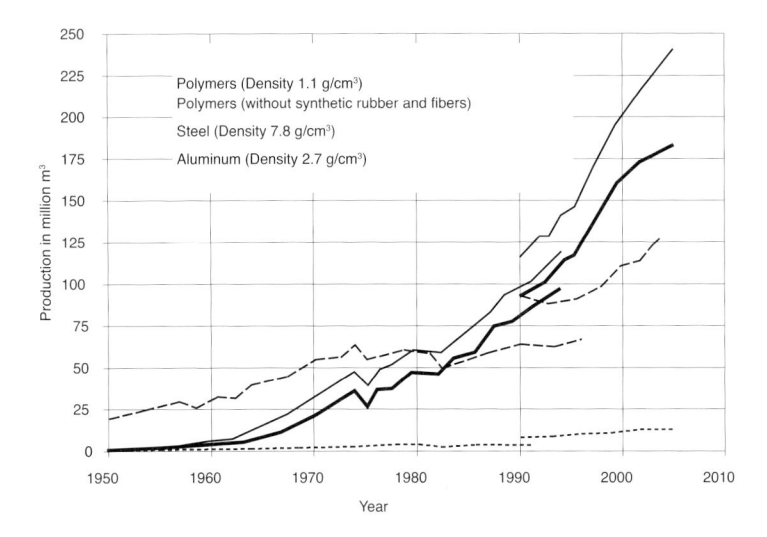

Figure 1.1: World production of raw materials (after Ehrenstein).

broken down into its most common types. Of the thermoplastics, polyethylenes are by far the most widely used polymeric material, accounting for 41% of the U.S. plastic production.

Table 1.1: Regional Break-Down of Per-Capita-Plastics Use in Kilograms

Region	1980	2002	2010	Annual % change (2002-2010)
Worldwide	10	26	37	4.5
US	45	105	146	4.2
Latin America	7.5	20.5	30.5	5.1
Europe	40	97	136	4.3
Eastern Europe	8.5	12.5	24	8.5
Japan	50	85	108	3.0
South East Asia	2	14.5	24	6.5
Africa Middle East	3	8	10	2.8

Today, the plastics industry implements polymers in a wide variety of applications as shown in Figs. 1.4 and 1.5 for thermoplastics and thermosets, respectively. As depicted in the figures, packaging accounts for over one-third of the captive use of thermoplastics, whereas construction accounts for about half that number, and transportation accounts for only 4% of the total captive use of thermoplastics. On the other hand, 69% of thermosets are used in building and construction, followed by 8% used in transportation.

1.1 Statistical data

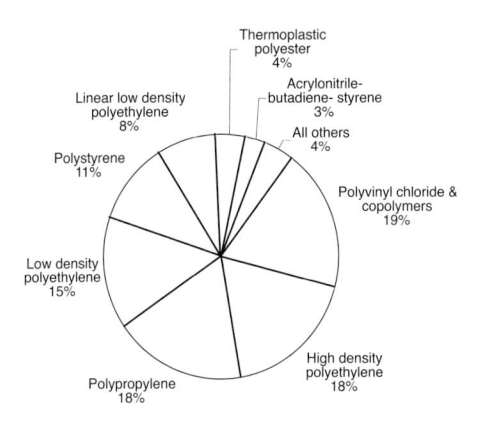

Figure 1.2: Break down of US thermoplastic production into common types.

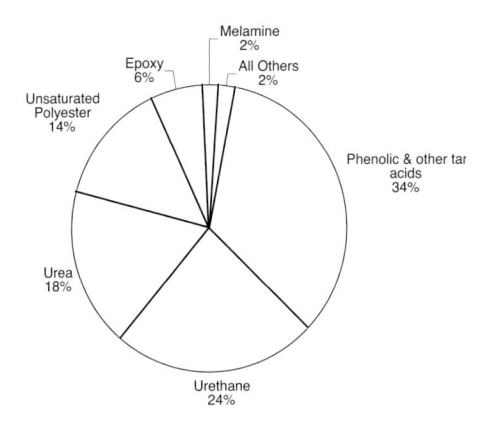

Figure 1.3: Break-down of US thermoset production into common types.

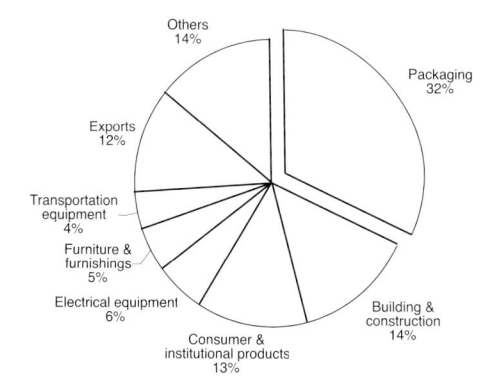

Figure 1.4: Break down of US thermoplastic applications into common areas.

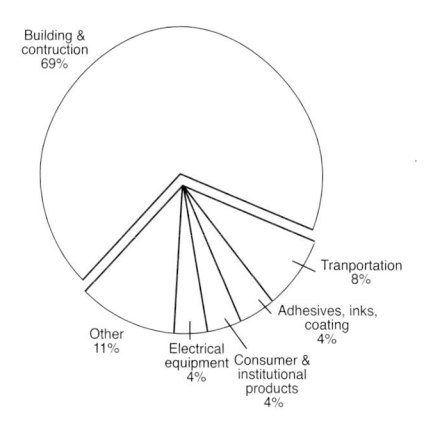

Figure 1.5: Break down of US thermoset applications into common areas.

Table 1.2: Regional Break-Down of Per-Capita-Plastics Use in Kilograms

Rank	State	No. of Employees
1	California	137,800
2	Ohio	112,100
3	Michigan	95,300
4	Texas	94,900
5	Illinois	89,100
6	Pennsylvania	74,400
7	Indiana	70,000
8	New York	52,800
9	North Carolina	51,700
10	Wisconsin	50,900

The transportation sector is one of the fastest growing areas of application for both thermoplastic and thermosetting resins. In the U.S. alone, plastics encompass a $310 billion industry that supplies 1.4 million jobs. Table 1.2 presents the top 10 states in the U.S. in terms of number of employees. These 10 states employ almost 60% of the US plastics workers.

1.2 POLYMER AND PLASTICS CATEGORIES

Plastics are organic and semi-organic materials that have as their main attribute a very large molecular weight. These very large molecules, or macromolecules, give them their distinct properties and material behavior, when compared to other materials used in manufacturing or found in nature.

Figure 1.6 presents the classification, break-down and provenance of plastics. As presented in the figure polymers can be placed into either a thermoset, thermoplastic or elastomer category. Thermoplastics in turn include a special family which is relatively new, called thermoplastic elastomers. However, all these materials have in common that they are made of huge molecules. Some of these molecules are uncrosslinked, which means that each molecule can move freely relative to its neighbors, and others are crosslinked, which means that "bridges", or physical links interconnect the polymer molecules. Thermoplastics and unvulcanized elastomers are uncrosslinked. Vulcanized rubber, or elastomers, and thermosets are cross-linked.

Thermoplastics are those polymers that solidify as they are cooled, no longer allowing the long molecules to move freely. When heated, these materials regain the ability to "flow", as the molecules are able to slide past each other with ease. Furthermore, thermoplastic polymers are divided into two classes: amorphous and semi-crystalline polymers. Amorphous thermoplastics are those with molecules that remain in disorder as they cool, leading to a material with a fairly random molecular structure. An amorphous polymer solidifies, or vitrifies, as

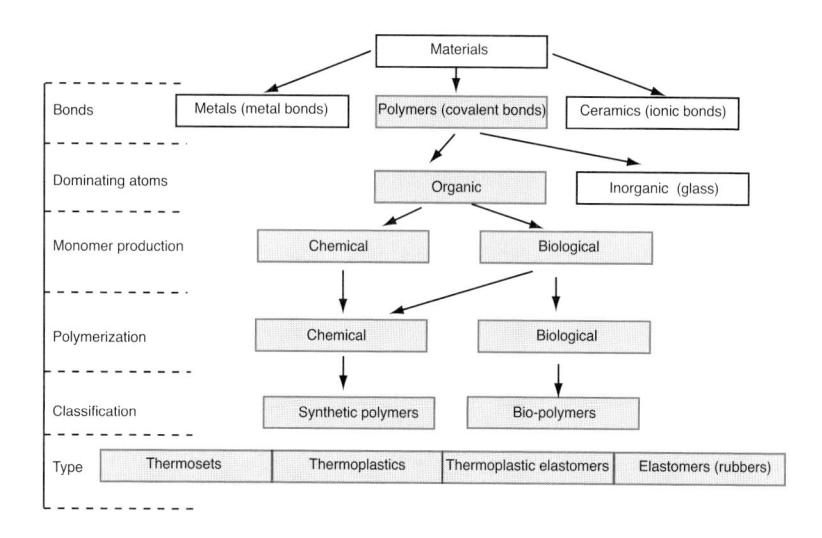

Figure 1.6: Classification, break-down and provenance of plastics in materials science.

it is cooled below its glass transition temperature. Semi-crystalline thermoplastics, on the other hand, solidify with a certain order in their molecular structure. Hence, as they are cooled, they harden when the molecules begin to arrange in a regular order below what is usually referred to as the melting temperature. The molecules in semi-crystalline polymers that are not transformed into ordered regions remain as small amorphous regions. These amorphous regions within the semi-crystalline domains lose their "flowability" below their glass transition temperature. Most semi-crystalline polymers have a glass transition temperature at subzero temperatures, hence, behaving at room temperature as rubbery or leathery materials. On the other hand, thermosetting polymers solidify by being chemically cured. Here, the long macromolecules cross-link with each other during cure, resulting in a network of molecules that cannot slide past each other. The formation of these networks causes the material to lose the ability to "flow" even after reheating. The high density of cross-linking between the molecules makes thermosetting material stiff and brittle. Thermosets also exhibit a glass transition temperature which is sometimes near or above thermal degradation temperatures. Compared to thermosets, elastomers are only lightly cross-linked which permits almost full extension of the molecules. However, the links across the molecules hinder them from sliding past each other, making even large deformations reversible. One common characteristic of elastomeric materials is that the glass transition temperature is much lower than room temperature. Their ability to "flow" is lost after they are vulcanized or cross-linked.

1.2 Polymer and plastics categories

Since cross-linked elastomers at room temperature are significantly above their glass transition temperature, they are very soft and very compliant elastic solids.

Although this handbook heavily concentrates on thermoplastic polymers, we have tried to incorporate thermosets as well as elastomers whenever necessary and fitting. Finally, the following generalizations can be made of plastics and should serve as a general guide to this Plastics Handbook:

- Unlike other materials such as metals, plastics have numerous grades and variations of every type of resin. These variations include different additives, fillers and reinforcing fibers, to name a few. Early on, plastics were lauded as the "material made to measure"; today, this has become reality and an everyday attribute that we take for granted.
- The strange molecular structure of polymers leads to peculiar behavior not observed with other materials. Such behavior includes viscoelasticity and other non-Newtonian effects during deformation, such as shear thinning. These characteristics not only affect how a final product may perform in its lifetime, controlling how we must approach design, but also the actual manufacturing process, such as mold filling, extrusion die flow, etc. This will often lead to residual stresses, as well as molecular and filler orientation, which causes anisotropy in the final part. General material science of polymers is covered in Chapter 2 of this handbook. Several standard tests are available to evaluate the performance of a material. These tests are described in detail in Chapter 3 of this handbook.
- During design and manufacturing of a product, material cost often becomes the most influential parameter. However, today we must also factor in ecological and environmental aspects. These include the effects of additives such as solvents or certain flame-retardants on the health of factory workers, as well as the environmental impact in general. In addition, the production of a product must keep in mind that the material used should be recyclable. Recycling issues are also introduced in Chapter 2 of this handbook.
- One of the great advantages of polymers is the low energy required during manufacturing. The melting, shaping and solidification all take place in an integrated fashion. Chapter 4 presents the various plastics manufacturing techniques, as well as material preparation and post-processing procedures.
- The design, performance and recyclability of a product is directly coupled to the choice of material and its additives as well as chosen processing technique and corresponding processing conditions. This can be referred to as the 5 P's: Polymer, Process, Product, Performance and Post-consumer life. Design aspects are covered in Chapter 5, plastics materials are presented in Chapter 6, and plastics additives are covered in Chapter 7.

1.3 PLASTICS ACRONYMS

In the plastics industry it is common to define a polymer by the chemical family it belongs to, and assign an abbreviation based on the chemistry. However, many times instead of using the standardized descriptive symbol, often engineers use the tradename given by the resin supplier.

This book uses the standardized notation presented in Table 1.3. The symbols which have been marked with an asterisk (*) have been designated by the ISO standards, in conjunction with the material data bank CAMPUS. The plastics presented in the table are presented in detail in Chapter 6 of this handbook. Furthermore, the acronyms presented in Table 1.3 may have additional symbols separated with a hyphen, such PE-LD for low density polyethylene, or PVC-P for plasticized PVC. The symbols for the most common characteristics are presented in Table 1.4.

Table 1.5 presents the most commonly used plasticizers and the symbols used to describe them. Plasticizers are also covered in detail in Chapter 6 of this handbook.

Table 1.3: Alphabetical overview of commonly used acronyms for plastics

Acronym	Chemical notation
ABS*	Acrylonitrile-butadiene-styrene
ACM	Acrylate rubber, (AEM, ANM)
ACS	Acrylonitrile-chlorinated polyethylene-styrene
AECM	Acrylic ester-ethylene rubber
AEM	Acrylate ethylene polymethylene rubber
AES	Acrylonitrile ethylene propylene diene styrene
AFMU	Nitroso rubber
AMMA	Acrylonitrile methylmethacrylate
ANBA	Acrylonitrile butadiene acrylate
ANMA	Acrylonitrile methacrylate
APE-CS	see ACS
ASA*	Acrylonitile styrene acrylic ester
AU	Polyesterurethane rubber
BIIR	Bromobutyl rubber
BR	Butadiene rubber
CA	Cellulose acetate
CAB	Cellulose acetobutyrate
CAP	Cellulose acetopropionate
CF	Cresol formaldehyde
CH	Hydratisierte cellulose, Zellglas
CIIR	Chloro butyl rubber
CM	Chlorinated polyethylene rubber
CMC	Carboxymethylcellulose
CN	Cellulose nitrate, Celluloid
CO	Epichlorhydrine rubber
COC*	Cyclopolyolefine-Copolymers

Continued on next page

1.3 Plastics Acronyms

Acronym	Chemical notation
COP	COC-Copolymer
CP	Cellulose propionate
CR	Chloroprene rubber
CSF	Casein formaldehyde, artificial horn
CSM	Chlorosulfonated polyethylene rubber
CTA	Cellulose triacetate
DPC	Diphenylene polycarbonate
E/P*	Ethylene-propylene
EAM	Ethylene vinylacetate rubber
EAMA	Ethylene acrylic acid ester-maleic acid anhydride-copoly
EB	Ethylene butene
EBA	Ethylene butylacrylate
EC	Ethylcellulose
ECB	Ethylene copolymer bitumen-blend
ECO	Epichlorohydrine rubber
ECTFE	Ethylene chlorotrifluoroethylene
EEAK	Ethylene ethylacrylate copolymer
EIM	Ionomer Copolymer
EMA	Ethylene methacrylic acid ester copolymer
EP*	Epoxy Resin
EP(D)M	see EPDM
EPDM	Ethylene propylene diene rubber
EPM	Ethylene propylene rubber
ET	Polyethylene oxide tetrasulfide rubber
ETER	Epichlorohydrin ethylene oxid rubber (terpolymer)
ETFE	Ethylene tetrafluoroethylene copolymer
EU	Polyetherurethane rubber
EVAC*	Ethylene vinylacetate
EVAL	Ethylene vinylalcohol, old acronym EVOH
FA	Furfurylalcohol resin
FEP	Polyfluoroethylene propylene
FF	Furan formaldehyde
FFKM	Perfluoro rubber
FKM	Fluoro rubber
FPM	Propylene tetrafluoroethylene rubber
FZ	Phosphazene rubber with fluoroalkyl- or fluoroxyalkyl gr
HIIR	Halogenated butyl rubber
HNBR	Hydrated NBR rubber
ICP	Intrinsically conductive polymers
IIR	Butyl rubber (CIIR, BIIR)
IR	Isoprene rubber
IRS	Styrene isoprene rubber
LCP*	Liquid crystal polymer
LSR	Liquid silicone rubber
MABS*	Methylmethacrylate acrylonitrile butadiene styrene
MBS*	Methacrylate butadiene styrene
MC	Methylcellulose (cellulose derivate)
MF*	Melamine formaldehyde
MFA	Tetrafluoroethylene perfluoromethyl vinyl ether copolyme
MFQ	Methylfluoro silicone rubber

Continued on next page

Acronym	Chemical notation
MMAEML	Methylmethacrylate-exo-methylene lactone
MPF*	Melamine phenolic formaldehyde
MPQ	Methylphenylene silicone rubber
MQ	Polydimethylsilicone rubber
MS	see PMS
MUF	Melamine urea formaldehyde
MUPF	Melamine urea phenolic formaldehyde
MVFQ	Fluoro silicone rubber
NBR	Acrylonitrile butadiene rubber
NCR	Acrylonitrile chloroprene rubber
NR	Natural rubber
PA	Polyamide (other notations see Section 6.7)
PA11*	Polyamide from aminoundecanoic acid
PA12*	Polyamide from dodecanoic acid
PA46*	Polyamide from polytetramethylene adipic acid
PA6*	Polyamide from e-caprolactam
PA610*	Polyamide from hexamethylene diamine sebatic acid
PA612*	Polyamide from hexamethylene diamine dodecanoic acid
PA66*	Polyamide from Hexamethylene diamine adipic acid
PA69*	Polyamide from hexamethylene diamine acelaic acid
PAA	Polyacrylic acid ester
PAC	Polyacetylene
PAE	Polyarylether
PAEK*	Polyarylether ketone
PAI	Polyamidimide
PAMI	Polyaminobismaleinimide
PAN*	Polyacrylonitrile
PANI	Polyaniline, polyphenylene amine
PAR	Polyarylate
PARA	Polyarylamide
PARI	Polyarylimide
PB	Polybutene
PBA	Polybutylacrylate
PBI	Polybenzimidazole
PBMI	Polybismaleinimide
PBN	Polybutylene naphthalate
PBO	Polyoxadiabenzimidazole
PBT*	Polybutylene terephthalate
PC*	Polycarbonate (from bisphenol-A)
PCPO	Poly-3,3-bis-chloromethylpropylene oxide
PCTFE	Polychlorotrifluoro ethylene
PDAP	Polydiallylphthalate resin
PDCPD	Polydicyclopentadiene
PE*	Polyethylene
PE-HD	Polyethylene-high density
PE-HMW	Polyethylene-high molecular weight
PE-LD	Polyethylene-low density
PE-LLD	Polyethylene-linear low density
PE-MD	Polyethylene medium density
PE-UHMW	Polyethylene-ultra high molecular weight

Continued on next page

1.3 Plastics Acronyms

Acronym	Chemical notation
PE-ULD	Polyethylene-ultra low density
PE-VLD	Polyethylene-very low density
PE-X	Polyethylene, crosslinked
PEA	Polyesteramide
PEDT	Polyethylenedioxythiophene
PEEEK	Polyetheretheretherketone
PEEK	Polyetheretherketone
PEEKEK	Polyetheretherketoneetherketone
PEEKK	Polyetheretherketoneketone
PEI*	Polyetherimide
PEK	Polyetherketone
PEKEEK	Polyetherketoneetheretherketone
PEKK	Polyetherketoneketone
PEN*	Polyethylenenaphthalate
PEOX	Polyethylene oxide
PESI	Polyesterimide
PES*	Polyethersulfone
PET*	Polyethylene terephthalate
PET-G*	Polyethylene terephthalate, glycol modified
PF*	Phenolic formaldehyde resin
PFMT	Polyperfluorotrimethyltriazine rubber
PFU	Polyfuran
PHA	Polyhydroxyalkanoate
PHB	Polyhydroxybutyrate
PHFP	Polyhexafluoropropylene
PI*	Polyimide
PIB	Polyisobutylene
PISO	Polyimidsulfone
PK*	Polyketone
PLA	Polylactide
PMA	Polymethylacrylate
PMI	Polymethacrylimide
PMMA*	Polymethylmethacrylate
PMMI	Polymethacrylmethylimide
PMP	Poly-4-methylpentene-1
PMPI	Poly-m-phenylene-isophthalamide
PMS	Poly-a-methylstyrene
PNF	Fluoro-phosphazene rubber
PNR	Polynorbornene rubber
PO	Polypropylene oxide rubber
PO	General notation for polyolefins, polyolefin-derivates u
POM*	Polyoxymethylene (polyacetal resin, polyformaldehyde)
PP*	Polypropylene
PPA	Polyphthalamide
PPB	Polyphenylenebutadiene
PPC	Polyphthalate carbonate
PPE*	Polyphenylene ether, old notation PPO
PPI	Polydiphenyloxide pyromellitimide
PPMS	Poly-para-methylstyrene
PPOX	Polypropylene oxide

Continued on next page

Acronym	Chemical notation
PPP	Poly-para-phenylene
PPQ	Polyphenylchinoxaline
PPS*	Polyphenylene sulfide
PPSU*	Polyphenylene sulfone
PPTA	Poly-p-phenyleneterephthalamide
PPV	Polyphenylene vinylene
PPY	Polypyrrol
PPYR	Polyparapyridine
PPYV	Polyparapyridine vinylene
PS*	Polystyrene
PSAC	Polysaccharide, starch
PSIOA	Polysilicooxoaluminate
PSS	Polystyrenesulfonate
PSU*	Polysulfone
PT	Polythiophene
PTFE*	Polytetrafluoroethylene
PTHF	Polytetrahydrofuran
PTT	Polytrimethyleneterephthalate
PUR*	Polyurethane
PVAC	Polyvinylacetate
PVAL	Polyvinylalcohol
PVB	Polyvinyl butyral
PVBE	Polyvinyl isobutylether
PVC*	Polyvinyl chloride
PVC/EVA	Polyvinyl chloride-ethylene vinylacetate
PVDC*	Polyvinylidene chloride
PVDF	Polyvinylidene fluoride
PVF	Polyvinyl fluoride
PVFM	Polyvinyl formal
PVK	Polyvinyl carbazole
PVME	Polyvinyl methylether
PVMQ	Polymethylsiloxane phenyl vinyl rubber
PVP	Polyvinyl pyrrolidone
PVZH	Polyvinyl cyclohexane
PZ	Phosphazene rubber with phenoxy groups
RF	Resorcin formaldehyde resin
SAN*	Styrene acrylonitrile
SB*	Styrene butadiene
SBMMA	Styrene butadiene methylmethacrylate
SBR	Styrene butadiene rubber
SBS	Styrene butadiene styrene
SCR	Styrene chloroprene rubber
SEBS	Styrene ethene butene styrene
SEPS	Styrene ethene propene styrene
SEPDM	Styrene ethylene propylene diene rubber
SI	Silicone, Silicone resin
SIMA	Styrene isoprene maleic acid anhydride
SIR	Styrene isoprene rubber
SIS	Styrene isoprene styrene block copolymer
SMAB	Styrene maleic acid anhydride butadiene

Continued on next page

1.3 Plastics Acronyms

Acronym	Chemical notation
SMAH*	Styrene maleic acid anhydride
SP	Aromatic (saturated) polyester
TCF	Thiocarbonyldifluoride copolymer rubber
TFEHFPVDF	Tetrafluoroethylene hexafluoropropylene vinylidene fluor
TFEP	Tetrafluoroethylene hexafluoropropylene
TM	Thioplastics
TOR	Polyoctenamer
TPA*	Thermoplastic elastomers based on polyamide
TPC*	Thermoplastic elastomers based on copolyester
TPE	Thermoplastic elastomers
TPE-A	see TPA
TPE-C	see TPC
TPE-O	see TPO
TPE-S	see TPS
TPE-U	see TPU
TPE-V	see TPV
TPO*	Thermoplastic elastomers based on olefins
TPS*	Thermoplastic elastomers based on styrene
TPU*	Thermoplastic elastomers based on polyurethane
TPV*	Thermoplastic elastomers based on crosslinked rubber
TPZ*	Other thermoplastic elastomers
UF	Urea formaldehyde resin
UP*	Unsaturated polyester resin
VCE	Vinylchloride ethylene
VCEMAK	Vinylchloride ethylene ethylmethacrylate
VCEVAC	Vinylchloride ethylene vinylacetate
VCMAAN	Vinylchloride maleic acid anhydride acrylonitrile
VCMAH	Vinylchloride maleic acid anhydride
VCMAI	Vinylchloride maleinimide
VCMAK	Vinylchloride methacrylate
VCMMA	Vinylchloride methylmethacrylate
VCOAK	Vinylchloride octylacrylate
VCPAEAN	Vinylchloride acrylate rubber acrylonitrile
VCPE-C	Vinylchloride-chlorinated ethylene
VCVAC	Vinylchloride vinylacetate
VCVDC	Vinylchloride vinylidenechloride
VCVDCAN	Vinylchloride vinylidenechloride acrylonitrile
VDFHFP	Vinylidenechloride hexafluoropropylene
VF	Vulcanized fiber
VMQ	Polymethylsiloxane vinyl rubber
VU	Vinylesterurethane
XBR	Butadiene rubber, containing carboxylic groups
XCR	Chloroprene rubber, containing carboxylic groups
XF	Xylenol formaldehyde resin
XNBR	Acrylonitrile butadiene rubber, containing carboxylic gr
XSBR	Styrene butadiene rubber, containing carboxylic groups

Table 1.4: Commonly Used Symbols Describing Polymer Characteristics

Symbol	Material characteristic
A	Amorphous
B	Block-copolymer
BO	Biaxially oriented
C	Chlorinated
CO	Copolymer
E	Expanded (foamed)
G	Grafted
H	Homopolymer
HC	Highly crystalline
HD	High density
HI	High impact
HMW	High molecular weight
I	Impact
LD	Low density
LLD	Linear low density
(M)	Metallocene catalyzed
MD	Medium density
O	Oriented
P	Plasticized
R	Randomly polymerized
U	Unplasticized
UHMW	Ultra high molecular weight
ULD	Ultra low density
VLD	Very low density
X	Cross-linked
XA	Peroxide cross-linked
XC	Electrically cross-linked

Table 1.5: Commonly Used Plasticizers and Their Acronyms

Acronym	Chemical notation
DODP	Dioctyldecylphthalate
ASE	Alkylsulfone acid ester
BBP	Benzylbutylphthalate
DBA	Dibutyladipate
DBP	Dibutylphthalate
DBS	Dibutylsebacate
DCHP	Dicyclohexylphthalate
DEP	Diethylphthalate
DHXP	Dihexylphthalate
DIBP	Diisobutylphthalate
DIDP	Diisodecylphthalate
DINA	Diisononyladipate
DMP	Dimethylphthalate
DMS	Dimethylsebazate
DNA	Dinonyladipate
DNODP	Di-n-octyl-n-decylphthalate

Continued on next page

1.3 Plastics Acronyms

Acronym	Chemical notation
DNOP	Di-n-octylphthalate
DNP	Dinonylphthalate
DOA (DEHA)	Dioctyladipate, also diethylhexyladipate, DEHA no longer used
DOP, DEHP, DODP	Dioctylphthalate, dioctyldecylphthalate
DOS	Dioctylsebacate
DOZ	Dioctylazelate
DPCF	Diphenylkresylphosphate
DPOF	Diphenyloctylphosphate
DPP	Dipropylphthalate
ELO	Epoxidized linseed oil
ESO	Epoxidized soy bean oil
ODA	Octyldecyladipate
ODP	Octyldecylphthalate
PO	Paraffin oil
TBP	Tributylphosphate
TCEF	Trichlorethylphosphate
TCF	Trikresylphosphate
TIOTM	Triisooctyltrimellitate
TOF	Trioctylphosphate
TPP	Triphenylphosphate

CHAPTER 2

MATERIALS SCIENCE OF POLYMERS

This chapter is intended to give the reader a general overview of polymer materials science. It presents the general chemistry, structure and morphology of polymers as well as common modifications done on polymeric materials to enhance their properties. At the end of the chapter the topic of recycling is briefly discussed.

2.1 POLYMER STRUCTURE

The material behavior of polymers is totally controlled by their molecular structure. In fact, this is true for all polymers; synthetically generated polymers as well as polymers found in nature (bio-polymers), such as natural rubber, ivory, amber, protein-based polymers or cellulose-based materials. To understand the basic aspects of material behavior and its relation to the molecular structure of polymers, in this chapter we attempt to introduce the fundamental concepts in a compact and simple way, leaving out material that has academic and scientific interest, but that is not needed to understand the basic technological aspects of plastics. Further information is presented in subsequent chapters. For more details on specific plastics and polymers the reader should consult Chapter 6 of this handbook.

2.1.1 Chemistry

As the word itself suggests, polymers are materials composed of molecules of very high molecular weight. These large molecules are generally referred to as *macromolecules*. Polymers are macromolecular structures that are generated synthetically or through natural processes. Historically, it has been always said that synthetic polymers are generated through *addition* or *chain growth polymerization*, and *condensation* or *radical initiated polymerization*. In addition polymerization, the final molecule is a repeating sequence of blocks with a chemical formula of the monomers. Condensation polymerization processes occur when the resulting polymers have fewer atoms than those present in the monomers from which they are generated. However, since many additional polymerization processes result in condensates, and various condensation polymerization processes are chain growth polymerization processes that resemble addition polymerization, today we rather break-down polymerization processes into *step polymerization* and *chain polymerization*. Table 2.1 shows a break-down of polymerization into step and chain polymerization, and presents examples for the various types of polymerization processes.

Table 2.1: Polymerization Classification

Classification	Polymerization	Examples
Step linear	Polycondensation	Polyamides
		Polycarbonate
		Polyesters
		Polyethers
		Polyimide
		Siloxanes
	Polyaddition	Polyureas
		Polyurethanes
Step non-linear	Network polymers	Epoxy resins
		Melamine
		Phenolic
		Polyurethanes
		Urea
Chain	Free radical	Polybutadiene
		Polyethylene (branched)
		Polyisoprene
		Polymethylmethacrylate
		Polyvinyl acetate
		Polystyrene
	Cationic	Polyethylene
		Polyisobutylene
		Polystyrene
		Vinyl esters
	Anionic	Polybutadiene
		Polyisoprene
		Polymethylmethacrylate
		Polystyrene
	Ring opening	Polyamide 6
		Polycaprolactone
		Polyethylene oxide
		Polypropylene oxide
	Ziegler-Natta	Polyethylene
		Polypropylene
		Polyvinyl chloride
		Other vinyl polymers
	Metallocene	Polyethylene
		Polypropylene
		Polyvinyl chloride
		Other vinyl polymers

2.1 Polymer Structure

Linear and non-linear step growth polymerization are processes where the polymerization occurs with more than one molecular species. On the other hand, chain growth polymerization processes occur with monomers with a reactive end group. Chain growth polymerization processes include *free-radical polymerization, ionic polymerization, cationic polymerization, ring opening polymerization, Ziegler-Natta polymerization and metallocene catalysis polymerization*. Free-radical polymerization is the most widely used polymerization process and it is used to polymerize monomers with the general structure $CH_2 = CR_1R_2$. Here, the polymer molecules grow by addition of a monomer with a free-radical reactive site called an active site. A chain polymerization process can also take place when the active site has an ionic charge. When the active site is positively charged, the polymerization process is called a *cationic polymerization*, and when the active site is negatively charged it is called *ionic polymerization*. Finally, monomers with a cyclic or ring structure such as caprolactam can be polymerized using the ring-opening polymerization process. Caprolactam is polymerized into polycaprolactam or polyamide 6.

The atomic composition of polymers encompasses primarily non-metallic elements such as carbon (C), hydrogen (H) and oxygen (O). In addition, recurrent elements are nitrogen (N), chlorine (Cl), fluoride (F) and sulfur (S). The so-called semi-organic polymers contain other non-metallic elements such as silicon (Si) in silicone or polysiloxane, as well as bor or beryllium (B). Although other elements can sometime be found in polymers, because of their very specific nature, we will not mention them here. The properties of the above elements lead to specific properties that are common of all polymers. These are:

- Polymers have very low electric conductance (electric insulators)
- Polymers have a very low thermal conductance (thermal insulators)
- Polymers have a very low density (between 0.8 and 2.2 g/cm^3)
- Polymers have a low thermal resistance and will easily irreversibly thermally degrade

There are various ways that the monomers can arrange during polymerization, however, we can break them down into two general categories: uncross-linked and cross-linked. Furthermore, the uncross-linked polymers can be subdivided into linear and branched polymers. The most common example of uncross-linked polymers that present the various degrees of branching is polyethylene (PE), as schematically depicted in Figure 2.1. Another important family of uncrosslinked polymers are copolymers. Copolymers are polymeric materials with two or more monomer types in the same chain. A copolymer that is composed of two monomer types is referred to as a *bipolymer* (e.g., PS-HI), and one that is formed by three different monomer groups is called a *terpolymer* (e.g., ABS). Depending on how

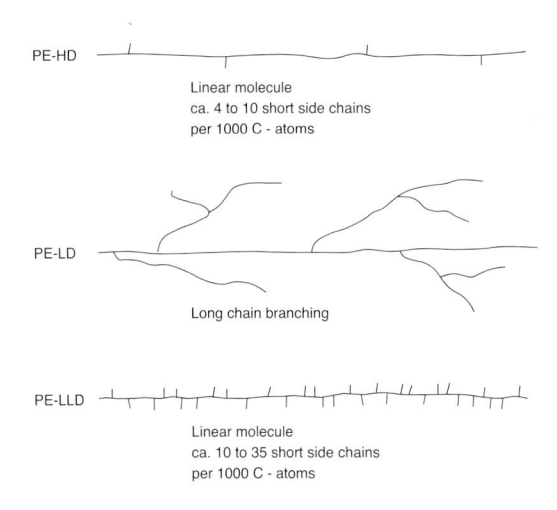

Figure 2.1: Schematic of the molecular structure of different polyethylenes.

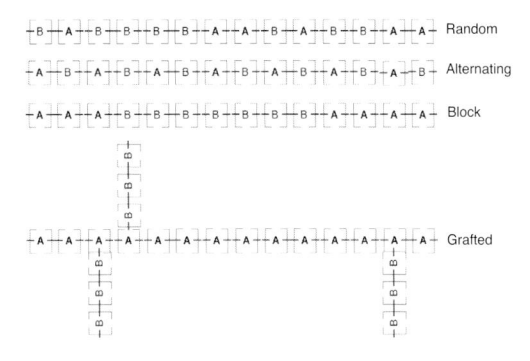

Figure 2.2: Schematic representation of different copolymers.

the different monomers are arranged in the polymer chain, one distinguishes between *random*, *alternating*, *block*, or *graft* copolymers. The four types of copolymers are schematically represented in Fig. 2.2.

Although thermoplastics can cross-link under specific conditions, such as gel formation when PE is exposed to high temperatures for prolonged periods of time, thermosets, and some elastomers, are polymeric materials that have the ability to cross-link. The cross-linking causes the material to become heat resistant after it has solidified. The cross-linking usually is a result of the presence of double bonds that break, allowing the molecules to link with their neighbors. One of

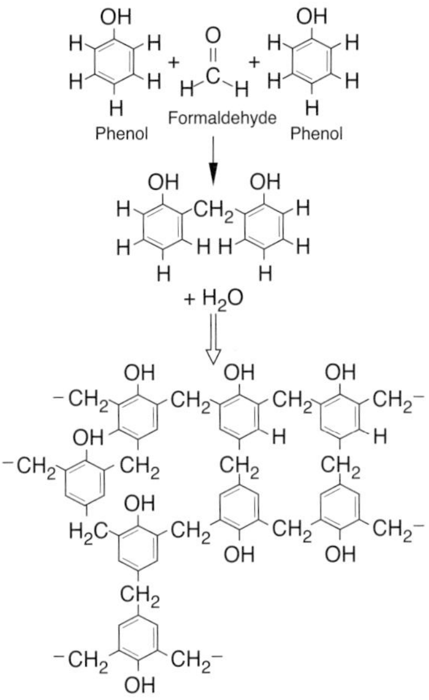

2.1 Polymer Structure

Figure 2.3: Symbolic representation of the condensation polymerization of phenol-formaldehyde resins.

the oldest thermosetting polymers is phenol-formaldehyde, or phenolic. Figure 2.3 shows the chemical symbol representation of the reaction where the phenol molecules react with formaldehyde molecules to create a three-dimensional cross-linked network that is stiff and strong, and leaving water as the by-product of this chemical reaction. This type of chemical reaction is called *condensation polymerization*.

Molecular Weight The size of the resulting macromolecules is the primary factor resulting from a polymerization reaction. After such a reaction, a polymeric material will consist of polymer chains of various lengths or repeat units. Hence, the molecular weight is determined by the average, or mean, molecular weight, which is defined by $M = W/N$. Here, W is the weight of the sample and N the number of moles in the sample. Figure 2.4 presents a schematic of a molecular weight distribution. The figure presents three molecular weight distributions. These are the *number avarage*, $\bar{M}_n$, defined by,

$$\bar{M}_n = \frac{\sum n_i M_i}{\sum n_i} \qquad (2.1)$$

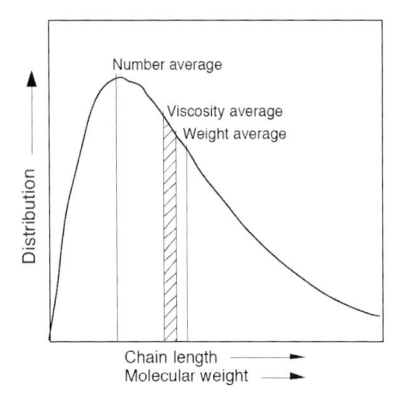

Figure 2.4: Molecular weight distribution of a typical thermoplastic.

the weight average, $\bar{M}_w$, defined by,

$$\bar{M}_w = \frac{\sum n_i M_i^2}{\sum n_i M_i} \tag{2.2}$$

and the *viscosity average*, $\bar{M}_v$. The viscosity average is a function of the viscosity of the polymers,

$$[\eta] = k \bar{M}_v^\alpha \tag{2.3}$$

where α is a material dependent parameter, which varies from $\alpha = 1$ for short molecules to $\alpha = 3.4$ for long molecules. The linear relation for short molecules is directly related to the intra-molecular friction, while the power relation of long molecules is related to the molecular entanglement.

Figure 2.5 presents the viscosity of various polymers as a function of molecular weight. The figure shows how for all these polymers the viscosity goes from the linear to the power dependence at some critical molecular weight. The linear relation is sometimes referred to as Staudinger's rule and applies for a perfectly monodispersed polymer. In a monodispersed polymer most molecules have the same molecular weight. A measure of the broadness of a polymer's molecular weight distribution is the polydispersity index defined by,

$$PI = \frac{\bar{M}_w}{\bar{M}_n} \tag{2.4}$$

Figure 2.6 presents a plot of flexural strength versus melt flow index for polystyrene samples with three different polydispersity indices. The figure shows that low

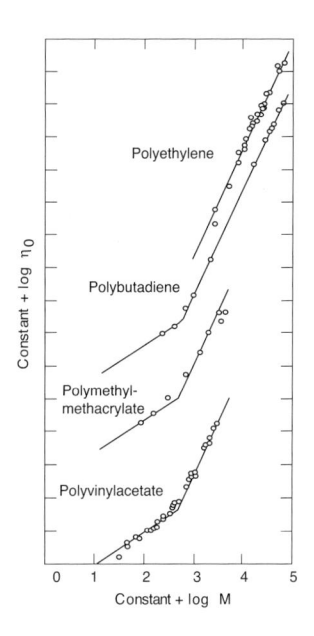

Figure 2.5: Zero shear rate viscosity for various polymers as a function of molecular weight.

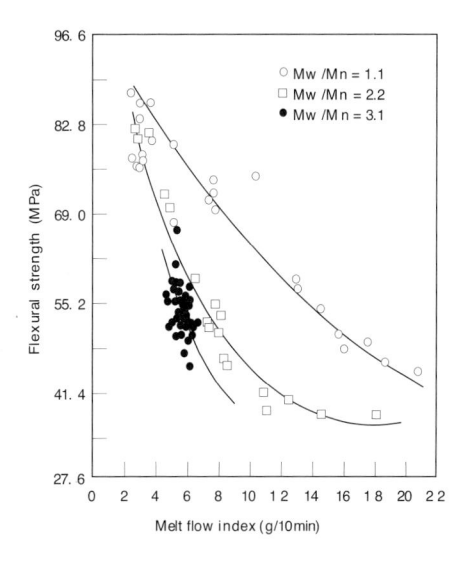

Figure 2.6: Effect of molecular weight on the strength-MFI interrelationship of polystyrene for three polydispersity indeces.

polydispersity index grade materials render higher strength properties and flowability, or processing ease, than high polydispersity index grades. Table 2.2 summarizes the various techniques used to measure molecular weight of polymers and the molecular weight range at which they can adequately perform the measurement.

Table 2.2: Various techniques used to measure molecular weight of polymers and oligomers

Method	Measurement	Range
Sedimentation and diffusion with centrifuge (a)	$\bar{M}_w$	Up to 10^8
Light scattering (a)	$\bar{M}_w$	Up to 10^7
Electron microscopy (a)	$\bar{M}_i, \bar{M}_n$	*
Gel-Permeation-Chromatography (r)	$\bar{M}_w, \bar{M}_n, \bar{M}_v$	*
Solvent viscometry (r)	$\bar{M}_v$	*
Melt viscometry (r)	$\bar{M}_w$	‡
Membrane osmometry (a)	$\bar{M}_n$	$2 \cdot 10^4$ to 10^6
Vapor pressure osmometry (a)	$\bar{M}_n$	Up to 10^5
End-group determination (e)	$\bar{M}_n$	Up to $5 \cdot 10^4$
Cryoscopy (a)	$\bar{M}_n$	Up to $5 \cdot 10^4$
Ebullioscopy (a)	$\bar{M}_n$	Up to 10^4

(a) Absolute method, (r) Relative method, and (e) Equivalent method.
* Dependent on solubility.
‡ Dependent on melting point.

Physically, the molecules can have rather large dimensions. For example, each repeat unit of a carbon backbone molecule, such as polyethylene, measures 0.252 nm in length. If completely stretched out, a high molecular weight molecule with say 10,000 repeat units can measure over 2 mm in length. Figure 2.7 serves to illustrate the range in dimensions associated with polymers as well as which microscopic devices are used to capture the detail at various orders of magnitude. If we go from the atomic structure to the part geometry, we easily travel between 0.1 nm and 1 mm, covering 8 orders of magnitude.

Conformation and Configuration The conformation and configuration of the polymer molecules have a great influence on the properties of the polymer component. The conformation describes the preferential spatial positions of the atoms in a molecule, which is described by the polarity flexibility and regularity of the macromolecule. Typically, carbon atoms are tetravalent, which means that they are surrounded by four substituents in a symmetric tetrahedral geometry. The configuration gives information about the distribution and spatial organization of the molecule. The general structure and symmetry of a molecule can be greatly influenced during polymer synthesis. This is illustrated in Fig. 2.8. Due to energy issues, linear polymers such as polyethylene, polypropylene, polystyrene and polyvinyl chloride prefer to polymerize through 1,2-addition. For technical polydienes, such as natural rubber (polyisoprene), polybutadiene and poly-

2.1 Polymer Structure

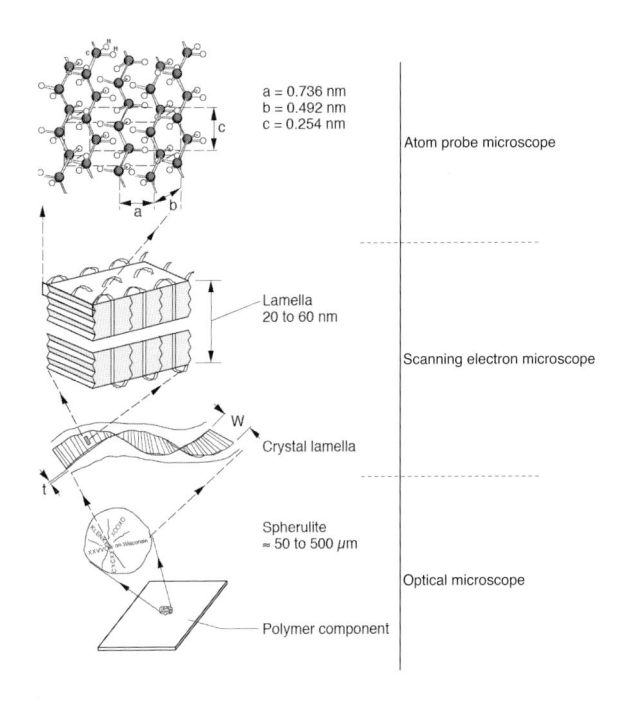

Figure 2.7: Schematic representation of the general molecular structure of semi-crystalline polymers and magnitudes as well as microscopic devices associated with such structures.

chloroprene, the preferred mode of polymerization is 1,4-addition. In effect, one can control the mode of polymerization by use of stereocatalysts as well as methalocene catalysts.

During polymerization it is possible to place the X groups on the carbon-carbon backbone in different directions. The order in which they are arranged is called the tacticity. The polymers with side groups placed in a random matter are called atactic. The polymers whose side groups are all on the same side are called isotactic, and those molecules with regularly alternating side groups are called syndiotactic. Figure 2.9 shows the three different tacticity cases for polypropylene. The tacticity in a polymer determines the degree of crystallinity that a polymer can reach. For example, polypropylene with a high isotactic content will reach a high degree of crystallinity and as a result will be stiff, strong, and hard.

Another type of geometric arrangement arises with polymers that have double bonds between carbon atoms. Double bonds restrict the rotation of the carbon atoms about the backbone axis. These polymers are sometimes referred to as

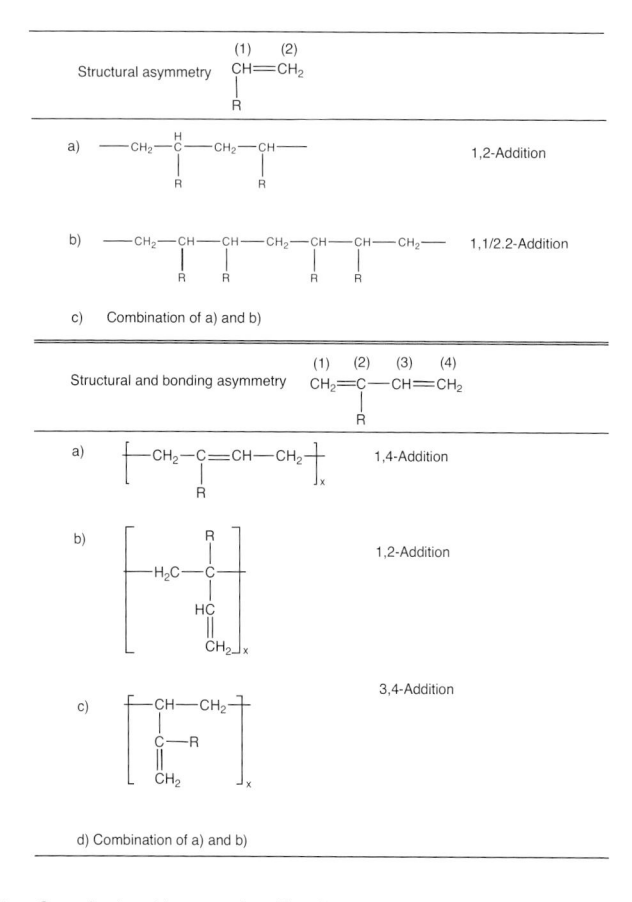

Figure 2.8: Constitutional isomer classification.

geometric isomers. The X groups may be on the same side (cis-) or on opposite sides (trans-) of the chain as schematically shown for polybutadiene in Fig. 2.10. The arrangement in a cis-1,4- polybutadiene results in a very elastic rubbery material, whereas the structure of the trans-1,4- polybutadiene results in a leathery and tough material. A cis-1,4- polybutadiene can be used to manufacture the outer tread of an automotive tire. A trans-1,4- polybutadiene can be used to make the outer skin of a golf ball. The same geometric arrangement is found in natural rubber, polyisoprene. The cis-1,4- polyisoprene is the elastic natural rubber used for the body of a tire, and the latex used to manufacture "rubber" gloves and condoms. The trans-1,4- polyisoprene, or the so-called gutta percha or ebony, was used to make dentures, statues, and other decorative items in the 1800s.

2.1 Polymer Structure

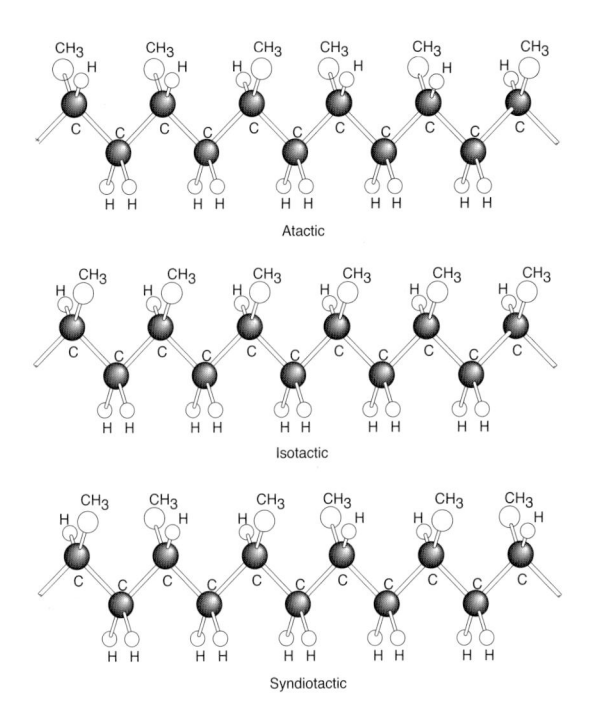

Figure 2.9: Different polypropylene structures.

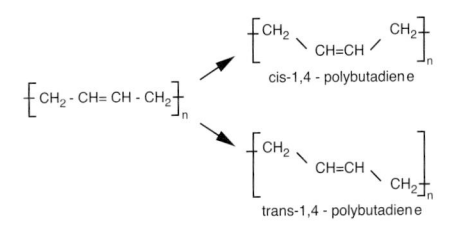

Figure 2.10: Symbolic representation of cis-1,4- and trans-1,4-polybutadiene molecules.

Intramolecular Forces The various intermolecular forces, generally called van der Waals forces, between macromolecules are of importance because of the size of the molecules. These forces are often the cause of the unique behavior of polymers. The so-called dispersion forces, the weakest of the intermolecular forces, are caused by the instantaneous dipoles that form as the charge in the molecules fluctuates. Very large molecules, such as ultra high molecular weight polyethylene, can have significant dispersion forces. Dipole-dipole forces are those intermolecular forces that result from the attraction between polar groups. Hydrogen bonding intermolecular forces, the largest of them all, take place when a polymer molecule contains -OH or -NH groups. The degree of polarity within a polymer determines how strongly it is attracted to other molecules. If a polymer is composed of atoms with different electronegativity (EN), it has a high degree of polarity and it is usually called a polar molecule. A non-polar molecule is composed of atoms with equal or similar electronegativity. For example, polyethylene, which is formed of carbon (C) and hydrogen (H) alone, is considered a non-polar material. An increase in polarity is to be expected when elements such as chlorine, fluorine, oxygen or nitrogen are present in a macromolecule.

Table 2.3: Electronegativity Number EN for Various Elements (After Pauling)

Element	Electronegativity Number
Fluorine (F)	4.0
Oxygen (O)	3.5
Chlorine (Cl)	3.0
Nitrogen (N)	3.0
Carbon (C)	2.5
Sulfur (S)	2.5
Hydrogen (H)	2.1
Silicone (Si)	1.8
Zink (Zn)	1.6
Sodium (Na)	0.9

Table 2.3 presents the electronegativity of common elements found in polymers. The intramolecular forces affect almost every property that is important when processing a polymer, including the effect that low molecular weight additives, such as solvents, plasticizers and permeabilizers, have on the properties of a polymer. Similarly, the intramolecular forces affect the miscibility of various polymers when making blends.

Polymer Groups In order to increase the understanding of existing polymer groups, it is important now to present these in form of tables that break them down into different categories. Chapter 6 of this handbook presents all these families of polymers in much more detail.

2.1 Polymer Structure

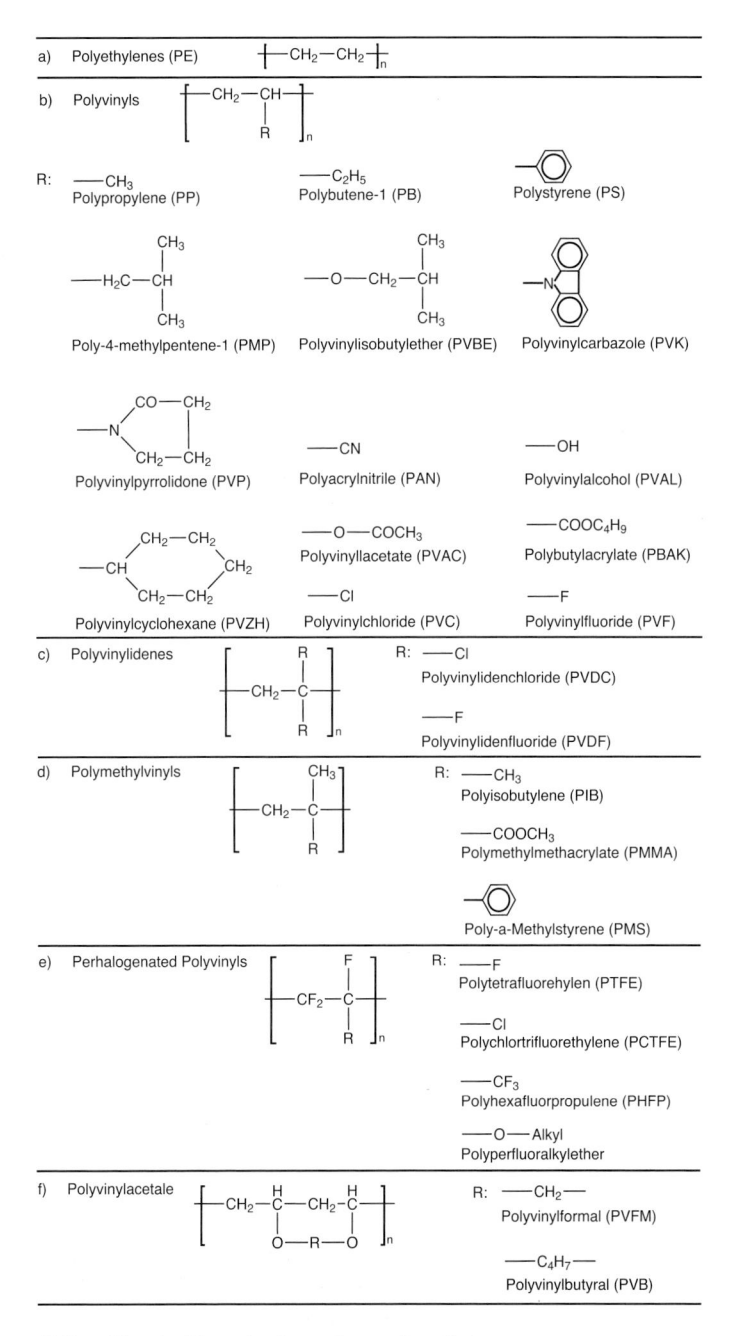

a) Polyethylenes (PE) — $\text{[—CH}_2\text{—CH}_2\text{—]}_n$

b) Polyvinyls — $\text{[—CH}_2\text{—CHR—]}_n$

R:
- —CH$_3$ Polypropylene (PP)
- —C$_2$H$_5$ Polybutene-1 (PB)
- —C$_6$H$_5$ (phenyl) Polystyrene (PS)
- —H$_2$C—CH(CH$_3$)$_2$ Poly-4-methylpentene-1 (PMP)
- —O—CH$_2$—CH(CH$_3$)$_2$ Polyvinylisobutylether (PVBE)
- —N-carbazole Polyvinylcarbazole (PVK)
- —N-pyrrolidone (CO—CH$_2$—CH$_2$—CH$_2$) Polyvinylpyrrolidone (PVP)
- —CN Polyacrylnitrile (PAN)
- —OH Polyvinylalcohol (PVAL)
- —cyclohexyl Polyvinylcyclohexane (PVZH)
- —O—COCH$_3$ Polyvinyllacetate (PVAC)
- —COOC$_4$H$_9$ Polybutylacrylate (PBAK)
- —Cl Polyvinylchloride (PVC)
- —F Polyvinylfluoride (PVF)

c) Polyvinylidenes — $\text{[—CH}_2\text{—CR}_2\text{—]}_n$

R:
- —Cl Polyvinylidenchloride (PVDC)
- —F Polyvinylidenfluoride (PVDF)

d) Polymethylvinyls — $\text{[—CH}_2\text{—C(CH}_3\text{)R—]}_n$

R:
- —CH$_3$ Polyisobutylene (PIB)
- —COOCH$_3$ Polymethylmethacrylate (PMMA)
- —C$_6$H$_5$ Poly-a-Methylstyrene (PMS)

e) Perhalogenated Polyvinyls — $\text{[—CF}_2\text{—CFR—]}_n$

R:
- —F Polytetrafluorehylen (PTFE)
- —Cl Polychlortrifluorethylene (PCTFE)
- —CF$_3$ Polyhexafluorpropulene (PHFP)
- —O—Alkyl Polyperfluoralkylether

f) Polyvinylacetale — $\text{[—CH}_2\text{—CH(O—R—O)—CH}_2\text{—CH—]}_n$

R:
- —CH$_2$— Polyvinylformal (PVFM)
- —C$_4$H$_7$— Polyvinylbutyral (PVB)

Figure 2.11: Chemical formulas for various polymethylenes.

30   2 Materials Science of Polymers

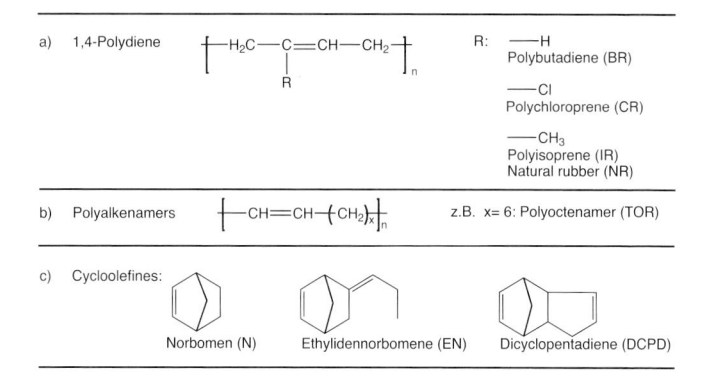

a) 1,4-Polydiene $\left[\!\!\begin{array}{c}H_2C-C=CH-CH_2\\|\\R\end{array}\!\!\right]_n$ R: —H Polybutadiene (BR)

—Cl Polychloroprene (CR)

—CH_3 Polyisoprene (IR) Natural rubber (NR)

b) Polyalkenamers $\left[\!-CH=CH-(CH_2)_x-\!\right]_n$ z.B. x= 6: Polyoctenamer (TOR)

c) Cycloolefines:

Norbomen (N)   Ethylidennorbomene (EN)   Dicyclopentadiene (DCPD)

Figure 2.12: Chemical formulas for various polymers with double bonds.

a) Polyacetals $\left[\!-CH_2-O-\!\right]_n$ $\left[\!\!\begin{array}{c}-CH-O-\\|\\CH_3\end{array}\!\!\right]_n$

Polyformaldehyde, Polyoxymethylene (POM)   Polyacetaldehyde

b) Polyethers $\left[\!-CH_2-CH_2-O-\!\right]_n$

Polyethylenoxide (PEOX)

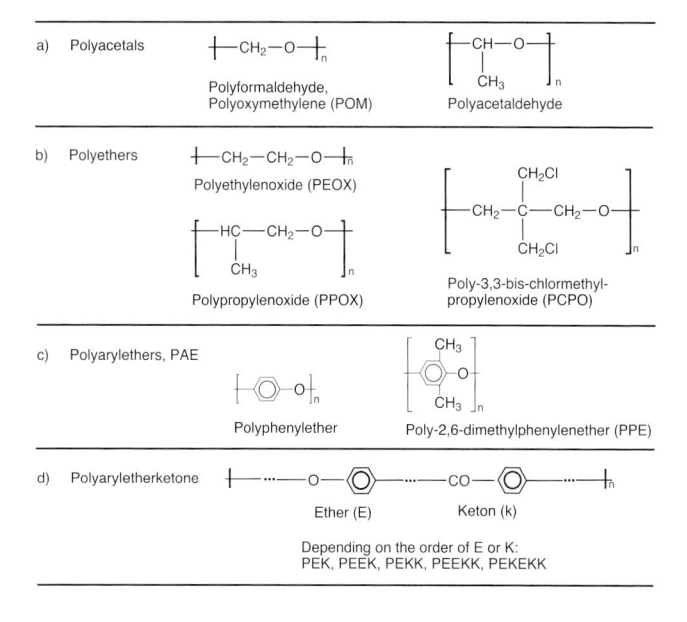

$\left[\!\!\begin{array}{c}-CH_2-C-CH_2-O-\\|\\CH_2Cl\\|\\CH_2Cl\end{array}\!\!\right]_n$

$\left[\!\!\begin{array}{c}-HC-CH_2-O-\\|\\CH_3\end{array}\!\!\right]_n$

Polypropylenoxide (PPOX)   Poly-3,3-bis-chlormethyl-propylenoxide (PCPO)

c) Polyarylethers, PAE

$\left[\!-\!\bigcirc\!-O-\!\right]_n$ $\left[\!\!\begin{array}{c}CH_3\\\bigcirc\!-O\\CH_3\end{array}\!\!\right]_n$

Polyphenylether   Poly-2,6-dimethylphenylenether (PPE)

d) Polyaryletherketone $\left[\!\cdots-O-\bigcirc-\cdots-CO-\bigcirc-\cdots\!\right]_n$

Ether (E)   Keton (k)

Depending on the order of E or K: PEK, PEEK, PEKK, PEEKK, PEKEKK

Figure 2.13: Chemical formulas for polymers with an oxygen in the backbone, such as polyacetals and polyethers.

Figure 2.11 presents linear homopolymers polymerized from either unsaturated monomers or from the transformation of analog polymers. This group of polymers include the four major commodity thermoplastics: polyethylene, polypropylene, polystyrene and polyvinyl chloride, as well as many others.

2.1 Polymer Structure

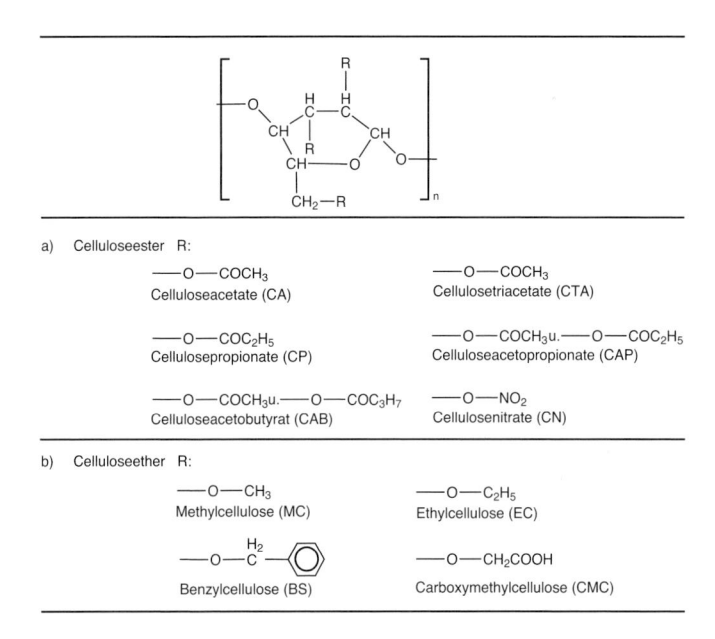

Figure 2.14: Chemical formulas for cellulose polymers.

a) Celluloseester R:

—O—COCH$_3$
Celluloseacetate (CA)

—O—COCH$_3$
Cellulosetriacetate (CTA)

—O—COC$_2$H$_5$
Cellulosepropionate (CP)

—O—COCH$_3$u.—O—COC$_2$H$_5$
Celluloseacetopropionate (CAP)

—O—COCH$_3$u.—O—COC$_3$H$_7$
Celluloseacetobutyrat (CAB)

—O—NO$_2$
Cellulosenitrate (CN)

b) Celluloseether R:

—O—CH$_3$
Methylcellulose (MC)

—O—C$_2$H$_5$
Ethylcellulose (EC)

—O—CH$_2$—C$_6$H$_5$
Benzylcellulose (BS)

—O—CH$_2$COOH
Carboxymethylcellulose (CMC)

Another group of polymers which exhibit double bonds and belong to the family of elastomers are presented in Fig. 2.12. The existence of a double bond allows these polymers to undergo a vulcanization or cross-linking process. Also presented in the figure is the special category of polycyclo-olefins, which are gaining significant attention as high temperature resistant thermoplastics. Polyacetals and polyethers are polymers that have a heteroatom (oxygen) present in the backbone of the molecule. This family of polymers is depicted in Fig.refethers. The heat resistance of these polymers is significantly increased by introducing bulky aromatic building blocks into the molecules, significantly reducing the intra-molecular mobility. The oldest thermoplastics which are still in use today are cellulose polymers, depicted in Fig. 2.14. The rigid cellulose molecules makes them difficult to process, requiring the use of plasticizers or other polymers. The high polarity of cellulose molecules makes them highly hydrophilic, and sometimes water soluble.

Polyamides are linear polycondensates which are among the most important engineering thermoplastics, used in engineering applications as well as fibers and films. Their general structure and notation is presented in Fig. 2.15. Due to hydrogen bonds and aromatic components, some of these materials are high temperature resistant. However, they are hydrophilic polymers, known for their

a) Homopolyamides

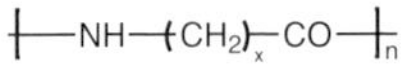

$+\!\!-\!\!NH\!\!-\!\!(CH_2)_x\!\!-\!\!CO\!\!-\!\!+_n$

Examples	PA(x+1)
Polycaprolactam	PA6
Poly-11-aminoundecanamide	PA11
Polylaurinlactam	PA12

b) Aliphatic homopolyamides made of brached diamines and dicarbon acids

$+\!\!-\!\!NH\!\!-\!\!(CH_2)_x\!\!-\!\!NHCO\!\!-\!\!(CH_2)_y\!\!-\!\!CO\!\!-\!\!+_n$

Examples	PA(x)(y+2)
Polytetramethylenadipinamid	PA46
Polyhexamethylenadipinamid	PA66
Polyhexamethylenacelainamid	PA69
Polyhexamethylensebacinamid	PA610
Polyhexamethylendodecanamid	PA612

c) Aromatic Homopolyamides (Polyarylamides)

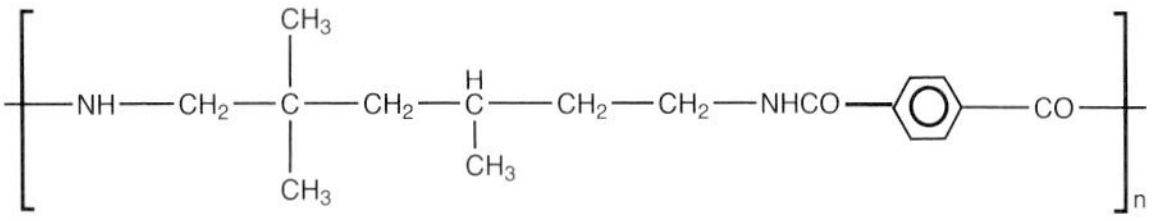

Polytrimethylhexamethylenterephthalamide (PANDT) (PA6-3-T)

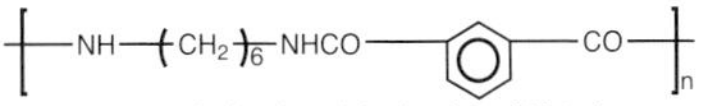

Polyhexamethylenisophthalamide (PA61) 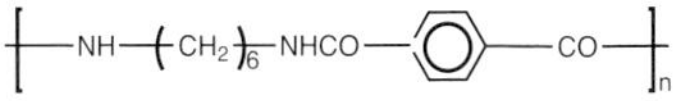   Polyhezamethylenterephthalamide (PA6T)

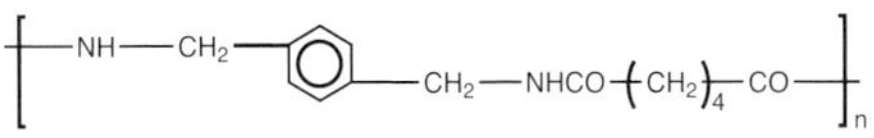

Poly-m-xylylenadipinamide (PAMXD6)

d) Polyurethane made of aliphatic diisocyanates and dioles

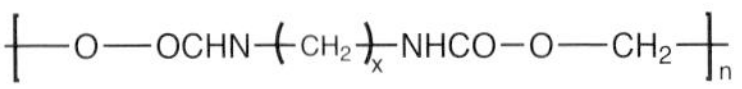

$+\!\!-\!\!O\!\!-\!\!OCHN\!\!-\!\!(CH_2)_x\!\!-\!\!NHCO\!\!-\!\!O\!\!-\!\!CH_2\!\!-\!\!+_n$

x= 6: 1,6-Hexamethyleendiisocyanate
x= 4: 1,4-Butandiol

Figure 2.15: Chemical formulas for common polyamides.

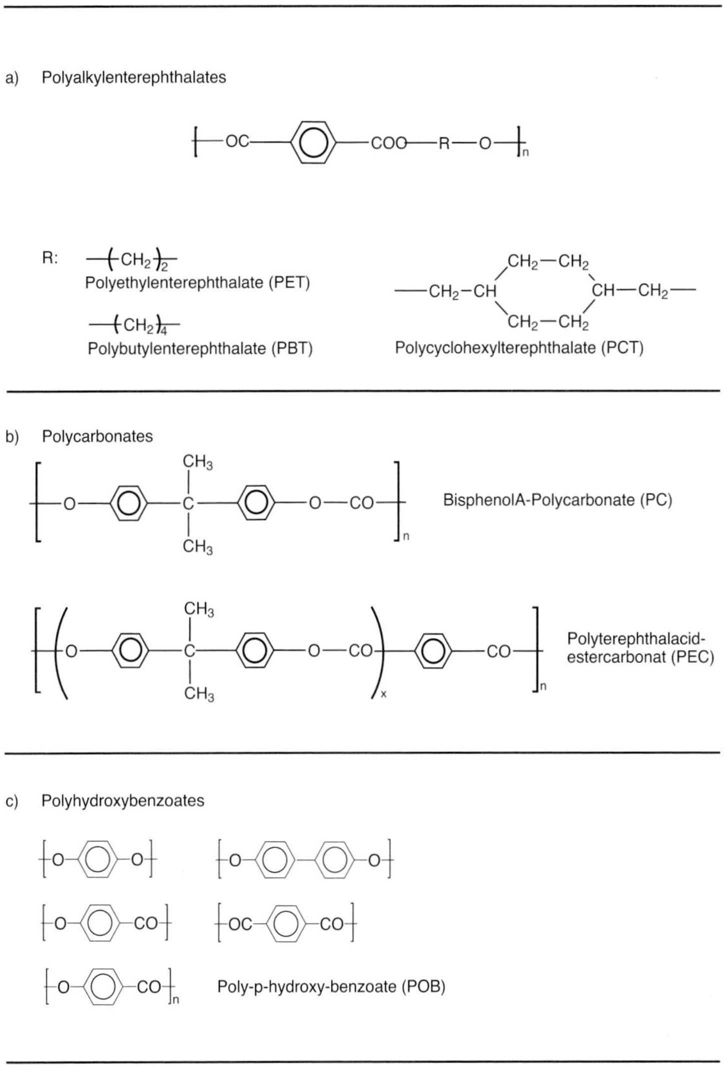

2.1 Polymer Structure

a) Polyalkylenterephthalates

$$\left[-OC-\bigcirc-COO-R-O-\right]_n$$

R: $-(CH_2)_2-$ Polyethylenterephthalate (PET)

$-(CH_2)_4-$ Polybutylenterephthalate (PBT)

$-CH_2-CH\begin{smallmatrix}CH_2-CH_2\\ \\CH_2-CH_2\end{smallmatrix}CH-CH_2-$ Polycyclohexylterephthalate (PCT)

b) Polycarbonates

BisphenolA-Polycarbonate (PC)

Polyterephthalacid-estercarbonat (PEC)

c) Polyhydroxybenzoates

Poly-p-hydroxy-benzoate (POB)

Figure 2.16: Chemical formulas for common polyesters.

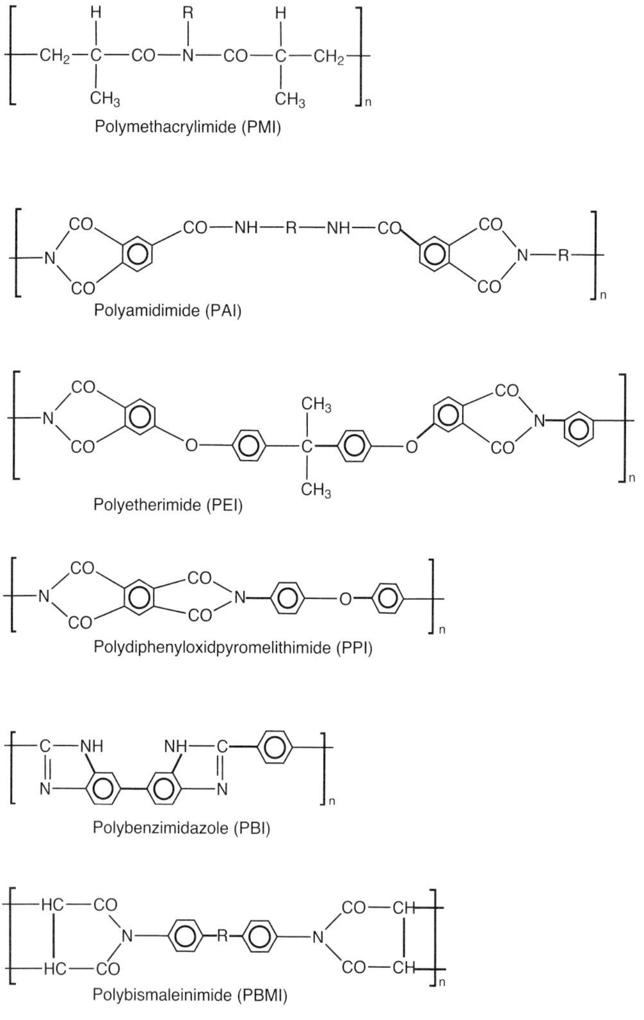

Figure 2.17: Chemical formulas for common polyimides.

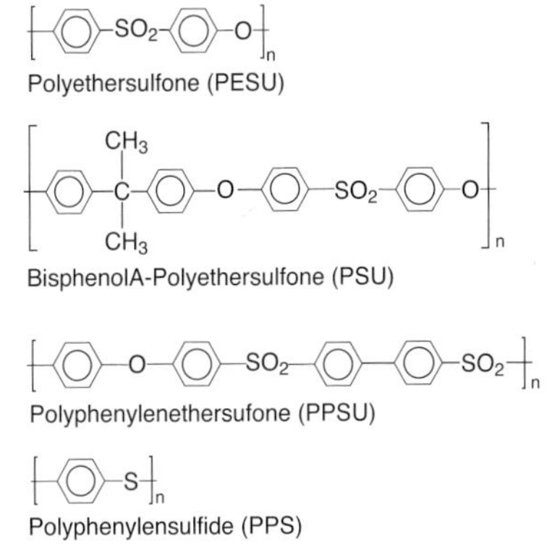

2.1 Polymer Structure

Figure 2.18: Chemical formulas for common polyarylethersulfones and polyarylsulfides.

high water absorption. Linear polyesters, as shown in Fig. 2.16, are also polycondensates which have only gained notariety in the form of polyarylesters. Special copolyesters have such stiff molecular segments (mesogenes) that they exhibit anisotropic behavior even in the melt state. Polymers with this behavior are termed liquid crystalline polymers (LCP). The stiffest molecules are realized by the molecular structure of polyimides, depicted in Fig. 2.17. Some of them, such as PPI, PBI and PBMI, are no longer processable thermoplastics. They are conductive and semi-conductive polymers that are high temperature resistant with a very high heat deflection temperature.

An important group of technical thermoplastics are the polyarylethersulfones and the polyarylsulfides presented in Fig. 2.18. Their unique molecular structure, composed of aromatic and sulfur based building blocks, make them high strength design plastics.

2.1.2 Morphological Structure

Morphology is the order or arrangement of the polymer structure. The possible "order" between a molecule or molecule segment and its neighbors can vary from a very ordered highly crystalline polymeric structure to an amorphous structure (i.e., a structure in greatest disorder or random). The possible range of order and disorder is clearly depicted on the left side of Fig. 2.19. For example, a purely amorphous polymer is formed only by the non-crystalline or amorphous chain structure, whereas the semi-crystalline polymer is formed by a combination of all the structures represented in Fig. 2.19. The semi-crystalline arrangement that certain polymer molecules take during cooling is in great part due to intramolecular

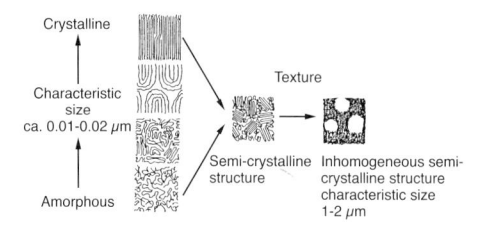

Figure 2.19: Schematic diagram of possible molecular structures that occur in thermoplastic polymers.

forces. As the temperature of a polymer melt is lowered, the free volume between the molecules is reduced, causing an increase in the intramolecular forces. As the free volume is reduced further, the intermolecular forces cause the molecules to arrange in a manner that brings them to a lower state of energy, as for example the folded chain structure of polyethylene molecules shown in Fig. 2.7. This folded chain structure, which starts at a nucleus, grows into the spherulitic structure shown in Fig. 2.7 and in the middle of Fig. 2.19, an image captured with an electron microscope. A macroscopic structure, shown in the right hand side of the figure, can be captured with an optical microscope. An optical microscope can capture the coarser macro-morphological structure such, as the spherulites in semi-crystalline polymers. Figure 2.7, presented earlier, shows a schematic of the spherulitic structure of polyethylene with the various microscopic devices that can be used to observe different levels of the formed morphology.

An amorphous polymer is defined as having a purely random structure. However, it is not quite clear if a "purely amorphous" polymer as such exists. Electron microscopic observations have shown amorphous polymers that are composed of relatively stiff chains, show a certain degree of macromolecular structure and order, for example, globular regions or fibrilitic structures. Nevertheless, these types of amorphous polymers are still found to be optically isotropic. Even polymers with soft and flexible macromolecules, such as polyisoprene which was first considered to be random, sometimes show band-like and globular regions. These bundle-like structures are relatively weak and short-lived when the material experiences stresses. The shear thinning viscosity effect of polymers sometimes is attributed to the breaking of such macromolecular structures.

Early on, before the existence of macromolecules had been recognized, the presence of highly crystalline structures had been suspected. Such structures were discovered when undercooling or when stretching cellulose and natural rubber. Later, it was found that a crystalline order also existed in synthetic macromolecular materials such as polyamides, polyethylenes, and polyvinyls. Because of the polymolecularity of macromolecular materials, a 100% degree

Figure 2.20: Polarized microscopic image of the spherulitic structure in polypropylene (After Wagner).

of crystallization cannot be achieved. Hence, these polymers are referred to as semi-crystalline. It is common to assume that the semi-crystalline structures are formed by small regions of alignment or crystallites connected by random or amorphous polymer molecules. With the use of electron microscopes and sophisticated optical microscopes the various existing crystalline structures are now well recognized. They can be listed as follows:

- Single crystals. These can form in solutions and help in the study of crystal formation. Here, plate-like crystals and sometimes whiskers are generated.
- Spherulites. As a polymer melt solidifies, several folded chain lamellae spherulites form which are up to 0.1 mm in diameter. A typical example of a spherulitic structure is shown in Fig. 2.20. The spherulitic growth in a polypropylene melt is shown in Fig. 2.21.
- Deformed crystals. If a semi-crystalline polymer is deformed while undergoing crystallization, oriented lamellae form instead of spherulites.
- Shish-kebab. In addition to spherulitic crystals, which are formed by plate- and ribbo - like structures, there are also shish-kebab crystals which are formed by circular plates and whiskers. Shish-kebab structures are generated when the melt undergoes a shear deformation during solidification. A typical example of a shish-kebab crystal is shown in Fig. 2.22.

The speed at which crystalline structures grow depends on the type of polymer and on the temperature conditions. Table 2.4 shows the maximum growth rate for common semi-crystalline thermoplastics as well the maximum achievable degree of crystallinity. Experimental evidence has demonstrated that the growth rate of the crystallized layer in semi-crystalline polymers is finite. This is mainly due to the fact that at the beginning the nucleation occurs at a finite rate. This is clearly demonstrated in Fig. 3.114 which presents the measured thickness of crystallized

Figure 2.21: Development of the spherulitic structure in polypropylene. Images were taken at 30 s intervals (After Menges and Winkel).

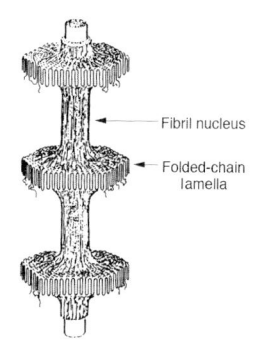

Figure 2.22: Model of the shish-kebab morphology (After Tadmor and Gogos).

2.1 Polymer Structure

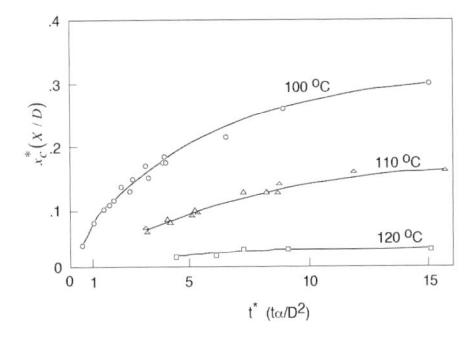

Figure 2.23: Dimensionless thickness of the crystallized layers of polypropylene as a function of dimensionless time for various temperatures of the quenching surfaces (After Eder and Janeschitz-Kriegl).

layers as a function of time for polypropylene plates quenched at three different temperatures.

Table 2.4: Maximum Crystalline Growth Rate and Maximum Degree of Crystallinity for Various Thermoplastics

Polymer	Growth rate (μ/min)	Maximum crystallinity (%)
Polyethylene	>1000	80
Polyamide 66	1000	70
Polyamide 6	200	35
Isotactic polypropylene	20	63
Polyethylene teraphthalate	7	50
Isotactic polystyrene	0.3	32
Polycarbonate	0.01	25

2.1.3 Thermal Transitions

A phase change or a thermal transition occurs with polymers when they undergo a significant change in material behavior. The phase change occurs as a result of either a reduction in material temperature or a chemical curing reaction. A thermoplastic polymer hardens as the temperature of the material is lowered below either the melting temperature for a semi-crystalline polymer, the glass transition temperature for an amorphous thermoplastic or the crystalline and glass transition temperatures in liquid crystalline polymers. A thermoplastic has the ability to soften again as the temperature of the material is raised above the solidification temperature. On the other hand, the solidification of thermosets leads to cross-linking of molecules. The effects of cross-linkage are irreversible and result in a network that hinders the free movement of the polymer chains independent of the material temperature. With thermoplastics, the term "solidification" is often misused to describe the hardening of amorphous thermoplastics.

Amorphous Thermoplastics The solidification of most materials is defined at a discrete temperature, whereas amorphous polymers do not exhibit a sharp transition between the liquid and the solid states. Instead, an amorphous thermoplastic polymer vitrifies as the material temperature drops below the glass transition temperature, T_g. Amorphous thermoplastics exhibit a "useful" behavior below their glass transition temperature. Due to their random structure, the characteristic size of the largest ordered region is on the order of a carbon-carbon bond. This dimension is much smaller than the wavelength of visible light and so generally makes amorphous thermoplastics transparent. The shear modulus versus temperature diagram is a very useful description of the mechanical behavior of certain materials. It is possible to generate a general shear modulus versus temperature diagram for all polymers by using a reduced temperature described by

$$T_{red} = \frac{293\text{K}}{T_g} \quad (2.5)$$

Figure 2.24 shows this diagram with the shear modulus of several polymers. The upper left side of the curve represents the stiff and brittle cross-linked materials, and the upper right side represents the semi-crystalline thermoplastics whose glass transition temperature is below room temperature. The lower right side of the curve represents elastomers, positioned accordingly on the curve depending on their degree of cross-linkage.

Figure 2.25 shows the shear modulus, G', versus temperature for polystyrene, one of the most common amorphous thermoplastics. The figure shows two general regions: one where the modulus appears fairly constant, and one where the modulus drops significantly with increasing temperature.

With decreasing temperatures, the material enters the glassy region where the slope of the modulus approaches zero. At high temperatures, the modulus is negligible and the material is soft enough to flow. Although there is not a clear transition between "solid" and "liquid", the temperature at which the slope is highest is T_g. For the polystyrene in Fig. 2.25, the glass transition temperature is approximately 110°C. Although data are usually presented in the form shown in Fig. 2.25, it should be mentioned here that the curve shown in the figure was measured at a constant frequency. If the frequency of the test is increased - reducing the time scale - the curve is shifted to the right since higher temperatures are required to achieve movement of the molecules at the new frequency. A similar effect is observed if the molecular weight of the material is increased. The longer molecules have more difficulty sliding past each other, thus requiring higher temperatures to achieve "flow".

2.1 Polymer Structure

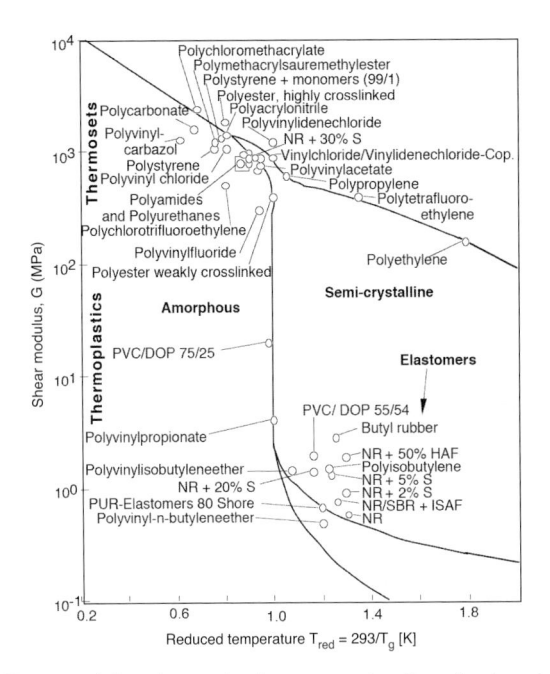

Figure 2.24: Shear modulus of several polymers as a function of reduced glass transition temperature (After Thimm).

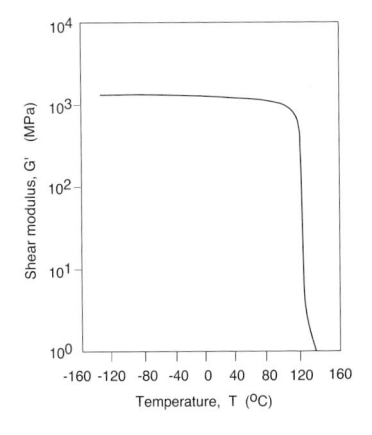

Figure 2.25: Shear modulus of polystyrene as a function of temperature.

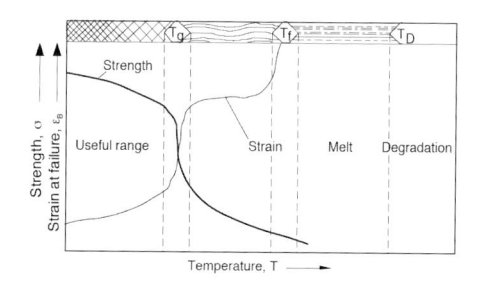

Figure 2.26: Tensile strength and strain at failure as a function of temperature for an amorphous thermoplastic.

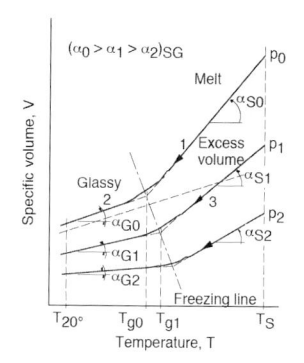

Figure 2.27: Schematic of a pvT diagram for amorphous thermoplastics.

2.1 Polymer Structure

Often, the tensile stress and strain at failure are plotted as a function of temperature. Figure 2.26 shows this for a typical amorphous thermoplastic. The figure shows how the material is brittle below the glass transition temperature and, therefore, fails at low strains. As the temperature increases, the stiffness of the amorphous thermoplastic decreases, as it becomes leathery in texture and is able to withstand larger deformations. Above T_g, the strength decreases significantly, as the maximum strain continues to increase, until the flow properties have been reached, at which point the mechanical strength is negligible. This occurs around the "flow temperature" marked as T_f in the diagram. If the temperature is further increased, the material will eventually thermally degrade at the degradation temperature.

Higher pressures reduce the free volume between the molecules which restricts their movement. This requires higher temperatures to increase the free volume sufficiently to allow molecular movement. This is clearly depicted in Fig. 2.27, which schematically presents the pressure-volume-temperature (pvT) behavior of amorphous polymers.

Semi-Crystalline Thermoplastics Semi-crystalline thermoplastic polymers show more order than amorphous thermoplastics. The molecules align in an ordered crystalline form as shown for polyethylene in Fig. 2.7. The size of the crystals or spherulites is much larger than the wavelength of visible light, making semi-crystalline materials translucent rather than transparent. However, the crystalline regions are very small with molecular chains comprised of both crystalline and amorphous regions. The degree of crystallinity in a typical thermoplastic will vary from grade to grade as for example in polyethylene, where the degree of crystallinity depends on the branching and the cooling rate. Because of the existence of amorphous as well as crystalline regions, a semi-crystalline polymer has two distinct transition temperatures, the glass transition temperature, T_g, and the melting temperature, T_m. Figure 2.28 shows the dynamic shear modulus versus temperature for a high density polyethylene, the most common semi-crystalline thermoplastic. Again, this curve presents data measured at one test frequency. The figure clearly shows two distinct transitions: one at about -110°C, the glass transition temperature, and another near 140°C, the melting temperature. Above the melting temperature, the shear modulus is negligible and the material will flow. Crystalline arrangement begins to develop as the temperature decreases below the melting point. Between the melting and glass transition temperatures, the material behaves as a leathery solid.

The tensile stress and the strain at failure for a common semi-crystalline thermoplastic is shown in Fig. 2.29. The figure shows an increase in toughness between the glass transition temperature and the melting temperature. The range between T_g and T_m applies to most semi-crystalline thermoplastics.

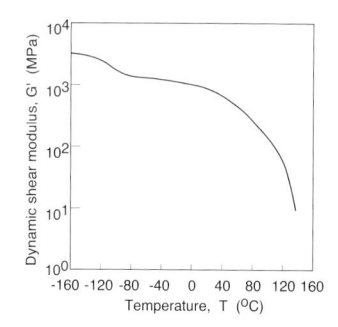

Figure 2.28: Shear modulus of a high density polyethylene as a function of temperature.

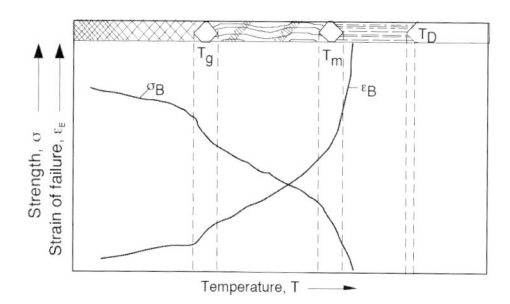

Figure 2.29: Tensile strength and strain at failure as a function of temperature for a semi-crystalline thermoplastic.

2.1 Polymer Structure

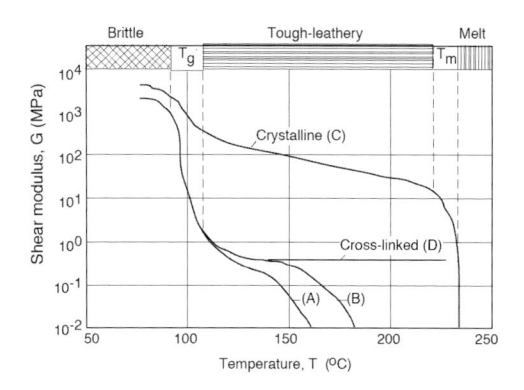

Figure 2.30: Shear modulus curves for amorphous, semi-crystalline and cross-linked polystyrene. (A) low molecular weight amorphous, (B) high molecular weight amorphous, (C) semi-crystalline, (D) cross-linked.

Strictly speaking, all polymers are semi-crystalline. However, for some polymers the speed of crystallization is so slow that for all practical purposes they render an amorphous structure. An interesting example is shown in Fig. 3.142, which presents plots of shear modulus versus temperature of a polystyrene with different molecular structures after having gone through different stereo-specific polymerization techniques: low molecular weight PS (A), a high molecular weight PS (B), a semi-crystalline PS (C), and a cross-linked PS (D). In Fig. 3.142 we can see that the low molecular weight material flows before the high molecular weight one, simply due to the fact that the shorter chains can slide past each other more easily - reflected in the lower viscosity of the low molecular weight polymer. The semi-crystalline PS shows a certain amount of stiffness between its glass transition temperature at around 100°C and its melting temperature at 230°C. Since a semi-crystalline polystyrene is still brittle at room temperature, it is not very useful to the polymer industry. Figure 3.142 also demonstrates that a cross-linked polystyrene will not melt. Semi-crystalline thermoplastics are leathery and tough at room temperature since their atactic and amorphous regions vitrify at much lower temperatures. While the amorphous regions ÒsolidifyÓ at a specific temperature, the melting temperature goes up significantly with increasing degree of crystallinity.

Figure 2.31 summarizes the property behavior of amorphous, crystalline, and semi-crystalline materials using schematic diagrams of material properties plotted as functions of temperature. Again, pressures affect the transition temperatures as schematically depicted in the pressure-volume-temperature diagram Fig. 2.32 for a semi-crystalline polymer.

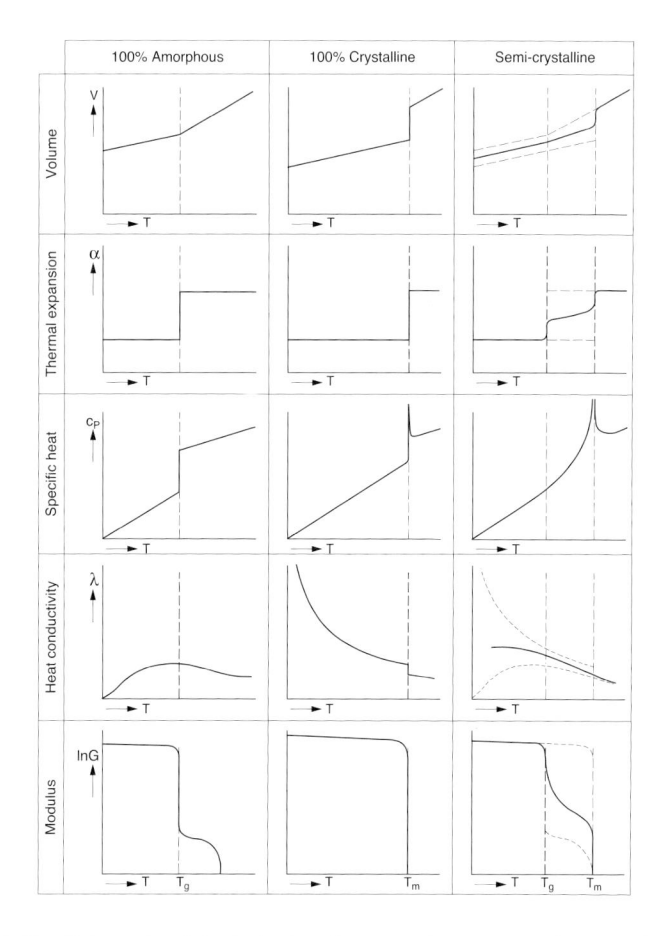

Figure 2.31: Schematic of the behavior of some polymer properties as a function of temperature for different thermoplastics (Hypothetical 100% amorphous and crystalline, as well as semi-crystalline polymers).

2.1 Polymer Structure

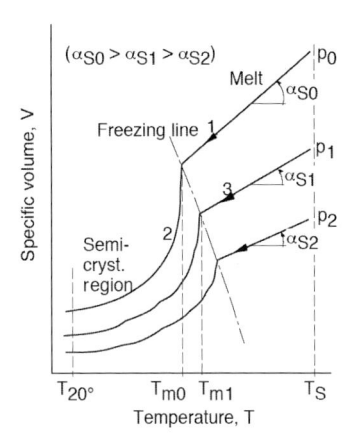

Figure 2.32: Schematic of a pvT diagram for semi-crystalline thermoplastics.

Liquid Crystalline Polymers The transition regions in liquid crystalline polymers or mesogenic polymers is much more complex. These transitions are referred to as mesomorphic transitions, and occur when one goes from a crystal to a liquid crystal, from a liquid crystal to another liquid crystal, and from a liquid crystal to an isotropic fluid. The volume-temperature diagram for a liquid crystalline polymer is presented in Fig. 2.33. The figure shows the various phases present in a liquid crystalline polymer. From a lower temperature to a high temperature, these are the glassy phase, partially crystalline phase, the smectic phase, the nematic or cholesteric phase and the isotropic phase. The smectic phase is where the molecules have all distinct orientation and where all their "centers of gravity" align with each other, giving them a highly organized structure. The nematic phase is where the axis of the molecules are aligned, giving them a high degree of orientation, but where the "centers of gravity" of the molecules are not aligned. During cooling, both the nematic and the smectic phases can be maintained, leading to nematic glass and smectic glass, respectively.

Oriented Thermoplastics If a thermoplastic is deformed at a temperature high enough so that the polymer chains can slide past each other but low enough such that the relaxation time is much longer than the time it takes to stretch the material, the orientation generated during stretching is retained within the polymer component. We note that the amount of stretching, L/L_0, is not always proportional to the degree of orientation within the component; for example, if the temperature is too high during stretching, the molecules may have a chance to fully relax, resulting in a component with little or no orientation. Any degree of orientation results in property variations within thermoplastic polymers. The

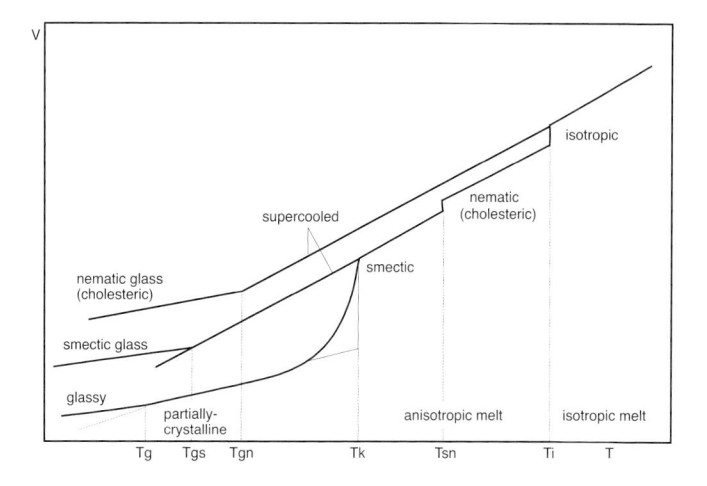

Figure 2.33: Schematic diagram of the specific volume of liquid crystalline polymers as a function of temperature (After van Krevelen).

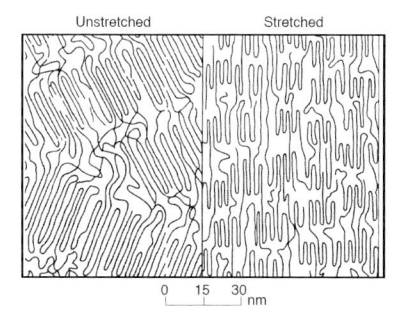

Figure 2.34: Schematic of the sliding and re-orientation of crystalline blocks in semi-crystalline thermoplastics (after Hosemann).

influence stretching has on various properties of common amorphous thermoplastics is presented in the next chapter in Figs. 3.12 and 3.32. Additionally, the stretching will lead to decreased strength and stiffness properties perpendicular to the orientation and increased properties parallel to the direction of deformation. In addition, highly oriented materials tend to split along the orientation direction under small loads. In amorphous thermoplastics the stretching that leads to permanent property changes occurs between 20 and 40°C above the glass transition temperature, T_g, whereas with semi-crystalline thermoplastics, they occur between 10 and 20°C below the melting temperature, T_m. After having stretched a semi-crystalline polymer, one must anneal it at temperatures high enough that the amorphous regions relax. During stretching, the spherulites break up as whole blocks of lamellae slide out, shown schematically in Fig. 2.34. Whole lamellae can also rotate such that by sufficiently high stretching, all molecules are oriented in the same direction. The lamellae blocks are now interconnected by what is generally called tie molecules. If this material is annealed in a fixed position, a very regular, oriented structure can be generated. This highly oriented material becomes dimensionally stable at elevated temperatures, including temperatures slightly below the annealing or fixing temperature. However, if the component is not fixed during the annealing process, the structure before stretching would be recovered. Figure 2.35 shows stress-strain plots for polyethylene with various morphological structures. If the material is stretched such that a needle-like or fibrilic morphological structure results, the resulting stiffness of the material is very high. Obviously, a more realistic structure that would result from stretching would lead to a stacked plate - like structure with lower stiffness and ultimate strength. An unstretched morphological structure would be composed of spherulites and exhibit much lower stiffness and ultimate strength. The strength of fibrilic structures is taken advantage of when making synthetic fibers. Figure 2.36 shows theoretical and achievable elastic moduli of various synthetic fiber materials.

Table 2.5: Mechanical Properties of Selected Fibers

Fiber	Tensile strength (MPa)	Tensile modulus (GPa)	Elongation to break (%)	Specific gravity
Polyethylene	3000	172	2.7	0.97
Aramid	2760	124	2.5	1.44
Graphite	2410	379	0.6	1.81
S-glass	4585	90	2.75	2.50

High-stiffness and high-strength synthetic fibers are becoming increasingly important for lightweight high-strength applications. Extended-chain ultra- high molecular weight polyethylene fibers have only been available commercially

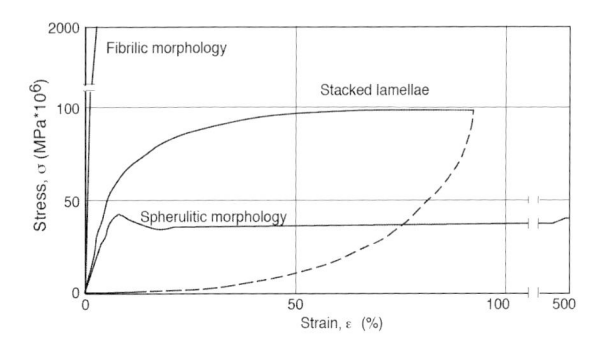

Figure 2.35: Stress-strain behavior of polyethylene with various morphologies.

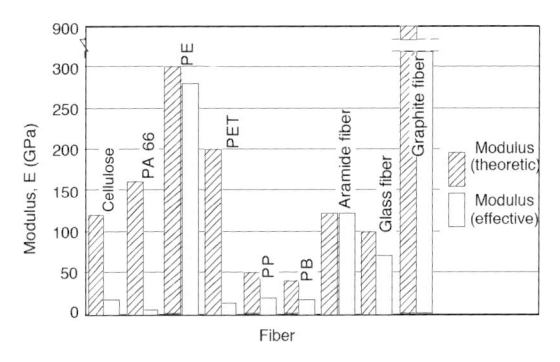

Figure 2.36: Tensile modulus for various fibers.

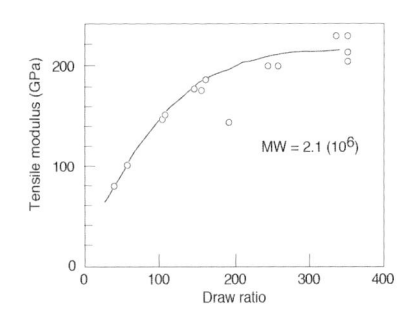

Figure 2.37: Tensile modulus as a function of draw ratio for a UHMWPE for molecular weight of approximately 2×10^6 (After Zachariades and Kanamoto).

2.1 Polymer Structure

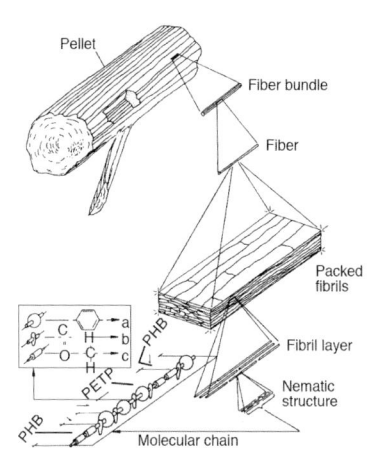

Figure 2.38: Schematic of the structure of a LC-PET (After Becker).

since the mid 1980s. The fibers are manufactured by drawing or extending fibers of small diameters at temperatures below the melting point. The modulus and strength of the fiber increase with the drawing ratio or stretch. Due to intermolecular entanglement, the natural draw ratio of high molecular weight high-density polyethylene is only five[1]. To increase the draw ratio by a factor of 10 or 100, polyethylene must be processed in a solvent such as paraffin oil or paraffin wax. Figure 2.37 presents the tensile modulus of super-drawn ultra-high molecular weight high-density polyethylene fibers as a function of draw ratio. It can be seen that at draw ratios of 250, a maximum modulus of 200 GPa is reached. In addition to amorphous and semi-crystalline thermoplastics, there is a whole family of thermoplastic materials whose molecules do not relax and, thus, retain their orientation even in the molten state. This class of thermoplastics is the *liquid crystalline polymers*. One such material is the aramid fiber, most commonly known by its tradename, Kevlar®, which has been available on the market for several years. To demonstrate the structure of liquid crystalline polymers, successive enlargement of an aramid pellet is shown in Fig. 2.38. For comparison, Table 2.5 presents mechanical properties of aramid and polyethylene fibers and compares them to other materials.

---

[1]It is interesting that a semi-crystalline thermoplastic stretches more at low molecular weights than at high molecular weights. This contradicts what we expect from theory that longer molecules allow the component to stretch following the relation $\lambda_{max} \approx M^{0.5}$. An explanation for this may be the *trapped entanglements* found in high molecular weight, semi-crystalline polymers that act as semi-permanent cross-links which rip at smaller deformations.

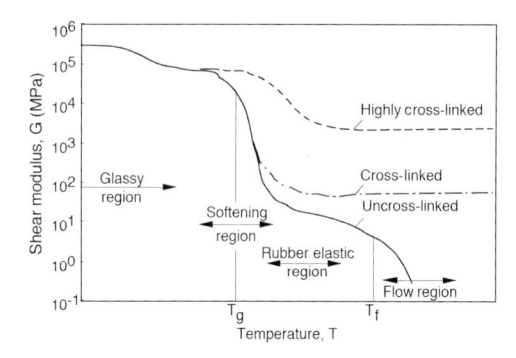

Figure 2.39: Shear modulus as a function of temperature for uncross-linked and cross-linked polymers.

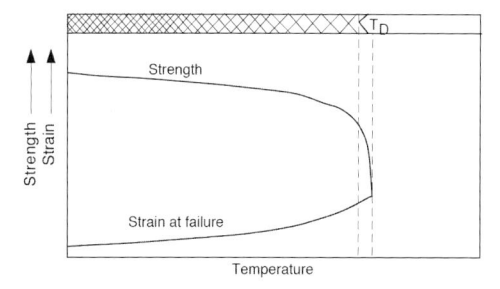

Figure 2.40: Tensile strength and strain at failure as a function of temperature for typical thermosets.

Cross-Linked Polymers Cross-linked polymers, such as thermosets and elastomers, behave completely differently than their counterparts, thermoplastic polymers. In cross-linked systems, the mechanical behavior is also best reflected by the plot of the shear modulus versus temperature. Figure 2.39 compares the shear modulus between highly cross-liked, cross-linked, and uncross-linked polymers. The coarse cross-linked system, typical of elastomers, has a low modulus above the glass transition temperature.

The glass transition temperature of these materials is usually below -50°C, so they are soft and flexible at room temperature. On the other hand, highly cross-linked systems, typical in thermosets, show a smaller decrease in stiffness as the material is raised above the glass transition temperature; the decrease in properties becomes smaller as the degree of cross-linking increases. Figure 2.40 shows the ultimate strength and strain curves plotted versus temperature. With thermosetting polymers, strength remains fairly constant up to the thermal degradation temperature of the material.

2.2 MATERIAL MODIFICATION OF PLASTICS

There are many ways a polymer or a plastic material can be modified in order to influence its material properties and behavior. The most common form of modification is through additives. Additives, such as plasticizers, flame retardants, stabilizers, antistatic agents and blowing agents, to name a few, can sometimes have a profound effect on the structure, material behavior and performance of the polymer. However, because of their importance and their variety, we dedicate Chapter 7 of this handbook to additives. This chapter will cover modifications that are more fundamental, and that lead to different families of plastics, such as polymer blends and polymer composites.

2.2.1 Polymer Blends

Polymer blends belong to another family of polymeric materials which are made by mixing or blending two or more polymers to enhance the physical properties of each individual component. Common polymer blends include PP-PC, PC-PBT, PVC-ABS, PE-PTFE, to name a few. The process of polymer blending is accomplished by distributing and dispersing a minor or secondary polymer within a major polymer that serves as a matrix. The major component can be thought of as the continuous phase, and the minor components as distributed or dispersed phases in the form of droplets or filaments. When creating a polymer blend, one must always keep in mind that the blend will probably be remelted in subsequent processing or shaping processes. For example, a rapidly cooled system, frozen as a homogenous mixture, can separate into phases because of coalescence when re-heated. For all practical purposes, such a blend is not processable. To avoid this problem, compatibilizers, which are macromolecules used to ensure compatibility in the boundary layers between the two phases, are common.

Blending can also be called mixing, which is subdivided into distributive and dispersive mixing. The morphology development of polymer blends is determined by three competing mechanisms: distributive mixing, dispersive mixing, and coalescence. Figure 2.41 presents a model that helps us visualize the mechanisms governing morphology development in polymer blends. The process begins when a thin tape of polymer is melted away from the pellet. As the tape is stretched, surface tension causes it to rip and to form into threads. These threads stretch and reduce in radius, until surface tension becomes significant

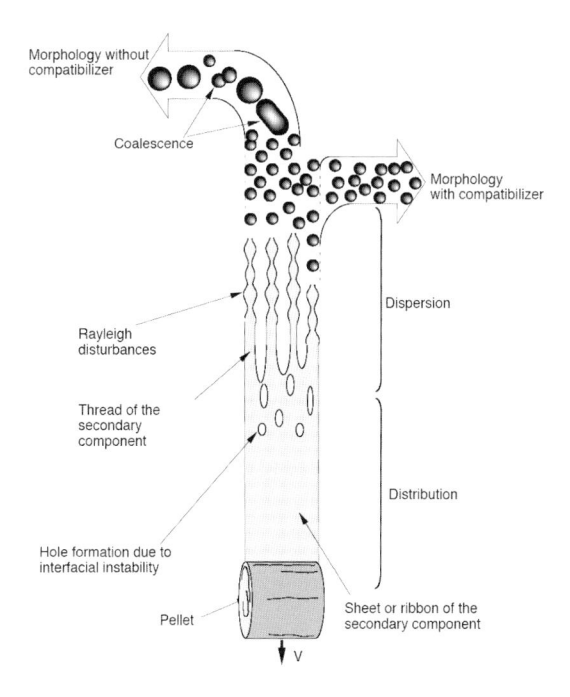

Figure 2.41: Mechanisms of morphology development in polymer blends.

enough leading to Rayleigh disturbances. These disturbances grow and cause the threads to break down into small droplets.

2.2.2 Filled Polymers and Reinforced Composites

When we talk about fillers, we refer to materials that are intentionally placed in polymers to make them stronger, lighter, electrically conductive, or cheaper. Any filler will affect the mechanical behavior of a polymeric material. For example, long fibers will make it stiffer but usually denser, whereas foaming will make it more compliant but much lighter. On the other hand, a filler such as calcium carbonate will decrease the polymer's toughness while making it considerably cheaper. Figure 2.42 shows a schematic plot of the change in stiffness as a function of filler content for several types of filler materials. Figure 2.43 shows the increase in dynamic shear modulus for polybutylene terephthalate with 10% and 30% glass fiber content. However, fillers often decrease the strength properties of polymers. When we refer to reinforced plastics, we talk about polymers (matrix) whose properties have been enhanced by introducing a reinforcement

2.2 Material Modification of Plastics

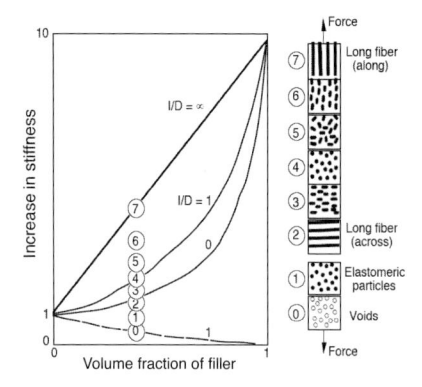

Figure 2.42: Relation between stiffness, filler and orientation in polymeric materials.

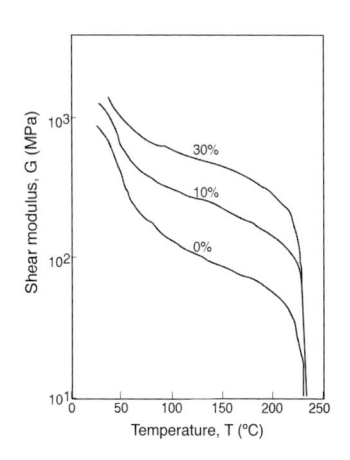

Figure 2.43: Shear modulus for a polybutylene terephthalate with various degrees of glass fiber content by weight.

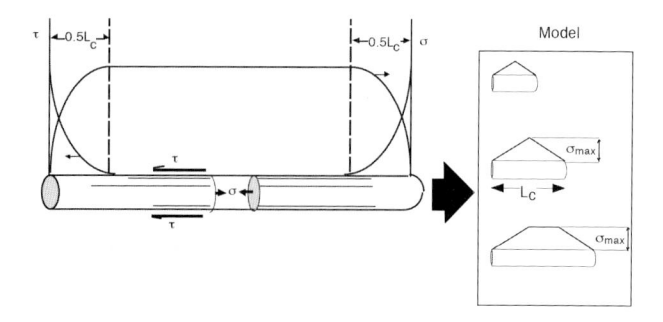

Figure 2.44: Schematic diagram of load transfer from matrix to fiber in a composite.

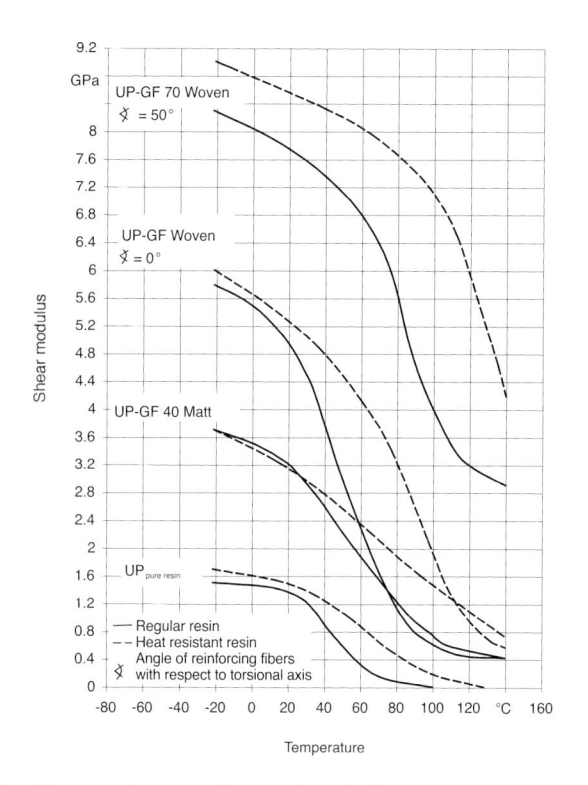

Figure 2.45: Shear modulus of a fiber reinforced UP resin.

2.2 Material Modification of Plastics

(fibers) of higher stiffness and strength. Such a material is usually called a fiber reinforced polymer (FRP) or a fiber reinforced composite (FRC). The purpose of introducing a fiber into a matrix is to transfer the load from the weaker material to the stronger one. This load transfer occurs over the length of the fiber as schematically represented in Fig. 2.44. The length it takes to complete the load transfer from the matrix to the fiber, without fiber or matrix fracture, is usually referred to as critical length, L_c. For the specific case where there is perfect adhesion between fiber and matrix, the critical length can be computed using

$$L_c = D \frac{\sigma_{uf}}{2\tau_{um}} \qquad (2.6)$$

where D is the diameter of the fiber, σ_{uf} is the ultimate tensile strength of the fiber, and τ_{um} is the ultimate shear strength of the matrix. Although this equation predicts as low as 10, experimental evidence suggests that aspect ratios of 100 or higher are required to achieve maximum strength. If composites have fibers that are shorter than their critical length, they are referred to as short fiber composites, and if the fibers are longer, they are referred to as long fiber composites.

In addition to the orientation of the fibers, how these reinforcing fibers are arranged plays a significant role on the properties of composites. This is illustrated in Fig. 2.45 that presents the shear modulus of fiber reinforced UP with different forms of glass fiber reinforcements.

2.2.3 Other Modifications

The properties of thermoplastics can be modified by adding plasticizing agents. This is shown in Fig. 2.46, where the glass transition temperature shifts to lower temperatures and consequently the shear modulus starts to drop at much lower temperatures when a plasticizing agent is added. In essence, the plasticizer replaces a temperature rise, by wedging these typically low molecular weight materials between the polymer's molecular structure, generating free volume between the chains. This additional free volume is equivalent to a temperature rise, consequently lowering the glass transition temperature, making a normally brittle and hard thermoplastic more rubbery and compliant.

Some amorphous thermoplastics can be made high impact resistant (less brittle) through copolymerization. The most common example is acrylonitrile-butadiene-styrene, also known as ABS. Since butadiene chains vitrify at temperatures below -50°C, ABS is very tough at room temperature in contrast to polystyrene and acrylics by themselves. Due to the different glass transition temperatures present in the materials which form the blend, ABS shows two general transition regions, one around -50°C and the other at 110°C, visible in both the logarithmic decrement and the shear modulus.

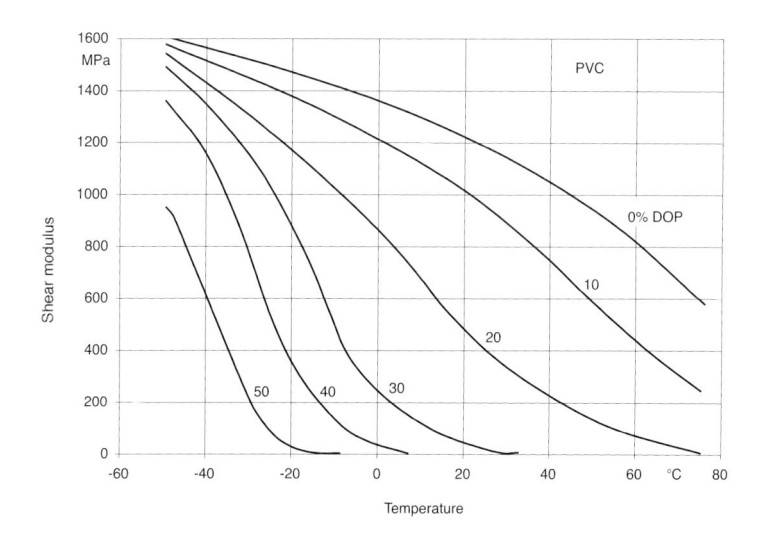

Figure 2.46: Shear modulus of PVC as a function of temperature and plasticizer (DOP) content).

2.3 PLASTICS RECYCLING

We can divide plastics recycling into two major categories: industrial and post-consumer plastic scrap recycling. Industrial scrap is rather easy to recycle and re-introduce into the manufacturing stream, either within the same company as a regrind or sold to third parties as a homogeneous, reliable and uncontaminated source of resin. Post-consumer plastic scrap recycling requires the material to go through a full life cycle prior to being reclaimed. This life cycle can be from a few days for packaging material to several years for electronic equipment housing material. The post-consumer plastic scrap can come from commercial, agricultural, and municipal waste. Municipal plastic scarp primarily consists of packaging waste, but also plastics from de-manufactured retired appliances and electronic equipment.

Post-consumer plastic scrap recycling requires collecting, handling, cleaning, sorting and grinding. Availability and collection of post-consumer plastic scrap is perhaps one of the most critical aspects. Today, the demand for recycled plastics is higher than the availability of these materials. Although the availability of HDPE from bottles has seen a slight increase, the availability of recycled PET bottles has decreased in the past two years. One of the main reasons for the decrease of PET is the fact that single serving PET bottles are primarily consumed outside

2.3 Plastics Recycling

| 1 PETE | 2 HDPE | 3 V | 4 LDPE | 5 PP | 6 PS | 7 OTHER |

Figure 2.47: SPI resin identification codes.

of the home, making recycling and collection more difficult. On the other hand, HDPE bottles, which come from milk containers, soap and cleaning bottles, are consumed in the home and are therefore thrown into the recycling bin by the consumer. A crucial issue when collecting plastic waste is identifying the type of plastic used to manufacture the product. Packaging is often identified with the standard SPI identification symbol, which contains the triangular-shaped recycling arrows and a number between 1 and 7. Often, this is accompanied by the abbreviated name of the plastic. Table 2.6 and Fig. 2.47 present the seven commonly recycled plastics in the United States along with the characteristics of each plastic, the main sources or packaging applications and the common applications for the recycled materials. Electronic housings are often identified with a molded-in name of the polymer used, such as ABS, as well as an identifier that reveals if a flame retardant was used, such as ABS-FR. When a product is not identified, various simple techniques exist, such as the water or burning tests. The water test simply consists in determining if a piece of plastic floats or sinks after having added a drop of soap to the a container filled with water. If a part floats, it is either a polyethylene, a polypropylene, or an expanded or foamed plastic. Most of the remaining polymers will likely sink. Tables 6.2- 6.5 in Chapter 6 of this handbook summarizes various tests that can be performed to quickly recognize a type of plastic material. Through simple observation, a burn test and experience, an engineer is often able to identify most plastics.

Table 2.7 presents a relation between sold and recycled plastic bottles in the United States in 2004. The numbers presented in the table had remained fairly constant until 2003, when the rate of recycled resins increased due to the high cost of virgin material. This increase continues, and itt is expected that this trend will continue in the years to come.

Table 2.6: Plastics, Characteristics, Applications and Use After Recycling

Codes	Characteristics	Packaging Applications	Recycled Products
(1) PET	Clarity, strength, toughness, barrier to gas and moisture, resistance to heat.	Plastic soft drink, water, sports drink, beer, mouthwash, catsup and salad dressing bottles; peanut butter, pickle, jelly and jam jars; heatable film and food trays.	Fiber, tote bags, clothing, film and sheet, food and beverage containers, carpet, strapping, fleece wear, luggage and bottles.
(2) PE-HD	Stiffness, strength, toughness, resistance to chemicals and moisture, permeability to gas, ease of processing, and ease of forming.	Milk, water, juice, shampoo, dish and laundry detergent bottles; yogurt and margarine tubs; cereal box liners; grocery, trash and retail bags.	Liquid laundry detergent, shampoo, conditioner and motor oil bottles; pipe, buckets, crates, flower pots, garden edging, film and sheet, recycling bins, benches, dog houses, plastic lumber, floor tiles, picnic tables, fencing.
(3) PVC	Versatility, clarity, ease of blending, strength, toughness, resistance to grease, oil and chemicals.	Clear food and non-food packaging, medical tubing, wire and cable insulation, film and sheet, construction products such as pipes, fittings, siding, floor tiles, carpet backing and window frames.	Packaging, loose-leaf binders, decking, paneling, gutters, mud flaps, film and sheet, floor tiles and mats, resilient flooring, cassette trays, electrical boxes, cables, traffic cones, garden hose, mobile home skirting.
(4) PE-LD	Ease of processing, strength, toughness, flexibility, ease of sealing, barrier to moisture.	Dry cleaning, bread and frozen food bags, squeezable bottles, e.g., honey, mustard.	Shipping envelopes, garbage can liners, floor tile, furniture, film and sheet, compost bins, paneling, trash cans, landscape timber, lumber.
(5) PP	Strength, toughness, resistance to heat, chemicals, grease and oil, versatile, barrier to moisture.	Catsup bottles, yogurt containers and margarine tubs, medicine bottles.	Automobile battery cases, signal lights, battery cables, brooms, brushes, ice scrapers, oil funnels, bicycle racks, rakes, bins, pallets, sheeting, trays.

Continued on next page

Codes	Characteristics	Packaging Applications	Recycled Products
(6) PS	Versatility, insulation, clarity, easily formed.	Compact disc jackets, food service applications, grocery store meat trays, egg cartons, aspirin bottles, cups, plates, cutlery.	Thermometers, light switch plates, thermal insulation, egg cartons, vents, desk trays, rulers, license plate frames, foam packing, foam plates, cups, utensils.
(7) Other	Dependent on resin or combination of resins.	Three and five gallon reusable water bottles, some citrus juice and catsup bottles.	Bottles, plastic lumber applications.

To achieve this, equipment that performs differential scanning calorimetry, infrared spectroscopy, Raman spectroscopy and dynamic mechanical analysis, to name a few, is available. Most process and design engineers do not have these measuring devices at hand, nor do they have the analytical experience to run them and interpret the resulting data. Once properly identified, either before or after cleaning, the plastic part is chopped down in size or ground. The ground clean plastic scrap is often directly used for processing. For some applications, where additives are needed or homogenization is required, the ground flakes are extruded and pelletized. However, this step adds to the cost of the recycled plastic.

Reprocessing of plastics has an effect on the flow and mechanical properties of the material, as the molecular weight is reduced each time the material is heated and sheared during the pelletizing and manufacturing process. The reduction in molecular weight is reflected by increases in the melt flow index, a common technique used to detect degradation. If a recycled polymer was in contact with a corrosive environment in its previous use, infrared spectroscopy is used to reveal penetration of the environment into the polymer's molecular structure.

Table 2.7: Plastic Bottles Recycling Statistics in 2004 (American Plastics Council)

Plastic bottle type	Resin sold (Million lb)	Recycled plastic	Recycling rate
PET soft drink	1,722	579.4	33.7%
PET custom	2,915	424.0	14.5%
Total PET bottles	4,637	1,003.4	21.6%
PE-HD natural	1,621	450.3	27.8%
PE-HD pigmented	1,865	453.9	24.3%
Total PE-HD bottles	3,486	904.2	25.9%
PVC	113	0.9	0.7%
PE-LD/PE-LLD	63	0.3	0.5%
PP	190	6.0	3.2%
Total	8,489	1,914.8	22.6%

CHAPTER 3

PROPERTIES AND TESTING

This chapter presents the basic properties of interest to the engineer, and the testing techniques used to measure them. Furthermore, each set of properties is presented with sets of graphs that illustrate how various conditions affect it. Additionally, the graphs compare the most important plastics to each other, and often to other materials such as steel, wood, glass, etc. The chapter is broken down into the following categories of properties

- Thermal properties
- Curing behavior
- Rheological properties
- Mechanical properties
- Permeability properties
- Friction and wear
- Environmental effects
- Electrical properties
- Optical properties
- Acoustic properties

3.1 COMPARABILITY OF MATERIAL PROPERTIES

Material properties should be considered the most important aspect of product design. Properties are used in the first stages of design, when comparing one product to another, and ultimately to evaluate the performance of an existing product. To be able to make fair comparisons between the various products and the materials used to make them, the properties used should fulfill the following requirements:

- They must be comparable to one another
- They must contain sufficient and accurate information

- They must be measured using reasonable efforts

Today's material data banks introduce approximately 200 different properties. It is clear that all that information is overwhelming and confusion to the designer and plastics engineer. To help overcome this problem, an international group of engineers and material scientists have developed a set of standards[1] upon which modern material data banks such as CAMPUS® are based on. For example, CAMPUS® is used by 50 large resin suppliers, and is distributed free through the world wide web.

This handbook adheres to the above philosophy of material data comparability. While the standards mentioned above use ISO tests, for completeness we also present the equivalent ASTM tests.

Tables 3.1, 7.6 and 3.3 summarize the most common standardized tests used to measure plastics technology relevant properties[2]. Table 3.1 presents the most fundamental or key testing techniques. Table 7.6 presents the most important testing techniques to measure design properties, and Table 3.3 presents the measuring techniques used to acquire process simulation relevant material properties.

Table 3.1: Key Properties of Plastics Reported in Datasheets and Common Test Methods (after Shastri)

Property	ISO method	ASTM method
Specific Gravity/Density	ISO 1183	D 792
		D 1505 (polyolefins)
Water absorption	ISO 62	D 570
Melt flow rate (MFR)	ISO 1133	D 1238
Mold shrinkage	ISO 294-4 (thermoplastics)	D 955
	ISO 2577 (thermosets)	
Tensile properties	ISO 527-1	2&D 638
Flexural properties	ISO 178	D 790
Notched Izod impact strength	ISO 180	D 256
Instrumented dart impact strength	ISO 6603-2&D 3763	
Deflection temperatures under load	ISO 75-1	2&D 648
Vicat softening temperature	ISO 306	D 1525
Coefficient of thermal expansion	ISO 11359-2	E 831

---

[1] The actual standards are described in the ISO 1035 (Plastics-acquisition and presentation of comparable single-point data) and ISO 11403 (Plastics-acquisition and presentation of comparable multipoint data) standards.
[2] This handbook presents the ISO and ASTM testing techniques based on tables developed by Dr. R. Shastri and published in *Modern Plastics Handbook*, Eds. Modern Plastics and C.A. Harper, McGraw-Hill, (2000).

3.1 Comparability of Material Properties

Table 3.2: Material Property Testing for Structural Analysis (after Shastri)

Property	ISO method	ASTM method	Suggested Conditions
Properties in tension	527-1,2,& 4	D 638	at 23°C, at least three elevated temperatures, and one temperature below standard laboratory conditions at standard strain rate; at 3 additional strain rates at 23°C.
Poisson's ratio	527-1, 2	D 638	at 23°C, at least one elevated temperature, and one temperature below standard laboratory conditions.
Properties in compression	604	D 695	at 23°C, two additional elevated temperatures, and one temperature below standard laboratory conditions.
Shear modulus	6721-2, 5	D 5279-150	°C to Tg+20 °C or Tm+10 °C @ 1 Hz.
Creep in tension	899-1	D 2990	at 23°C and at least two elevated temperatures for 1000 hrs at 3 stress levels.
Fatigue in tension			a. S - N curves at 3 Hz at 23°C; 80%, 70%, 60%, 55%, 50% and 40% of tensile stress at yield; R =0.5; 1 million cycles run out. b. a - N curves at 3 Hz at 23°C; Single Edge Notched specimens; 3 stress levels; R = 0.5.
Coefficient of friction	8295	D 3028	at 23°C against itself and steel
Application specific			
Creep in bending	899-2	D 2990	at 23°C and at least two elevated temperatures for 1000 hrs and at least 3 stress levels.
Creep in compression		D 2990	at 23°C and at least two elevated temperatures for 1000 hrs and at east 3 stress levels.
Fatigue in bending			at 23°C; fully reversed; 80%, 70%, 60%, 55%, 50% and 40% of tensile stress at yield @ 3 Hz.
Fracture toughness	13586-1	D 5045	

Table 3.3: Material Property Characterizations for Processing Simulations (after Shastri)

Property	ISO method	ASTM method	Suggested Conditions
Melt viscosity-Shear rate data	11443	D 3835	at three temperatures, over shear rate range 10-10000 s^{-1}.
Reactive viscosity of thermosets	6721-10		a slit die rheometer according to ISO 11443 can also be used.
Uniaxial extensional viscosity			
Biaxial extensional viscosity			
First normal stress difference	6721-10		
Melt density		D 3835	at 0 MPa and processing temperature.
Bulk density	61	D 1895	
Density - reacted system	1183	D 792	
pvT data			at a cooling rate of 2.5 °C/min at 40, 80, 120, 160 and 200 MPa with an estimation at 1 MPa
Thermal conductivity*		D 5930	23°C to processing temperature.
Specific heat	11357-4	D 3418	DSC cooling scan @ 10°C/min from processing temperature to 23°C.
No flow temperature*			DSC cooling scan @ 10°C/min from processing temperature to 23°C (11357-3).
Ejection temperature*			DSC cooling scan @ 10°C/min from processing temperature to 23°C (11357-3).
Glass transition temperature	11357-2	D 3418	DSC cooling scan @ 10°C/min from processing temperature to 23°C.
Crystallization temperature	11357-3	D 3418	DSC cooling scan @ 10°C/min from processing temperature to 23°C.
Degree of crystallinity	11357-3	D 3418	DSC cooling scan @ 10°C/min from processing temperature to 23°C.
Enthalpy of fusion	11357-3	D 3417	DSC heating scan @ 10°C/min from 23°C to processing temperature.
Enthalpy of crystallization	11357-3	D 3417	Cooling scan @ 10°C/min, 50°C/min, 100°C/min and 200°C/min from processing temperature to 23°C.

Continued on next page

3.2 Thermal Properties

Property	ISO method	ASTM method	Suggested Conditions
Crystallization kinetics	11357-7	D 3417	Isothermal scans at different cooling rates at 3 temperatures in the crystallization range.
Heat of Reaction of thermosets	11357-5	D 4473	Heating scan @ 10°C/min from 23°C to reaction temperature.
Reaction kinetics of thermosets	11357-5	D 4473	Isothermal DSC runs at 3 temperatures in the reaction temperature range.
Gelation conversion Isothermal induction time	11357-5		Heating scan @ 10°C/min.
Coefficient of Linear Thermal Expansion	11359-2	E 831	with specimens cut from ISO 294-3 plate over the range -40°C to 100°C.
Mold shrinkage - thermoplastics	294-4	D 955	at 1 mm, 1.5 mm and 2 mm thickness with cavity pressure of 25, 50, 75 and 100 MPa.
Mold shrinkage - thermosets	2577		
In-plane shear modulus	6721-2 or 7		

3.2 THERMAL PROPERTIES

The heat flow through a material is controlled by heat conduction, determined by the thermal conductivity, k. However, at the onset of heating, the polymer responds solely as a heat sink, and the amount of energy per unit volume stored in the material before reaching steady state conditions is controlled by the density, ρ, and the specific heat, C_p, of the material. These three thermal properties can be combined to form the thermal diffussivity, α, using

$$\alpha = \frac{k}{\rho C_p} \quad (3.1)$$

3.2.1 Thermal Conductivity

When analyzing thermal processes, the thermal conductivity, k, is the most commonly used property that helps quantify the transport of heat through a material. By definition, energy is transported proportionally to the speed of sound. Accordingly, thermal conductivity follows the relation

$$k \approx C_p \rho u l \quad (3.2)$$

where u is the speed of sound and l the molecular separation. Amorphous polymers show an increase in thermal conductivity with increasing temperature, up to the glass transition temperature, T_g. Above T_g, the thermal conductivity decreases with increasing temperature. Figure 3.1 presents the thermal conductivity, below the glass transition temperature, for various amorphous thermoplastics as a function of temperature.

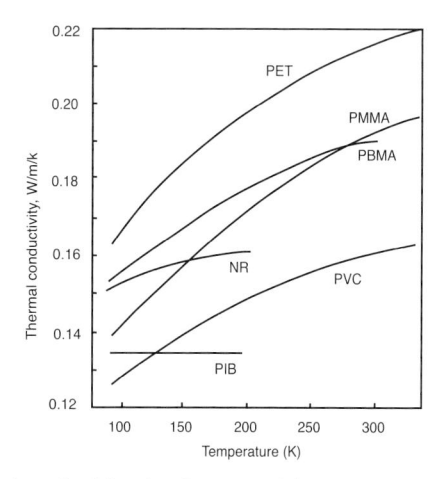

Figure 3.1: Thermal conductivity of various materials.

Due to the increase in density upon solidification of semi-crystalline thermoplastics, the thermal conductivity is higher in the solid state than in the molten. In the molten state, however, the thermal conductivity of semi-crystalline polymers reduces to that of amorphous polymers as can be seen in Fig. 3.2.

Furthermore, it is not surprising that the thermal conductivity of melts increases with hydrostatic pressure. This effect is clearly shown in Fig. 3.3. Figures 5.23 and 5.26 present the combined effect that pressure and temperature have on the thermal conductivity of PS and PE, respectively.

As long as thermosets are unfilled, their thermal conductivity is very similar to amorphous thermoplastics. Anisotropy in thermoplastic polymers also plays a significant role in the thermal conductivity. Highly drawn semi-crystalline polymer samples can have a much higher thermal conductivity as a result of the orientation of the polymer chains in the direction of the draw. For amorphous polymers, the increase in thermal conductivity in the direction of the draw is usually not higher than two. Figure 3.6 presents the thermal conductivity in the directions parallel and perpendicular to the draw for high density polyethylene, polypropylene, and polymethyl methacrylate. Figures 5.22 and 5.21 present the combined effect that draw and temperature have on thermal conductivity of PMMA and PVC-U, respectively. The higher thermal conductivity of inorganic fillers increases the thermal conductivity of filled polymers. Nevertheless, a sharp

3.2 Thermal Properties

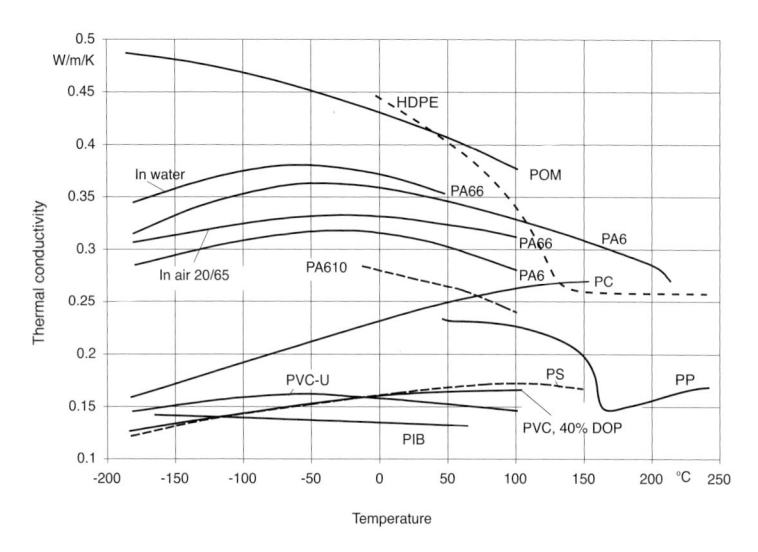

Figure 3.2: Thermal conductivity of various thermoplastics.

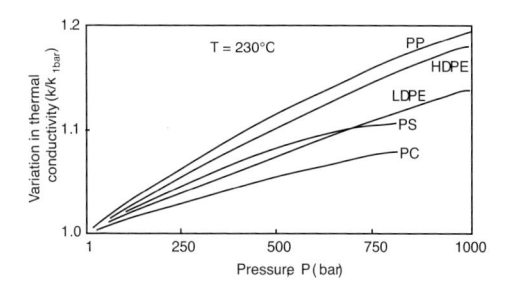

Figure 3.3: Influence of pressure on thermal conductivity of various thermoplastics.

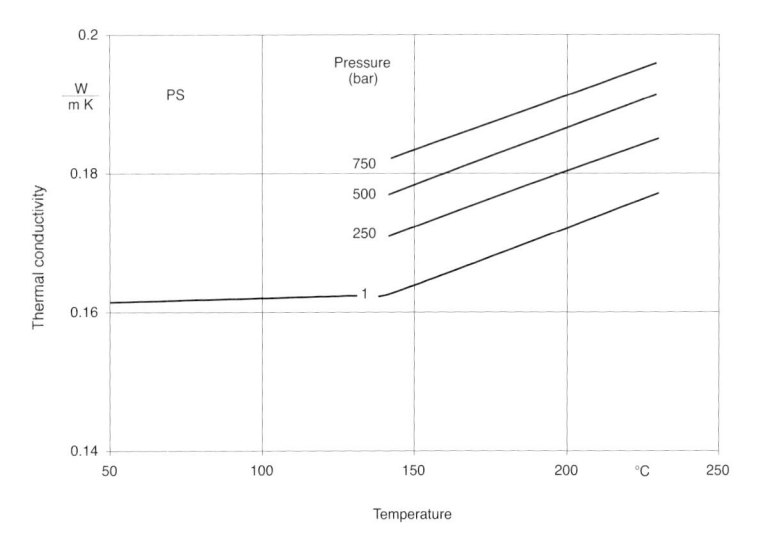

Figure 3.4: Influence of pressure and temperature on the thermal conductivity of PS.

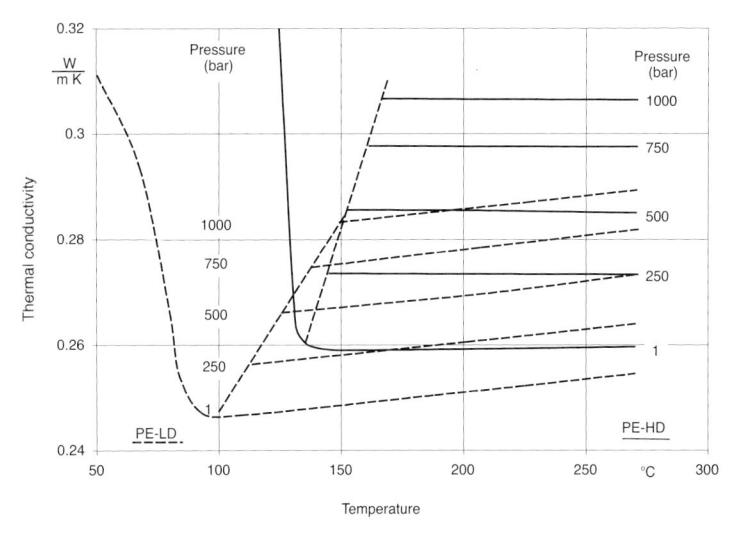

Figure 3.5: Influence of pressure and temperature on the thermal conductivity of PE-LD and PE-HD.

3.2 Thermal Properties

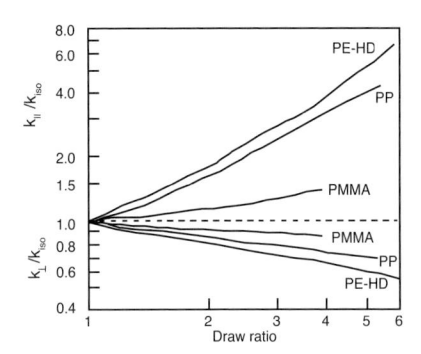

Figure 3.6: Thermal conductivity as a function of draw ratio in the directions perpendicular and parallel to the stretch for various oriented thermo-plastics.

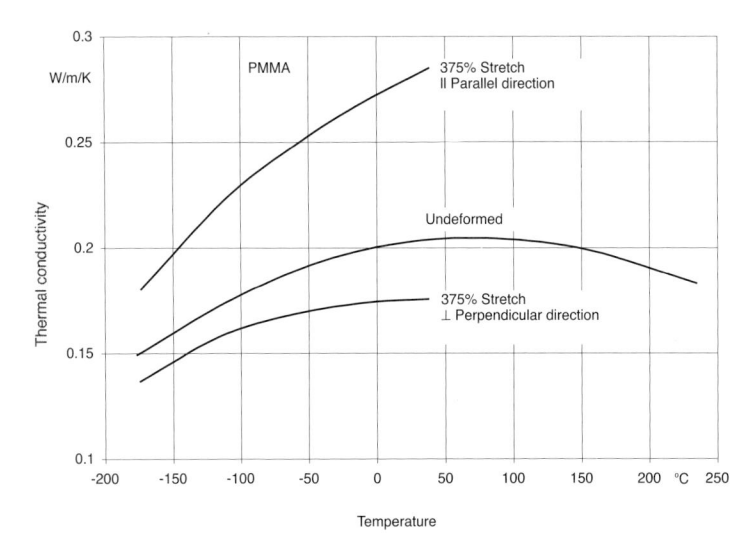

Figure 3.7: Influence of draw and temperature on the thermal conductivity of PMMA.

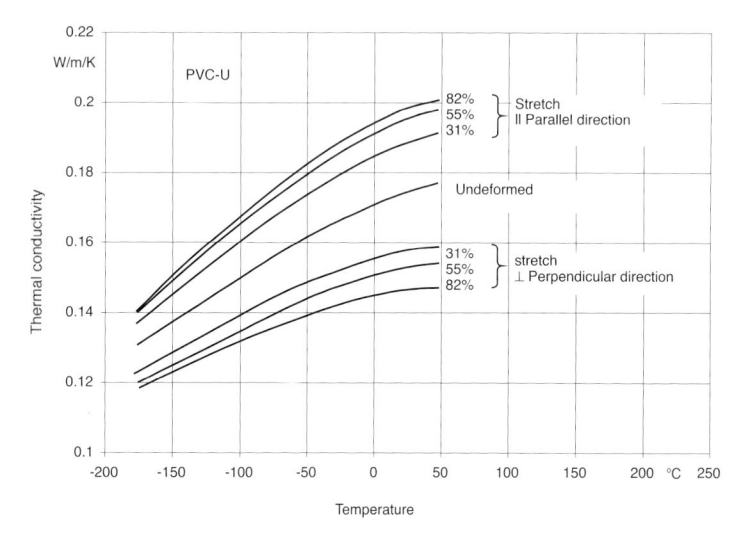

Figure 3.8: Influence of draw and temperature on the thermal conductivity of PVC-U.

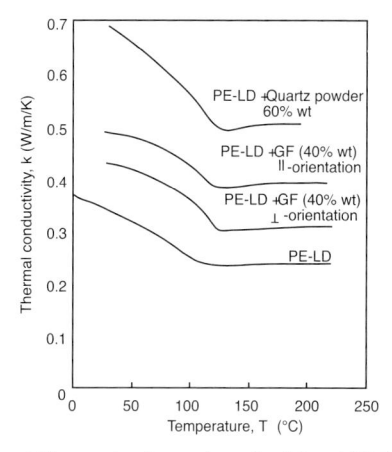

Figure 3.9: Influence of filler on the thermal conductivity of PE-LD.

decrease in thermal conductivity around the melting temperature of crystalline polymers can still be seen with filled materials. The effect of fillers on thermal conductivity for PE-LD is shown in Fig. 3.9 . Where the effect of fiber orientation as well as the effect of quartz powder on the thermal conductivity of low density polyethylene are shown.

3.2 Thermal Properties

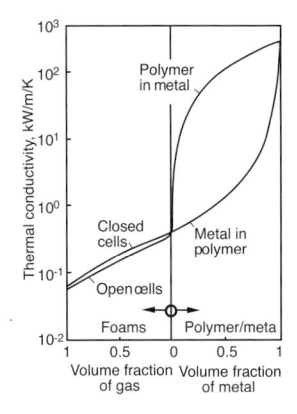

Figure 3.10: Thermal conductivity of plastics filled with glass or metal.

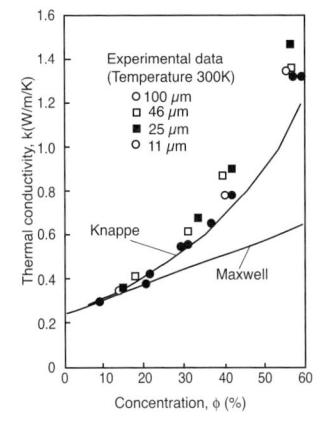

Figure 3.11: Thermal conductivity versus volume concentration of metallic particles of an epoxy resin. Solid lines represent predictions using Maxwell and Knappe models.

Figure 3.10 demonstrates the influence of gas content on expanded or foamed polymers, and the influence of mineral content on filled polymers.

Figure 3.11 compares theory with experimental data for an epoxy filled with copper particles of various diameters. The figure also compares the data to the classic model given by Maxwell.

With fiber reinforced systems, one must differentiate between the direction longitudinal to the fibers and that transverse to them. For high fiber content, one can approximate the thermal conductivity of the composite by the thermal conductivity of the fiber. Similar to fiber orientation of reinforced polymers, the molecular orientation plays a significant role in the thermal properties of a

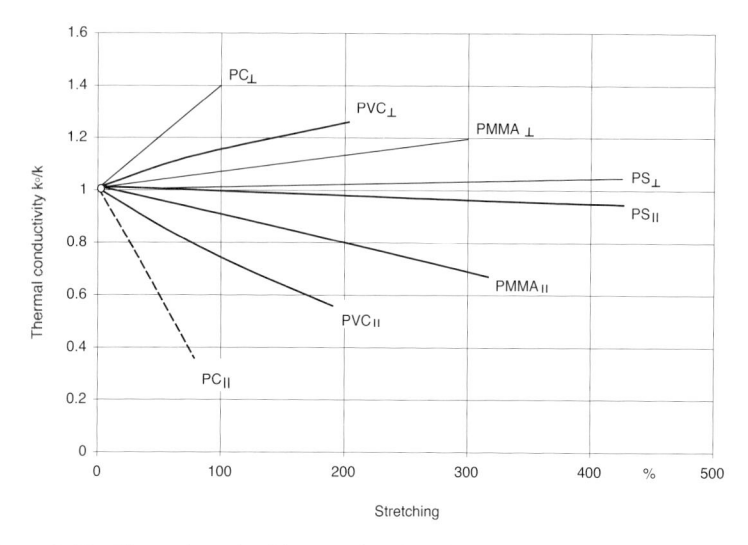

Figure 3.12: Thermal conductivity as a function of % stretch, parallel and perpendicular to the direction of stretch.

polymer. Figure 3.12 presents the effect of stretch on the thermal conductivity of various amorphous polymers.

The thermal conductivity can be measured using the standard tests ASTM C177 and DIN 52612. A new method currently being balloted (ASTM D20.30) is preferred by most people today.

3.2.2 Specific Heat and Specific Enthalpy

The *specific heat*, C, represents the energy required to change the temperature of a unit mass of material by one degree. The *specific enthalpy* represents the energy of a material per unit mass at a specific temperature. The specific heat can be measured at either constant pressure, C_p, or constant volume, C_v. Since the specific heat at constant pressure includes the effect of volumetric change, it is larger than the specific heat at constant volume. However, the volume changes of a polymer with changing temperatures have a negligible effect on the specific heat. Hence, one can usually assume that the specific heat at constant volume or constant pressure are the same. It is usually true that specific heat only changes modestly in the range of practical processing and design temperatures of polymers. However, semi-crystalline thermoplastics display a discontinuity in the specific heat at the melting point of the crystallites. This jump or discontinuity in specific heat includes the heat that is required to melt the crystallites which is

3.2 Thermal Properties

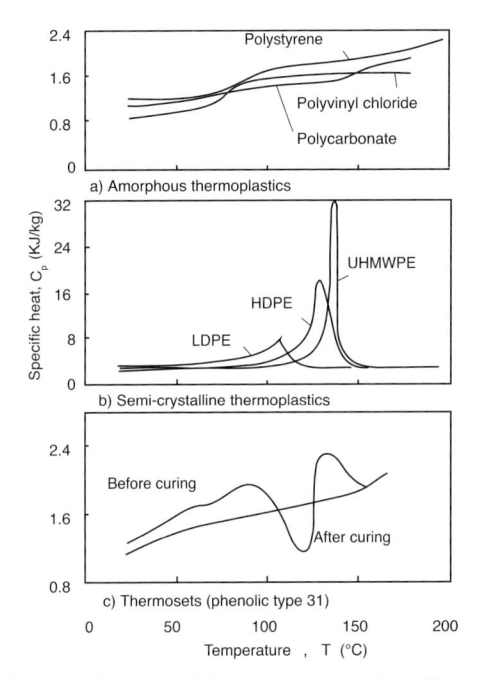

Figure 3.13: Specific Heat Curves for Selected Polymers of the Three General Polymer Categories.

usually called the heat of fusion. Hence, specific heat is dependent on the degree of crystallinity. Values of heat of fusion for typical semi-crystalline polymers are shown in Table 3.4.

Table 3.4: Heat of Fusion of Various Thermoplastic Polymers

Polymer	λ (kJ/kg)	$T_m(°C)$
Polyamide 6	193-208	223
Polyamide 66	205	265
Polyethylene	268-300	141
Polypropylene	209-259	183
Polyvinyl chloride	181	285

The chemical reaction that takes place during solidification of thermosets also leads to considerable thermal effects. In a hardened state, their thermal data are similar to the ones of amorphous thermoplastics. Figure 3.13 shows specific heat graphs for the three polymer categories. The graphs that pertain to polyethylene

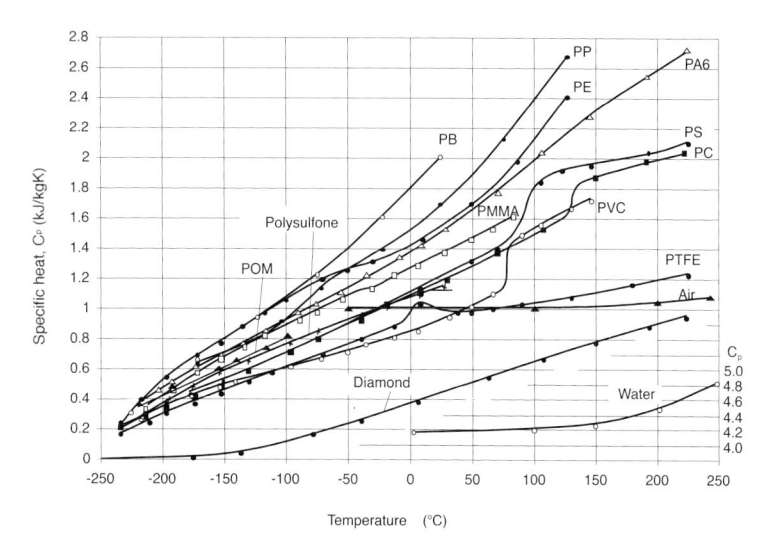

Figure 3.14: Specific heat curves for selected polymers.

have the heat of of fusion lumped into the specific property curve. Similarly, the graph for phenolic presents the effect of cure.

Curing effects of thermosetting plastics are discussed in more detail later in this chapter. Figure 3.14 presents the specific heat of several thermoplasticsas a function of temperature and compares them to the curves for air, diamond, and water.

Specific enthalpy curves for selected amorphous thermoplastics are presented in Fig. 3.15. Figure 3.16 presents the specific enthalpy for various polyamides, while Fig. 3.17 shows the curves for other semi-crystalline polymers. The specific heat is also affected by the filler and reinforcements. Figure 3.18 shows a specific heat curve of an unfilled polycarbonate and its corresponding computed specific heat curves for 10%, 20%, and 30% glass fiber content. In most cases, temperature dependence of C_p on inorganic fillers is minimal and need not be taken into consideration.

3.2.3 Density

The density or its reciprocal, the specific volume, is a commonly used property for polymeric materials. As with other properties, the specific volume is greatly affected by the temperature. Figure 5.27 shows the specific volume at atmospheric pressure for several plastics as a function temperature. However,

3.2 Thermal Properties

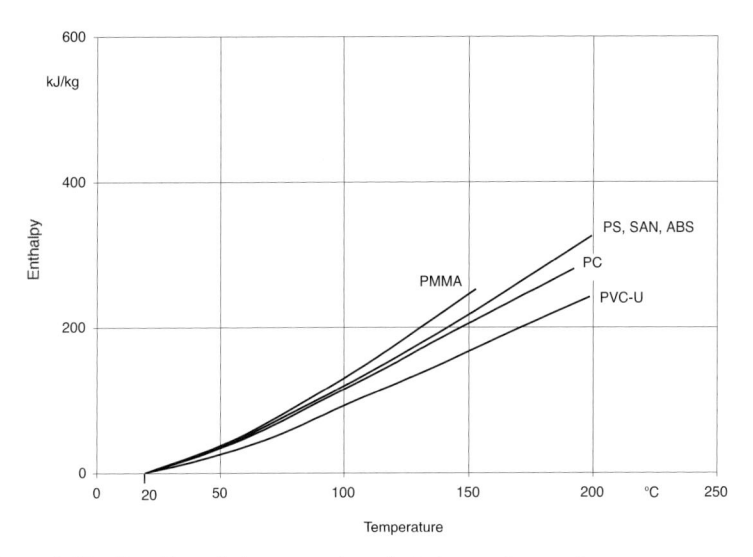

Figure 3.15: Specific enthalpy curves for selected amorphous polymers.

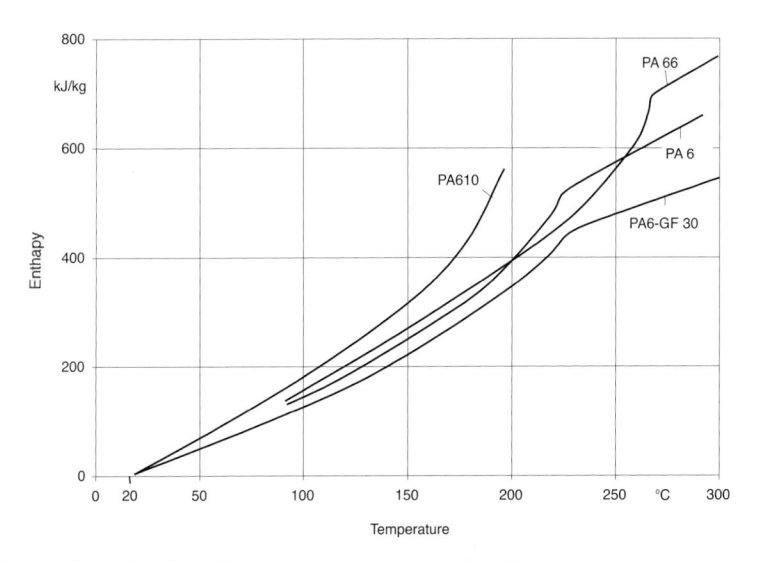

Figure 3.16: Specific enthalpy curves for various polyamides.

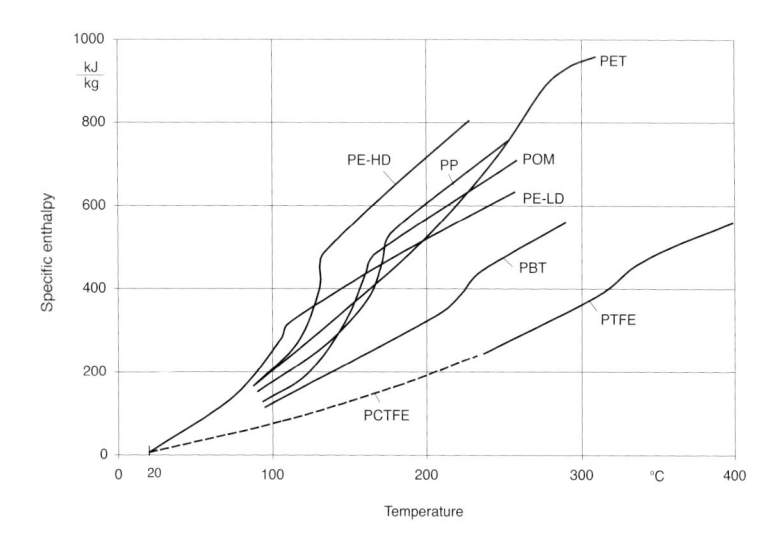

Figure 3.17: Specific enthalpy curves for selected semi-crystalline polymers.

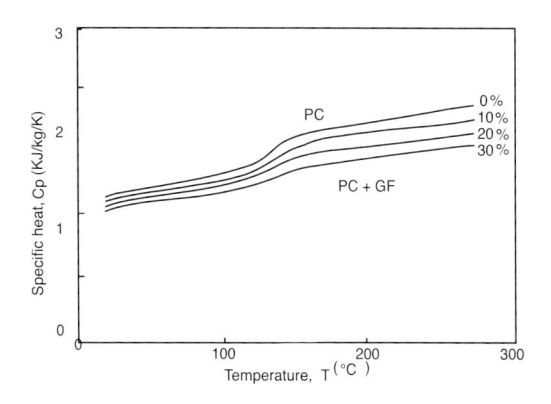

Figure 3.18: Generated specific heat curves for a filled and unfilled polycarbonate. (Courtesy Bayer AG.)

3.2 Thermal Properties

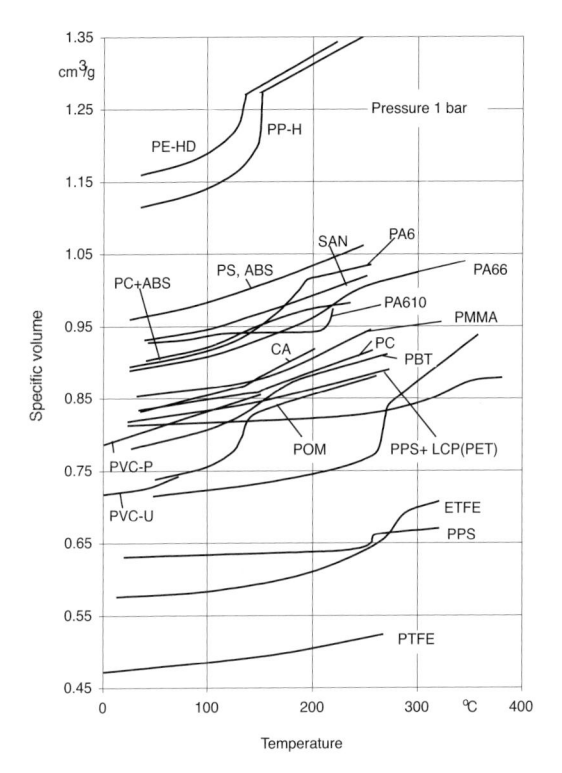

Figure 3.19: Specific volume as a function of temperature at 1 bar for selected thermoplastic polymers.

the specific volume is also affected by the processing pressure and is therefore often plotted as a function of pressure, as well as temperature, in what is known as a pvT diagram. A typical pvT diagram for an unfilled and filled amorphous polymer is shown, using polycarbonate as an example, in Figs. 3.20 and 3.21. The two slopes in the curves represent the specific volume of the melt and of the glassy amorphous polycarbonate, separated by the glass transition temperature. Figure 3.22 presents the pvT diagram for polyamide 66 as an example of a typical semi-crystalline polymer. Figure 3.23 shows the pvT diagram for polyamide 66 filled with 30% glass fiber. The curves clearly show the melting temperature (i.e., $T_m \approx 250°C$ for the unfilled PA66 cooled at 1 bar, which marks the beginning of crystallization as the material cools). It should also come as no surprise that the glass transition temperatures are the same for the filled and unfilled materials.

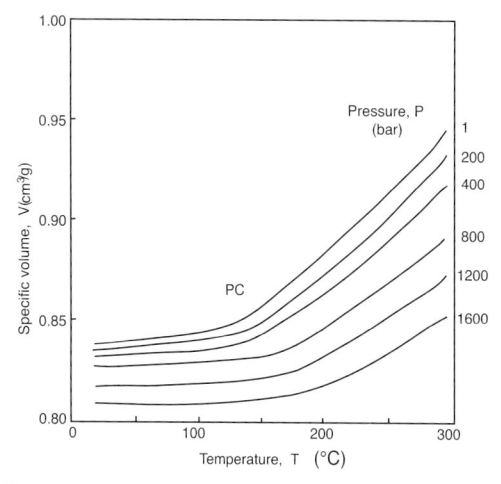

Figure 3.20: pvT diagram for a polycarbonate. (Courtesy Bayer AG.)

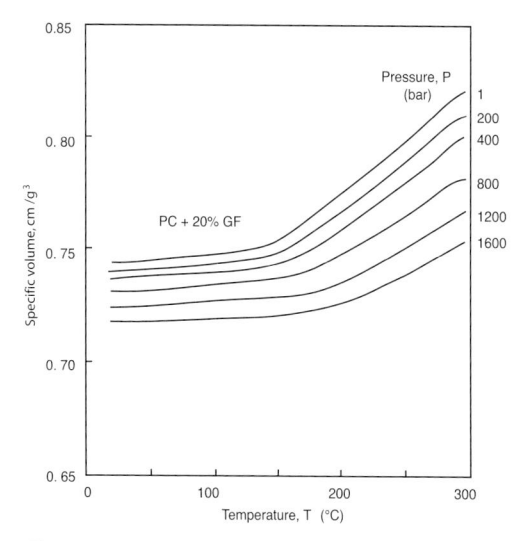

Figure 3.21: pvT diagram for a polycarbonate filled with 20% glass fiber. (Courtesy Bayer AG.)

3.2 Thermal Properties

The density of polymers filled with inorganic materials can be computed at any temperature using the rule of mixtures. Density measurements can be done using standard tests ISO 1183 and ASTM D792, presented in Table 3.5.

Table 3.5: Standard Methods of Measuring Density of Polymers (after Shastri)

Standard	ISO 1183 : 87 Methods A, B, C, or D	D 792 - 98 Method A or B, or D1505 - 88 (Reapproved in 1990)
Specimen geometry	The specimen shall be of convenient size to give adequate clearance between the specimen and the beaker (a mass of 1-5 g is often convenient). This specimen shall be taken from the center portion of the ISO 3167 multipurpose test specimen.	Single piece of material of any size or shape that can be conveniently prepared, provided that its volume shall be not less than 1 cm^3 and its surface and edges made smooth, with a thickness of at least 1 mm for each 1 g weight (D792 requirements). The test specimens should have dimensions that permit the most accurate position measurement of the center of volume of the suspended specimen. If it is suspected that interfacial tension affects the equilibrium position of the specimens in the thickness range from 0.025 - 0.051 mm, then films not less than 0.127 mm thick should be tested (D 1505 requirements)
Conditioning	Specimen conditioning, including any post molding treatment, shall be carried out at 23°C ± 2°C and 50 ± 5 % R.H. for a minimum length of time of 88h, except where special conditioning is required as specified by the appropriate material standard.	At 23 ± 2 °C and 50 ± 5 % relative humidity for not less than 40h according to D 618 - 95.
Test procedures	Any one of the 4 methods - immersion in liquid, pyknometer, titration, or density gradient column. Three specimens are required for Method D (density gradient column)	D 792 - displacement in water (Method A) or other liquids (Method B) D1505 - density gradient column. Several specimens are required for D792. Three specimens are required for D1505.
Values and units	Density $\Rightarrow$ kg/m^3	Density $\Rightarrow$ kg/m^3

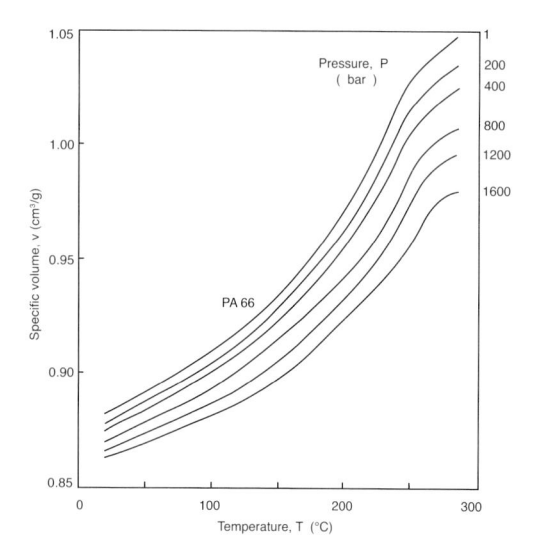

Figure 3.22: pvT diagram for a polyamide 66. (Courtesy Bayer AG.)

3.2.4 Thermal Diffusivity

Thermal diffusivity, defined in eqn. (3.1), is the material property that governs the process of thermal diffusion over time. The thermal diffusivity in amorphous thermoplastics decreases with temperature. A small jump is observed around the glass transition temperature due to the decrease in heat capacity at T_g. Figure 3.24 presents the thermal diffusivity for selected amorphous thermoplastics.

A decrease in thermal diffusivity, with increasing temperature, is also observed in semi-crystalline thermoplastics. These materials show a minimum in the thermal diffusivity curve at the melting temperature, as demonstrated in Fig. 3.25 for a selected number of semi-crystalline thermoplastics. It has also been observed that the thermal diffusivity increases with increasing degree of crystallinity and that it depends on the rate of crystalline growth, hence, on the cooling speed.

3.2.5 Linear Coefficient of Thermal Expansion

The linear coefficient of thermal expansion is related to volume changes that occur in a polymer due to temperature variations and is well represented in the pvT diagram. For many materials, thermal expansion is related to the melting

3.2 Thermal Properties

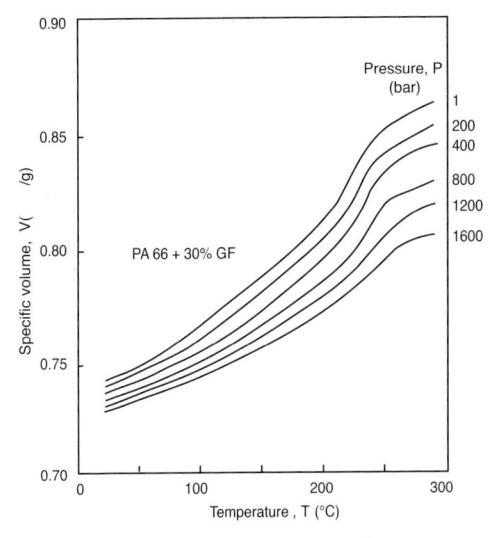

Figure 3.23: pvT diagram for a polyamide 66 filled with 30% glass fiber. (Courtesy Bayer AG.)

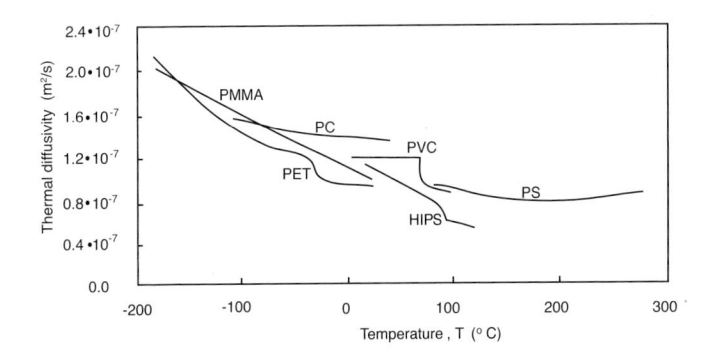

Figure 3.24: Thermal diffusivity as a function of temperature for various amorphous thermoplastics.

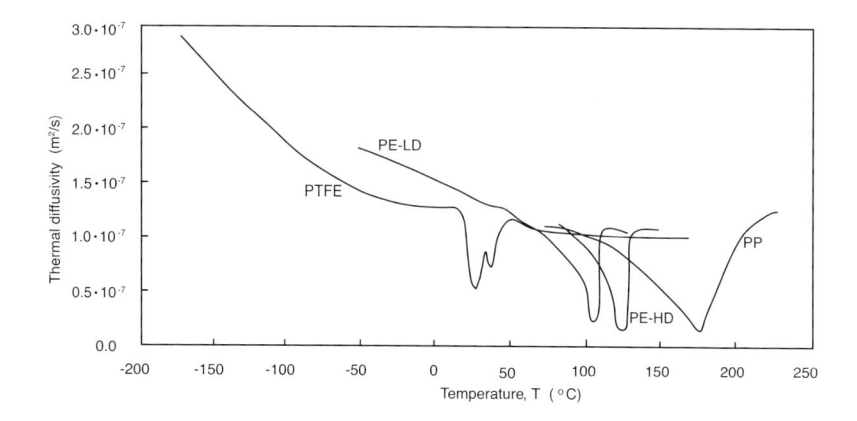

Figure 3.25: Thermal diffusivity as a function of temperature for various semi-crystalline thermoplastics.

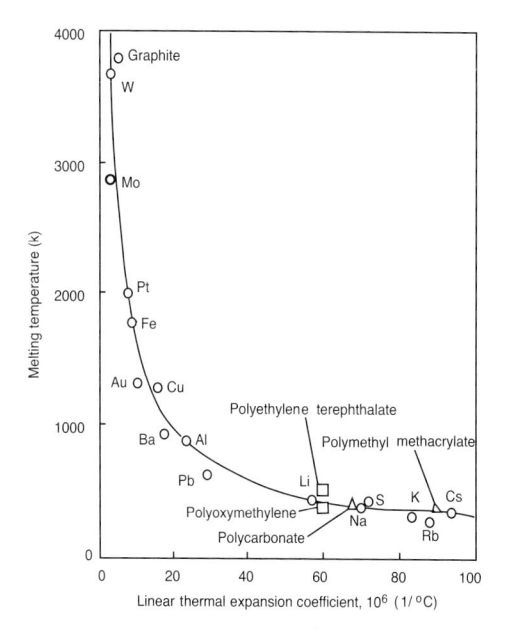

Figure 3.26: Relation between thermal expansion of some metals and plastics at 20°C and their melting temperature.

3.2 Thermal Properties

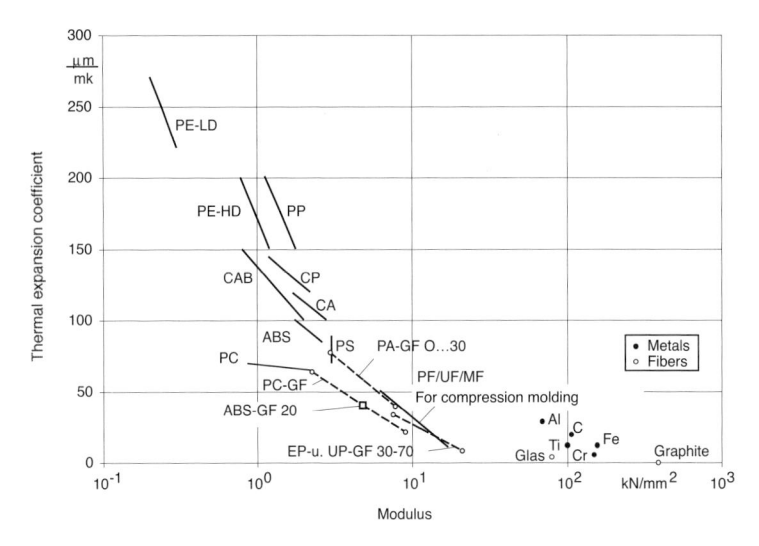

Figure 3.27: Relation between thermal expansion of some metals and plastics and their elastic modulus.

temperature of that material, shown for some important polymers in Fig. 3.26. Similarly, there is also a relation between the thermal expansion coefficient of polymers and their elastic modulus, as depicted in Fig 3.27.

Although the linear coefficient of thermal expansion varies with temperature, it is often considered constant within typical design and processing conditions. It is especially high for polyolefins, where it ranges from $1.5 \times 10^{-4} K^{-1}$ to $2 \times 10^{-4} K^{-1}$; however, fibers and other fillers significantly reduce thermal expansion. The linear coefficient of thermal expansion is a function of temperature. Figure 3.28 presents the thermal expansion coefficient for a selected number of thermoplastic polymers. The figure also presents thermal expansion perpendicular and parallel to the fiber orientation for a fiber filled POM. The percent thermal expansion of several fiber filled plastics is presented in Figs. 3.29 and 3.30. The figures present the effect of orientation and fiber arrangement, and are all plotted as a function of temperature.

A rule of mixtures is sufficient to calculate the thermal expansion coefficient of polymers that are filled with powdery or small particles as well as with short fibers. In case of continuous fiber reinforcement, the rule of mixtures applies for the coefficient perpendicular to the reinforcing fibers. In the fiber direction, however, the thermal expansion of the fibers determines the linear coefficient of thermal expansion of the composite. Extensive calculations are necessary to

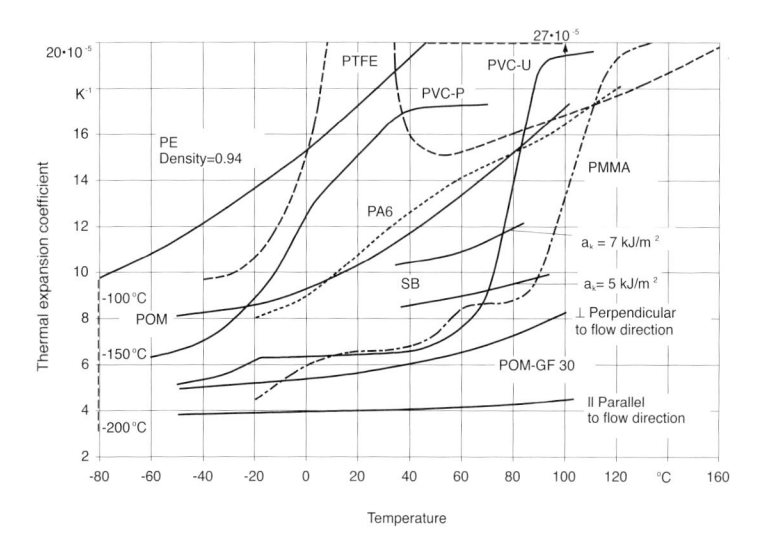

Figure 3.28: Coefficient of thermal expansion of various plastics as a function of temperature.

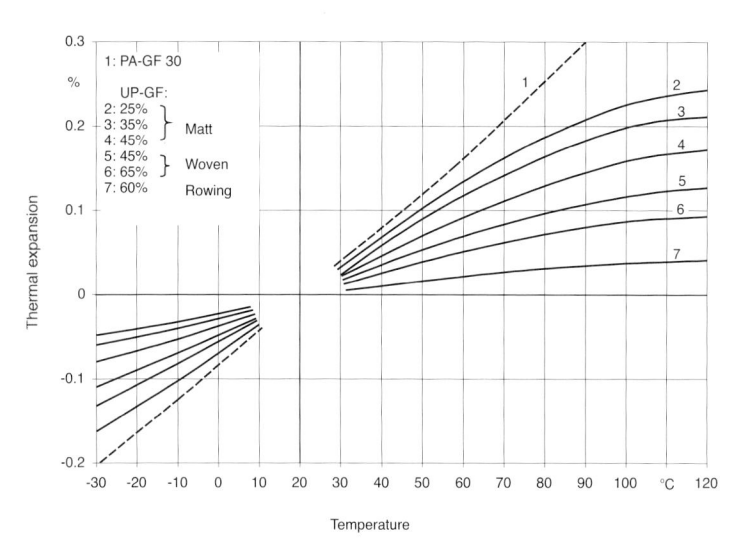

Figure 3.29: Thermal expansion of various fiber reinforced plastics as a function of temperature.

3.2 Thermal Properties

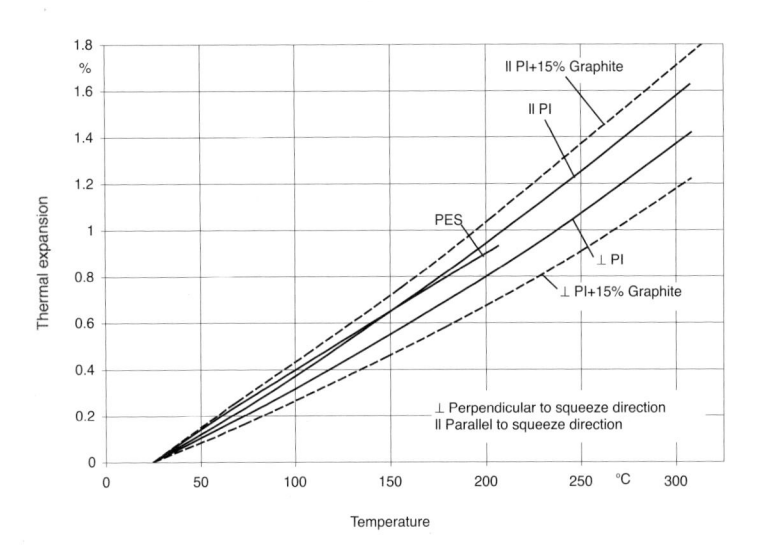

Figure 3.30: Thermal expansion of various fiber reinforced plastics as a function of temperature.

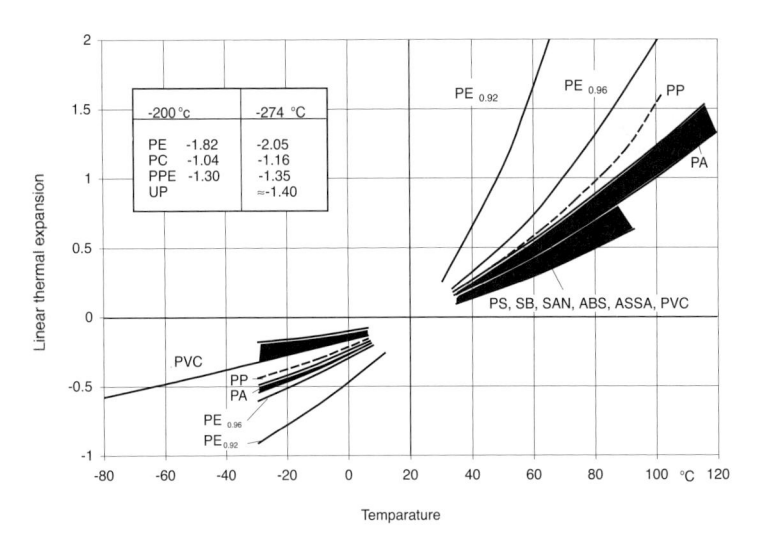

Figure 3.31: Length change as a function of temperature for various thermoplastics. Index for PE denotes specific gravity.

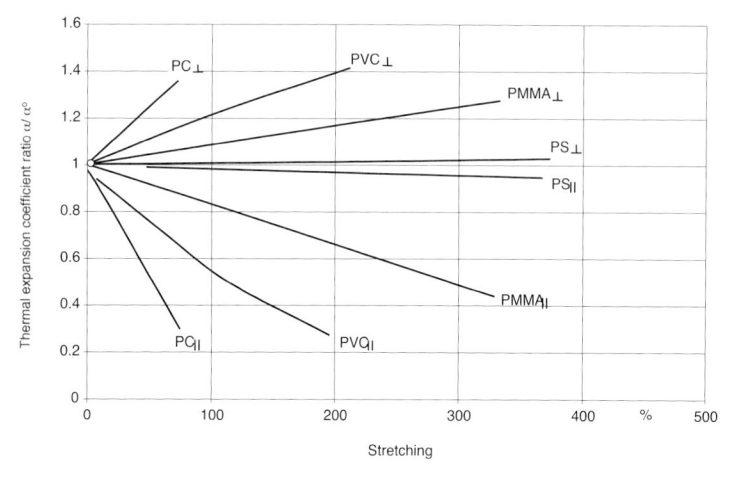

Figure 3.32: Thermal expansion coefficient for various thermoplastics, parallel and perpendicular to direction of molecular orientation, as a function of % stretch.

determine coefficients in layered laminated composites and in fiber reinforced polymers with varying fiber orientation distribution.

The temperature variation of the thermal expansion coefficients can be seen by the shape of the curves in Fig. 3.31 which present % length change as a function of temperature for several thermoplastics.

Molecular orientation also affects the thermal expansion of plastics. Figure 3.32 presents the linear expansion coefficient as a function of % stretch for various polymers in the direction parallel and perpendicular to the orientation.

The thermal expansion is often affected by the cooling time during processing. This is especially true with semi-crystalline polymers whose crystallization process requires time. For example, a thin part that cools fast will have a lower degree of crystallinity and will therefore shrink less. This is illustrated in Fig. 3.33 where presents the range of wall shrinkage as a function of wall thickness for two semi-crystalline polymers (PA6 and PBT) and two amorphous polymers (PC and ABS)is presened. The figures show how the semi-crystalline plastics are more affected by the increase in wall thickness. The coefficient of linear thermal expansion is measured using the standard tests ISO 11359 and ASTM E831, which are both presented in Table 3.6.

3.2 Thermal Properties

Table 3.6: Standard Methods of Measuring Coefficient of Linear Thermal Expansion (CLTE) (after Shastri)

Standard	ISO 11359 - 2	E 831 - 93
Specimen	Prepared from ISO 3167 multi-purpose test specimen cut from the specimen taken from the middle parallel region.	Specimen shall be between 2 and 10 mm in length and have flat and parallel ends to with ± 25 micrometers. Lateral dimensions shall not exceed 10 mm.
Conditioning	No conditioning requirements given. If the specimens are heated or mechanically treated before testing, it should be noted in the report.	No conditioning requirements given. If the specimens are heated or mechanically treated before testing, it should be noted in the report.
Apparatus	TMA	TMA
Test procedures	Three specimens are required. Measure the initial specimen length in the direction of the expansion test to ± 25 micrometers at room temperature. Place the specimen in the specimen holder in the furnace. If measurements at sub-ambient temperatures are to be made, then cool the specimen to at least 20°C below the lowest temperature of interest. Heat the specimen at a constant heating rate of 5°C /min over the desired temperature range and record changes in specimen length and temperature to all available decimal places. Determine the measurement instrument baseline by repeating the two steps above without a specimen present. The measured change in expansion length of the specimen should be corrected for the instrument baseline. Record the secant value of the expansion vs. temperature over the temperature range of 23°C to 55°C.	Three specimens are required. Measure the initial specimen length in the direction of the expansion test to ± 25 micrometers at room temperature. Place the specimen in the specimen holder in the furnace. If measurements at sub-ambient temperatures are to be made, then cool the specimen to at least 20°C below the lowest temperature of interest. Heat the specimen at a constant heating rate of 5°C /min over the desired temperature range and record changes in specimen length and temperature to all available decimal places. Determine the measurement instrument baseline by repeating the two steps above without a specimen present. The measured change in expansion length of the specimen should be corrected for the instrument baseline. Select a temperature range from a smooth portion of the thermal curves in the desired temperature range, then obtain the change in expansion length over that temperature range.
Values and units	Coefficient of linear thermal expansion micrometers/(m.°C)	Coefficient of linear thermal expansion micrometers/(m.°C)

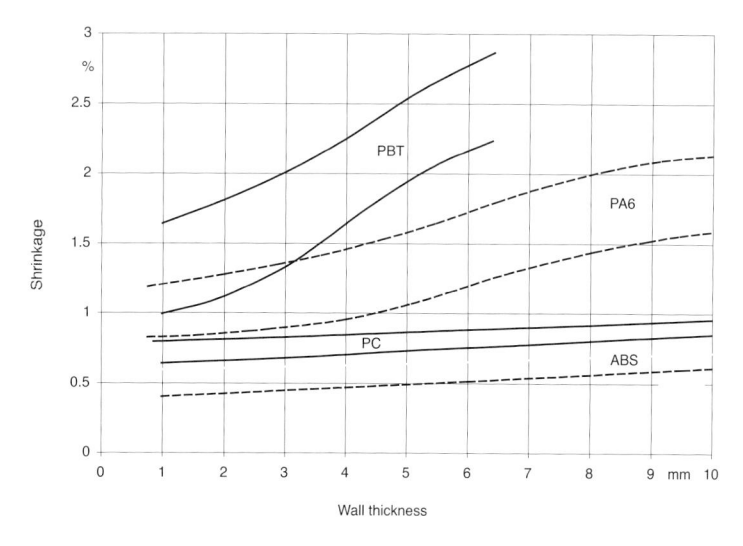

Figure 3.33: Wall shrinkage as a function of wall thickness for a selected number of thermoplastics.

Thermal expansion data is often used to predict shrinkage in injection molded parts. The injection molding shrinkage data is measured using ASTM D 955 and ISO 294 - 4 tests, presented in Table 3.7.

Table 3.7: Standard Methods of Measuring Injection Molded Shrinkage (after Shastri)

Standard	ISO 294 - 4	D 955 - 89 (Reapproved 1996)
Conditioning	At 23 ± 2°C between 16h to 24h. Materials which show marked difference in mold shrinkage if stored in a humid or dry atmosphere, must be stored in dry atmosphere.	At 23 ± 2°C and 50 ± 5 % relative humidity for 1-2 hours for "initial molding shrinkage" (optional), 16-24 hours for "24-h shrinkage" (optional), and 40-48 hours for "48-h or normal shrinkage"
Test procedures	Mold at least five specimens, using a 2-cavity ISO 294-3 Type D2 mold, equipped with cavity pressure sensor. Molding equipment complies with the relevant 4.2 clauses in ISO 294-1 and ISO 294-3. In addition, accuracy of the cavity pressure sensor must be ± 5%.	Molding in accordance with the Practice D 3641 such that the molding equipment is operated without exceeding 50-75% of its rated Molding in accordance with the Practice D 3641 such that the molding equipment is operated without exceeding 50-75% of its rated shot capacity.

Continued on next page

3.2 Thermal Properties

Standard	ISO 294 - 4	D 955 - 89 (Reapproved 1996)
	The machine is operated such that the ratio of the molding volume to the screw-stroke volume is between 20-80%, when using the injection molding conditions specified in Part 2 of the relevant material standard.	
	Perform mold shrinkage measurements on specimens which have been molded such that one or more of the preferred "cavity pressure at pressure at hold pch" of 20, 40, 60, 80 and/or 100 MPa is achieved.	No cavity pressure requirements are given.
	Allow molded specimens to cool to room temperature by placing them on a material of low thermoconductivity with an appropriate load to prevent warping. Any specimen that has warpage > 3% of its length is discarded.	Allow molded specimens to cool at 23 ± 2°C and 50 ± 5% relative humidity. No warpage limits are specified.
	Measure the length and width of the cavity and the corresponding molded specimens to within 0.02 mm at 23 ± 2°C.	Measure the length or diameter (both parallel and normal to the flow) of the cavity and the corresponding molded specimens to within 0.02 mm. Temperature requirement of the mold while measuring the cavity dimensions is not specified.
Values and units	Molding shrinkage (16-24h): % reported as mean value of the five specimens measured	Initial molding shrinkage: mm/mm (optional) 24h shrinkage: mm/mm (optional) 48h or normal shrinkage: mm/mm reported as mean value of the five specimens measured

3.2.6 Thermal Penetration

In addition to thermal diffusivity, the thermal penetration number is of considerable practical interest. It is given by

$$b = \sqrt{kC_p\rho} \tag{3.3}$$

If the thermal penetration number is known, the contact temperature T_C which results when two bodies A and B touch, that are at different temperatures, can easily be computed using

$$T_C = \frac{b_A T_A + b_B T_B}{b_A + b_B} \tag{3.4}$$

where T_A and T_B are the temperatures of the touching bodies and b_A and b_B are the thermal penetrations for both materials. The contact temperature is very important for many objects in daily use (e.g., from the handles of heated objects or drinking cups made of plastic, to the heat insulation of space crafts). It is also very important for the calculation of temperatures in tools and molds during polymer processing. The constants used to compute temperature dependent thermal penetration numbers for common thermoplastics are given by

$$b_i = a_1 T + a_2 \tag{3.5}$$

The coefficients in the above equation are found in Table. 3.8.

Table 3.8: Heat of Fusion of Various Thermoplastic Polymers

Polymer	a_1	a_2
PE-HD	1.41	441.7
PE-LD	0.0836	615.1
PMMA	0.891	286.4
POM	0.674	699.6
PP	0.846	366.8
PS	0.909	188.9
PVC	0.649	257.8

3.2.7 Thermal Data Measuring Devices

Thanks to modern analytical instruments it is possible to measure thermal data with a high degree of accuracy. These data allow a good insight into chemical and manufacturing processes. Accurate thermal data or properties are necessary for everyday calculations and computer simulations of thermal processes. Such analyses are used to design polymer processing installations and to determine and optimize processing conditions. In the last twenty years, several physical thermal measuring devices have been developed to determine thermal data used to analyze processing and polymer component behavior.

Differential Thermal Analysis (DTA) The differential thermal analysis test serves to examine transitions and reactions which occur on the order between seconds and minutes, and involve a measurable energy differential of less than 0.04 J/g. Usually, the measuring is done dynamically (i.e., with linear temperature variations in time). However, in some cases isothermal measurements are also done. DTA is mainly used to determine the transition temperatures. The principle is shown schematically in Fig. 3.34. Here, the sample, S, and an inert substance, I, are placed in an oven that has the ability to raise its temperature linearly.

3.2 Thermal Properties

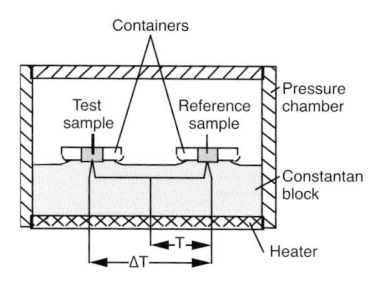

Figure 3.34: Schematic of a differential thermal analysis test.

Two thermocouples that monitor the samples are connected opposite to one another such that no voltage is measured as long as S and I are at the same temperature:

$$\Delta T = T_S - T_I = 0 \qquad (3.6)$$

However, if a transition or a reaction occurs in the sample at a temperature, T_C, then heat is consumed or released, in which case $\Delta T \neq 0$. This thermal disturbance in time can be recorded and used to interpret possible information about the reaction temperature, T_C, the heat of transition or reaction, ΔH, or simply about the existence of a transition or reaction.

Figure 3.35 shows the temperature history in a sample with an endothermic melting point (i.e., such as the one that occurs during melting of semi-crystalline polymers). The figure also shows the functions $\Delta T(T_I)$ and $\Delta T(T_S)$ which result from such a test. A comparison between Figs. 3.35a, b and c, demonstrates that it is very important to record the sample temperature, T_S, to determine a transition temperature, such as the melting or glass transition temperature.

Differential Scanning Calorimeter (DSC) The differential scanning calorimeter permits us to determine thermal transitions of polymers in a range of temperatures between -180 and +600°C. Unlike the DTA cell, in the DSC device, thermocouples are not placed directly inside the sample or the reference substance. Instead, they are embedded in the specimen holder or stage on which the sample and reference pans are placed; the thermocouples make contact with the containers from the outside. A schematic diagram of a differential scanning calorimeter is very similar to the one shown in Fig. 3.34. Materials that do not show or undergo transition or react in the measuring range (e.g., air, glass powder, etc.) are placed inside the reference container. For standardization, one generally uses mercury, tin, or zinc, whose properties are exactly known. In contrast to the DTA test, where samples larger than 10 g are needed, the DSC test requires samples that are in the mg range (<10 mg). Although DSC tests are less sensitive

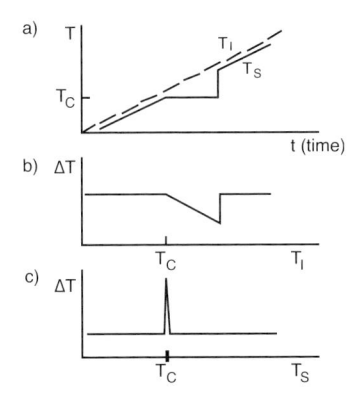

Figure 3.35: Temperature and temperature differences measured during melting of a semi-crystalline polymer sample.

than the DTA tests, they are the most widely used tests for thermal analysis. In fact, DTA tests are rarely used in the polymer industry.

Figure 3.36 shows a typical DSC curve measured using a partly crystalline polymer sample. In the figure, the area that is enclosed between the trend line and the base line is a direct measurement for the amount of heat, ΔH, needed for transition. In this case, the transition is melting and the area corresponds to the heat of fusion.

The degree of crystallinity, $\mathcal{X}$, is determined from the ratio of the heat of fusion of a polymer sample, ΔH_{SC}, and the enthalpy of fusion of a 100% crystalline sample ΔH_C.

$$\mathcal{X} = \frac{\Delta H_{SC}}{\Delta H_C} \tag{3.7}$$

In a DSC analysis of a semi-crystalline polymer, a jump in the specific heat curve, as shown in Fig. 3.36, becomes visible. The glass transition temperature, T_g, is determined at the inflection point of the specific heat curve. The release of residual stresses as a material's temperature is raised above the glass transition temperature is often observed in a DSC analysis.

Specific heat, C_p, is one of the many material properties that can be measured with DSC. During a DSC temperature sweep, the sample pan and the reference pan are maintained at the same temperature. This allows the measurement of the differential energy required to maintain identical temperatures. The sample with the higher heat capacity will absorb a larger amount of heat, which is proportional to the difference between the heat capacity of the measuring sample and the reference sample. It is also possible to determine the purity of a polymer sample when additional peaks or curve shifts are detected in a DSC measurement.

3.2 Thermal Properties

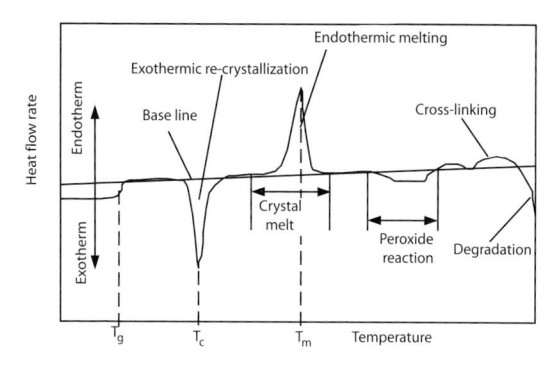

Figure 3.36: Typical DSC heat flow for a semi-crystalline polymer.

Thermal degradation is generally accompanied by an exothermic reaction which may result from oxidation. Such a reaction can easily be detected in a DSC output. By further warming of the test sample, cross-linking may take place and, finally, chain breakage, as shown in Fig. 3.36.

An important aspect in DSC data interpretation is the finite heat flow resistance between the sample pan and the furnace surface. Recent studies by Janeschitz-Kriegl, Eder and co-workers have demonstrated that the heat transfer coefficient between the sample pan and furnace is of finite value, and cannot be disregarded when interpreting the data. In fact, with materials that have a low thermal conductivity, such as polymers, the finite heat transfer coefficient will significantly influence the temperature profiles of the samples.

The differential scanning calorimeter is used to measure the melting, T_m, and the glass transition temperatures of polymers using the ISO 11357 and ASTM 3418 tests. Tables 3.9 and 3.10 present the tests for melting and glass transitions, respectively.

Thermomechanical Analysis (TMA) Thermomechanical analysis (TMA) measures shape stability of a material at elevated temperatures by physically penetrating it with a metal rod. A schematic diagram of TMA equipment is shown in Fig. 3.37. In TMA, the test specimen's temperature is raised at a constant rate, the sample is placed inside the measuring device, and a rod with a specified weight is placed on top of it. To allow for measurements at low temperatures, the sample, oven, and rod can be cooled with liquid nitrogen. Most instruments are so precise that they can be used to measure the melting temperature of the material and, by using linear dilatometry, to measure the thermal expansion coefficients. The thermal expansion coefficient can be measured using

$$\alpha_t = \frac{1}{L_0} \frac{\Delta L}{\Delta T} \tag{3.8}$$

where L_0 is the initial dimension of the test specimen, ΔL the change in size and ΔT the temperature difference. For isotropic materials a common relation between the linear and the volumetric thermal expansion coefficient can be used:

$$\gamma = 3\alpha_t \tag{3.9}$$

Table 3.9: Standard Methods of Measuring Melting Temperature (after Shastri)

Standard	ISO 11357 - 3	D 3418 - 97
Specimen	Molding compound.	Powders; granules; pellets or molded part cut with a microtome, razor blade, hypodermic punch, paper punch or cork borer; slivers cut from films and sheets.
Apparatus	DSC Calibrate the temperature measuring system periodically over the temperature range used for the test.	DSC or DTA Using the same heating rate to be used for specimen, calibrate the temperature scale with the appropriate reference materials covering the materials of interest.
Test procedures	Sample mass of up to 50 mg is recommended. Perform and record a preliminary thermal cycle by heating the specimen at a rate of 10 K/min under inert gas from ambient to 30 K above the melting point to erase previous thermal history. Hold for 10 min at temperature. Cool to 50°C below the peak crystallization temperature at a rate of 10 K/min.	Sample weight of 5 mg is recommended. An appropriate sample will result in 25 to 95 % of scale deflection. Immediately repeat heating under inert gas at rate of 10 K/min. Perform and record a preliminary thermal cycle by heating the specimen at a rate of 10°C/min under nitrogen from ambient to 30°C above the melting point to erase previous thermal history. Hold for 10 min at temperature. Cool to 50°C below the peak crystallization temperature at a rate of 10°C/min. Repeat heating as soon as possible under N_2 at rate of 10°C/min.
Values and units	T_p- peak melting point(s) from the second heat cycle $\Rightarrow$ °C or K	T_m-melting point(s) from the second heat cycle $\Rightarrow$ °C

Thermogravimetry (TGA) A thermogravimetric analyzer can measure weight changes of less than 10 μg as a function of temperature and time. This measurement technique, typically used for thermal stability, works on the principle of a beam balance. The testing chamber can be heated (up to approximately 1,200°C) and rinsed with gases (inert or reactive). Measurements are performed on isothermal reactions or at temperatures sweeps of less than 100 K/min. The maximum sample weight used in thermogravimetric analyses is 500 mg. Thermogravimetry is often used to identify the components in a blend or a compound based on the thermal stability of each component.

Table 3.10: Standard Methods of Measuring Glass Transition Temperature (after Shastri)

Standard	ISO 11357 - 2	D 3418 - 97
Specimen	Molding compound	Powders; granules; pellets or molded part cut with a microtome, razor blade, hypodermic punch, paper punch or cork borer; slivers cut from films and sheets.
Apparatus	DSC Calibrate the temperature measuring system periodically over the temperature range used for the test.	DSC or DTA Using the same heating rate to be used for specimen, calibrate the temperature scale with the appropriate reference materials covering the materials of interest.
Test procedures	Sample mass of 10 - 20 mg is satisfactory. Perform and record an initial thermal cycle up to a temperature high enough to erase previous thermal history, by using a heating rate of 20°C ± 1 K/min in 99.9% pure nitrogen or other inert gas. Hold temperature until a steady state is achieved (usually 5-10 min) Quench cool at a rate of at least (20±1) K/min to well below the T_g (usually 50 K below). Reheat at a rate (20 ± 1) K/min and record heating curve until all desired transitions are recorded.	Sample weight of 10 - 20 mg is recommended Perform and record a preliminary thermal cycle by heating the specimen at a rate of 20°C /min in air or nitrogen from ambient to 30°C above the melting point to erase previous thermal history. Hold for 10 min at temperature. Quench cool to 50°C below the transition peak of interest. Hold temperature until a steady state is reached (usually 5-10 min). Repeat heating as soon as possible at a rate of 20°C /min until all desired transitions have been completed.
Values and units	T_{mg} midpoint temperature $\Rightarrow$ °C	T_m (T_g) midpoint temperature $\Rightarrow$ °C T_f (T_g) extrapolated onset temperature $\Rightarrow$ °C For most applications the T_f is more meaningful than T_m and may be designated as T_g in place of the midpoint of the T_g curve.

Figure 3.38 shows results from a TGA analysis on a PVC fabric. The figure shows the transitions at which the various components of the compound decompose. The percent of the original sample weight is recorded along with the change

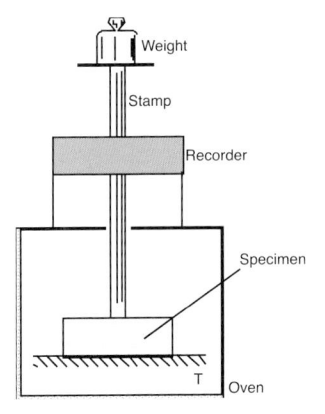

Figure 3.37: Schematic diagram of the thermomechanical analysis (TMA) device.

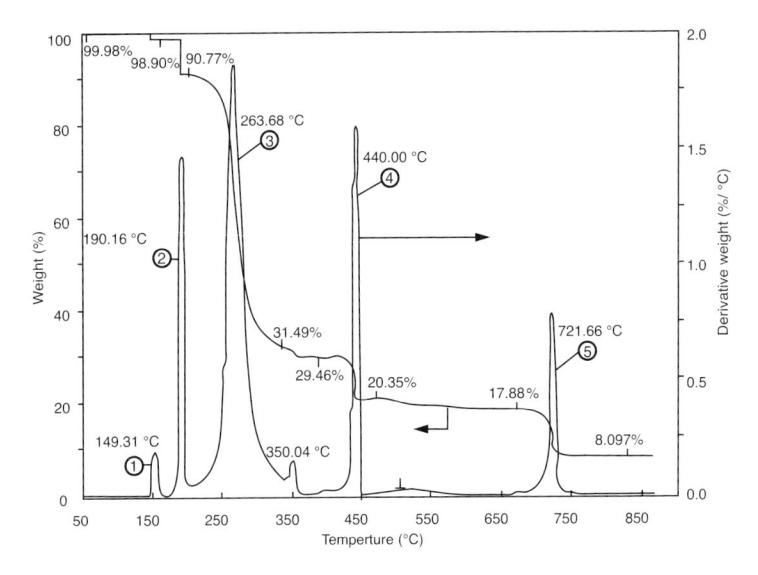

Figure 3.38: TGA analysis of a PVC fabric. (1) volatiles: humidity, monomers, solvents etc., (2) DOP plasticizer, (3) HCl formation, (4) carbon-carbon scission, and (5) CO_2 formation.

of the weight with respect to temperature. Five transitions representing (1) the decomposition of volatile components, (2) decomposition of the DOP plasticizer, (3) formation of HCl, (4) carbon-carbon scission, and (5) the forming of CO_2, are clearly visible.

Density Measurements One simple form of calculating the density of a polymer sample is to first weigh the sample immersed in water. Assuming the density of water to be 1.0g/cm^3, we can use the relation

$$\rho = \frac{m}{(m - m_i)(1 \text{ cm}^3/\text{g})} \quad (3.10)$$

where m is mass of the specimen, m_i is the mass of the immersed specimen and $(m - m_i)$ is the mass of the displaced body of water.

Some common ways of determining density of polymeric materials are described by ASTM D792, ISO 1183, and DIN 53 479 test methods. Another common way of measuring density is the *through flow density meter*. Here, the density of water is changed to that of the polymer by adding ethanol until the plastic shavings are suspended in the solution. The density of the solution is then measured in a device that pumps the liquid through a U-pipe, where it is measured using ultrasound techniques. A density gradient technique is described by the standard ASTM D1505 test method.

3.3 CURING BEHAVIOR

Both, thermosets and elastomeric materials undergo a reaction process during processing. They can be classified in two general processing categories: heat activated cure and mixing activated cure thermosets. However, no matter which category a reactive polymer belongs to, its curing reaction can be described by the reaction between two chemical groups denoted by A and B which link two segments of a polymer chain. The reaction can be followed by tracing the concentration of unreacted As or Bs, C_A or C_B. If the initial concentration of A and B is defined as C_{A_0} and C_{B_0}, the degree of cure can be described by

$$c = \frac{C_{A_0} - C_{B_0}}{C_{A_0}} \quad (3.11)$$

The degree of cure or conversion, c, equals zero when there has been no reaction and equals one when all As have reacted and the reaction is complete. However, it is impossible to monitor reacted and unreacted As and Bs during the curing reaction of a thermoset polymer. It is known that the exothermic heat released during curing can be used to monitor the conversion, c. When a small sample of an unreacted thermoset polymer is placed in a differential scanning calorimeter (DSC), the sample will release a certain amount of heat, Q_T. This occurs because every cross-link that forms during a reaction releases a small amount of energy in the form of heat. For example, Fig. 3.39 shows the heat rate released during isothermal cure of a vinyl ester at various temperatures.

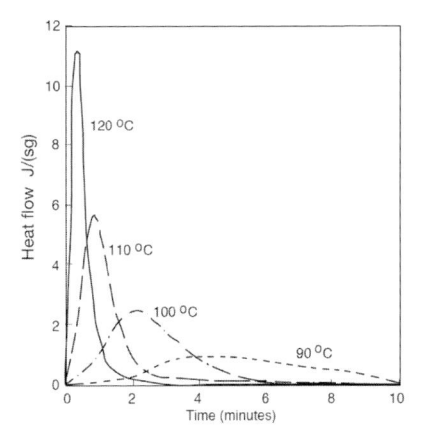

Figure 3.39: DSC scans of the isothermal curing reaction of vinyl ester at various temperatures.

Using the exothermic heat as a measure of cure, the degree of cure can be defined by the following relation

$$c = \frac{Q}{Q_T} \quad (3.12)$$

where Q is the heat released up to an arbitrary time t, and is defined by

$$Q = \int_0^t \dot{Q} dt \quad (3.13)$$

DSC data is commonly fitted to semi-empirical models that accurately describe the curing reaction. Hence, the rate of cure can be described by the exotherm, $\dot{Q}$, and the total heat released during the curing reaction, Q_T, as

$$\frac{dc}{dt} = \frac{\dot{Q}}{Q_T} \quad (3.14)$$

With the use of eqn. (3.14), it is now easy to take the DSC data and find the models for $\frac{dc}{dt}$ that best describe the curing reaction.

During cure, thermoset resins exhibit three distinct phases; viscous liquid, gel, and solid. Each of these three stages is marked by dramatic changes in the thermomechanical properties of the resin. The transformation of a reactive thermosetting liquid to a glassy solid generally involves two distinct macroscopic transitions: molecular gelation and vitrification. Molecular gelation is defined as

3.3 Curing Behavior

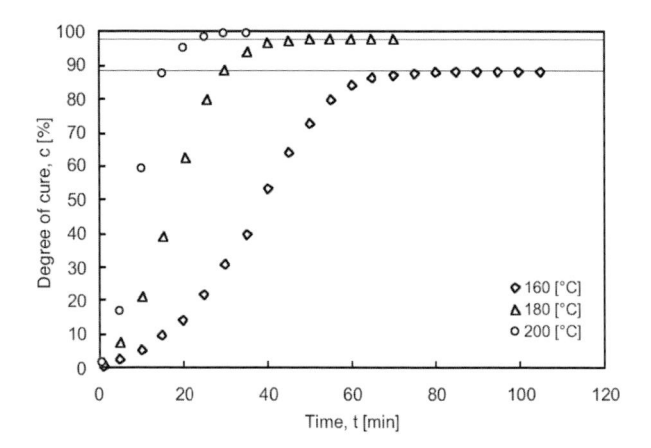

Figure 3.40: Degree of cure as a function time for an epoxy resin measured using isothermal DSC.

the time or temperature at which covalent bonds connect across the resin to form a three-dimensional network which gives rise to long range elastic behavior in the macroscopic fluid. This point is also referred to as the gel point, where $c = c_g$. As a thermosetting resin cures, the cross-linking begins to hinder molecular movement, leading to a rise in the glass transition temperature. Eventually, when T_g nears the processing temperature, the rate of curing reduces significantly, and becomes dominated by diffusion. At this point, the resin has reached its vitrification point. Figure 3.40, which shows the degree of cure as a function of time, illustrates how an epoxy resin reaches a maximum degree of cure at various processing temperatures.

The resin processed at 200°C reaches 100%, cure because the glass transition temperature of fully cured epoxy is 190°C, less than the processing temperature. On the other hand, the sample processed at 180°C reaches 97% cure and the one processed at 160°C only reaches 87% cure. Figures 3.39 and 3.40 also illustrate how the curing reaction is accelerated as the processing temperature is increased. The curing reaction of thermally cured thermoset resins is not immediate, thus the blend can be stored in a refrigerator for a short period of time without having any significant curing reaction.

The behavior of curing thermosetting resins can be represented with the generalized time-temperature-transformation (TTT) cure diagram developed by Enns and Gillham; it can be used to relate the material properties of thermosets as a function of time and the processing temperature as shown in Fig. 3.41.

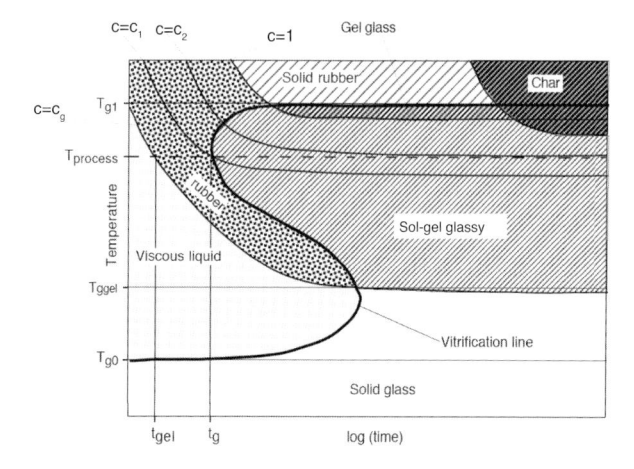

Figure 3.41: Time-temperature-transformation (TTT) diagram for a thermoset.

The diagram presents various lines that represent constant degrees of cure. The curve labeled $c = c_g$ represents the gel point and $c = 1$ the fully cured resin. Both curves have their corresponding glass transition temperatures, T_{g_1} and $T_{g_{gel}}$, for the glass transition temperature of the fully cured resin and at its gel point, respectively. The glass transition temperature of the uncured resin, T_{g_0}, and an S-shaped curve labeled *vitrification line*, are also depicted. The vitrification line represents the boundary where the glass transition temperature becomes the processing temperature. Hence, to the left of the vitrification curve, the curing process is controlled by a very slow diffusion process. The TTT-diagram shows an arbitrary process temperature. The material being processed reaches the gel point at $t = t_{gel}$ and the vitrification line at $t = t_g$. At this point, the material has reached a degree of cure of c_1 and the glass transition temperature of the resin is equal to the processing temperature. The material continues to cure very slowly (diffusion controlled) until it reaches a degree of cure just below c_2. There are also various regions labeled in the diagram. The one labeled *viscous liquid* is the one where the resin is found from the beginning of processing until the gel point has been reached. The flow and deformation that occurs during processing or shaping must occur within this region. The region labeled *char* must be avoided during processing, since at high processing temperatures the polymer will eventually undergo thermal degradation.

3.4 RHEOLOGICAL PROPERTIES

Rheology is the field of science that studies fluid behavior during flow-induced deformation. From the variety of materials that rheologists study, polymers have been found to be the most interesting and complex. Polymer melts are shear thinning, viscoelastic, and their flow properties are temperature dependent. Viscosity is the most widely used material parameter when determining the behavior of polymers during processing. Since the majority of polymer processes are shear rate dominated, the viscosity of the melt is commonly measured using shear deformation measurement devices. However, there are polymer processes, such as blow molding, thermoforming, and fiber spinning, that are dominated by either elongational deformation or by a combination of shear and elongational deformation. In addition, some polymer melts exhibit significant elastic effects during deformation. Modeling and simulation of polymer flows will be presented in Chapter 5 of this handbook.

3.4.1 Flow Phenomena

There are three important phenomena seen is polymeric liquids that make them different from simple fluids: a non-Newtonian viscosity, normal stresses in shear flow, and elastic effects. All these effects are a result of the complex molecular structure of polymer macromolecules.

Non-Newtonian Viscosity In a Newtonian fluid the deviatoric stresses that occur during deformation, τ, are directly proportional to the rate of deformation tensor, $\dot{\gamma}$,

$$\tau = \eta \dot{\gamma} \tag{3.15}$$

For Newtonian liquids the proportionality constant, η, is considered to be only dependent on temperature. However, the viscosity of most polymer melts is *shear thinning* in addition to being temperature dependent. The shear thinning effect is the reduction in viscosity at high rates of deformation. This phenomenon occurs because, at high rates of deformation, the molecules are stretched out and disentangled, enabling them to slide past each other with more ease, hence, lowering the bulk viscosity of the melt. This is clearly demonstrated in Figs. 3.42-3.46 shows the shear thinning behavior and temperature dependance of the viscosity of a selected number of thermoplastics.

Normal Stresses in Shear Flow The tendency of polymer molecules to *curl-up* while they are being stretched in shear flow results in normal stresses in the fluid. Figure 3.47 presents the first normal stress difference coefficient for the low density polyethylene melt of Fig. 3.48 at a reference temperature of 150°C.

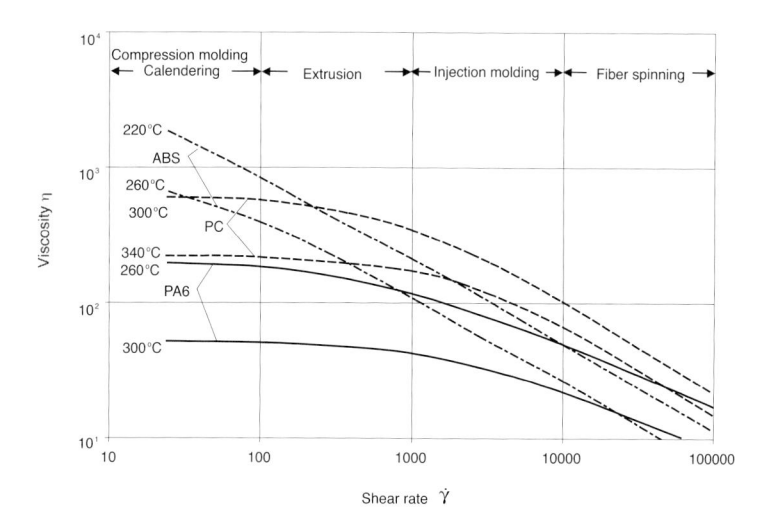

Figure 3.42: Viscosity curves for a selected number of thermoplastics.

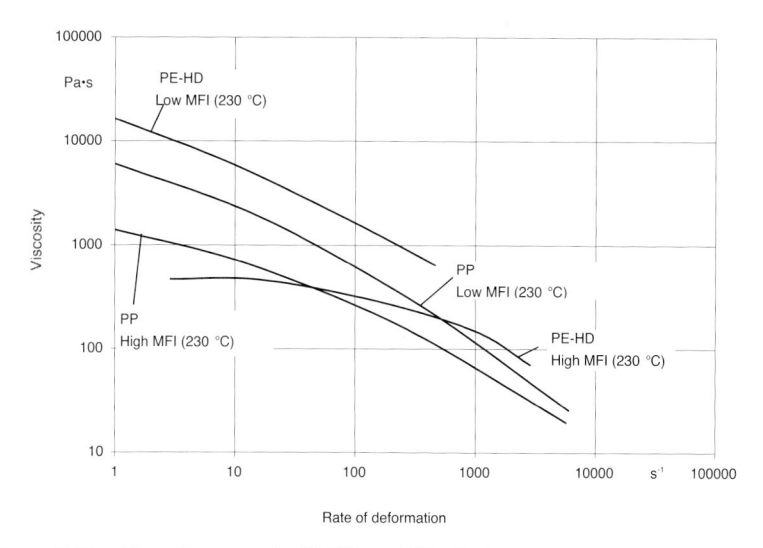

Figure 3.43: Viscosity curves for PE-HD and PP with low and high MFI.

3.4 Rheological Properties

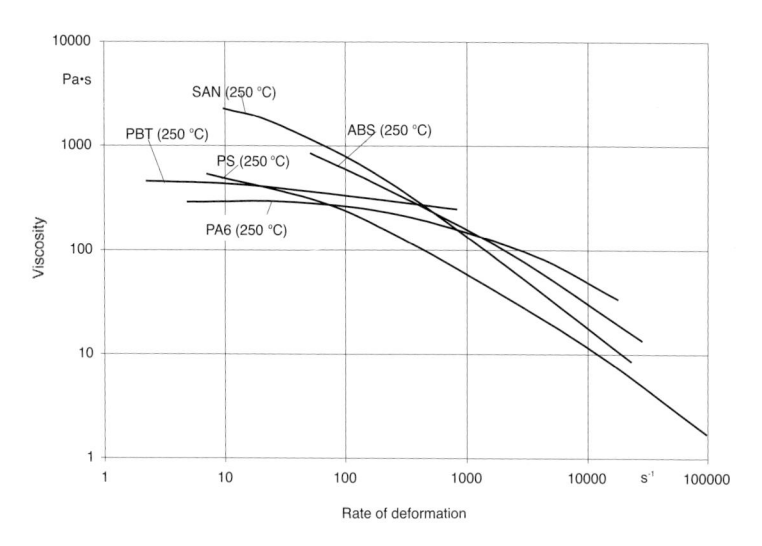

Figure 3.44: Viscosity curves for PE-HD and PP with low and high MFI at 230°C.

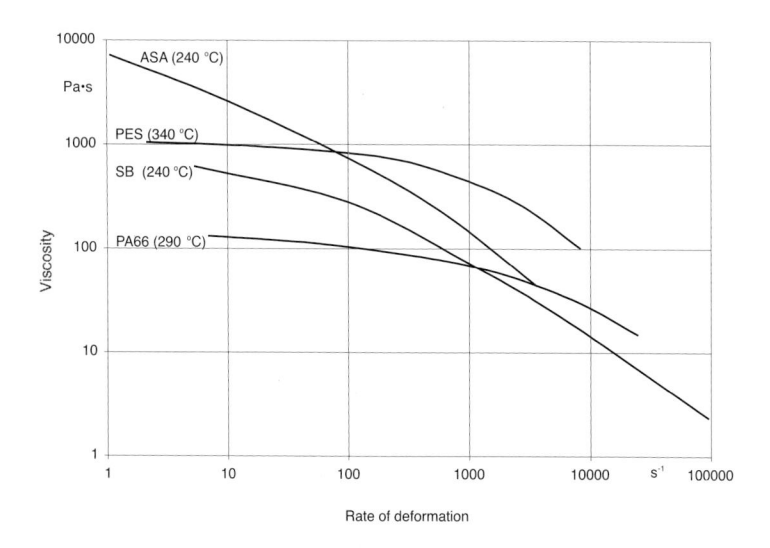

Figure 3.45: Viscosity curves for selected thermoplastics.

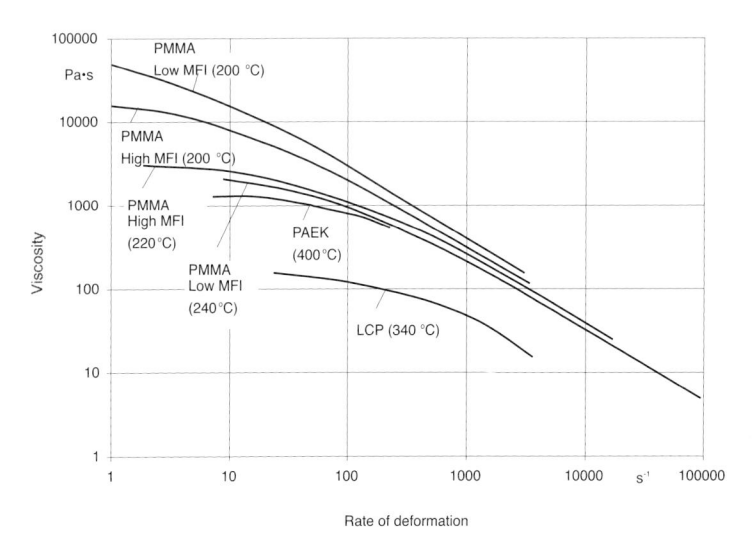

Figure 3.46: Viscosity curves for PMMA, PAEK and LCP.

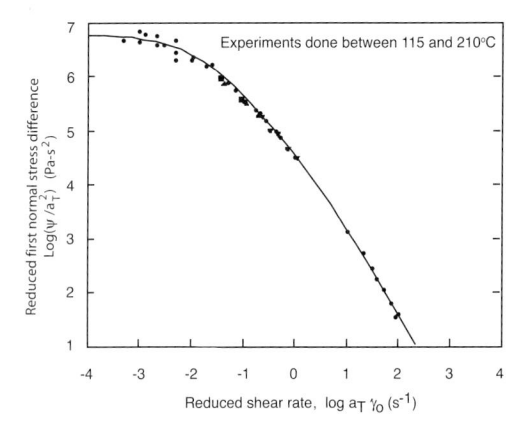

Figure 3.47: Reduced first normal stress difference coefficient for a low density polyethylene melt at a reference temperature of $150^\circ C$.

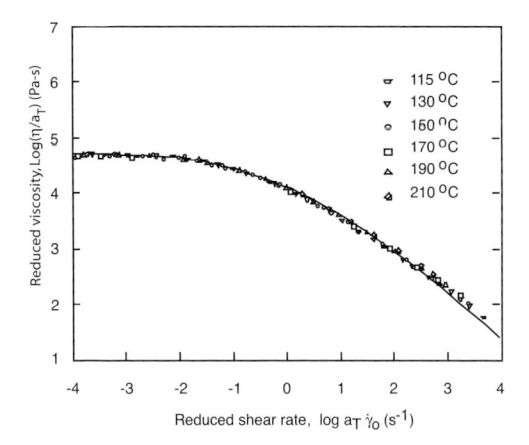

Figure 3.48: Reduced viscosity for a low density polyethylene melt at a reference temperature of 150°C.

The second normal stress difference is much smaller than the first normal stress difference and is therefore difficult to measure.

Viscoelastic Memory Effects or Stress Relaxation When a polymer melt is deformed, either by stretching, shearing or often by a combination of the above, the polymer molecules are stretched and untangled. In time, the molecules try to recover their initial shape, in essence getting used to their new state of deformation. If the deformation is maintained for a short period of time, the molecules may go back to their initial position, and the shape of the melt is fully restored to its initial shape. Here, it is said that the molecules *remembered* their initial position. However, if the shearing or stretching goes on for an extended period of time, the polymer cannot recover its starting shape, in essence *forgetting* the initial positions of the molecules. The time it takes for a molecule to fully relax and get used to its new state of deformation is referred to as the relaxation time, λ.

A useful parameter often used to estimate the elastic effects during flow is the Deborah[3] number, De. The Deborah number is defined by

$$De = \frac{\lambda}{t_{\text{process}}} \qquad (3.16)$$

where t_{process} is a characteristic process time. For example, in an extrusion die, a characteristic process time can be defined by the ratio of characteristic die dimen-

---

[3]From the Song of Deborah, Judges 5:5 - "The mountains flowed before the Lord." M. Rainer is credited for naming the Deborah number; Physics Today, 1, (1964).

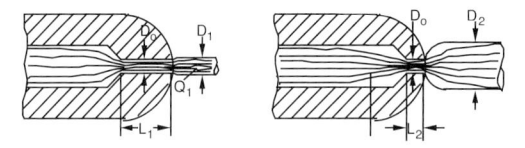

Figure 3.49: Schematic diagram of extrudate swell during extrusion.

sion in the flow direction and average speed through the die. A Deborah number of zero represents a viscous fluid and a Deborah number of ∞ an elastic solid. As the Deborah number becomes > 1, the polymer does not have enough time to relax during the process, resulting in possible extrudate dimension deviations or irregularities, such as extrudate swell, shark skin, or even melt fracture.

Although many factors affect the amount of extrudate swell, fluid *memory* and normal stress effects are the most significant ones. However, abrupt changes in boundary conditions, such as the separation point of the extrudate from the die, also play a role in the swelling or cross section reduction of the extrudate. In practice, the fluid memory contribution to die swell can be mitigated by lengthening the land length of the die. This is schematically depicted in Fig. 3.49 a long die land separates the polymer from the manifold long enough to allow it to *forget* its past shape.

Waves in the extrudate may also appear as a result of high speeds during extrusion, where the polymer is not allowed to relax. This phenomenon is generally referred to as shark skin and is shown for a high density polyethylene in Fig. 3.50a. It is possible to extrude at such high speeds that an intermittent separation of melt and inner die walls occurs as shown in Fig. 3.50b. This phenomenon is often referred to as the stick-slip effect or spurt flow and is attributed to high shear stresses between the polymer and the die wall. This phenomenon occurs when the shear stress is near the critical value of 0.1 MPa. If the speed is further increased, a helical geometry is extruded as shown for a polypropylene extrudate in Fig. 3.50c. Eventually, the speeds are so high that a chaotic pattern develops, such as the one shown in Fig. 3.50d. This well known phenomenon is called melt fracture. The shark skin effect is frequently absent and spurt flow seems to occur only with linear polymers.

The critical shear stress has been reported to be independent of the melt temperature but to be inversely proportional to the weight average molecular weight.

To summarize, the Deborah number and the size of the deformation imposed on the material during processing determine how the system can most accurately be modeled. Figure 3.51 helps visualize the relation between time scale, deformation and applicable model. At small Deborah numbers, the polymer can be modeled as a Newtonian fluid, and at very high Deborah numbers the material can be modeled as a Hookean solid. In between, the viscoelastic region is divided in

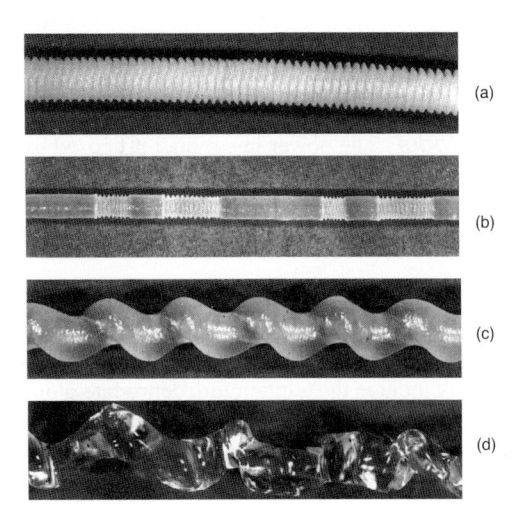

Figure 3.50: Various shapes of extrudates under melt fracture.

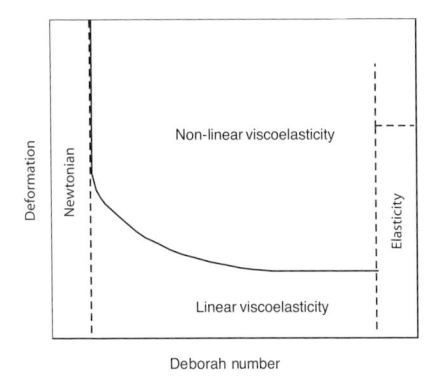

Figure 3.51: Schematic of Newtonian, elastic, linear, and non-linear viscoelastic regimes as a function of deformation and Deborah number during deformation of polymeric materials.

two: the linear viscoelastic region for small deformations, and the non-linear viscoelastic region for large deformations.

3.4.2 Viscous Flow Models

Strictly speaking, the viscosity η, measured with shear deformation viscometers, should not be used to represent the elongational terms located on the diagonal of the stress and strain rate tensors. Elongational flows are briefly discussed later in this chapter. A rheologist's task is to find the models that best fit the measured viscosity data. Some of the models used by polymer processors on a day-to-day basis to represent the viscosity of industrial polymers are presented in this section.

The Power Law Model The power law model proposed by Ostwald and de Waale is a simple model that accurately represents the shear thinning region in the viscosity versus strain rate curve but neglects the Newtonian plateau present at small strain rates. The power law model can be written as follows:

$$\eta = m(T)\dot{\gamma}^{n-1} \tag{3.17}$$

where m is referred to as the consistency index and n the power law index. The consistency index may include the temperature dependence of the viscosity, and the power law index represents the shear thinning behavior of the polymer melt. It should be noted that the limits of this model are

$$\eta \longrightarrow 0 \text{ as } \dot{\gamma} \longrightarrow \infty$$

and

$$\eta \longrightarrow \infty \text{ as } \dot{\gamma} \longrightarrow 0$$

The infinite viscosity at zero strain rates leads to an erroneous result when there is a region of zero shear rate, such as at the center of a tube. This results in a predicted velocity distribution that is flatter at the center than the experimental profile. In computer simulation of polymer flows, this problem is often overcome by using a truncated model such as

$$\eta = m_0(T)\dot{\gamma}^{n-1} \quad \text{for} \quad \dot{\gamma} > \dot{\gamma}_0 \tag{3.18}$$

and

$$\eta = m_0(T) \quad \text{for} \quad \dot{\gamma} \leqslant \dot{\gamma}_0 \tag{3.19}$$

where η_0 represents a zero-shear-rate viscosity ($\dot{\gamma}_0$). Table 3.11 presents a list of typical power law and consistency indices for common thermoplastics.

Table 3.11: Power Law and Consistency Indices for Common Thermoplastics

Polymer	m (**Pa-s**^{n})	n	T (°**C**)
High density polyethylene	2.0×10^4	0.41	180
Low density polyethylene	6.0×10^3	0.39	160
Polyamide 66	6.0×10^2	0.66	290
Polycarbonate	6.0×10^2	0.98	300
Polypropylene	7.5×10^3	0.38	200
Polystyrene	2.8×10^4	0.28	170
Polyvinyl chloride	1.7×10^4	0.26	180

The Bird-Carreau-Yasuda Model A model that fits the whole range of strain rates was developed by Bird and Carreau and Yasuda and contains five parameters:

$$\frac{\eta - \eta_0}{\eta_0 - \eta_\infty} = [1 + |\lambda\dot\gamma|^a]^{(n-1)/a} \quad (3.20)$$

where η_0 is the zero-shear-rate viscosity, η_∞ is an infinite-shear-rate viscosity, λ is a time constant and n is the power law index. In the original Bird-Carreau model, the constant $a = 2$. In many cases, the infinite-shear-rate viscosity is negligible, reducing eqn. (3.20) to a three parameter model. Equation (3.20) was modified by Menges, Wortberg and Michaeli to include a temperature dependence using a WLF relation. The modified model, which is used in commercial polymer data banks, is written as follows:

$$\eta = \frac{k_1 a_T}{[1 + k_2 \dot\gamma a_T]^{k_3}} \quad (3.21)$$

where the shift a_T applies well for amorphous thermoplastics and is written as

$$a_T = \frac{8.86(k_4 - k_5)}{101.6 + k_4 - k_5} - \frac{8.86(T - k_5)}{101.6 + T - k_5} \quad (3.22)$$

Table 3.12 presents constants for Carreau-WLF (amorphous) and Carreau-Arrhenius models (semi-crystalline) for various common thermoplastics. In addition to the temperature shift, Menges, Wortberg and Michaeli measured a pressure dependence of the viscosity and proposed the following model, which includes both temperature and pressure viscosity shifts:

$$\log \eta(T, p) - \log \eta_0 = \frac{8.86(T - T_0)}{101.6 + T - T_0} - \frac{8.86(T - T_0 + 0.02p)}{101.6 + (T - T_0 + 0.02p)} \quad (3.23)$$

where p is in bar, and the constant 0.02 represents a 2°C shift per bar.

Table 3.12: Constants for Carreau-WLF (Amorphous) and Carreau-Arrhenius (Semi-Crystalline) Models for Various Common Thermoplastic

Polymer	k_1 Pa-s	k_2 s	k_3	k_4 °C	k_5 °C	T_0 °C	E_0 J/mol
High density polyethylene	24,198	1.38	0.60	-	-	200	22,272
Low density polyethylene	317	0.015	0.61	-	-	189	43,694
Polyamide 66	44	0.00059	0.40	-	-	300	123,058
Polycarbonate	305	0.00046	0.48	320	153	-	-
Polypropylene	1,386	0.091	0.68	-	-	220	427,198
Polystyrene	1,777	0.064	0.73	200	123	-	-
Polyvinyl chloride	1,786	0.054	0.73	185	88	-	-

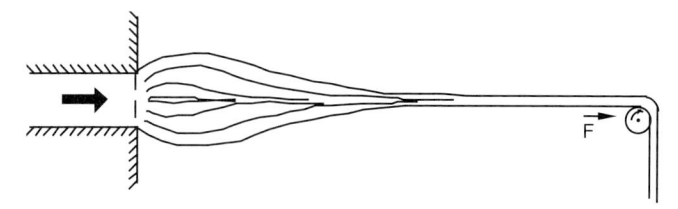

Figure 3.52: Schematic diagram of a fiber spinning process.

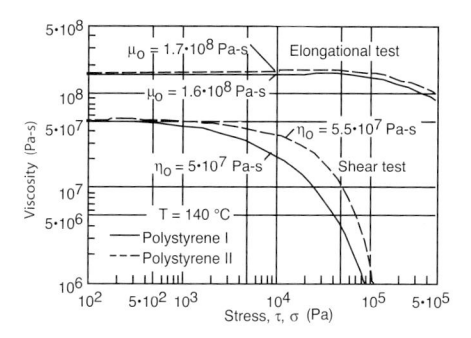

Figure 3.53: Shear and elongational viscosity curves for two types of polystyrene (After Muenstedt).

Elongational Viscosity In polymer processes such as blow molding, fiber spinning (Fig.3.52), thermoforming, foaming, certain extrusion die flows, and compression molding with specific processing conditions, the major mode of deformation is elongational.

Figure 3.53 compares shear and elongational viscosities for two types of polystyrene. It is clear that the elengational viscosities are much larger than the shear viscosities. In the region of the Newtonian plateau, a limit of 3 is quite clear. Figure 3.54 presents plots of elongational viscosities as a function of stress for various thermoplastics at common processing conditions. It should be emphasized that measuring elongational or extensional viscosity is an extremely difficult task. For example, in order to maintain a constant strain rate, the specimen must be deformed uniformly exponentially. In addition, a molten polymer must be tested completely submerged in a heated neutrally buoyant liquid at constant temperature.

Rheology of Curing Thermosets A curing thermoset polymer has a conversion or cure dependent viscosity that increases as the molecular weight of the reacting polymer increases. For vinyl ester, whose curing history is shown in Fig. 3.55, the viscosity behaves as shown in Fig. 3.56.

3.4 Rheological Properties 113

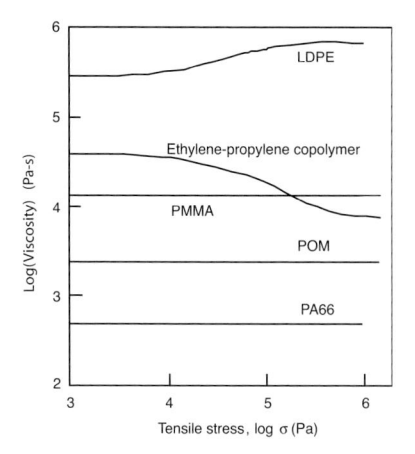

Figure 3.54: Elongational viscosity curves as a function of tensile stress for several thermoplastics.

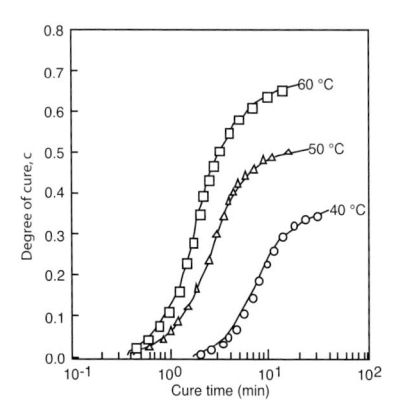

Figure 3.55: Degree of cure as a function of time for a vinyl ester at various isothermal cure temperatures.

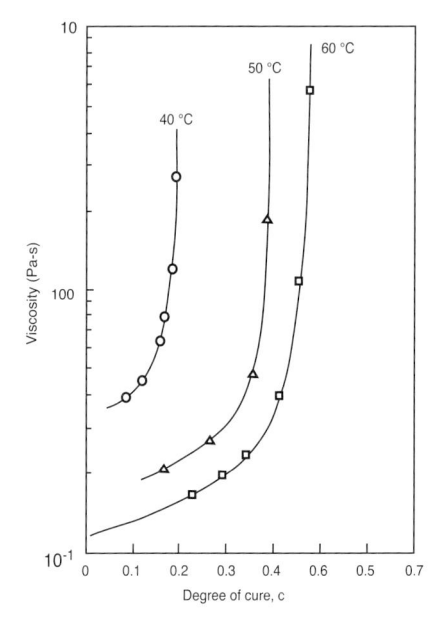

Figure 3.56: Viscosity as a function of degree of cure for a vinyl ester at various isothermal cure temperatures.

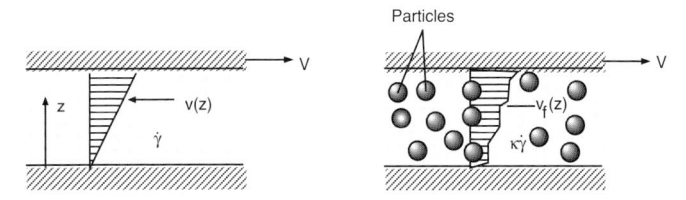

Figure 3.57: Schematic diagram of strain rate increase in a filled system.

3.4 Rheological Properties

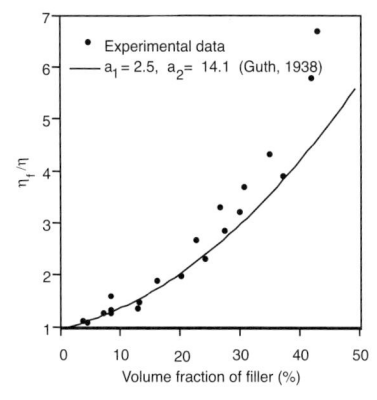

Figure 3.58: Viscosity increase as a function of volume fraction of filler for polystyrene and low density polyethylene containing spherical glass particles with diameters ranging from 36 to 99.8μm (After Geisüsch).

At the gel point, the cross-linking forms a closed network, at which point it is said that the molecular weight and the viscosity go to infinity.

Suspension Rheology Particles suspended in a material, such as in filled or reinforced polymers, have a direct effect on the properties of the final article and on the viscosity during processing. Numerous models have been proposed to estimate the viscosity of filled liquids. Most models proposed are a power series of the form

$$\frac{\eta_f}{\eta} = 1 + 2.5\phi + 14.1\phi^2 \tag{3.24}$$

The linear term in eqn. (3.24) represents the narrowing of the flow passage caused by the filler that is passively entrained by the fluid and sustains no deformation as shown in Fig. 3.57.

For instance, Einstein's model, which only includes the linear term with $a_1 = 2.5$, was derived based on a viscous dissipation balance. For high deformation stresses, which are typical in polymer processing, the yield stress in the filled polymer melt can be neglected. Figure 3.58 compares experimental data to eqn. (3.24).

3.4.3 Rheometry

There are various ways to qualify and quantify the properties of the polymer melt in industry. The techniques range from simple analyses for checking the consistency of the material at certain conditions, to more complex measurements

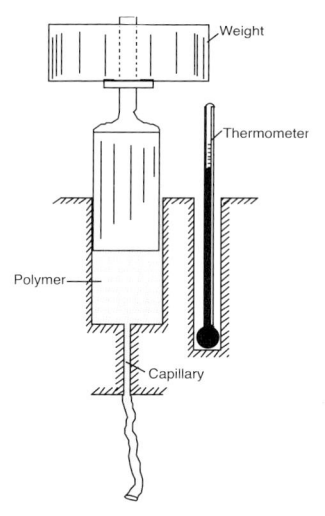

Figure 3.59: Schematic diagram of an extrusion plastometer used to measure the melt flow index.

to evaluate viscosity and normal stress differences. This section includes three such techniques, to give the reader a general idea of current measuring techniques.

The melt flow indexer The melt flow indexer is often used in industry to characterize a polymer melt and as a simple and quick means of quality control. It takes a single point measurement using standard testing conditions specific to each polymer class on a ram type extruder or extrusion plastometer as shown in Fig. 3.59.

The standard procedure for testing the flow rate of thermoplastics using a extrusion plastometer is described in the ASTM D1238 test as presented in Table 3.13. During the test, a sample is heated in the barrel and extruded from a short cylindrical die using a piston actuated by a weight. The weight of the polymer in grams extruded during the 10-minute test is the melt flow index (MFI) of the polymer.

The capillary viscometer The most common and simplest device for measuring viscosity is the capillary viscometer. Its main component is a straight tube or capillary, and it was first used to measure the viscosity of water by Hagen and Poiseuille. A capillary rheometer has a pressure driven flow for which the velocity gradient or strain rate and also the shear rate will be maximum at the wall and zero at the center of the flow, making it a non-homogeneous flow. Since pressure driven viscometers employ non-homogeneous flows, they can only measure steady shear functions such as viscosity, $\eta(\dot{\gamma})$. However, they are widely used

3.4 Rheological Properties

because they are relatively inexpensive to build and simple to operate. Despite their simplicity, long capillary viscometers give the most accurate viscosity data available. Another major advantage is that the capillary rheometer has no free surfaces in the test region, unlike other types of rheometers, such as the cone and plate rheometers, which we will discuss in the next section. When the strain rate dependent viscosity of polymer melts is measured, capillary rheometers may provide the only satisfactory method of obtaining such data at shear rates >10 s^{-1}. This is important for processes with higher rates of deformation such as mixing, extrusion, and injection molding.

Table 3.13: Standard Methods of Measuring Melt Flow Index (MFI) - Melt Flow Rate (MFR), Melt Volume Rate (MVR), and Flow Rate Ratio (FRR)(after Shastri)

Standard	ISO 1133	ASTM D 1238 - 98
Specimen	Powder, pellets, granules or strips of films	Powder, pellets, granules, strips of films, or molded slugs
Conditioning	In accordance with the material standard, if necessary	Check the applicable material specification for any conditioning requirements before using this test. See practice D 618 for appropriate conditioning practices.
Apparatus	Extrusion plastometer with a steel cylinder (115 - 180) mm (L) x 9.55 ± 0.025 mm (D), and a die with an orifice of 8.000 ± 0.025 mm (L) x 2.095 ± 0.005 mm (D)	Extrusion plastometer with a steel cylinder 162 mm (L) x 9.55 ± 0.008 mm (D), and a die with an orifice of 8.000 ± 0.025 mm (L) x 2.0955 ± 0.0051 mm (D)
Test procedures	Test temperature and test load as specified in Part 2 of the material designation standards, or as listed in ISO 1133. Some examples are: PC (300°C /1.2 kg) ABS (220°C /10 kg) PS (200°C /5 kg) PS-HI (200°C /5 kg) SAN (220°C /10 kg) PP (230°C /2.16 kg) PE (190°C / 2.16 kg) POM (190°C /2.16 kg) PMMA (230°C /3.8 kg) Charge $\Rightarrow$ within 1 min Preheat $\Rightarrow$ 4 min Test time $\Rightarrow$ last measurement not to exceed 25 min from charging. Procedure A - manual operation using the mass and cut-time intervals shown below:	Test temperature and test load as specified in the applicable material specification, or as listed in D1238. Some examples are: PC (300°C /1.2 kg) ABS (230°C /10 kg) PS (200/5 kg) PS-HI (200°C /5 kg) SAN (220/10 kg) PP (230°C /2.16 kg) PE (190°C / 2.16 kg) POM (190°C /2.16 kg) Acrylics (230°C /3.8 kg) Charge $\Rightarrow$ within 1 min Preheat $\Rightarrow$ 6.5 min Test time $\Rightarrow$ 7.0 ± 0.5 min from initial charging. Procedure A - manual operation using the mass and cut-time intervals shown below:

Continued on next page

Standard	ISO 1133	ASTM D 1238 - 98
	MFR ; Mass ; Time	**MFR ; Mass ; Time**
	0.1 to 0.5g/10min; 3-5g ; 4min	0.15 to 1g/10min; 2.5-3g; 6 min
	>0.5 to 1g/10min; 4-5g; 2min	>1 to 3.5g/10min; 3-5g; 3 min
	>1 to 3.5g/10min; 4-5g; 1min	>3.5 to 10g/10min; 4-8g; 1 min
	>3.5 to 10 g/10 min 6-8g; 30s > 10g/10min; 6-8g; 5-15s	>10 to 25g/10min; 4-8g; 30s > 25g/10min; 4-8g; 15s
	Procedure B - automated time or travel indicator is used to calculate the MFR (MVR) using the mass as specified above in Procedure A for the predicted MFR	Procedure B - MFR (MVR) is calculated from automated time measurement based on specified travel distances, < 10 MFR $\Rightarrow$ 6.35 ± 0.25 mm. > 10 MFR $\Rightarrow$ 25.4 ± 0.25 mm. and using the mass as specified above for the predicted MFR
Values and units	MFR $\Rightarrow$ g/10min MVR $\Rightarrow$ cm^3/10min	MFR $\Rightarrow$ g/10min MVR $\Rightarrow$ cm^3/10min FRR $\Rightarrow$ Ratio of the MFR (190/10) by MFR (190/2.16) (used specifically for PE)

Because its design is basic and it only needs a pressure head at its entrance, capillary rheometers can easily be attached to the end of a screw- or ram-type extruder for on-line measurements. This makes the capillary viscometer an efficient tool for industry. The basic features of the capillary rheometer are shown in Fig. 3.60. A capillary tube of a specified radius, R, and length, L, is connected to the bottom of a reservoir. Pressure drop and flow rate through this tube are used to determine the viscosity.

The cone-plate rheometer The cone-plate rheometer is often used when measuring the viscosity and the primary and secondary normal stress coefficient functions as a function of shear rate and temperature. The geometry of a cone-plate rheometer is shown in Fig. 3.61. Since the angle Θ_0 is very small, typically $< 5^o$, the shear rate can be considered constant throughout the material confined within the cone and plate. Although it is also possible to determine the secondary stress coefficient function from the normal stress distribution across the plate, it is very difficult to get accurate data.

The Couette rheometer Another rheometer commonly used in industry is the concentric cylinder or Couette flow rheometer schematically depicted in Fig. 3.62. The torque, T, and rotational speed, Ω, can easily be measured. The torque is related to the shear stress that acts on the inner cylinder wall and the rate of deformation in that region is related to the rotational speed.

The major sources of error in a concentric cylinder rheometer are the end-effects. One way of minimizing these effects is to provide a large gap between the inner cylinder end and the bottom of the closed end of the outer cylinder.

3.4 Rheological Properties

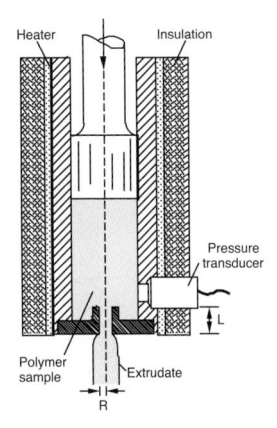

Figure 3.60: Schematic diagram of a capillary viscometer.

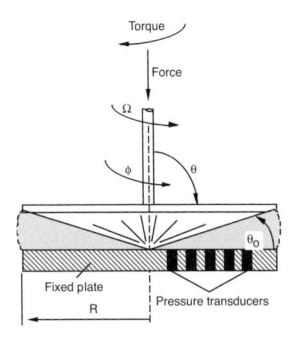

Figure 3.61: Schematic diagram of a cone-plate rheometer.

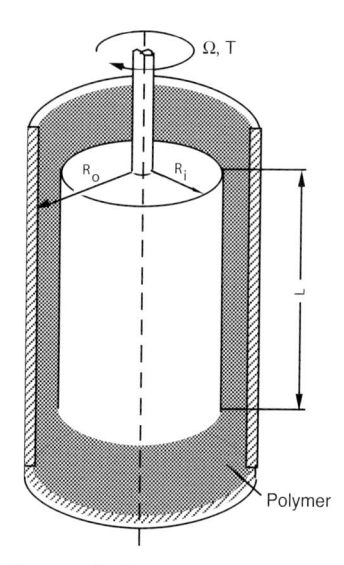

Figure 3.62: Schematic diagram of a Couette rheometer.

Extensional rheometry It should be emphasized that the shear behavior of polymers measured with the equipment described in the previous sections cannot be used to deduce the extensional behavior of polymer melts. Extensional rheometry is the least understood field of rheology. The simplest way to measure extensional viscosities is to stretch a polymer rod held at elevated temperatures at a speed that maintains a constant strain rate as the rod reduces its cross-sectional area. The viscosity can easily be computed as the ratio of instantaneous axial stress to elongational strain rate. The biggest problem when trying to perform this measurement is to grab the rod at its ends as it is pulled apart. The most common way to grab the specimen is with toothed rotary clamps to maintain a constant specimen length. A schematic of Meissner's extensional rheometer incorporating rotary clamps is shown in Fig. 3.63.

Another set-up that can be used to measure extensional properties without clamping problems and without generating orientation during the measurement is the lubricating squeezing flow, which generates an equibiaxial deformation. A schematic of this apparatus is shown in Fig. 3.64.

It is clear from the apparatus description in Fig. 3.63 that carrying out tests to measure extensional rheometry is a very difficult task. One of the major problems arises because of the fact that, unlike shear tests, it is not possible to achieve steady state condition with elongational rheometry tests. This is simply because the cross-sectional area of the test specimen is constantly diminishing.

3.4 Rheological Properties

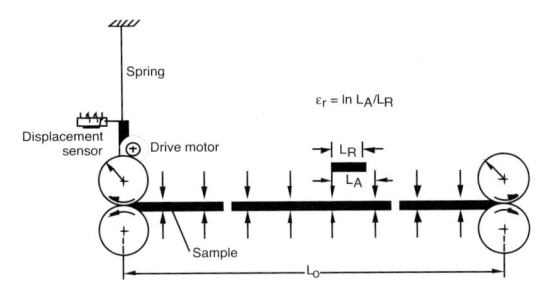

Figure 3.63: Schematic diagram of an extensional rheometer.

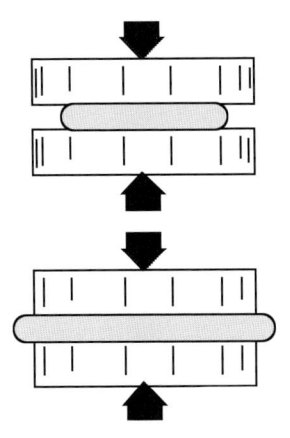

Figure 3.64: Schematic diagram of squeezing flow.

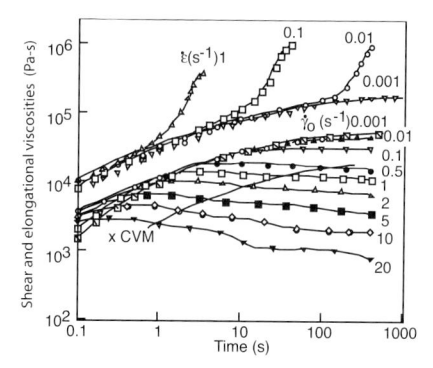

Figure 3.65: Development of elongational and shear viscosities during deformation for polyethylene samples (after Meissner).

Figure 3.65 shows this effect by comparing shear and elongational rheometry data of polyethylene.

Finally, another equibiaxial deformation test is carried out by blowing a bubble and measuring the pressure required to blow the bubble and the size of the bubble during the test, as schematically depicted in Fig. 3.66. This test has been successfully used to measure extensional properties of polymer membranes for blow molding and thermoforming applications. Here, a sheet is clamped between two plates with circular holes and a pressure differential is introduced to deform it. The pressure applied and deformation of the sheet are monitored over time and related to extensional properties of the material.

3.4.4 Surface Tension

Surface tension plays a significant role in the deformation of polymers during flow, especially in dispersive mixing of polymer blends. Surface tension, σ_S, between two materials appears as a result of different intermolecular interactions. In a liquid-liquid system, surface tension manifests itself as a force that tends to maintain the surface between the two materials to a minimum. Thus, the equilibrium shape of a droplet inside a matrix, which is at rest, is a sphere. When three phases touch, such as liquid, gas, and solid, we get different contact angles depending on the surface tension between the three phases.

Figure 3.67 schematically depicts three different cases. In case 1, the liquid perfectly wets the surface with a continuous spread, leading to a wetting angle of zero. Case 2, with moderate surface tension effects, shows a liquid that has a tendency to flow over the surface with a contact angle between zero and $\pi/2$. Case 3, with a high surface tension effect, is where the liquid does not wet the surface which results in a contact angle greater than $\pi/2$. In Fig. 3.67, σ_S denotes

3.4 Rheological Properties

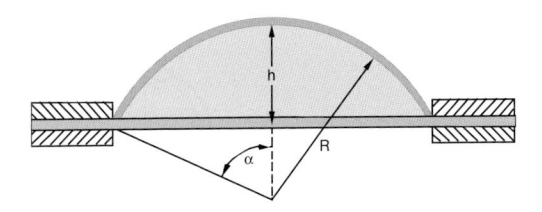

Figure 3.66: Schematic diagram of sheet inflation.

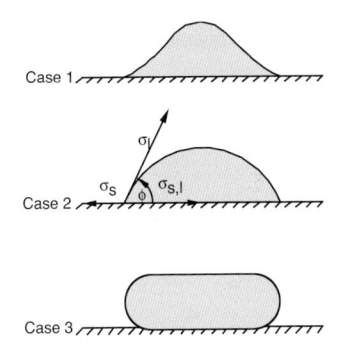

Figure 3.67: Schematic diagram of contact between liquids and solids with various surface tension effects.

the surface tension between the gas and the solid, σ_l the surface tension between the liquid and the gas, and σ_{sl} the surface tension between the solid and liquid.

The wetting angle can be measured using simple techniques such as a projector, as shown schematically in Fig. 3.68. This technique, originally developed by Zisman, can be used in the ASTM D2578 standard test. Here, surface tension, σ_l are applied to a film. The measured values of $\cos\phi$ are plotted as a function of surface tension σ_l, as shown in Fig. 3.69, and extrapolated to find the critical surface tension, σ_c, required for wetting.

For liquids of low viscosity, a useful measurement technique is the tensiometer, schematically represented in Fig. 3.70. Here, the surface tension is related to the force it takes to pull a platinum ring from a solution. Surface tension for selected polymers are listed in Table 3.14, for some solvents in Table 3.15 and between polymer-polymer systems in Table 3.16.

There are many areas in polymer processing and in engineering design with polymers where surface tension plays a significant role. These are mixing of polymer blends, adhesion, treatment of surfaces to make them non-adhesive and sintering. During manufacturing, it is often necessary to coat and crosslink a

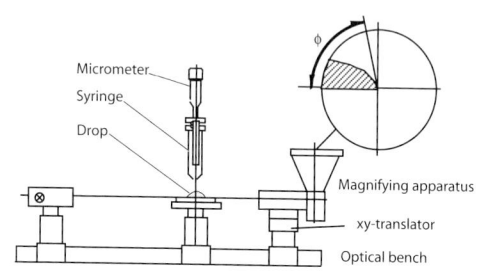

Figure 3.68: Schematic diagram of apparatus to measure contact angle between liquids and solids.

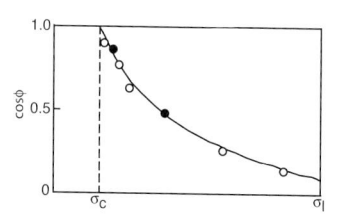

Figure 3.69: Contact angle as a function of surface tension.

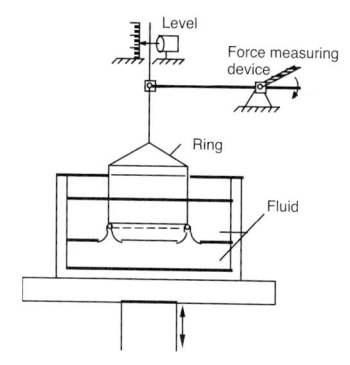

Figure 3.70: Schematic diagram of a tensiometer used to measure surface tension of liquids.

3.4 Rheological Properties

surface with a liquid adhesive or bonding material. To enhance adhesion it is often necessary to raise surface tension by oxidizing the surface, by creating COOH-groups, using flames, etching or releasing electrical discharges.

Table 3.14: Typical Surface Tension Values of Selected Polymers at $180^{\circ}C$

Polymer	σ_s (N/m)	$\partial\sigma_s/\partial T$ (N/m/K)
Polyamide resins ($290^{\circ}C$)	0.0290	-
Polyethylene (linear)	0.035	-5.7×10^{-5}
Polyethylene teraphthalate ($290^{\circ}C$)	0.045	-6.5×10^{-5}
Polyisobutylene	0.0234	-6×10^{-5}
Polymethyl methacrylate	0.0289	-7.6×10^{-5}
Polypropylene	0.030	-5.8×10^{-5}
Polystyrene	0.0292	-7.2×10^{-5}
Polytetrafluoroethylene	0.0094	-6.2×10^{-5}

Table 3.15: Surface Tension for Several Solvents

Solvent	Surface Tension - σ_s (N/m)
n-Hexane	0.0184
Formamide	0.0582
Glycerin	0.0634
Water	0.0728

Table 3.16: Surface Tension Between Polymers

Polymers	σ_s (N/m)	$\partial\sigma_s/\partial T$ (N/m/K)	T ($^{\circ}C$)
PE-PP	0.0011	-	140
PE-PS	0.0051	2.0×10^{-5}	180
PE-PMMA	0.0090	1.8×10^{-5}	180
PP-PS	0.0051	-	140
PS-PMMA	0.0016	1.3×10^{-5}	140

This is also the case when enhancing the adhesion properties of a surface before painting. On the other hand, it is often necessary to reduce adhesiveness of a surface such as required when releasing a product from the mold cavity or when coating a pan to give it nonstick properties. A material that is often used for this purpose is polytetrafluoroethylene (PTFE), mostly known by its tradename of teflon.

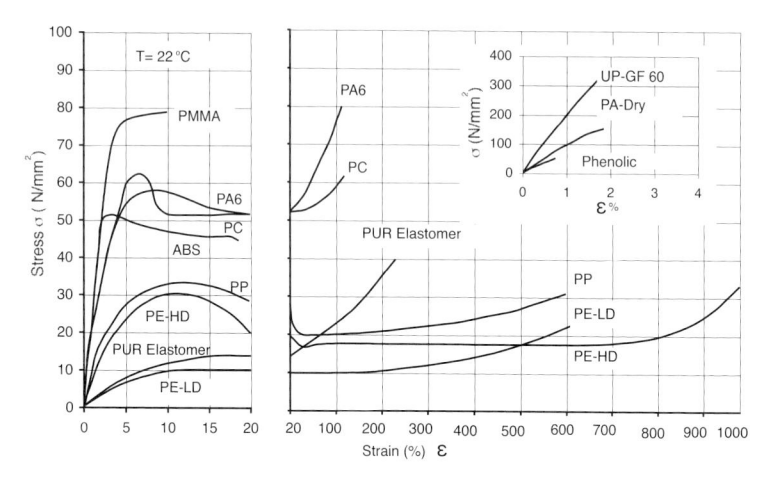

Figure 3.71: Tensile stress-strain curves for several materials.

3.5 MECHANICAL PROPERTIES

Polymeric materials are implemented into various designs because of their low cost, processability, and desirable material properties. Of interest to the design engineer are the short and long-term responses of a loaded component. Properties for short-term responses are usually acquired through short-term tensile tests and impact tests, whereas long-term responses depend on properties measured using techniques such as the creep and the dynamic test.

3.5.1 The Short-Term Tensile Test

The most commonly used mechanical test is the short-term stress-strain tensile test. Stress-strain curves for selected polymers are displayed in Fig. 3.71.

The next two sections discuss the short-term tensile test for elastomers and thermoplastic polymers separately. The main reason for identifying two separate topics is that the deformation of a cross-linked elastomer and an uncross-linked thermoplastic vary greatly. The deformation in a cross-linked polymer is in general reversible, whereas the deformation in typical uncross-linked polymers is associated with molecular chain relaxation, which makes the process time-dependent, and sometimes irreversible.

Rubber Elasticity The main feature of elastomeric materials is that they can undergo very large and reversible deformations. This is because the curled-up polymer chains stretch during deformation but are hindered in sliding past each

3.5 Mechanical Properties

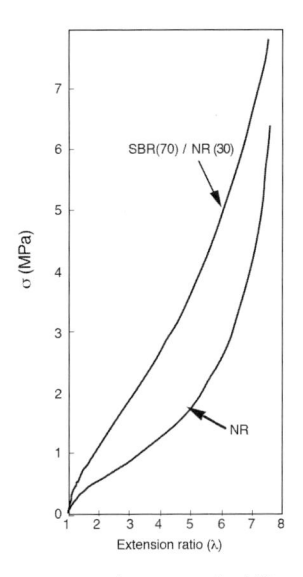

Figure 3.72: Experimental stress-extension curves for NR and a SBR/NR compound.

other by the cross-links between the molecules. Once a load is released, most of the molecules return to their coiled shape. As an elastomeric polymer component is deformed, the slope of the stress-strain curve drops significantly as the uncurled molecules provide less resistance and entanglement, allowing them to move more freely. Eventually, at deformations of about 400%, the slope starts to increase since the polymer chains are fully stretched. This is followed by polymer chain breakage or crystallization which ends with fracture of the component. Stress-deformation curves for natural rubber (NR) and a rubber compound [3] composed of 70 parts of styrene-butadiene-rubber (SBR) and 30 parts of natural rubber are presented in Fig. 3.72. Because of the large deformations, typically several hundred percent, the stress-strain data are usually expressed in terms of extension ratio, λ defined by

$$\lambda = \frac{L}{L_0} \qquad (3.25)$$

where L represents the instantaneous length and L_0 the initial length of the specimen.

Finally, it should be noted that the stiffness and strength of rubber is increased by filling with carbon black. The most common expression for describing the effect of carbon black content on the modulus of rubber was originally derived by Guth and Simha [8] for the viscosity of particle suspensions, and later used

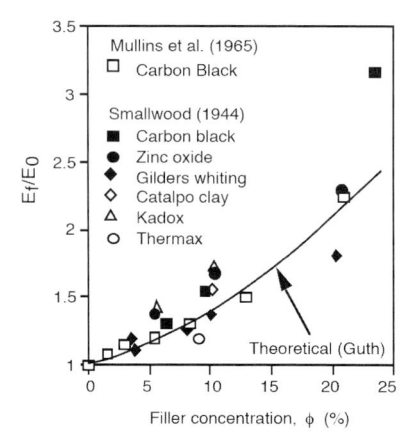

Figure 3.73: Effect of filler on modulus of natural rubber.

by Guth [9] to predict the modulus of filled polymers. The Guth equation can be written as

$$\frac{G_f}{G_0} = 1 + 2.5\phi + 14.1\phi^2 \qquad (3.26)$$

where G_f is the shear modulus of the filled material, G_0 is the shear modulus of the unfilled material and ϕ the volume fraction of particulate filler. The above expression is compared to experiments in Fig. 3.73.

The Tensile Test and Thermoplastic Polymers Of all the mechanical tests done on thermoplastic polymers, the tensile test is the least understood, and the results are often misinterpreted and misused. Since the test was inherited from other materials that have linear elastic stress-strain responses, it is often inappropriate for testing polymers. However, standardized tests such as DIN 53457 and ASTM D638 are available to evaluate the stress-strain behavior of polymeric materials. The DIN 53457, for example, is performed at a constant elongational strain rate of 1% per minute, and the resulting data are used to determine the short-term modulus. The ASTM D638 test also uses one rate of deformation per material to measure the modulus; a slow speed for brittle materials and a fast speed for ductile ones. However, these tests do not reflect the actual rate of deformation experienced by the narrow portion of the test specimen, making it difficult to maintain a constant speed within the region of interest. The standard tests ASTM D638 and ISO 527-1 are presented in Table 3.17.

3.5 Mechanical Properties

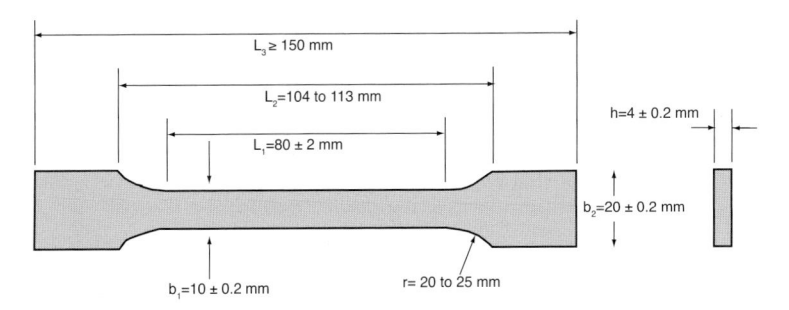

Figure 3.74: Standard ISO-3167 tensile bar.

Table 3.17: Standard Methods of Measuring Tensile Properties (after Shastri)

Standard	ISO 527-1:93 and 527-2:93	D 638-98
Specimen	ISO 3167 (Type A or B*) multipurpose test specimens (Figure 3.74). * Type A is recommended for directly molded specimens, so the 80 mm x 10 mm x 4 mm specimens required for most tests in ISO 10350 - 1 can be cut from the center of these specimens. Type B is recommended for machined specimens. **Dimensions for ISO 3167 specimens are:** Overall Length $\Rightarrow$ >150 mm Width $\Rightarrow$ 10 mm Thickness $\Rightarrow$ 4 mm Fillet radius $\Rightarrow$ 20-25 mm (Type A) or >60 mm (Type B)	For rigid/semirigid plastics: D 638 Type I specimens (Figure 3.75) are the preferred specimen and shall be used when sufficient material having a thickness of 7 mm or less is available. **Dimensions for D 638 Type I specimens are:** Overall Length $\Rightarrow$ 165 mm Width $\Rightarrow$ 12.7 mm Thickness $\Rightarrow$ 3.2 mm Fillet radius $\Rightarrow$ 76 mm Length of parallel narrow section $\Rightarrow$ 57 mm Length of parallel narrow section $\Rightarrow$ 80 mm (Type A) or 60 mm (Type B)
Conditioning	Specimen conditioning, including any post molding treatment, shall be carried out at 23°C ± 2°C and 50 ± 5 % R.H. for a minimum length of time of 88h, except where special conditioning is required as specified by the appropriate material standard.	At 23 ± 2 °C and 50 ± 5 % relative humidity for not less than 40h prior to testing in accordance with D 618 Procedure A for those tests where conditioning is required. For hygroscopic materials, the material specification takes precedence over the above routine preconditioning requirements.

Continued on next page

Standard	ISO 527-1:93 and 527-2:93	D 638-98
Test procedures	A minimum of five specimens shall be prepared in accordance with the relevant material standard. When none exists, or unless otherwise specified specimens shall be directly compression or injection molded in accordance with ISO 293 or ISO 294-1. Test speed for ductile failure (defined as yielding or with a strain at break >10%) is 50 mm/min and for a brittle failure (defined as rupture without yielding or strain at break < 10%) is 5 mm/min. For modulus determinations the test speed is not specified in ISO 10350; however, in ISO 527-2 it is specified for molding and extrusion plastics that the test speed is 1 mm/min. Extensometers are required for determining strain at yield and tensile modulus. The specified initial gauge length is 50 mm. The extensometer shall be essentially free of inertia lag at the specified speed of testing and capable of measuring the change in gauge with an accuracy of 1% of the relevant value or better. This corresponds to ± 1 micrometer for the measurement of modulus on a gauge length of 50 mm. The reported tensile modulus is a chord modulus determined by drawing a straight line that connects the stress at 0.05% strain and the stress at 0.25% strain. There is no requirement for toe compensation in determining a corrected zero point, if necessary.	A minimum of five test specimens shall be prepared by machining operations or die cutting the materials in sheet, plate, slab or similar form. Specimens can also be prepared by injection or compression molding the material to be tested. Test speed is specified in the specification for the material being tested. If no speed is specified, then use the lowest speed given in Table 1 (5, 50, or 500 mm/min) which gives rupture within 0.5 to 5.0 minutes. Modulus testing may be conducted at the same speed as the other tensile properties provided that recorder response and resolution are adequate. Extensometers are required for determining strain at yield and tensile modulus. The specified initial gauge length is 50 mm. For modulus determinations, an extensometer which meets Class B-2 (Practice E-38) is required, for low extensions (<20%) the extensometer must at least meet Class C (Practice E38) requirements, for high extensions (>20%) any measurement technique which has an error no greater than ± 10% can be used. Tangent modulus is determined by drawing a tangent to the steepest initial straight line portion of the load-deflection curve and then dividing the difference in stress on any section of this line by the corresponding difference in strain. Secant modulus is the ratio of stress to corresponding strain at any given point on the stress-strain curve, or the slope of the straight line that joins the zero point or corrected zero point and the selected point corresponding to the strain selected on the actual stress-strain curve. Toe compensation, if applicable as defined, is mandatory.

Continued on next page

3.5 Mechanical Properties

Standard	ISO 527-1:93 and 527-2:93	D 638-98
Values and units	For ductile materials: Stress at yield ⇒ MPa Strain at yield ⇒ % Stress at 50% strain* ⇒ MPa Nominal strain at break** ⇒ % Tensile modulus ⇒ MPa * If the material does not yield before 50% strain, report stress at 50% strain. ** Nominal strain at break based on initial and final grip separations, if rupture occurs above 50% nominal strain one can either report the strain at break or simply > 50%. Stress at break ⇒ MPa Strain at break ⇒ % Chord modulus (0.5-0.25% strain) ⇒ MPa	For ductile materials: Stress at yield ⇒ MPa Strain at yield ⇒ % Stress at break ⇒ MPa Strain at break ⇒ % Tangent modulus or ⇒ MPa Secant modulus ⇒ MPa Stress at break ⇒ MPa Strain at break ⇒ % Modulus ⇒ MPa

However, the rate of deformation has a great impact on the measured results. A typical test performed on PMMA at various strain rates at room temperature is shown in Fig. 3.76. The increased curvature in the results with slow elongational speeds suggests that stress relaxation plays a significant role during the test. Similarly, Fig. 3.77 reflects the effect of rate of deformation on the stress-strain behavior of a typical semi-crystalline polymer. The ultimate strength is also affected by the deformation rate, and the trend depends on the polymer, as depicted in Fig. 3.78. Again, the effect is due to the relaxation behavior of the polymer. The relaxation behavior and memory effects of polymers is illustrated in Fig. 3.79 which shows the strain one minute after the specimen failed, for tests performed at different rates of deformation.

It can be shown that for small strains the secant modulus, described by

$$E_s = \frac{\sigma}{\epsilon} \tag{3.27}$$

and the tangent modulus, defined by

$$E_t = \frac{d\sigma}{d\epsilon} \tag{3.28}$$

are independent of strain rate and are functions only of time and temperature. This is schematically shown in Fig. 3.80.

The figure shows two stress-strain responses: one at a slow elongational strain rate, $\dot{\epsilon}_1$, and one at twice the speed, defined by $\dot{\epsilon}_2$. The tangent modulus at ϵ_1 in the curve with $\dot{\epsilon}_1$ is identical to the tangent modulus at ϵ_2 in the curve with $\dot{\epsilon}_2$, where ϵ_1 and ϵ_2 occurred at the same time. For small strains the tangent

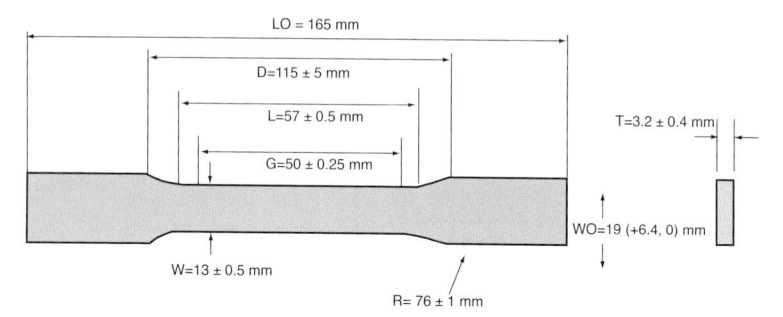

Figure 3.75: Standard ASTM-D638 tensile bar.

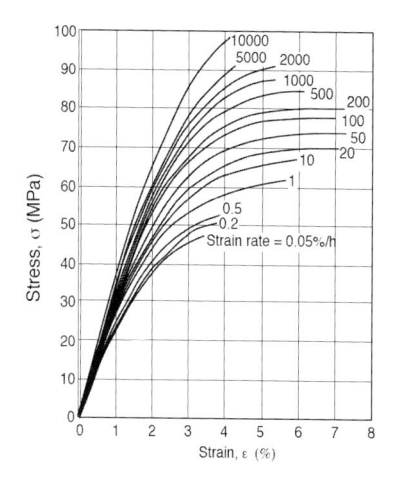

Figure 3.76: Stress-strain behavior of PMMA at various strain rates.

3.5 Mechanical Properties

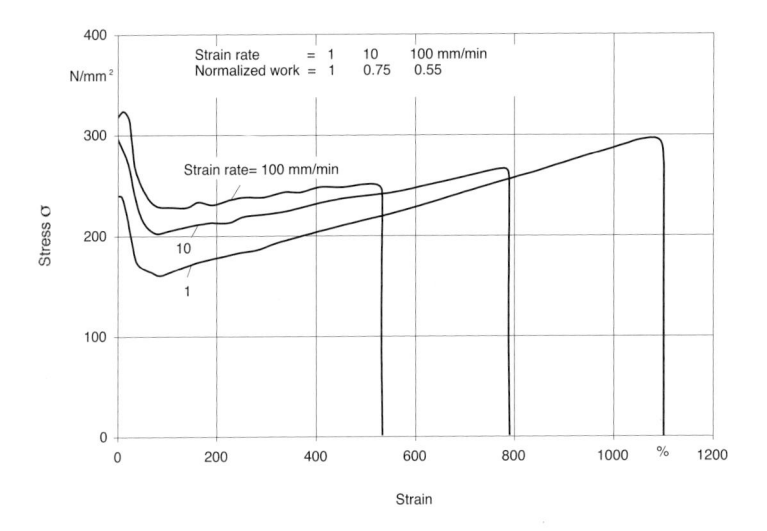

Figure 3.77: Stress-strain behavior of PE at various rates of deformation.

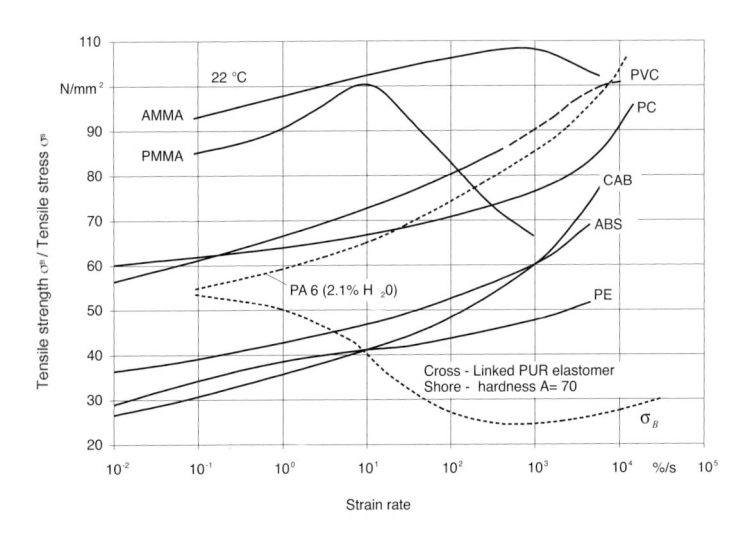

Figure 3.78: Rate of deformation dependance of strength for various thermoplastics.

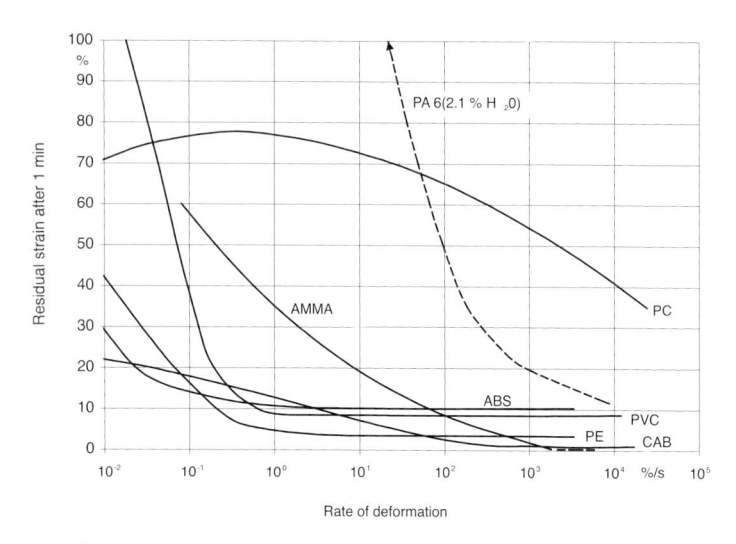

Figure 3.79: Residual strain in the test specimen as a function of strain rate for various thermoplastics.

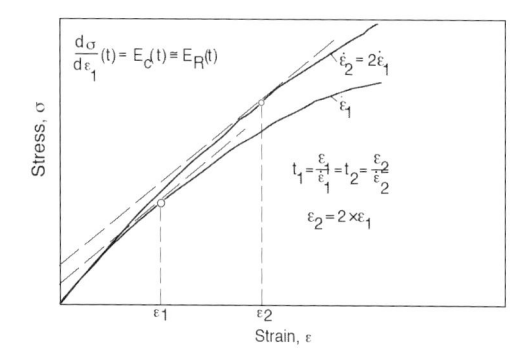

Figure 3.80: Schematic of the stress-strain behavior of a viscoelastic material at two rates of deformation.

modulus, E_t, is identical to the relaxation modulus, E_r, measured with a stress relaxation test. This is important since the complex stress relaxation test can be replaced by the relatively simple short-term tensile test by plotting the tangent modulus versus time. Generic stress-strain curves and stiffness and compliance plots for amorphous and semi-crystalline thermoplastics are shown in Fig. 3.81. The stress-strain behavior for thermoplastic polymers can be written in a general

3.5 Mechanical Properties

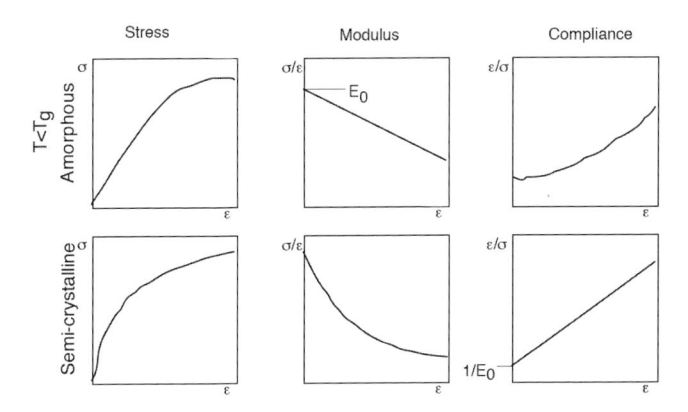

Figure 3.81: Schematic of the stress-strain response, modulus and compliance of amorphous and semi-crystalline thermoplastics at constant rates of deformation.

form as

$$\sigma = E_0 \epsilon \frac{1 - D_1 \epsilon}{1 + D_2 \epsilon} \tag{3.29}$$

where E_0, D_1 and D_2 are time- and temperature-dependent material properties. The constant $D_1 = 0$ for semi-crystaline polymers and $D_2 = 0$ for amorphous plastics.

Figure 3.82 shows E_0 and D_2 for a high density polyethylene at 23 °C as a function of strain rate. The values of E_0, D_1 and D_2 can be easily calculated for each strain rate from the stress-strain diagram [16]. The modulus E_0 simply corresponds to the tangent modulus at small deformations where

$$\sigma = E_0 \epsilon \tag{3.30}$$

Assuming that for amorphous thermoplastics $D_2 \approx 0$ when $T \ll T_g$ and for semi-crystalline thermoplastics $D_1 \approx 0$ when $T \gg T_g$, we can compute D_1 from

$$D_1 = \frac{\sigma_2 \epsilon_1 - \sigma_1 \epsilon_2}{\sigma_2 \epsilon_1^2 - \sigma_1 \epsilon_2^2} \tag{3.31}$$

and D_2 from

$$D_2 = \frac{\sigma_1 \epsilon_2 - \sigma_2 \epsilon_1}{\epsilon_1 \epsilon_2 (\sigma_2 - \sigma_1)} \tag{3.32}$$

Depending on the time scale of the experiment, a property that also varies considerably during testing is Poisson's ratio, ν. Figure 3.83 shows Poisson's ratio for PMMA deformed at rates (%/h) between 10^{-2} (creep) and 10^3 (impact).

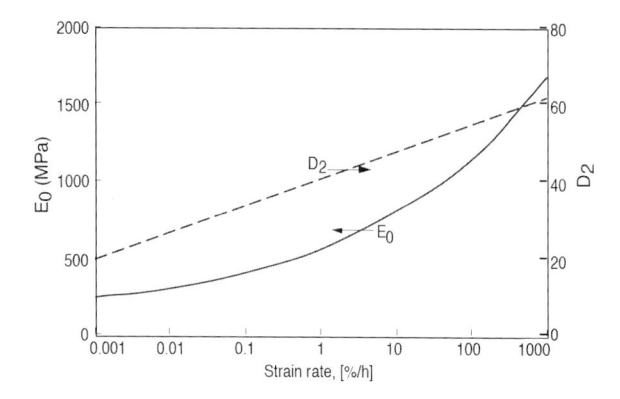

Figure 3.82: Coefficients E_0 and D_2 for a high density polyethylene at 23 °C.

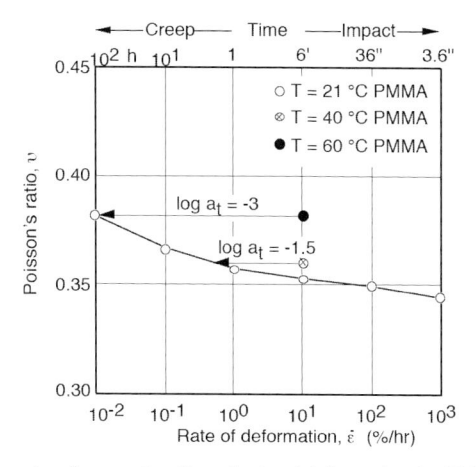

Figure 3.83: Poisson's ratio as a function of rate of deformation for PMMA.

Temperature affects Poisson's ratio in a similar way, as depicted in Fig. 3.84 for several thermoplastics. The limits are $\nu=0.5$ (fluid) for high temperatures or very slow deformation speeds and $\nu=0.33$ (solid) at low temperatures or high deformation speeds. In fiber filled plastics, Poisson's ratio is affected by the fiber content and the orientation of the reinforcing fibers. This is demonstrated in Fig. 3.85 for fiber filled thermosets.

Flexular Test The flexural test is a widely accepted test in the plastics industry, because it portrays well bending load cases, which often reflects realistic situations.

3.5 Mechanical Properties

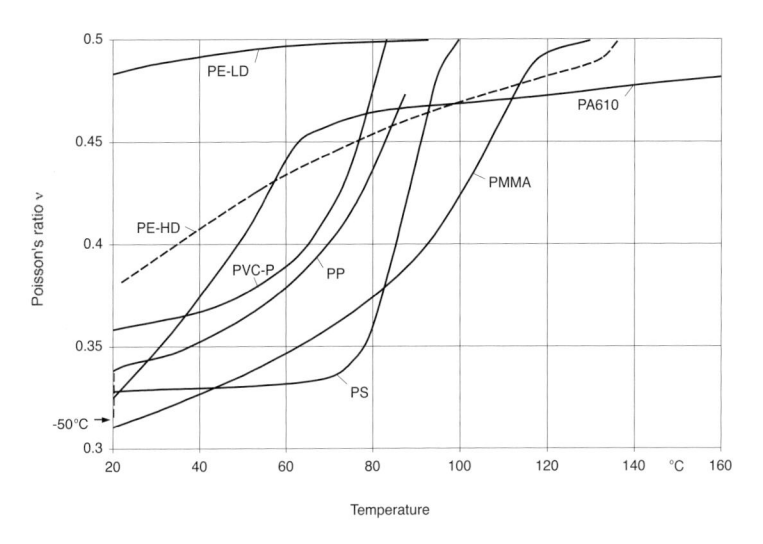

Figure 3.84: Poisson's ratio as a function of temperature for various temperatures.

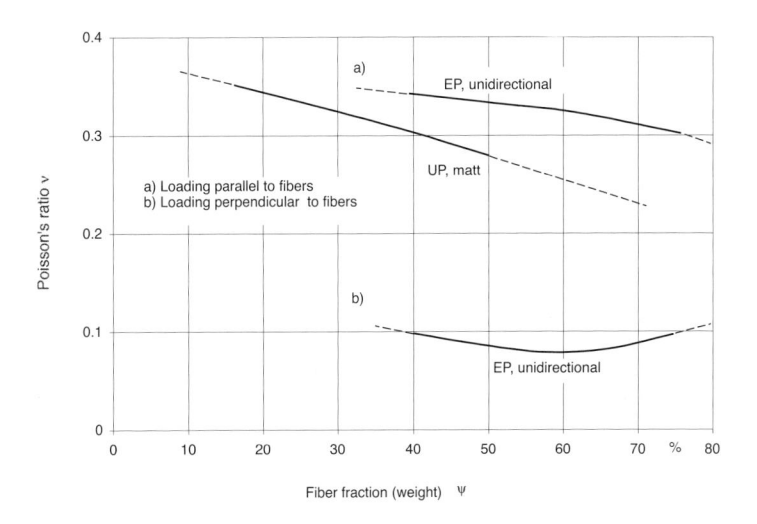

Figure 3.85: Poisson's ratio as a function of fiber content for fiber filled thermosets.

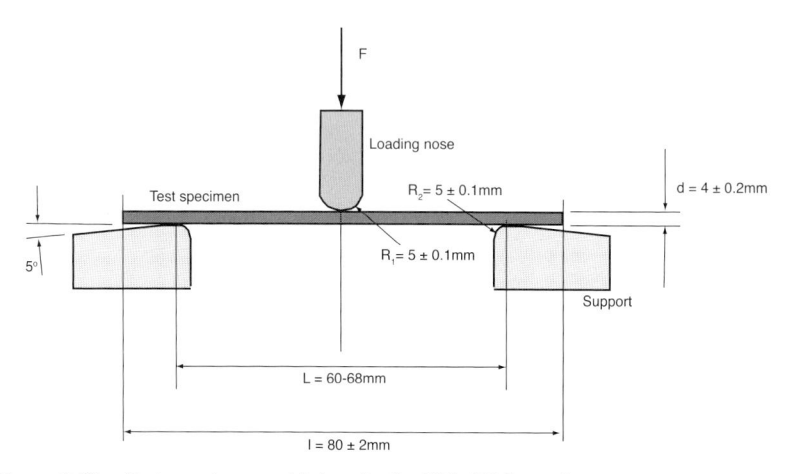

Figure 3.86: Test specimen and fixture for the ISO 178 flexural test.

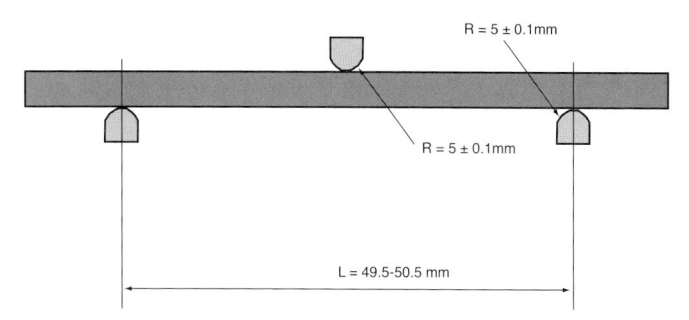

Figure 3.87: Test specimen and fixture for the ASTM D790 flexural test.

3.5 Mechanical Properties

However, due to the combined tensile and compressive stresses, encountered in bending, it is a test which renders properties that should be regarded with caution. The test is summarized for ISO and ASTM standards in Table 3.18.

Table 3.18: Standard Methods of Measuring Flexural Properties (after Shastri)

Standard	ISO 178	D 790 - 98
Specimen	80 mm x 10 mm x 4 mm cut from the center of an ISO 3167 Type A specimen. In any one specimen the thickness within the central one-third of length shall not deviate by more than 0.08 mm from its mean value, and the corresponding allowable deviation in the width is 0.3 mm from its mean value.	Specimens may be cut from sheets, plates, molded shapes or molded to the desired finished dimensions. The recommended specimen for molding materials is 127 mm x 12.7 mm x 3.2 mm.
Conditioning	Specimen conditioning, including any post molding treatment, shall be carried out at $23°C \pm 2°C$ and 50 ± 5 % R.H. for a minimum length of time of 88h, except where special conditioning is required as specified by the appropriate material standard.	At 23 ± 2 °C and 50 ± 5 % relative humidity for not less than 40h prior to testing in accordance with to D 618 Procedure A for those tests where conditioning is required. For hygroscopic materials, the material specification takes precedence over the above routine preconditioning requirements.
Apparatus	Support and loading nose radius 5.0 ± 0.1 mm (Fig. 3.86) Parallel alignment of the support and loading nose must be less than or equal to 0.02 mm.	Support and loading nose radius 5.0 ± 0.1 mm (Fig. 3.87) Parallel alignment of the support and loading noses may be checked by means of a jig with parallel grooves into which the loading nose and supports will fit if properly aligned.
	ISO/IEC Methods as specified by ISO 10350 - 1	**ASTM Methods**
	Support span length 60 - 68 mm (Adjust the length of the span to within 0.5%, which is 0.3 mm for the span length specified above) Support span to specimen depth ratio 16 ± 1; 1 mm/mm	Support span length* 49.5 -50.5 mm (Measure the span accurately to the nearest 0.1 mm for spans less than 63 mm. Use the measured span length for all calculations). Support span to specimen depth ratio 16 (+ 4, -1); 1 mm/mm (Specimens with a thickness exceeding the tolerance of ± 0.5%).

Continued on next page

Standard	ISO 178	D 790 - 98
Test procedures		Testing conditions indicated in material specifications take precedence; therefore, it is advisable to refer to the material specification before using the following procedures.
	Test speed ⇒ mm/min	Procedure A crosshead speed* ⇒ 1.3 mm/min Procedure B crosshead speed* ⇒ 13 mm/min * *Procedure A must be used for modulus determinations, Procedure B may be used for flexural strength determination only*
	A minimum of five specimens shall be prepared in accordance with the relevant material standard. When none exists, or unless otherwise specified, specimens shall be directly compression or injection molded in accordance with ISO 293 or ISO 294-1. Test specimens that rupture outside the central one-third of the span length shall be discarded and new specimen shall be tested in their place.	A minimum of five test specimens are required. No specimen preparation conditions are given.
	Measure the width of the test specimen to the nearest 0.1 mm and the thickness to the nearest 0.01 mm in the center of the test specimen.	Measure the width and depth of the test specimen to the nearest 0.03 mm at the center of the support span.
	The reported flexural modulus is a chord modulus determined by drawing a straight line that connects the stress at 0.05% strain and the stress at 0.25% strain. There is no requirement for toe compensation in determining a corrected zero point, if necessary.	Tangent modulus is determined by drawing a tangent to the steepest initial straight line portion of the load-deflection curve and then dividing the difference in stress on any section of this line by the corresponding difference in strain. Secant modulus is the ratio of stress to corresponding strain at any given point on the stress-strain curve, or the slope of the straight line that joins the zero point and a selected point on the actual stress-strain curve. Toe compensation, if applicable, as defined is mandatory.

Continued on next page

3.5 Mechanical Properties

Standard	ISO 178	D 790 - 98
Values and units	Flexural modulus $\Rightarrow$ MPa Flexural strength, at rupture $\Rightarrow$ MPa Flexural strength, at maximum strain* $\Rightarrow$ MPa *At conventional deflection which is 1.5 x height: therefore 4 mm specimens would have a maximum strain at 3.5%.	Tangent modulus or $\Rightarrow$ MPa Secant modulus $\Rightarrow$ MPa Flexural strength, (at rupture*) $\Rightarrow$ MPa Flexural yield strength** $\Rightarrow$ MPa * Maximum allowable strain in the outer fibers is 0.05 mm/mm **The point where the load does not increase with increased deflection, provided it occurs before the maximum strain rate*

3.5.2 Impact Strength

In practice, nearly all polymer components are subjected to impact loads. Since many polymers are tough and ductile, they are often well suited for this type of loading. However, under specific conditions even the most ductile materials, such as polypropylene, can fail in a brittle manner at very low strains. These types of failure are prone to occur at low temperatures and at very high deformation rates. As the rate of deformation increases, the polymer has less time to relax. The limiting point is when the test is so fast that the polymer behaves as a linear elastic material. At this point, fracture occurs at a minimum value of strain, ϵ_{min} and its corresponding stress, σ_{max}. During impact, one should always assume that if this minimum strain value is exceeded at any point in the component, initial fracture has already occurred. Table 3.19 presents minimum elongation at break and their corresponding stresses for selected thermoplastics during impact loading.

Table 3.19: Minimum Elongation at Break and corresponding stress on Impact Loading

Polymers	ϵ_{min} (%)	σ_{max} (MPa)
HMW-PMMA	2.2	135
PA6+25% SFR	1.8	175
PVC-U	2.0	125
POM	4.0	>130
PC+20% SFR	4.0	>110
PC	6.0	>70

Figure 3.88 summarizes the stress-strain and fracture behavior of a HMW-PMMA tested at various rates of deformation. The area under the stress-strain curves represents the *volume-specific energy to fracture* (w). For impact, the elongation at break of 2.2% and the stress at break of 135 MPa represent a minimum of volume-specific energy because the stress increases with higher

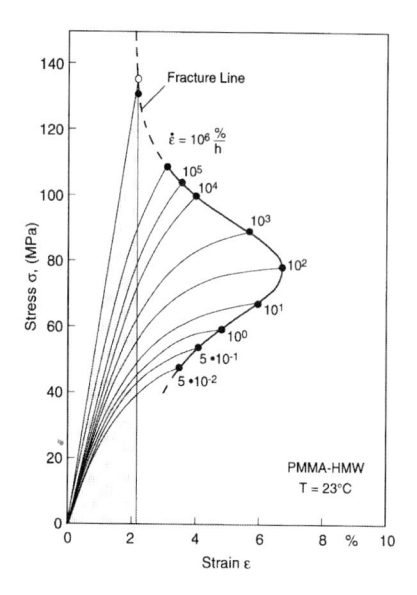

Figure 3.88: Stress-strain behavior of HMW-PMMA at various rates of deformation.

rates of deformation, but the elongation at break remains constant. Hence, if we assume a linear behavior, the *minimum volume-specific energy absorption* up to fracture can be calculated using

$$w_{min} = \frac{1}{2}\sigma_{max}\epsilon_{min} \qquad (3.33)$$

The impact strength of a copolymer and polymer blend of the same materials can be quite different, as shown in Fig. 3.89. From the figure it is clear that the propylene-ethylene copolymer, which is an elastomer, has a much higher impact resistance than the basic polypropylene-polyethylene blend. It should be pointed out here that elastomers usually fail by ripping. The ripping or tear strength of elastomers can be tested using the ASTM D1004, ASTM D1938, or DIN 53507 test methods. The latter two methods make use of rectangular test specimens with clean slits cut along the center. The tear strength of elastomers can be increased by introducing certain types of particulate fillers. For example, a well dispersed carbon black filler can double the ripping strength of a typical elastomer. Figure 3.90 shows the effect that different types of fillers have on the ripping strength of a polychloroprene elastomer.

In general, one can say if the filler particles are well dispersed and have diameters between 20 nm and 80 nm, they will reinforce the matrix. Larger particles will act as microscopic stress concentrators and will lower the strength

3.5 Mechanical Properties

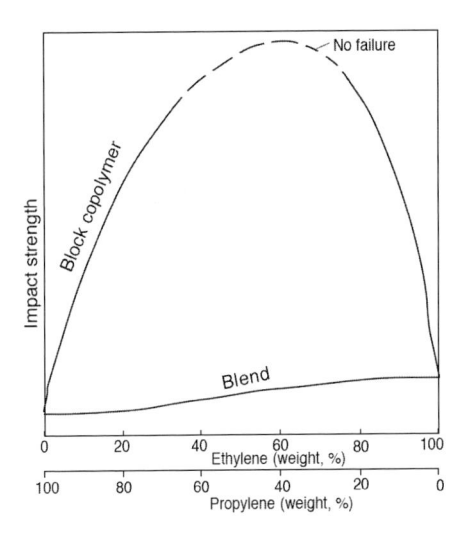

Figure 3.89: Impact strength of a propylene-ethylene copolymer and a polypropylene-polyethylene polymer blend.

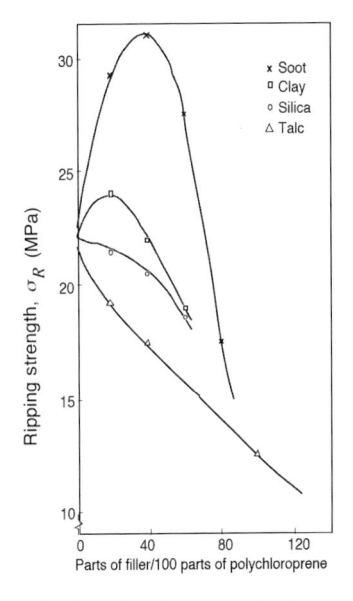

Figure 3.90: Ripping strength of a polychloroprene elastomer as a function of filler content for different types of fillers (after Menges).

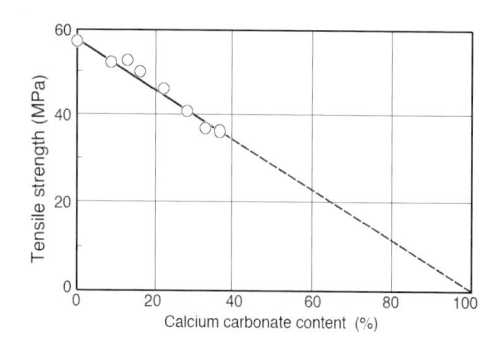

Figure 3.91: Tensile strength of PVC as a function of calcium carbonate content (after Menges).

of the polymer component. A case where the filler adversely affects the polymer matrix is presented in Fig. 3.91, where the strength of PVC is lowered with the addition of a calcium carbonate powder.

Impact Test The most common impact tests used to evaluate the strength of polymers are the *Izod* and the *Charpy* tests.

The Charpy test evaluates the bending impact strength of a small notched or unnotched simply supported specimen that is struck by a swinging hammer. There are notched and unnotched Charpy impact tests. The standard unnotched Charpy impact test is given by the ISO 179 test, however, ASTM does not offer such a test. The ISO 179 test is presented in Table 3.20. The notched Charpy test is done such that the notch faces away from the swinging hammer creating tensile stresses within the notch, as shown in Fig. 3.96. The standard ISO 179 also describes the notched Charpy test, as well as the ASTM D256 and DIN 53453 tests. The standard Charpy notched tests ISO 179 and ASTM D256 are presented in Table 3.21.

The Izod test evaluates the impact resistance of a cantilevered notched bending specimen as it is struck by a swinging hammer. Figure 3.94 shows a typical Izod-type impact machine, and Fig. 3.95 shows a detailed view of the specimen, the clamp, and the striking hammer. The standard test method that describes the Izod impact test is also the ASTM-D 256 test.

The Izod and Charpy impact tests impose bending loads on the test specimens. For tensile impact loading one uses the standard tensile impact tests prescribed by tests ISO 8256 and ASTM D1822 presented in Table 3.22.

3.5 Mechanical Properties

Table 3.20: Standard Methods of Measuring Unnotched Charpy Impact Strength (after Shastri)

Standard	ISO 179 - 1 and ISO 179 - 2	No ASTM equivalent
Specimen	80 mm x 10 mm x 4 mm cut from the center of an ISO 3167 Type A specimen, also referred to as an ISO 179/1eU specimen	
Conditioning	Specimen conditioning, including any post molding treatment, shall be carried out at $23°C \pm 2°C$ and 50 ± 5 % R.H. for a minimum length of time of 88h, except where special conditioning is required as specified by the appropriate material standard.	
Apparatus	The machine shall be securely fixed to a foundation having a mass at least 20 times that of the heaviest pendulum in use and be capable of being leveled. Striking edge of the hardened steel pendulums is to be tapered to an included angle of $30 \pm 1°$ and rounded to a radius of 2.0 ± 0.5 mm The striking edge of the pendulum shall pass midway, to within ± 0.2 mm, between the specimen supports. The line of contact shall be within $\pm 2°$ of perpendicular to the longitudinal axis of the test specimen. Pendulums with specified nominal energies shall be used: 0.5, 1.0 2.0, 4.0, 5.0, 7.5, 15.0, 25.0, and 50.0 J. Velocity at impact is 2.9 + 10% m/s for the 0.5 to 5.0 J pendulums and 3.8 ± 10% m/s for pendulums with energies from 7.5 to 50.0 J. The support anviles line of contact with the specimen shall be 62.0 (+0.5, -0.0) mm.	

Continued on next page

Standard	ISO 179 - 1 and ISO 179 - 2	No ASTM equivalent
Test procedures	A minimum of ten specimens shall be prepared in accordance with the relevant material standard. When none exists, or unless otherwise specified, specimens shall be directly compression or injection molded in accordance with ISO 293 or ISO 294-1. Edgewise impact is specified. Consumed energy is 10 to 80% of the pendulum energy, at the corresponding specified velocity of impact. If more than one pendulum satisfies these conditions, the pendulum having the highest energy is used. (It is not advisable to compare results obtained using different pendulums) Maximum permissible frictional loss without specimen: 0.02% for 0.5 to 5.0 J pendulum 0.04% for 7.5 J pendulum 0.05% for 15.0 J pendulum 0.10% for 25.0 J pendulum 0.20% for 50.0 J pendulum Permissible error after correction with specimen: 0.01 J for 0.5, 1.0, and 2.0 J pendulums. No correction applicable for pendulums with energies > 2.0 J. Four types of failure are defined as: C- Complete break; specimen separates into one or more pieces. H- Hinge break; an incomplete break such that both parts of the specimen are only held together by a thin peripheral layer in the form of a hinge. P- Partial break; an incomplete break which does meet the definition for a hinge break. NB-Non-break; in the case of the non-break, the specimen is only bent and passed through, possibly combined with stress whitening.	

Continued on next page

3.5 Mechanical Properties

Standard	ISO 179 - 1 and ISO 179 - 2	No ASTM equivalent
Values and units	The measured values of complete and hinged breaks can be used for a common mean value with remark. If in the case of partial breaks a value is required, it shall be assigned with the letter P. In case of non-breaks, no figures are to be reported. (If within one sample the test specimens show different types of failures, the mean value for each failure type shall be reported). Unnotched Charpy impact strength $\Rightarrow$ kJ/m^2.	

Table 3.21: Standard Methods of Measuring Notched Charpy Impact Strength (after Shastri)

Standard	ISO 179 - 1 and ISO 179 - 2	D 256 - 97
Specimen	80 mm x 10 mm x 4 mm cut from the center of an ISO 3167 Type A specimen with a single notch A, also referred to as an ISO 179/1eA specimen. (See Figure 3.92)	124.5 to 127 mm x 12.7 mm x (*) mm specimen, * The width of the specimens shall be between 3.0 and 12.7 mm as specified in the material specification, or as agreed upon as representative of the crosssection in which the particular material may be used. (Figure 3.93)
	Notch A has a 45° ± 1° included angle with a notch base radius of 0.25 ± 0.05 mm. The notch should be at a right angle to the principal axis of the specimen. The specimens shall have a remaining width of 8.0 ± 0.2 mm after notching. These machined notches shall be prepared in accordance with ISO 2818.	A single notch with 45° ±1° included angle with a radius of curvature at the apex 0.25 ± 0.05 mm. The plane bisecting the notch angle shall be perpendicular to the face of the test specimen within 2° The depth of the plastic material remaining in the bar under the notch shall be 10.16 ± 0.05 mm. The notches are to be machined.
Conditioning	Specimen conditioning, including any post molding treatment, shall be carried out at 23°C ± 2°C and 50 ± 5 % R.H. for a minimum length of time of 88h, except where special conditioning is required as specified by the appropriate material standard.	At 23°C ± 2 °C and 50 ± 5 % relative humidity for not less than 40h prior to testing in accordance with D 618 Procedure A for those tests where conditioning is required. For hygroscopic materials, the material specification takes precedence over the above routine preconditioning requirements.

Continued on next page

Standard	ISO 179 - 1 and ISO 179 - 2	D 256 - 97
Apparatus	The machine shall be securely fixed to a foundation having a mass at least 20 times that of the heaviest pendulum in use and be capable of being leveled.	The machine shall consist of a massive base.
	Striking edge of the hardened steel pendulums is to be tapered to an included angle of 30° ± 1° and rounded to a radius of 2.0 ± 0.5 mm.	Striking edge of hardened steel pendulums is to be tapered to an included angle of 45° ± 2° and rounded to a radius of 3.17 ± 0.12 mm.
	Pendulums with the specified nominal energies shall be used: 0.5, 1.0, 2.0, 4.0, 5.0, 7.5, 15.0, 25.0, and 50.0 J.	Pendulum with an energy of 2.710 ± 0.135 J is specified for all specimens that extract up to 85% of this energy. Heavier pendulums are to be used for specimens that require more energy; however, no specific levels of energy pendulums are specified.
	Velocity at impact is 2.9 ± 10% m/s for the 0.5 to 5.0 J pendulums and 3.8 ± 10% m/s for pendulums with energies from 7.5 to 50.0 J. The support anvils line of contact with the specimen shall be 62.0 (+0.5, -0.0) mm.	Velocity at impact is approximately 3.46 m/s, based on the vertical height of fall of the striking nose specified at 610 + 2 mm. The anvils line of contact with the specimen shall be 101.6 ± 0.5 mm.
Test procedures	A minimum of ten specimens shall be prepared in accordance with the relevant material standard. When none exists, or unless otherwise specified, specimens shall be directly compression or injection molded in accordance with ISO 293 or ISO 294-1. Edgewise impact is specified (Figure 3.92).	At least five, preferably 10 specimens shall be prepared from sheets, composites (not recommended), or molded specimen. Specific specimen preparations are not given or referenced. Edgewise impact is specified (Figures 3.95 and 3.96).
	Consumed energy is 10 to 80% of the pendulum energy, at the corresponding specified velocity of impact. If more than one pendulum satisfies these conditions, the pendulum having the highest energy is used. (It is not advisable to compare results obtained using different pendulum)	

Continued on next page

3.5 Mechanical Properties

Standard	ISO 179 - 1 and ISO 179 - 2	D 256 - 97
Test procedures	Maximum permissible frictional loss without specimen: 0.02% for 0.5 to 5.0 J pendulum 0.04% for 7.5 J pendulum 0.05% for 15.0 J pendulum 0.10% for 25.0 J pendulum 0.20% for 50.0 J pendulum Permissible error after correction with specimen: 0.01 J for 0.5, 1.0, and 2.0 J pendulums. No correction applicable for pendulums with energies > 2.0 J. Four types of failure are defined as: C- Complete break; specimen separates into two or more pieces. H- Hinge break; an incomplete break such that both parts of the specimen are only held together by a thin peripheral layer in the form of a hinge. P- Partial break; an incomplete break which does not meet the definition for a hinge break. NB-Non-break; in the case of the non-break, the specimen is only bent and passed through, possibly combined with stress whitening.	Windage and friction correction are not mandatory; however, a method of determining these values is given. Four types of failure are specified: C- Complete break; specimen separates into two or more pieces. H- Hinge break; an incomplete break such that one part of the specimen cannot support itself above the horizontal when the other part is held vertically (less than 90o included angle). P- Partial break; an incomplete break which does not meet the definition for a hinge break, but has fractured at least 90% of the distance between the vertex of the notch and the opposite side. NB-Non-break; an incomplete break where the fracture extends less than 90% of the distance between the vertex of the notch and the opposite side.
Values and units	The measured values of complete and hinged breaks can be used for a common mean value with remark. If in the case of partial breaks a value is required, it shall be signed with the letter P. In case of non-breaks, no figures are to be reported. (If within one sample the test specimens show different types of failures, the mean value for each failure type shall be reported) Notched Charpy impact strength $\Rightarrow$ kJ/m	Only measured values for complete breaks can be reported. (If more than one type of failure is observed for a sample material, then report the average impact value for the complete breaks, followed by the number and percent of the specimen failing in that manner suffixed by the letter code). Notched Charpy impact strength $\Rightarrow$ J/m

Table 3.22: Standard Methods of Measuring Tensile Impact Strength (after Shastri)

Standard	ISO 8256 : 90	D 1822 - 93
Specimen	80 mm x 10 mm x 4 mm, cut from the center of an ISO 3167 Type A specimen, with a double notch. Also referred to as an ISO 8256 Type 1 specimen (Fig. 3.97). Type S or L specimen as specified by this standard (Fig. 3.98). 63.50 mm length x 9.53 or 12.71 mm tab width x 3.2 mm (preferred thickness).	Type S has a non-linear narrow portion width of 3.18 mm, whereas Type L has a 9.53 mm length linear narrow portion width of 3.18 mm.
Conditioning	Specimen conditioning, including any post molding treatment, shall be carried out at 23 °C ± 2 °C and 50 ± 5 % R.H. for a minimum length of time of 88h, except where special conditioning is required as specified by the appropriate material standard.	At 23 ± 2 °C and 50 ± 5 % relative humidity for not less than 40h, prior to testing in accordance with Practice D618, procedure A. Material specification conditioning requirements take precedence.
Apparatus	The machine shall be securely fixed to a foundation having a mass at least 20 times that of the heaviest pendulum in use and be capable of being leveled. Pendulums with the specified initial potential energies shall be used: 2.0, 4.0, 7.5, 15.0, 25.0, and 50.0 J. Velocity at impact is 2.6 to 3.2 m/s for the 2.0 to 4.0 J pendulums and 3.4 to 4.1 m/s for pendulums with energies from 7.5 to 50.0 J. Free length between grips is 30 ± 2 mm. The edges of the serrated grips in close proximity to the test region shall have a radius such that they cut across the edges of the first serrations. Unless otherwise specified in the relevant material standard, a minimum of ten specimens shall be prepared in accordance with that same material standard.	The base and suspending frame shall be of sufficiently rigid and massive construction to prevent or minimize energy losses to or through the base and frame. No pendulums specified Velocity at impact is approximately 3.444 m/s, based on the vertical height of fall of the striking nose specified at 610 ± 2 mm. Jaw separation is 25.4 mm. The edge of the serrated jaws in close proximity to the test region shall have a 0.40 mm radius to break the edge of the first serrations. Material specification testing conditions take precedence; therefore, it is advisable to refer to the material specification before using the following procedures.

Continued on next page

3.5 Mechanical Properties

Standard	ISO 8256 : 90	D 1822 - 93
	When none exists, or unless otherwise specified, specimens shall be directly compression or injection molded in accordance with ISO 293 or ISO 294-1.	At least five, preferably 10, sanded, machined, die cut or molded in a mold with the dimensions specified for Type S and L specimen.
Test procedures	Notches shall be machined in accordance with ISO 2818. The radius of the notch base shall be 1.0 ± 0.02 mm, with an angle of 45° ± 1°.	Specimens are unnotched.
	The two notches shall be at right angles to its principal axis on opposite sides with a distance between the two notches of 6 ± 0.2 mm. The two lines drawn perpendicular to the length direction of the specimen through the apex of each notch shall be within 0.02 mm of each other.	
	The selected pendulum shall consume at least 20%, but not more than 80% of its stored energy in breaking the specimens. If more than one pendulum satisfies these conditions, the pendulum having highest energy is used.	Use the lowest capacity pendulum available, unless the impact values go beyond the 85% scale reading. If this occurs, use a higher capacity pendulum.
	Run three blank tests to calculate the mean frictional loss. The loss should not exceed 1% for a 2.0 J pendulum and 0.5% for those specified pendulums with a 4.0 J or greater energy pendulum.	A friction and windage correction may be applied. A non-mandatory appendix provides the necessary calculations to determine the amount of this type of correction.
	Determine the energy correction, using Method A or B, before one can determine the notched tensile impact strength, E_n. Method A- Energy correction due to the plastic deformation and kinetic energy of the crosshead, E_q. Method B-Crosshead-bounce energy, E_b.	The bounce correction factor may be applied. A non-mandatory appendix provides the necessary calculations to determine the amount of this correction factor. (A curve must be calculated for the cross head and pendulum used before applying in bounce correction factors).
	Calculate the notched tensile impact strength, E_n by dividing the the corrected energy(Method A or B) by the cross sectional area between the two notches.	Calculate the corrected impact energy to break by subtracting the friction and windage correction and/or the bounce correction factor from the scale reading of energy to break.
Values and units	Notched tensile impact strength, $E_n \Rightarrow kJ/m^2$	Tensile-impact energy $\Rightarrow$ J.

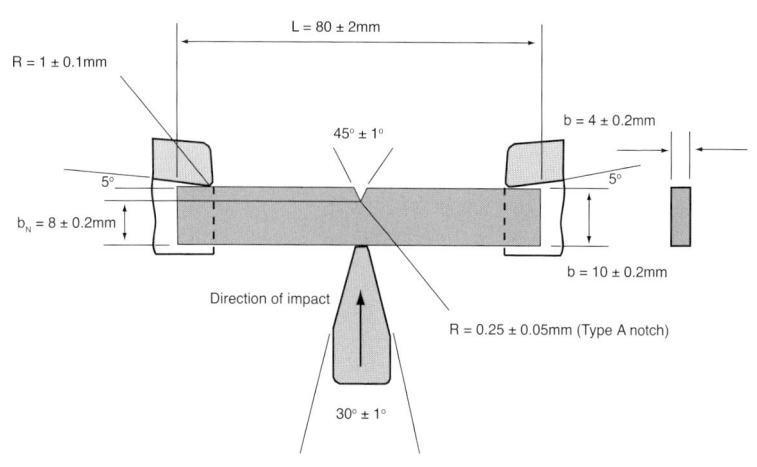

Figure 3.92: Dimensions of Charpy impact test with support and striking edge for ISO 179.

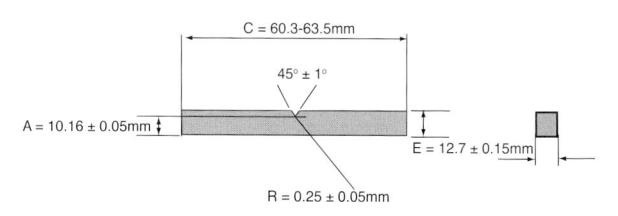

Figure 3.93: Dimensions of Charpy impact test specimen ASTM D256.

Depending on the type of material, the notch tip radius may significantly influence the impact resistance of the specimen. Figure 3.99 presents impact strengths for various thermoplastics as a function of notch tip radius. As expected, impact strength is significantly reduced with decreasing notch radius. Another factor that influences the impact resistance of polymeric materials is the temperature. This is clearly demonstrated in Fig. 3.100, in which PVC specimens with several notch radii are tested at various temperatures. In addition, the impact test sometimes brings out brittle failure in materials that undergo a ductile breakage in a short-term tensile test.

Similar to a small notch radius, brittle behavior is sometimes developed by lowering the temperature of the specimen. Figure 3.101 shows the brittle to

3.5 Mechanical Properties

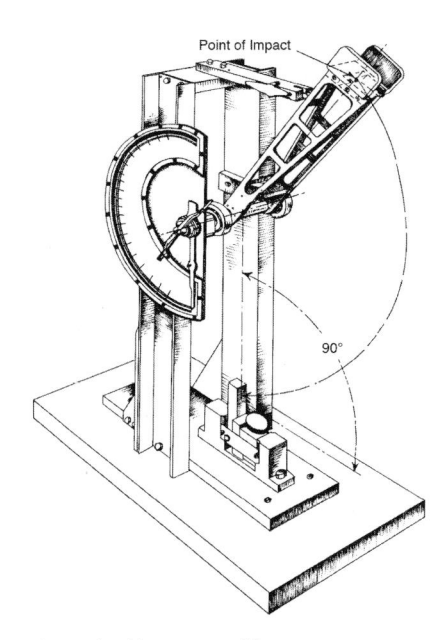

Figure 3.94: Cantilever beam Izod impact machine.

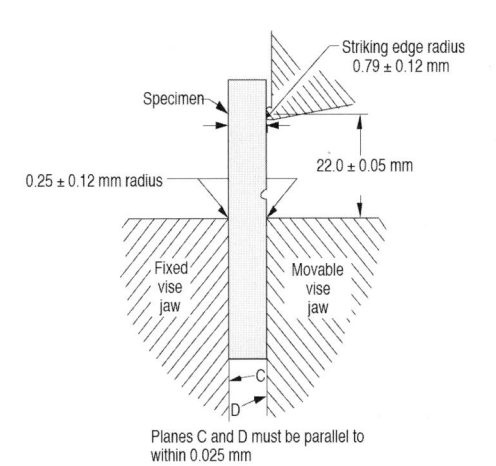

Figure 3.95: Schematic of the clamp, specimen and striking hammer in an Izod impact test.

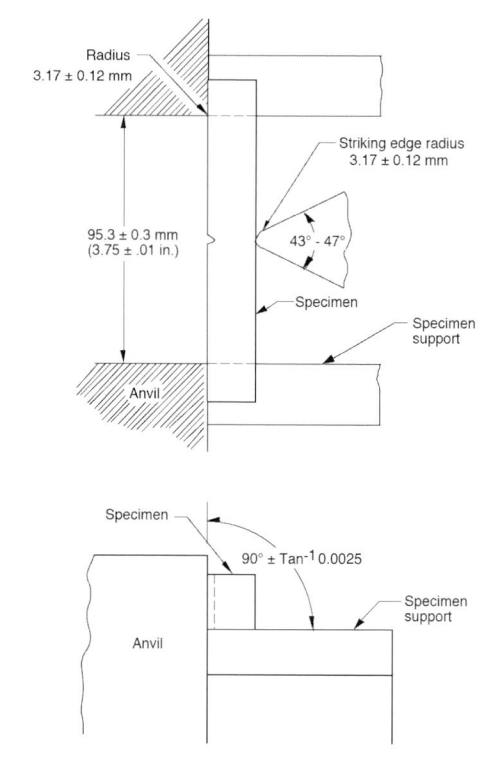

Figure 3.96: Schematic of the clamp, specimen, and striking hammer in a Charpy impact test.

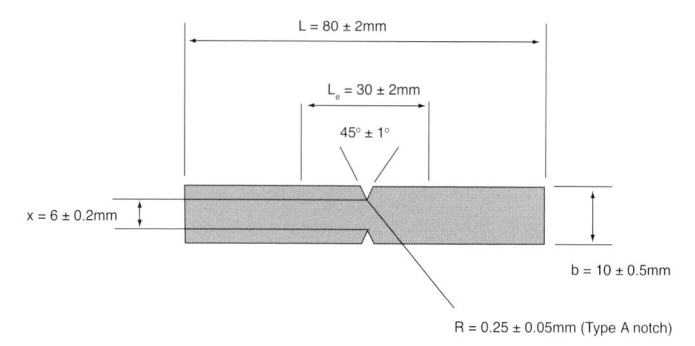

Figure 3.97: Tensile impact specimen (Type 1) for ISO 8256.

3.5 Mechanical Properties

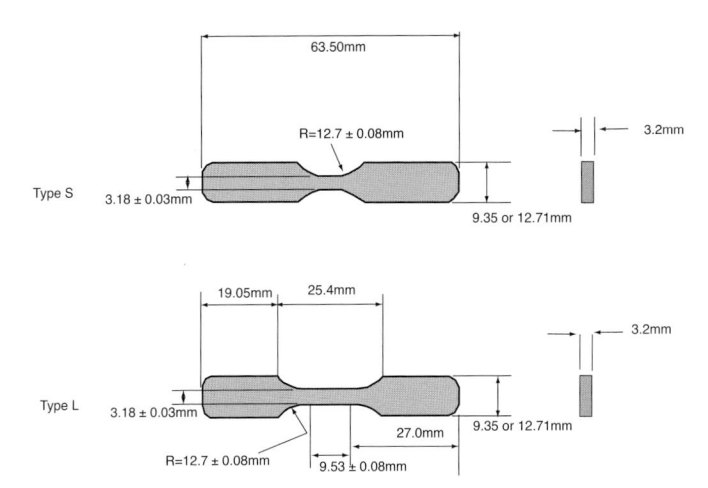

Figure 3.98: Type S and L tensile impact test specimens (ASTM D1822).

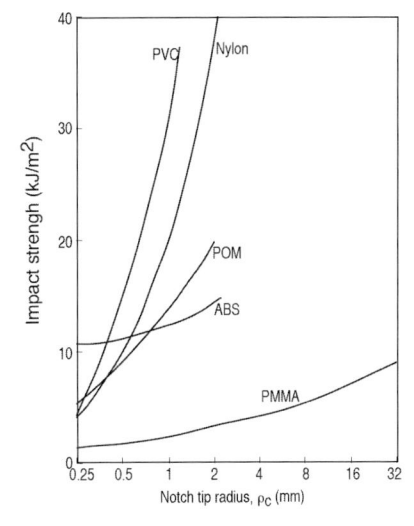

Figure 3.99: Impact strength as a function of notch tip radius for various polymers (after Kinloch and Young).

ductile behavior regimes as a function of temperature for several thermoplastic polymers.

Finally, processing conditions such as barrel temperature during injection molding or extrusion and residence time inside the barrel, can also affect the impact properties of a plastic component. Higher processing temperatures as

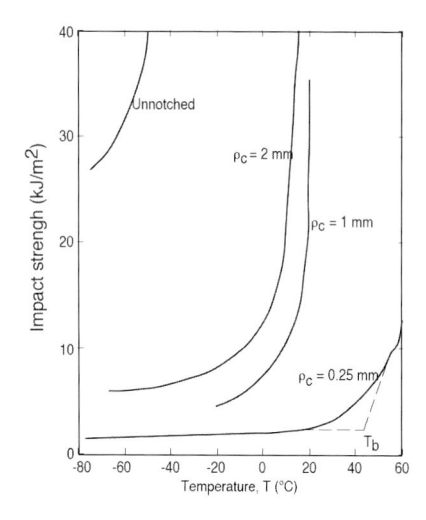

Figure 3.100: Impact strength of PVC as a function of temperature for various notch tip radii (after Kinloch and Young).

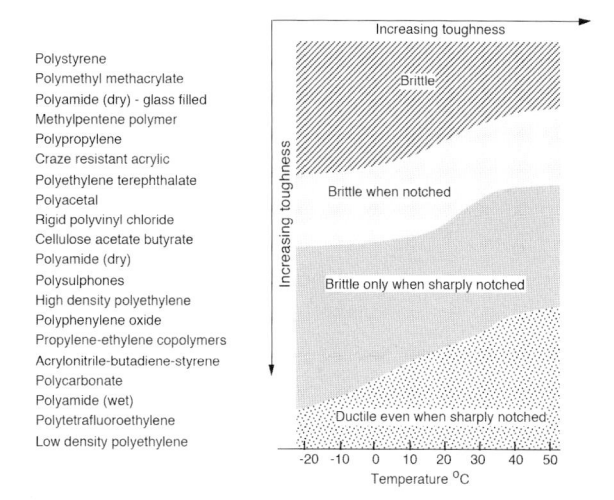

Polystyrene
Polymethyl methacrylate
Polyamide (dry) - glass filled
Methylpentene polymer
Polypropylene
Craze resistant acrylic
Polyethylene terephthalate
Polyacetal
Rigid polyvinyl chloride
Cellulose acetate butyrate
Polyamide (dry)
Polysulphones
High density polyethylene
Polyphenylene oxide
Propylene-ethylene copolymers
Acrylonitrile-butadiene-styrene
Polycarbonate
Polyamide (wet)
Polytetrafluoroethylene
Low density polyethylene

Figure 3.101: Brittle to ductile behavior regimes as a function of temperature for several thermoplastic polymers (after Crawford).

3.5 Mechanical Properties

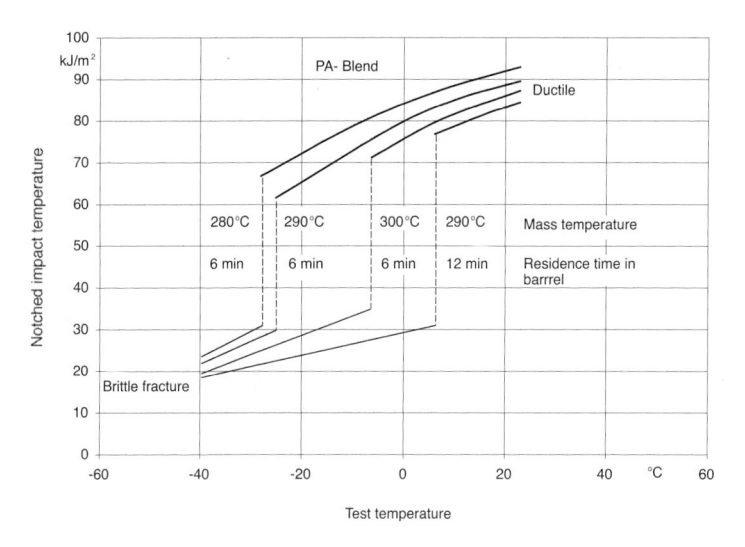

Figure 3.102: Notched impact strength of a PA blend as a function of test temperature, barrel temperature and barrel residencde time.

well as higher residence times will have an adverse effect on impact properties, as depicted for a PA blend in Fig. 3.102.

Another impact test worth mentioning is the *falling dart* test. This test, described by the ASTM 3029 and DIN 53 453 standard methods, is well suited for specimens that are too thin or flexible to be tested using the Charpy and Izod tests and when the fracture toughness of a finished product with large surfaces is sought. Figure 3.103 shows a schematic of a typical falling dart test set-up.

Analysis of Impact Data Although the most common interpretation of impact tests is qualitative, it is possible to use linear elastic fracture mechanics to quantitatively evaluate impact test results. Using LEFM, it is common to compute the material's fracture toughness G_{IC} from impact test results. Obviously, LEFM is only valid if the Izod or Charpy test specimen is assumed to follow linear elastic behavior and contains a sharp notch. The Izod or Charpy test specimen absorbs a certain amount of energy, U_e, during impact. This energy can be related to the fracture toughness using,

$$U_e = G_{IC} t w \widetilde{a} \qquad (3.34)$$

where t and w are the specimens thickness and width, respectively. The parameter $\widetilde{a}$ is a *geomettric crack factor* found in Table 3.23 for various Charpy impact test specimens and in Table 3.24 for various Izod impact test specimens. The elastic

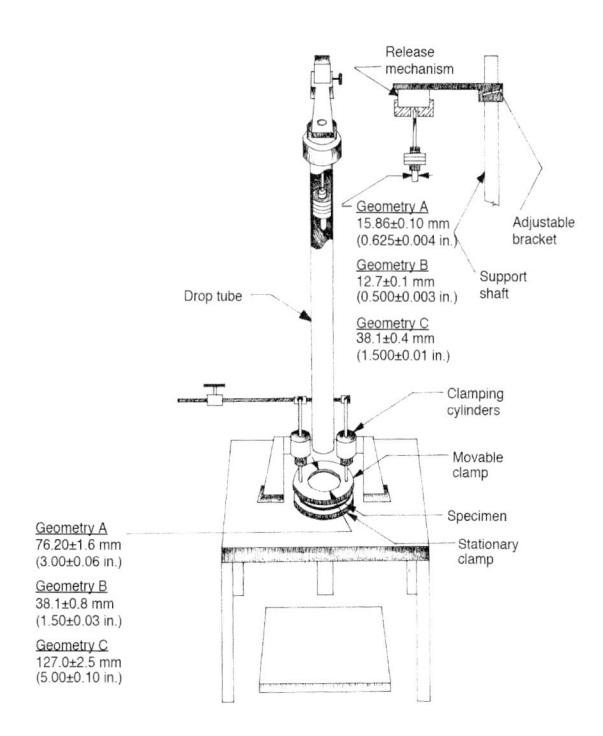

Figure 3.103: Schematic of a drop weight impact tester.

energy absorbed by the test specimen during fracture can be represented with energy lost by the pendulum during the test. This allows the test engineer to relate impact test results with the fracture toughness of a material. Figure 3.104 contains both Charpy and Izod test result data for a medium density polyethylene as plots of U_e versus $tw\widetilde{a}$ with kinetic energy corrections. The fracture toughness is the slope of the curve.

Table 3.23: Charpy Impact Test Geometric Crack Factors $\widetilde{a}$

a/w	$2L/w = 4$	$2L/w = 6$	$2L/w = 8$	$2L/w = 10$	$2L/w = 12$
			$\widetilde{a}$		
0.04	1.681	2.456	3.197	3.904	4.580
0.06	1.183	1.715	2..220	2.700	3.155
0.08	0.933	1.340	1.725	2.089	2.432
0.10	0.781	1.112	1.423	1.716	1.990
0.12	0.680	0.957	1.217	1.461	1.688
0.14	0.605	0.844	1.067	1.274	1.467

Continued on next page

3.5 Mechanical Properties

a/w	2L/w = 4	2L/w = 6	2L/w = 8	2L/w = 10	2L/w = 12
			$\widetilde{a}$		
0.16	0.550	0.757	0.950	1.130	1.297
0.18	0.505	0.688	0.858	1.015	1.161
0.20	0.468	0.631	0.781	0.921	1.050
0.22	0.438	0.584	0.718	0.842	0.956
0.24	0.413	0.543	0.664	0.775	0.877
0.26	0.391	0.508	0.616	0.716	0.808
0.28	0.371	0.477	0.575	0.665	0.748
0.30	0.354	0.450	0.538	0.619	0.694
0.32	0.339	0.425	0.505	0.578	0.647
0.34	0.324	0.403	0.475	0.542	0.603
0.36	0.311	0.382	0.447	0.508	0.564
0.38	0.299	0.363	0.422	0.477	0.527
0.42	0.276	0.328	0.376	0.421	0.462
0.44	0.265	0.311	0.355	0.395	0.433
0.46	0.254	0.296	0.335	0.371	0.405
0.48	0.244	0.281	0.316	0.349	0.379
0.50	0.233	0.267	0.298	0.327	0.355
0.52	0.224	0.253	0.281	0.307	0.332
0.54	0.214	0.240	0.265	0.88	0.310
0.56	0.205	0.228	0.249	0.270	0.290
0.58	0.196	0.216	0.235	0.253	0.271
0.60	0.187	0.205	0.222	0.238	0.253

Figure 3.105 compares plots of impact-absorbed energy as a function of $tw\widetilde{a}$ for unfilled epoxy and epoxies filled with irregular-shaped silica with weight percents of 55% and 64%.

Table 3.24: Izod Impact Test Geometric Crack Factors $\widetilde{a}$

a/w	2L/w = 4	2L/w = 6	2L/w = 8	2L/w = 10	2L/w = 12
			$\widetilde{a}$		
0.06	1.540	1.744	1.850	2.040	-
0.08	1.273	1.400	1.485	1.675	1.906
0.10	1.060	1.165	1.230	1.360	1.570
0.12	0.911	1.008	1.056	1.153	1.294
0.14	0.795	0.890	0.932	1.010	1.114
0.16	0.708	0.788	0.830	0.900	0.990
0.18	0.650	0.706	0.741	0.809	0.890
0.20	0.600	0.642	0.670	0.730	0.810
0.22	0.560	0.595	0.614	0.669	0.750
0.24	0.529	0.555	0.572	0.617	0.697
0.26	0.500	0.525	0.538	0.577	0.656
0.28	0.473	0.500	0.510	0.545	0.618
0.30	0.452	0.480	0.489	0.519	0.587
0.32	0.434	0.463	0.470	0.500	0.561
0.34	0.420	0.446	0.454	0.481	0.538
0.36	0.410	0.432	0.440	0.468	0.514

Continued on next page

	$2L/w=4$	$2L/w=6$	$2L/w=8$	$2L/w=10$	$2L/w=12$
a/w			$\widetilde{a}$		
0.38	0.397	0.420	0.430	0.454	0.494
0.40	0.387	0.410	0.420	0.441	0.478
0.42	0.380	0.400	0.411	0.431	0.460
0.44	0.375	0.396	0.402	0.423	0.454
0.46	0.369	0.390	0.395	0.415	0.434
0.48	0.364	0.385	0.390	0.408	0.422
0.50	0.360	0.379	0.385	0.399	0.411

Table 3.25 presents values for *stress intesnsity factor* and *fracture toughness* for several plastics and other materials.

Table 3.25: Values of Plane Stress Intensity Factor and Strain Toughness for Various Materials

Material	$K_{IC}(MN/m^{3/2})$	$G_{IC}(kJ/m^2)$
ABS	2-4	5
POM	4	1.2-2
EP	0.3-0.5	0.1-0.3
PE-LD	1	6.5
PE-MD and PE-HD	0.5-5	3.5-6.5
PA66	3	0.25-4
PC	1-2.6	5
UPE-glass reinforced	5-7	5-7
PP-co	3-4.5	8
PS	0.7-1.1	0.3-0.8
PMMA	1.1	1.3
PVC-U	1-4	1.3-1.4
Aluminum-alloy	37	20
Glass	0.75	0.01-0.02
Steel-mild	50	12
Steel-alloy	150	107
Wood	0.5	0.12

3.5.3 Creep Behavior

The stress relaxation and the creep test are well-known long-term tests. The stress relaxation test is difficult to perform and is therefore often approximated by data acquired through the more commonly used creep test. The stress relaxation of a polymer is often thought of as the inverse of creep. The creep test, which can be performed either in shear, compression, or tension, measures the flow of a polymer component under a constant load. It is a common test that measures the strain, ϵ, as a function of stress, time, and temperature. Standard creep tests such as ISO 899, ASTM D2990 and DIN 53 444 can be used. The ISO 899 and ASTM D2990 standard creep tests are presented in Table 3.26.

3.5 Mechanical Properties

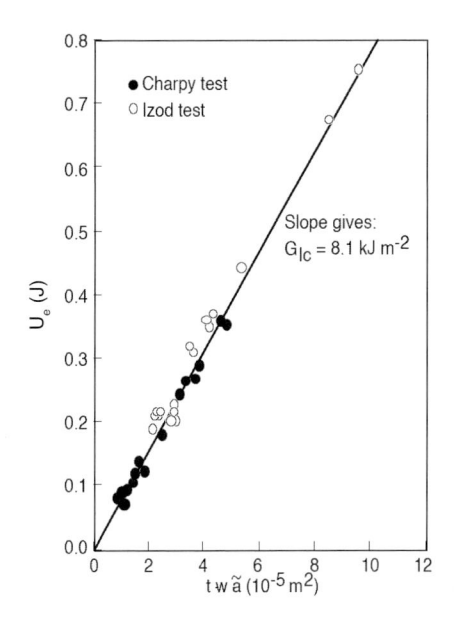

Figure 3.104: Elastic energy absorbed at impact fracture as a function of test specimen cross-sectional geometry for a medium-density polyethylene (after Plati and Williams).

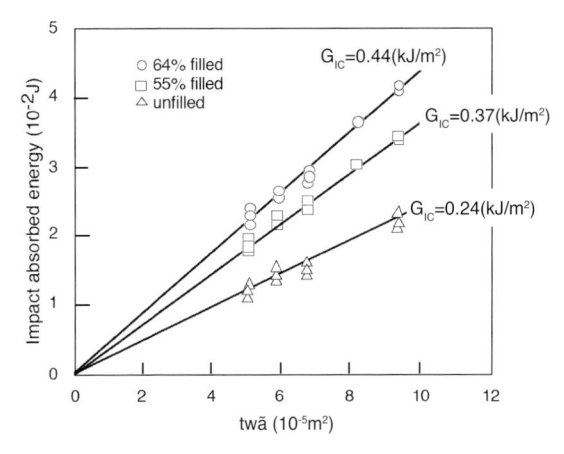

Figure 3.105: Impact absorbed energy as a function of specimen size for unfilled epoxy and epoxies filled with irregular-shaped silica with weight percents of 55% and 64%.

Table 3.26: Standard Methods of Measuring Tensile Creep Modulus (after Shastri)

Standard	ISO 899 - 1	D 2990 - 95
Specimen	ISO 3167 Type A specimen	D 638 Type I specimens may be prepared by injection or compression molding or by machining from sheets or other fabricated forms.
Conditioning	Specimen conditioning, including any post molding treatment, shall be carried out at 23°C ± 2°C and 50 ± 5 % R.H. for a minimum length of time of 88h, except where special conditioning is required as specified by the appropriate material standard.	At 23 ± 2 °C and 50 ± 5 % relative humidity for not less than 40h, prior to testing in accordance with D 618 Procedure A. The specimens shall be preconditioned in the test environment for at least 48 h prior to testing. Those materials whose creep properties are suspected to be affected by moisture content shall be brought to moisture equilibrium appropriate to the test conditions prior to testing.
Test procedures	Conduct the test in the same atmosphere as used for conditioning, unless otherwise agreed upon by the interested parties, e.g., for testing at elevated or low temperatures. Select appropriate stress levels to produce data for the application requirements. Where it is necessary to preload the test specimen prior to loading, preloading shall not be applied until the temperature and humidity of the test specimen (finally gripped in the testing apparatus) correspond to the test conditions, and the total load (including preload) shall be taken as the test load.	For material characterization, select two or more test temperatures to cover the useful temperature range. For simple material comparisons, select the test temperatures from the following: 23, 50, 70, 90, 120, and 155°C . For simple material comparisons, determine the stress to produce 1% strain in 1000 h. Select several loads to produce strains in the approximate range of 1% strain and plot the 1000-h isochronous stress-strain curve* from which the stress to produce 1% strain may be determined by interpolation. *Since only one point of an isochronous plot is obtained from each creep test, it is usually necessary to run at least three stress levels (preferably more) to obtain an isochronous plot. For creep testing at a single temperature, the minimum number of test specimens at each stress shall be two if four or more stress levels are used or three if fewer than four levels are used.

Continued on next page

3.5 Mechanical Properties

Standard	ISO 899 - 1	D 2990 - 95
	Unless the elongation is automatically and/or continuously measured, record the elongations at the following time schedule: 1, 3, 6, 12 and 30 min; 1, 2, 5, 10, 20, 50, 100, 200, 500, 1000 h.	Measure the extension of the specimens in accordance with the approximate time schedule: 1, 6, 12, and 30 min; 1, 2, 5, 50, 100, 200, 500, 700, and 1000 h.
Units	Tensile creep modulus at 1h and at a strain < 0.5% $\Rightarrow$ MPa Tensile creep modulus at 1000h and at a strain < 0.5% $\Rightarrow$ MPa	Tensile creep modulus in MPa plotted vs. time in h.

Figure 5.41 presents the creep responses of a polybutylene teraphthalate for a range of stresses in a graph with a log scale for time. When plotting creep data in a log-log graph, in the majority of the cases, the creep curves reduce to straight lines as shown for polypropylene in Fig. 3.107. Hence, the creep behavior of most polymers can be approximated with a power-law model, sometimes referred to a as the *Norton model*, represented by

$$\epsilon(t) = k(T)\sigma^n t^m \tag{3.35}$$

where k, n and m are material-dependent properties.

Isochronous and Isometric Creep Curves Typical creep test data, as shown in Fig. 5.41, can be manipulated to be displayed as short-term stress-strain tests or as stress relaxation tests. These manipulated creep-test-data curves are called *isochronous* and *isometric* graphs.

An isochronous plot of the creep data is generated by cutting sections through the creep curves at constant times and plotting the stress as a function of strain. The isochronous curves of the creep data displayed in Fig. 5.41 are presented in Fig. 5.42. Similar curves can also be generated by performing a series of short creep tests, where a specimen is loaded at a specific stress for a short period of time, typically around 100 s. The load is then removed, and the specimen is allowed to relax for a period of 4 times greater than the time of the creep test. The specimen is then reloaded at a different stress, and the test is repeated until a sufficient number of points exists to plot an isochronous graph. This procedure is less time-consuming than the regular creep test and is often used to predict the short-term behavior of polymers. However, it should be pointed out that the short-term tests described in the previous section are more accurate and are less time consuming and cheaper to perform. The isometric or "equal size" plots of the creep data are generated by taking constant strain sections of the creep curves and by plotting the stress as a function of time. Isometric curves of the polypropylene creep data presented in Fig. 5.41 are shown in Fig. 5.43. This plot resembles the stress relaxation test results and is often used in the same

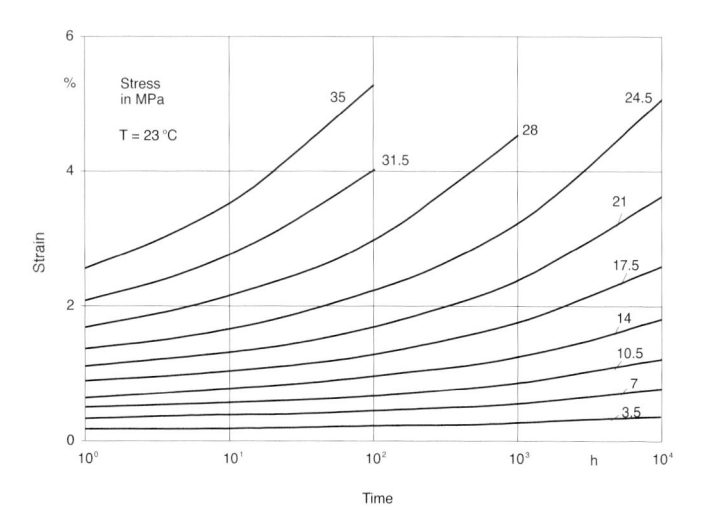

Figure 3.106: Creep response of a PBT at 23°C.

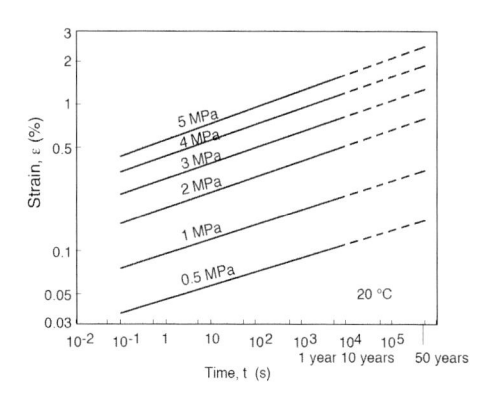

Figure 3.107: Creep response of a polypropylene plotted on a log-log scale.

3.5 Mechanical Properties

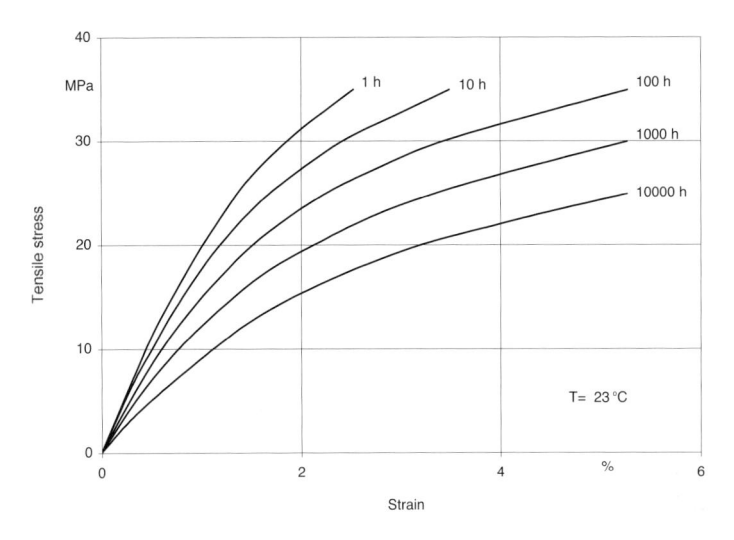

Figure 3.108: Isochronous stress-strain curves for the PBT at 23°C creep responses shown in Fig. 5.41.

manner. When we divide the stress axis by the strain, we can also plot the modulus versus time. Creep data can sometimes be presented in terms of secant creep modulus. For this, the data can be generated for a given stress as presented in Fig. 3.110. For specific applications, plastics should also be tested at higher temperatures. To further illustrate the effect temperature has on the mechanical behavior of thermoplastics, Figs. 3.111 and 3.112 present 1000h isochronous curves for a selected number of thermoplstics at 23°C and 60°C, respectively. Creep of thermoplastic polymers can be mitigated by the use of fiber reinforcements. Figures 3.113 and 3.114 show 1000h isochronous curves for fiber reinforced thermoplastics at 23°C and 60°C, respectively.

Creep Rupture During creep, a loaded polymer component will gradually increase in length until fracture or failure occurs. This phenomenon is usually referred to as creep rupture or, sometimes, as static fatigue. During creep, a component is loaded under a constant stress, constantly straining until the material cannot withstand further deformation, causing it to rupture. At high stresses, the rupture occurs sooner than at lower stresses. However, at low enough stresses, failure may never occur. The time it takes for a component or test specimen to fail depends on temperature, load, manufacturing process, environment, etc. It is important to point out that damage is often present and visible before creep rupture occurs. This is clearly demonstrated in Fig. 3.115, which presents

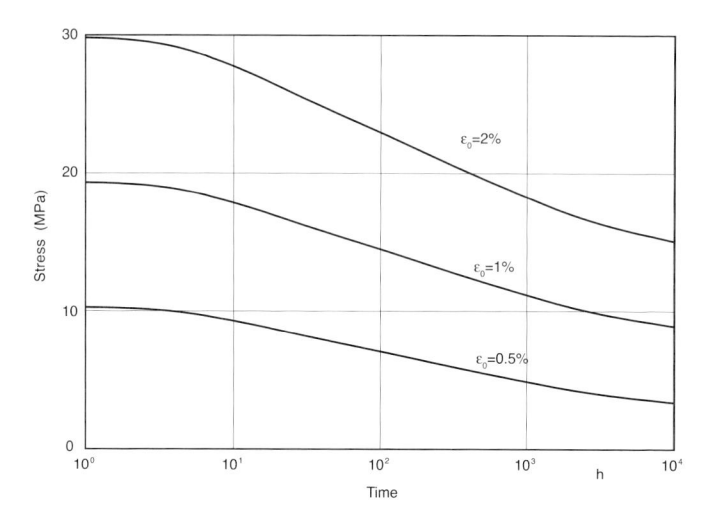

Figure 3.109: Isometric stress-time curves for the PBT at 23°C creep responses shown in Fig. 5.41.

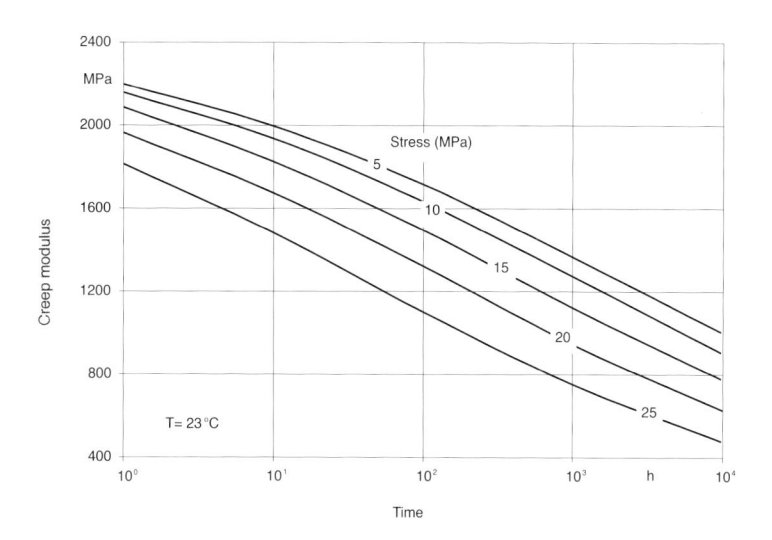

Figure 3.110: Secant creep modulus curves as a function of time for the PBT at 23°C creep responses shown in Fig. 5.41.

3.5 Mechanical Properties 167

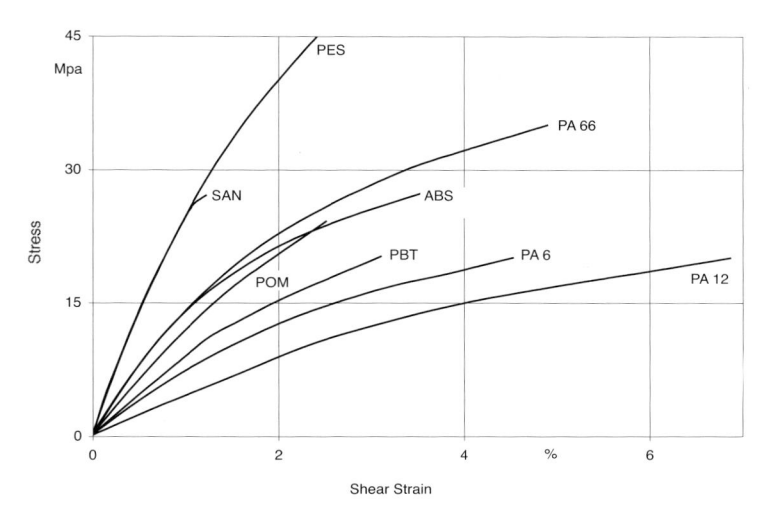

Figure 3.111: Isochronous (1000h) stress-strain curves for selected thermoplastics at 23°C.

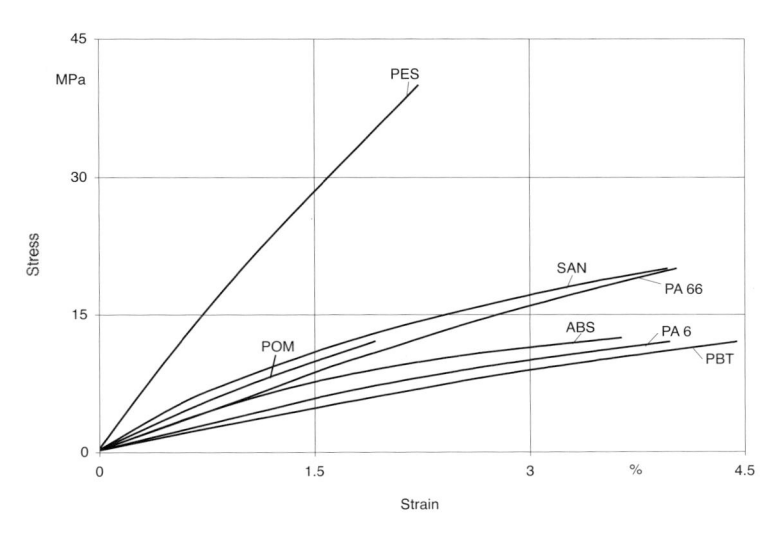

Figure 3.112: Isochronous (1000h) stress-strain curves for selected thermoplastics at 60°C.

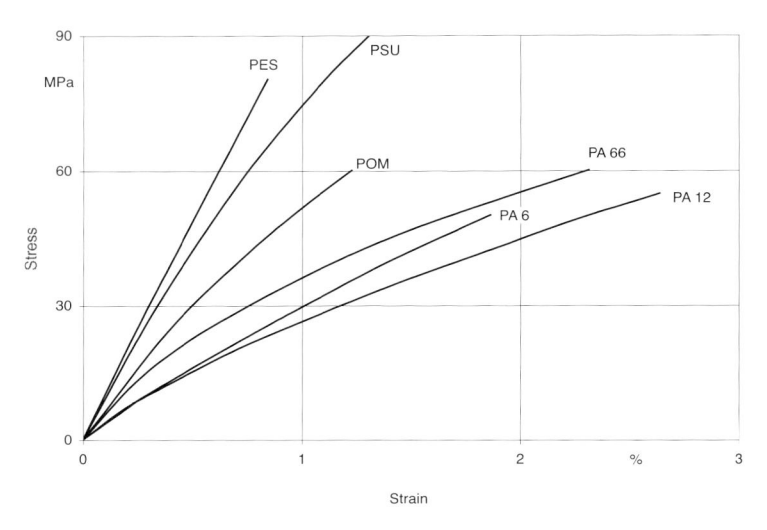

Figure 3.113: Isochronous (1000h) stress-strain curves for selected fiber reinforced (25-35 volume %) thermoplastics at 23°C.

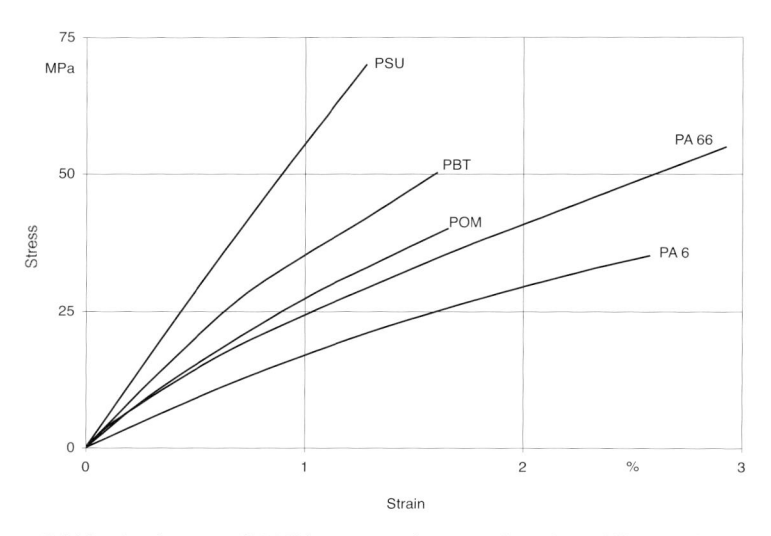

Figure 3.114: Isochronous (1000h) stress-strain curves for selected fiber reinforced (25-35 volume %) thermoplastics at 60°C.

3.5 Mechanical Properties 169

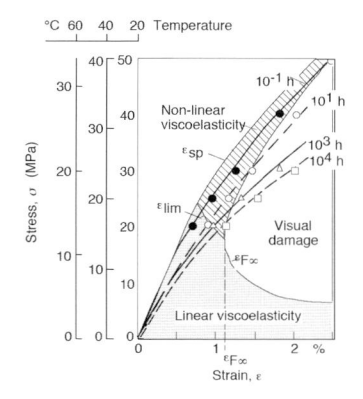

Figure 3.115: Isochronous creep curves for PMMA at three different temperatures (after Menges).

isochronous creep curves for polymethyl methacrylate at three different temperatures. The regions of linear and non-linear viscoelasticity and of visual damage are highlighted in the figure.

The standard test to measure creep rupture is the same as the creep test. Results from creep rupture tests are usually presented in graphs of applied stress versus the logarithm of time to rupture. An example of a creep rupture test that ran for 10 years is shown in Fig. 3.116. Here, the creep rupture of high density polyethylene pipes under internal pressures was tested at different temperatures. Two general regions with different slopes become obvious in the plots. The points to the left of the knee represent pipes that underwent a ductile failure, whereas those points to the right represent the pipes that had a brittle failure. As pointed out, generating a graph such as the one presented in Fig. 3.116, is an extremely involved and lengthy task, that takes several years of testing[4]. Figures 3.117 and 3.118 compare the static fatigue or creep rupture life curves of several thermoplastics at 20°C and 60°C, respectively. Since these tests are so time consuming, they are usually only carried out to 1,000 h (6 weeks) and in some cases to 10,000 h (60 weeks). Once the steeper slope, which is typical of the brittle fracture has been reached, the line can be extrapolated with some degree of confidence to estimate values of creep rupture at future times.

Although the creep test is considered a long - term test, in principle it is difficult to actually distinguish it from monotonic stress strain tests or even impact tests. In fact, one can plot the full behavior of the material, from impact to creep, on the same graph as shown for PMMA under tensile loads at room temperature in

---

[4]These tests where done between 1958 and 1968 at Hoechst AG, Germany.

170　　　　　　　　　　　　　　　　　　　　　　　　　　3 Properties and Testing

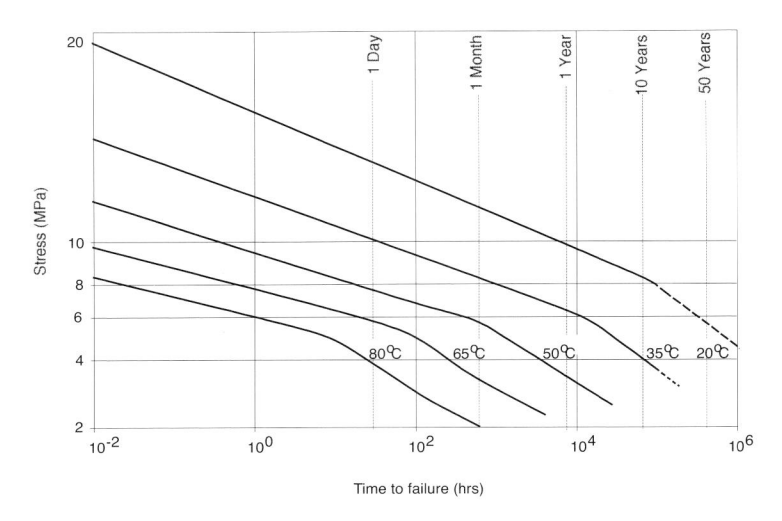

Figure 3.116: Creep rupture behavior as a function of temperature for a high density polyethylene (after Gaube and Kausch).

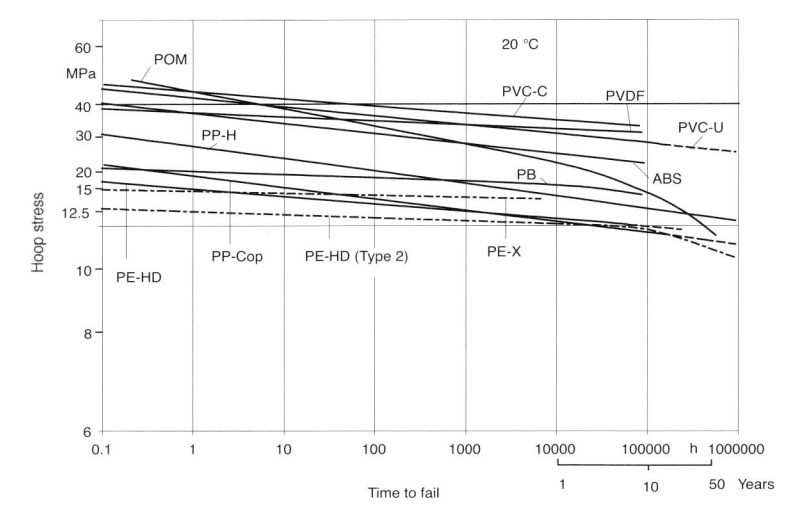

Figure 3.117: Creep rupture behavior of a several thermoplastics at $20\,°C$.

3.5 Mechanical Properties

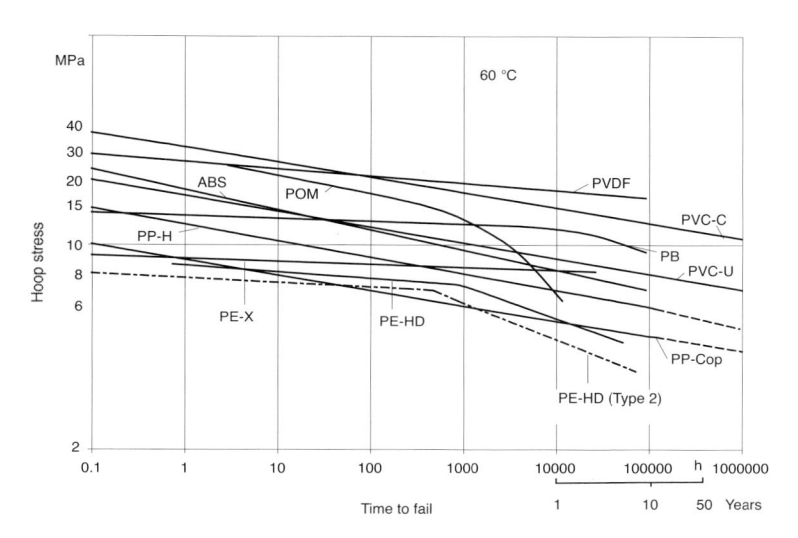

Figure 3.118: Creep rupture behavior of a several thermoplastics at 60°C.

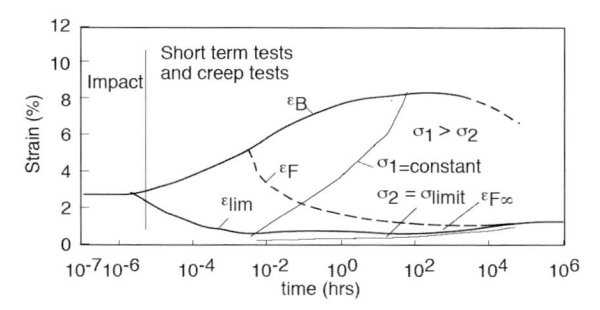

Figure 3.119: Plot of material behavior at room temperature, from impact to creep, for a PMMA under tensile loads (after Menges).

Fig. 3.119. The figure represents strain as a function of the logarithm of time. The strain line that represents rupture is denoted by ϵ_B. This line represents the maximum attainable strain before failure as a function of time. Obviously, a material tested under an impact tensile loading will strain much less than the same material tested in a creep test. Of interest in Fig. 3.119 are the two constant stress lines denoted by σ_1 and σ_2. For example, it can be seen that a PMMA specimen loaded to a hypothetical stress of σ_1 will behave as a linear viscoelastic material up to a strain of 1%, at which point the first microcracks start forming or the craze nucleation begins. The crazing appears a little later after the specimen's

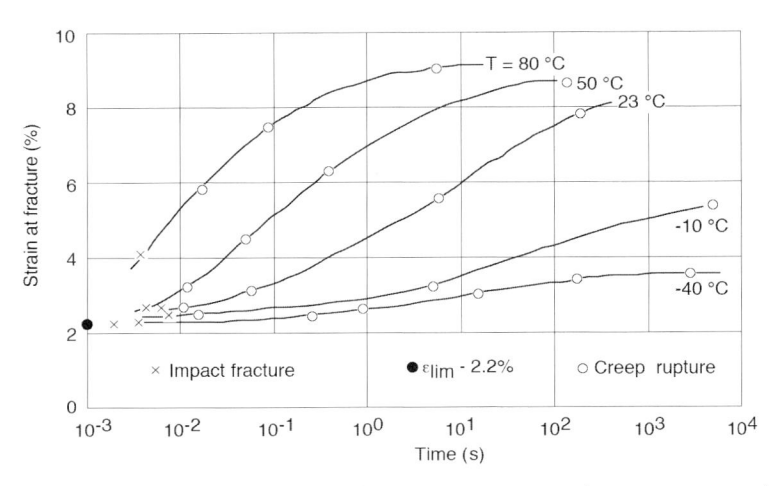

Figure 3.120: Strain at fracture for a PMMA in creep tests at various temperatures (after Menges).

deformation is slightly over 2%. The test specimen continues to strain for the next 100 h until it ruptures at a strain of about 8%. From the figure it can be deduced that the first signs of crazing can occur days and perhaps months or years before the material actually fractures. The stress line denoted by σ_2, where $\sigma_1 > \sigma_2$, is a limiting stress under which the component will not craze. Figure 3.119 also demonstrates that a component loaded at high speeds (i.e., impact) will craze and fail at the same strain. A limiting strain of 2.2% is shown. Since these tests take a long time to perform, it is often useful to test the material at higher temperatures, where a similar behavior occurs in a shorter period of time. Figure 3.120 shows tests performed on PMMA samples at five different temperatures. When comparing the results in Fig. 3.120 to the curve presented in Fig. 3.119, a clear time-temperature superposition becomes visible. In the applied stress versus logarithm of time to rupture curves, such as the one shown in Fig. 3.116, the time-temperature superposition is also evident.

3.5.4 Dynamic Mechanical Tests

The simplest dynamic mechanical test is the torsion pendulum. The standard procedure for the torsional pendulum, schematically shown in Fig. 3.121, is described in DIN 53445 and ASTM D2236. The technique is applicable to virtually all plastics, through a wide range of temperatures; from the temperature of liquid nitrogen, -180°C to 50-80°C above the glass transition temperature in

3.5 Mechanical Properties

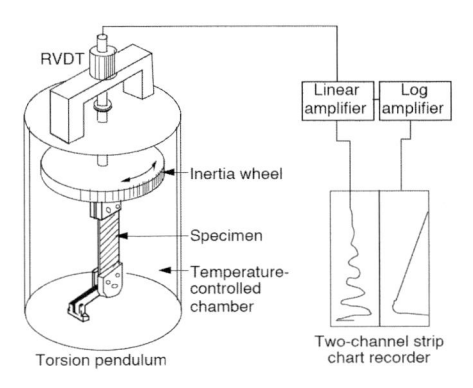

Figure 3.121: Schematic diagram of the torsion pendulum test equipment.

amorphous thermoplastics and up to the melting temperature in semi-crystalline thermoplastics. With thermoset polymers one can apply torsional tests up to the degradation temperatures of the material. The torsion pendulum apparatus consist of an inertia wheel, grips, and the specimen contained in a temperature-controlled chamber. The rectangular test specimen can be cut from a polymer sheet or part, or it can be made by injection molding. To execute the test, the inertia wheel is deflected, then released and allowed to oscillate freely. The angular displacement or twist of the specimen is recorded over time. The frequency of the oscillations is directly related to the elastic shear modulus of the specimen, G', and the decay of the amplitude is related to the damping or logarithmic decrement, Δ, of the material. The elastic shear modulus (in Pascals) can be computed using the relation

$$G' = \frac{6.4\pi^2 I L f^2}{\mu b t^3} \quad (3.36)$$

where I is the polar moment of inertia (g/cm^2), L the specimen length (cm), the frequency (Hz), b the width of the specimen, t the thickness of the specimen, and μ a shape factor which depends on the width-to-thickness ratio. Values of μ vary between 5.0 for $b/t = 10$ and 5.333 for $b/t = $ inf. The logarithmic decrement can be computed using

$$\Delta = \text{Ln}\left(\frac{A_n}{A_{n+1}}\right) \quad (3.37)$$

where A_n represents the amplitude of the nth oscillation.[5] Although the elastic shear modulus, G', and the logarithmic decrement, Δ, are sufficient to characterize a material, one can also compute the loss modulus G'' by using

$$G'' = \left(\frac{G'\Delta}{\pi}\right) \qquad (3.38)$$

The logarithmic decrement can also be written in terms of loss tangent, $\tan\delta$, where δ is the out-of-phase angle between the strain and stress responses. The loss tangent is defined as

$$\tan\delta = \frac{G''}{G'} = \frac{\Delta}{\pi} \qquad (3.39)$$

Since the frequency in the torsional pendulum test depends on the stiffness of the material under consideration, the test's rate of deformation is also material dependent, and can therefore not be controlled. To overcome this problem, the *dynamic mechanical analysis* (DM) test, or *sinusoidal oscillatory test* was developed. In the sinusoidal oscillatory test, a specimen is excited with a predetermined low frequency stress input, which is recorded along with the strain response. The shapes of the test specimen and the testing procedure vary significantly from test to test. The various tests and their corresponding specimens are described by ASTM D4065 and the terminology, such as the one already used in the above equations, is described by ASTM D4092. If the test specimen in a sinusoidal oscillatory test is perfectly elastic, the stress input and strain response would be in phase, as

$$\tau(t) = \tau_0 = \cos\Omega t \qquad (3.40)$$

and

$$\gamma(t) = \gamma_0 = \cos\Omega t \qquad (3.41)$$

For an ideally viscous test specimen, the strain response would lag $\pi/2$ radians behind the stress input as,

$$\gamma(t) = \gamma_0 = \cos\left(\Omega t - \frac{\pi}{2}\right) \qquad (3.42)$$

Polymers behave somewhere in between the perfectly elastic and the perfectly viscous materials and their response is described by

$$\gamma(t) = \gamma_0 = \cos(\Omega t - \delta) \qquad (3.43)$$

The shear modulus takes a complex form of

$$G^* = \frac{\tau(t)}{\gamma(t)} = \frac{\tau_0}{\gamma_0}(\cos\delta + i\sin\delta) = G' + G'' \qquad (3.44)$$

---

[5] When $\Delta > 1$, a correction factor must be used to compute G'.

3.5 Mechanical Properties

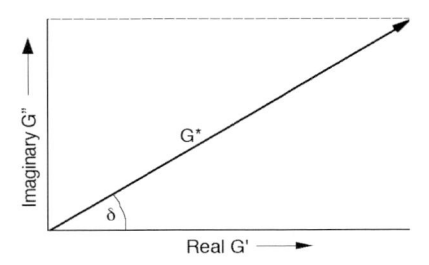

Figure 3.122: Vector representation of the complex shear modulus.

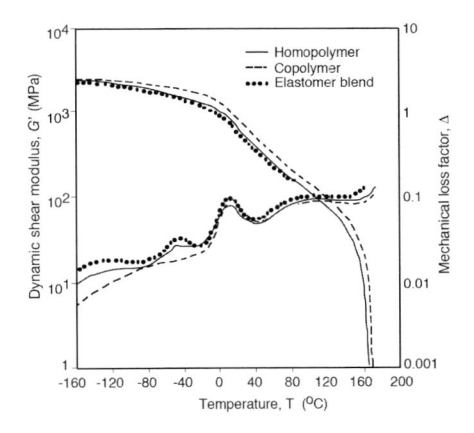

Figure 3.123: Elastic shear modulus and loss factor for various polypropylene grades.

which is graphically represented in Fig. 3.122. G' is usually referred to as storage modulus and G'' as loss modulus. The ratio of loss modulus to storage modulus is referred to as loss tangent.

Figure 3.123 shows the elastic shear modulus and the loss tangent for various polypropylene grades. In the graph, the glass transition temperatures and the melting temperatures can be seen. The vertical scale in plots such as Fig. 3.123 is usually a logarithmic scale. However, a linear scale better describes the mechanical behavior of polymers in design aspects. Figures 3.124-3.127 present the elastic shear modulus on a linear scale for several thermoplastic polymers as a function of temperature. The shear modulus of high temperature application plastics are presented in Fig. 3.128.

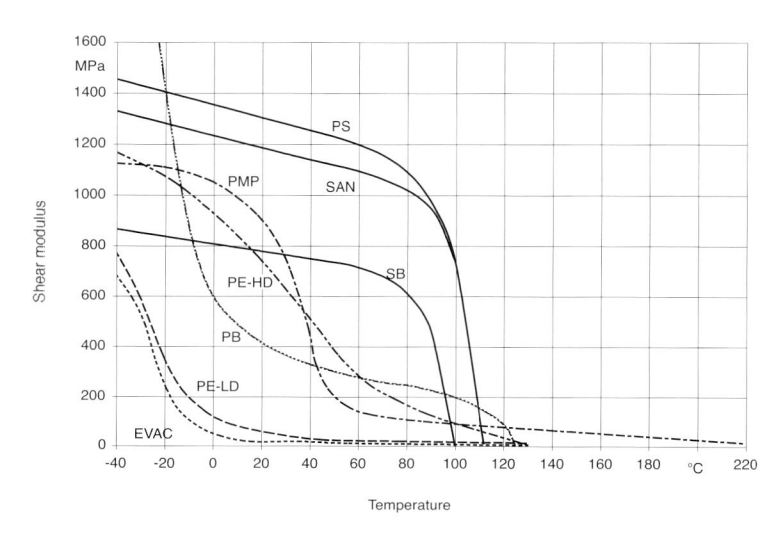

Figure 3.124: Elastic shear modulus for several thermoplastics.

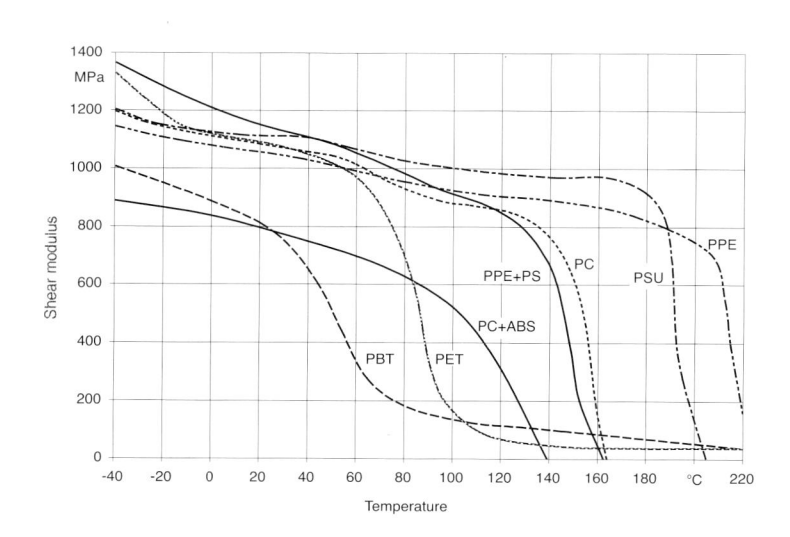

Figure 3.125: Elastic shear modulus for several thermoplastics.

3.5 Mechanical Properties

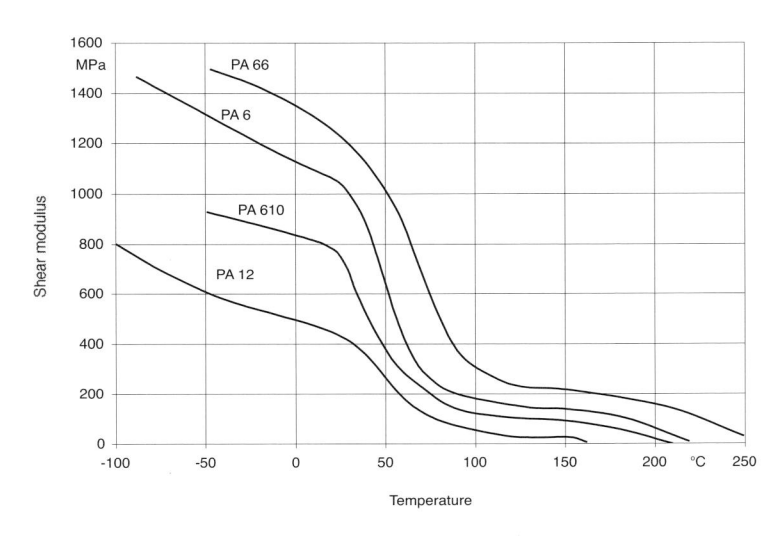

Figure 3.126: Elastic shear modulus for several thermoplastics.

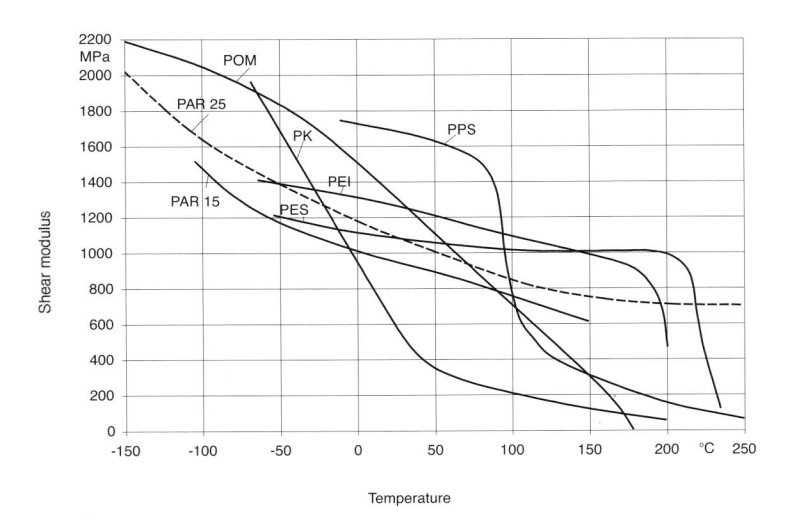

Figure 3.127: Elastic shear modulus for several thermoplastics.

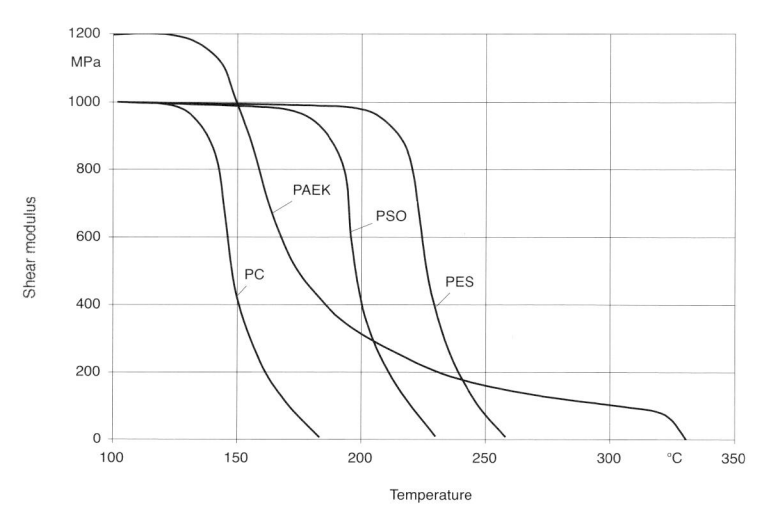

Figure 3.128: Elastic shear modulus for several high temperature application thermoplastics.

3.5.5 Fatigue Tests

Dynamic loading of any material that leads to failure after a certain number of cycles is called *fatigue* or *dynamic fatigue*. Dynamic fatigue is of extreme importance since a cyclic or fluctuating load will cause a component to fail at much lower stresses than it does under monotonic loads. Fatigue testing results are plotted as stress amplitude versus number of cycles to failure. These graphs are usually called *S-N curves*, a term inherited from metal fatigue testing. Figure 3.129 presents S-N curves for several thermoplastic and thermoset polymers tested at a 30-Hz frequency and about a zero mean stress, σ_m.

We must point out here, that most fatigue data presented in the literature and in resin supplier data sheets do not present the frequency, specimen geometry or environmental conditions at which the tests were performed. Hence, such data is not suitable for use in design. The data we present in this section is only intended to illustrate the various problems that arise when measuring fatigue life of a polymer. The information should also serve to reflect trends and as a comparison between various materials and conditions. Fatigue in plastics is strongly dependent on the environment, the temperature, the frequency of loading, the surface, etc. For example, due to surface irregularities and scratches, crack initiation at the surface is more likely in a polymer component that has been machined than in one that was injection molded. An injection molded article is

3.5 Mechanical Properties

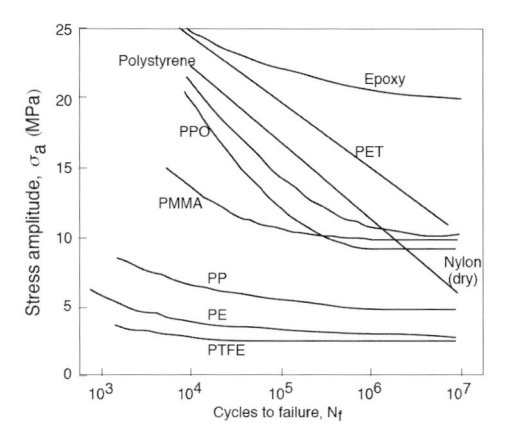

Figure 3.129: Stress-life (S-N) curves for several thermoplastic and thermoset polymers tested at a 30-Hz frequency about a zero mean stress (after Riddell).

formed by several layers of different orientation. In such parts, the outer layers act as a protective skin that inhibits crack initiation. In an injection molded article, cracks are more likely to initiate inside the component by defects such as weld lines and filler particles. The gate region is also a prime initiator of fatigue cracks. Corrosive environments also accelerate crack initiation and failure via fatigue. Corrosive environments and weathering will be discussed in more detail later in this chapter. It is interesting to point out in Fig. 3.129 that thermoset polymers show a higher fatigue strength than thermoplastics. An obvious cause for this is their greater rigidity. However, more important is the lower internal damping or friction, which reduces temperature rise during testing. Temperature rise during testing is one of the main factors that lead to failure when experimentally testing thermoplastic polymers under cyclic loads. The heat generation during testing is caused by the combination of internal frictional or hysteretic heating and low thermal conductivity. At a low frequency and low stress level, the temperature inside the polymer specimen will rise and eventually reach thermal equilibrium when the heat generated by hysteretic heating equals the heat removed from the specimen by conduction. As the frequency is increased, viscous heat is generated faster, causing the temperature to rise even further. This phenomenon is shown in Fig. 3.130, in which the temperature rise during uniaxial cyclic testing of polyacetal is plotted. After thermal equilibrium has been reached, a specimen eventually fails by conventional brittle fatigue, assuming the stress is above the endurance limit. However, if the frequency or stress level is increased even further, the temperature will rise to the point at which the test specimen

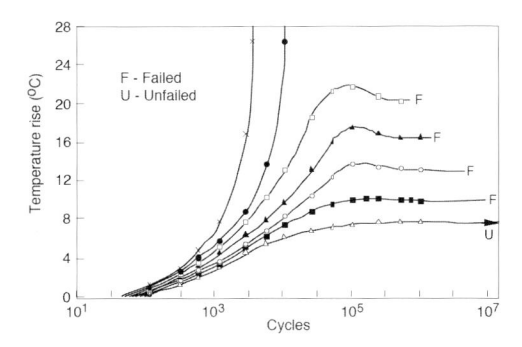

Figure 3.130: Temperature rise during uniaxial cycling under various stresses at 5-Hz (after Crawford).

softens and ruptures before reaching thermal equilibrium. This mode of failure is usually referred to as thermal fatigue. This effect is clearly demonstrated in Fig. 3.131. The points marked T denote those specimens that failed due to thermal fatigue. The other points represent the specimens that failed by conventional mechanical fatigue. A better picture of how frequency plays a significant role in fatigue testing of polymeric materials is generated by plotting results such as those shown in Fig. 3.131 for several frequencies (Fig. 3.132). The temperature rise in the component depends on the geometry and size of test specimen. For example, thicker specimens will cool slower and are less likely to reach thermal equilibrium. Similarly, material around a stress concentrator will be subjected to higher stresses which will result in temperatures that are higher than the rest of the specimen leading to crack initiation caused by localized thermal fatigue. To neglect the effect of thermal fatigue, cyclic tests with polymers must be performed at very low frequencies that make them much lengthier than those performed with metals and other materials which have high thermal conductivity. It is important to understand that although most fatigue data curves state the testing temperature, the resultant data points all have their corresponding temperature at failure. For example, the curves presented in Fig. 3.133 were tested at 23°C; however, each specimen failed at a different temperature. The curves also illustrate how the shape of the imposed stress cycles affect the fatigue life of the polymer.

Stress concentrations have a great impact on the fatigue life of a component. Figures 3.134 and 3.135 compare S-N curves for PVC-U and PA 66, respectively, for specimens with and without a 3 mm circular hole acting as a stress concentrator. Material irregularities caused by filler particles or by weld lines also affect the fatigue of a component. Figures 3.136 and 3.137 compare S-N curves for

3.5 Mechanical Properties

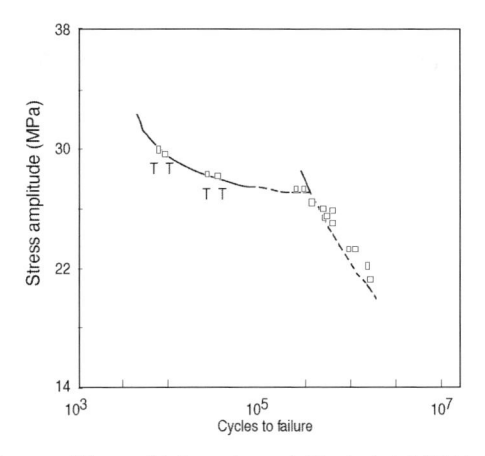

Figure 3.131: Fatigue and thermal failures in acetal tested at 1.67 Hz (after Crawford).

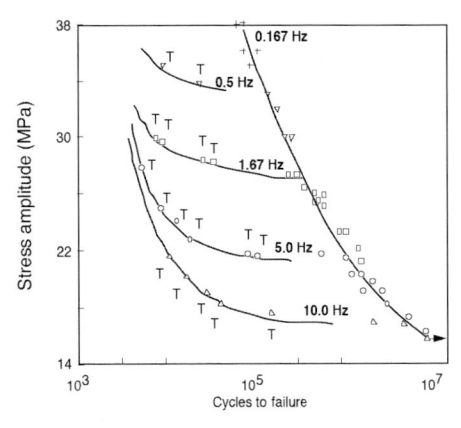

Figure 3.132: Fatigue and thermal failures in acetal tested at various frequencies (after Crawford).

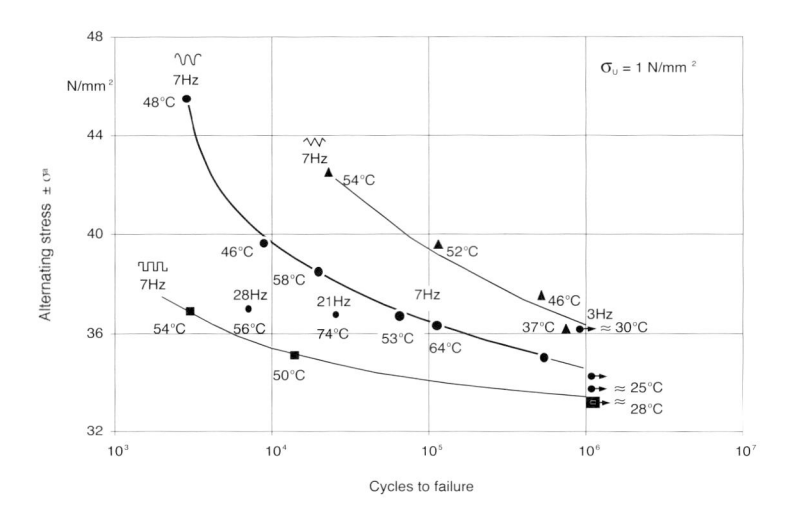

Figure 3.133: Fatigue curves for a glass fiber reinforced PA6 tested with three different imposed stress cycles (23°C.)

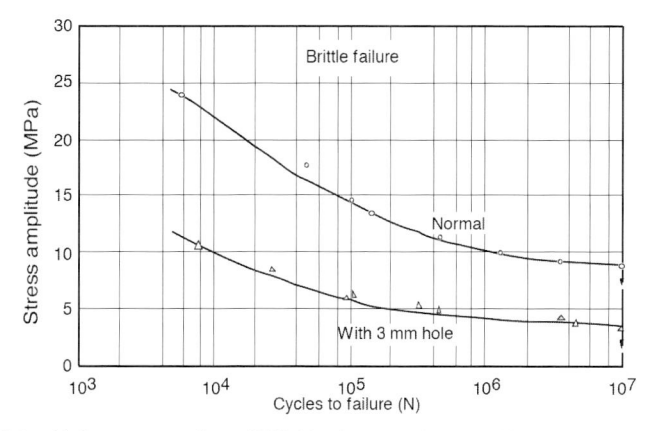

Figure 3.134: Fatigue curves for a PVC-U using specimens with and without 3 mm hole stress concentrators tested at 23°C and 7-Hz with a zero mean stress.

regular PC and ABS test specimens to fatigue behavior of specimens with a weld line and specimens with a 3-mm circular hole.

The previous fatigue graphs pertained to tests with zero mean stress, σ_m. However, many polymer components that are subjected to cyclic loading have

3.5 Mechanical Properties 183

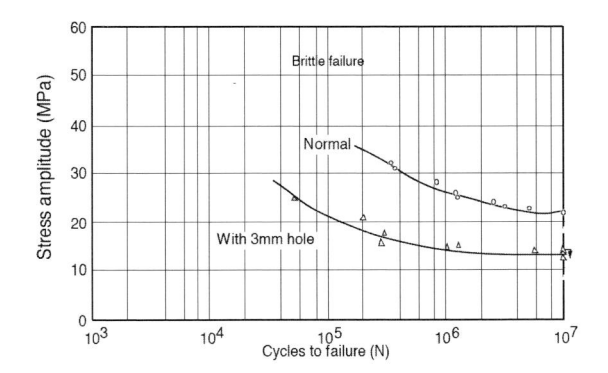

Figure 3.135: Fatigue curves for a PA66 using specimens with and without 3 mm hole stress concentrators tested at 23°C and 7-Hz with a zero mean stress.

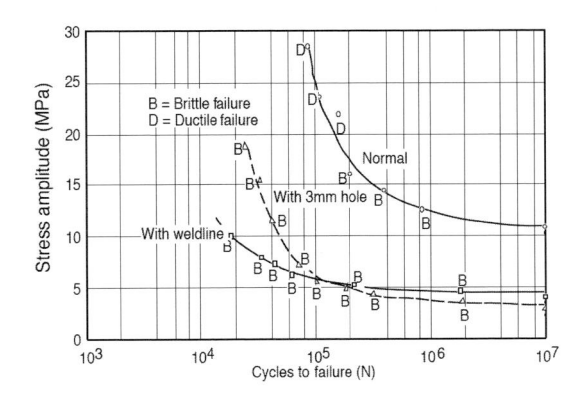

Figure 3.136: Fatigue curves for a PC using regular specimens and specimens with 3 mm hole stress concentrators and weldlines tested at 23°C and 7-Hz with a zero mean stress.

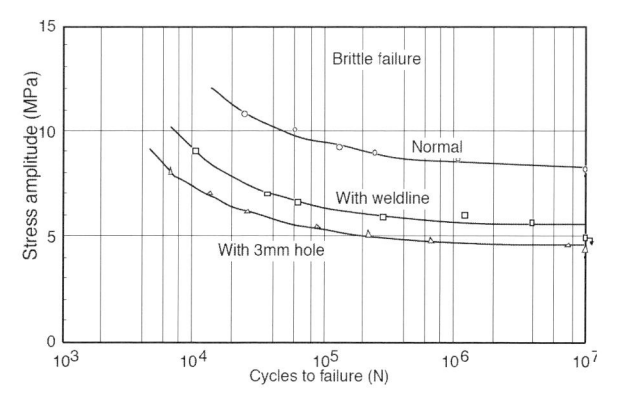

Figure 3.137: Fatigue curves for ABS (Novodur PH/AT) using regular specimens and specimens with 3 mm hole stress concentrators and weldlines tested at 23°C and 7-Hz with a zero mean stress.

other loads and stresses applied to them, leading to non-zero mean stress values. This superposition of two types of loading will lead to a combination of creep, caused by the mean stress, and fatigue, caused by the cyclic stress, σ_a. Test results from experiments with cyclic loading and non-zero mean stresses are complicated by the fact that some specimens fail due to creep and others due to conventional brittle fatigue. Figure 3.138 illustrates this phenomenon for both cases with and without thermal fatigue, comparing them to experiments in which a simple static loading is applied. For cases with two or more dynamic loadings with different stress or strain amplitudes, a similar strain deformation progression is observed. The strain progression, $\Delta\epsilon$, is the added creep per cycle caused by different loadings, similar to ratcheting effects in metal components where different loadings are combined.

Fiber-reinforced composite polymers are stiffer and less susceptible to fatigue failure. Reinforced plastics have also been found to have lower hysteretic heating effects, making them less likely to fail by thermal fatigue. Figure 3.139 presents the flexural fatigue behavior for glass fiber filled and unfilled PA66 tested at 20°C and a 0.5 Hz frequency with a zero mean stress. Parallel to the fiber orientation, the fatigue life was greater than the life of the specimens tested perpendicular to the orientation direction and the unfilled material specimens. The fatigue life of the unfilled specimen and the behavior perpendicular to the orientation direction were similar. However, the unfilled material failed by thermal fatigue at high stresses, whereas both the specimens tested perpendicular and parallel to the orientation direction failed by conventional fatigue at high stress levels. Fiber reinforced systems generally follow a sequence of events during failure consisting

3.5 Mechanical Properties

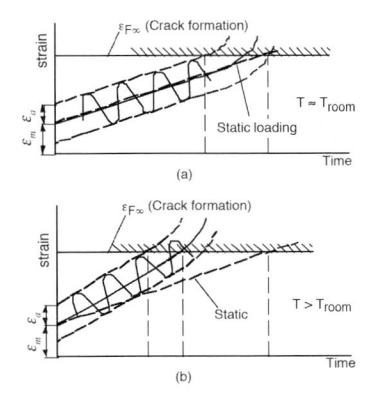

Figure 3.138: Creep and thermal fatigue effects during cyclic loading.

of debonding, cracking, and separation. Figure 3.140 clearly demonstrates this sequence of events with a glass-filled polyester mat tested at 20°C and a frequency of 1.67 Hz. In most composites, debonding occurs after just a few cycles. It should be pointed out that often reinforced polymer composites do not exhibit an endurance limit, making it necessary to use factors of safety between 3 and 4. The fracture by fatigue is generally preceded by cracking of the matrix material, which gives a visual warning of imminent failure. It is important to mention that the fatigue life of thermoset composites is also affected by temperature. Figure 3.141 shows the tensile strength versus number of cycles to failure for a 50% glass fiber filled unsaturated polyester tested at 23°C and 93°C. At ambient temperature, the material exhibits an endurance limit of about 65 MPa, which is reduced to 52 MPa at 93°C.

3.5.6 Strength Stability Under Heat

Polymers soften and eventually flow as they are heated. It is, therefore, important to know what the limiting temperatures are at which a polymer component can still be loaded with moderate deformations. Figure 3.142 presents the shear modulus as a function of temperature for various thermoplastics, with the region of maximum temperature.

Three tests are commonly performed on polymer specimens to determine this limiting temperature for a specific material. They are the Vicat temperature test (ISO 306, ASTM D648 and DIN 53460), shown in Fig. 3.143, the heat-distortion temperature (HDT) test (ISO 75 and ASTM D 648) shown in Fig. 3.144 and the Martens temperature test (DIN 53458 or 53462). In the Vicat temperature test, a needle loaded with weights is pushed against a plastic specimen inside a glycol

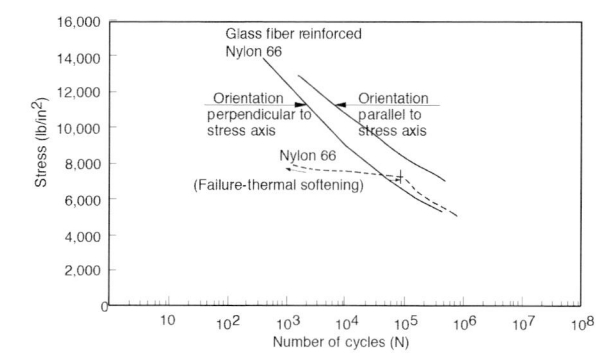

Figure 3.139: Flexural fatigue curves for a PA66 and a glass fiber filled polyamide 66 tested at 20°C and 0.5 Hz with a zero mean stress (after Bucknall, Gotham and Vincent).

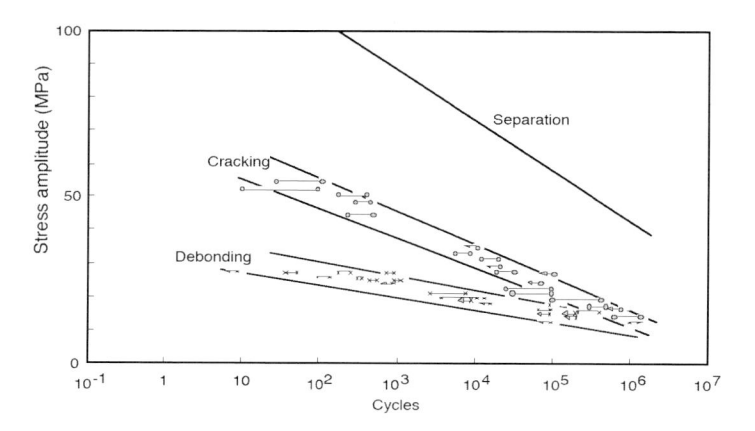

Figure 3.140: Fatigue curves for a glass filled polyester mat tested at 20°C and a frequency of 1.67 Hz (after Hertzberg and Mason).

3.5 Mechanical Properties

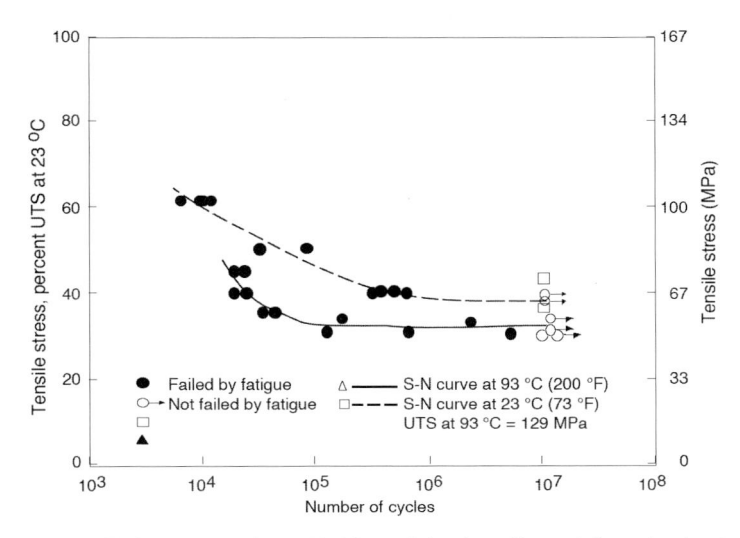

Figure 3.141: Fatigue curves for a 50% by weight glass fiber reinforced polyester resin sheet molding compound tested at 23°C and 93°C and 10 Hz (after Denton).

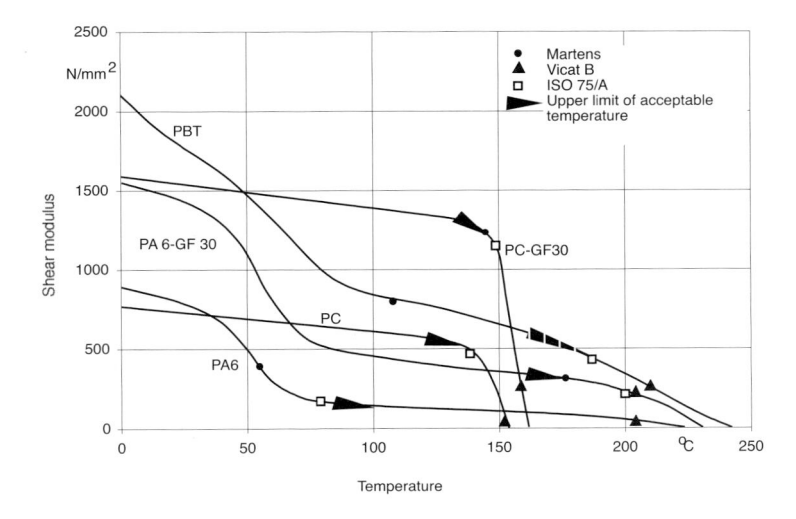

Figure 3.142: Shear modulus as a function of temperature for several thermoplastics.

bath. This is shown schematically in Fig. 3.143. The uniformly heated glycol bath rises in temperature during the test. The Vicat number or Vicat temperature is measured when the needle has penetrated the polymer by 1 mm. The advantage of this test method is that the test results are not influenced by the part geometry or manufacturing technique. The practical limit for thermoplastics, such that the finished part does not deform under its own weight, lies around 15K below the Vicat temperature. To determine the heat distortion temperature, the standard specimen lies in a fluid bath on two knife edges separated by a 10 cm distance. A bending force is applied on the center of the specimen. The standard Vicat temperature tests ISO 306 and ASTM D648 are presented in Table 3.27.

Table 3.27: Standard Methods of Measuring Vicat Softening Temperature (after Shastri)

Standard	ISO 306	D 1525 - 98
Specimen	10 mm x 10 mm x 4 mm, from middle region of the ISO 3167 multipurpose test specimen.	Use at least two specimens to test each sample. The specimen shall be flat, between 3 and 6.5 mm thick, and at least 10 mm x 10 mm in area or 10 mm in diameter.
Conditioning	Specimen conditioning, including any post molding treatment, shall be carried out at 23°C ± 2°C and 50 ± 5 % R.H. for a minimum length of time of 88h, except where special conditioning is required as specified by the appropriate material standard.	If conditioning of the test specimens is required, then condition at 23°C ± 2 °C and 50 ± 5 % relative humidity for not less than 40 h prior to testing in accordance with Test Method D618.
Apparatus	The indenting tip shall preferably be of hardened steel 3 mm long, of circular cross section 1.000 ± 0.015 mm^2 fixed at the bottom of the rod. The lower surface of the indenting tip shall be plane and perpendicular to the axis of the rod and free from burrs.	A flat-tipped hardened steel needle with a cross-sectional area of 1.000 ± 0.015 mm^2 shall be used. The needle shall protruded at least 2 mm from the end of the loading rod.
	Heating bath containing a suitable liquid (e.g., liquid paraffin, glycerol, transformer and silicone oil) which is stable at the temperature used and does not affect the material under test (e.g., swelling or cracking) in which the test specimen can be immersed to a depth of at least 35 mm is used. An efficient stirrer shall be provided.	Immersion bath containing the heat transfer medium (eg., silicone oil, glycerine, ethylene glycol, and mineral oil) that will allow the specimens to be submerged at least 35 mm below the surface.

Continued on next page

3.5 Mechanical Properties

Standard	ISO 306	D 1525 - 98
Test procedures	At least 2 specimens to test each sample.	Use at least two specimens to test each sample. Molding conditions shall be in accordance with the applicable material specification or should be agreed upon by the co-operating laboratories. Specimens shall be annealed only if required in the material specification.
	Specimens tested flatwise. The temperature of the heating equipment should be 20 to 23°C at the start of each test, unless previous tests have shown that, for the material under test, no error is caused by starting at another temperature.	Specimens tested flatwise. The bath temperature shall be 20 to 23 °C at the start of the test unless previous tests have shown that, for a particular material, no error is introduced by starting at a higher temperature.
	Mount the test specimen horizontally under the indenting tip of the unloaded rod. The indenting tip shall at no point be nearer than 3 mm to the edge of the test specimen.	Place the specimen on the support so that it is approximately centered under the needle. The needle should not be nearer than 3 mm to the edge of the test specimen.
	Put the assembly in the heating equipment.	Lower the needle rod (without extra load) and then lower the assembly into the bath.
	After 5 min, with the indenting tip still in position, add the weights to the load carrying plate so that the total thrust on the test specimen is 50 ± 1 N.	Apply the extra mass required to increase the load on the specimen to 10 ± 0.2 N (Loading 1) or 50 ± 1.0 N (Loading 2)
	Set the micrometer dial-gauge reading to zero.	After waiting five minutes, set the penetration indicator to zero.
	Increase the temperature of the heating equipment at a uniform rate: Heating rate ⇒ 50 ± 5 °C /h	Start the temperature rise at one of these rates: 50 ± 5 °C /h (Rate A) or 120 ± 12 °C /h (Rate B) The rate selection shall be agreed upon by the interested parties.
	Note the temperature at which the indenting tip has penetrated into the test specimen by 1 ± 0.01 mm beyond the starting position, and record it as the Vicat softening temperature of the test specimen.	Record the temperature at which the penetration depth is 1 mm. If the range of the temperatures recorded for each specimen exceeds 2°C , then record the individual temperatures and rerun the test.
Values and units	Vicat softening temperature ⇒ °C	Vicat softening temperature ⇒ °C

Similar to the Vicat temperature test, the bath's temperature is increased during the test. The HDT is the temperature at which the rod has bent 0.2 to 0.3 mm

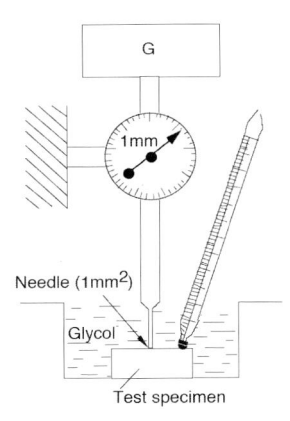

Figure 3.143: Apparatus to determine a material's shape stability under heat using the Vicat temperature test.

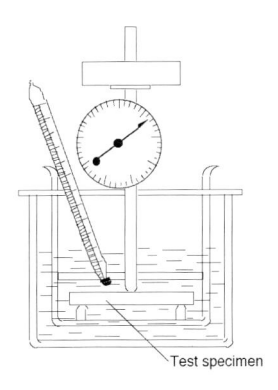

Figure 3.144: Apparatus to determine a material's shape stability under heat using the heat-distortion-temperature test (HDT).

3.5 Mechanical Properties

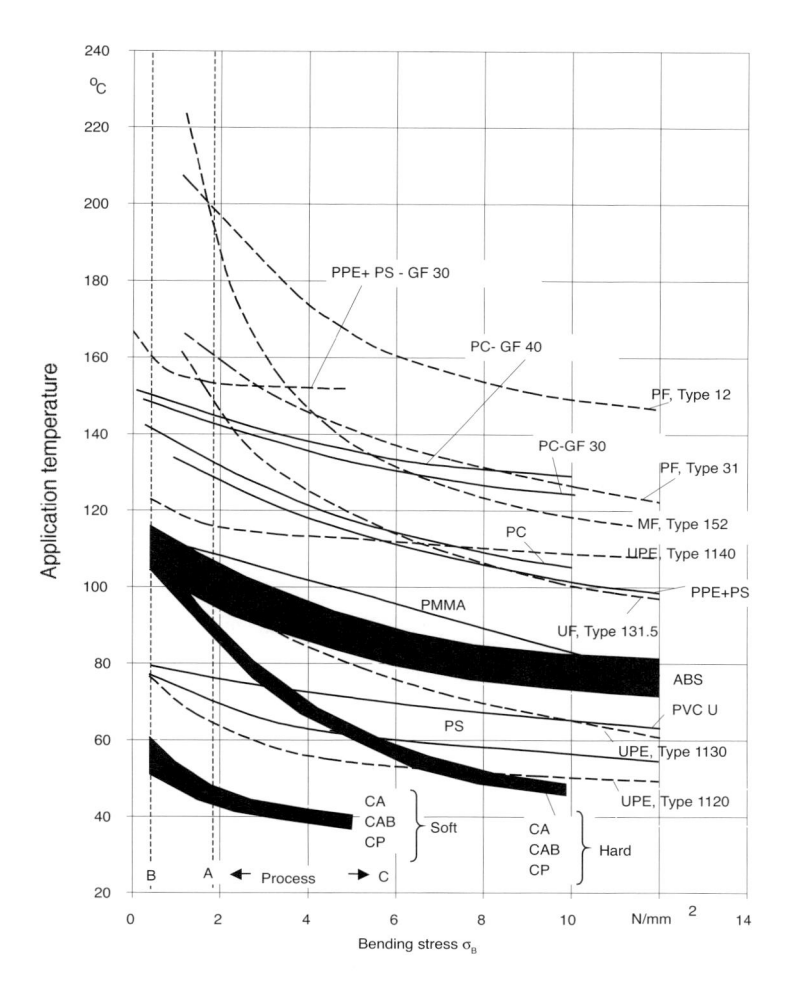

Figure 3.145: Heat distortion temperature for selected thermoplastics as a function of bending stress.

(see Fig. 3.144). The Vicat temperature is relatively independent of the shape and type of part, whereas the heat-distortion-data are influenced by the shaping and pretreatment of the test sample.

Figure 3.145 presents the heat distortion temperature for selected thermoplastics and thermosets as a function of bending stress, measured using ISO 75 and Table 3.28 presents HDT for selected thermoplastics measured using ASTM

D648. The standard HDT tests ISO 75 and ASTM D648 are presented in Table 3.29.

In the Martens temperature test, the temperature at which a cantilevered beam has bent 6 mm is recorded. The test sample is placed in a convection oven with a constantly rising temperature. In Europe, the HDT test has replaced the Martens temperature test.

Table 3.28: Heat Distortion Temperature for Selected Thermoplastics

Material	HDT(°C)1.86 MPa	HDT(°C)0.45 MPa
HDPE	50	50
PP	45	120
uPVC	60	82
PMMA	60	100
PA66	105	200
PC	130	145

Table 3.29: Standard Methods of Measuring Temperature of Deflection Under Load (after Shastri)

Standard	ISO 75 - 1 and 75 - 2	D 648 - 98c
Specimen	Flatwise ⇒ 80 mm x 10mm x 4 mm, cut from the ISO 3167 Type A specimen.	Edgewise ⇒ 120 ± 10 mm x (3 - 13) mm x 12.7 ± 0.3 mm
Conditioning	Specimen conditioning, including any post molding treatment, shall be carried out at 23°C ± 2°C and 50 ± 5 % R.H. for a minimum length of time of 88h, except where special conditioning is required as specified by the appropriate material standard.	At 23°C ± 2°C and 50 ± 5 % relative humidity for not less than 40 h prior to testing in accordance with Procedure A of Method D618.
Apparatus	The contact edges of the supports and the loading nose radius are rounded to a radius of 3.0 ± 0.2 mm and shall be longer than the width of the test specimen. Specimen supports should be about 100 mm apart (edgewise specimens).	The contact edges of the supports and loading nose shall be rounded to a radius of 3.0 ± 0.2 mm. Specimen supports shall be 100 ± 2 mm apart, or 64 mm apart (flatwise specimens).

Continued on next page

3.5 Mechanical Properties 193

Standard	ISO 75 - 1 and 75 - 2	D 648 - 98c
	Heating bath shall contain a suitable liquid (e.g. liquid paraffin, glycerol, transformer oil, and silicone oils) which is stable at the temperature used and does not affect the material under the test (e.g., swelling, softening, or cracking). An efficient stirrer shall be provided with a means of control so that the temperature can be raised at a uniform rate of 120 K/h ± 10 K/h. This heating rate shall be considered to be met if over every 6 min interval during the test, the temperature change is 12 K ± 1 K	Immersion bath shall have a suitable heat-transfer medium (e.g. mineral or silicone oils) which will not affect the specimen and which is safe at the temperatures used. It should be well stirred during the test and provided with means of raising the temperature at a uniform rate of $2°C \pm 0.2\,°C$. This heating rate is met if over every 5 min interval the temperature of the bath shall rise $10°C \pm 1°C$ at each specimen location.
	A calibrated micrometer dial-gauge or other suitable measuring instrument capable of measuring to an accuracy of 0.01 mm deflection at the mid point of the test specimen shall be used.	The deflection measuring device shall be capable of measuring specimen deflection to at least 0.25 mm and is readable to 0.01 mm or better.
Test procedures	At least 2 unannealed specimens	At least 2 specimens shall be used to test each sample at each fiber stress of 0.455 MPa ± 2.5% or 1.820 MPa ± 2.5%.
	The temperature of the heating bath shall be 20 to 23 °C at the start of each test, unless previous tests have shown that, for the particular materials under test, no error is introduced by starting at other temperatures.	The bath temperature shall be about room temperature at the start of the test unless previous tests have shown that, for a particular material, no error is introduced by starting at a higher temperature.
	Apply the calculated force to give the desired nominal surface stress.	Apply the desired load to obtain the desired maximum fiber stress of 0.455 MPa or 1.82 MPa to the specimen.
	Allow the force to to act for 5 min, to compensate partially for the creep exhibited at room temperature when subjected to the specified nominal surface stress. Set the reading of the deflection measuring instrument to zero.	Five minutes after applying load, adjust the deflection measuring device to zero/ starting position.
	Heating rate $\Rightarrow 120 \pm 10\,°C$/h Deflections $\Rightarrow 0.32$ mm (edgewise) for 10.0 to 10.3 mm height 0.34 mm (flatwise) for height equal to 4 mm	Heating rate $\Rightarrow 2.0 \pm 0.2°C$/min The deflection when the specimen is positioned edgewise is: 0.25 for a specimen with a depth of 12.7 mm.

Continued on next page

Standard	ISO 75 - 1 and 75 - 2	D 648 - 98c
	Note the temperature at which the test specimen reaches the deflection corresponding to height of the test specimen, as the temperature of deflection under load for the applied nominal surface stress.	Record the temperature at which the specimen has deflected the specific amount, as the deflection temperature at either 0.455 MPa or 1.820 MPa.
Values and units	HDT at 1.8 MPa and (0.45 MPa or 8 MPa) $\Rightarrow$ °C	HDT at 0.455 MPa or 1.820 MPa $\Rightarrow$ °C

It is important to point out that these test methods do not give enough information to determine the allowable operating temperature of molded plastic components subjected to a stress. Heat distortion data is excellent when comparing the performance of different materials and should only be used as a reference not as a direct design criterion.

3.6 PERMEABILITY PROPERTIES

Because of their low density, polymers are relatively permeable by gases and liquids. A more in-depth knowledge of permeability is necessary when dealing with packaging applications and with corrosive protection coatings. The material transport of gases and liquids through polymers consists of various steps. They are:

- Absorption of the diffusing material at the interface of the polymer, a process also known as adsorption,
- *Diffusion* of the attacking medium through the polymer, and
- Delivery or secretion of the diffused material through the polymer interface, also known as desorption.

With polymeric materials, these processes can occur only if the following conditions are fulfilled:

- The molecules of the permeating materials are inert,
- The polymer represents a homogeneous continuum, and
- The polymer has no cracks or voids which channel the permeating material.

In practical cases, such conditions are often not present. Nevertheless, this chapter shall start with these *ideal cases,* since they allow for useful estimates and serve as learning tools for these processes.

3.6.1 Sorption

We talk about adsorption when environmental materials are deposited on the surface of solids. Interface forces retain colliding molecules for a certain time. Possible causes include van der Waals' forces in the case of physical adsorption, chemical affinity (chemical sorption), or electrostatic forces. With polymers, we have to take into account all of these possibilities.

A gradient in concentration of the permeating substance inside the material results in a transport of that substance which we call molecular diffusion. The cause of molecular diffusion is the thermal motion of molecules that permit the foreign molecule to move along the concentration gradient using the intermolecular and intramolecular spaces. However, the possibility to migrate essentially depends on the size of the migrating molecule.

The rate of permeation for the case shown schematically in Fig. 3.146 is defined as the mass of penetrating gas or liquid that passes through a polymer membrane per unit time. The rate of permeation, $\dot{m}$, can be defined using Fick's first law of diffusion as

$$\dot{m} = -DA\rho \frac{dc}{dx} \qquad (3.45)$$

where D is defined as the *diffusion coefficient*, A is the area and ρ the density. If the diffusion coefficient is constant, eqn. (3.45) can be easily integrated to give

$$\dot{m} = -DA\rho \frac{c_1 - c_2}{L} \qquad (3.46)$$

The equilibrium concentrations c_1 and c_2 can be calculated using the pressure, p, and the *sorption equilibrium parameter*, S:

$$c = Sp \qquad (3.47)$$

which is often referred to as Henry's law.Henry's law The sorption equilibrium constant, also referred to as solubility constant, is almost the same for all polymer materials. However, it does depend largely on the type of gas and on the boiling, T_b, or critical temperatures, T_{cr}, of the gas, as shown in Fig. 3.147

3.6.2 Diffusion and Permeation

Diffusion, however, is only one part of permeation. First, the permeating substance has to infiltrate the surface of the membrane; it has to be absorbed by the membrane. Similarly, the permeating substance has to be desorbed on the opposite side of the membrane. Combining eqn. (3.46) and (3.47), we can calculate the sorption equilibrium using

$$\dot{m} = -DS\rho A \frac{p_1 - p_2}{L} \qquad (3.48)$$

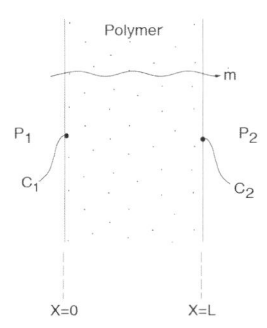

Figure 3.146: Schematic diagram of permeability through a film.

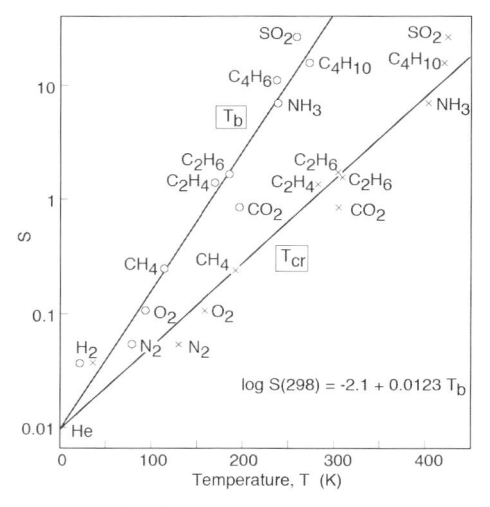

Figure 3.147: Solubility (cm^3/cm^3) of gas in natural rubber at 25°C and 1 bar as a function of the critical and the boiling temperatures.

3.6 Permeability properties

where the product of the sorption equilibrium parameter and the diffusion coefficient is defined as the *permeability* of a material

$$P = -DS = \frac{\dot{m}L}{A\Delta p \rho} \tag{3.49}$$

Equation (3.49) does not take into account the influence of pressure on the permeability of the material and is only valid for dilute solutions. The Henry-Langmuir model takes into account the influence of pressure and works very well for amorphous thermoplastics. It is written as

$$P = -DS\left(1 + \frac{KR'}{1 + b\Delta p}\right) \tag{3.50}$$

where $K = c'_H b/S$, with c'_H being a saturation capacity constant and b an affinity coefficient. The constant R' represents the degree of mobility, where $R' = 0$ for complete immobility and $R' = 1$ for total mobility. Table 3.30 presents permeability of various gases at room temperature through several polymer films. In the case of multi-layered films commonly used as packaging material, we can calculate the permeation coefficient P_C for the composite membrane using

$$\frac{1}{P_C} = \frac{1}{L_C} \sum_{i=1}^{n} \frac{L_i}{P_i} \tag{3.51}$$

Table 3.30: Permeability of Various Gases Through Several Polymer Films. Permeability units are in cm^3-mil/100in^2/24h/atm (after Rosato)

Polymer	CO_2	O_2	H_2O
PET	12-20	5-10	2-4
OPET	6	3	1
PVC	4.75-40	8-15	2-3
PE-HD	300	100	0.5
PE-LD	-	425	1-1.5
PP	450	150	0.5
EVOH	0.05-0.4	0.05-0.2	1-5
PVDC	1	0.15	0.1

Sorption, diffusion, and permeation are processes activated by heat and as expected follow an Arrhenius type behavior. Thus, we can write

$$S = S_0 e^{-\Delta H_s/RT} \tag{3.52}$$
$$D = D_0 e^{-E_D/RT} \tag{3.53}$$
$$P = P_0 e^{-E_P/RT} \tag{3.54}$$

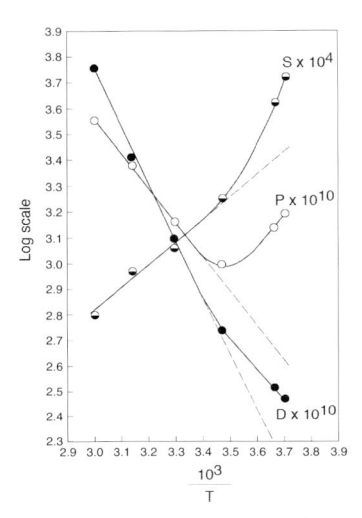

Figure 3.148: Sorption, diffusion, and permeability coefficients, as a function of temperature for polyethylene and methyl bromine at 600 mm of Hg (after Knappe).

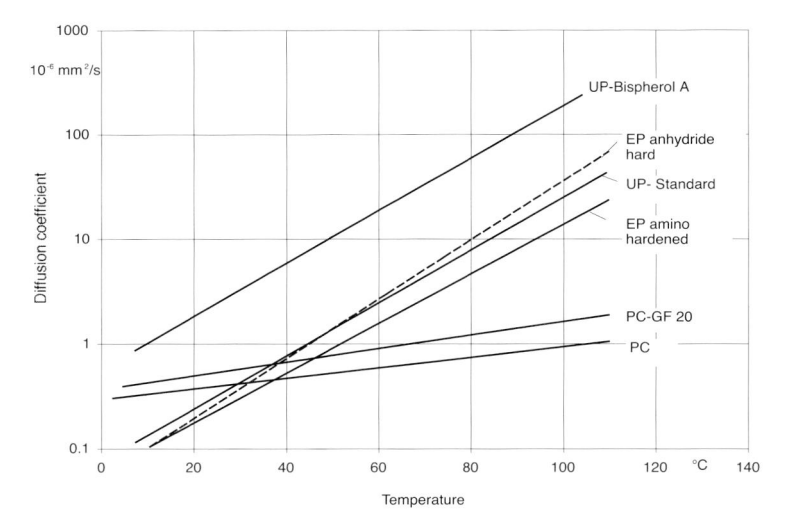

Figure 3.149: Diffusion coefficients as a function of temperature for PC, EP and UP.

3.6 Permeability properties 199

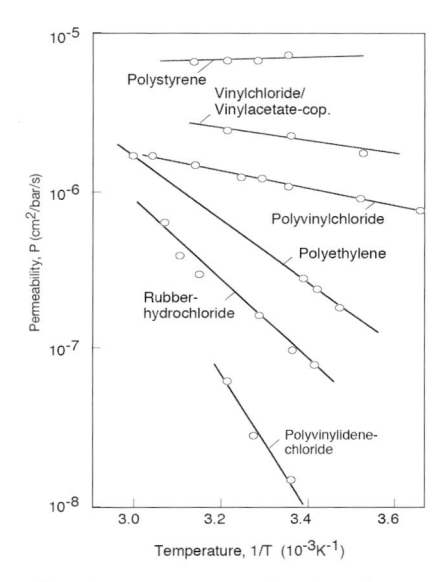

Figure 3.150: Permeability of water vapor as a function of temperature through various polymer films.

where ΔH_S is the enthalpy of sorption, E_D and E_P are diffusion and permeation activation energies, R is the ideal gas constant, and T is the absolute temperature. The Arrhenius behavior of sorption, diffusion and permeability coefficients, as a function of temperature for polyethylene and methyl bromine at 600 mm of Hg are shown in Fig. 3.148. Figure 3.149 presents the effect that temperature has on the diffusion coefficient of a selected number of plastics. Figure 3.150 presents the permeability of water vapor through several polymers as a function of temperature. It should be noted that permeability properties drastically change once the temperature increases above the glass transition temperature. This is demonstrated in Table 3.31, which presents Arrhenius constants for diffusion of selected polymers and CH_3OH.

The diffusion activation energy E_D depends on the temperature, the size of the gas molecule d, and the glass transition temperature of the polymer. This relationship is well represented in Fig. 3.151 with the size of nitrogen molecules, d_{N_2} as a reference. Table 3.31 contains values of the effective cross section size of important gas molecules. Using Fig. 3.151 with the values from Table 3.30 and using the equations presented in Table 3.32, the diffusion coefficient, D, for several polymers and gases can be calculated.

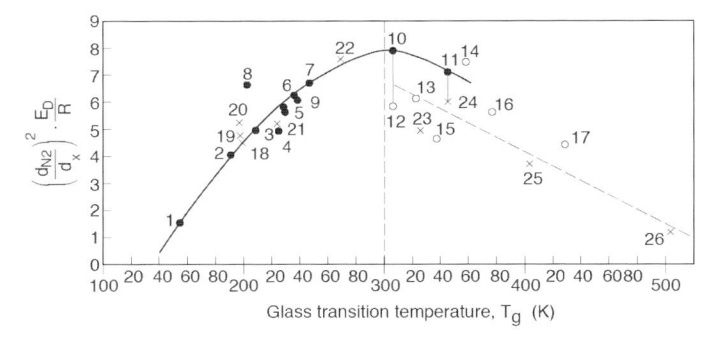

Figure 3.151: Graph to determine the diffusion activation energy E_D as a function of glass transition temperature and size of the gas molecule d_x, using the size of a nitrogen molecule, d_{N2}, as a reference. Rubbery polymers (•): 1 =Silicone rubber, 2 =Polybutadiene, 3 =Natural rubber, 4 =Butadiene/Acrylonitrile K $80/20$, 5 =Butadiene/Acrylonitrile K $73/27$, 6 =Butadiene/Acrylonitrile K $68/32$, 7 =Butadiene/Acrylonitrile K $61/39$, 8 =Butyl rubber, 9 =Polyurethane rubber, 10 =Polyvinyl acetate (r), 11 =Polyethylene terephthalate (r). Glassy polymers (○): 12 =Polyvinyl acetate (g), 13 =Vinylchloride/vinyl acetate copolymer, 14 =Polyvinyl chloride, 15 =Polymethyl methacrylate, 16 =Polystyrene, 17 =Polycarbonate. Semi-crystalline polymers (×): 18 =High-density polyethylene, 19 =Low density polyethylene, 20 =Polymethylene oxide, 21 =Gutta percha, 22 =Polypropylene, 23 =Polychlorotrifluoroethylene, 24 =Polyethyleneterephthalate, 25 =Polytetraflourethylene, 26 =Poly(2,6-diphenylphenyleneoxide)(after Rosato).

Table 3.31: Diffusion Constants Below and Above the Glass Transition Temperature (after van Krevelen)

Polymer	T_g (°C)	$D_0(H_2O)$ (cm²/s)		E_D (kcal/mol)	
		$T < T_g$	$T > T_g$	$T < T_g$	$T > T_g$
Polymethylmethacrylate	90	0.37	110	12.4	21.6
Polystyrene	88	0.33	37	9.7	17.5
Polyvinyl acetate	30	0.02	300	7.6	20.5

Table 3.32: Important Properties of Gases

Gas	d (nm)	V_{cr} (cm³)	T_b (K)	T_{cr} (K)	dN_2/dx
He	0.255	58	4.3	5.3	0.67
H₂O	0.370	56	373	647	0.97
H₂	0.282	65	20	33	0.74
Ne	0.282	42	27	44.5	0.74
NH₃	0.290	72.5	240	406	0.76

Continued on next page

3.6 Permeability properties

Gas	d (nm)	V_{cr} (cm^3)	T_b (K)	T_{cr} (K)	dN_2/dx
O_2	0.347	74	90	55	0.91
Ar	0.354	75	87.5	151	0.93
CH_3OH	0.393	118	338	513	0.96
Kr	0.366	92	121	209	0.96
CO	0.369	93	82	133	0.97
CH_4	0.376	99.5	112	191	0.99
N_2	0.380	90	77	126	1.00
CO_2	0.380	94	195	304	1.00
Xe	0.405	119	164	290	1.06
SO_2	0.411	122	263	431	1.08
C_2H_4	0.416	124	175	283	1.09
CH_3Cl	0.418	143	249	416	1.10
C_2H_6	0.444	148	185	305	1.17
CH_2Cl_2	0.490	193	313	510	1.28
C_3H_8	0.512	200	231	370	1.34
C_6H_6	0.535	260	353	562	1.41

Table 3.33 also demonstrates that permeability properties are dependent on the degree of crystallinity. Figure 3.153 presents the permeability of polyethylene films of different densities as a function of temperature. Again, the Arrhenius relation becomes evident.

Table 3.33: Equations to Compute D Using Data from Table 3.30 and Table 3.31[a]

Elastomers	$\log D = \dfrac{E_D}{2.3R}\left(\dfrac{1}{T} - \dfrac{1}{T_R}\right) - 4$
Amorphous thermoplastics	$\log D = \dfrac{E_D}{2.3R}\left(\dfrac{1}{T} - \dfrac{1}{T_R}\right) - 5$
Semi-crystalline thermoplastics	$\log D = \left(\dfrac{E_D}{2.3R}\left(\dfrac{1}{T} - \dfrac{1}{T_R}\right) - 5\right)(1 - \mathcal{X})$

Figures 3.154 and 3.155 present the permeability of water vapor through several polymers as a function of film thickness.

3.6.3 Measuring S, D, and P

The permeability P of a gas through a polymer can be measured directly by determining the transport of mass through a membrane per unit time. The sorption constant S can be measured by placing a saturated sample into an environment, which allows the sample to desorb and measure the loss of weight. As shown in Fig. 3.156, it is common to plot the ratio of concentration of absorbed substance $c(t)$ to saturation coefficient c_∞ with respect to the root of time.

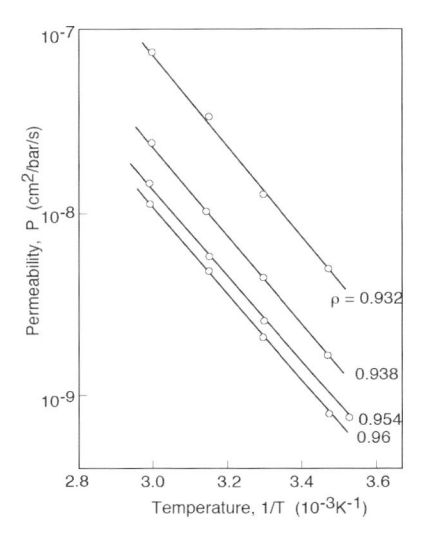

Figure 3.152: Permeation of nitrogen through polyethylene films of various densities.

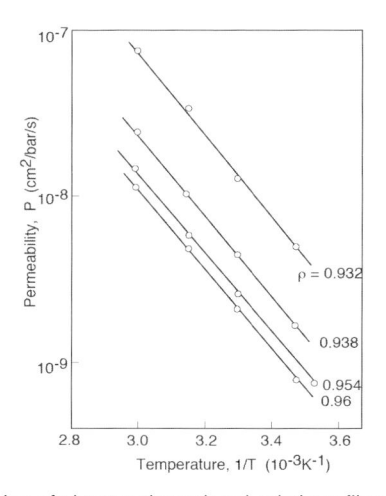

Figure 3.153: Permeation of nitrogen through polyethylene films of various densities.

3.6 Permeability properties

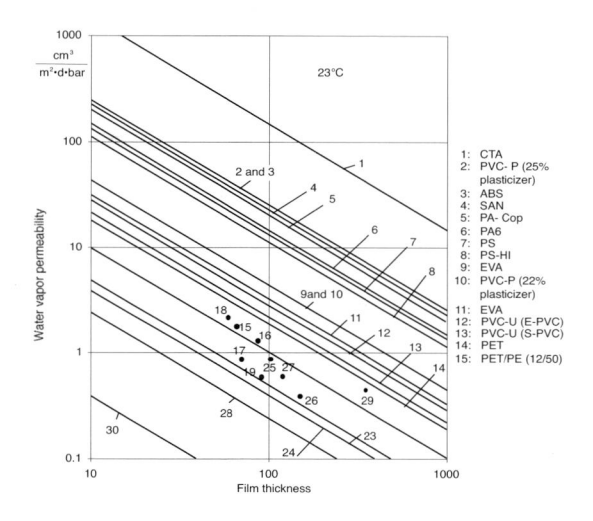

Figure 3.154: Permeability of water vapor through polymer films as a function of film thickness (d=24h).

The diffusion coefficient D is determined using sorption curves as the one shown in Fig. 3.156. Using the slope of the curve, a, we can compute the diffusion coefficient as

$$D = \frac{\pi}{16}L^2 a^2 \qquad (3.55)$$

where L is the thickness of the membrane.

Another method uses the lag time, t_0, from the beginning of the permeation process until the equilibrium permeation has occurred, as shown in Fig. 3.157. Here, the diffusion coefficient is calculated using

$$D = \frac{L^2}{6t_0} \qquad (3.56)$$

The most important techniques used to determine gas permeability of polymers are the ISO 2556, DIN 53 380 and ASTM D 1434 standard tests.

3.6.4 Diffusion of Polymer Molecules and Self-Diffusion

The ability to infiltrate the surface of a host material decreases with molecular size. Molecules of $M > 5 \times 10^3$ can hardly diffuse through a porous-free

204　　　　　　　　　　　　　　　　　　　　　　　　　　　　3 Properties and Testing

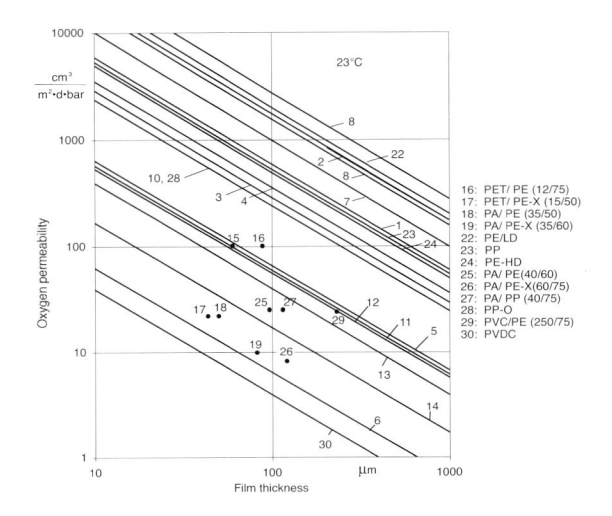

Figure 3.155: Permeability of water vapor through polymer films as a function of film thickness (d=24h).

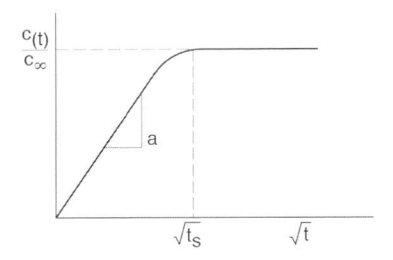

Figure 3.156: Schematic diagram of sorption as a function of time.

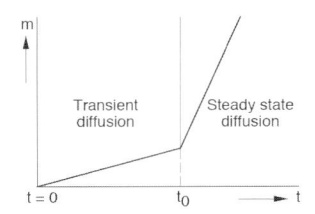

Figure 3.157: Schematic diagram of diffusion as a function of time.

membrane. Self-diffusion occurs when a molecule moves, say in the melt, during crystallization. Also, when bonding rubber, the so-called tack is explained by the self-diffusion of the molecules. The diffusion coefficient for self-diffusion is of the order of

$$D \sim \frac{T}{\eta} \tag{3.57}$$

where T is the temperature and η the viscosity of the melt.

3.7 FRICTION AND WEAR

Friction is the resistance that two surfaces experience as they slide or try to slide past each other. Friction can be dry (i.e., direct surface-surface interaction) or lubricated, where the surfaces are separated by a thin film of a lubricating fluid.

The force that arises in a dry friction environment can be computed using Coulomb's law of friction as

$$F = \mu N \tag{3.58}$$

where F is the force in surface or sliding direction, N the normal force, and μ the coefficient of friction.

Coefficients of friction between several polymers and different surfaces are listed in Table 3.34. However, when dealing with polymers, the process of two surfaces sliding past each other is complicated by the fact that enormous amounts of frictional heat can be generated and stored near the surface due to the low thermal conductivity of the material. The analysis of friction between polymer surfaces is complicated further by environmental effects such as relative humidity and by the likeliness of a polymer surface to deform when stressed, as shown in Fig. 3.158. The top two figures illustrate metal-metal friction, wheareas the bottom figures illustrate metal-polymer friction.

Temperature plays a significant role for the coefficient of friction μ, as demonstrated in Fig. 3.159 for polyamide 66 and polyethylene. In the case of polyethylene, the friction first decreases with temperature. At 100°C, the friction increases because the polymer surface becomes tacky. The friction coefficient starts to drop as the melt temperature is approached. A similar behavior can be seen in the polyamide curve.

Wear is also affected by the temperature of the environment. Figure 3.160 shows how wear rates increase dramatically as the surface temperature of the polymer increases, causing it to become tacky.

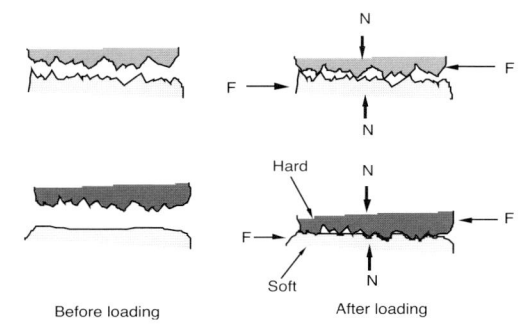

Figure 3.158: Effect of surface finish and hardness on frictional force build-up.

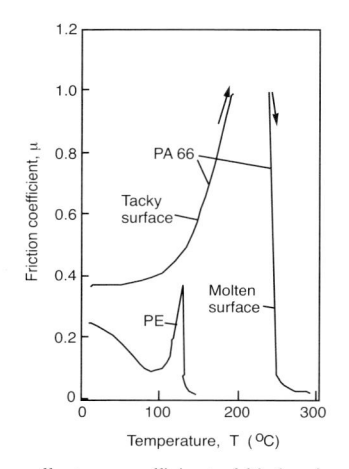

Figure 3.159: Temperature effect on coefficient of friction for a polyamide 66 and a high density polyethylene.

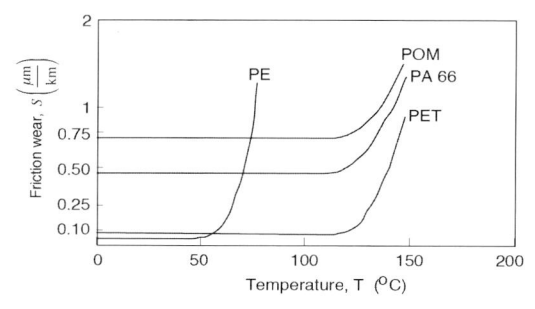

Figure 3.160: Wear as a function of temperature for various thermoplastics.(Courtesy BASF).

3.8 Environmental effects

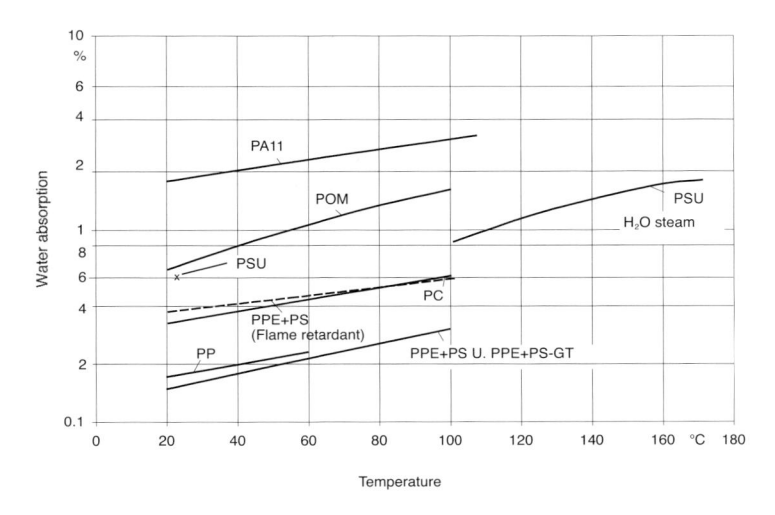

Figure 3.161: Temperature dependance of the water saturation point for various thermoplastics.

Table 3.34: Coefficient of Friction for Various Polymers (superscripts: i=injection molded, s=sandblasted, and m=machined)

Specimen	Partner	Velocity		(mm/s)			
		0.03	0.1	0.4	0.8	3.0	10.6
PP^i	PP^s	0.54	0.65	0.71	0.77	0.77	0.71
PA^i	PA^i	0.63	-	0.69	0.70	0.70	0.65
PP^s	PP^s	0.26	0.29	0.22	0.21	0.31	0.27
PA^m	PA^m	0.42	-	0.44	0.46	0.46	0.47
Steel	PP^s	0.24	0.26	0.27	0.29	0.30	0.31
Steel	PA^m	0.33	-	0.33	0.33	0.30	0.30
PP^s	Steel	0.33	0.34	0.37	0.37	0.38	0.38
PA^m	Steel	0.30	-	0.41	0.41	0.40	0.40

3.8 ENVIRONMENTAL EFFECTS

The environment or the media in contact with a loaded or unloaded component plays a significant role on its properties, life span, and mode of failure. The environment can be a natural one, such as rain, hail, solar ultra-violet radiation, extreme temperatures, etc., or an artificially created one, such as solvents, oils, detergents, high temperature environments, etc. Damage in a polymer component due to natural environmental influences is usually referred to as *weathering*.

3.8.1 Water Absorption

While all polymers absorb water to some degree, some are sufficently hydrophilic that they absorb large quatities water to significantly affect their performance. Water will cause the polymer to swell and serves as a platicizer, consequently lowering its performance, such as electrical and mechanical behavior. Figure 3.161 presents the water saturation point for a selected number of thermoplastics. Increases in temperature result in an increase of free volume between the molecules, allowing the polymer to absorb more water. The standard tests ISO 62 and ASTM D570, presented in Table 3.35, are used to measure the water absorption of polymers.

Table 3.35: Standard Methods of Measuring Water Absorption (after Shastri

Standard	ISO 62	D 570 - 98
Specimen geometry	50 ± 1 mm square or diameter disks x 3 ± 0.2 mm thick for 24h immersion and <1 mm thick for saturation values.	50.8 mm diameter x 3.2 mm disk for molded plastics. The thickness shall be measured to the nearest 0.025 mm.
Conditioning	Dry specimens in an oven for 24 ± 1h at 50 ± 2 °C, allow to cool to ambient temperature in the desiccator and weigh to the nearest 1 mg.	Specimens of a material whose water absorption value is appreciably affected by temperatures close to 110°C, shall be dried in an oven for 24 h at 50 ± 3 °C, cooled in a desiccator, and immediately weighed to the nearest 0.001 g. Specimens of a material whose water absorption value is not appreciably affected by temperatures up to 110°C, shall be dried in an oven for 1 h at 105 to 110 °C. (No weighing requirement is given in the method; however the authors assume that the specimen should be weighed immediately to the nearest 0.001 g.) When data comparisons with other plastics are desired, the specimens shall be dried in oven for 24h at 50 ± 3 °C, cooled in a desiccator, and immediately weighed to the nearest 0.001 g.
Apparatus	Three specimens shall be prepared in accordance with the relevant material standard. When none exists, specimens shall be directly compression or injection molded in accordance with ISO 293 or ISO 294-1.	Three specimens shall be tested. No specimen preparation conditions are given.

Continued on next page

3.8 Environmental effects

Standard	ISO 62	D 570 - 98
	The volume of water shall be at least 8 ml per cm^2 of the total surface area of the test specimen.	No specifics given on the volume of water required.
Test procedures	Place the conditioned specimens in a container of distilled water, controlled at 23°C with a tolerance of ± 0.5 or ± 2.0 °C according to the relevant material standard. In absence of such standard, the tolerance shall be ± 0.5 °C After immersion for 24 ± 1 h, take the specimens from the water and remove all surface water with a clean, dry cloth or with filter paper. Re-weigh the specimens to the nearest 1 mg within 1 min of taking them out of the water. (Method 1). Saturation values in water or air at 50% relative humidity at 23 °C.	The conditioned specimens shall be placed in a container of distilled water maintained at 23 ± 1°C, and shall rest on edge and be entirely immersed. At 24 (+0.5, -0) h, the specimens shall be removed one at a time and wiped off with a dry cloth and weighed to the nearest 0.0001 g immediately. Long-term immersion - To determine the saturation value, the specimens are tested according to the 24 h procedure, except after weighing the specimen are replaced in the water. The weighings shall be repeated at the end of the first week and every two weeks thereafter until the increase in weight per two-week period, as shown by three consecutive weighings, averages less than 1% of the total increase in weight or 5 mg, whichever is greater. The difference between the saturated and dry weight shall be considered the water absorbed when substantially saturated.
	If it is desired to allow for the presence of water soluble matter, dry the test specimens again for 24 ± 1 hr in the oven controlled at 50 ± 2 °C, after completion of Method 1. Allow the specimen to cool to ambient temperature in the desiccator and reweigh to the nearest 1 mg. (Method 2). The percentage of water absorbed is a total of the % weight increase after immersion either by Method 1 or 2.	Materials that known or suspected to contain appreciable amounts of water-soluble ingredients, shall be reconditioned for the same time and at the same temperature as used for conditioning the specimen originally. If the weight of the specimen is less than the original conditioned weight, then that difference in weight shall be considered as water-soluble matter lost during the immersion test. The percentage of water absorbed is a total of the % weight increase (to be noted whether it is 24h or saturation) and the % soluble matter lost.
Values and units	Water absorption (24 h) ⇒wt	Water absorption (24 h)(24 h or saturation) ⇒wt

Figures 3.162 and 3.163 present the % water absorption as a function of relative humidity for a selected number of thermoplastics.

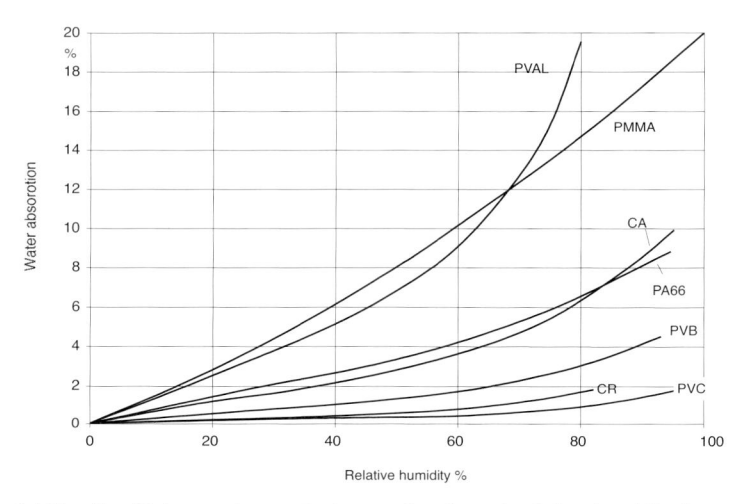

Figure 3.162: Equilibrium water content as a function of relative humidity for several thermoplastics.

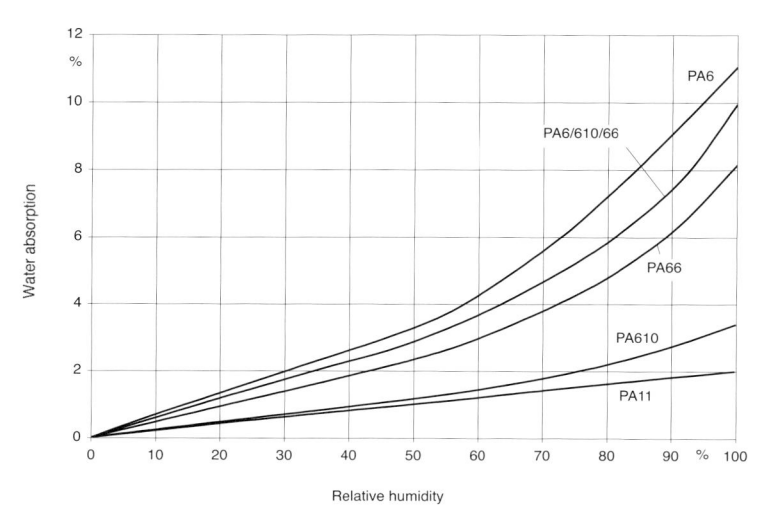

Figure 3.163: Equilibrium water content as a function of relative humidity for various polyamides.

3.8 Environmental effects 211

Figure 3.164: Electron micrograph of the surface of a high density polyethylene beer crate after nine years of use and exposure to weathering (after Ehrenstein).

3.8.2 Weathering

When exposed to the elements, polymeric materials begin to exhibit environmental cracks, which lead to early failure at stress levels significantly lower than those in the absence of these environments. Figure 3.164 shows an electron micrograph of the surface of a high-density polyethylene beer crate after nine years of use and exposure to weathering. The surface of the PE-HD exhibits brittle cracks, which resulted from ultra violet rays, moisture, and extremes in temperature. Standard tests such as the DIN 53486 test are available to evaluate effects of weathering on properties of polymeric materials. It is often unclear which weathering aspects or what combinations of aspects influence material decay the most. Hence, laboratory tests are often done to isolate individual weathering factors such as ultra-violet radiation. For example, Fig. 3.165 shows the surface of a POM specimen irradiated with ultra violet light for 100 h in a laboratory environment. The DIN 53487 xenotest is a standard test to expose polymer test specimens to UV radiation in a controlled environment. Figure 3.166 is a plot of impact strength of notched PMMA specimens as a function of hours of UV radiation exposure in a controlled DIN 53487 test and years of weathering under standard DIN 53486 conditions. The correlation between the two tests is clear. The ASTM-D 4674 test evaluates the color stability of polymer specimens exposed to ultra violet radiation. Standard tests also exist to test materials for specific applications, such as the ASTM-D 2561 test, which evaluates the environmental stress cracking resistance of blow molded polyethylene containers. As can be seen, the effect of ultra violet radiation, moisture, and extreme temperature is detrimental to the mechanical properties of plastic parts. One example in which

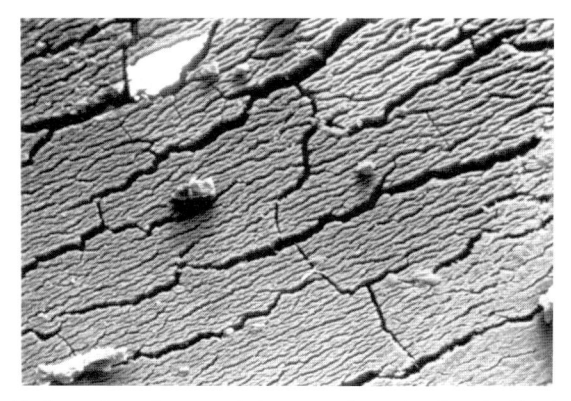

Figure 3.165: Surface of a polyoxymethylene specimen irradiated with ultra violet light for 100 h in a laboratory environment (after Ehrenstein).

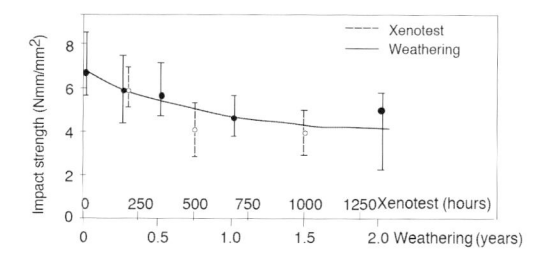

Figure 3.166: Impact strength of notched PMMA specimens as a function of hours of UV radiation exposure in a controlled test and weathering exposure time (after Ehrenstein).

3.8 Environmental effects

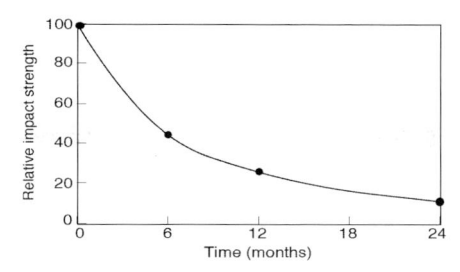

Figure 3.167: Impact strength of PVC pipe as a function weathering exposure time in the United Kingdom (after Davis and Sims).

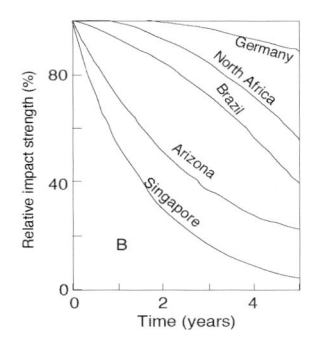

Figure 3.168: Impact strength as a function of weathering time of uPVC exposed in different geographic locations (after Davis and Sims).

weathering completely destroys the strength properties of a material is shown for PVC in Fig. 3.167. The figure shows the decay of the impact strength of PVC pipes exposed to weathering in the United Kingdom. As can be seen, the impact strength rapidly decreases in the first six months and is only 11% of its original value after only two years. The location and climate of a region can play a significant role on the weathering of polymer components. Figure 3.168 shows the decrease in impact strength of rigid PVC as a function of time at five different sites. After five years of weathering, the PVC exposed in Germany still has 95% of its original impact strength, whereas the samples exposed in Singapore have less than 5% of their initial strength.

The strength losses and discoloration in a weathering process are mainly attributed to the ultra-violet rays received from sunshine. This can be demonstrated by plotting properties as a function of actual sunshine received instead of time exposed. Figure 3.169 is a plot of percent of initial impact strength for an ABS as a function of total hours of exposure to sun light in three different locations:

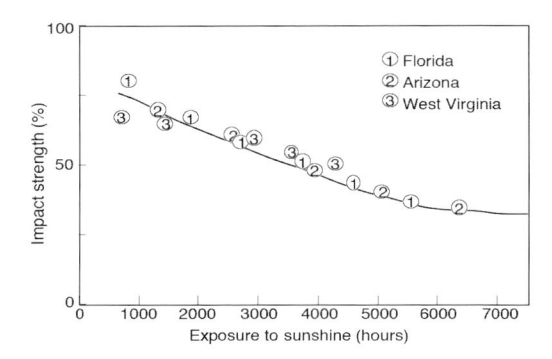

Figure 3.169: Impact strength of an ABS as a function of hours to actual sunshine exposure (after Ruhnke and Biritz).

Florida, Arizona, and West Virginia. The curve reveals the fact that by "normalizing" the curves with respect to exposure to actual sunshine, the three different sites with three completely different weather conditions[6] lead to the same relation between impact strength and total sunshine.

The effect of weathering can often be mitigated with the use of pigments, such as TiO_2 or soot, which absorb ultra violet radiation, making it more difficult to penetrate the surface of a polymer component. The most important one is soot. For example, ABS with white and black pigments exhibit a noticeable improvement in properties after exposure to ultra violet radiation. Figure 3.170 shows the reduction of impact strength in ABS samples as a function of exposure time to sunshine for four pigment concentrations: 0.5%, 0.7%, 1%, and 2%. It is clear that the optimal pigment concentration is around 1%. Beyond 1% of pigmentation there is little improvement in the properties.

3.8.3 Chemical Degradation

Liquid environments can have positive and negative effects on the properties of polymeric materials. Some chemicals or solvents can have detrimental effects on a polymer component. Figure 3.171 shows results of creep rupture tests done on PVC tubes as a function of the hoop stress. It can be seen that the life span of the tubes in contact with the iso-octane and isopropanol has been significantly reduced as compared to the tubes in contact with water. The measured data for the pipes that contained iso-octane clearly show a slope reduction with a visible

---

[6]Florida has a subtropical coastal climate with a yearly rainfall of 952 mm and sunshine of 2750 hours. Arizona has a hot dry climate with 116 mm of rainfall and 3850 hours of sunshine. West Virginia has a milder climate with 992 mm of rainfall and 2150 hours of sunshine

3.8 Environmental effects

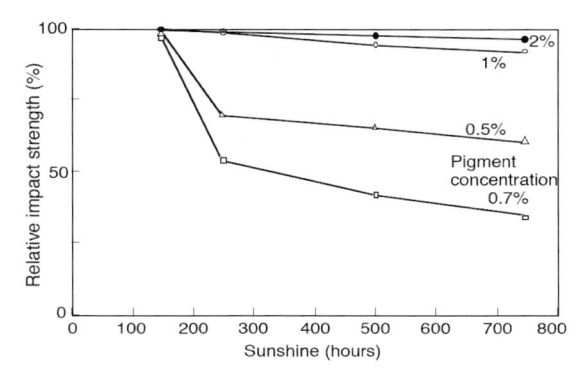

Figure 3.170: Influence of pigment concentration on the impact strength reduction of ABS specimens exposed to weathering (after Ruhnke and Biritz).

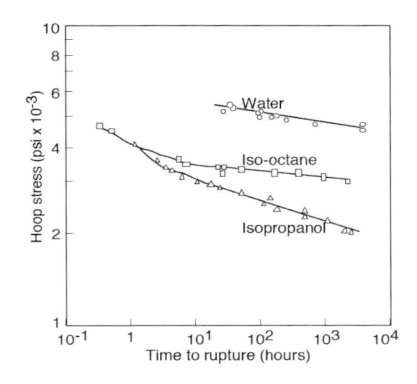

Figure 3.171: Effect of different environments on the stress rupture life of PVC pipe at 23°C (after Riddell).

endurance limit, making it possible to do long-life predictions. On the other hand, the isopropanol samples do not exhibit such a slope reduction, suggesting that isopropanol is a harmful environment which acts as a solving agent and leads to gradual degradation of the PVC surface.

The question of whether a chemical is harmful to a specific polymeric material needs to be addressed if the polymer component is to be placed in a possibly threatening environment. Similar to polymer solutions, a chemical reaction between a polymer and another substance is governed by Gibbs free energy equation. If the change in enthalpy, ΔH, is negative, a chemical reaction will occur between the polymer and the solvent.

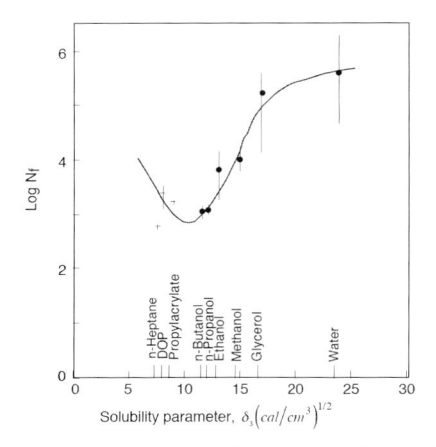

Figure 3.172: Effect of solubility parameter of the surrounding media on the fatigue life of polystyrene specimens (after Hertzberg and Mason).

The effect of the solubility parameter of several solvents on the fatigue response of polystyrene samples is presented in Fig. 3.172. When the absolute difference between the solubility parameter of polystyrene, which is 9.1 (cal/cm^3)$^{1/2}$, and the solubility parameter of the solvent decreases, the fatigue life drops significantly.

It should be pointed out again that some substances are more likely to be absorbed by the polymer than others. A polymer that is in a soluble environment is more likely to generate stress cracks and fail. This is illustrated in Fig. 3.173, which shows the strain for crack formation in polyphenylene oxide samples as a function of solubility parameter of various solutions. The specimens in solutions that were ± 1 (cal/cm^3)$^{1/2}$ away from the solubility parameter of the polymer generated cracks at fairly low strains, whereas those specimens in solutions with a solubility parameter further away from the solubility of the polymer formed crazes at much higher strains.

Environmental stress cracking or stress corrosion in a polymer component only occurs if crazes or microcracks exist.

3.8.4 Thermal Degradation of Polymers

Because plastics are organic materials, they are threatened by chain breaking, splitting off of substituents, and oxidation. This degradation generally follows a reaction which can be described by the Arrhenius principle. The period of dwell

3.8 Environmental effects

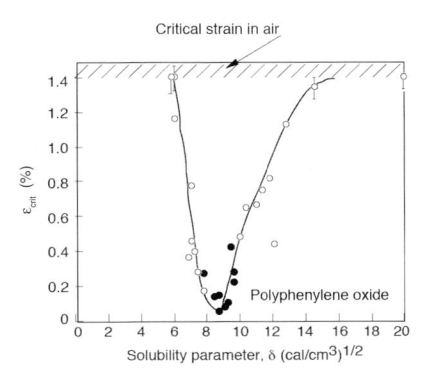

Figure 3.173: Strains at failure as a function of solubility parameter for polyphenylene oxide specimens: (filled circle) cracking, (open circle) crazing(after Bernier and Kambour).

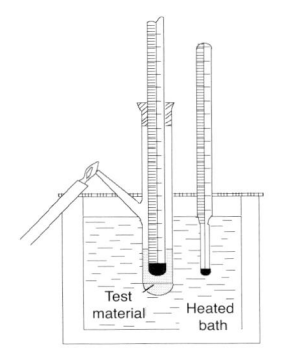

Figure 3.174: Test procedure to determine flash point of polymers.

or residence time permitted before thermal degradation occurs is given by

$$t_{permited} \approx e^{\frac{\Delta}{RT}} \qquad (3.59)$$

where Δ is the activation energy of the polymer, R the gas constant and T the absolute temperature.

A material that is especially sensitive to thermal degradation is PVC; in addition, the hydrogen chloride that results during degradation attacks metal parts. Ferrous metals act as a catalyzer and accelerate degradation. An easy method for determining the flash point of molding batches is by burning the hydrocarbons which are released at certain temperatures. This is shown schematically in Fig.3.174. For PVC a vial with soda lye, instead of a flame, should be used to determine the conversion of chlorine. Figure 3.175 presents the allowable expo-

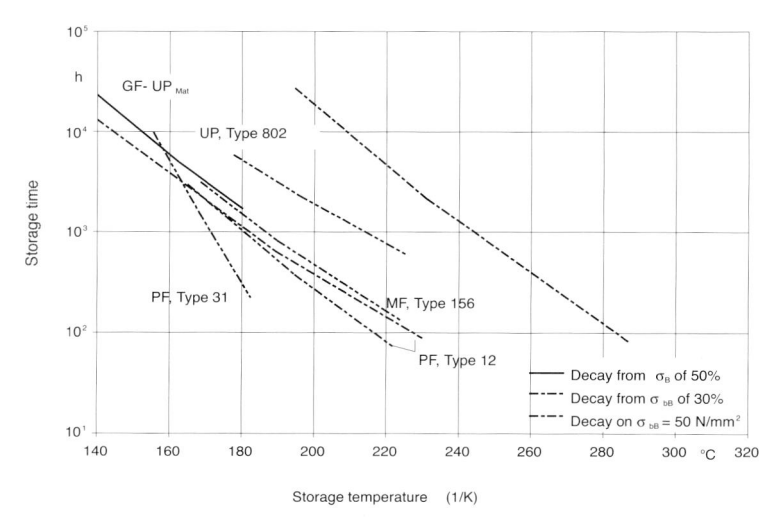

Figure 3.175: Allowable thermal loading times as a function of temperature for various thermosets.

sure time as a function of the environment's temperature for various thermosets.

Thermal degradation can be critical during processing. Figure 3.176 schematically depicts processing windows for injection molding with respect to optimal melt conditions and thermal degradation due to thermal loading. An actual thermal degradation versus thermal loading time is presented for two types of PBT in Fig. 3.177. The upper bound of the processing window is controlled by a 10% drop in viscosity criteria.

Contact to flames is considered extreme thermal loading of polymers. The flammability of polymers is critical in many products such as clothing, construction, electric and electronic equipment, under-the-hood automotive applications, to name a few. There are various standard tests with which one can determine and assess the flammability of polymers. The flammability of horizontally oriented specimens is measured using the standard tests ISO 1210 and ASTM D635 presented in Table 3.36. ISO 1210 also prescribes the flammability of vertical specimens along with the ASTM D3801 test. The vertical specimen flammability tests are presented in Table 3.37.

3.8 Environmental effects

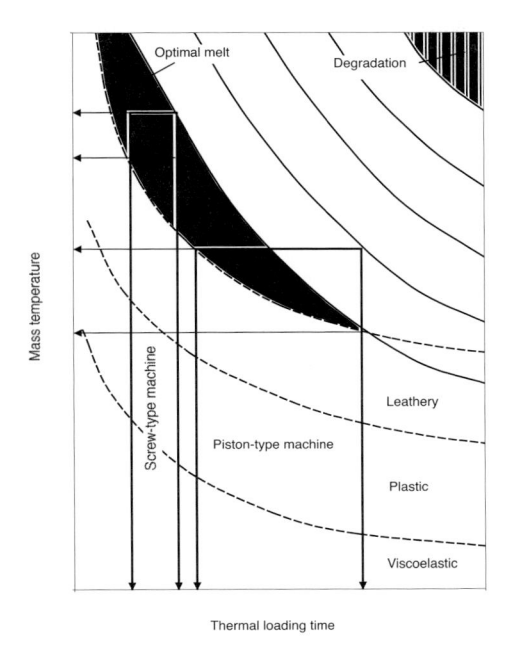

Figure 3.176: Schematic of thermal loading of polymer melts for injection molding processes (after Albers).

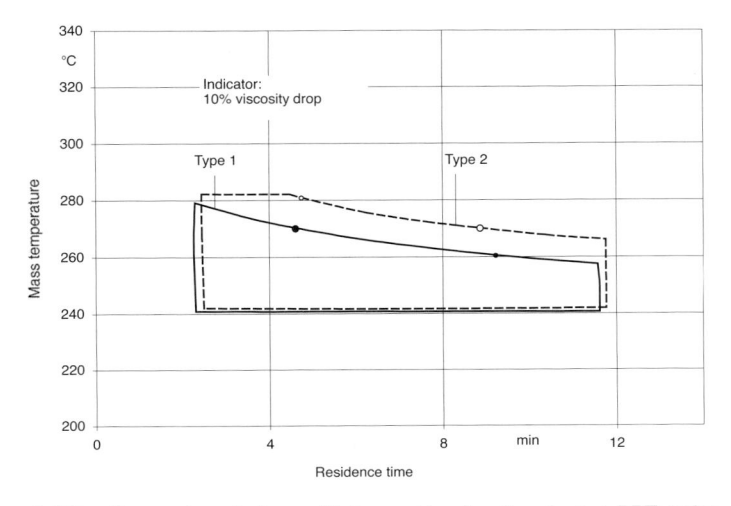

Figure 3.177: Processing windows with thermal loading time for two PBT resins.

Table 3.36: Standard Methods of Measuring Flammability of Polymers (Linear Burning Rate of Horizontal Specimens) (after Shastri)

Standard	ISO 1210 , Method A	D 635 - 98
Specimen	125 mm x 13 mm x 3 mm (Additional specimen thickness <3 mm may be used)	125 mm x 12.5 mm in thickness normally supplied (3 - 12 mm cut from sheet or molded).
Conditioning	Specimen conditioning, including any post molding treatment, shall be carried out at 23°C ± 2°C and 50 ± 5 % R.H. for a minimum length of time of 88h, except where special conditioning is required as specified by the appropriate material standard.	As received, unless otherwise specified.
Apparatus	Laboratory burner in accordance with ISO 10093	Laboratory burner in accordance with D 5025 - 94.
Test procedures	Three specimens	At least 10 specimens
Values and units	Burning rate $\Rightarrow$ mm/min *(If additional specimens with thicknesses < 3 mm are tested, the specimen thickness must also be reported)* Average time of burning $\Rightarrow$ s	Average extent of burning $\Rightarrow$ mm

Table 3.37: Standard Methods of Measuring Flammability (After flame and after-glow times of vertical specimens) (after Shastri)

Standard	ISO 1210 Method B	D 3801 - 96
Specimen	125 mm x 13 mm x 3 mm (Additional specimen thickness <13 mm may be used)	125 mm x 13 mm x 3 mm (Additional specimen thickness <13 mm may be used)
Conditioning	Two sets of five specimens at 23 ± 2 2°C and 50 ± 5 % relative humidity for at least 48h; two sets of five specimens at 70 ± 1/2°C for 168h ± 2 h	One set of five specimens at 23 ± 2 2°C and 50 ± 5 % relative humidity for at least 48h; second set of five specimens at 70 ± 1/2°C for 168h.
Apparatus	Laboratory burner in accordance with ISO 10093. Barrel length is 100 ± 10 mm and inside diameter of 9.5 ± 0.3 mm	Bunsen Tirrill type burner of tube length 95 ± 6 mm and inside diameter 9.5 (+1.6 mm, - 0.0 mm).
Test procedures	Technical grade methane gas or natural gas having a heat content of approximately 37 MJ/m^3.	Technical grade methane gas or natural gas having energy density approximately 37 MJ/m^3.
Values and units	Flame classification	Flame classification

To asses the flammability of polymers the UL Subject 94 of the *Underwriters Laboratories* has been adopted throughout the world. In the tests horizontal and

… 3.8 Environmental effects

vertical specimens are exposed to flames from a Bunsen burner. The evaluation includes burning speed, burned distance, after burn, and dripping. The evaluation is done in steps of increasing rigor: "no," "HB," "V-2," "V-1," "V-0," "5VA," or "5VB." The UL Subject 94 has been adopted by the CAMPUS data bank.

The ignitability of polymers is measured using the standard tests ISO 4589 and ASTM D2863 presented in Table 3.38. The ISO 4589 has also been adopted by the CAMPUS data bank.

Table 3.38: Standard Methods of Measuring Ignitability (after Shastri)

Standard	ISO 4589 - 2 , Procedure A	D 2863 - 97
Specimen	80 mm x 10 mm x 4 mm cut from the center of the ISO 3167 multipurpose specimen (ISO 4589-2, Form 1)	70 - 150 mm x 6.5 mm x 3 mm
Conditioning	Specimen conditioning, including any post molding treatment, shall be carried out at $23°C \pm 2°C$ and 50 ± 5 % R.H. for a minimum length of time of 88h, except where special conditioning is required as specified by the appropriate material standard.	As received, unless otherwise agreed upon
Apparatus	Test chimney dimensions of 450 mm in height x 75 mm minimum diameter cylindrical bore. The upper outlet should be restricted as necessary to produce an exhaust velocity of at least 90 mm/s from a flow rate within the chimney of 30 mm/s. The base of the chimney will preferably have a layer of glass beads (3-5 mm in diameter) between 80 and 100 mm deep. The specimen shall be held by a small clamp which is a least 15 mm away from the nearest point at which the specimen may burn. The moisture content of the gas entering the chimney shall be < 0.1% (m/m) and the variation in oxygen concentration rising in the chimney, below the level of the test specimen, is <0.2% (V/V)	Test column of heat-resistant glass tube 450 mm in height x 75 mm minimum inside diameter. The base of the column contain a non- combustible material which can evenly distribute the gas mixture. A layer of glass beads (3-5 mm in diameter) between 80 and 100 mm deep has been found suitable. The specimen shall be held by a small clamp that will support the specimen at its base and hold it vertically in the center of the column.

The flow control and measuring devices shall be such that the volumetric flow of each gas into the column is within 1% of the range being used. |

Continued on next page

Standard	ISO 4589 - 2 , Procedure A	D 2863 - 97
	The flame ignitor is a tube with an outlet of 2 ± 1 mm diameter which projects a 16 ± 4 mm flame vertically downward from the outlet when the tube is vertical within the chimney. The flame fuel shall be propane without premixed air.	The flame ignitor is a tube with a small orifice 1-3 mm in diameter, which projects a flame 6 to 25 mm long. The flame fuel can be propane, hydrogen or other gas flame.
Test procedures	A minimum of 15 specimens shall be prepared in accordance with the relevant material standard. When none exists, or unless otherwise specified, specimens shall be directly compression or injection molded in accordance with ISO 293 or ISO 294-1.	A sufficient number of specimens (normally 5 to 10)
	Test specimens shall be marked 50 mm from the end to be ignited.	No marking indicated, author's would assume that some indication is needed to know when the burn length of 50 mm is achieved.
	Select the initial concentration of oxygen to be used based on experience with similar materials, or ignite the specimen in air and note the burning behavior. Select an initial concentration ca. 18%, 21% , or 25% (V/V) depending on the burning behavior.	Select the initial concentration of oxygen to be used based on experience with similar materials, or ignite the specimen in air and note the burning behavior. Select an initial concentration ca. 18%, or 25% depending on the burning behavior.
	Specimen is mounted such that the top of the specimen is at least 100 mm below the top of the chimney, and the lowest exposed part of the specimen is 100 m above the top of the gas distribution device.	Specimen is mounted vertically in approximate center of the column with the top of specimen at least 100 mm below the top of the column.
	Gas flow rate of 40 ± 10 mm/s, must flow at least 30 min prior to ignition.	Gas flow rate of 40 ± 10 mm/s, must flow at least 30 min prior to ignition.
	Apply the flame, with a sweeping motion to the top of the specimen for up to 30 s, removing it every 5s to determine if the top is burning.	Ignite the entire top of the specimen so that the specimen is well lit, remove the flame.

Continued on next page

Standard	ISO 4589 - 2 , Procedure A	D 2863 - 97
	Commence timing the period of burning. If the burning ceases but spontaneous combustion occurs in < 1s, continue timing. If period and extent of burning does not exceed 180s and 50 mm, the oxygen concentration would need to be incrementally increased. Adjust the oxygen concentration either up or down until there are two concentrations which differ by <1.0% and in which one specimen met the criteria and the other did not. Repeat the test on four more specimens.	Start timing. If the burning time and the extent of burning does not exceed 180s and 50 mm, the oxygen concentration would need to be incrementally increased. Adjust the oxygen either up or down until the critical concentration of oxygen is determined. This is the lowest level which meets the 180 s/50 mm criteria. At the next lower oxygen concentration that will give a difference in oxygen index of 0.2% or less, the specimen should not meet the 180 s/50 mm criteria. Repeat the test on three more specimens, but starting at a slightly different flow rate, yet in the criteria of 30-50% (V/V)
Values and units	Average Oxygen Index ⇒ % (V/V)	Average Oxygen Index ⇒ % (V/V)

3.9 ELECTRICAL PROPERTIES

In contrast to metals, common polymers are poor electron conductors. Similar to mechanical properties, their electric properties depend to a great extend on the flexibility of the polymer's molecular blocks. The intent of this chapter is to familiarize the reader with electrical properties of polymers by discussing the dielectric, conductive and magnetic properties.

3.9.1 Dielectric Behavior

Dielectric Coefficient The most commonly used electrical property is the dielectric coefficient, ϵ_r, also known as the electric permittivity. This property describes the ability of a material to store an electric charge. Table 3.39 lists the relative dielectric coefficients of important polymers. The measurements were conducted using the standard test DIN 53 483 in condensers of different geometries which, in turn depended on the sample type. The ASTM standard test is described by ASTM D150 and the ISO test. Both the ASTM D150 and the IEC 60250 tests are presented in Table 3.40. Figures 3.178 and 3.179 present the dielectric coefficient for selected polymers as a function of temperature and frequency, respectively.

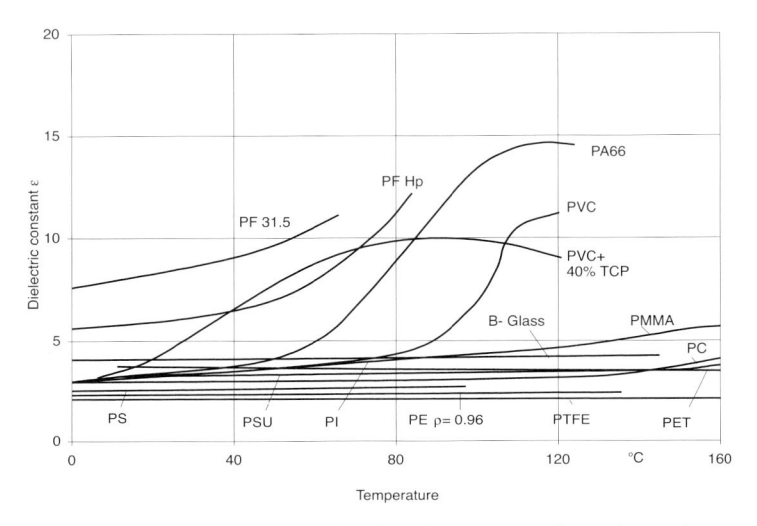

Figure 3.178: Dielectric constant as a function of temperature for various polymers (after Domininghaus).

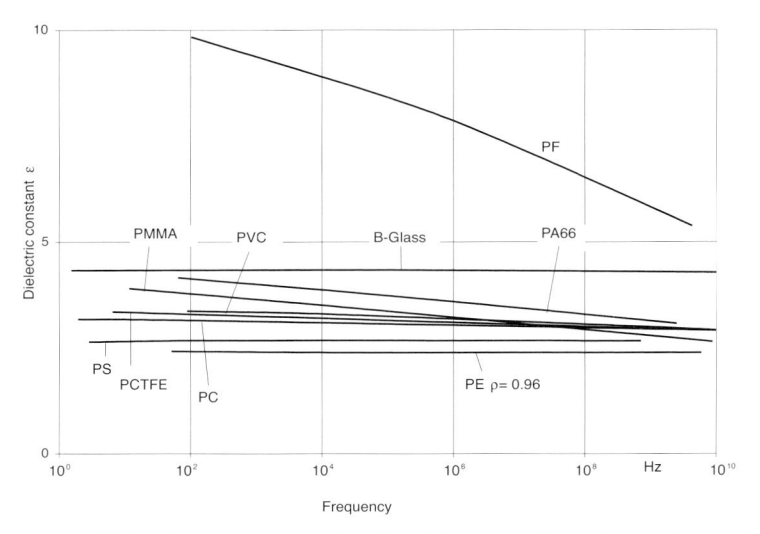

Figure 3.179: Dielectric constant as a function of frequency for various polymers (after Domininghaus).

3.9 Electrical Properties

Table 3.39: Relative Dielectric Coefficient, ϵ_r, of Various Polymers

Polymer	Relative dielectric coefficient, ϵ_r	
	800 Hz	106 Hz
ABS	4.6	3.4
CA, type 433	5.3	4.6
EP (unfilled)		2.5-5.4
Expanded PS	1.05	1.05
PA6 (moisture content dependent)	3.7-7.0	
PA66 (moisture content dependent)	3.6-5.0	
PE (density dependent)	2.3-2.4	2.3-2.4
PC	3.0	3.0
PET	3.0-4.0	3.0-4.0
MF type 154	5.0	10.0
PF type 31.5	6.0-9.0	6.0
PF type 74	6.0-10.0	4.0-7.0
PP	2.3	2.3
PPE	2.7	2.7
PS	2.5	2.5
PTFE	2.05	2.05
UF type 131.5	6.0-7.0	6.0-8.0

Table 3.40: Standard Methods of Measuring Relative Permittivity of Polymers (after Shastri)

Standard	IEC 60250 : 69	D 150 - 95
Specimen geometry	> 80 mm x > 80 mm x 1 mm (greater thickness may be used for those materials that cannot be molded reliably at 1 mm thickness)	Test specimens are of suitable shape and thickness determined by the material specification or by the accuracy of measurement required, and the frequency at which the measurements are to be made.
Conditioning	Specimen conditioning, including any post molding treatment, shall be carried out at 23°C ± 2°C and 50 ± 5 % R.H. for a minimum length of time of 88h, except where special conditioning is required as specified by the appropriate material standard.	Clean the test specimen with a suitable solvent or as prescribed in the material specification. Use Recommended Practice D1371 as a guide to the choice of suitable cleaning procedures.
Test procedures	100 Hz and 1 MHz.	Frequency not specified but recorded.
	Null methods are used at frequencies up to 50 MHz and results are compensated for electrode edge effects.	Null method with resistive or inductive ratio arm capacitance bridge suggested for frequencies of < 1Hz to a few MHz.
Values and units	Relative permittivity $\Rightarrow$ unitless	Relative permittivity $\Rightarrow$ unitless

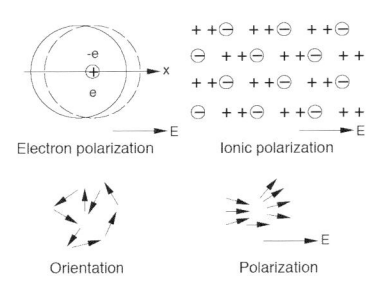

Figure 3.180: Polarization processes.

Dielectric Polarization The two most important molecular types for the polarization of a dielectric in an electric field are *displacement polarization* and *orientation polarization*. Under the influence of an electric field, the charges deform in field direction by aligning with the atomic nucleus (electron polarization) or with the ions (ionic polarization). This is usually called *displacement polarization* and is shown in Fig. 3.180. Because of their structure, some molecules possess a dipole moment in the spaces that are free of an electric field. Hence, when these molecules enter an electric field, they will orient according to the strength of the field. This is generally referred to as orientation polarization and is schematically shown in Fig. 3.180.

It takes some time to displace or deform the molecular dipoles in the field direction and even longer time for the orientation polarization. The more viscous the surrounding medium is, the longer it takes. In alternating fields of high frequency, the dipole movement can lag behind at certain frequencies. This is called dielectric relaxation, which leads to dielectric losses that appear as dielectric heating of the polar molecules. In contrast to this, the changes in the displacement polarization happen so quickly that it can even follow a lightwave. Hence, the refractive index, n, of light is determined by the displacement contribution, ϵ_v, of the dielectric constant. The relation between n and ϵ_v is given by

$$n = \sqrt{\epsilon_v} \qquad (3.60)$$

Hence, we have a way of measuring polarization properties since the polarization of electrons determines the refractive index of polymers. It should be noted that ion or molecular segments of polymers are mainly stimulated in the middle of the infrared spectrum. A number of polymers have permanent dipoles. The best known polar polymer is polyvinyl chloride, and C=O groups also represent a permanent dipole. Therefore, polymers with that kind of building block suffer dielectric losses in alternating fields of certain frequencies. For example, Fig. 3.181 shows the frequency dependence of susceptibility.

3.9 Electrical Properties

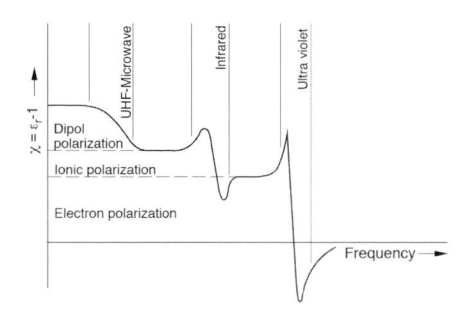

Figure 3.181: Frequency dependence of different polarization cases.

In addition, the influence of fillers on the relative dielectric coefficient is of considerable practical interest. The rule of mixtures can be used to calculate the effective dielectric coefficient of a matrix with assumingly spherically shaped fillers. Materials with air entrapments such as foams, have a filler dielectric coefficient of $\epsilon_{air} = 1$. Whether a molecule is stimulated to its resonant frequency in alternating fields or not depends on its relaxation time. The relaxation time, in turn, depends on viscosity, temperature, and radius of the molecule.

Dielectric Dissipation Factor Typical ranges for the dielectric dissipation factor of various polymer groups are:

- Non - polar polymers (PS, PE, PFE): $\tan\delta < 0.005$
- Polar polymers: $\tan\delta = 0.001 - 0.02$
- Thermosets resins filled with glass, paper or cellulose: $\tan\delta = 0.02 - 0.5$

Figures 3.182 and 3.183 present the dissipation factor $\tan\delta$ as a function of temperature and frequency, respectively.

The standard tests to measure the dielectric dissipation factor of polymers are the IEC 60250 and ASTM D150, presented in Table 3.41.

Electrical and Thermal Loss in a Dielectric The electric losses through wire insulation running high frequency currents must be kept as small as possible. Insulators are encountered in transmission lines or in high frequency fields such as the housings of radar antennas. Hence, we would select materials that have low electrical losses for these types of applications. On the other hand, in some cases we want to generate heat at high frequencies. Heat sealing of polar polymers at high frequencies is an important technique used in the manufacturing of soft PVC sheets such as the ones encountered in automobile vinyl seat covers. To assess whether a material is suitable for either application, one must know the loss

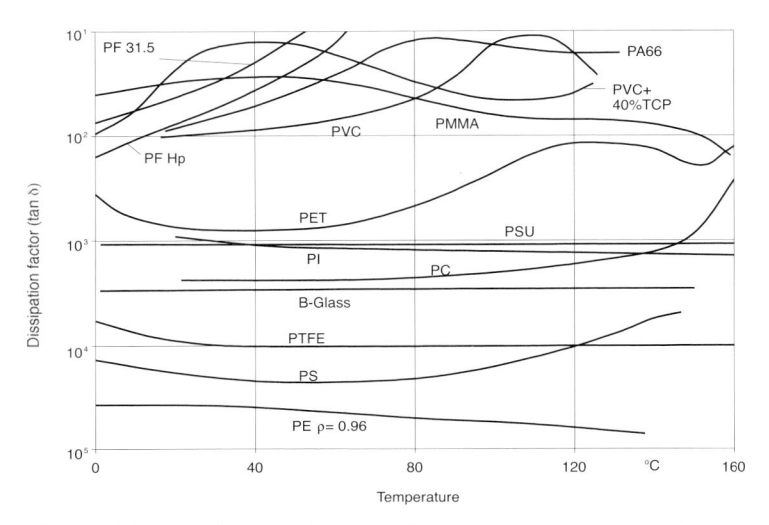

Figure 3.182: Dielectric dissipation factor as a function of temperature for various polymers (after Domininghaus).

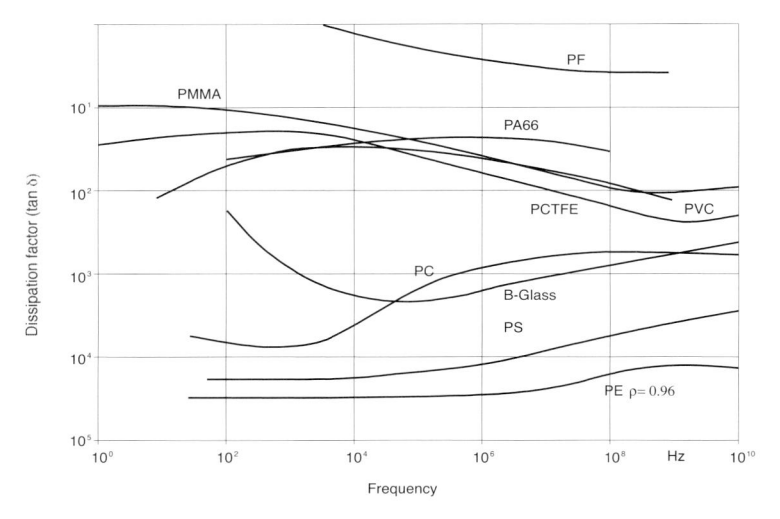

Figure 3.183: Dielectric dissipation factor as a function of frequency for various polymers (after Domininghaus).

3.9 Electrical Properties

properties of the material and calculate the actual electrical loss. Polyethylene and polystyrene are perfectly suitable as insulators in high frequency applications. To measure the necessary properties of the dielectric, the standard DIN 53 483 and ASTM D 150 tests are recommended.

Table 3.41: Standard Methods of Measuring Dissipation Factor of Polymers (after Shastri)

Standard	IEC 60250 : 69	D 150 - 95
Specimen geometry	> 80 mm x > 80 mm x 1 mm (greater thickness may be used for those materials that cannot be molded reliably at 1 mm thickness)	Test specimens are of suitable shape and thickness determined by the material specification or by the accuracy of measurement required, and the frequency at which the measurements are to be made.
Conditioning	Specimen conditioning, including any post molding treatment, shall be carried out at 23°C ± 2°C and 50 ± 5 % R.H. for a minimum length of time of 88h, except where special conditioning is required as specified by the appropriate material standard.	Clean the test specimen with a suitable solvent or as prescribed in the material specification. Use Recommended Practice D1371 as a guide to the choice of suitable cleaning procedures.
Test procedures	100 Hz and 1 MHz.	Frequency not specified but recorded.
	Null methods are used at frequencies up to 50 MHz and results are compensated for electrode edge effects.	Null method with resistive or inductive ratio arm capacitance bridge suggested for frequencies of < 1Hz to a few MHz.
Values and units	Dissipation Factor ⇒ unitless	Dissipation Factor ⇒ unitless

3.9.2 Electric Conductivity

Electric Resistance The current flow resistance, R, in a plate-shaped sample in a direct voltage field is defined by Ohm's law as

$$R = \frac{U}{I} \tag{3.61}$$

where U is the consumed voltage and I the current. The resistance is often described as the inverse of the conductance, G,

$$R = \frac{1}{G} \tag{3.62}$$

The simple relationship found in the above equations is seldom encountered since the voltage, U, is rarely steady and usually varies in a cyclic fashion between 10^{-1} to 10^{11} Hertz. Figures 3.184-3.186 compare the specific resistance of

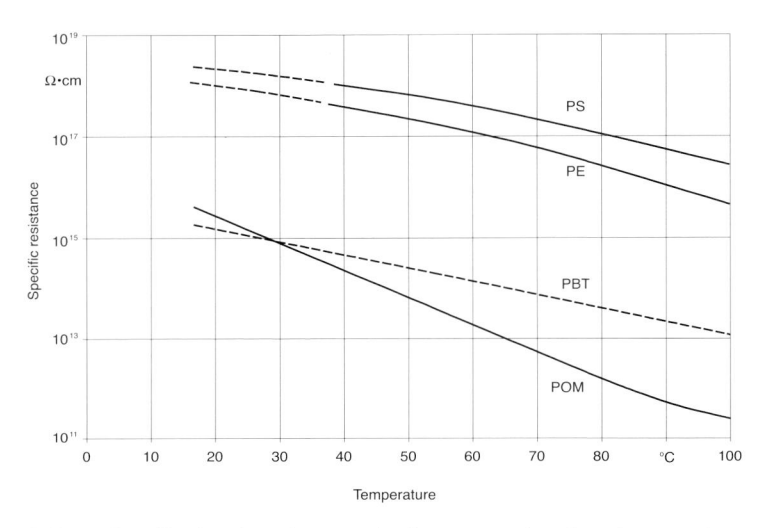

Figure 3.184: Specific electric resistance of polymers as a function of temperature.

various polymers and show its dependence on temperature. Here, we can see that similar to other polymer properties, such as the relaxation modulus, the specific resistance not only decreases with time but also with temperature. The surface of polymer parts often shows different electric direct-current resistance values than their volume. The main cause is surface contamination (e.g., dust and moisture). We therefore have to measure the surface resistance using a different technique. Other tests often used to measure surface resistance are the ASTM D257, DIN 53 480 and DIN 53 482 tests. The surface resistance of various filled thermoplastics is presented in Fig. 3.187 as a function of filler weight percentage. Figure 3.188 shows the effect of moisture content on the specific electric resistance of polyamides.

Physical Causes of Volume Conductivity Polymers with a homopolar atomic bond, which leads to pairing of electrons, do not have free electrons and are not considered to be conductive. Conductive polymers - still in the state of development - in contrast, allow for movement of electrons along the molecular cluster, since they are polymer salts. The classification of these polymers with different materials is given in Fig. 3.189. Potential uses of electric conductive polymers in electrical engineering include flexible electric conductors of low density, strip heaters, anti-static equipment, high frequency shields and housings. In semi-conductor engineering, some applications include semi-conductor devices (Schottky-Barriers) and solar cells. In electrochemistry, applications in-

3.9 Electrical Properties

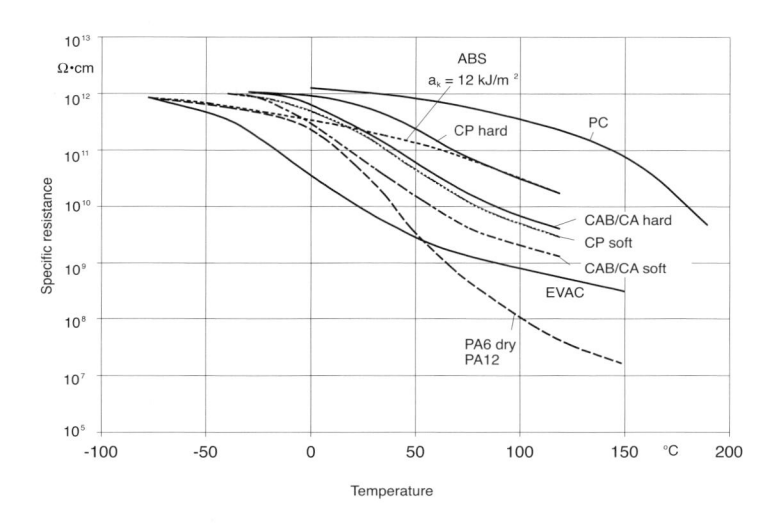

Figure 3.185: Specific electric resistance of polymers as a function of temperature.

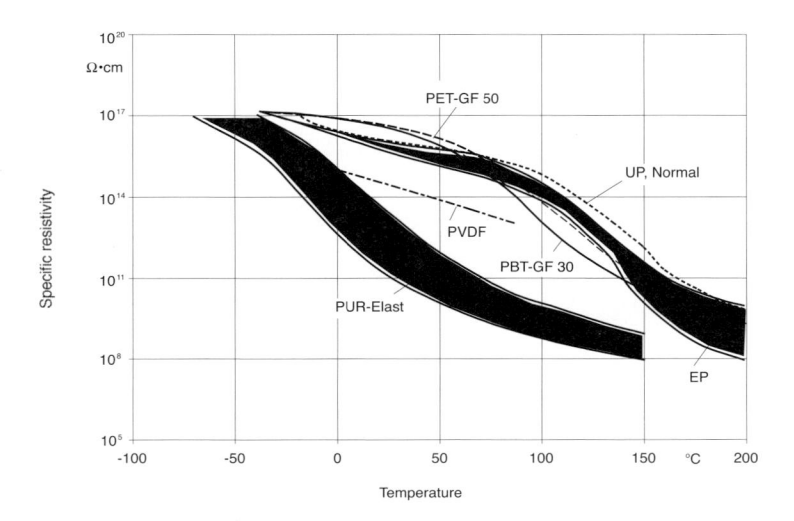

Figure 3.186: Specific electric resistance of polymers as a function of temperature.

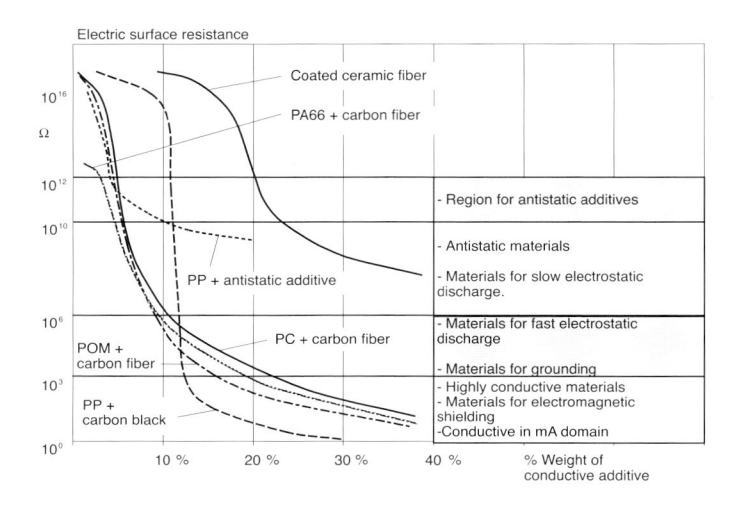

Figure 3.187: Surface resistance for several thermoplastics as a function of filler content.

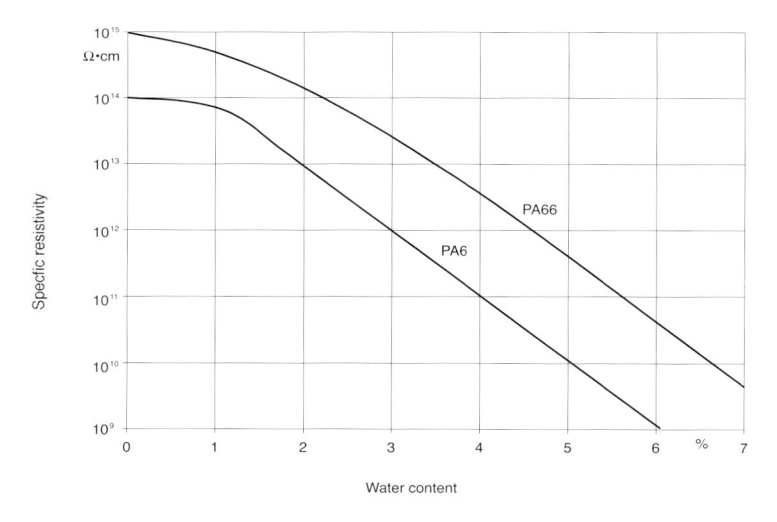

Figure 3.188: Specific electric resistance for PA6 and PA66 as a function of moisture content.

3.9 Electrical Properties

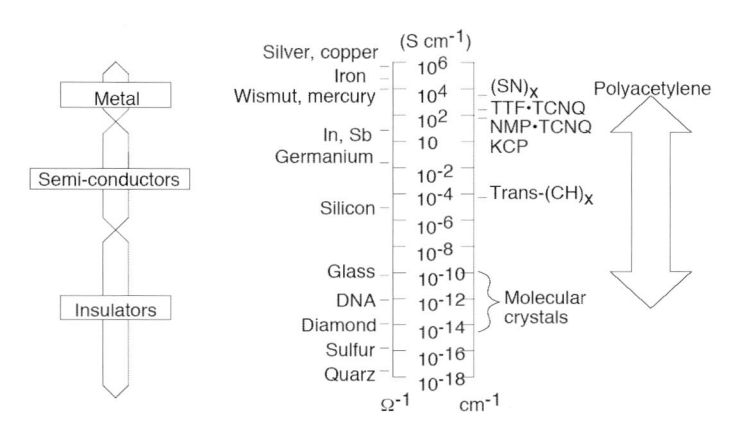

Figure 3.189: Electric conductivity of Polyacetylene (trans-$(CH)_x$) in comparison to other materials.

clude batteries with high energy and power density, electrodes for electrochemical processes and electrochrome instruments. Because of their structure, polymers cannot be expected to conduct ions. Yet the extremely weak electric conductivity of polymers at room temperature and the fast decrease of conductivity with increasing temperatures is an indication that ions do move. They move because engineering polymers always contain a certain amount of added low molecular constituents which act as moveable charge carriers. This is a diffusion process which acts in field direction and across the field. The ions "jump" from potential hole to potential hole as activated by higher temperatures. At the same time, the lower density speeds up this diffusion process. The strong decrease of specific resistance with the absorption of moisture is caused by ion conductivity.

Conducting polymers are useful for certain purposes. When we insulate high energy cables, for example, as a first transition layer we use a polyethylene filled with conductive filler particles such as soot. Figure 3.190 demonstrates the relationship between filler content and resistance. When contact tracks develop, resistance drops spontaneously. The number of inter-particle contacts, M, determines the resistance of a composite. At M_1 or $M = 1$ there is one contact per particle. At this point, the resistance starts dropping. When two contacts per particle exist, practically all particles participate in setting up contact and the resistance levels off. The sudden drop in the resistance curve indicates why it is difficult to obtain a medium specific resistance by filling a polymer. Figure 3.191 presents the resistance in metal flakes or powder filled epoxy resins and Fig. 3.192 presents the specific resistance of carbon black filled polypropylenes. Figure 3.191 shows how the *critical volume concentration* for the epoxy systems

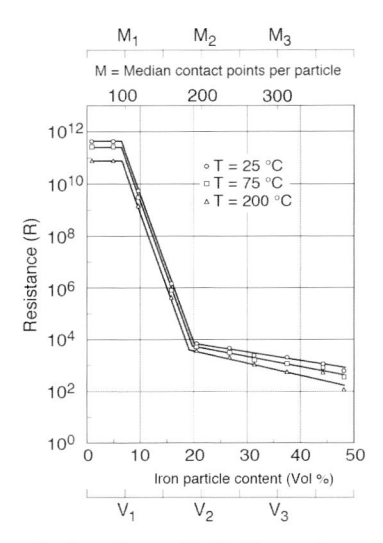

Figure 3.190: Resistance R of a polymer filled with metal powder (iron).

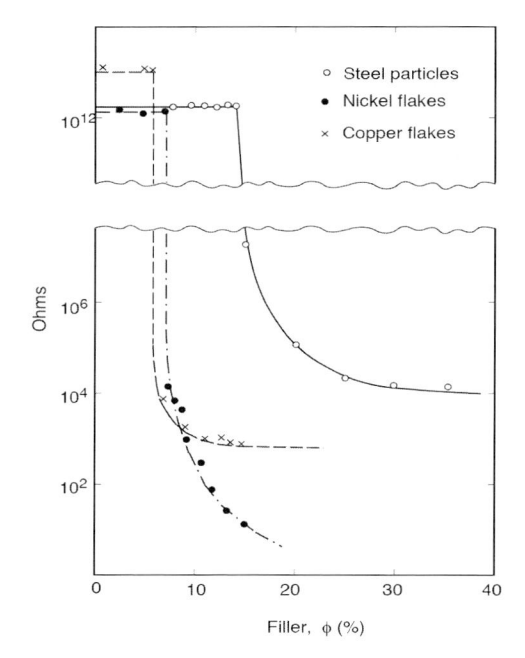

Figure 3.191: Resistance in metal flakes and powder filled epoxy resins (after Reboul).

3.9 Electrical Properties

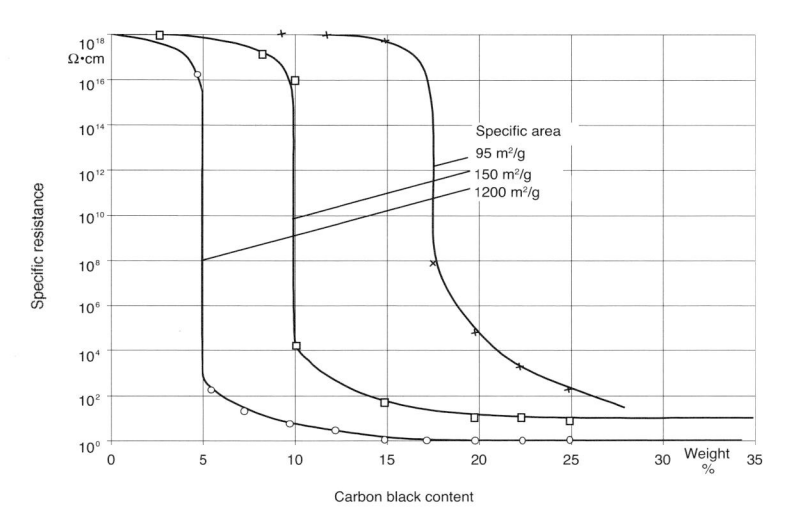

Figure 3.192: Specific resistance in carbon black filled polypropylene.

filled with copper or nickel flakes is about 7% concentration of filler, and the critical volume concentration for the epoxy filled with steel powder is around 15%. A similar effect is seen in the carbon black filled polypropylene. In addition, the effect of carbon black surface area is significant.

3.9.3 Application Problems

Electric Breakdown Since the electric breakdown of insulation may lead to failure of an electric component or may endanger people handling the component, it must be prevented. Hence, we have to know the critical load of the insulating material to design the insulation for long continuous use and with a great degree of confidence. The standard tests used to generate this important material property data for plate or block-shaped specimens are the ASTM D149 and DIN 53 481 tests. From the properties already described, we know that the electric breakdown resistance or dielectric strength must depend on time, temperature, material condition, load application rate and frequency. It is furthermore dependent on electrode shape and sample thickness as is demonstrated in Figs. 3.193 and 3.194 which present the dielectric strength of various thermoplastics as a function of test specimen thickness. In practice, however, it is very important that the upper limits measured on the experimental specimens in the laboratory are never reached. The rule of thumb is to use long term - load values of only

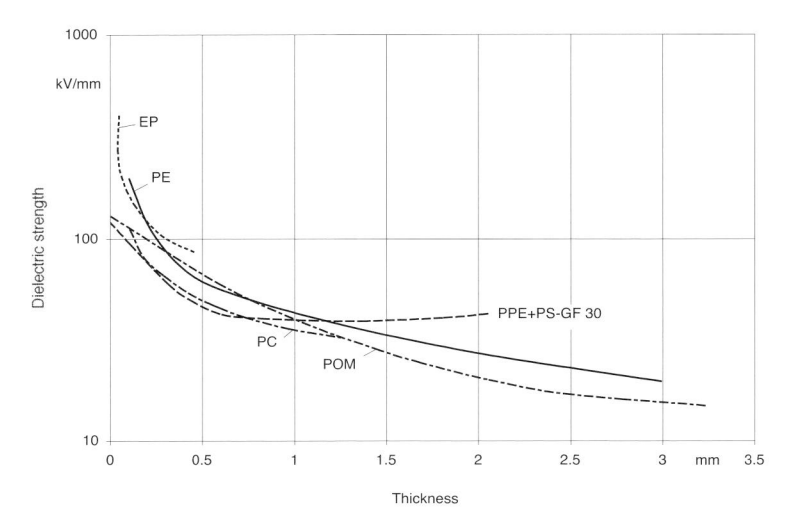

Figure 3.193: Dielectric strength of various plastics as a function of test specimen thickness.

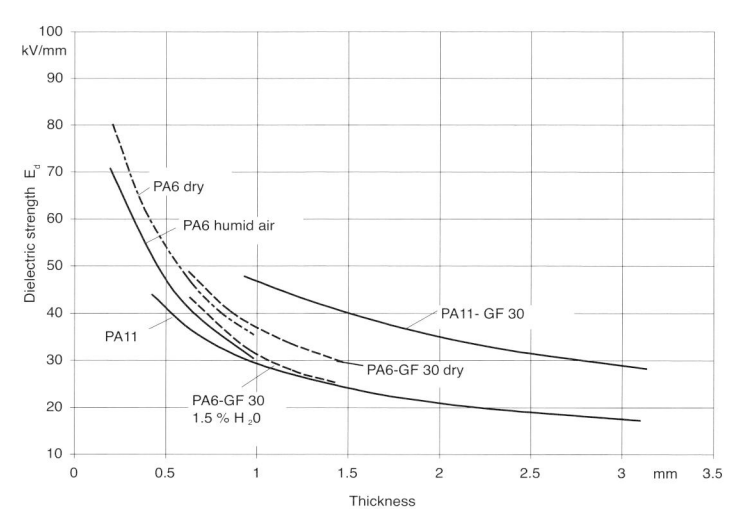

Figure 3.194: Dielectric strength of various polyamides as a function of test specimen thickness.

3.9 Electrical Properties

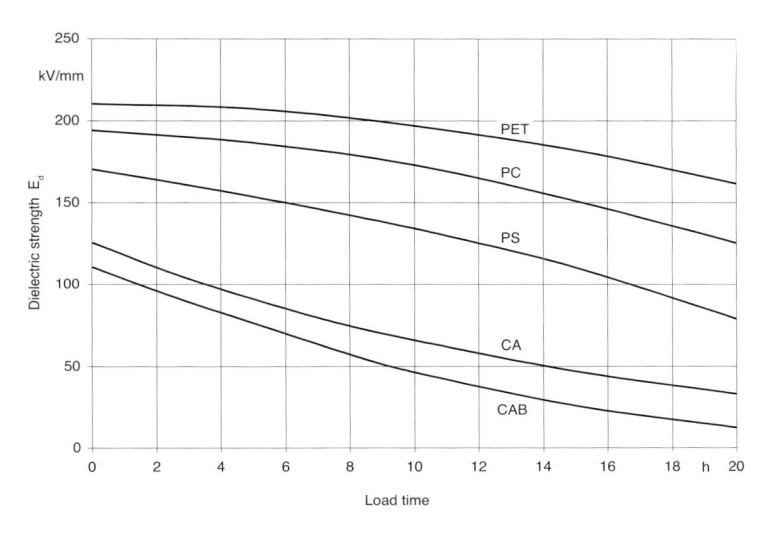

Figure 3.195: Dielectric strength of various thermoplastics as a function of load time.

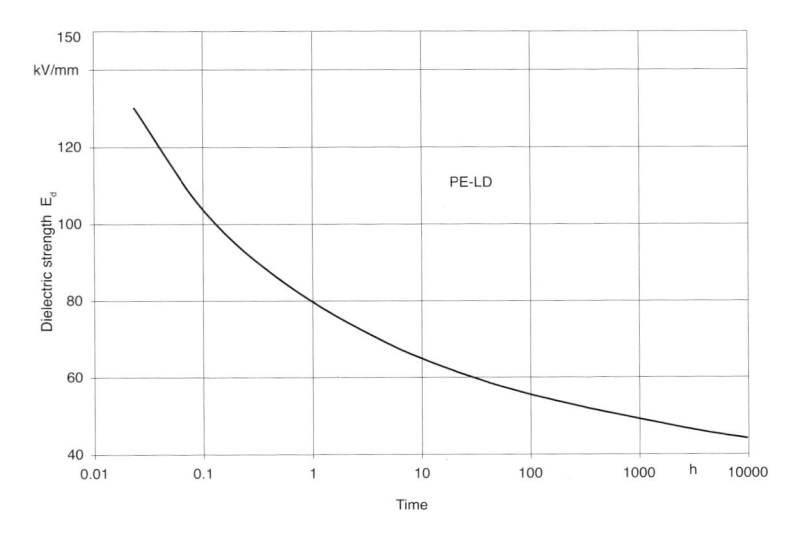

Figure 3.196: Dielectric strength of PE-LD as a function of load time.

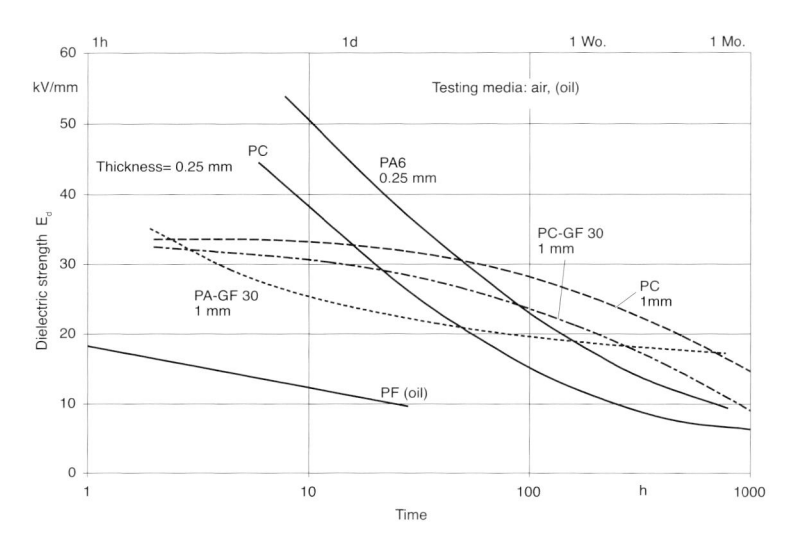

Figure 3.197: Dielectric strength of various thermoplastics as a function of load time.

10% of the short - term laboratory data. Figures 3.195-3.197 present dielectric strength as a function of load time for several thermoplastics.

Figure 3.198 shows the progression of dielectric breakthrough during time; starting with dielectric breakthrough, and followed by heat breakthrough and erosion.

The temperature and frequency also significantly affect the dielectric strength of polymers. Figures 5.28 and 5.38 show the effect of temperature on dielectric strength of selected thermoplastics, and Fig. 3.201 represents the dielectric strength as a function of frequency. Figure 3.202 shows the combined effect of frequency and temperature on the dielectric strength of PF paper. As with other properties, additives such as plasticizers can significantly influence the dielectric strength of polymer sheets, as is illustrated in Figs. 3.203 and 3.204 for a PVC-P with 25 and 35% plasticizers, respectively.

Experimental evidence shows that the dielectric strength decreases as soon as crazes form in a specimen under strain and continues to decrease with increasing strain. This is demonstrated in Fig. 3.205. On the other hand, Fig. 3.206 demonstrates how the *dielectric dissipation factor*, $\tan \delta$, rises with strain. Hence, one can easily determine the beginning of the viscoelastic region (begin of crazing) by noting the starting point of the change in $\tan \delta$. It is also known that amorphous polymers act more favorably to electric breakdown resistance than partly crystalline polymers. Semi-crystalline polymers are more susceptible to

3.9 Electrical Properties 239

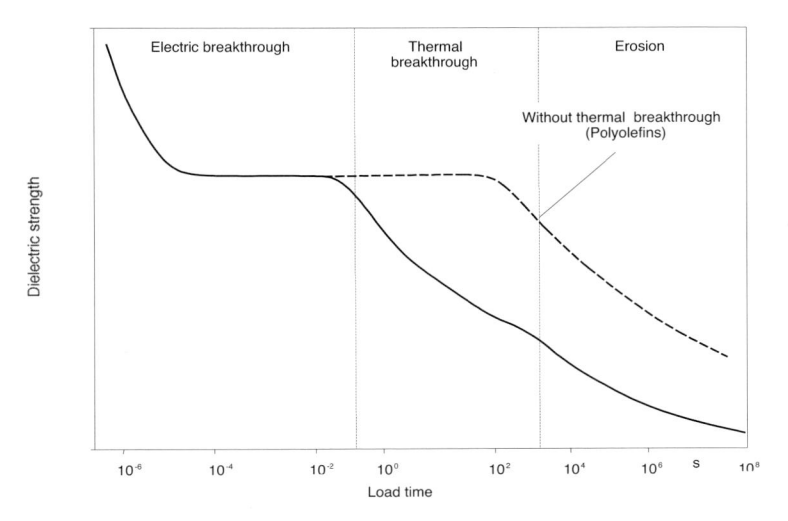

Figure 3.198: Dielectric strength breakthrough progression in time.

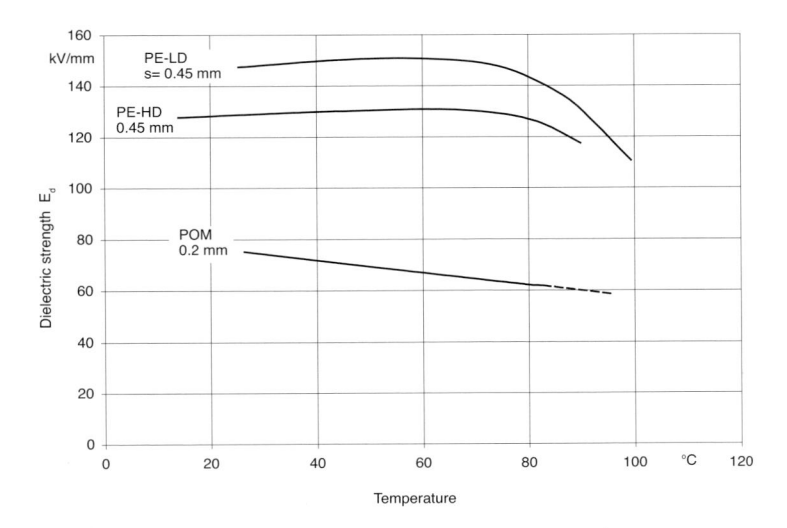

Figure 3.199: Dielectric strength of PE and POM as a function of temperature.

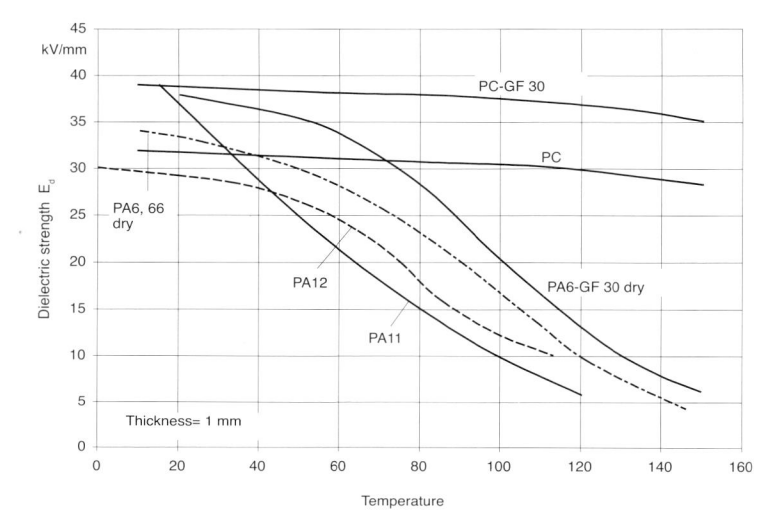

Figure 3.200: Dielectric strength of selected plastics as a function of temperature.

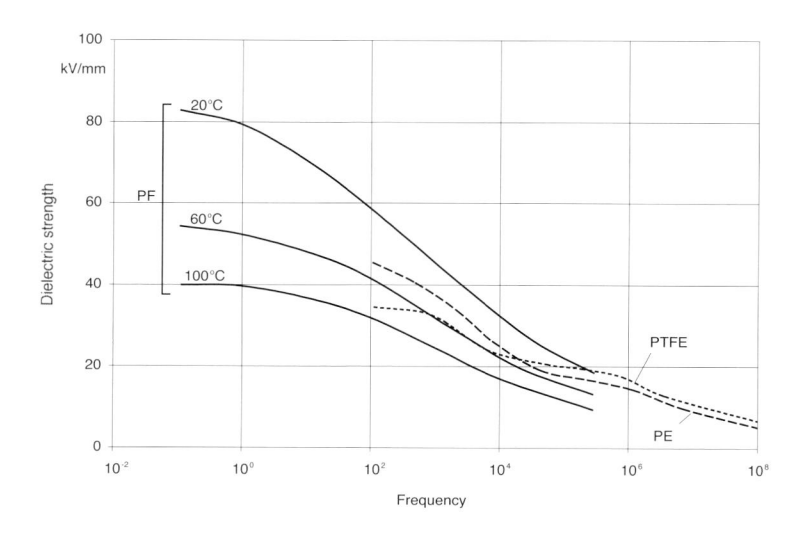

Figure 3.201: Dielectric strength of selected plastics as a function of frequency.

3.9 Electrical Properties

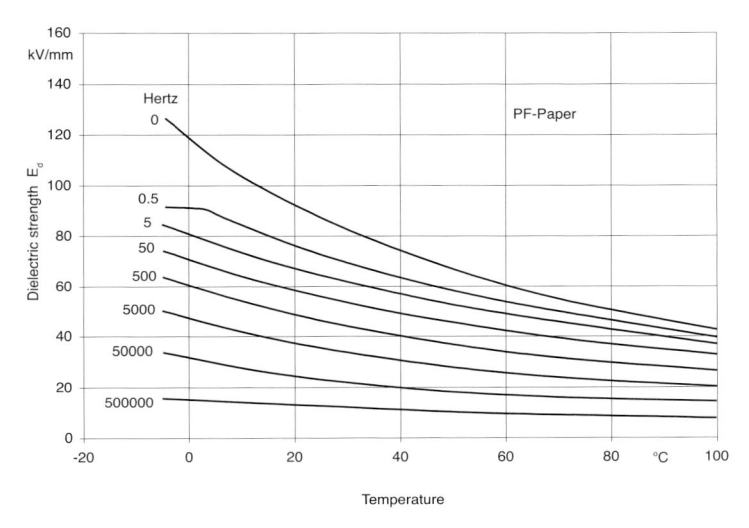

Figure 3.202: Dielectric strength of PF paper measured at several frequencies as a function of temperature.

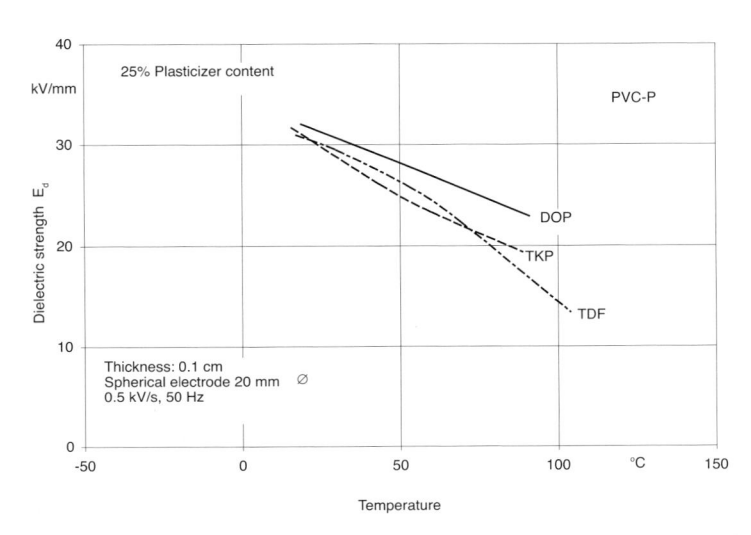

Figure 3.203: Dielectric strength of a PVC-P, with various plasticizers (25%), as a function of temperature.

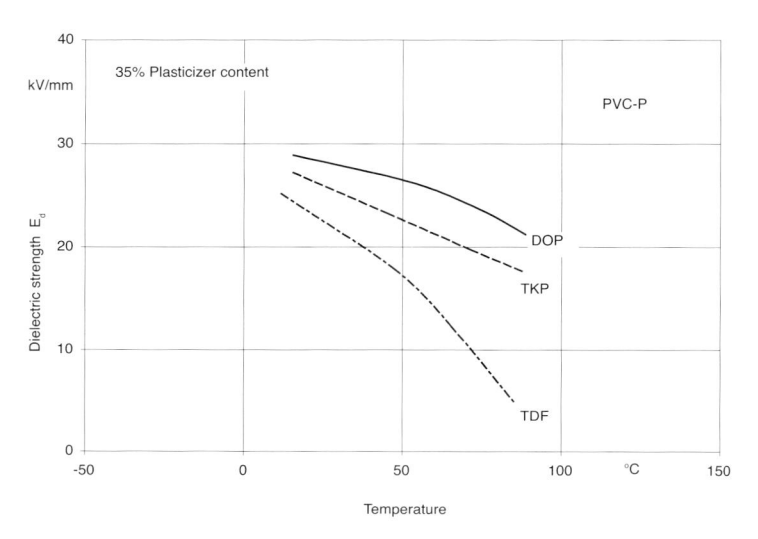

Figure 3.204: Dielectric strength of a PVC-P, with various plasticizers (35%), as a function of temperature.

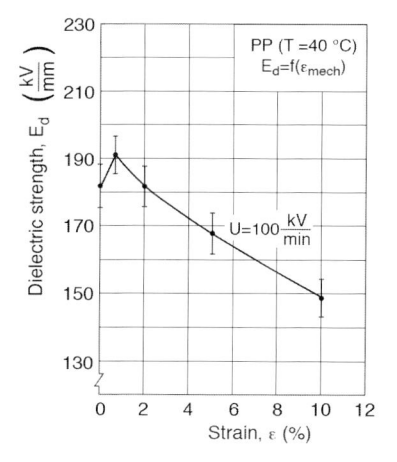

Figure 3.205: Drop of the dielectric strength of PP films with increasing strain (after Menges and Berg).

3.9 Electrical Properties

electric breakdown as a result of breakdown along inter-spherulitic boundaries. Long-term breakdown of semi-crystalline polymers is either linked to "treeing", or occurs as a heat breakdown, burning a hole into the insulation. In general, with rising temperature and frequency, the dielectric strength continuously drops. Insulation materials- mostly LDPE- are especially pure and contain voltage stabilizers. These stabilizers are low molecular cyclic aromatic hydrocarbons. Presumably, they diffuse into small imperfections or failures, fill the empty space and thereby protect them from breakdown. Table 3.44 gives dielectric strength and resistivity for selected polymeric materials. The standard tests used to measure dielectric strength of polymers are the IEC 60243 and ASTM D149 tests, presented in Table 3.42. Another test to measure the dielectric strength of polymers is the electric tracking test where the voltage required to cause tracking is recorded. The standard IEC 60112 and ASTM D3638 tests, described in Table 3.43, are used to evaluate the Comparative Tracking Index (CTI).

Table 3.42: Standard Methods of Measuring Electric Strength of Polymers (after Shastri)

Standard	IEC 60243-1 : 88	D 149 - 95a, Method B
Specimen geometry	> 80 mm x > 80 mm x 1 mm or 3 mm, sufficiently wide to prevent discharge along the surface.	Thickness not specified but measured. It shall be of sufficient size to prevent flashover under the conditions of the test.
Conditioning	Specimen conditioning, including any post molding treatment, shall be carried out at 23°C ± 2°C and 50 ± 5 % R.H. for a minimum length of time of 88h, except where special conditioning is required as specified by the appropriate material standard.	If not specified in the applicable material specification, follow the procedures in Practice D 618.
Apparatus	Two coaxial cylinder electrodes (25 mm diameter x 25 mm and 75mm diameter x 15 mm) with edges rounded to 3 mm radius.	Electrode Type 6: Two coaxial cylinder electrodes (25 mm diameter x 25 mm and 75mm diameter x 15 mm) with edges rounded to 3 mm radius.
Test procedures	Immersion in transformer oil in accordance with IEC 296. Power frequencies between 48 - 62 Hz 20s step-by-step.	Immersion in mineral oil, meeting D3487 Type I or II requirements. 60 Hz, unless otherwise specified 60 ± 5s step-by-step.
Values and units	Dielectric Strength $\Rightarrow$ kV/mm	Dielectric Strength $\Rightarrow$ kV/mm

Table 3.43: Standard Methods of Measuring Comparative Tracking Index (CTI) of Polymers (after Shastri)

Standard	IEC 60112 : 79	D 3638 - 93
Specimen geometry	> 15 mm x > 15 mm x 4 mm from the shoulder of the ISO 3167 multipurpose test specimen.	Sample size is 50 mm or 100 mm disk with minimum thickness of 2.5 mm. Thin samples are to be clamped together to get minimum thickness.
Conditioning	Specimen conditioning, including any post molding treatment, shall be carried out at 23°C ± 2°C and 50 ± 5 % R.H. for a minimum length of time of 88h, except where special conditioning is required as specified by the appropriate material standard.	In accordance with Procedure A of Practice D618.
Apparatus	Two platinum electrodes of rectangular cross-section 5 mm x 2 mm with one end chisel edged with an angle of 30° and slightly rounded. Electrodes are symmetrically arranged in a vertical plane, the total angle between them being 60° and with opposing faces vertical and 4.0 ± 0.1 mm apart on the specimen surface. Force exerted on the surface by the electrode is 1.0 ± 0.05 N.	Two platinum electrodes of rectangular cross-section 5 mm x 2 mm with one end chisel edged with an angle of 30° and slightly rounded. Position the electrodes so that the chisel edges contact the specimen at a 60° angle and the chisel faces are parallel in the vertical plane and are separated by 4 ± 0.2 mm.
Test procedures	0.1 ± 0.002% by mass ammonium chloride in distilled or deionized water (Solution A) with a resistivity of 395 ± 5 Ohm-cm at 23 ± 1°C. Voltage between 100V and 600 V at frequency between 46-60 Hz. Determine maximum voltage at which no failure occurs at 50 drops in the test on five sites. This is the CTI provided no failure occurs below 100 drops when the voltage is dropped by 25V. At lerast five test sites (can be on one specimen).	0.1 ± 0.002% by mass ammonium chloride in distilled or deionized water (Solution A) with a resistivity of 395 ± 5 Ohm-cm at 23 ± 1°C. Voltage should be limited to 600 V at a frequency of 60 Hz. Plot the number of drops of electrolyte at breakdown vs. voltage. The voltage which corresponds to 50 drops is the CTI. At least five specimen of each sample shall be tested.
Values and units	CTI ⇒ V	CTI ⇒ V

3.9 Electrical Properties

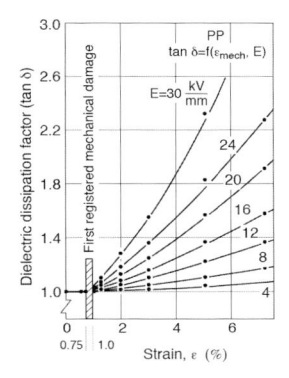

Figure 3.206: Increase of dielectric dissipation with increased strain in PP foils (after Menges and Berg).

Table 3.44: Dielectric Strength and Resistivity for Selected Polymers (after Crawford)

Polymer	Dielectric strength MV/m	Resistivity Ohm-m
ABS	25	10^{14}
Acrylic	11	10^{13}
CA	11	10^{9}
CAB	10	10^{9}
Epoxy	16	10^{13}
Modified PPO	22	10^{15}
PA66	8	10^{13}
PA66 + 30% GF	15	10^{12}
PEEK	19	10^{14}
PET	17	10^{13}
PET + 36% GF	50	10^{14}
PF (mineral filled)	12	10^{9}
PC	23	10^{15}
PE-PE	27	10^{14}
PE-PE	22	10^{15}
POM (homopolymer)	20	10^{13}
POM (copolymer)acrylic	20	10^{13}
PP	28	10^{15}
PS	20	10^{14}
PTFE	45	10^{16}
PVC-U	14	10^{12}
PVC-P	30	10^{11}
SAN	25	10^{14}

Electrostatic Charge An electrostatic charge is often a result of the excellent insulation properties of polymers - the very high surface resistance and current-

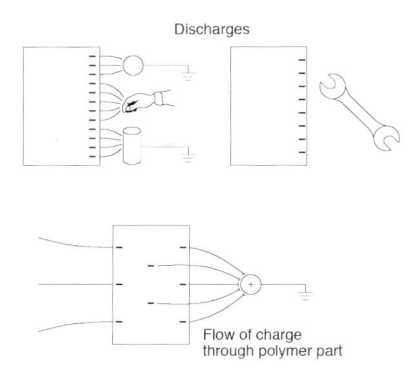

Figure 3.207: Electrostatic charges in polymers.

flow resistance. Since polymers are bad conductors, the charge displacement of rubbing bodies, which develops with mechanical friction, cannot equalize. This charge displacement results from a surplus of electrons on one surface and a lack of electrons on the other. Electrons are charged positively or negatively up to hundreds of volts. They release their surface charge only when they touch another conductive body or a body which is inversely charged. Often, the discharge occurs without contact, as the charge arches through the air to the close - by conductive or inversely charged body, as demonstrated in Fig. 3.207. The currents of these breakdowns are low. For example, there is no danger when a person suffers an electric shock caused by a charge from friction of synthetic carpets or vinyls. There is danger of explosion, though, when the sparks ignite flammable liquids or gases.

As the current-flow resistance of air is generally about 10^9 Ωcm, charges and flashovers only occur if the polymer has a current-flow resistance of $> 10^9$ to 10^{10} Ωcm. Another effect of electrostatic charges is that they attract dust particles on polymer surfaces. Electrostatic charges can be reduced or prevented by the following means: Reduce current-flow resistance to values of $< 10^9$ Ωcm, for example, by using conductive fillers such as graphite. Make the surfaces conductive by using hygroscopic fillers that are incompatible with the polymer and surface. It can also be achieved by mixing in hygroscopic materials, such as strong soap solutions. In both cases, the water absorbed from the air acts as a conductive layer. It should be pointed out that this treatment loses its effect over time. Especially, the rubbing in of hygroscopic materials has to be repeated over time. Reduce air resistance by ionization through discharge or radioactive radiation.

3.9 Electrical Properties   247

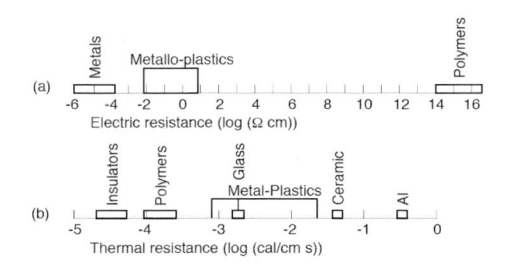

Figure 3.208: Comparison of conductive polymers with other materials: a) Electric resistance ρ of metal-plastics compared to resistance of metals and polymers b) Thermal resistance λ of metal-plastics compared to other materials.

Electrets An electret is a solid dielectric body that exhibits permanent dielectric polarization. One can manufacture electrets out of some polymers when they are solidified under the influence of an electric field, when bombarded by electrons, or sometimes through mechanical forming processes. Applications include films for condensers (polyester, polycarbonate or fluoropolymers).

Electromagnetic Interference Shielding (EMI Shielding) Electric fields surge through polymers as shown schematically in Fig. 3.207. Since we always have to deal with the influence of interference fields, signal sensitive equipment, such as computers, cannot operate in polymer housings. Such housings must therefore have the function of Faradayic shields. Preferably, a multilayered structure is used - the simplest solution is to use one metallic layer. Figure 3.207 classifies several materials in a scale of resistances. We need at least 10^2 Ωcm for a material to fulfill the shielding purpose. With carbon fibers or nitrate coated carbon fibers used as a filler, one achieves the best protective properties. The shielding properties are determined using the standard ASTM ES 7-83 test. Figures 3.209 and 3.210 presents the magnetic shielding as a function of frequency of aluminum coated polymers and steel fiber filled plastics, respectively.

3.9.4 Magnetic Properties

External magnetic fields have an impact on substances that are subordinate to them, because the external field interacts with the internal fields of electrons and atomic nuclei.

Magnetizability Pure polymers are diamagnetic; that is, the external magnetic field induces magnetic moments. However, permanent magnetic moments, which are induced on ferromagnetic or paramagnetic substances, do not exist in

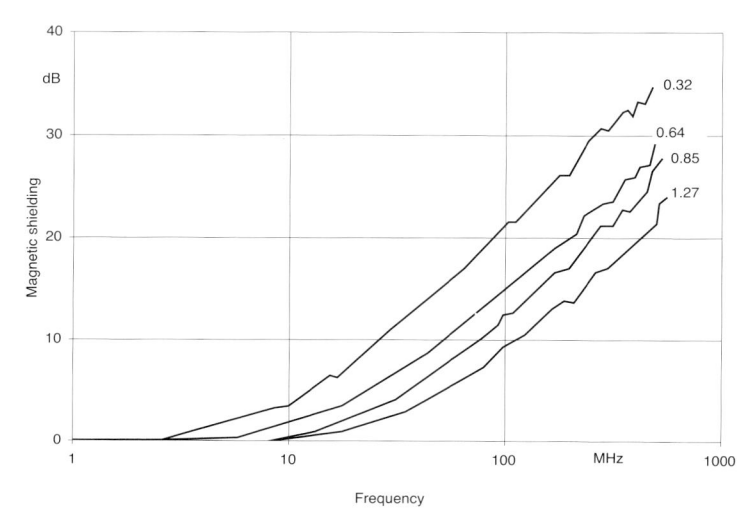

Figure 3.209: Electromagnetic shielding of aluminum - coated plastics as a function of frequency and the square resistance of the plastic.

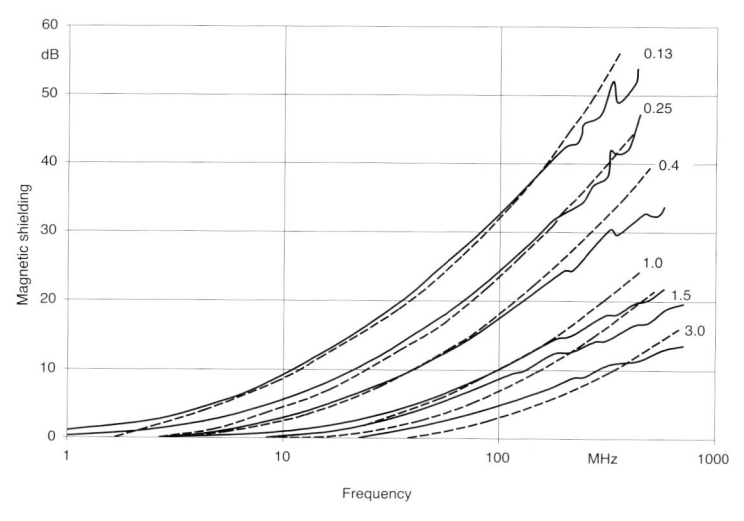

Figure 3.210: Electromagnetic shielding of steel fiber filled (0.7-1.4 vol %) plastics as a function of frequency and the square resistance of the plastic.

3.10 Optical Properties

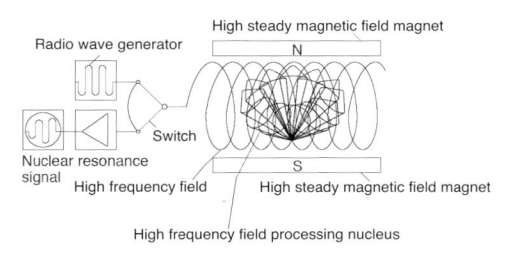

Figure 3.211: Schematic of the operating method of a nuclear spin tomograph.

polymers. This magnetizability M of a substance in a magnetic field with a field intensity H is computed with the *magnetic susceptibility*, X, as

$$M = XH \qquad (3.63)$$

The susceptibility of pure polymers as *diamagnetic substances* has a very small and negative value. However, in some cases, we make use of the fact that fillers can alter the magnetic character of a polymer completely. The magnetic properties of polymers are often changed using magnetic fillers. Well-known applications are injection molded or extruded magnets or magnetic profiles, and all forms of electronic storage such as recording tape, floppy or magnetic disks.

Magnetic Resonance Magnetic resonance occurs when a substance, in a permanent magnetic field, absorbs energy from an oscillating magnetic field. This absorption develops as a result of small paramagnetic molecular particles stimulated to vibration. We use this phenomenon to a great extent to clarify structures in physical chemistry. Methods to achieve this include *electron spinning resonance* (ESR) and, above all, *nuclear magnetic resonance* (NMR) spectroscopy. Electron spinning resonance becomes noticeable when the field intensity of a static magnetic field is altered and the microwaves in a high frequency alternating field are absorbed. Since we can only detect unpaired electrons using this method, we use it to determine radical molecule groups. When atoms have an odd number of nuclei, protons and neutrons, the magnetic fields caused by self-motivated spin cannot equalize. The alignment of nuclear spins in an external magnetic field leads to a magnetization vector which can be measured macroscopically as is schematically demonstrated in Fig. 3.211. This method is of great importance for the polymer physicist to learn more about molecular structures.

3.10 OPTICAL PROPERTIES

Since some polymers have excellent optical properties and are easy to mold and form into any shape, they are often used to replace transparent materials such

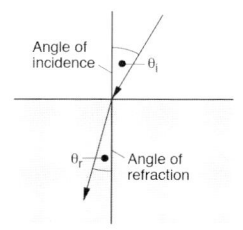

Figure 3.212: Schematic of light refraction.

as inorganic glass. Polymers have been introduced into a variety of applications such as automotive headlights, signal light covers, optical fibers, imitation jewelry, chandeliers, toys and home appliances. Organic materials such as polymers are also an excellent choice for high impact applications where inorganic materials such as glass would easily shatter. However, due to the difficulties encountered in maintaining dimensional stability, they are not apt for precision optical applications. Other drawbacks include lower scratch resistance, when compared to inorganic glasses, making them still impractical for applications such as automotive windshields. In this section, we will discuss basic optical properties which include the index of refraction, birefringence, transparency, transmittance, gloss, color and behavior of polymers in the infrared spectrum.

3.10.1 Index of Refraction

As rays of light pass through one material into another, the rays are bent due to the change in the speed of light from one medium to the other. The fundamental material property that controls the bending of the light rays is the *index of refraction*, N. The index of refraction for a specific material is defined as the ratio between the speed of light in a vacuum to the speed of light through the material under consideration. In more practical terms, the refractive index can also be computed as a function of the angle of incidence, θ_i, and the angle of refraction, θ_r, as follows

$$N = \frac{\sin \theta_i}{\sin \theta_r} \qquad (3.64)$$

where θ_i and θ_r are defined in Fig. 3.212.

The index of refraction for organic plastic materials can be measured using the standard ASTM D 542 test. It is important to mention that the index of refraction is dependent on the wavelength of the light under which it is being measured. Figure 3.213 shows plots of the refractive index for various organic and inorganic materials as a function of wavelength. One of the significant points of this plot

3.10 Optical Properties

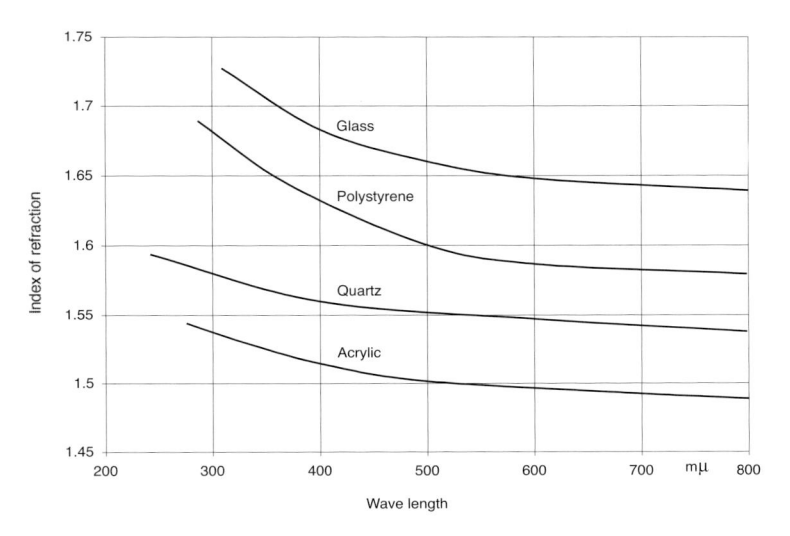

Figure 3.213: Index of refraction as a function of wavelength for various materials.

is that acrylic materials and polystyrene have similar refractive properties as inorganic glasses.

An important quantity that can be deduced from the light's wavelength dependence on the refractive index is the dispersion, D, which is defined by

$$D = \frac{dN}{d\lambda} \tag{3.65}$$

Figure 3.214 shows plots of dispersion as a function of wavelength for the same materials shown in Fig. 3.213. The plots show that polystyrene and glass have a high dispersion in the ultra-violet light domain.

It is also important to mention that since the index of refraction is a function of density, it is indirectly affected by temperature. Figure 3.215 shows how the refractive index of PMMA changes with temperature. A closer look at the plot reveals the glass transition temperature.

3.10.2 Photoelasticity and Birefringence

Photoelasticity and flow birefringence are applications of the optical anisotropy of transparent media. When a transparent material is subjected to a strain field or

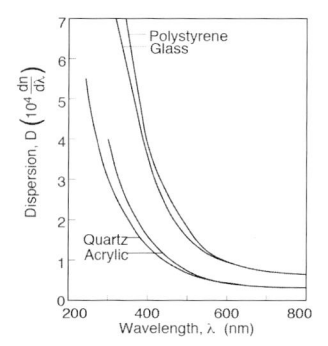

Figure 3.214: Dispersion as a function of wavelength for various materials.

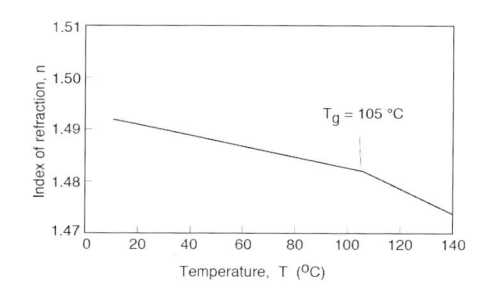

Figure 3.215: Index of refraction as a function of temperature for PMMA (λ= 589.3 nm).

3.10 Optical Properties

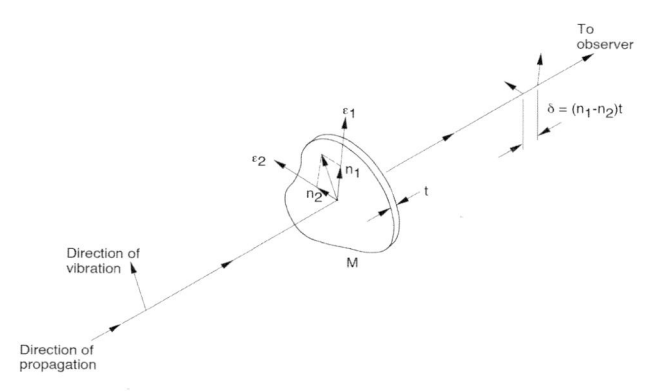

Figure 3.216: Propagation of light in a strained transparent media.

a molecular orientation, the index of refraction becomes directional; the principal strains N_1 and N_2 are associated with principal indices of refraction N1 and N2 in a two-dimensional system. The difference between the two principal indices of refraction (*birefringence*) can be related to the difference of the principal strains using the *strain-optical coefficient*, k, as

$$N_1 - N_2 = k(\epsilon_1 - \epsilon_2) \tag{3.66}$$

or, in terms of principal stress, and

$$N_1 - N_2 = C(\sigma_1 - \sigma_2) \tag{3.67}$$

where C is the stress-optical coefficient. *Double refractance* in a material is caused when a beam of light travels through a transparent media in a direction perpendicular to the plane that contains the principal directions of strain or refraction index, as shown schematically in Fig. 3.216. The incoming light waves split into two waves that oscillate along the two principal directions. These two waves are out of phase by a distance δ.

The out-of-phase distance, δ, between the oscillating light waves is usually referred to as the *retardation*. In photoelastic analysis, one measures the direction of the principal stresses or strains and the retardation to determine the magnitude of the stresses. The technique and apparatus used to performed such measurements is described in the ASTM D 4093 test. Figure 3.217 shows a schematic of such a set-up, called a *polariscope*, composed of a narrow wavelength band light source, two polarizers, two quaterwave plates, a compensator, and a monochromatic filter. The polarizers and quaterwave plates must be perpendicular to each other (90°). The compensator is used for measuring retardation, and the monochromatic filter is needed when white light is not sufficient to perform the photoelastic measurement. The parameter used to quantify the strain field in a specimen observed through a polariscope is the color. The retardation in a strained specimen is associated to a specific color. The sequence of colors and their respective retardation values and fringe order are shown in Table 3.45. The retardation and color can also be associated to a *fringe order* using

$$\text{fringe order} = \frac{\delta}{\lambda} \qquad (3.68)$$

Table 3.45: Retardation and Fringe Order Produced in a Polariscope

Color	Retardation (nm)	Fringe order
Black	0	0
Gray	160	0.28
White	260	0.45
Yellow	350	0.60
Orange	460	0.79
Red	520	0.90
Tint of passage	577	1.00
Blue	620	1.06
Blue-green	700	1.20
Green-yellow	800	1.38
Orange	940	1.62
Red	1050	1.81
Tint of passage	1150	2.00
Green	1350	2.33
Green-yellow	1450	2.50
Pink	1550	2.67
Tint of passage	1730	3.00
Green	1800	3.10
Pink	2100	3.60
Tint of passage	2300	4.00
Green	2400	4.13

A black body (fringe order zero) represents a strain free body, and closely spaced color bands represent a component with high strain gradients. The color bands are generally called the *isochromatics*. Figure 3.218 shows the isochromatic fringe pattern in a stressed notched bar. The fringe pattern can also be a result of molecular orientation and residual stresses in a molded transparent polymer component. Figure 3.219 shows the orientation induced fringe pattern in a molded part. The residual stress-induced birefringence is usually smaller than the orientation-induced pattern, making them more difficult to measure. Flow induced birefringence is an area explored by several researchers. Likewise, the flow induced principal stresses can be related to the principal refraction indices.

3.10 Optical Properties

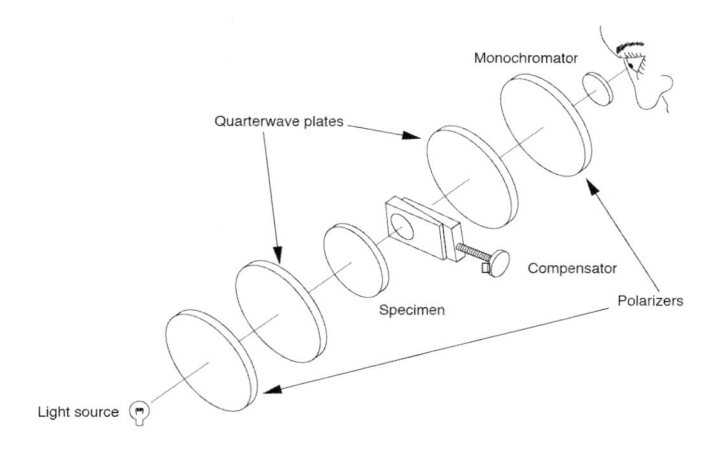

Figure 3.217: Schematic diagram of a polariscope.

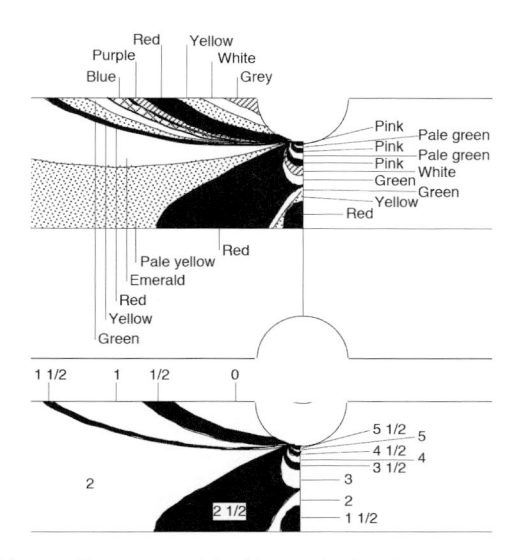

Figure 3.218: Fringe pattern on a notched bar under tension.

Figure 3.220 shows the birefringence pattern for the flow of linear low density polyethylene in a rectangular die.

Figure 3.219: Transparent injection molded part viewed through a polariscope.

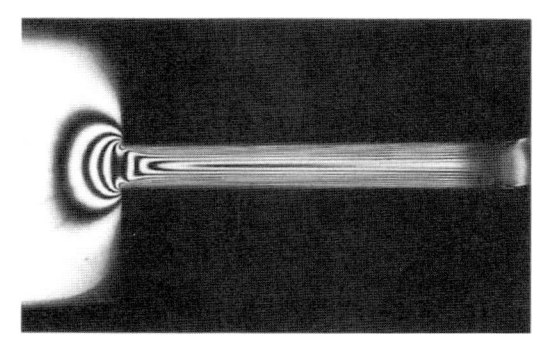

Figure 3.220: Birefringence pattern for flow of LLDPE in a rectangular die.

3.10 Optical Properties

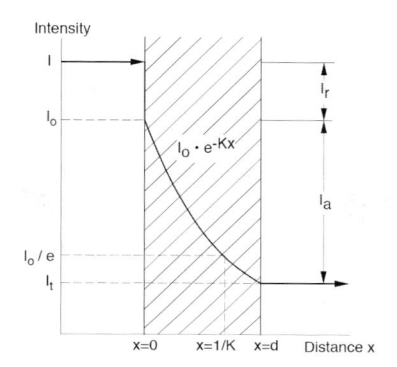

Figure 3.221: Schematic of light transmission through a plate.

3.10.3 Transparency, Reflection, Absorption and Transmittance

As rays of light pass through one media into another of a different refractive index, light will be scattered if the interface between the two materials shows discontinuities larger than the wavelength of visible light. Hence, the transparency in semi-crystalline polymers is directly related to the crystallinity of the polymer. Since the characteristic size of the crystalline domains are larger than the wavelengths of visible light, and since the refractive index of the denser crystalline domains is higher compared to the amorphous regions, semi-crystalline polymers are not transparent; they are opaque or translucent. Similarly, high impact polystyrene -which is actually formed by two amorphous components, polybutadiene rubber particles and polystyrene,- appears white and translucent due to the different indices of refraction of the two materials. However, filled polymers can be made transparent if the filler size is smaller than the wavelength of visible light. The concept of absorption and transmittance can be illustrated using the schematic and notation shown in Fig. 3.221. The figure plots the intensity of a light ray as it strikes and travels through an infinite plate of thickness d. For simplicity, the angle of incidence, θ_i, is $0°$. The initial intensity of the incoming light beam, I, drops to I_0 as a fraction ρ_0 of the incident beam is reflected out. The reflected light beam can be computed using

$$I_r = \rho_0 I \qquad (3.69)$$

The fraction of the beam that does penetrate into the material continues to drop due to absorption as it travels through the plate. However, as illustrated in Fig. 3.222, part of the beam is reflected back by the rear surface of the plate and is subsequently reflected and absorbed several times as it travels between the front and back surfaces of the plate. The fraction of incident beam absorbed by the

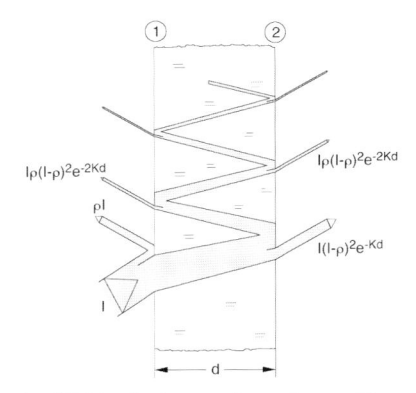

Figure 3.222: Schematic of light reflectance, absorption and transmission through a plate.

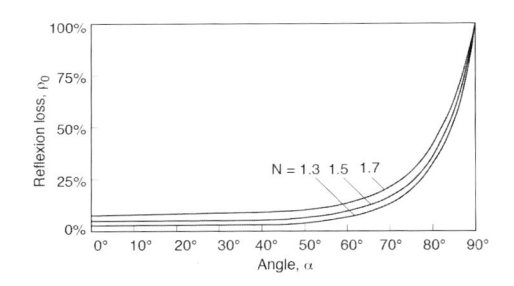

Figure 3.223: Influence of incidence angle on reflection losses.

material, α, is transformed into heat inside the material and be written as

$$\alpha = 1 - \tau - \rho \tag{3.70}$$

where τ and ρ is the fraction of transmitted and reflected light. Plots of reflection loss as a function of incidence angle are shown in Fig. 3.223 for various refraction indices.

The transmittance becomes less as the wavelength of the incident light decreases, as shown for PMMA in Fig. 3.224. The figure also demonstrates the higher absorption of the thicker sheet.

The transmissivity is generally measured in air and is plotted as a function of wavelength. Figure 3.225 presents plots of the transmissivity of CAB and PC and compares them to window glass. The transmissivity of polymers can be improved by altering their chemical composition. For example, the transmissivity of PMMA can be improved by substituting hydrogen atoms by fluorine atoms. The improvement is clearly demonstrated in Fig. 3.226. Such modifications

3.10 Optical Properties

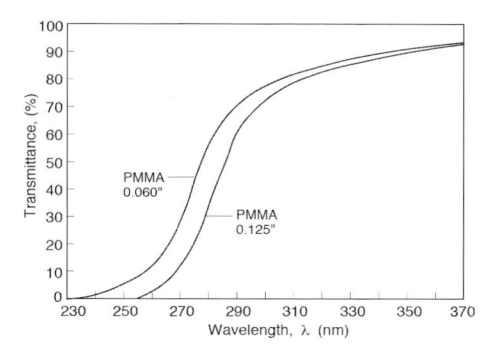

Figure 3.224: Ultraviolet light transmission through PMMA.

bring polymers a step closer to materials appropriate for usage in fiber optic applications. Their ability to withstand shock and vibration and cost savings during manufacturing make some amorphous polymers important materials for fiber optics applications. However, in unmodified polymer fibers, the initial light intensity drops to 50% after only 100 m, whereas when using glass fibers the intensity drops to 50% after 3000 m. Nucleating agents can also be used to improve the transmissivity of semi-crystalline polymers. A large number of nuclei will reduce the average spherulite size to values below the wavelength of visible light. The haziness or luminous transmittance of a transparent polymer is measured using the standard ASTM D 1003 test, and the transparency of a thin polymer film is measured using the ASTM D 1746 test. The *haze measurement* (ASTM D 1003) is the most popular measurement for film and sheet quality control and specification purposes.

3.10.4 Gloss

Strictly speaking, all of the above theory is valid only if the surface of the material is perfectly smooth. However, the reflectivity of a polymer component is greatly influenced by the quality of the surface of the mold or die used to make the part. Specular gloss can be measured using the ASTM D 2457 standard technique, which describes a part by the quality of its surface. A glossmeter or lustremeter is usually composed of a light source and a photometer as shown in schematic diagram in Fig. 3.227. These types of glossmeters are called *goniophotometers*. As shown in the figure, the specimen is illuminated with a light source from an angle α, and the photometer reads the light intensity from the specimen from a variable angle β. The angle α should be chosen according to the glossiness of the surface. For example, for transparent films, values for α are $20°$ for high

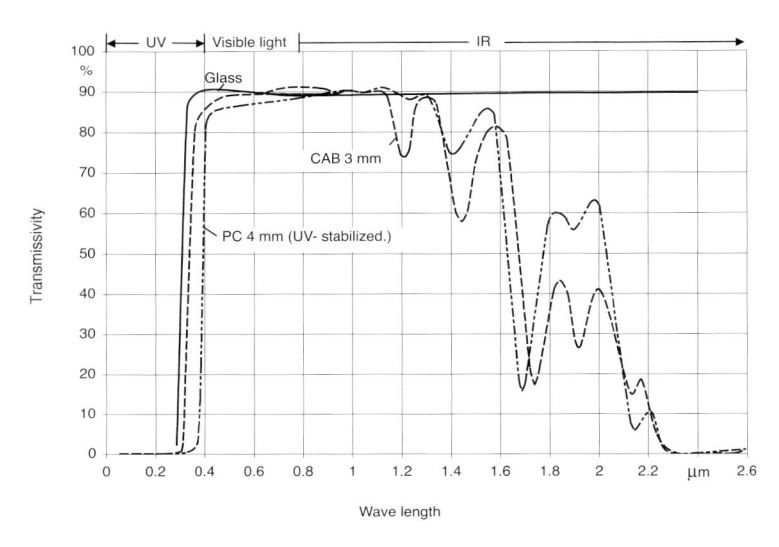

Figure 3.225: Transmissivity of CAB, PC and glass as a function of wavelength.

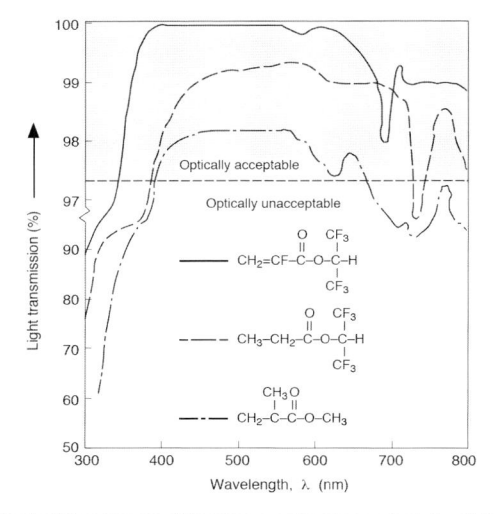

Figure 3.226: Effect of fluorine modification on the transmissivity of light through PMMA.

3.10 Optical Properties

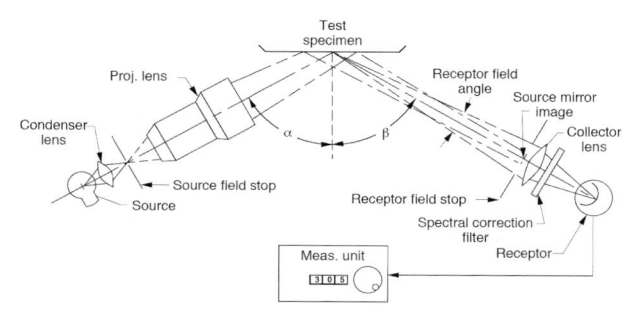

Figure 3.227: Schematic diagram of a glossmeter.

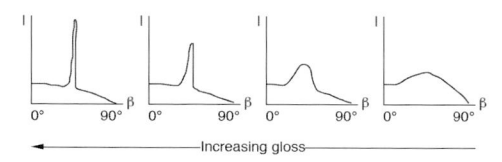

Figure 3.228: Reflective intensity as a function of photometer orientation for specimens with various degrees of surface gloss.

Figure 3.229: Reflective intensity as a function of photometer orientation for black and white specimens with equal surface gloss.

gloss, 45° for intermediate and 60° for low gloss. For opaque specimens ASTM test E 97 should be used. Figure 3.228 presents plots of reflective intensity as a function of photometer orientation for several surfaces with various degrees of gloss illuminated by a light source oriented at a 45° angle from the surface. The figure shows how the intensity distribution is narrow and sharp at 45° for a glossy surface, and the distribution becomes wider as the surface becomes matte. The color of the surface also plays a significant role on the intensity distribution read by the photometer as it sweeps through various angular positions. Figure 3.229 shows plots for a black and a white surface with the same degree of glossiness. The specular gloss is used as a measurement of the glossy appearance of films. However, gloss values of opaque and transparent films should not be compared with each other.

3.10.5 Color

The surface quality of a part is not only determined by how smooth or glossy it is, but also by its color. Color is often one of the most important specifications for a part. In the following discussion it will be assumed that the color is homogeneous throughout the surface. This assumption is linked to processing, where efficient mixing must take place to disperse and distribute the pigments that will give the part color. Color can always be described by combinations of basic red, green and blue. Hence, to quantitatively evaluate or measure a color, one must filter the intensity of the three basic colors. A schematic diagram of a color measurement device is shown in Fig. 3.230. Here, a specimen is lit in a diffuse manner using a photometric sphere, and the light reflected from the specimen is passed through red, green and blue filters. The intensity coming from the three filters are allocated the variables X, Y, and Z for red, green and blue, respectively. The variables X, Y and Z are usually referred to as *tristimulus values*. Another form of measuring color is to have an observer compare two surfaces. One surface is the sample under consideration illuminated with a white light. The other surface is a white screen illuminated by light coming from three basic red, green and blue sources. By varying the intensity of the three light sources, the color of the two surfaces are matched. This is shown schematically in Fig. 3.231. Here too, the intensities of red, green and blue are represented with X, Y and Z, respectively. The resulting data is better analyzed by normalizing the individual intensities as

$$x = \frac{X}{X+Y+Z} \tag{3.71}$$

$$y = \frac{Y}{X+Y+Z} \tag{3.72}$$

$$z = \frac{Z}{X+Y+Z} \tag{3.73}$$

The parameters x, y and z, usually termed *trichromatic coefficients*, are plotted on a three-dimensional graph that contains the whole spectrum of visible light, as shown in Fig. 3.232. This graph is usually referred to as a *chromaticity diagram*. The standard techniques that make use of the chromaticity diagram are the ASTM E 308-90 and the DIN 5033. Three points in the diagram have been standardized. These are:

- Radiation from a black body at 2848 K corresponding to a tungsten filament light, denoted by A in the diagram
- Sunlight, denoted by B
- North sky light, denoted by C

3.10 Optical Properties

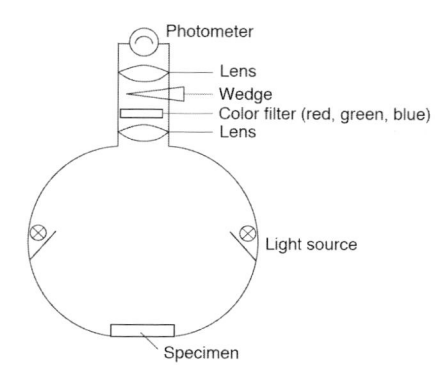

Figure 3.230: Schematic diagram of a colorimeter.

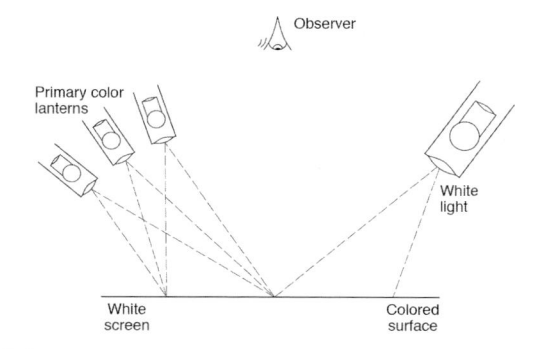

Figure 3.231: Schematic diagram of a visual colorimeter.

It is important to note that colors plotted on the chromaticity diagram are only described by their *hue* and *saturation*. The *luminance* factor is plotted in the z direction of the diagram. Hence, all neutral colors such as black, gray and white lie on point C of the diagram.

3.10.6 Infrared Spectroscopy

Infrared spectroscopy has developed into one of the most important techniques used to identify polymeric materials. It is based on the interaction between matter and electromagnetic radiation of wavelengths between 1 and 50 μm. The atoms in a molecule vibrate in a characteristic mode, which is usually called a fundamental frequency. Thus, each molecule has a set group of characteristic frequencies which can be used as a diagnostics tool to detect the presence of dis-

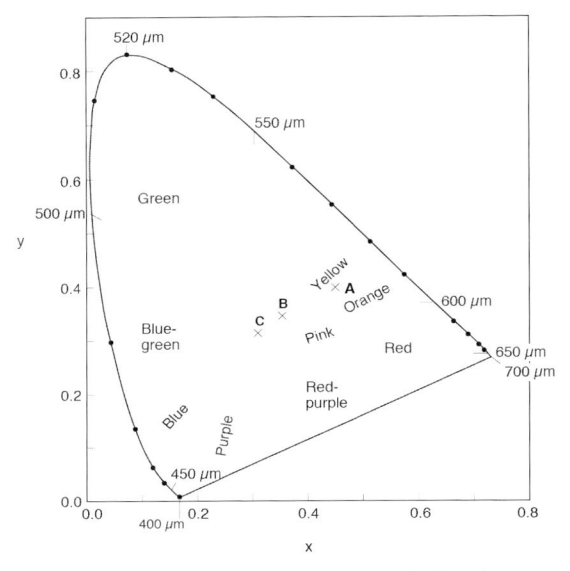

Figure 3.232: Chromaticity diagram with approximate color locations.

tinct groups. Table 3.46 presents the absorption wavelength for several chemical groups. The range for most commercially available infrared spectroscopes is between 2 and 25 µm. Hence, the spectrum taken between 2 and 25 µm serves as a fingerprint for that specific polymer, as shown in Fig. 3.233 for polycarbonate. An *infrared spectrometer* to measure the absorption spectrum of a material is schematically represented in Fig. 3.234. It consists of an infrared light source that can sweep through a certain wavelength range, and that is split in two beams: one that serves as a reference and the other that is passed through the test specimen. The comparison of the two gives the absorption spectrum, shown in Fig. 3.233. Using infrared spectroscopy can also help in quantitatively evaluating the effects of weathering (e.g., by measuring the increase of the absorption band of the COOH group, or by monitoring the water intake over time). One can also use the technique to follow reaction kinetics during polymerization.

3.10 Optical Properties

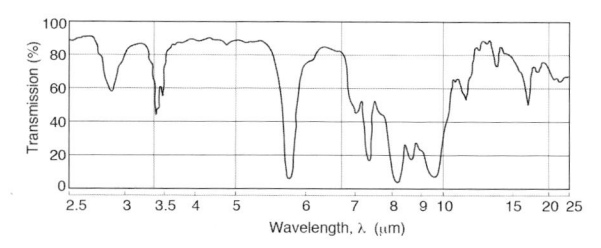

Figure 3.233: Infrared spectrum of a polycarbonate film.

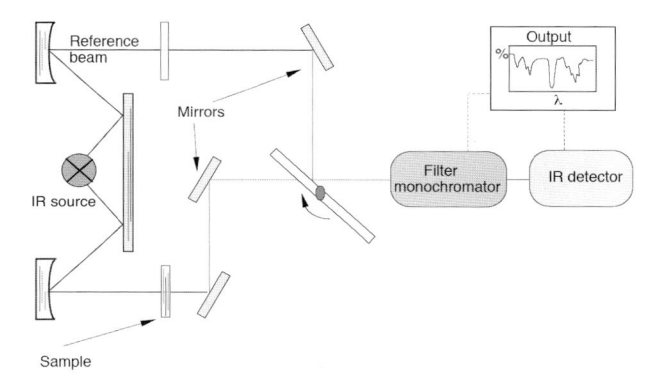

Figure 3.234: Schematic diagram of an infrared spectrometer.

Table 3.46: Absorption Wavelengths for Various Groups

Group	Wavelength region (μm)
O-H	2.74
N-H	3.00
C-H	3.36
C-O	9.67
C-C	11.49
C=O	5.80
C=N	5.94
C=C	6.07
C=S	6.57

3.11 ACOUSTIC PROPERTIES

Sound waves, similar to light waves and electromagnetic waves, can be reflected, absorbed and transmitted when they strike the surface of a body. The transmission of sound waves through polymeric parts is of particular interest to the design engineer. Of importance is the absorption of sound and the speed at which acoustic waves travel through a body, for example in a pipe, in the form of longitudinal, transversal and bending modes of deformation.

3.11.1 Speed of Sound

The speed at which sound is transmitted through a solid barrier is proportional to Young's modulus of the material, E, but inversely proportional to its density, ρ. For sound waves transmitted through a rod, in the longitudinal direction, the speed of sound can be computed as

$$C_L^{rod} = \sqrt{\frac{E}{\rho}} \qquad (3.74)$$

Similarly, the transmission speed of sound waves through a plate along its surface direction can be computed as

$$C_L^{plate} = \sqrt{\frac{E}{\rho(1-\nu^2)}} \qquad (3.75)$$

where ν is Poisson's ratio.

The speed of sound through a material is dependent on its state. For example, sound waves travel much slower through a polymer melt than through a polymer in the glassy state and the speed of sound through a polymer in the rubbery state is 100 times slower than that through a polymer in a glassy state. In the melt state, the speed of sound drops with increasing temperature due to density increase. Figure 3.235 presents plots of speed of sound through several polymer melts as a function of temperature. On the other hand, speed of sound increases with pressure as clearly shown in Fig. 3.236.

3.11.2 Sound Reflection

Sound reflection is an essential property for practical noise reduction. This can be illustrated using the schematic in Fig. 3.237. As the figure shows, sound waves

3.11 Acoustic Properties

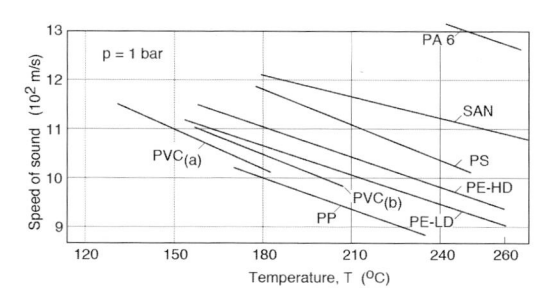

Figure 3.235: Speed of sound as a function of temperature through various polymers (after Offergeld and Menges).

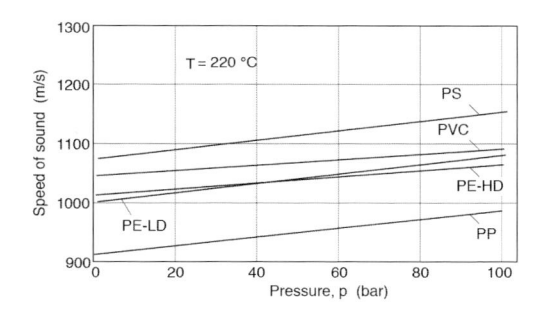

Figure 3.236: Speed of sound as a function of pressure through various polymers (after Offergeld and Menges).

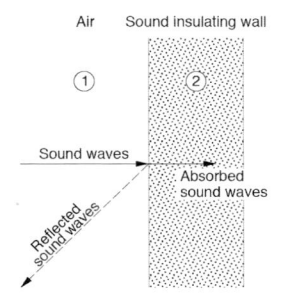

Figure 3.237: Schematic diagram of sound transmission through a plate.

that travel through media 1 strike the surface of media 2, and a fraction of the sound waves reflect back into media 1.

In order to obtain high sound reflection, the mass of the media 2 must be high compared to the mass of media 1. The mass of insulating sound walls can be increased with the use of fillers, such as plasticized PVC with barium sulfate or by spraying similar anti noise compounds on the insulating walls. It is common practice to use composite plates as insulating walls. This is only effective if the flexural resonance frequencies the walls do not coincide with the frequency of the sound waves.

3.11.3 Sound Absorption

Similar to sound reflection, sound absorption is an essential property for practical noise insulation. Materials that have the same characteristic impedance as air are the best sound-absorbent materials. The sound waves that are not reflected back out into media 1, penetrate media 2 or the sound insulating wall. Sound waves that penetrate a polymer medium are damped out similar to that of mechanical vibrations. Hence, sound absorption also depends on the magnitude of the loss tangent $\tan \delta$, or logarithmic decrement Δ, described earlier in this chapter. Table 3.47 presents orders of magnitude for the logarithmic decrement for several types of materials. As expected, elastomers and amorphous polymers have the highest sound absorption properties, whereas metals have the lowest.

Table 3.47: Damping Properties for Various Materials

Material	Temperature range	Logarithmic decrement Δ
Amorphous polymers	$T < T_g$	0.01-0.1
	$T > T_g$	0.1-1
Elastomers		0.1-1
Semi-crystalline polymers	$T_g < T < T_m$	≈ 0.1
Fiber reinforced polymers	$T_g < T < T_m$	< 0.01
Wood	$T < T_g$	0.01-0.02
Ceramic and glass	$T < T_g$	0.001-0.01
Metals	$T < T_m$	<0.0001

In a material, sound absorption takes place by transforming acoustic waves into heat. Since foamed polymers have an impedance of the same order as air, they are poor reflectors of acoustic waves. This makes them ideal for eliminating multiple reflections of sound waves in *acoustic* or sound proof rooms. Figure 3.238 presents the sound absorption coefficient for several foamed polymers as a function of the sound wave frequency. It should be noted that the speed at which sound travels in foamed materials is similar to that of the solid polymers,

3.11 Acoustic Properties

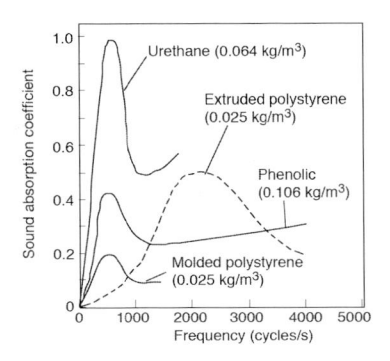

Figure 3.238: Sound absorption coefficients as a function of frequency for various foams (after Griffin and Skochdopole.

since foaming affects the stiffness and the density in the same proportion. When compared to wood, even semi-crystalline polymers are considered *sound-proof materials*. Materials with a glass transition temperature lower than room temperature are particularly suitable as damping materials. Commonly used for this purpose are thermoplastics and weakly cross-linked elastomers. Elastomer mats are often adhered on one or both sides of sheet metal, preventing resonance flexural vibrations of the sheet metal such as in automotive applications.

CHAPTER 4

PLASTICS PROCESSES

Manufacturing of plastic parts can involve one or several of the following steps:

- Shaping operations - This involves transforming a polymer pellet, powder or resin into a final product or into a preform using extrusion or molding processes such as injection, compression molding or roto molding.
- Secondary shaping operation - Here a preform such as a parison or sheet is transformed into a final product using thermoforming or blow molding.
- Material removal - This type of operation involves material removal using machining operations, stamping, laser, drilling, etc.
- Joining operations - Here, two or more parts are assembled physically or by bonding or welding operations.

Most plastic parts are manufactured using shaping operations. Here, the material is deformed into its final shape at temperatures between room temperature and 350°C, using wear resistant tools, dies and molds. For example, an injection mold would allow making between 10^6 and 10^7 parts without much wear of the tool, allowing for the high cost of the molds utilized. One of the many advantages of polymer molding processes is the accuracy, sometimes with features down to the micrometer scale, with which one can shape the finished product without the need of trimming or material removal operations. For example, when making compact discs by an injection-compression molding process, it is possible to accurately produce features, that contain digital information smaller than 1μm, on a disc with a thickness of less than 1mm and a diameter of several centimeters. The cycle time to produce such a part can be less than 3 seconds.

In the past few years, we have seen trends where more complex manufacturing systems are developed that manufacture parts which use various materials and components such as co-extrusion of multilayer films and sheets, multi-component injection molding, sandwiched parts, or hollow products.

Thermoplastics and thermoplastic elastomers are shaped and formed by heating them above glass transition or melting temperatures and then freezing them into their final shape by lowering the temperature. At that point, the crystallization, molecular or fiber orientation and residual stress distributions are an integral feature of the final part, dominating the material properties and performance of the finished product. Similarly, thermosetting polymers and vulcanizing elastomers solidify by a chemical reaction that results in a cross-linked molecular structure. Here too, the filler or fiber orientation as well as the residual stresses are frozen into the finished structure after cross-linking.

This chapter is intended to give an introduction to the most important polymer processes[1].

4.1 RAW MATERIAL PREPARATION

Raw material preparation is understood as the necessary steps taken before the polymeric material is processed into the finished product. Such steps include the addition of components or additives such as pigments, fillers, fibers, platicizers, lubricants, stabilizers, flame retardants, foaming agents, solvents, or other polymers, or the material's transformation into a powder, paste or pellet. The most important material preparation operations are mixing, kneading, disolving, granulating or pelletizing, and drying.

4.1.1 Mixing Processes

Today, most processes involve some form of mixing. For example, an integral part of a screw extruder is a mixing zone. In fact, most twin screw extruders are primarily used as mixing devices. Similarly, the plasticating unit of an injection molding machine often has a mixing zone. This is important because the quality of the finished product in almost all polymer processes depends in part on how well the material was mixed. Both the material properties and the formability of the compound into shaped parts are highly influenced by the mixing quality. Hence, a better understanding of the mixing process helps to optimize processing conditions and increase part quality.

The process of polymer blending or mixing is accomplished by distributing or dispersing a minor or secondary component within a major component serving as a matrix. The major component can be thought of as the continuous phase, and the minor components as distributed or dispersed phases in the form of droplets, filaments, or agglomerates. When creating a polymer blend, one must always keep in mind that the blend will probably be remelted in subsequent processing or shaping processes. For example, a rapidly cooled system, frozen as a homogenous mixture, can separate into phases because of coalescence when re-heated. For all practical purposes, such a blend is not processable. To avoid this problem,

---

[1] For further reading in the area of extrusion and injection molding we recommend that the reader consult the literature. A list of books is given in the appendices of this handbook.

compatibilizers, which are macromolecules used to ensure compatibility in the boundary layers between the two phases, are common. Mixing can be distributive or dispersive. For example, the morphology development of polymer blends is determined by three competing mechanisms: distributive mixing, dispersive mixing, and coalescence. There are three general categories of mixtures that can be created:

- Homogeneous mixtures of compatible polymers,
- Single phase mixtures of partly incompatible polymers, and
- Multi-phase mixtures of incompatible polymers.

Table 4.1 lists examples of compatible, partially incompatible, and incompatible polymer blends.

Table 4.1: Common Polymer Blends

Compatible polymer blends	Natural rubber and polybutadiene
	Polyamides (e.g., PA 6 and PA 66)
	Polyphenylene ether (PPE) and polystyrene
Partially incompatible polymer blends	Polyethylene and polyisobutylene
	Polyethylene and polypropylene (5% PE in PP)
	Polycarbonate and polyethylene terephthalate
Incompatible polymer blends	Polystyrene/polyethylene blends
	Polyamide/polyethylene blends
	Polypropylene/polystyrene blends

Distributive mixing or laminar mixing of compatible materials is usually characterized by the distribution of the secondary phase within the matrix. This distribution is achieved by imposing large strains on the system such that the interfacial area between the two or more phases increases and the local dimensions, or striation thicknesses, of the secondary phases decrease. Imposing large strains on the blend is not always sufficient to achieve a homogeneous mixture. The type of mixing device, initial orientation, and position of the two or more fluid components play a significant role in the quality of the mixture.

Dispersive mixing in polymer processing involves breaking a secondary immiscible fluid or an agglomerate of solid particles and dispersing them throughout the matrix. Here, the imposed strain is not as important as the imposed stress which causes the system to break-up. Hence, the type of flow inside a mixer plays a significant role on the break-up of solid particle clumps or fluid droplets when dispersing them throughout the matrix. The most common example of dispersive mixing of particulate solid agglomerates is the dispersion and mixing of carbon black into a rubber compound. Figure 4.1 relates the viscosity

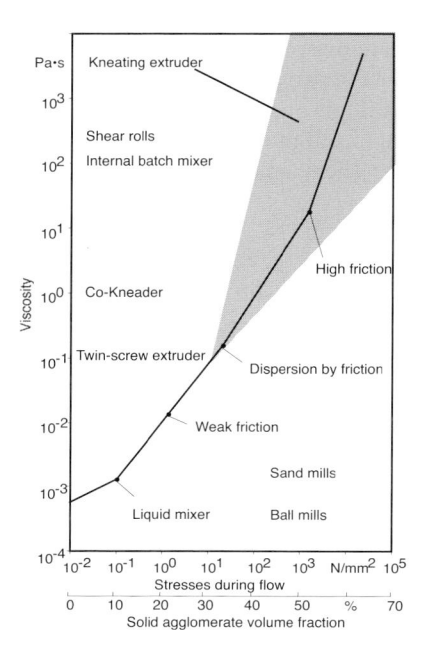

Figure 4.1: Overview of dispersive mixing equipment as a function of compound viscosity and volume fraction of solid agglomerates.

of the compound to the volume fraction of the solid agglomerates and the type of mixing device. When breaking up one polymer melt within a matrix, when compounding a polymer blend, the droplets inside the incompatible matrix tend to stay or become spherical due to the natural tendencies of the drop trying to maintain the lowest possible surface - to - volume ratio. However, a flow field within the mixer applies a stress on the droplets, causing them to deform. If this stress is high enough, it will eventually cause the drops to break up. The droplets will disperse when the surface tension can no longer maintain their shape in the flow field and the filaments break-up into smaller droplets. This phenomenon of dispersion and distribution continues to repeat itself until the stresses caused by the flow field can no longer overcome the surface tension of the small droplets that are formed. The mechanism of melt droplet break-up is similar in nature to solid agglomerate break-up in the sense that both rely on forces to disperse the particulates. It is well known that when breaking up solid agglomerates or melt droplets, the most effective type of flow is a stretching or elongational flow. Hence, devices that stretch the melt, instead of shearing it, lead to more effective mixers while significantly lowering energy consumption.

4.2 MIXING DEVICES

The final properties of a polymer component are heavily influenced by the blending or mixing process that takes place during processing or as a separate step in the manufacturing process. As mentioned earlier, when measuring the quality of mixing it is also necessary to evaluate the efficiency of mixing. For example, the amount of power required to achieve the highest mixing quality for a blend may be unrealistic or unachievable. This section presents some of the most commonly used mixing devices encountered in polymer processing.

In general, mixers can be classified in two categories: internal batch mixers and continuous mixers. Internal batch mixers, such as the Banbury type mixer, are the oldest type of mixing devices in polymer processing and are still widely used in the rubber compounding industry. Industry often also uses continous mixers because they combine mixing in addition to their normal processing tasks. Typical examples are single and twin screw extruders that often have mixing heads or kneading blocks incorporated into their system.

4.2.1 Mixing of Particulate Solids

There is a variety of mixing drums that are used to blend granulated solids. They range from internal impeller speeds of less than 2m/s to speeds up to 50m/s. As the speed increases, so does the energy input and degradation of the mixed components. These types of mixers can be continuous or discontinuous. Just as is the case with mixers involving viscous substances, with continuous systems, the mixing impellers impose a conveying action on the particulates or mixture. Typical types of particulate solids batch mixers are presented in Fig. 4.2, and continuous mixers are found in Fig. 4.3, where the various mixers are equipped with different types of mixing elements or impellers such as augers, paddles or spirals.

The conical hopper mixers with the rotating auger have a capacity of up to 30 m^3, while the silo mixers with an auger have a capacity of up to 100 m^3. The discontinuous mixers with horizontal impellers have a capacity of up to 30 m^3, and the continuous systems have a throughput of up to 450 m^3/h.

4.2.2 Screw-Type Mixers

Screw-type mixing devices are used to continuously compound plastics to processable materials. They are fed by premixed solid agglomerates or they are fed or metered into the mixers by dosing systems. There are single screw and twin screw mixing devices. Table 4.2 presents various screw compounders with characteristic sizes, power consumption, and throughputs.

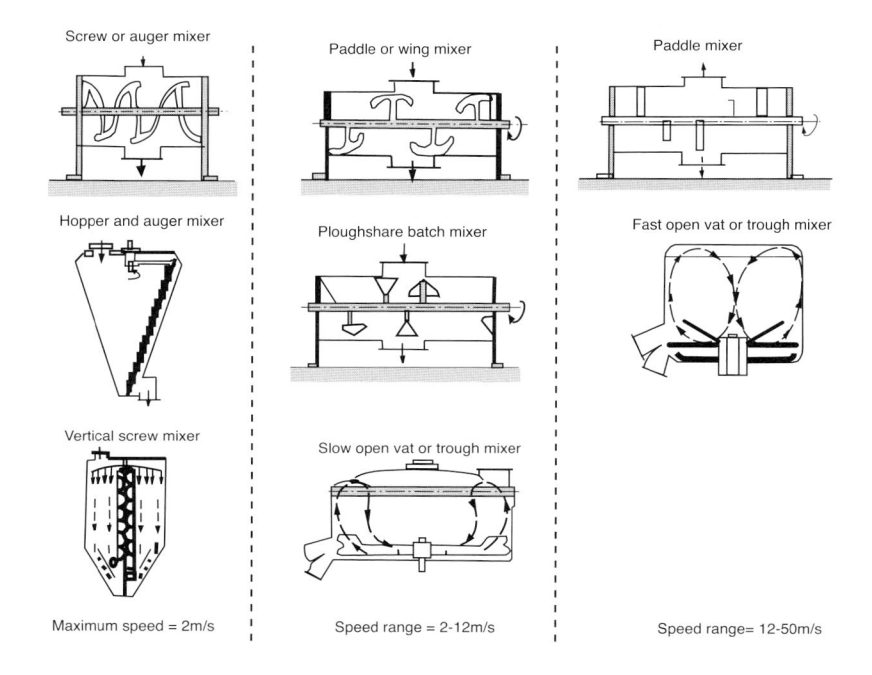

Figure 4.2: Solid particulate batch mixing devices.

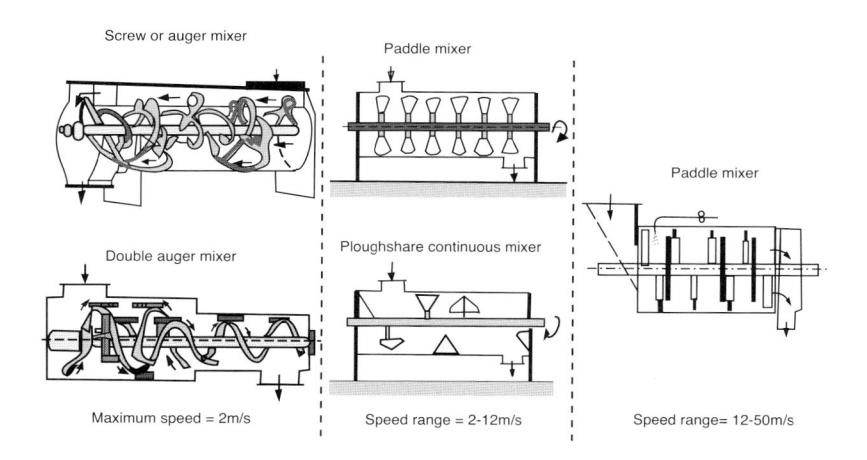

Figure 4.3: Solid particulate continuous mixing devices.

4.2 Mixing Devices

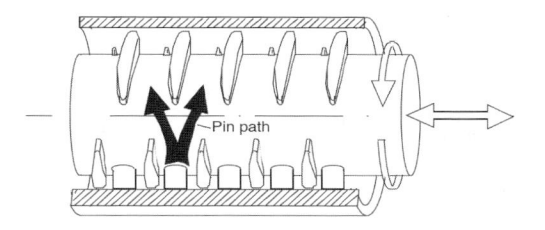

Figure 4.4: Schematic diagram of a co - kneader.

Table 4.2: Typical Sizes, Power Consumption, and Throughput of Screw Compounders

Compounder Type	Screw diameter mm	Power kW	Energy Input kW-h/kg	Throughput kg/h
Single screw type	250-800	500-6,700	0.07-0.13	4,000-74,000
Plasticizer	330-555	22-160	0.07-0.10	200-2,300
Co-Kneader	46-400	11-650	0.08-0.40	10-8,000
Twin screw type	25-380	75-20,000	0.10-0.40	10-75,000

Co - kneader The co - kneader is a single screw mixer with pins on the barrel and a screw that oscillates in the axial direction. Figure 4.4 shows a schematic diagram of a co - kneader. The pins on the barrel practically wipe the entire surface of the screw, making it the only self-cleaning single-screw extruder. This results in a reduced residence time, which makes it appropriate for processing thermally sensitive materials. The pins on the barrel also disrupt the solid bed creating a dispersed melting which improves the overall melting rate while reducing the overall temperature in the material.

A simplified analysis of a co - kneader gives a number of striations per L/D of

$$N_s = 2^{12} \tag{4.1}$$

which means that over a section of 4D, the number of striations is $2^{12}(4) = 28^{13}$.

Mixing In Single Screw Extruders Distributive mixing caused by the cross-channel flow component in single screw extruders can be enhanced by introducing pins (Fig. 4.5) or rhomboidal elements (Fig. 4.6) in the flow channel and slots on the screw flights. The pins can either sit on the screw or on the barrel as shown in Fig. 4.5. The extruder with the adjustable pins on the barrel is generally referred to as QSM-extruder[2]. In all cases, the pins disturb the flow by re-orienting

---
[2] QSM comes from the German words *Quer Strom Mischer* which translates into cross-flow mixing

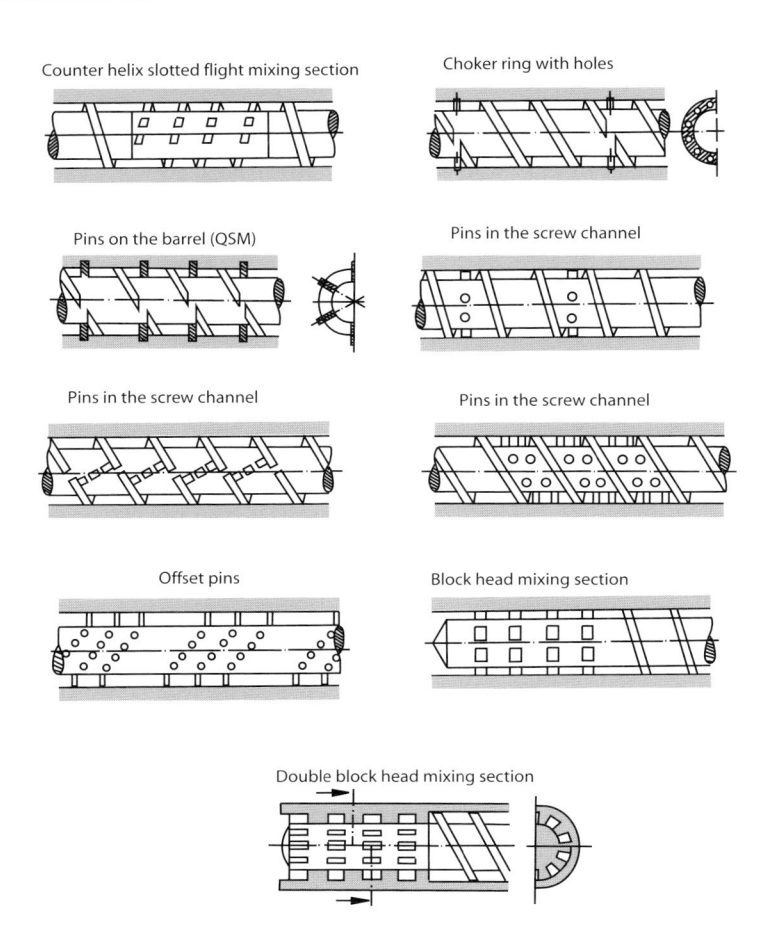

Figure 4.5: Typical pin and slotted flight type single screw extruder distributive mixing sections.

the surfaces between fluids and by creating new surfaces by splitting the flow. Figure 4.7 presents a photograph of the channel contents of a QSM-extruder. The photograph shows the re-orientation of the layers as the material flows past the pins. The pin-type extruder is especially useful for the mixing of high viscosity materials, such as rubber compounds; and is therefore, often referred to as cold feed rubber extruder. This machine is widely used in the production of rubber profiles of any shape and size. As mentioned earlier, dispersive mixing is required when breaking down particle agglomerates or when surface tension effects exist between primary and secondary fluids in the mixture. To disperse

4.2 Mixing Devices

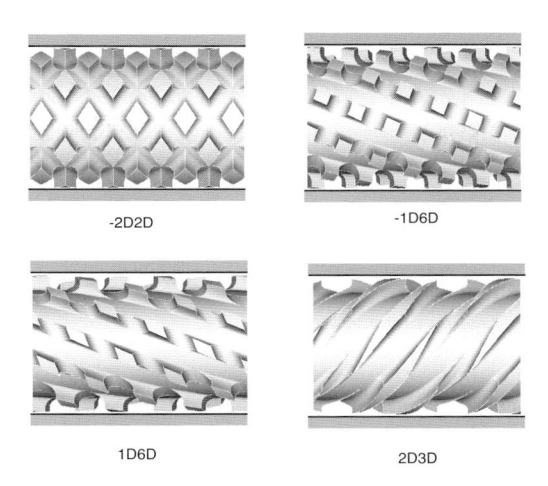

-2D2D

-1D6D

1D6D

2D3D

Figure 4.6: Various rhomboidal single screw extruder distributive mixing sections.

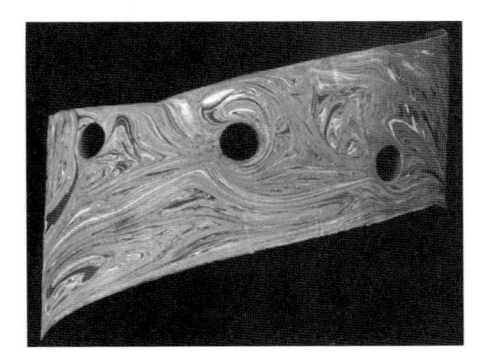

Figure 4.7: Photograph of the unwrapped channel contents of a pin barrel extruder. Courtesy of the Paul Troester Maschinenfabrik, Hannover, Germany.

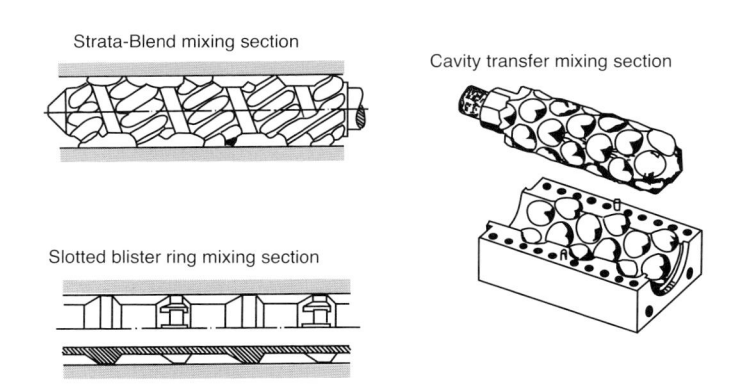

Figure 4.8: Typical single screw extruder distributive mixing sections with dispersive capabilities.

such systems, the mixture must be subjected to large stresses. Some distributive mixing heads, such as those depicted in Fig. 4.8, also have a strong dispersive component due to the stretching flows that they generate during mixing. A very popular one is the cavity transfer mixing (CTM) head depicted in Fig. 4.8. Barrier-type screws are often used in lieu of dispersive mixing heads, because they apply high shear stresses to the polymer melt when the molten material passes to the melt pool channel. However, more intensive mixing can be applied by using a mixing head intended for this purpose. When using barrier-type screws or a mixing head as shown in Fig. 4.9, the mixture is forced through narrow gaps, causing high stresses in the melt. It should be noted that dispersive as well as distributive mixing heads result in a resistance to the flow, which results in viscous heating and pressure losses during extrusion.

Twin Screw Extruders In the past two decades, twin screw extruders have developed into the best available continuous mixing devices. In general, they can be classified into intermeshing or non-intermeshing, and co-rotating or counter-rotating twin screw extruders. The intermeshing twin screw extruders render a self-cleaning effect which evens-out the residence time of the polymer in the extruder. The self-cleaning geometry for a co-rotating double flighted twin screw extruder is shown in Fig. 4.10 The main characteristic of this type of configuration is that the surfaces of the screws are sliding past each other, constantly removing the polymer that is stuck to the screw. In the last two decades, the co-rotating twin screw extruder systems have established themselves as efficient continuos mixers, including reactive extrusion. In essence, the co-rotating systems have a high pumping efficiency caused by the double transport action of the two screws. Counter-rotating systems generate high stresses because of the calendering action

4.2 Mixing Devices

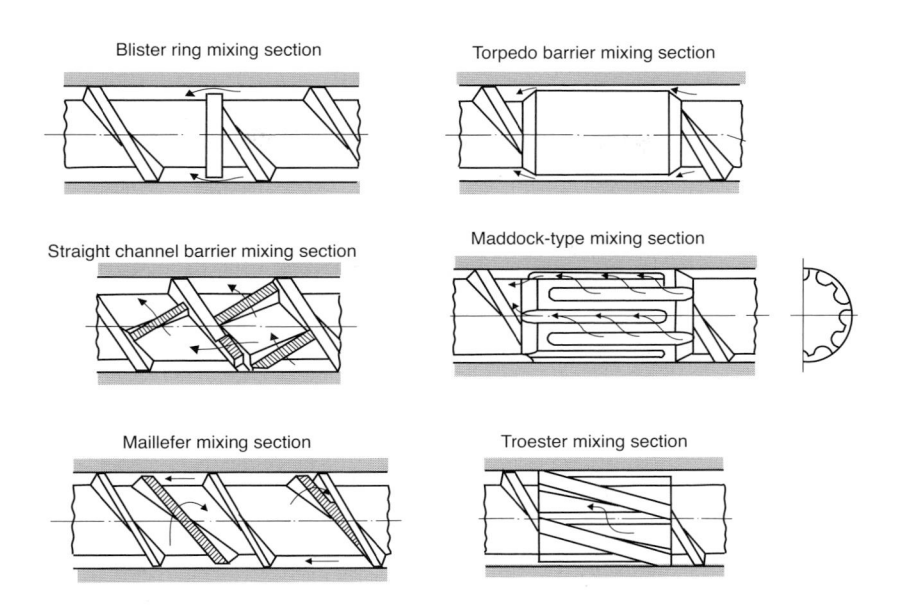

Figure 4.9: Commonly used single screw extruder dispersive mixing section.

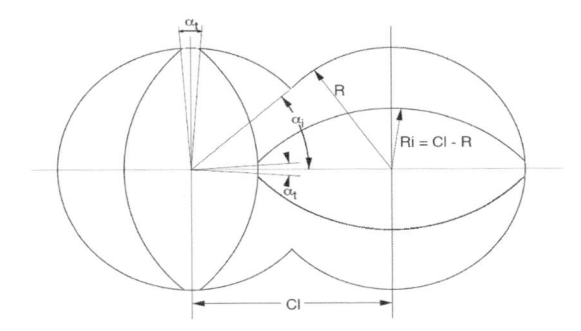

Figure 4.10: Geometry description of a double-flighted, co-rotating, self-cleaning twin screw extruder.

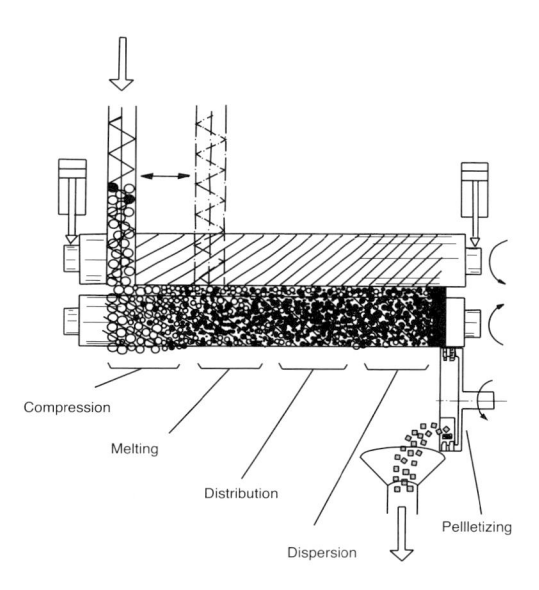

Figure 4.11: Schematic diagram of a shear roll mixer.

between the screws, making them efficient machines to disperse pigments and lubricants. However, there seems to be considerable disagreement about co - versus counter-rotating twin screw extruders between different groups in the polymer processing industry and academic community.

A special form of twin screw mixing devices are *shear roll mixers* schematically depicted in Fig. 4.11. Such a system, which is open to the atmosphere, presents several advantages over its enclosed counterpart. It is used to compact, melt, homogenize, disperse and pelletize materials from medium to high viscosity in temperature ranges between 20 and 280°C. Each one of the counter-rotating rolls has a series of counter-oriented grooves that force the material to convey from one side to the other of the horizontally aligned rolls. The additives, such as solid agglomerates or fibers are dosed into the rolls' nip region. The roll separation is adjusted and decreases from the beginning to the end of the rolls in order to enhance the dispersive mixing action. The finished material is either removed from the end of the rolls in form of a tape, or it is forced through small holes and pelletized as shown in Fig. 4.12.

Static Mixers Static mixers, or motionless mixers, are pressure-driven continuous mixing devices through which the melt is pumped, rotated, and divided, leading to effective mixing without the need for movable parts and mixing heads. One of the most commonly used static mixers is the twisted tape static mixer

4.2 Mixing Devices

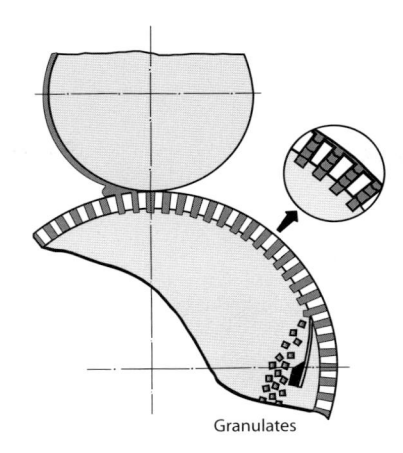

Figure 4.12: Pelletizing mechanism of a shear roll mixer.

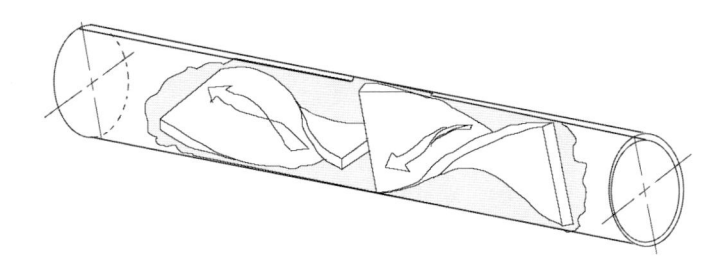

Figure 4.13: Schematic diagram of a Kenics static mixer.

schematically shown in Fig. 4.13 The polymer is sheared and then rotated by $90°$ by the dividing wall; the interfaces between the fluids increase. The interfaces are then re-oriented by $90°$ once the material enters a new section. The stretching-re-orientation sequence is repeated until the number of striations is so high that a seemingly homogeneous mixture is achieved. Figure 4.14 shows a sequence of cuts down a Kenics static mixer. From the figure it can be seen that the number of striations increases from section to section by 2, 4, 8, 16, 32, etc., which can be computed using

$$N = 2^n \qquad (4.2)$$

where N is the number of striations and n is the number of sections in the mixer.

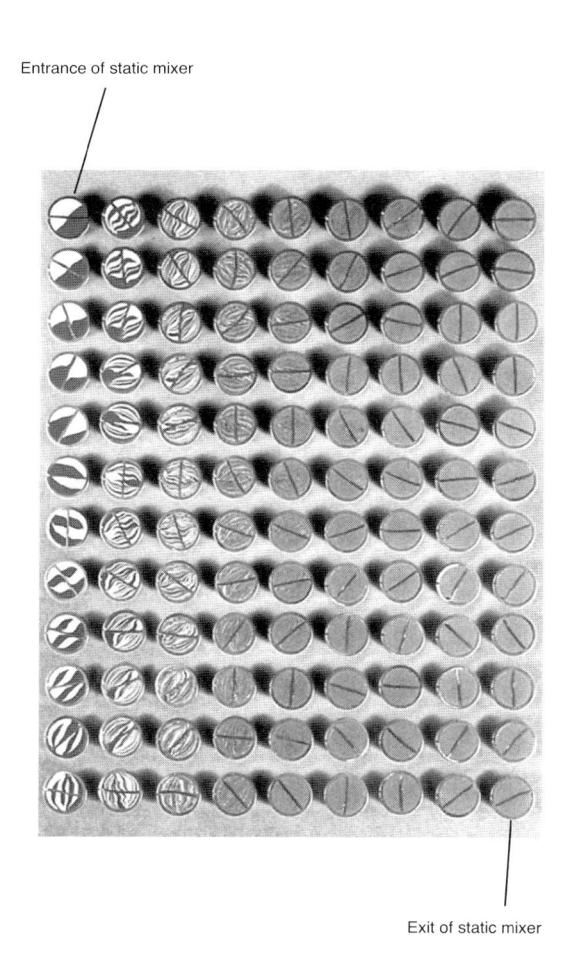

Figure 4.14: Experimental progression of the layering of colored resins in a Kenics static mixer. Courtesy Chemineer, Inc., North Andover, Massachusetts.

4.2 Mixing Devices

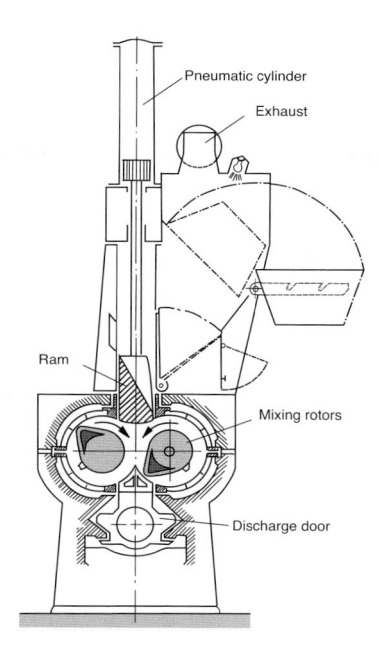

Figure 4.15: Schematic diagram of an internal batch mixer.

Internal Batch Mixer The internal batch or Banbury - type mixer, schematically shown in Fig. 4.15, is perhaps the most commonly used internal batch mixer. Internal batch mixers are high intensity mixers that generate complex shearing and elongational flows that work especially well in the dispersion of solid particle agglomerates within polymer matrices. One of the most common applications for high - intensity internal batch mixing is the break-up of carbon black agglomerates into rubber compounds, but have also been implemented for the compounding of ABS. The dispersion of agglomerates is strongly dependent on mixing time, rotor speed, temperature, and rotor blade geometry. Figure 4.16 shows the fraction of undispersed carbon black as a function of time in a Banbury mixer at 77rpm and $100^\circ C$. The broken line in the figure represents the fraction of particles smaller than 500 nm. An internal batch mixer can be equipped with rotors that are intermeshing or tangential to each other as schematically depicted in Fig. 4.17 The intermeshing system renders a more effective mixing process but it consumes significantly more power than its tangential counterpart. Instead of using a discharge door to remove the mixed contents from the internal batch mixer, some systems come equipped with a screw pump, such as the one shown in Fig. 4.18. These systems range between 10 liter (2 gallon), for laboratory

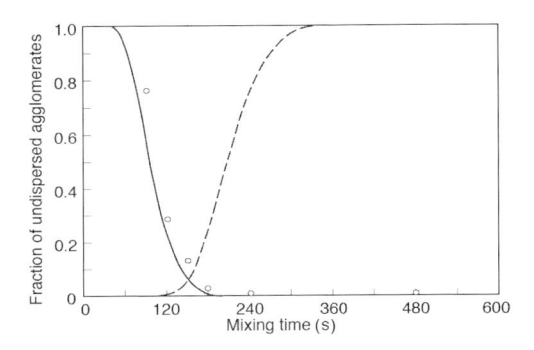

Figure 4.16: Fraction of undispersed carbon black of size above $9\mu m$, as a function of mixing time inside a Banbury mixer. The open circles denote experimental results and the solid line a theoretical prediction. Broken line denotes the fraction of aggregates of size below 500nm.

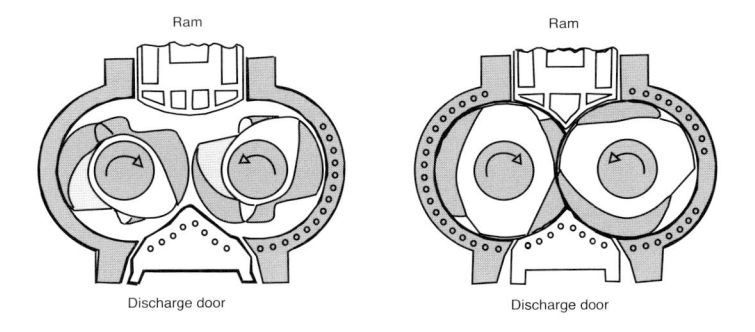

Figure 4.17: Schematic diagram of tangential (left) and intermeshing (right) internal batch mixers.

mixers, to 4,000 liters (800 gallon), for industrial batch mixers. The continuous kneader, schematically depicted in Fig. 4.19, is composed of a series of 6 to 12 kneading chambers. Such a system combines various steps, such as mixing, kneading, extrusion, and dosing, into a single processing equipment. The discharge is often metered using gear pumps or screw pumps, which can feed pelletizing equipment. Mixers or blenders, such as the double planetary mixer depicted in Fig. 4.20, are used to blend and mix solutions and pastes of relatively low viscosities, up to 150Pa-s.

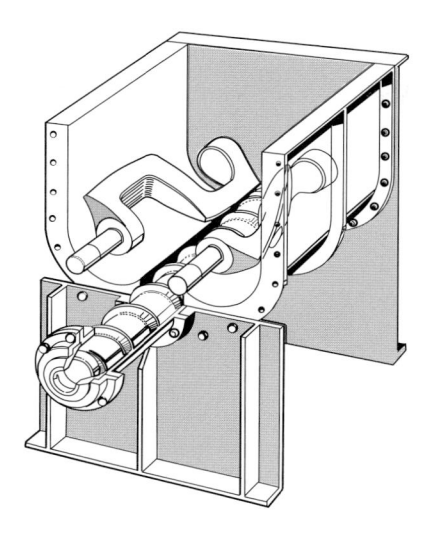

Figure 4.18: Schematic diagram of an internal batch mixer equipped with a continuous discharge screw.

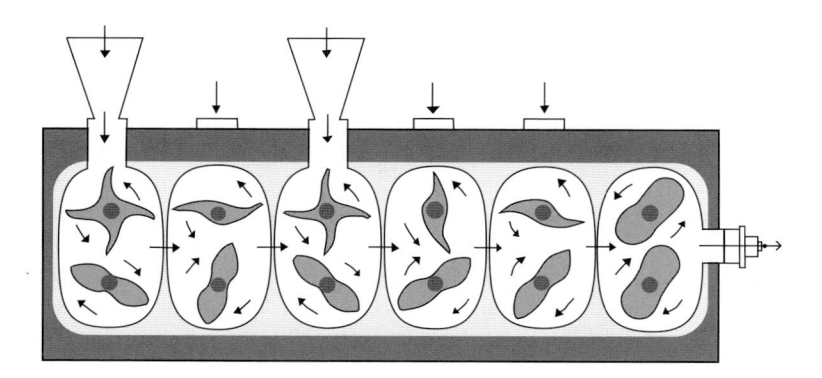

Figure 4.19: Schematic diagram of a continuous kneader.

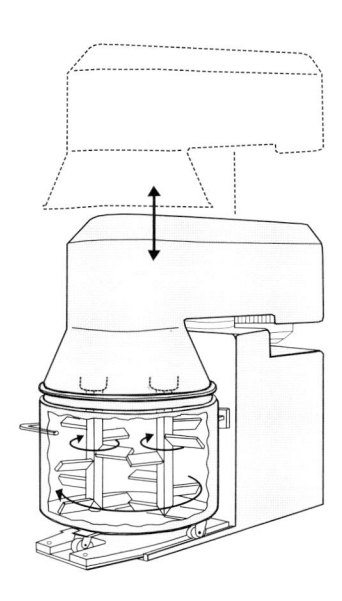

Figure 4.20: Schematic diagram of a planetary mixer.

4.2.3 Granulators and Pelletizers

Generally, thermoplasic resins are delivered in pellet form. There are two general forms of pelletizing: *hot plate* or *hot face pelletizing* and *cold cutting* systems. Figure 4.21 presents a *strand pelletizer*, where molten polymer is fed through holes on a circular die face. The polymer strands are cut into pellets by rotating knives and cooled by forced air. Figure 4.22 shows three different types of pelletizers: the *rotary knife pelletizer*, where the pellets are cooled with air and water, presented in Fig. 4.22(A), and the *water ring pelletizer* and the *under water pelletizer*, presented in Figs.4.22(B) and 4.22(C), respectively. Today, the throughput of hot face pelletizers range between a couple of pounds to 25 tons per hour. The underwater pelletizing systems render the highest throughput. When using cold cutting pelletizing systems, the polymer melt is extruded through a perforated plate into strands or tapes, which are cooled before the they are cut into pellets. Although cold cutting systems are generally less expensive and simpler than their hot face counterparts, they are more labor - intensive and since they are open systems, they are more prone to contamination. However, they are the systems of choice in smaller laboratory or pilot systems. Draw-backs of the cold cutting systems, which in essence cut solid polymer, are the high noise levels and the shorter life of the blades.

4.2 Mixing Devices

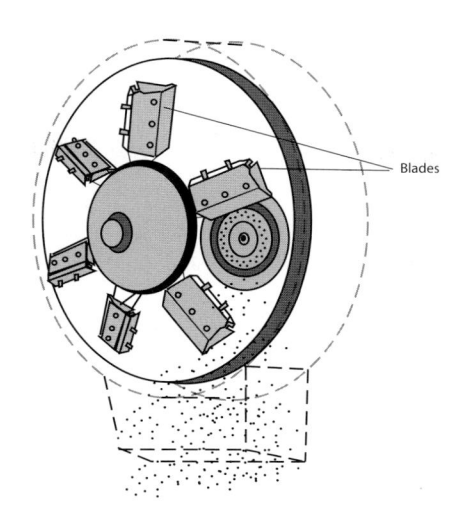

Figure 4.21: Schematic diagram of an air dried rotary knife pelletizer.

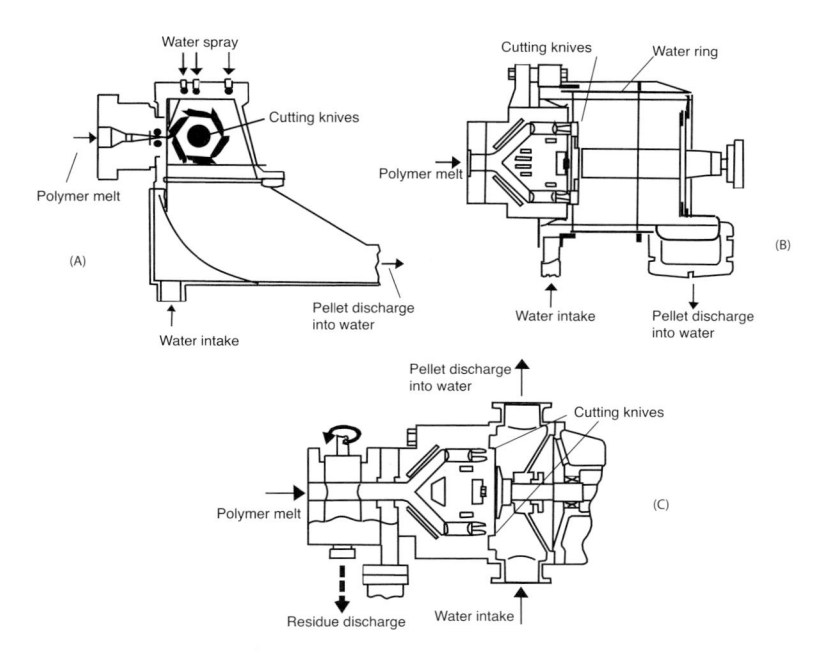

Figure 4.22: Schematic of various pelletizing systems. (A) Rotary knife pelletizer, (B) water ring pelletizer and (C) underwater pelletizer.

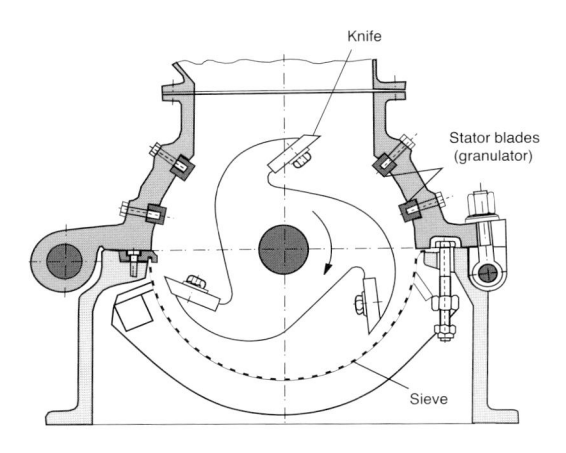

Figure 4.23: Schematic of an shear-type pulverizer.

Sometimes it is desired to transform polymer materials into fine granules or powders, such as those used for sintering, coating or roto-molding applications, or when processing or regrinding residues or left-overs from injection molding and extrusion processes. In some applications, regrind of up to 50% is re-introduced into the raw material stream. The systems used to transform plastics into granules or powders are referred to as granulators or pulverizers. Typically, these systems crush or shear the material. Figure 4.23 presents a system where the plastic residues are sheared and reduced between the static and rotating knives until they fall through the sieve. To avoid overheating due to frictional heating, the pulverzing is often done in steps, or the material is pre-cooled using liquid nitrogen. Such a system is presented in Fig. 4.24

4.2.4 Dryers

To avoid hydrolysis during the melt stage, many plastics must be dried before processing. Surface moisture can dissipate due to rising heat from the barrel of the process machine while in the hopper or via vented barrels, which is normally done with polyolefins and PVC. Removing moisture in these materials can easily be achieved by prewarming the granules with hot air in an oven fitted with removable baskets or trays. A more automated method uses a drying hopper that can be mobile, free-standing, or mounted directly on the throat of the processing machine. The capacity of the holding hopper is determined by the necessary drying time required at the temperature specified by the polymer supplier. For instance,

4.2 Mixing Devices

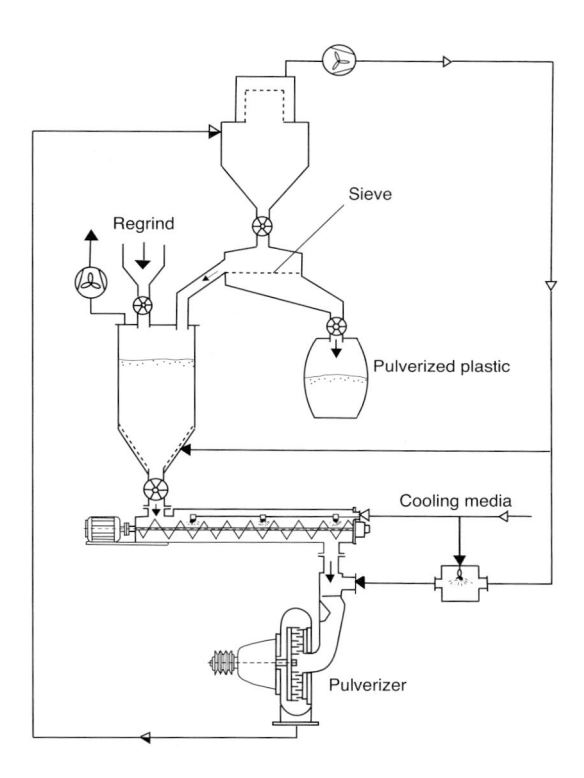

Figure 4.24: Schematic of a cooled pulverizing system.

an injection machine that processes 50 kg/hour, with a determined drying time of 2 hours, requires a hopper holding of at least 100 kg.

The main methods of drying use hot air or dehumidified air by various means. The simplest hot-air technique uses hot-air lances or probes that are easily fixed directly in the processor's existing hopper and are therefore portable between presses. These drying probes consist of a steel tube that contains a heater, thermocouple, and thermal overload through which atmospheric air is blown to the base of the filled hopper, and dissipated into the material through a diffuser head such as the one shown in Fig. 4.25 Another popular method is throat-mounted drying; here the insulated drying chamber with a flanged collar support replaces the process machine hopper. Drying at the machine throat eliminates the possibility of contamination or reabsorption of moisture, which can occur when transporting granules from a central drying position. Hot-air drying normally reduces moisture content to 0.2% or less. To remove in-built moisture requires more complex dehumidified air drying equipment, which injects super-dry heated air into the base of a cylindrical insulated container, up through an inverted diffuser cone supporting the granules. This super-dry air is generated by injecting compressed air through a molecular sieve desiccant, such as crystalline aluminum silicate, which is contained in a pressure-sealed cylinder. Compressed air is known to give up its moisture more rapidly than air at atmospheric pressure, so passing it through the desiccant lowers the relative humidity and dew point of the air supplied to the drying chamber, efficiently removing moisture [1 part per million (ppm) can be achieved]. Two or more cylinders containing the desiccant are alternatively charged with compressed air as each becomes saturated with the moisture-laden air. Over time, this dry air, when mixed with heated air at up to 165°C, draws moisture from the granules (see Fig. 4.26). Table 4.3 presents common thermoplastics with their recommended drying temperatures and times. Materials that have been stored for long periods in unsealed, unlined bags may need longer drying times. Microwave drying is prohibitively expensive commercially, but is used in laboratories for measuring moisture content.

Table 4.3: Recommended Drying Parameters for Common Thermoplastics

Polymer	Abb.	Temperature (°C)	Time (hours)
Polystyrene and Copolymers	PS	60-80	1-3
	SAN	70-90	1-4
	ABS	70-80	4
	ASA	80-85	2-4
Polymethylmethacrylate	PMMA	70-100	8
Polyoxymethylene	POM	80-120	3-6
Polycarbonate	PC	100-120	8
Polyamide	PA	80-100	16
Ethylen-Butyl-Acrylate	EBA	70-80	3
Polyether-Block-Amide	PEBA	70-80	2-4
Celluloseacetobutyrate	CAB	60-80	2-4
Polyphenylene oxide	PPO	100	2
Polysulfone	PSU	120-150	3-4
Thermoplastic polyester	PET	120-140	2-12
Ethylen-Vinylacetate	EVA	70-80	3
Polyarylate	PAR	120-150	4-8
Polyarylsulfone	PASU	135-180	3-6
Polyphenylene sulfide	PPS	140-250	3-6
Polyether-Etherketone	PEEK	150	8
Thermoplastic elastomers	TPE	120	3-4

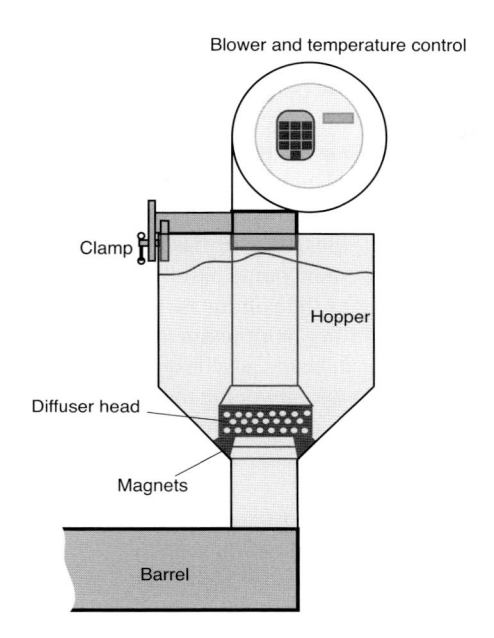

Figure 4.25: Schematic diagram of an in-hopper probe drying system.

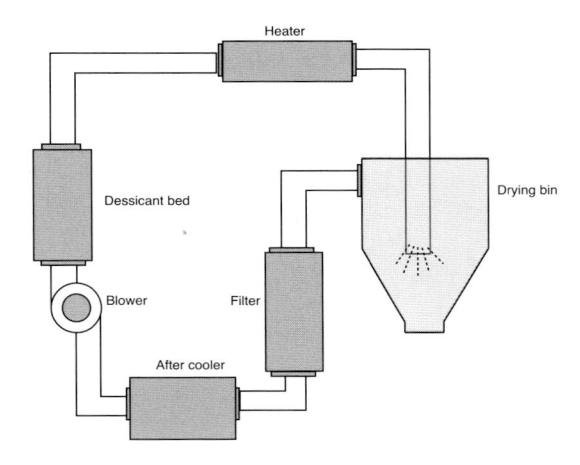

Figure 4.26: Schematic diagram of a dehumidifier drying system.

4.3 EXTRUSION

During extrusion, a polymer melt is pumped through a shaping die and formed into a profile. This profile can be a plate, a film, a tube, or have any shape for its cross section. The first extrusion of thermoplastic polymers was done at the Paul Troester Maschinenfabrik in Hannover, Germany in 1935. Although ram and screw extruders are both used to pump highly viscous polymer melts through passages to generate specified profiles, they are based on different principles. The ram extruder is a positive displacement pump based on the pressure gradient term of the equation of motion. Here, as the volume is reduced, the fluid is displaced from one point to the other, resulting in a pressure rise. The gear pump, widely used in the polymer processing industry, also works on this principle. On the other hand, a screw extruder is a viscosity pump that works based on the pressure gradient term generated by friction of the pellets against the barrel in the soilds section and by dragging the highly viscous polymer melt in the metering section of the extruder.

In today's polymer industry, the most commonly used extruder is the single screw extruder, schematically depicted in Fig. 4.27

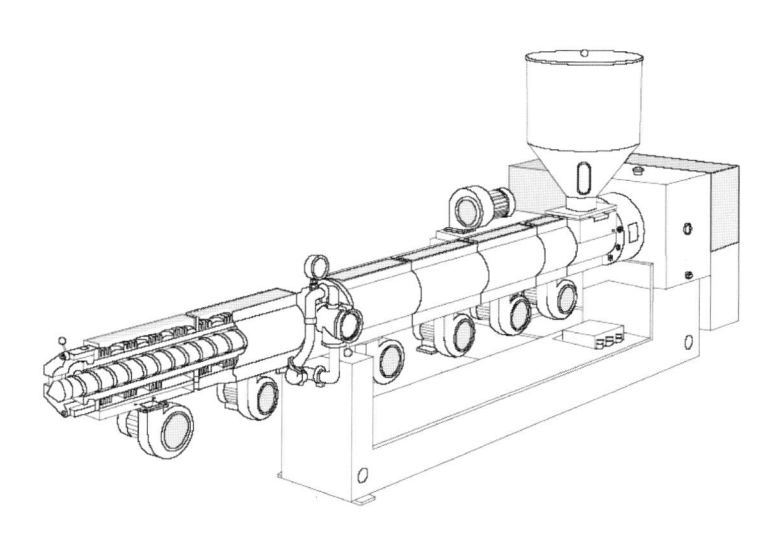

Figure 4.27: Schematic of a single screw extruder (Reifenhäuser).

4.3 Extrusion

The single screw extruder can either have a smooth inside barrel surface, called a conventional single screw extruder, or a grooved feed zone, called a grooved feed extruder. In some cases, an extruder can have a degassing zone, required to extract moisture, volatiles, and other gases that form during the extrusion process, as shown in Fig. 4.28.

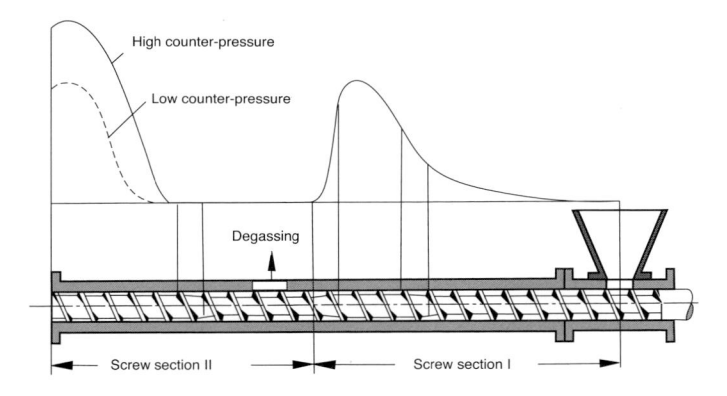

Figure 4.28: Schematic of a single screw extruder with a degassing section with the pressure distribution along the screw axis.

Other special systems of single screw extruders with degassing sections are the cascade-type single screw extruder system, schematically depicted in Fig. 4.29.

The cascade-type system is a twin screw extruder with the screws in series, with a plasticating screw, and a mixing and pressure build-up screw that pumps the melt through the die. Each screw has its own drive system that is used to adjust the rotational speed of each screw in order to synchronize the throughput through each system.

A whole other family of twin screw systems, where the screws are parallel to each other and rotate inside an interconnected twin-cylinder barrel system, are what we commonly refer to as twin screw extruders. Various types of twin screw extruder systems are schematically depicted in Figs. 4.30 and 4.31. Twin screw extruders can have co-rotating or counter-rotating screws, and the screws can be intermeshing or non-intermeshing. Twin screw extruders are primarily employed as mixing and compounding devices, as well as polymerization reactors. The *conical twin screw extruder* depicted in Fig. 4.31 is used to convey and compound PVC and PE-HD powders and other difficult to feed materials. The mixing aspects of single and twin screw extruders were detailed earlier in this chapter.

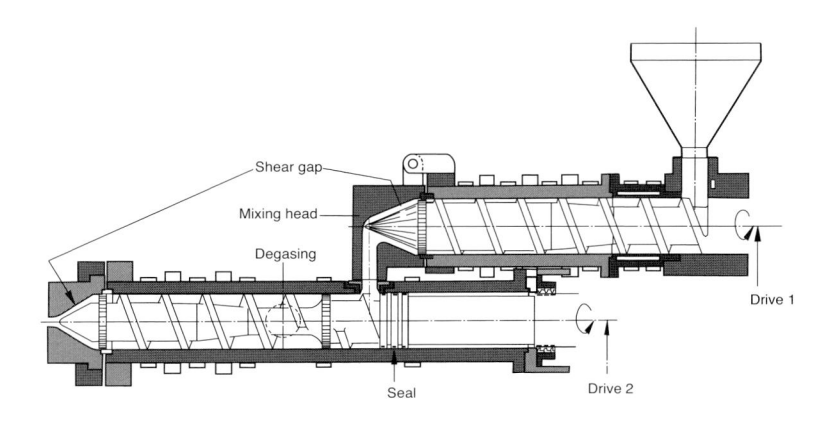

Figure 4.29: Schematic diagram of a cascade system of single screw extruders.

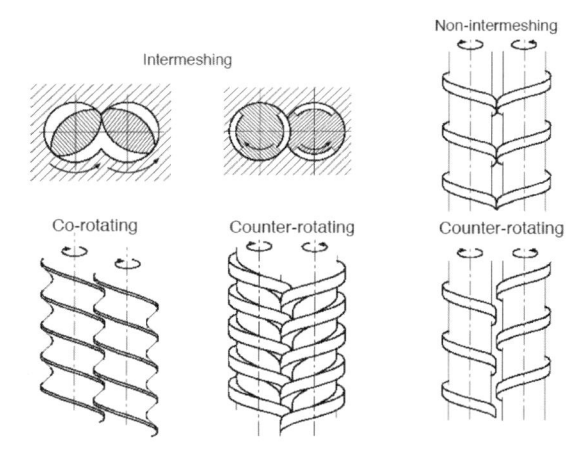

Figure 4.30: Schematic of different twin screw extruders.

4.3.1 The Plasticating Extruder

The plasticating single screw extruder is the most common equipment in the polymer industry. It can be part of a molding unit and found in numerous other extrusion processes, including blow molding, film blowing, and wire coating. A schematic of a plasticating or three-zone, single screw extruder, with its most

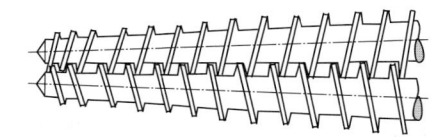

Figure 4.31: Schematic of conical twin screw extruder.

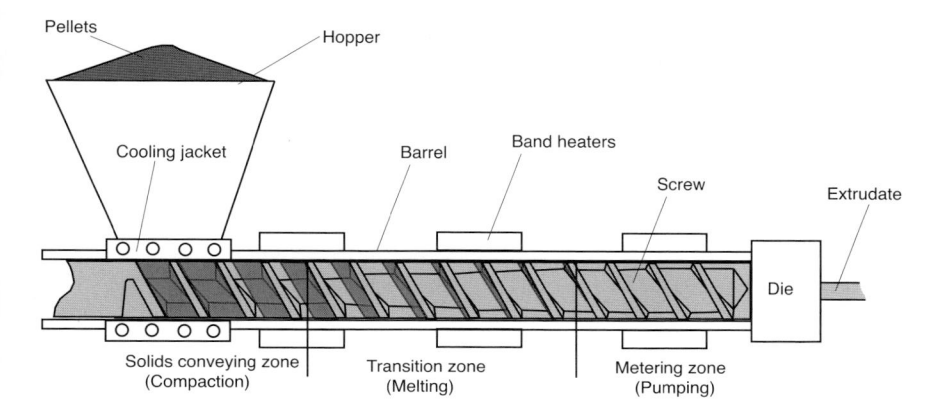

Figure 4.32: Schematic of a plasticating single screw extruder.

important elements is given in Fig. 4.32 Table 4.4 presents typical extruder dimensions and relationships common in single screw extruders, using the notation presented in Fig. 4.33

The plasticating extruder can be divided into three main zones:

- The solids conveying zone
- The melting or transition zone
- The metering or pumping zone

The tasks of a plasticating extruder are to:

- Transport the solid pellets or powder from the hopper to the screw channel
- Compact the pellets and move them down the channel
- Melt the pellets
- Mix the polymer into a homogeneous melt
- Pump the melt through the die

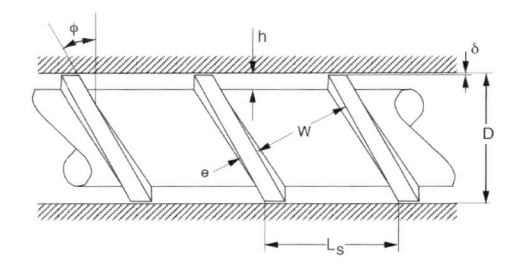

Figure 4.33: Schematic diagram of a screw section.

Table 4.4: Typical Extruder Dimensions and Relationships

L/D	Length - to - diameter ratio
	20 or less for feeding or melt extruders
	25 for blow molding, film blowing, and injection molding
	30 or higher for vented extruders or high output extruders
D	Standard diameter
US (inches)	0.75, 1.0, 1.5, 2, 2.5, 3.5, 4.5, 6, 8, 10, 12, 14, 16, 18, 20, and 24
Europe (mm)	20, 25, 30, 35, 40, 50, 60, 90, 120, 150, 200, 250, 300, 350, 400, 450, 500, and 600
ϕ	Helix angle
	17.65^o for a square pitch screw where $L_s = D$
	New trend: $0.8 < L_s/D < 1.2$
h	Channel depth in the metering section
	(0.05-0.07)D for D <30 mm
	(0.02-0.05)D for D >30 mm
β	Compression ratio
	$h_{feed} = \beta h$
	2 to 4
δ	Clearance between the screw flight and the barrel
	0.1 mm for D <30 mm
	0.15 mm for D >30 mm
N	Screw speed
	1-2 rev/s (60-120 rpm) for large extruders
	1-5 rev/s (60-300 rpm) for small extruders
V_b	Barrel velocity (relative to screw speed) $= \pi D N$
	0.5 m/s for most polymers
	0.2 m/s for unplasticized PVC
	1.0 m/s for LDPE

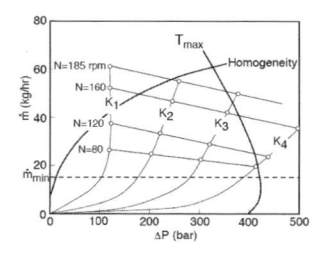

Figure 4.34: Screw and die characteristic curves for a 45mm diameter extruder with an LDPE.

The pumping capability and characteristic of an extruder can be represented with sets of die and screw characteristic curves. Figure 4.34 presents such curves for a conventional (smooth barrel) single screw extruder. The die characteristic curves are labelled K_1, K_2, K_3, and K_4 in ascending order of die restriction. Here, K_1 represents a low resistance die, such as for a thick plate, and K_4 represents a restrictive die, such as is used for film. The different screw characteristic curves represent different screw rotational speeds. In a screw characteristic curve, the point of maximum throughput and no pressure build-up is called the point of open discharge. This occurs when there is no die. The point of maximum pressure build-up and no throughput is called the point of closed discharge. This occurs when the extruder is plugged. Shown in Fig. 4.34 are also lines that represent critical aspects encountered during extrusion. The curve labeled T_{max} represents the conditions at which excessive temperatures are reached as a result of viscous heating. The feasibility line represents the throughput required to have an economically feasible system. The processing conditions to the right of the homogeneity line render a thermally and physically heterogeneous polymer melt.

The Solids Conveying Zone The task of the solids conveying zone is to move the polymer pellets or powders from the hopper to the screw channel. Once the material is in the screw channel, it is compacted and transported down the channel. The process to compact the pellets and to move them can only be accomplished if the friction at the barrel surface exceeds the friction at the screw surface. This can be visualized if one assumes the material inside the screw channel to be a nut sitting on a screw. As we rotate the screw without applying outside friction, the nut (polymer pellets) rotates with the screw without moving in the axial direction. As we apply outside forces (barrel friction), the rotational

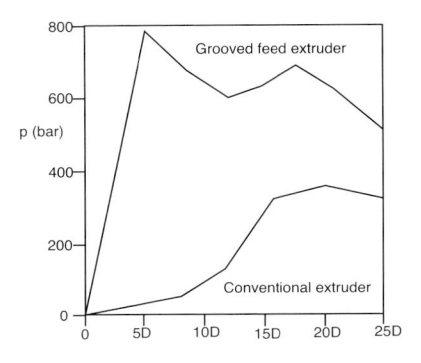

Figure 4.35: Typical conventional and grooved feed extruder pressure distributions in a 45mm diameter extruder.

speed of the nut is less than the speed of the screw, causing it to slide in the axial direction. Virtually, the solid polymer is then "unscrewed" from the screw. To maintain a high coefficient of friction between the barrel and the polymer, the feed section of the barrel must be cooled, usually with cold water cooling lines. The frictional forces also result in a pressure rise in the feed section. This pressure compresses the solids bed, which continues to travel down the channel as it melts in the transition zone. Figure 4.35 compares the pressure build-up in a conventional, smooth barrel extruder with that in a grooved feed extruder. In these extruders, most of the pressure required for pumping and mixing is generated in the metering section. The simplest mechanism for ensuring high friction between the polymer and the barrel surface is grooving its surface in the axial direction. Extruders with a grooved feed section where developed by Menges and Predöhl in 1969, and are called grooved feed extruders. To avoid excessive pressures that can lead to barrel or screw failure, the length of the grooved barrel section must not exceed $3.5D$. A schematic diagram of the grooved section in a single screw extruder is presented in Fig. 4.36. The key factors that propelled the development and refinement of the grooved feed extruder were the processing problems, excessive melt temperature, and reduced productivity, posed by high viscosity and low coefficients of friction typical of high molecular weight polyethylenes and polypropylenes. In a grooved feed extruder, the conveying and pressure build-up tasks are assigned to the feed section. The high pressures in the feed section (Fig. 4.35) lead to the main advantages over conventional systems. With grooved feed systems, there is a higher productivity and a higher melt flow stability and pressure invariance. This is demonstrated with the screw characteristic curves in Fig. 4.37, which presents

4.3 Extrusion

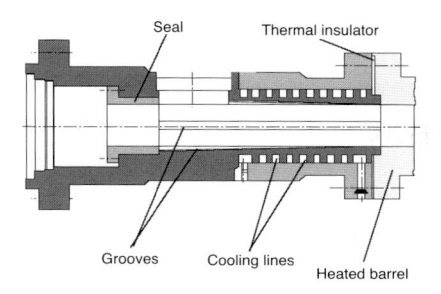

Figure 4.36: Schematic diagram of the grooved feed section of a single screw extruder.

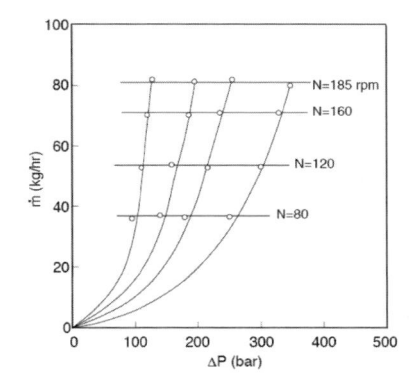

Figure 4.37: Screw and die characteristic curves for a grooved feed 45mm diameter extruder with an LDPE.

screw characteristic curves for a 45mm diameter grooved feed extruder with comparable mixing sections and die openings as shown in Fig. 4.34

The Melting or Transition Zone The melting or transition zone is the portion of the extruder were the material melts. The length of this zone is a function of the material properties, screw geometry, and processing conditions. During melting, the size of the solid bed shrinks as a melt pool forms at its side, as depicted in Fig. 4.38 which shows the polymer unwrapped from the screw channel.

Figure 4.38 also presents a cross section of the screw channel in the melting zone. The solid bed is pushed against the leading flight of the screw as freshly molten polymer is wiped from the melt film into the melt pool by the relative

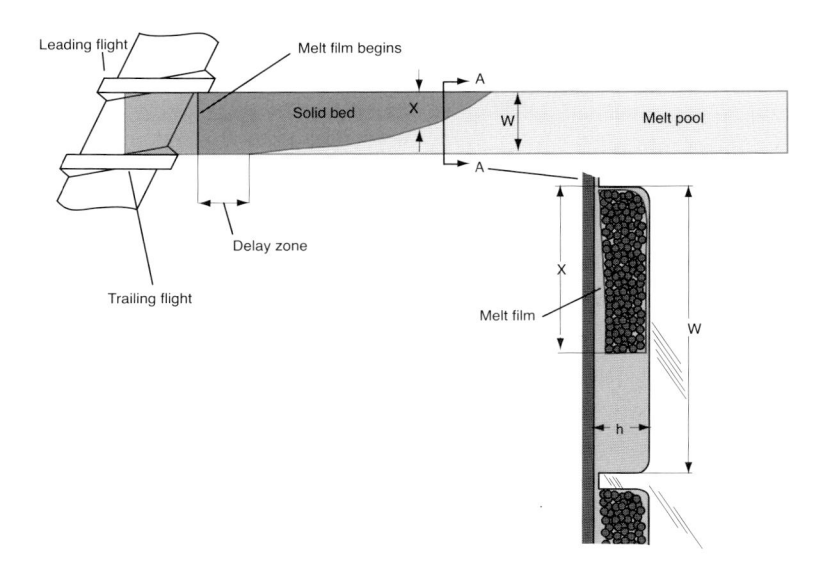

Figure 4.38: Solids bed in an unwrapped screw channel with a screw channel cross-section.

motion between the solids bed and the barrel surface. Knowing where the melt starts and ends is important when designing a screw for a specific application. The solid bed profile that develops during plastication remains one of the most important aspects of screw design.

From experiment to experiment there are always large variations in the experimental solids bed profiles. The variations in this section of the extruder are caused by slight variations in processing conditions and by the uncontrolled solids bed break-up towards the end of melting. This effect can be eliminated by introducing a screw with a barrier flight that separates the solids bed from the melt pool. The Maillefer screw and barrier screw in Fig. 4.39 are commonly used for high quality and reproducibility. The Maillefer screw maintains a constant solids bed width, using most effectively the melting with melt-removal mechanism, while the barrier screw uses a constant channel depth with a gradually decreasing solids bed width.

The Metering Zone The metering zone is the most important section in melt extruders and conventional single screw extruders that rely on it to generate pressures sufficient for pumping.

4.3 Extrusion

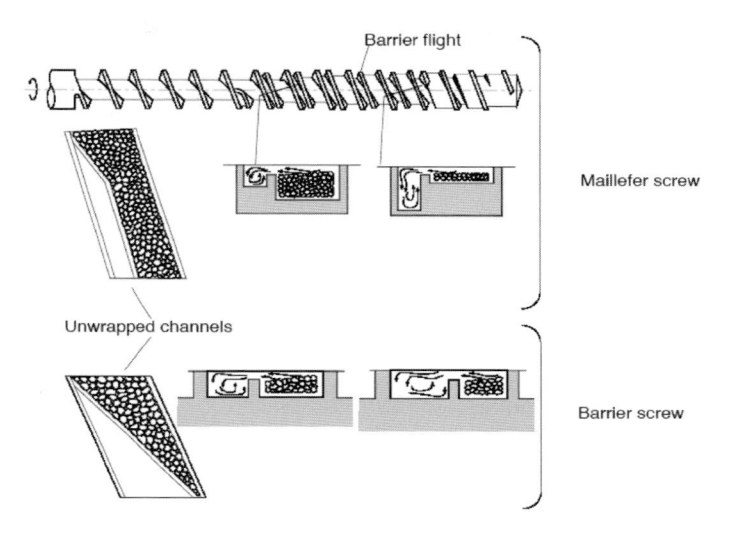

Figure 4.39: Schematic diagram of screws with different barrier flights.

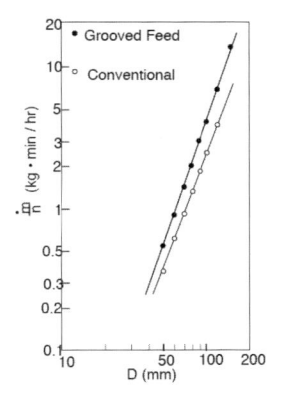

Figure 4.40: Throughput for conventional and grooved feed extruders.

In both the grooved barrel and the conventional extruder, the diameter of the screw determines the metering or pumping capacity of the extruder. Figure 4.40 presents typical normalized mass throughput as a function of screw diameter for both systems.

4.3.2 Troubleshooting Extrusion

This troubleshooting guide covers common problems encountered with extrusion processes. Table 4.5[3] presents several defects with possible causes and remedies.

Table 4.5: Troubleshooting Guide for Extrusion

Problem	Possible causes and recommendations
High drive motor amperage	**Low resin temperature.** Possible bad heater or heater temperatures too low. Raise the temperatures and check the electrical output of the heaters. **Resin.** Resin molecular weight may be too high (MFI too low). The resin may be cross-linking or degrading. **Plugged screens.** Change the screens. **Motor.** Maintenance needed. Motor speed too high. **Contamination.** A large contaminant could have dropped into the extruder. You may have to pull the screw to check.
Interrupted resin output	**Hopper.** Clumping in the hopper due to resin density too low or some contaminant. Heat conduction from the feed zone to the hopper causing the resin to stick together, especially with low molecular weight resin. Lower the feed-zone temperature. **Cooling jacket.** Chilled water may be turned off, leading to melt film formation in the solids section. **Bridging.** Temperatures in the feed zone are too high; lower the heater. Density of the resin too low; use a cram feeder or extrude the material into pellets in a separate operation. **Clogging.** Check the screen pack. Look for degraded or crosslinked resin.
No output	**Feed hopper slide valve closed.** Open slide valve. **Bridging in the feed hopper.** Use soft rod to dislodge the bridging. Put vibrating pads on feed hopper. Use stirrer in feed hopper. **Screw turning the wrong way.** Switch the terminals on the screw drive motor. **Screw broken.** Remove screw and install spare screw. Repair broken screw or have spare made. **Material blockage at feed opening.** Dislodge blockage. Adjust wall temperatures so they are low enough to avoid sticking. **Material stuck to the screw.** Clean screw. Review operating procedures. Use screw cooling. Use low friction screw coating. **Insufficient barrel friction.** Adjust barrel temperature. Use an extruder with a grooved feed section. **Blockage along the screw.** Pull screw and clean it. Avoid barrier sections along the screw.

Continued on next page

---

[3] Parts of this table were presented in C. Rauwendaal, *Polymer Extrusion*, Hanser Publishers, Munich (2001).

4.3 Extrusion

Problem	Possible causes and recommendations
	Excessive screen pack restriction. Replace with new screen pack. Use fewer screens in screen pack. Use lower mesh screens in screen pack. Increase die temperature. **Excessive die restriction.** Increase die temperature.
Uneven flow (surging)	**Temperatures.** Raise the temperatures in the heating zones, especially if the extrudate has a high viscosity. If an internal mixer is used, it may have to be removed. If partial bridging is suspected, lower the feed-zone temperature, which may require that the metering and, perhaps, the compression-zone temperatures be raised. **Melting problems.** Increase barrel temperatures at low rpm. Reduce barrel temperatures at high rpm. Change screw design. **Melt conveying problems.** Reduce barrel temperature in metering section. Increase screw temperature in metering section. Clean screen pack. Use less restrictive screen pack. Use less restrictive die. Change screw design in metering section.
	Barrel or die temperature variations. Check temperature measurement. Check temperature control system. Tune temperature control system. Insulate extruder from outside influences. **Cooling jacket.** Chilled water may be turned off, eventually leading to interrupted resin output. **Contamination/plugging.** Check the pressure across the screen pack and if high, change the screen pack; if the screens are not blinded, increase the screen openings. Check for plugging in the hopper. If no other solution is found, pull the screw and check for a large contaminant. **Equipment.** Extruder too long for the barrel or improperly placed in the thrust bearing, causing the screw to drag on the bottom of the barrel. This requires pulling the screw. Motor not functioning properly due to a need for maintenance or because it is undersized. The puller could be slipping - monitor its speed and the speed of the material going through it; if irregular, increase the pulling pressure on the part. **Feeding problems in extruder.** Adjust feed temperatures. Low friction screw coating. Change screw geometry. Used grooved barrel in feed section. **Resin.** Density of resin could be too low, thus requiring a cram feeder, starve feeding, or pelletization. **Low bulk density feed stock.** Densify feed stock. Use special extruder for low bulk density materials. **Feed stock variation.** Work with resin supplier. Set specifications on pellet size, bulk density, etc. Improve blending operation. Avoid segregation of ingredients. **Unstable flow in feed hopper.** Improve feed hopper design. Low friction screw coating. Crammer feed.
Unmelted particles in the extrudate	**Screen pack.** Hole in the screen pack. **Temperatures.** Raise the temperature in the compression and metering zones. Damaged heater. Test reheaters. **Contamination.** Crosslinked or degraded polymer, especially in the die. Lower the die temperature if the material seems off-color or if the particles won't melt if put on a hot plate. If the particles will melt, raise the die temperature to melt them. Streamline the die better.

Continued on next page

Problem	Possible causes and recommendations
Discolored extrudate	**Degraded polymer.** Heater temperatures or screw speed too high. **Poor mixing.** Pigments or dyes not well mixed; add a mixing head. Use concentrates. **Die design.** Die not streamlined. **Output control.** Screw speed too high, especially if resin is subjected to degradation from adiabatic heating. Extruder too large for the output.
Die pressure drop too high	**Plugging.** Plugged screen pack. Screen pack has openings that are too small or has too many screens in the pack. **Unfinished melting.** Temperatures too low.
Extrudate viscosity too low	**Resin.** Melt temperatures too high. Resin melt flow index too high. Resin molecular weight distribution too narrow. **Temperatures.** Die temperature too high.
Cross-section too small	**Equipment not synchronized.** Puller speed too high or extruder speed too low. Gap too large, so move the cooling tank closer to the die face. Melt temperature too low. **Equipment design.** Die opening too small or land too long. Sizing plates in cooling tank too small requires changing the plates. **Resin.** Molecular weight too high. *For larger than desired cross-sections do the opposite*
Rough surface/die lines/melt fracture	**Die.** Not streamlined properly. Land too short. Die temperature too low. **Resin.** Melt temperature too low. Molecular weight too high or molecular weight distribution too narrow.
Sharkskin	**Die.** Die temperature too low. Land too long. Resin gap in the die might be too small. **Resin.** Modulus of resin too high or too narrow molecular weight distribution. **Operation.** Extruder speed too high. Back pressure could be too high; change screen packs. Raise melt temperature by increasing heater temperatures.
Fish eyes	**Contamination.** The contaminant could be from degradation or from material introduced at the hopper. Check the screen pack for discolored material, which would indicate a hopper origin. Water could be the contaminant; this would require drying the resin. **Degradation.** Temperatures are probably too high, especially in the die, leading to cross-linking or gels.
Bubbles in the part	**Water.** Dry the resin. **Degradation.** Check for an odor and if present, lower the temperatures of the melt.
Warped part	**Die.** Spider mandrel not concentric in the die opening and therefore needs adjusting. Land is not even on all sides. Entry angle of the die is not uniform on all sides. **Cooling tank.** Part is being twisted or turned as it enters the cooling tank. Align the tanks to be parallel with the extruder outlet. **Part design.** Look for nonsymmetries and thickness differentials in the part, which may result in variable degree of crystallinity.

Continued on next page

Problem	Possible causes and recommendations
Sheet not uniform in thickness	**Die.** The coat-hanger die is not properly adjusted. Check by injecting a colored material just upstream of the die and see if it comes out uniformly. Die temperatures not uniform. Poorly designed die (unbalanced). **Downstream equipment.** The rollers could be nonparallel. Puller could be unaligned with extruder.
Blown film not uniform	**Die.** Mandrel not concentric. Air inlet off center. Flow channel in the die could be partially plugged. Temperatures in the die may not be uniform. **Downstream equipment.** Puller not aligned with the die. Sizing rollers/cage may not be aligned.
Excessively high melt temperatures	**Poor melt temperature measurement.** Do not use P/T probe. Do not use flush-mounted probe. Do use an immersion probe. **For amorphous plastic, is $T_{melt} \gg T_g + 100°C$** No, temperature probably okay. Yes, continue troubleshooting. **For semi-crystalline plastic, is $T_{melt} \gg T_{mp} + 50°C$** No, temperature probably okay. Yes, continue troubleshooting. **High restriction at screen pack.** Clean screen pack. Use less restrictive screen pack. **High die restriction.** Open up die restriction. Increase die temperatures. Change die design. **Blockage along the screw.** Pull the screw and clean. **Incorrect screw design.** Design screw for the correct plastic viscosity. **Cooling system not working correctly.** Fix cooling system. Use more efficient cooling system. Consider using screw cooling. **Barrel or die temperatures too high.** Check temperature measurement. Use correct temperature settings. Check temperature control system.
Excessive wear	**Wear occurred over a long time period (over one year).** Replace the worn parts. **Wear occurred over a short time period (less than one year).** Reduce the wear rate. **Corrosive wear.** Eliminate corrosive substance. Use corrosion resistant materials. **Abrasive wear by fillers.** Add abrasive fillers downstream. Use wear-resistant screw and barrel material. **Metal-to-metal wear.** Do not use high compression ratio. Do not use short compression section. Use compatible screw and barrel materials. Use double flighted screw, not single flighted. Make sure barrel temperatures are correct.
Gel formation	**Gels created in polymerization process.** Check gel level in incoming raw material. Set specifications on acceptable gel level. Consider other resin suppliers. **Gels created in extrusion process.** Reduce residence times in extruder. Minimize hangup of material in screw and die. Minimize stock temperatures in extruder. Fix scratches in screw and die. Use low friction coating on screw and die. Check startup and shutdown procedures. Use filter with good gel capture capability. **Contamination.** Clean storage bins and conveying system. Thoroughly clean extruder before startup. Avoid contamination at every point.

Continued on next page

Problem	Possible causes and recommendations
Poor extrudate appearance	**Thickness variation.** Adjust die bolts. Make sure die temperatures are stable. Improve mixing by using a mixing screw. **Melt fracture.** Improve streamlining of the die. Reduce shear stress at die land by: increasing die temperature; increasing die opening; increasing melt temperature; reducing throughput. Use processing aid. Use special material at die wall. Use "super extrusion". **Draw resonance.** Reduce draw ratio. Cool extrudate more rapidly. **Bubbles or voids in product.** Remove volatiles by predrying or venting. Reduce stock temperatures in the extrusion process. Reduce shrink voids by cooling more slowly. Reduce air entrapment by: using vented extruder; increasing particle size; using vacuum feed hopper system. Improve venting efficiency by: higher vacuum level; better screw design; better vent port design.
	Weld lines in extruded product. Change die design. Run at higher temperatures. Run at lower throughput. **Die lines.** Make sure the die is properly cleaned. Fix scratches in die. Polish internal die surfaces. Use low friction coating for die surfaces. **Die drool.** Remove incompatible component from compound. Use compatibilizer. Change compounding procedure. Adjust die temperatures. Use special die land material (e.g., ceramic). Use low friction coating. Use longer die land.

4.3.3 Extrusion Dies

The extrusion die shapes the polymer melt into its final profile. The extrusion die is located at the end of the extruder and it is used to extrude

- Flat films and sheets
- Pipes and tubular films for bags
- Filaments and strands
- Hollow profiles for window frames
- Open profiles

As shown in Fig. 4.41, depending on the functional needs of the product, several rules of thumb can be followed when designing an extruded plastic profile. These are:

- Avoid thick sections. Thick sections add to the material cost and increase sink marks caused by shrinkage.
- Minimize the number of hollow sections. Hollow sections add to die cost and make the die more difficult to clean.
- Generate profiles with constant wall thickness. Constant wall thickness in a profile makes it easier to control the thickness of the final profile and results in a more even crystallinity distribution in semi-crystalline profiles.

4.3 Extrusion

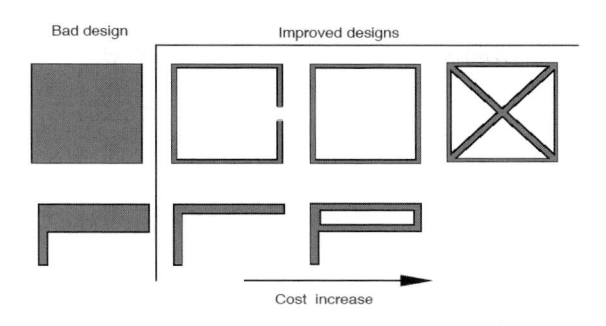

Figure 4.41: Extrusion profile design.

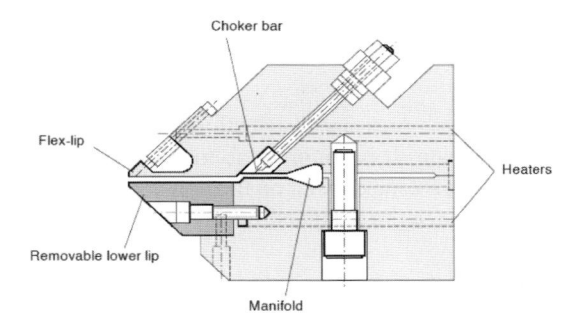

Figure 4.42: Cross-section of a coat-hanger die.

Sheeting Dies One of the most widely used extrusion dies is the coat-hanger sheeting die. A sheeting die, as depicted in Fig. 4.42, is formed by the following elements:

- Manifold. evenly distributes the melt to the approach or land region
- Approach or land. carries the melt from the manifold to the die lips
- Die lips. perform the final shaping of the melt
- Flex lips. for fine tuning when generating a uniform profile

To generate a uniform extrudate geometry at the die lips, the geometry of the manifold must be specified appropriately. Figure 4.43 presents the schematic

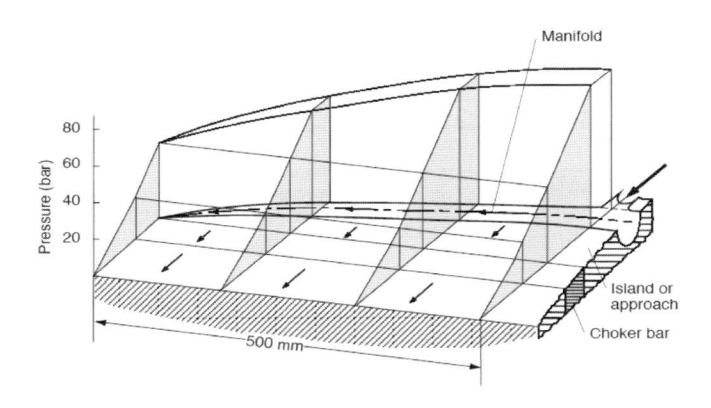

Figure 4.43: Pressure distribution in a coat-hanger die.

of a coat-hanger die with a pressure distribution that corresponds to a die that renders a uniform extrudate. It is important to mention that the flow though the manifold and the approach zone depend on the non-Newtonian properties of the polymer extruded. Hence, a die designed for one material does not necessarily work for another.

Tubular dies In a tubular die, the polymer melt exits through an annulus. These dies are used to extrude plastic pipes and tubular film. The film blowing operation is discussed in more detail later in this chapter. The simplest tubing die is the spider die, depicted in Fig. 4.44 Here, a symmetric mandrel is attached to the body of the die by several legs. The polymer must flow around the spider legs causing weld lines along the pipe or film. These weld lines, visible streaks along the extruded tube, are weaker regions.

To overcome weld line problems, the cross-head tubing die is often used. Here, the die design is similar to that of the coat-hanger die, but wrapped around a cylinder. This die is depicted in Fig. 4.45 Since the polymer melt must flow around the mandrel, the extruded tube exhibits one weld line. In addition, although the eccentricity of a mandrel can be controlled using adjustment screws, there is no flexibility to perform fine tuning such as in the coat-hanger die. This can result in tubes with uneven thickness distributions. The spiral die, commonly used to extrude tubular blown films, eliminates weld line effects and produces a thermally and geometrically homogeneous extrudate. The polymer melt in a spiral die flows through several feed ports into independent spiral channels wrapped around the

4.3 Extrusion

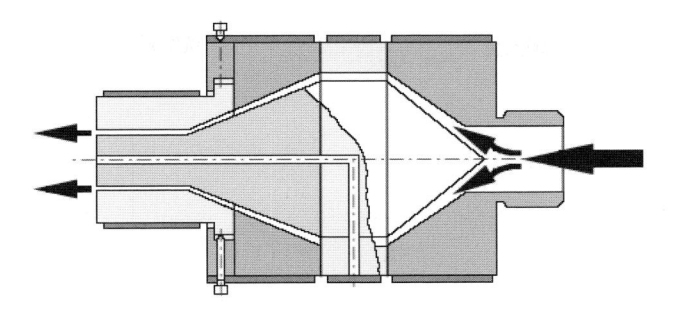

Figure 4.44: Schematic diagram of a spider leg tubing die.

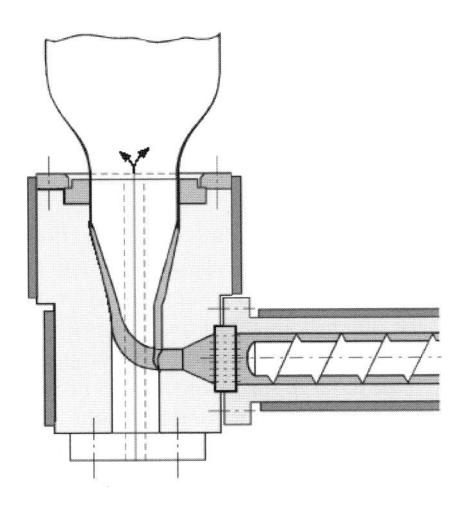

Figure 4.45: Schematic diagram of a cross-head tubing die used in film extrusion.

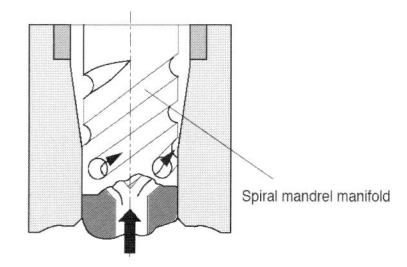

Figure 4.46: Schematic diagram of a spiral die.

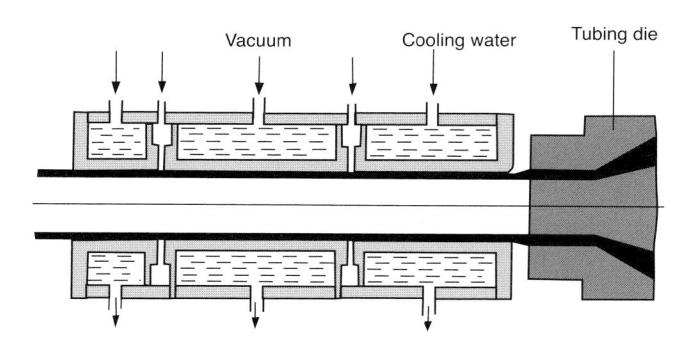

Figure 4.47: Schematic diagram of a vacuum sleeve pipe calibration system.

circumference of the mandrel. This type of die is schematically depicted in Fig. 4.46

One important aspect of pipe extrusion is the calibration or sizing of the pipe after extrusion. Often, a plug that controls air flow is tied to the mandrel with a chain, while the profile is cooled. In a vacuum calibration sleeve system, as depicted in Fig. 4.47, the outer diameter of the pipe is controlled, and therefore, an inner plug is not required. Although solid extrusion profiles are not desirable, they are often needed. Due to large shrinkage in such profiles, they often require calibration or sizing sleeves to reduce or control the shrinkage, as well as speed-up the cooling process. These sizing sleeves, such as the one depicted in Fig. 4.48 usually allow a back-pressure build-up to better control the dimensional stability of the extrusion profile.

Sizing sleeves are also required when extruding foamed profiles. These sizing sleeves also assure rapid cooling of the surface, creating a smooth rigid skin and a foamed core structure. Figures 4.49 and 4.50 present two different types of foamed profile extrusion with a sizing sleeve. Figure 4.49 shows the free

4.3 Extrusion

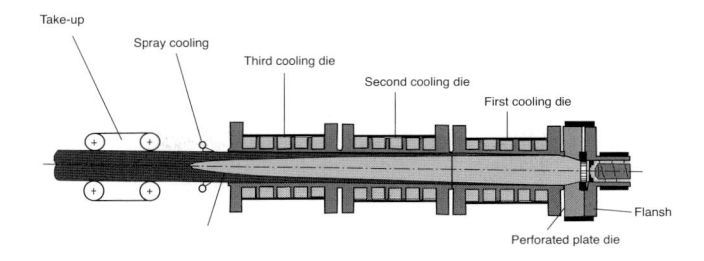

Figure 4.48: Schematic diagram of a solid extrusion profile calibration system.

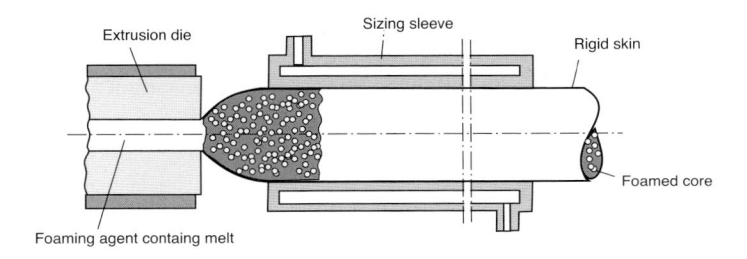

Figure 4.49: Schematic diagram of a calibration sleeve for a foamed extrusion profile.

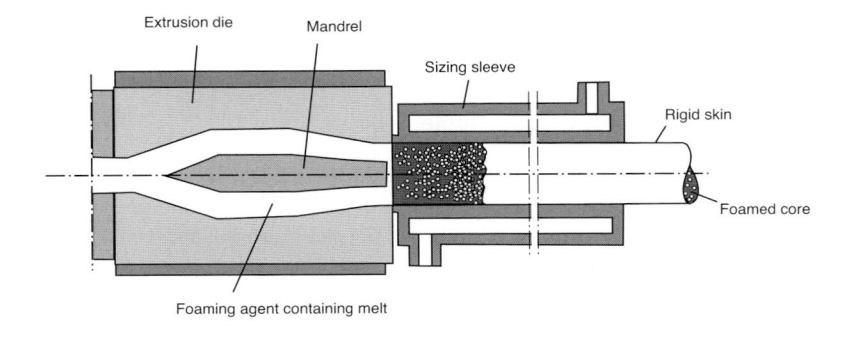

Figure 4.50: Schematic diagram of a calibration sleeve for a Celuka-type foamed extrusion profile.

extrusion, where a small profile is extruded and is let grow into the sizing sleeve. Figure 4.50 presents the so-called *Celuka process*, where a mandrel generates a hollow space inside the die, into which the foamed structure will expand. Here, the calibration sleeve is butted against the extrusion die.

4.4 INJECTION MOLDING

Injection molding is the most important process used to manufacture plastic products. Today, more than one-third of all thermoplastic materials are injection molded and more than half of all polymer processing equipment is for injection molding. The injection molding process is ideally suited to manufacture mass-produced parts of complex shapes requiring precise dimensions. The process goes back to 1872, when the Hyatt brothers patented their stuffing machine to inject cellulose into molds. However, today's injection molding machines are mainly related to the reciprocating screw injection molding machine patented in 1956. A modern injection molding machine with its most important elements is shown in Fig. 4.51 The components of the injection molding machine are the plasticating unit, clamping unit, and the mold. Today, injection molding machines are classified by the following international convention[4]

$$\text{MANUFACTURER } T/P$$

where T is the clamping force in metric tons and P is defined as

$$P = \frac{V_{max} P_{max}}{1000} \qquad (4.3)$$

where V_{max} is the maximum shot size in cm^3 and P_{max} is the maximum injection pressure in bar. The clamping forced T can be as low as 1metric ton for small machines, and as high as 11,000 tons.

4.4.1 The Injection Molding Cycle

The sequence of events during the injection molding of a plastic part, as shown in Fig. 4.52, is called the injection molding cycle. The cycle begins when the mold closes, followed by the injection of the polymer into the mold cavity. Once the cavity is filled, a holding pressure is maintained to compensate for material shrinkage. In the next step, the screw turns, feeding the next shot to the front of the screw. This causes the screw to retract as the next shot is prepared. Once the part is sufficiently cool, the mold opens and the part is ejected. Figure 4.53 presents the sequence of events during the injection molding cycle. The figure shows that the cycle time is dominated by the cooling of the part inside the mold cavity. The total cycle time can be calculated using

$$t_{\text{cycle}} = t_{\text{closing}} + t_{\text{cooling}} + t_{\text{ejection}} \qquad (4.4)$$

---

[4]The old US convention uses MANUFACTURER T/V where T is the clamping force in British tons and V the shot size in ounces of polystyrene.

4.4 Injection Molding 315

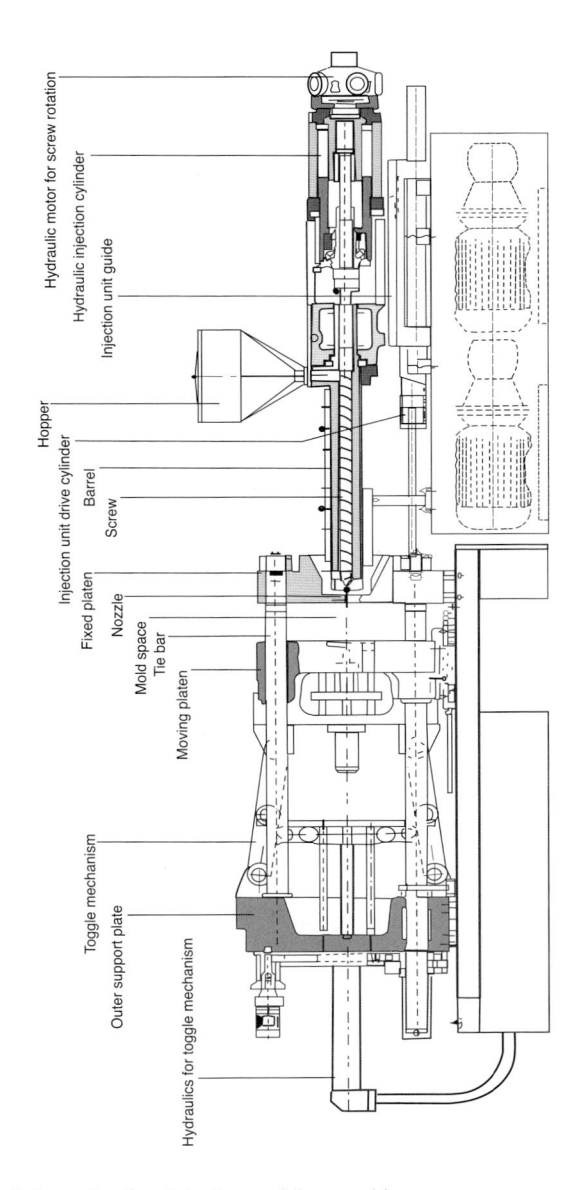

Figure 4.51: Schematic of an injection molding machine.

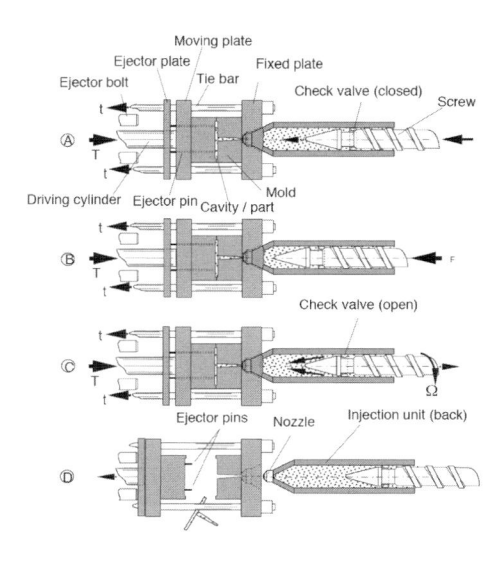

Figure 4.52: Sequence of events during an injection molding cycle.

where the closing and ejection times, t_{closing} and t_{ejection}, can last from a fraction of second to a few seconds, depending on the size of the mold and machine. The cooling times, which dominate the process, depend on the maximum thickness of the part. Using the average part temperature history and the cavity pressure history, the process can be followed and assessed using the pvT diagram as depicted in Fig. 4.54. To follow the process on the pvT diagram, we must transfer both the temperature and the pressure at matching times. The diagram reveals four basic processes: an isothermal injection (0-1) with pressure rising to the holding pressure (1-2), an isobaric cooling process during the holding cycle (2-3), an isochoric cooling after the gate freezes with a pressure drop to atmospheric (3-4), and then isobaric cooling to room temperature (4-5). The point on the pvT diagram at which the final isobaric cooling begins (4) controls the total part shrinkage. This point is influenced by the two main processing conditions – the melt temperature and the holding pressure as depicted in Fig. 4.55 Here, the process in Fig. 4.54 is compared to one with a higher holding pressure. Of course, there is an infinite combination of conditions that render acceptable parts, bound by minimum and maximum temperatures and pressures. Figure 4.56 presents the molding diagram with all limiting conditions. The melt temperature is bound by a low temperature that results in a short shot or unfilled cavity and a high temperature that leads to material degradation. The hold pressure is bound by

4.4 Injection Molding

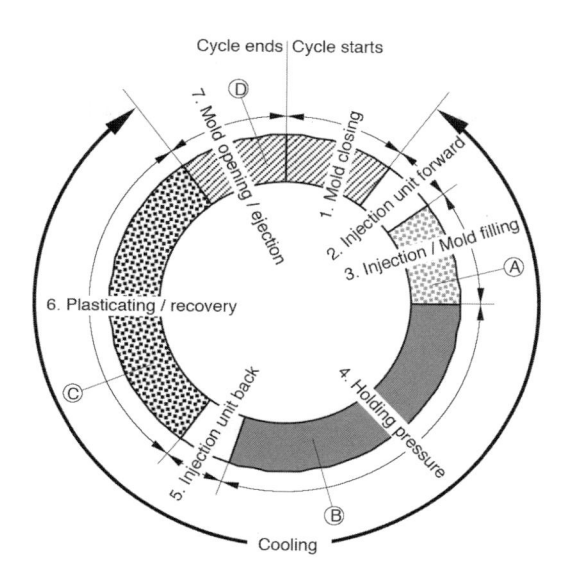

Figure 4.53: Injection molding cycle.

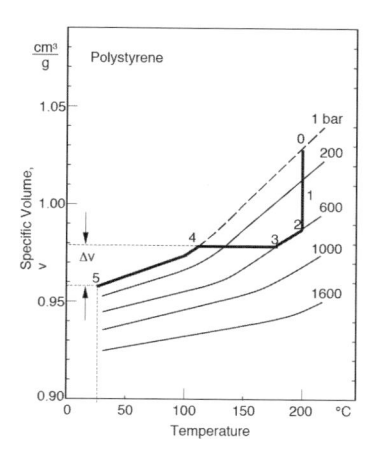

Figure 4.54: Trace of an injection molding cycle in a pvT diagram.

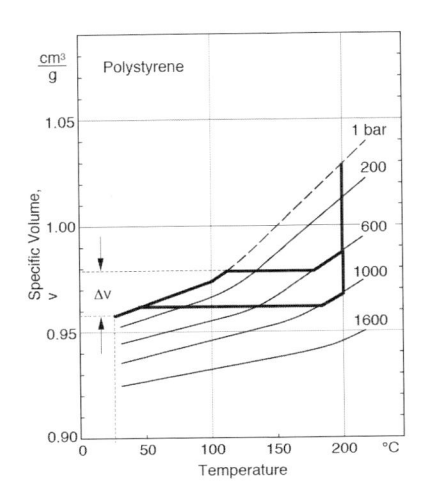

Figure 4.55: Trace of two different injection molding cycles in a pvT diagram.

a low pressure that leads to excessive shrinkage or low part weight, and a high pressure that results in flash. Flash results when the cavity pressure force exceeds the machine clamping force, leading to melt flow across the mold parting line. The holding pressure determines the corresponding clamping force required to size the injection molding machine. An experienced polymer processing engineer can usually determine which injection molding machine is appropriate for a specific application. For the untrained polymer processing engineer, finding this appropriate holding pressure and its corresponding mold clamping force can be difficult. With difficulty one can control and predict the component's shape and residual stresses at room temperature. For example, sink marks in the final product are caused by material shrinkage during cooling, and residual stresses can lead to environmental stress cracking under certain conditions. Warpage in the final product is often caused by processing conditions that lead to asymmetric residual stress distributions through the part thickness. The formation of residual stresses in injection molded parts is attributed to two major coupled factors: cooling and flow stresses. The first and most important is the residual stress formed as a result of rapid cooling which leads to large temperature variations.

4.4.2 The Injection Molding Machine

The Plasticating and Injection Unit A plasticating and injection unit is shown in Fig. 4.57 The major tasks of the plasticating unit are to melt the polymer,

4.4 Injection Molding

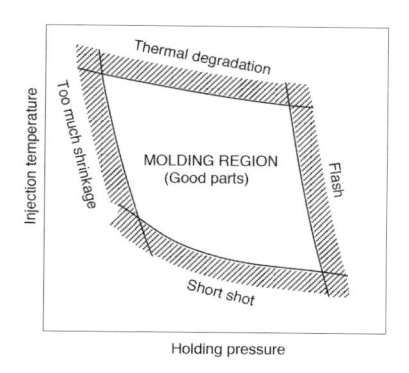

Figure 4.56: The molding diagram.

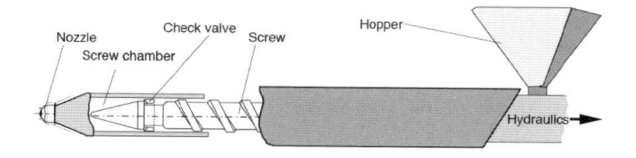

Figure 4.57: Schematic of a plasticating unit.

to accumulate the melt in the screw chamber, to inject the melt into the cavity, and to maintain the holding pressure during cooling. The main elements of the plasticating are:

- Hopper
- Screw
- Heater bands
- Check valve
- Nozzle

The hopper, heating bands, and the screw are similar to a plasticating single screw extruder, except that the screw in an injection molding machine can slide back and forth to allow for melt accumulation and injection. This characteristic gives it the name reciprocating screw. For quality purposes, the maximum stroke in a reciprocating screw should be set smaller than $3D$. Although the most common screw used in injection molding machines is the three-zone plasticating screw, as

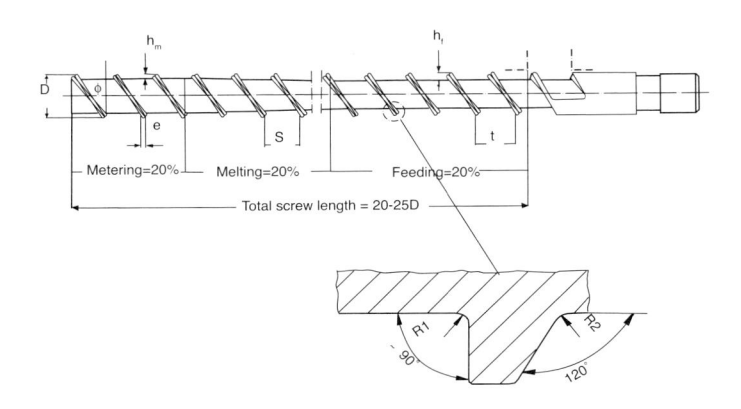

Figure 4.58: Schematic of a plasticating screw.

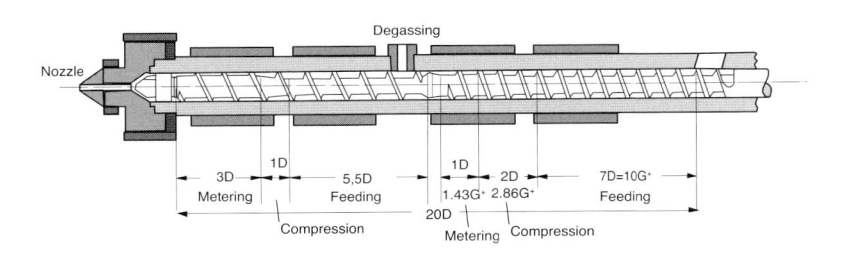

Figure 4.59: Schematic of a two stage degassing screw used in injection molding.

shown in Fig. 4.58, two - stage vented screws are often used to extract moisture and monomer gases just after the melting stage. A typical two - stage screw is schematically depicted in Fig. 4.59. The check valve, or non-return valve, is at the end of the screw and enables it to work as a plunger during injection and packing without allowing polymer melt to flow back into the screw channel. A check valve and its function during operation is depicted in Fig. 4.52, and in Fig. 4.57 A high quality check valve allows less then 5% of the melt back into the screw channel during injection and packing. The nozzle is at the end of the plasticating unit and fits tightly against the sprue bushing during injection. The

4.4 Injection Molding

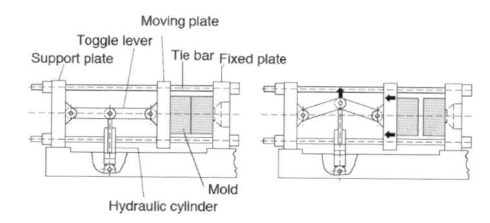

Figure 4.60: Clamping unit with a toggle mechanism.

nozzle type is either open or shut-off. The open nozzle is the simplest, rendering the lowest pressure consumption.

The Clamping Unit The job of a clamping unit in an injection molding machine is to open and close the mold, and to close the mold tightly to avoid flash during the filling and holding. Modern injection molding machines have two predominant clamping types: mechanical and hydraulic. Figure 4.60 presents a toggle mechanism in the open and closed mold positions. Although the toggle is essentially a mechanical device, it is actuated by a hydraulic cylinder. The advantage of using a toggle mechanism is that, as the mold approaches closure, the available closing force increases and the closing decelerates significantly. However, the toggle mechanism only transmits its maximum closing force when the system is fully extended. Figure 4.61 presents a schematic of a hydraulic clamping unit in the open and closed positions. The advantages of the hydraulic system is that a maximum clamping force is attained at any mold closing position and that the system can take different mold sizes without major system adjustments.

The Mold Cavity The central element in an injection molding machine is the mold. The mold distributes polymer melt into and throughout the cavities, shapes the part, cools the melt and, ejects the finished product. As depicted in Fig. 4.62, the mold is custom-made and consists of the following elements:

- Sprue and runner system
- Gate
- Mold cavity
- Cooling system (thermoplastics)
- Ejector system

During mold filling, the melt flows through the sprue and is distributed into the cavities by the runners, as seen in Fig. 4.63 The runner system in Fig. 4.63 (a)

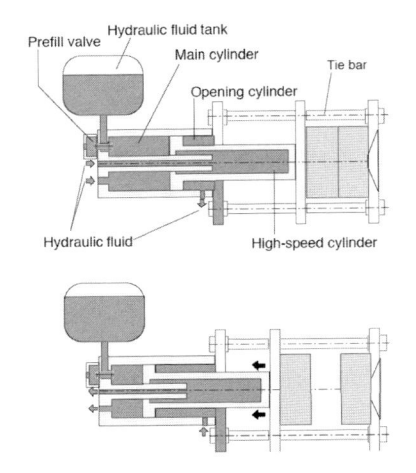

Figure 4.61: Hydraulic clamping unit.

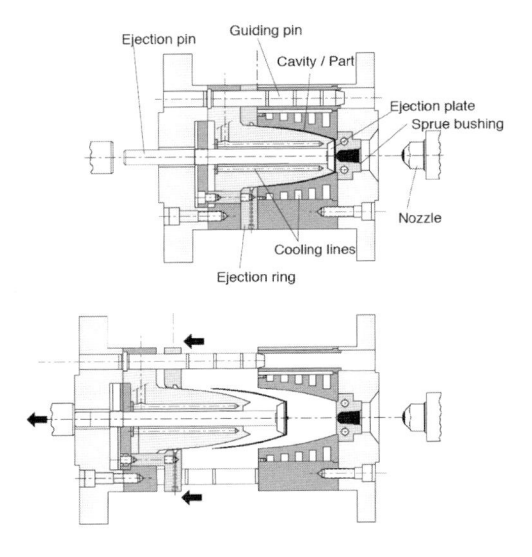

Figure 4.62: An injection mold.

4.4 Injection Molding

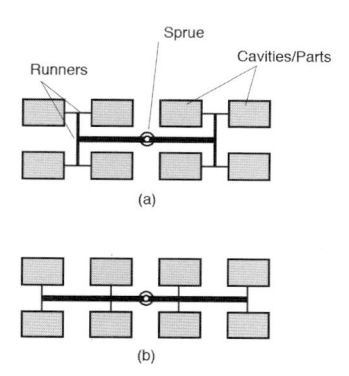

Figure 4.63: Schematic of different runner system arrangements.

is symmetric where all cavities fill at the same time, causing the polymer to fill all cavities in the same way. The disadvantage of this balanced runner system is that the flow paths are long, leading to high material and pressure consumption. On the other hand, the asymmetric runner system shown in Fig. 4.63 (b) leads to parts of different quality. Equal filling of the mold cavities can also be achieved by varying runner diameters. There are two types of runner systems – cold and hot. Cold runners are ejected with the part, and are trimmed after mold removal. The advantage of the cold runner is lower mold cost. The hot runner keeps the polymer at its melt temperature. The material stays in the runner system after ejection, and is injected into the cavity in the following cycle. There are two types of hot runner system: externally and internally heated. The externally heated runners have a heating element surrounding the runner that keeps the polymer isothermal. The internally heated runners have a heating element running along the center of the runner, maintaining a polymer melt that is warmer at its center and possibly solidified along the outer runner surface. Although a hot runner system considerably increases mold cost, its advantages include elimination of trim and lower pressures for injection. Various arrangements of hot runners are schematically depicted in Fig. 4.64. It should be noted that there are two parting lines in a hot runner cavity system, and that the second parting line is only opened during maintance of the molds.

When large items are injection molded, the sprue sometimes serves as the gate, as shown in Fig. 4.65. The sprue must be subsequently trimmed, often requiring further surface finishing. On the other hand, a pin-type or pinpoint gate (Fig. 4.65) is a small orifice that connects the sprue or the runners to the mold cavity. The part is easily broken off from such a gate, leaving only a small mark that usually does not require finishing. Other types of gates, also shown

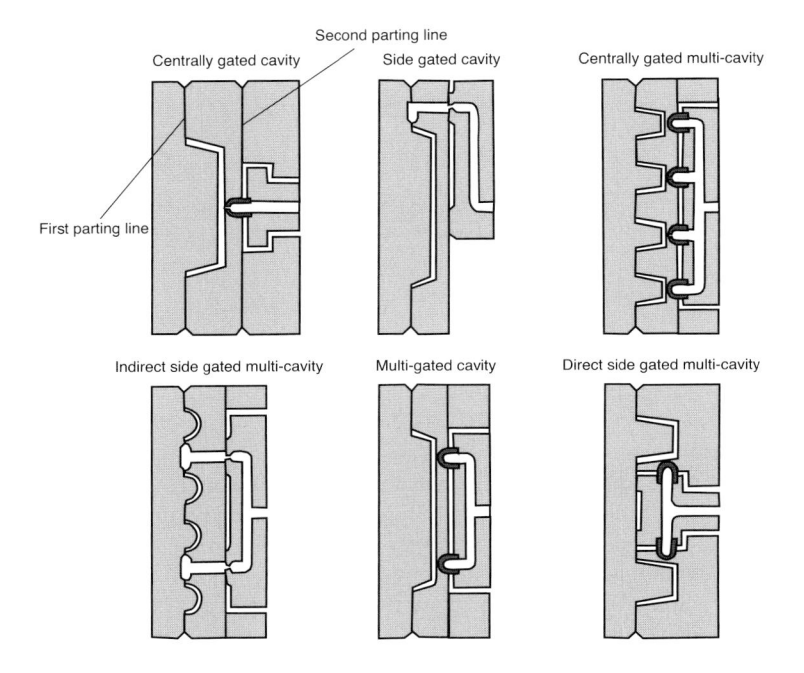

Figure 4.64: Various hot runner system arrangements.

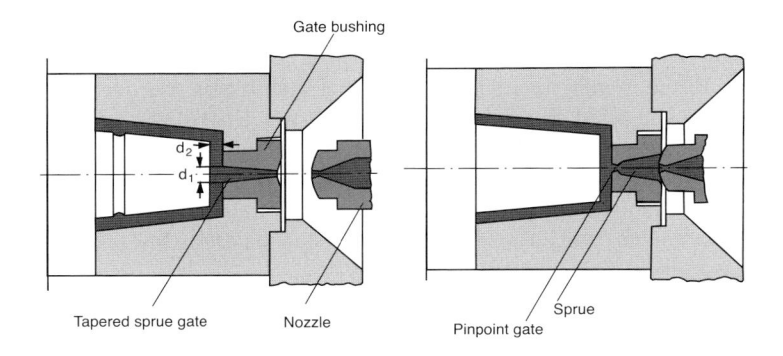

Figure 4.65: Tapered sprue and pinpoint gates used for centrally gated injection molded parts.

4.4 Injection Molding

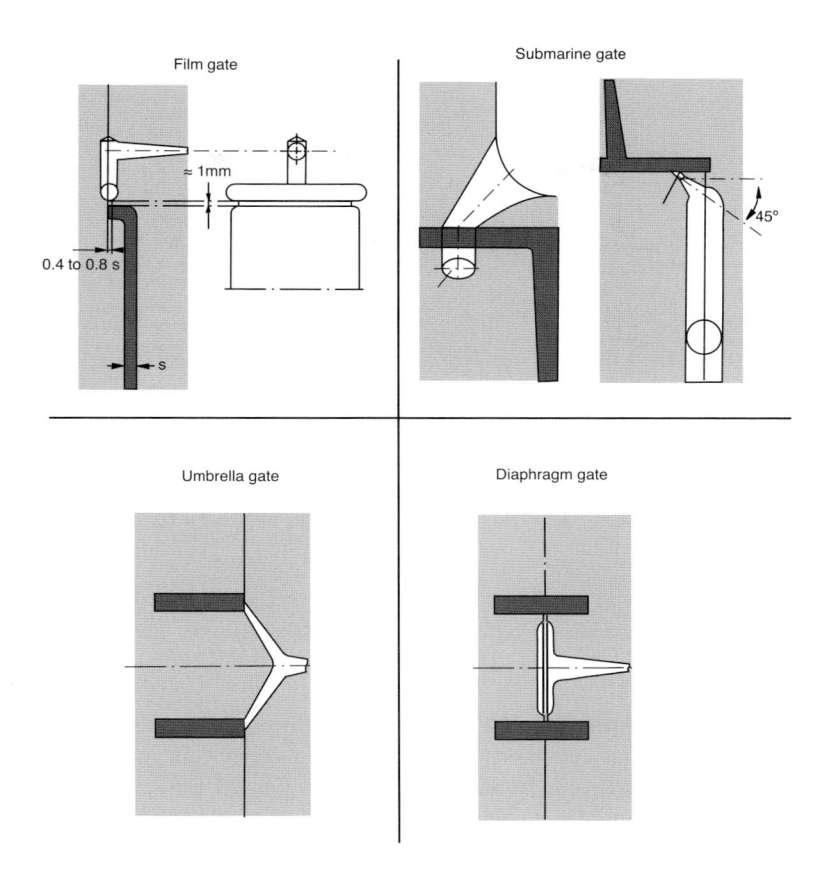

Figure 4.66: Schematic film, submarine, umbrella and diaphragm gating systems.

in Fig. 4.66, are film gates, used to eliminate orientation, submarine gates, and disk or diaphragm gates, as well as umbrella gates, for symmetric parts such as compact discs.

4.4.3 Special Injection Molding Processes

There are numerous variations of injection molding processes, many of which are still under development. Furthermore, due to the diversified nature of these special injection molding processes, there is no unique method to categorize them. Figure 4.67 attempts to schematically categorize special injection molding processes for thermoplastics. This section presents the most notable special

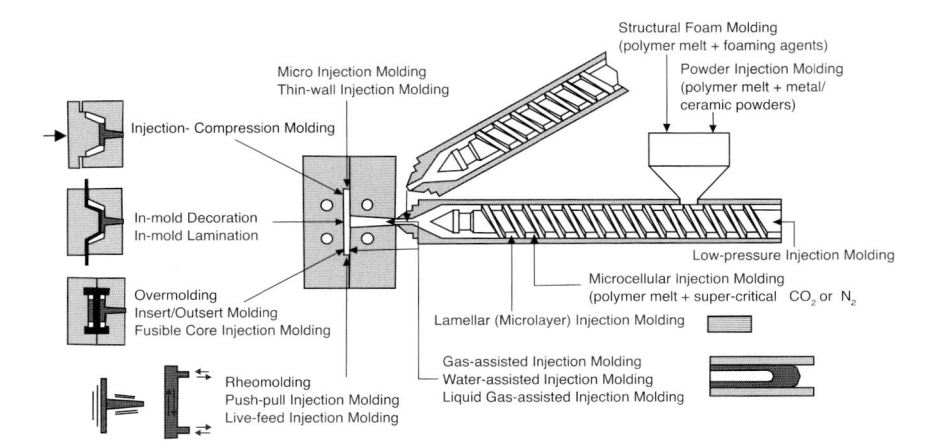

Figure 4.67: Schematic classification of special injection molding processes for thermoplastics.

injection molding processes, with their general descriptions, advantages, and applications.

Co-Injection or Sandwich Molding Co-injection molding, sometimes called sandwich molding, comprises sequential and/or concurrent injection of an outer skin material and a dissimilar but compatible central core material into a cavity. This process offers the inherent flexibility of using the optimal properties of each material to reduce the material cost, injection pressure, clamping tonnage, and residual stresses to modify the property of the molded part, and/or to achieve particular engineering effects.

Co-injection is one of the two-component or multi-component injection molding processes available today. Unlike other multi-component molding processes, however, the co-injection molding process is characterized by its ability to encapsulate an inner core material with an outer skin material completely. Figure 4.68 illustrates the typical sequences of the co-injection molding process using the *one-channel technique* and the resulting flow of skin and core materials inside the cavity. This is accomplished with the use of a machine that has two separate, individually controllable injection units and a common injection nozzle block with a switching head. Due to the flow behavior of the polymer melts and the solidification of skin material, a frozen layer of polymer starts to grow from the colder mold walls. The polymer flowing in the center of the cavity remains molten. As the core material is injected, it flows within the frozen skin layers, pushing the molten skin material at the hot core to the extremities of the cavity.

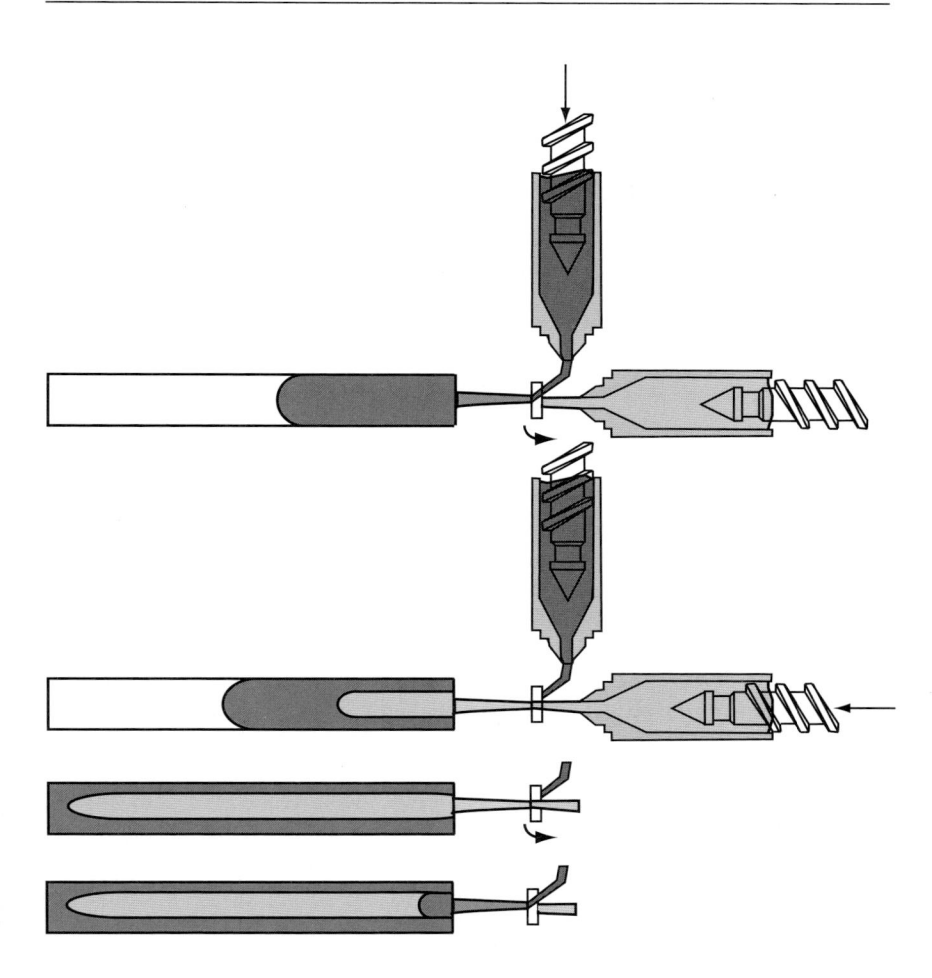

Figure 4.68: Sequential co-injection molding process.

Because of the fountain-flow effect at the advancing melt front, the skin material at the melt front will show up at the region adjacent to the mold walls. This process continues until the cavity is nearly filled, with skin material appearing on the surface and the end of the part. Finally, a small additional amount of skin material is injected again to purge the core material away from the sprue so that it will not appear on the part surface in the next shot. When there is not enough skin material injected prior to the injection of core material, the skin material may sometimes eventually be depleted during the filling process and the core

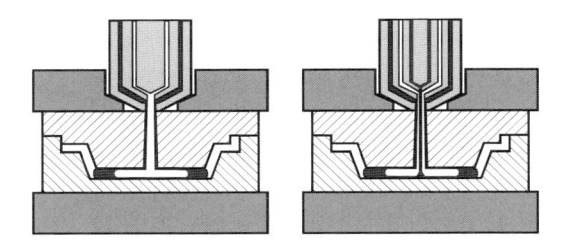

Figure 4.69: Two - and three - channel co-injection molding process.

material will show up on portions of the surface and the end of the part that is last filled. This is referrred to as *core surfacing* or *core breakthrough*. There are other variations to the sequential (namely, skin-core-skin, or A-B-A) co-injection molding process. In particular, one can start to inject the core material while the skin material is being injected (i.e., A-AB-B-A). That is, a majority of skin material is injected into a cavity, followed by a combination of both skin and core materials flowing into the same cavity, and then followed by the balance of the core material to fill the cavity. Again, an additional small amount of skin injection will cap the end of the sequence, as described previously. In addition to the one-channel technique configuration, two- and three-channel techniques have been developed that use nozzles with concentric flow channels to allow simultaneous injection of skin and core materials (Fig.4.69). More recently, a new version of the co-injection molding process that employs a multi-gate co-injection hot-runner system has become available. Such a system moves the joining of skin and core materials into the mold, as shown in Fig. 4.70. In particular, this hot-runner system has separate flow channels for the skin and core materials. The two flow streams are joined at each hot runner co-injection nozzle. In addition to all the benefits associated with conventional hot-runner molding, this system allows an optimum ratio of skin and core in multicavity or single-cavity molds.

Despite all the potential benefits of co-injection molding, the process has been slow to gain widespread acceptance because the equipment can cost 50 to 100% more than standard injection molding equipment. Co-injection molding offers a technically and economically viable solution for a wide range of commercial applications in emerging markets, including automotive, business machines, packaging, electronic components, leisure, agriculture, and *soft touch* products. Example applications include canoe paddles, toilet seats and cisterns, computer housings, copier parts, cash register covers, television escutcheons, audio cabinets, circuitry and electronics enclosures, garden chairs, boxes and containers, shoes and soles, paint brush handles, metal hand-rims on wheelchairs, thin-wall

4.4 Injection Molding

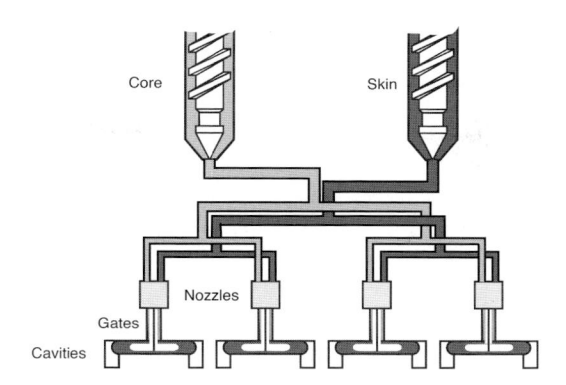

Figure 4.70: Multi-gate co-injection hot runner system with separate flow channels for skin and core materials, which join at each hot runner nozzle.

containers and beverage bottles, automotive parts such as exterior mirror housings and interior door handles and knobs, components for high-end ovens, audio speaker housings, and music center mainframes.

Fusible, Lost or Soluble Core Injection Molding The fusible, lost or soluble core injection molding process produces complicated, hollow components with complex and smooth internal geometry in a single molding operation. This process is a form of insert molding in which plastic is injected around a temporary core of low melting-point material, such as tin-bismuth alloy, wax, or a thermoplastic. An alloy core used to injection mold an intake manifold is presented in Fig. 4.71. After molding, the core is physically melted (or chemically dissolved), leaving its outer geometry as the internal shape of the plastic part. This process reduces the number of components required to make a final assembly or substitutes plastic for metal castings to boost performance (e.g., corrosion resistance) while saving weight, machining, and cost. Different techniques such as fusible core, soluble core and salt core techniques, are available to produce single-piece components featuring complex, smooth internal geometry and a high dimensional stability, which cannot be obtained through the conventional injection molding process. Among these lost core processes, the fusible core technique is the most energy-intensive method. Nevertheless, this drawback is offset by the low core losses, smoother internal surface requiring low finishing cost, faster heat dissipation by using a stronger and highly conductive metal core. The primary advantage of fusible core injection molding is the ability to produce single-piece plastic parts with highly complex, smooth internal shapes without a large number of secondary operations for assembly. Compared with options of

Figure 4.71: Bismuth-tin alloy core for used to manufature an intake manifold.

aluminum casting and machining, cost savings with fusible core injection molding are claimed to be up to 45% and weight savings can be as high as 75%. Plastic injection molding tools also have a much longer life than metal casting tools due to the absence of chemical corrosion and heavy wear. The main disadvantage of fusible core injection molding is cost. Another disadvantage is mold and machine development for casting and injection molding, which take time due to complex design and prototyping requirements. Because of this cost factor, the fusible core injection molding process for nylon air intake manifolds has been tempered by a shift to less-capital-intensive methods such as standard injection molding, where part halves are welded (twin-shell welding) or mechanically fastened.

Fusible core injection molding is generally employed for engineering thermoplastics, primarily glass fiber reinforced polyamide (PA) 6 and polyamide 66. Other materials used are glass fiber reinforced polyphenylene sulfide (PPS) and polyaryletherketone (PAEK); glass reinforced polypropylene (PP); and glass reinforced polyoxymethylene (POM). Low-profile thermosets have reportedly been used in fusible core injection molding, which require a core with a higher melting point. A number of factors must be considered in selecting the plastic materials. For example, the plastic component must be compatible with the core alloy and be able to withstand melting bath temperatures, which range from 100 to 180°C, in addition to typical considerations of the operation environment and mechanical property requirements.

4.4 Injection Molding

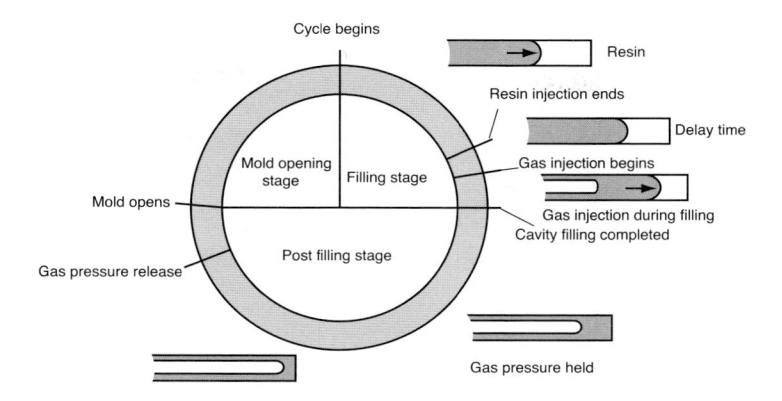

Figure 4.72: Gas-assisted injection molding cycle.

Gas-Assisted Injection Molding The gas-assisted injection molding (GAIM) process consists of a partial or nearly full injection of polymer melt into the mold cavity, followed by injection of an inert gas (typically nitrogen) into the core of the polymer melt through the nozzle, sprue, runner, or directly into the cavity. The compressed gas takes the path of the least resistance, flowing toward the melt front where the pressure is lowest. As a result, the gas penetrates and hollows out a network of predesigned, thick-sectioned gas channels, displacing molten polymer at the hot core to fill and pack out the entire cavity. As depicted in Fig. 4.73, gas assisted injection, as well as other fluid assisted injection molding, work based on several variations of two principles. The first principle is based on partially filling a mold cavity, and completing the mold filling by displacing the melt with a pressurized fluid. Figure 4.72 presents the gas-assisted injection molding process cycle based on this principle. with the second principle, the cavity is nearly or completely filled and the molten core is evacuated into a secondary cavity. This secondary cavity can be either a side cavity that will be scrapped after demolding, a side cavity that will result in an actual part, or the melt shot cavity in front of the screw in the plasticating unit of the injection molding machine. In the latter the melt is reused in the next molding cycle. In the so-called gas-pressure control process, the compressed gas is injected with a regulated gas pressure profile, either constant, ramped, or stepped. In the *gas-volume control process*, gas is initially metered into a compression cylinder at preset volume and pressure; then, it is injected under pressure generated from reducing the gas volume by movement of the plunger. Conventional injection molding machines with precise shot volume control can be adapted for gas-assisted injection molding with add-on conversion equipment,

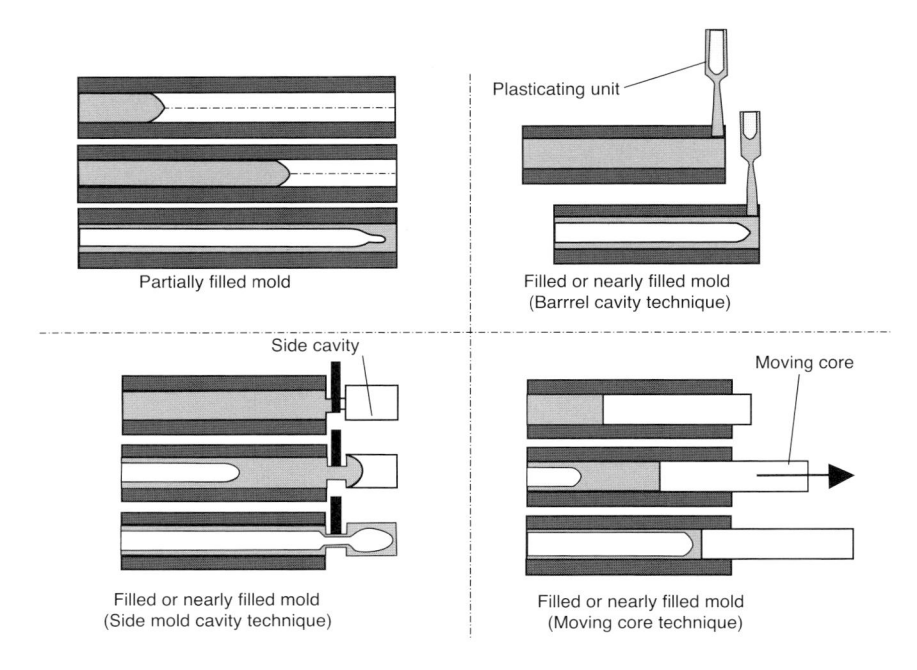

Figure 4.73: Schematic classification of fluid assisted injection molding processes.

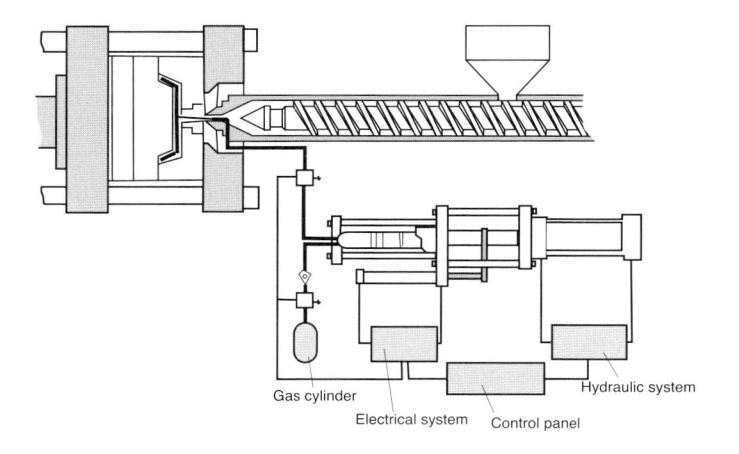

Figure 4.74: Schematic diagram of a typical injection molding machine adapted for gas-assisted injection molding with an add-on gas-compression cylinder and accessory equipment.

4.4 Injection Molding 333

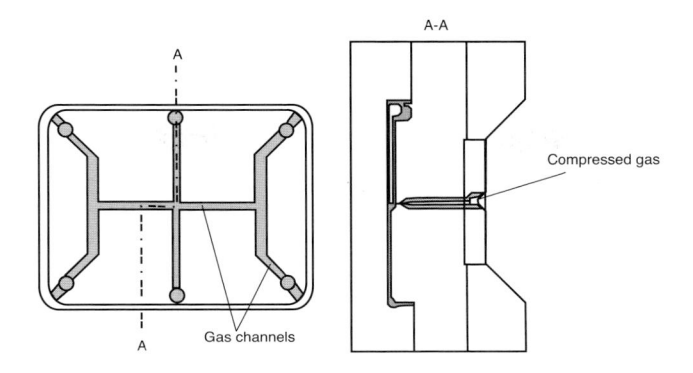

Figure 4.75: Schematic showing injection of gas through the nozzle and the gas distribution in the network of thicker ribs that serve as gas channels.

a gas source, and a control device for gas injection, as schematically depicted in Fig. 4.74. Gas-assisted injection molding, however, requires a different approach to product, tool, and process design due to the need for control of additional gas injection and the layout and sizing of gas channels to guide the gas penetration in a desirable fashion. As an illustration, Fig. 4.75 shows the schematic of gas injecting through the nozzle and the gas distribution in the rib of the molding with a network of thick-sectioned gas channels.

Because of the versatile and promising capabilities of this process, some alternative gas-assisted injection molding processes have been developed and become commercially available. For example, instead of using compressed nitrogen, the *liquid gas-assist process* injects a proprietary liquid into the melt stream. This liquid is converted to a gas in a compressed state with the heat of the polymer melt. After the filling and packing of the cavity, the gas is absorbed as the part cools down, thereby eliminating the need to vent the gas pressure. In the so-called *water-assisted injection molding* process, water, which does not evaporate during displacement of the melt, is injected after resin injection. Compared with conventional gas-assisted injection molding, water facilitates superior cooling effects and thus a shorter cooling time, as well as thinner residual wall thickness and larger component diameters. The *partial frame process* injects compressed gas into the strategically selected thick sections to form small voids of 1 to 2 mm (0.04 to 0.08 in) in diameter. Such a process can be employed to reduce sink marks and residual stresses over the thick sections. Finally, the *external gas molding process* is based on injecting gas in localized, sealed locations (typically on the ejector side) between the plastic material and the mold wall. The gas pressure maintains contact of the cooling plastic part with the opposite mold wall while providing a uniform gas pressure on the part surface supplementing or substituting for conventional holding pressure. This kind of process is suitable for parts with one visible surface where demands on surface finish are high and conventional use of gas channels is not feasible.

Other advantages of gas-assisted injection molding include:

- Increased part rigidity due to enlarged cross-sections at hollowed gas-channels, especially for large structural parts (also called open channel flow parts)
- Reduced material consumption, especially for rod - like parts (also called contained channel gas flow parts)
- Reduced cooling time since the thick sections are cored out
- Greater design flexibility to incorporate both thick and thin sections in the same part for part consolidation

Perhaps the main disadvantage of gas assisted injection molding is the uncertainty that still exists within the industry about patent rights involving the process. Most of the thermoplastic materials, technically speaking, can be used for gas-assisted injection molding. Gas-assisted injection molding has been extended to thermosetting polyurethane and bulk molding compounds, as well as powder injection molding. With bulk molding compound, the technique is referred to as *gas evacuation technique*. Here, the mold is completely filled and the outer layers are allowed to cure somewhat before gas is injected into the cavity, expelling the uncured central core of the compound. This material can be reused in the next cycle.

Typical applications for the gas-assisted injection molding process can be classified into three categories, or some combination thereof:

- Tube- and rod - like parts, where the process is used primarily for saving material, reducing the cycle time by coring out the part, and incorporating the hollowed section with product function. Examples are clothes hangers, grab handles (Fig. 4.76), chair armrests, shower heads, and water faucet spouts.
- Large, sheet - like, structural parts with a built-in gas-channel network, where the process is used primarily for reducing part warpage and clamp tonnage as well as to enhance rigidity and surface quality. Examples are automotive panels, business machine housings, outdoor furniture, and satellite dishes.
- Complex parts consisting of both thin and thick sections, where the process is used primarily to reduce manufacturing cost by consolidating several assembled parts into one single design. Examples are automotive door modules, television cabinets, computer printer housing bezels, and automotive parts.

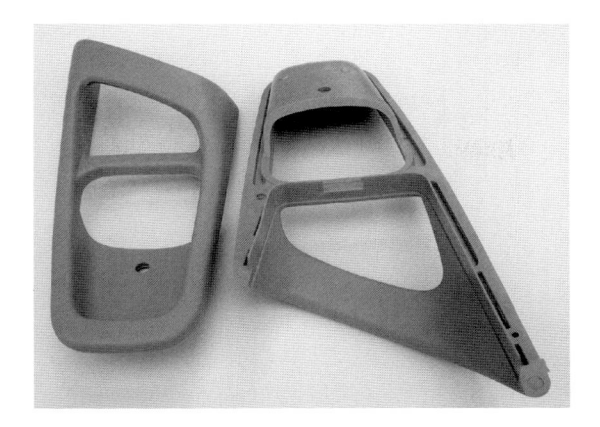

Figure 4.76: Gas-assisted injection molded door handles (Coutesy of Battendeld of America).

Injection-Compression Molding   The injection-compression molding (ICM) is an extension of conventional injection molding by incorporating a mold compression action to compact the polymer material for producing parts with dimensional stability and surface accuracy. In this process, the mold cavity has an enlarged cross-section initially, which allows polymer melt to proceed readily to the extremities of the cavity under relatively low pressure. At some time during or after filling, the mold cavity thickness is reduced by a mold closing movement, which forces the melt to fill and pack out the entire cavity. This mold compression action results in a more uniform pressure distribution across the cavity, leading to more homogenous physical properties and less shrinkage, warpage, and molded-in stresses than are possible with conventional injection molding. The injection-compression molding process is schematicallly depicted in Fig. 4.77.

The *two-stage sequential ICM* consists of separate injection stage and compression stage. During the injection stage, resin is injected into a cavity whose thickness is initially oversized by 0.5 to 10 mm greater than the nominal thickness. At the end of the resin injection, the compression stage begins during which the mold cavity thickness is reduced to its final value. The mold compression action forces the resin to fill the rest of the cavity and, after the cavity is completely filled, provides packing to compensate for the material shrinkage due to cooling. A potential drawback associated with the two-stage sequential ICM is the *hesitation* or *witness* mark resulting from flow stagnation during injection-compression transition. To avoid this surface defect and to facilitate continuous flow of the polymer melt, the simultaneous ICM activates mold compression while resin is

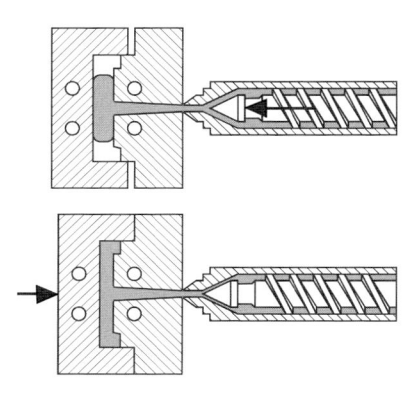

Figure 4.77: Schematic of the injection-compression molding process.

being injected. The *selective ICM* method starts with the cross-section at the final nominal value. During the injection stage, melt pressure drives the mold back toward the cylinder, which is mounted to an unpressurized movable core. Based on cavity pressure or time, the compression stage is activated by pressurizing the cylinder to force the part-forming surface on the movable core to compress the melt. A typical injection-molding machine with precise shot-volume control can be adapted for ICM; however, an additional control module is required for the mold compression stage. The primary advantage of ICM is the ability to produce dimensionally stable, relatively stress-free parts, at a low pressure, clamp tonnage (typically 20 to 50% lower), and reduced cycle time. For thin-wall applications, difficult-to-flow materials, such as polycarbonate and polyetherimide, have been molded to 0.5 mm. Additionally, the compression of a relatively circular charge significantly lowers molecular orientation, consequently leading to reduced birefringence, improving the optical properties of a finished part. The ICM is the most suitable technology for the production of high-quality and cost-effective CD-audio/ROMs as well as many types of optical lenses.

In-Mold Decoration and In-Mold Lamination *In-mold decoration (IMD)* comprises insertion of a predecorated film or foil into the cavity, followed by injection of polymer melts on the inner side of the insert to produce a part with the final finish defined by the decorative film/foil. *In-mold lamination process* resembles IMD except that a multilayered textile laminate is used. These two processes provide a cost-effective way to enhance and/or modify the appearance of a product for marking, coding, product differentiation, and model change without costly retooling. The laminated component can have desirable attributes (e.g., fabric or plastic skin finish with soft-touch) or properties (e.g., electromagnetic

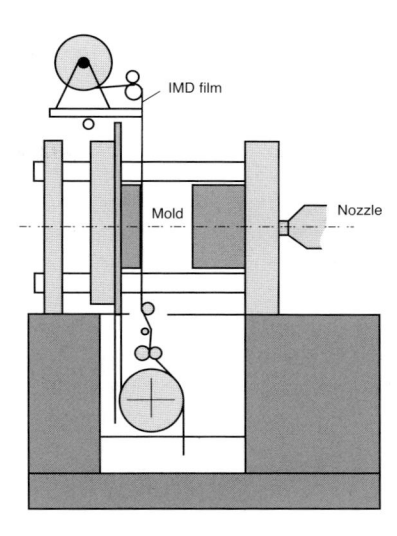

Figure 4.78: Schematic of an in-mold decorating set-up with a foil feedinf device built into the injection molding machine.

interference, EMI, or radiofrequency interference, RFI, shielding). During the molding stage, the polymer melt contacts the film and fuses with it so that the decoration can be lifted off from the carrier film and strongly attach to the surface of the molding. An injection molding machine with a typical IMD set up is shown in Fig. 4.78.

In one of the IMD techniques, the so-called paintless film molding (PFM) or laminate painting process, a three-layer coextrusion film with pigment incorporated into layers of clear-coat cap layers and a core layer is first thermoformed into the shape of the finished part and then inserted into the cavity and overmolded with thermoplastics to produce a final part. With this technique, it is possible to obtain a high-quality, extremely smooth paint finish on thermoplastic exterior body claddings and moldings ready for assembly without subsequent spray painting or finishing. Unlike in-mold coating and mold-in-color, this process provides high-gloss metallic and nonmetallic finishes. It provides unique patterns and designs that are not feasible with paint. The paint laminate finish provides superior weatherability, acid etch resistance, and a safe worker environment because it is virtually pollution-free. In addition to injection molding, IMD can be used with a variety of other processes, such as structural foam injection molding, ICM, compression molding, blow molding, thermoforming, resin transfer molding, and rotational molding. Instead of using a thin film/foil as does IMD, the in-mold lamination process employs a multilayered textile laminate positioned in the parting plane to be overmolded by the polymer melt on the inner side. The decorative laminate can be placed in the mold as a cut sheet, pulled from the roll with needle grippers, or, by means of a clamping frame method. By means of a thermoforming operation, the clamping frame method allows defined predeformation of the decorative laminate during the mold-closing operation. In-mold lamination is also known as *laminate insert molding* or *fabric molding* for manufacturing automotive instrument panels and interior panels, respectively. For in-mold lamination, the outer, visible layer of the decorative laminate can be made of polyester, PA, PP, PVC, or ABS film, cotton textile (woven, knitted, tufted, or looped fabrics), or leather. This outer layer typically comes with a variety of features to create a specific appearance or feel. In general, low surface texture is preferred because the ironing effect typically occurs with high pile or fiber loops. To provide the product with a soft-touch effect, there is typically an intermediate layer of polyurethane PU, PP, PVC, or PES foam between the top layer and the liner layer. Underneath the foam layer is the liner layer, which is used to stabilize the visible layer against shear and displacement, prevent the penetration of polymer melt into the intermediate layer, and provide thermal insulation against the polymer melt. To avoid damage or undesirable folding of the laminate during molding, low injection pressure and low temperature are desirable. This makes low-pressure injection molding, compression molding, and cascade injection molding with sequential valve-gate opening and closing suitable candidates for in-mold lamination. Although existing tools can often be adapted for IMD and in-mold lamination, it is generally recommended to design the tool specifically for these processes. In-mold decoration has been used to produce rooftops, bumper fascia, and exterior mirror housings, as well as automotive lenses and body-side molding with mold-in colors to help automakers eliminate costs and environmental concerns associated with painting. In-mold lamination has been used widely for automotive interior panels and other applications such as automotive interior panels and shell chairs, both featuring textile surfaces.

Microinjection Molding Microinjection molding (also called micromolding) produces parts that have overall dimension, functional features, or tolerance requirements that are expressed in terms of milli- or even micrometers. Due to the miniature characteristics of the molded parts, it requires a special molding machine and auxiliary equipment to perform tasks such as shot volume control, evacuation of mold (vacuum), injection, ejection, inspection, separation, handling, deposition, orientation, and packaging of molded parts. Special techniques are also being used to make the mold inserts and cavities. The demand for miniature injection molded parts and the equipment and processing capability to produce them with desirable precision began around 1985 and has been growing ever since. Although there is no clear way to define microinjection molding, ap-

4.4 Injection Molding

plications of this process can be broadly categorized into three types of products or components:

- Microinjection molded parts (micromolding) that weigh a few milligrams to a fraction of a gram, and possibly have dimensions on the micrometer (μm) scale (e.g., micro-gearwheels, micro operating pins (Fig. 9.37))
- Injection molded parts of conventional size but exhibiting microstructured regions or functional features (e.g., compact disc with data pits, optical lenses with microsurface features, and wafers for making microgearwheels with Plastic Wafer Technology)
- Microprecision parts that can have any dimensions, but have tolerances on the micrometer scale (e.g., connectors for optical fiber technology)

Although modern conventional injection molding machines can achieve impressive results, these machines must be adapted to meet the special requirements of microinjection molded parts such as:

- Small plasticating units with screw diameters ranging from 12 to 18 mm and a shorter screw length with L/D ratio around 15 to avoid material degradation from prolonged residence (dwell) time
- Precise shot volume control and desirable injection rate
- Repeatable control, such as fill to pack switchover, based on screw position or cavity pressure (preferred)
- Capability to raise the mold-wall temperature to such a level (sometimes slightly above the melting temperature of the polymer) to avoid premature solidification of ultrathin sections
- Mold evacuation if the wall thickness of the microinjection molded part is down to 5 μm, which is the same order as the dimension of vents for air to escape
- Shutoff nozzle to avoid drooling from the nozzle due to the high processing temperature of the melt
- Precise alignment and gentle mold-closing and -opening speeds to avoid deformation of delicate microinjection molded parts
- Special handling technique to remove and molded parts for inspection and packaging
- Possible local clean room enclosure or laminar flow boxes to avoid contamination of microinjection molded parts

In addition to the various features described earlier, these machines are sometimes equipped with separate metering and injection ram (pistons) and screw design. These features are used to meter the shot volume accurately and to eliminate problems associated with material degradation resulting from by-pass and dead corners with the conventional injection screw. Because the size and weight of the microinjection molded parts differ significantly from the conventional parts, certain steps have to be taken to ensure proper part ejection. For example, a vision system can be installed in the molding machine to confirm ejection of the miniature parts. In addition, part removal can be performed by suction pads, which keep parts separate and oriented for quality control and packaging, or by electrostatic charging or blowing out. Traditional quality control methods, such as measuring the part weight, becomes impractical with microinjection molding. New quality control techniques employ a video inspection system for sorting acceptable and unacceptable parts. Because microinjection molded parts are usually part of an assembly, they are packaged in an oriented way ready for assembly.

Microcellular Molding Microcellular molding (also commercially known as MuCell process) blends supercritical gas (usually nitrogen or carbon dioxide) with polymer melt in the machine barrel using a selenoid metering valve to create a single-phase solution. A specially designed screw is used to create the gas-polymer solution for both the microcellular molding process as well as for the solid plastics injection molding. During the molding process, the gas forms highly uniform microscale cells (bubbles) of 0.1 to 10 μm in diameter, and the internal pressure arising from the foaming eliminates the need of packing pressure while improving the dimensional stability. The original rationale is to reduce the amount of plastic used without sacrificing the mechanical properties. Typical mucell structures are shown in the fotographs of Fig. 4.79. Because the gas fills the interstitial sites between polymer molecules, it effectively reduces the viscosity and glass transition temperature of polymer melt; therefore, the part can be injection molded with lower temperatures and pressures, which leads to significant reduction of clamp tonnage requirement and cycle time. Microcellular plastics (MCPs) are single-phase, polymer-gas solutions by dissolving or saturating a polymer in a supercritical fluid (CO_2 or N_2) and then precipitating it by triggering polymer nucleation by adjusting process conditions, such as pressure and temperature. The underlying rationale of the MuCell process is to create enough voids smaller than the pre-existing flaw in polymers so that the amount of plastic used could be reduced without compromising the mechanical properties. Even though realizing a part weight reduction of 5 to 95% by replacing plastics with gas, the microcells also serve as crack arrestors by blunting crack tips, thereby greatly enhancing part toughness. When properly prepared, microcellular PS has five times the impact strength of its unfoamed counterpart. The fatigue life of microcellular PC with a relative foam density of 0.97 is four times that of its solid counterpart. Furthermore, because the gas fills the interstitial sites between polymer molecules, it effectively reduces the viscos-

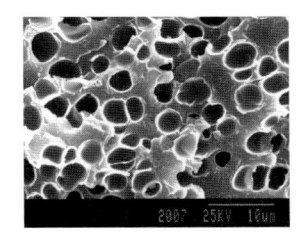

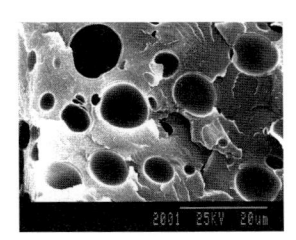

Figure 4.79: Micrographs of microcellular polystyrenes prepared under different conditions.

ity and the glass transition temperature of the polymer melt, which results in a substantial decrease in processing temperatures (as much as 78°C), 30 to 50% reduction in hydraulic injection pressure, and 30% or more reduction in clamp tonnage. The material, therefore, can be processed at much lower pressures and temperatures. For injection molding applications, MCPs has potential in housing and construction, sporting goods, automotive, electrical and electronic products, electronic encapsulation, and chemical and biochemical applications. Microcellular molding also has a potential in in-mold decoration and lamination due to its low pressure and temperature characteristics that reduce the damage to the film or fabric overlay during molding. Specific application examples include fuse boxes, medical handles, air intake manifold gaskets, automotive trims, and plastic housings.

Multicomponent Injection Molding (Overmolding) Multicomponent injection molding, sometimes also referred to as *overmolding* is a versatile and increasingly popular injection molding process that provides increased design flexibility for making multicolor or multifunctional products at reduced cost. By adopting multiple mold designs and shot transfer strategy, this process consists of injecting a polymer over another molded plastic insert to marry the best features of different materials while reducing or eliminating postmolding assembly, bonding, or welding operations. Special machine equipment with multiple injection units, a rotating mold base, and/or moving cores is available to complete the overmolding in one cycle. Multicomponent injection molding is a special process used in the plastic processing industry, and is synonymous with overmolding, multishot injection molding, and the *in-mold assembly* process. This process, however, is different from co-injection molding, which also incorporates two different materials into a single molded part. To be specific, co-injection molding involves sequential and/or concurrent injection of two dissimilar but compatible materials into a cavity to produce parts that have a sandwich structure, with the core material embedded between the layers of the skin material. On the other hand, for multicomponent injection molding, different polymer melts are injected at different stages of the process using different cavities or cavity geometry. In particular, a plastic insert is first molded and then transferred to a different cavity to be overmolded by the second polymer filling inside a cavity defined by the surfaces of the insert and the tool. The adhesion between the two different materials can be mechanical bonding, thermal bonding, or chemical bonding. Multicomponent injection molding date back several decades ago, as the multicolor molding of typewriter keys to produce permanent characters on the keys. This process has since advanced to allow consistent, cost-effective production of multicolor or multifunctional products in a variety of innovative and commonly used methods. The decision of choosing an appropriate molding technique depends on the production volume, quality requirements, and the molder's capabilities and preference. For example, without any additional equipment investment, one can use two separate molds and the conventional injection molding machine for producing multicomponent injection molded parts. In this approach, the insert is first molded and then transferred to a second mold, where it is overmolded with a second polymer. The disadvantage of this approach, however, is that it involves additional steps to transfer and load the prefabricated insert into the second mold. Nevertheless, the loading and pick up of the insert and molded system can be accelerated and precisely controlled by using robots or automated machine systems. Another commonly used method of multicomponent injection molding employs a rotating mold and multiple injection units, as shown in Fig. 4.80. Once the insert is molded, a hydraulic or electric servo drive rotates the core and the part by 180 degrees (or 120 degrees for a three-shot part), allowing alternating polymers to be injected. This is the fastest and most common method because two or more parts can be molded every cycle. Utilities for the rotating mold (i.e., cooling water, compressed air, or special heating) are connected through a central rotary union. If there is a family of parts that requires rotational transfers, it is more economical to have the frictional capability built into the machine (and pay for it once) than it is to buy a family of molds with the capability in each.

Another variation of multicomponent injection molding involves automatically expanding the original cavity geometry using retractable (movable) cores or slides while the insert is still in the mold. This process is called core-pull or core-back, as shown in Fig. 4.81. To be specific, the core retracts after the insert has solidified to create open volume to be filled by the second material within the same mold.

4.4 Injection Molding

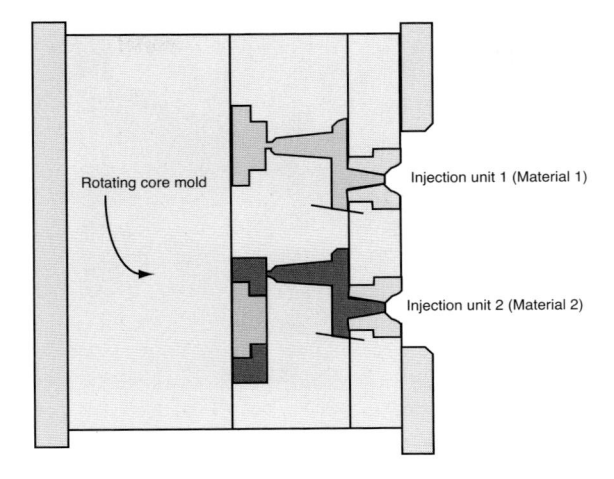

Figure 4.80: Schematic diagram of a rotating mold used to produce multi-component injection molded parts.

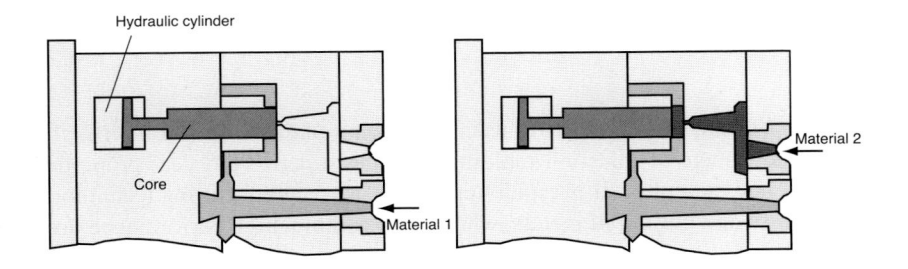

Figure 4.81: Schematic diagram of multi-component injection molding using a *core pull* or *a core back technique*.

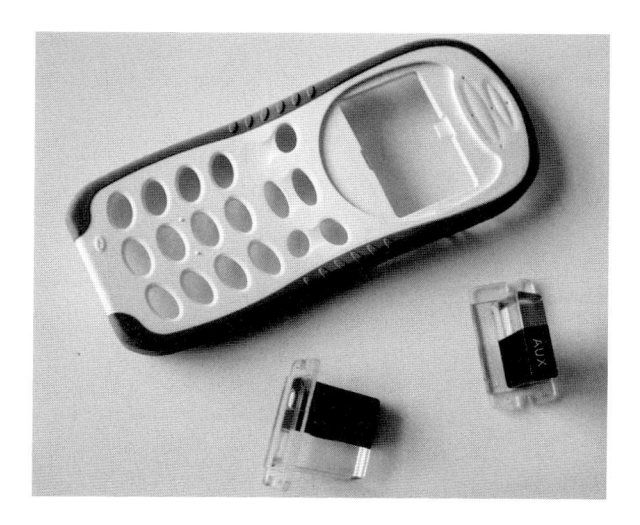

Figure 4.82: TPE (thermoplastic elastomer) overmolded on ABS/PC for a mobile phone to provide soft touch feeling and drop protection (Top) and black color ABS with transparent acrylic, used on audio push button for light transmission (Bottom) (Courtesy of Yomura Company Ltd., Taiwan).

Material selection is vital for multicomponent injection molding. A thorough analysis needs to be conducted to determine material compatibility, chemical- and wear-resistance, environmental performance, and other program-specific requirements. One of the most popular applications is the overmolding of a flexible thermoplastics elastomer (TPE) onto a rigid substrate to create the soft-touch feel and improved handling in a finished product such as shown in the photograph in Fig. 4.82.

Powder Injection Molding Powder injection molding (PIM) combines the shaping advantage of injection molding with the superior physical properties of metals and ceramics. It is rapidly emerging as a manufacturing technology for ceramics and powdered metals, where high volume, high performance, low cost, and complex shape are required. In these processes, a custom formulated mixture of metal or ceramic powder and polymer binder is injected into a mold and formed into the desired shape (the *green part*), which is akin to conventional injection molding and high-pressure die casting. After the molding process, the binder is removed and the resulting *brown part* undergoes a sintering process to create metallurgical bonds between the powder particles imparting the necessary mechanical and physical properties to the final part.

4.4 Injection Molding

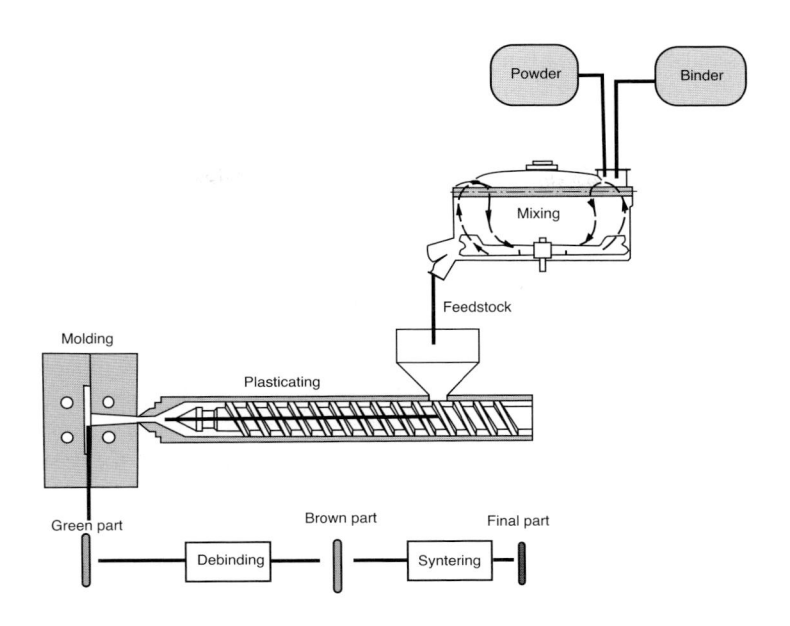

Figure 4.83: Schematic of the steps involved in the powder injection molding process.

Powdered metal injection molding and ceramic injection molding are special injection molding processes used to produce highly complex metal and ceramic parts that otherwise would require extensive finish machining or assembly operations. Figure 4.83 illustrates the basic steps involved in PIM: mixing for forming feedstock, molding, debinding (binder extraction), and sintering (densification). The last two steps can be combined into a single thermal cycle.

Feedstock is the mixture of powder and binder in forms of pellets and granules. Its formation involves mixing selected ceramic raw materials or powdered metal alloys and polymeric binders to form a moldable formulation. Powders are small particles that usually have sizes between 0.1 and 20 μm with a nearly spherical shape to ensure desirable densification. In principle, any metal that can be produced as a powder can be processed using this method. One exception is aluminum, which forms an oxide layer on the surface, thus preventing sintering. Commonly used particles are stainless steels, steels, tool steels, alumina, iron, silicates, zirconia, and silicon nitride. Binder is usually based on thermoplastics such as wax or polyethylene, but cellulose, gels, silanes, water, and various inorganic substances are also in use. The binder usually consists of two or three components: thermoplastics and additives for lubrication, viscosity control, wetting, and debinding. A typical feedstock is composed of approximately 60 volume %

powders and 40 volume % binders. Twin-screw or high-shear cam action extruders are often used to mix the materials at elevated temperatures (100 to 200°C). Upon exiting the extruder, the mixture is chopped into pellets and granules to be used for injection molding. There are five factors that determine the attributes of the feedstock: powder characteristics, binder composition, powder-binder ratio, mixing method, and pelletization technique. An ideal feedstock is easy to mold and easily attains the final product dimensions.

The tooling for powder injection molding is similar to that used in conventional polymer injection molding except that molding is oversized to account for sintering shrinkage. During the molding stage, feedstock pellets are heated to temperatures of 150 to 200°C and injected into the mold at an injection pressure range of 35 to 140 MPa (5000 to 20,000 psi) and with cycle times of 30 to 60 seconds, depending on the part dimension and actual molding formulation employed. The mold can be either heated to control the viscosity of the mixture or be kept at room temperature to facilitate cooling. At the end of the cycle, the molded components (so-called green parts) have a strong waxlike consistency and are suitable for autoejection from the mold via push-out pins or stripper rings. Debinding is the most critical step in the powder injection molding process. Debinding is typically accomplished by one of the three general procedures: thermal binder removal method, catalytic binder removal agents, or solvent extraction method. The debinding process, which removes as much as 30 to 98% of the binder, can be a long process. It typically takes many hours or even days, depending on part thickness and powder grain size. The last step in the process is the sintering, which is a thermal treatment for bonding the particles into a solid mass. Furnaces are used at this stage that permit sintering at temperatures close to the material melting temperature under a controlled atmosphere or in a vacuum. During the sintering process, intergranular pores within the molded part are removed, resulting in significant shrinkage (typically on the order of 10 to 20%). Meanwhile, the finished part achieves a density of around 97%, at which the mechanical properties differ only slightly, if at all, from those parts manufactured by alternative methods. PIM is applicable to a variety of automotive, consumer, electronic, computer peripherals, medical, industrial, and aerospace components. Automotive applications include turbocharger, brake, and ignition components, as well as oxygen sensors.

Reaction Injection Molding Reaction injection molding (RIM) involves mixing of two reacting liquids in a mixing head before injecting the low-viscosity mixture into mold cavities at relatively high injection speeds. The liquids react in the mold to form a cross-linked solid part. Figure 4.84 presents a schematic of a high pressure polyurethane injection system. The mixing of the two components occurs at high speeds in *impingment mixing heads* as schematically depicted in Fig. 4.85. Low pressure polyurethane systems, such as the one schematically

4.4 Injection Molding

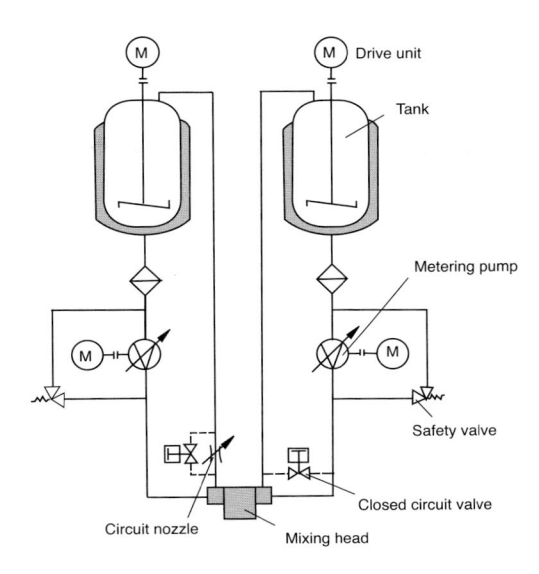

Figure 4.84: Schematic diagram of a high pressure polyurethane injection system.

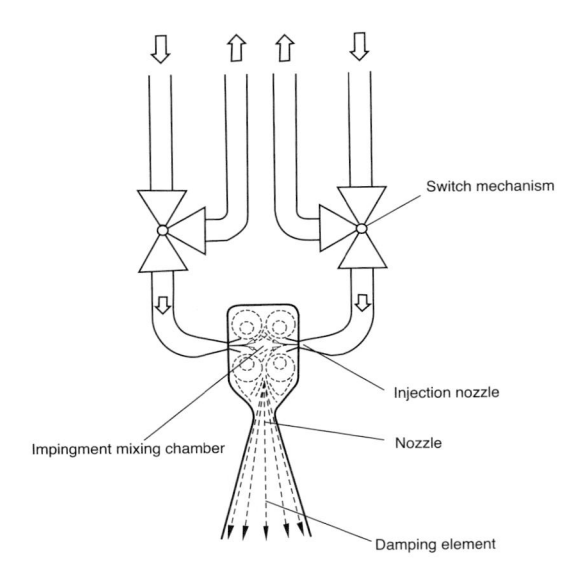

Figure 4.85: Schematic diagram of an impingment mixing head.

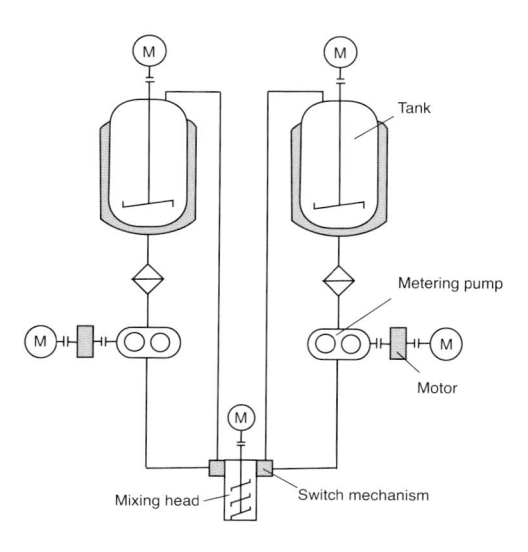

Figure 4.86: Schematic diagram of a low pressure polyurethane injection system.

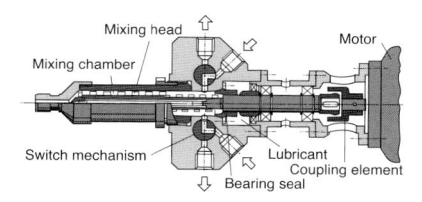

Figure 4.87: Schematic diagram of a mechanical mixing head used with low pressure polyurethane systems.

presented in Fig. 4.86, require mixing heads with a mechanical stirring device as presented in Fig. 4.87. The short cycle times, low injection pressures, and clamping forces, coupled with superior part strength and heat and chemical resistance of the molded part make RIM well suited for the rapid production of large, complex parts, such as automotive bumper covers and body panels. Reaction injection molding is a process for rapid production of complex parts directly from monomers or oligomers. Unlike thermoplastic injection molding, the shaping of solid RIM parts occurs through polymerization (cross-linking or phase separation) in the mold rather than solidification of the polymer melts. RIM is also different from thermoset injection molding in that the polymerization in RIM is activated via chemical mixing rather than thermally activated by the warm mold.

During the RIM process, the two liquid reactants (e.g., polyol and an isocyanate, which were the precursors for polyurethanes) are metered in the correct proportion into a mixing chamber where the streams impinge at a high velocity and start to polymerize prior to being injected into the mold. Due to the low-viscosity of the reactants, the injection pressures are typically very low even though the injection speed is fairly high. Because of the fast reaction rate, the final parts can be de-molded in typically less than one minute. There are a number of RIM variants. For example, in the so-called reinforced reaction injection molding (RRIM) process, fillers, such as short glass fibers or glass flakes, have been used to enhance the stiffness, maintain dimensional stability, and reduce material cost of the part. As another modification of RIM, structural reaction injection molding (SRIM) is used to produce composite parts by impregnating a reinforcing glass fibermat (preform) preplaced inside the mold with the curing resin. On the other hand, resin transfer molding (RTM) is very similar to SRIM in that it also employs reinforcing glass fibermats to produce composite parts; however, the resins used in RTM are formulated to react slower, and the reaction is thermally activated as it is in thermoset injection molding. The process advantages of RIM stem from the low pressure, temperature, and clamp force requirements that make RIM suitable for producing large, complex parts with lower-cost molds. In addition, the capital investment on molding equipment for RIM is lower compared with that of injection molding machines. Finally, RIM parts generally possess greater mechanical and heat-resistant properties due to the resulting cross-linking structure. The mold and process designs for RIM become generally more complex because of the chemical reaction during processing. For example, slow filling may cause premature gelling, which results in short shots, whereas fast filling may induce turbulent flow, creating internal porosity. Improper control of mold-wall temperature and/or inadequate part thickness will either give rise to moldability problem or cause scorching of the materials. Moreover, the low viscosity of the material tends to cause flash that requires trimming. Another disadvantage of RIM is that the reaction with isocyanate requires special environmental precaution due to health issues. Finally, like many other thermosetting materials, the recycling of RIM parts is not as easy as that of thermoplastics. Polyurethane materials (rigid, foamed, or elastomeric) have traditionally been synonymous with RIM as they and ureaurethanes account for more than 95% of RIM production. Alternative RIM materials include nylon (NYRIM), dicyclopentadiene (DCPD-RIM), acrylamate/acrylesterol, epoxies, unsaturated polyesters, phenolics, thermoset resins, and modified polyisocyanurate.

Structural Foam Injection Molding Structural foam injection molding is an extension of injection molding for producing parts with cellular (or foam) core sandwiched by solid external skins as shown in the photograph in Fig. 4.88. This process is suitable for large, thick parts that have enough strength-to-weight ratio to be used in load- or bending-bearing in their end-use applications. Despite the surface swirling pattern, which can be eliminated with special molding techniques or the postmolding finishing operations, this process permits molding of large parts at low pressure and with no sink marks and warpage problems. Structural foam parts can be produced with physical blowing agents such as nitrogen gas or chemical blowing agents. There are several variants of this process. The most common one is the low-pressure structural foam injection molding. Other additional processes are gas counter pressure, high pressure, co-injection, and expanding mold techniques. The low-pressure structural foam molding process involves a short-shot injection accompanied by expansion of melt during molding. The melt expansion is created from gas (typically nitrogen) dissolved in the polymer melt prior to injection. It can also be assisted by gas released from the chemical blowing agents in a resin-compatible carrier blended with the polymer pellets. The melt expansion from the foaming process generates an internal gas pressure of 21 to 34 bars (2.1 to 3.4 MPa or 300 to 500 psi), which is sufficient to drive the polymer melt continuously to fill the extremities of the cavity. Note that this process requires only 10% of the pressure normally required to fill and pack out the mold with the conventional injection molding process. In addition, structural foam injection molding realizes part weight reduction by replacing plastics with gas and eliminates the typical shrinkage associated with thick parts with the melt expansion. The foam structure provides an insulation effect so that the part cools down at a slower rate. If the part is removed from the mold before the material is sufficiently cooled, the internal gas pressure will cause blisters on the part surface, especially at the thick sections. Moreover, if the part is painted before the internal gas pressure reaches the ambient, atmospheric pressure, blisters will also occur. Due to the low injection pressure and slow cooling rate (which allows material relaxation), the process produces parts with very low molded-in residual stresses.

As indicated earlier, both nitrogen and chemical blowing agents can be added to the polymer melt to create a foam structure. To add nitrogen into the polymer melt and to hold the polymer-gas mixture under pressure until it is injected into the mold, it requires special structural-foam molding equipment with an accumulator. When the external pressure of the melt is relieved as the material enters the mold cavity, the gas begins to emerge from the solution. At the end of a short-shot injection, the pressure decreases rapidly as the material relaxes. Thus, gas expands immediately at this point, pushing the material to fill the rest of the mold. A gas counterpressure technique has been developed and employed to produce parts with smooth, thicker, and nonporous skin surfaces, and more uniform cell density in the core. In this process, the mold is pressurized with an inert gas to create a pressure level of 14 to 34 bars (1.4 to 3.4 MPa or 200 to 500 psi) during filling. The pressure in the mold must be high enough to prevent the melt from

4.4 Injection Molding

Figure 4.88: Cross section of a typical structural foam molded part showing the integral skin layers and the foam core (from Malloy).

foaming. To maintain the pressure level, the mold must be sealed with an O-ring. It may also be necessary to seal ejector pins, sprue bushing, and other moving slides or cores. The control of a sufficient short-shot weight and a critical timing for the counterpressure to be released are also critical. After the injection, the gas is vented and the mold is depressurized to allow the expansion of polymer. Because of the counterpressure, the flow length of the material may be reduced by 10 to 20%, and the density reduction is limited to 8 to 10%, as compared with a level of 10 to 15% with the low-pressure structural foam injection molding. Gas counter pressure using a hot runner mold has been used to mold a polystyrene part with surface smoothness substantially better than that of the parts molded under the same processing conditions using low-pressure structural injection molding. In addition, the surface roughness was found to increase progressively in the direction of melt flow from the cold sprue toward the end of the part. The processing conditions found to enhance the surface smoothness of the part were a higher melt temperature, higher mold temperatures, higher short-shot weight, higher injection speed, and lower blowing agent concentration.

To improve surface appearance and quality, secondary operations, such as sanding, priming, and painting, are required. Parts with substantially improved surface quality have successfully been produced using gas counterpressure structural foam injection molding and co-injection structural foam, as well as expanding, high-pressure methods. It should also be noted that the selection of blowing agents should avoid the possible material degradation from the by-products of the blowing agents during their decomposition. Finally, due to the characteristic cellular structure, the mechanical properties of structural foamed parts can be significantly lower than the solid parts, even though thicker foamed parts are more rigid than are their solid counterpart on a weight-to-weight basis. Among the materials used for structural foam injection molding, the most common are

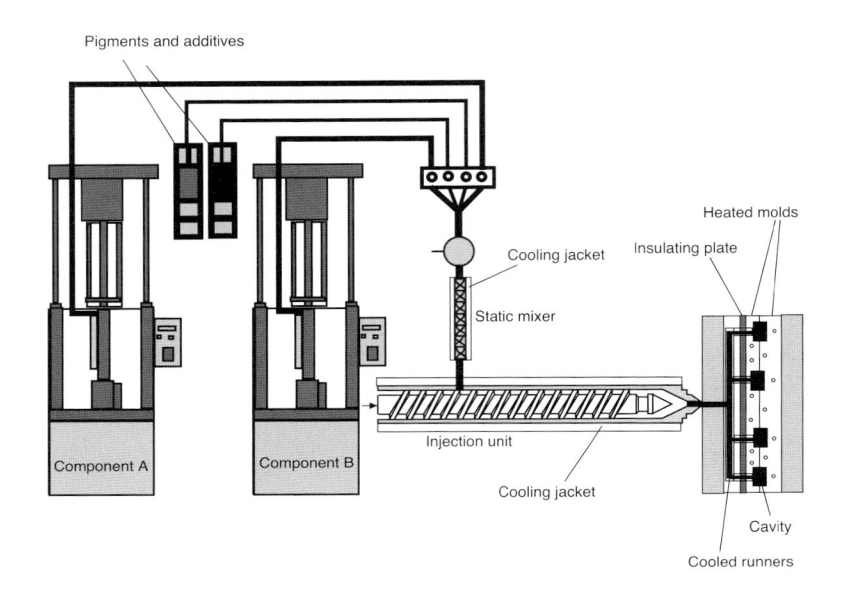

Figure 4.89: Schematic diagram of a liquid silicone rubber injection molding system.

low-density and high-density polyethylenes for their low cost, ease of processing, chemical resistance, and low-temperature impact strength. Polypropylene is an attractive material for its increased stiffness and chemical resistance, and its reinforced grades exhibit increased temperature resistance. High-impact polystyrene and modified polyphenylene ether (m-PPE), modified polyphenylene oxide (m-PPO), and polycarbonate are commonly found in high-performance applications. The versatility inherent in polyurethane chemistry and the advantages of the RIM process makes the utilization of polyurethane structural foams possible in many applications. Because structural foam injection molding enables molding of large, complex parts with rigidity, dimensional stability, and load-bearing capability, it finds many suitable applications in material handling, business machines, automotive components, and medical analysis equipment.

Injection Molding of Liquid Silicone Rubber Injection molding of liquid silicone rubber (LSR) has existed and evolved in the past 25 years. Due to the thermosetting nature of the material, liquid silicone injection molding requires special treatment, such as intensive distributive mixing, while maintaining the material cool before it is pushed into the heated cavity and vulcanized. Figure 4.89 schematically depicts an LSR injection molding process.

4.4 Injection Molding

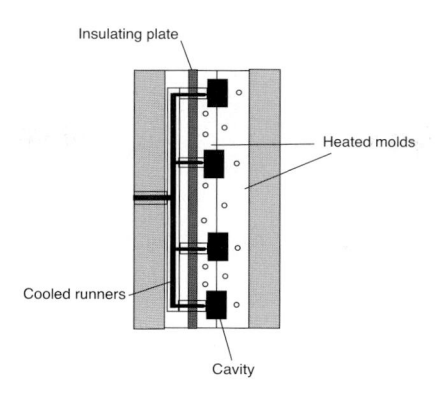

Figure 4.90: Schematic diagram of a liquid silicone rubber injection mold.

Liquid silicone rubbers are supplied in barrels or hobbocks. Because of their low viscosity, these rubbers can be pumped through pipelines and tubes to the vulcanization equipment. The two components (labeled A and B component in the figure) are pumped through a static mixer by a metering pump. One of the components contains the catalyst, typically platinum based. A coloring paste as well as other additives can also be added before the material enters the static mixer section. In the static mixer the components are well mixed and are transferred to the cooled metering section of the injection molding machine.

The static mixer renders a very homogeneous material that results in products that are not only very consistent throughout the part, but also from part to part. This is in contrast to solid silicone rubber materials that is purchased pre-mixed and partially vulcanized. In contrast, hard silicone rubbers are processed by transfer molding and result in less material consistency and control, leading to higher part variability. Additionally, solid silicone rubber materials are processed at higher temperatures and require longer vulcanization times.

From the metering section of the injection molding machine, the compound is pushed through cooled sprue and runner systems into a heated cavity where the vulcanization takes place. The cold runner and general cooling results in no loss of material in the feed lines. A schematic of a liquid silicone rubber mold with cooled runners is depicted in Fig. 4.90. The cooling allows production of LSR parts with nearly zero material waste, eliminating trimming operations and yielding significant savings in material cost.

Silicone rubber is a family of thermoset elastomers that have a backbone of alternating silicone and oxygen atoms and methyl or vinyl side groups. Silicone rubbers constitute about 30% of the silicone family, making them the largest group of that family. Silicone rubbers maintain their mechanical properties over a wide

Figure 4.91: Photographs of sample injection molded liquid silicone rubber parts. (a) Single wire seal, (b) Peripheral seal. These parts were molded with an *oil bleeding formulation*, a silicone rubber-silicone oil blend for a continuous lubrication effect. (Courtesy Simtec Silicone Parts, LLC, Madison, WI, USA).

range of temperatures and the presence of methyl-groups in silicone rubbers makes these materials extremely hydrophobic. Typical applications for liquid silicone rubber are products that require high precision such as seals, sealing membranes, electric connectors, multi-pin connectors, infant products where smooth surfaces are desired, such as bottle nipples, medical applications as well kitchen goods such as baking pans, spatulas, etc. Figure 4.91 presents examples of liquid silicone injection molded parts.

Table 4.6: Troubleshooting Guide for Injection Molding

Problem	Possible causes and recommendations
Black specks or flakes in the part	**Poor screw or nozzle design.** Check for dead-spots within the screw channel (Most general purpose screws). Check nonreturn valve, nozzle end-cap and nozzle tip for abrupt changes in flow path. **Excessive nozzle length.** Make sure the nozzle length is as short as possible and it is PID temperature controlled. **Faulty or poorly designed hot runner flow channels.** Repair and replace any that are working on % or variacs. Redesign hot runner. Check location of thermocouples. **High melt temperature; hot spots in screw or barrel.** Check melt temperature with IR sensor. Check screw and barrel for hot spots. Are any zones overriding their temperature setting? Lower back pressure. Slow screw speed RPM. **Contamination in virgin resin.** Check virgin resin using a white pan, spread 1.5 kg of resin over a 1500 cm^2 area, and inspect for 5 minutes under appropriate lighting.

Continued on next page

4.4 Injection Molding

Problem	Possible causes and recommendations
	Contamination in reprocessed or reground resin. Check screw design, material grinding and handling procedures. **Excessive, high temperatures or long residence times.** Use 25-65% of the barrel capacity. Using lower than 25% provides long residence time for the resin or additive to degrade. Use smaller barrel. Check heater band function, nozzle should be PID controlled, check screw for high shear dispersion mixing elements. **Excessive fines.** Remove fines before processing; a must for lens applications. **Special note:** Purging compound usually does not solve a black specking. Most of the time it will be wiser to pull the screw and clean it. Screws should have a high polish. **Vented barrel.** Check for dead spot or hang-up area. Difficult to access, but must be cleaned. **Sharp angle or corner at the gate.** Polish gate area, radius sharp corners, change gate type, or enlarge gate. **Hot runners.** Check for proper temperature control, burnt out heating elements, open thermocouples. Check depth of probe relative to gate surface. Check for hang up areas or dead spots. **Poor nozzle temperature control.** Placement of the thermocouple should be one third the distance of the nozzle from its tip if he nozzle is more than 75 mm long. For 75 mm lengths and shorter it is OK to have the thermocouple imbedded into the hex on the nozzle body. Do not allow the thermocouple to be attached via the screws on the clamp for the heater band. Clean out the nozzle, inspect for blockages. **Sharp angle on screw tip or broken nonreturn valve.** Check entire flow path of plastic looking for burrs, sharp corners, grooves, and the like, in any of the plasticating components: screw, tip, check ring and flights. **Contamination.** Check for foreign material resin.
Blisters on the part's surface	**Gas traveling across surface during fill or pack.** Check for moisture, trapped air, or excessive volatiles. Also, check for excessive decompression or suck back. **Trapped air due to inadequate venting.** Check number of vents and check vent depth. **Trapped air due to flow pattern.** Perform a short shot sequence changing transfer position or shot size to make various-sized short shots. Check trapped air in blind ribs. Jetting. Simulate mold filling to predict filling pattern. **Trapped air due to decompression or suck back.** Minimize decompression or screw suck back, especially with hot runners or hot sprues. Maintain proper nonreturn valve function. **Trapped air due to low L/D screw (L/D$\leq$ 18).** Raise back pressure to 1000 to 1500 psi melt pressure. **Degraded resin or additive package.** Check melt temperature. Minimize residence time. Try virgin resin.
Blooming	**Additive migrating to the surface of the part.** Try different lots or grade. Try slower injection rate. Seek a different additive or formulation. **Inadequate venting.** Clean vents. Add more vents. **White powder on part or mold surface.** Polyacetal may be depositing formaldehyde on mold. Dry the acetal before processing. *See also Mold build up, Surface finish, and Blush*

Continued on next page

Problem	Possible causes and recommendations
Blush	**Improper temperature control of nozzle.** PID temperature control nozzle and tip. Minimize length of nozzle. **Sharp corners at the gate.** Provide a minimal radius to reduce shear at the sharp corner. **Improper gate location.** Change gate location **Size of the gate.** Decrease land, increase area of the gate. This will change gate seal time! **Fast injections speeds.** Change injection velocity. Slow fill as plastic enters gate area then increase velocity. **Too high or low melt temperature.** Target middle of melt range specified by resin producer. **Too low or high a mold temperature.** Try high end and low end of supplier recommended mold temperature. **Trapped gas.** If blush is near an area of non-fill or short, check venting. Pull a vacuum on the mold cavity during fill. **If just in a small area** Look for excessive ejector speed, part sticking remedies or hot spot on the mold surface. See Sticking.
Brittleness	**Defining the brittle areas.** Determine if brittleness is throughout the part. Check molecular weight degradation via MFI. If brittle at lower temperatures check T_g of material for proper resin selection. **Molecular weight degradation due to excessive temperatures.** Check melt temperature, via the hot probe IR sensor. Check barrel heats and duty cycles. Check for temperature zone override. Check for screw and barrel wear. Check for damaged nonreturn valve. Check residence time in the barrel. **Too much degraded regrind or contamination.** Check quality of regrind. Run virgin resin. **Part not conditioned properly.** PA needs to regain moisture to achieve properties. Parts not conditioned according to resin specifications. **Too much molecular orientation.** Check part with cross-polarizers. Gate location. Sharp corners. Wall uniformity.
Bubbles	**Trapped gas.** Generate short shoots. Simulate mold filling. Check for race-tracking, jetting, or if ribs covered before filling. Check for moisture. Change gate location. Pull a vacuum on the mold cavity during fill. **Vacuum voids.** Voids occur upon cooling in thick sections. Increase packing. Bottoming out screw. Open the gate for longer gate seal times. Increase runner diameter. Raise the mold temperature. Eject the part sooner to allow the outside walls to collapse more upon cooling. Reduce melt temperature.
Burns	**Trapped air or volatiles dieseling.** Air trapped with gases in a mold. Check for venting. Core pins not vented. **Excess volatiles.** Large percentage of mold release agents, lubes, or other additives in resin. Lower melt temperature if allowed by the resin supplier. Change resin. **Excessive decompression.** Decompression pulls air into the nozzle. Reduce amount of decompression.
Burns at the gate.	See *Burns*; see also *Black streaks*.
Check ring.	See *Nonreturn valve*.

Continued on next page

4.4 Injection Molding

Problem	Possible causes and recommendations
Cloudiness or haze in clear parts.	**Excess additive or noncompatible additive package.** Try a different lot, change material supplier. **Contamination.** Clean barrel and screw. Check resin for contamination. **Nonuniform melt.** Check melt quality by adding a small amount of compatible color granules. **Incorrect mold temperature.** Raise mold temperature. Lower mold temperature for amorphous PETG. **Moisture in resin.** Check moisture content. Check dryer. Check dew point. **Worn or improper mold texture or surface.** Check mold surface for deposits, plate-out. **Crazing.** Check for stress in the area; for example stress whitening near ejector pins or places where the part sticks. Check for chemical attack.
Color Mixing	**Back pressure incorrect.** Increase back pressure to 1000-1500 psi. **Melt temperature.** Check resin supplier's recommendation. Use the hot probe technique or IR sensor. **Color concentrate carrier incompatible with base resin.** Concentrate should be of same resin family. Ask supplier for compatibility. **Incorrect colorant let down ratio.** Assure correct ratio of colorant to base resin. **Melt uniformity.** Check melt uniformity with few colored pellets. If you have a melt uniformity issue, specify a new screw; see "Screw design" below. Worn barrels and screws provide better mixing due to back flow. **Low L/D screw.** Try higher back pressure as above. Use precolored resin or get a melt uniformity screw with mixing pins. **Fast screw RPMs.** Slow screw rotation. **Screw design.** Standard general-purpose screws are known to provide unmelted solids within the melt. Do the melt uniformity test above. Use melt uniformity screws. Minimum recommended L/D is 20. **Miscellaneous.** Use the maximum amount of regrind allowed. Regrind acts like additional coloring agent.
Color variation, instability, or color shift	**Melt temperature too high.** Check resin supplier's recommendation. Check via hot probe or IR sensor. **Colorant thermally unstable.** Run at the minimum temperature allowed by supplier. Run the colorant at the maximum allowed by supplier. View parts in a light booth. Thermally stable colors will look the same. **Incorrect coloring agent.** Check the coloring agent for correct color, type, and base resin. Consult supplier. **Long residence time.** Use 25 to 65% of the barrel capacity. Minimize cycle time. **Virgin resin color shift or instability.** Check color of different virgin lots. Check if resin is thermally stable. **Gloss differential between parts.** Check gloss level for mating parts. **Surface finish of the mold.** Check finish of mold cavities for mating parts. **View angle different.** Disassemble the components and look at them at the same angle. **Colorant let down ratio not constant.** Check blend ratio and test for color accuracy. Recalibrate feeder. Check virgin only samples for color accuracy. Check level and quality of regrind.

Continued on next page

Problem	Possible causes and recommendations
Core pin bending or core shift	**Nonuniform pressure on the core or core pin during first-stage of filling.** Lock or support the core during fill. Change gate location. Perform a 99% full short shot and check core shift: problem may occur during second-stage. **Nonuniform pressure on the core or core pin during second-stage or pack, hold and cooling.** Do a gate seal analysis. Inspect parts for core shift with and without gate seal. Set second-stage time from the best results of the gate seal study. Choose pressure that provides least core shift.
Cracking (If troughout the part)	**Molecular weight degradation.** Check part and resin for proper MFI. Check for contamination. **Improper resin.** Check grade and type of resin. **Improper type or amount of colorant.** Check let down ratio and type of colorant carrier.
Cracking (If localized).	**Solvent, surfactant, or chemical attack.** Inspect mold and part handling for contamination by oils, solvents, fingerprints, mold sprays, soaps and cleaners. **Contamination.** Check localized area for off color or foreign material. Check regrind. **Exposure to radiation.** Check the resin for UV, sunlight, or gamma radiation stability.
Crazing	**Solvent, surfactant, or chemical attack.** Inspect mold and part handling for contamination by oils, solvents, fingerprints, mold sprays, soaps and cleaners. **Exposure to radiation.** Check the resin for UV, sunlight, or gamma radiation stability. **Part distorting upon ejection.** If craze is near an ejector pin the part may be sticking in the mold; see *Sticking*.
Cycle time too long	**Thick or nonuniform nominal wall.** Keep wall uniform within design guidelines. Maximum variation for amorphous is 20 to 25%; for semicrystalline it is 10 to 15%. **Slow filling.** Increase injection rate. First stage must be velocity controlled filling 95 to 99% of the cavity. **Improper cooling.** Cooling temperature entrance and exit difference is 4°F or 2°C. Assure water flow is turbulent, Re >5000. **Robot movement too slow.** Optimize or update robotics. **Mold movement too slow.** Optimize mold opening speed. **Long or excessive ejection strokes.** Optimize speed of ejection, but avoid pin marks. **Screw recovery too long.** Set rear zone to lowest recommended temperature. Note recovery time. Raise rear zone temperature 15°F (8°C) and note recovery time. Repeat until rear zone is at maximum recommended temperature setting of resin manufacturer. Plot data and pick best temperature for minimum recovery time. Keep backpressure and rpm constant. **Not enough ejector pin area.** Add or enlarge ejector pins. Larger surface area will allow ejection earlier. **Excessive second stage time.** Perform gate seal experiment to optimize second-stage time setting.
Dark streaks	See *Black streaks, Color mixing, and Black specks*.
Deformed or pulled parts	See *Sticking and Warp*.
Degradation	See *Black specks*.

Continued on next page

Problem	Possible causes and recommendations
Delamination	**Contamination.** Incompatible resin contamination. Run virgin resin. Check regrind. **High molecular orientation at the surface.** Try a hotter mold, slow injection rate down for longer fill times. Change gate location.
Dimensional variation	**First stage not consistent.** Take second-stage pressure to minimum. If minimum is above 1000 psi (70 bar) plastic pressure take second-stage time to 0 seconds. Do not reduce first stage pressure to do this test (velocity control). Make parts 99% full and run 20 shots. If inconsistent fix the nonreturn valve. **Pressure gradient variations.** Do a gate seal experiment. It is critical to know if you are using the correct amount of second-stage time. This time should not vary run to run. Gate seal, (longer second-stage times) packs the most plastic into a part and holds it in. Check resin viscosity and percentage of water if hygroscopic. **Too little second-stage pressure.** Increase the amount of second-stage pressure (avoid flash). Switchover from first to second stage is less than 0.1 seconds. **Nonreturn valve malfunction or screw and barrel wear.** Check nonreturn valve function, check cushion repeatability. Try adding more decompression to help seat the check ring. Watch out for splay development with too much decompression. Repair if more than 0.003 inches clearance between barrel wall and flight. Check machines repeatability of second-stage pressure, it should be < +/- 5-10 psi. **Degree of crystallinity.** Check cooling rate by measuring mold temperature. Hot mold yields higher crystallinity, more shrinkage and a smaller part. Colors affect crystallinity and therefore size of parts. Try nucleating agent. **Too much post mold shrinkage.** Longer cooling time may hold the part to size but is costly. Rapid cooling prevents some shrinkage. **Low melt temperature.** Increase melt temperature. This may lower density and allow for more shrink. **Gate too small, gate seal occurs too quickly.** Enlarge gate. Add a filler such as glass fibers to reduce shrinkage. **Improper measurements.** Check for identical conditioning of parts before measurements are taken. Have the parts been cooled the same and for the identical times after molding. Check measuring device for calibration. **Too much second-stage pressure.** Reduce second-stage pressure. **Degree of crystallinity.** Check cooling rate by measuring mold temperature. Cold mold yields lower crystallinity, less shrinkage and a bigger part. Colors affect crystallinity and therefore size of parts. Try nucleating agent. **Not enough post mold shrinkage.** Reduce cooling time to eject the part hotter to allow for more shrinkage. **Part sticking to the ejector side.** See sticking in mold, especially gate seal analysis, mold issues, and cooling or mold temperature. **Ejector velocity is to fast.** Slow velocity of ejection. **Not enough ejection pin surface area.** Add ejector pins or use larger diameter ejector pins. **Ejector plate cocking.** Check the length of the knock out bars, they should be identical in length, maximum difference is 0.003 inches. **Ejector pins not all the same length.** Check ejector pins for proper length.

Continued on next page

Problem	Possible causes and recommendations
Flash	**Parting line mismatch or mold damage.** If minimum second stage pressure is above 1000 psi (70 bar) take second-stage time to 0 seconds (velocity control). If part is short, inspect part for flash. If flash is present, clean mold surfaces and carefully inspect for any material on the surface or in the cavity that would prevent proper clamping at the parting line. Mold another shot under same conditions. If flash is still present, there is a parting line mismatch or tool damage.
	Clamp pressure too high. If flash is concentrated in the center of the tool, lower the clamp tonnage. Small molds in large platens tend to cause platen warp. Molds should take up 70% of the distance between the tie bars. **Melt temperature too high.** Take melt temperature via hot probe or IR sensor. Adjust if necessary. Reduce residence time if degradation is possible. Check MFI before and after molding. **Viscosity too low.** If resin is hygroscopic check moisture content. Check for correct resin. Try different lot. Check for degradation. **Mold improperly supported.** Check number, placement and length of support pillars in mold.
Flow lines	See *Weldlines*.
Gate blush	See *Blush*. **Gate design.** Break any sharp corners on the gate. Change gate type. **Gate location.** Change gate location. **Angle of impingement from gate.** Change the angle at which the plastic enters the cavity. **Injection velocity too fast.** Slow injection rate to avoid blush. **Wrong mold temperature.** Raise or lower the mold temperature significantly. **Wrong melt temperature.** Check nozzle for proper size and temperature control. Nozzle and tip should be as short as possible.
Gloss	**Mold surface finish.** Clean and inspect mold surface finish. See Mold build up. **Incorrect mold temperature.** A hotter mold provides higher gloss. **Pressure in the cavity.** Check second-stage pressure. Higher pressures increase gloss. **Injection velocity.** Increase injection velocity for higher gloss. **Resin type.** Certain ABS and PC/ABS resins are made specifically for high or low gloss applications. For high gloss you want emulsion-type ABS. For low gloss use mass-type ABS. **Venting.** If gloss is different in small areas, check venting.
Hot tip stringing	**Residual melt pressure.** Plastic is compressible, during screw rotation with 1000 to 1500 psi of pressure, the plastic is compressed. Use screw decompress to relieve this pressure; be careful not to pull air into the nozzle. **Volatiles.** Moisture and low molecular weight additives can turn to gas under heat. This gas formation can provide enough pressure to push molten plastic out the tip. Trapped air in the hot manifold or runner system will do the same. **Poor hot tip design or geometry.** Check tip design and clearance with gate surface. Check thermocouple placement and heater energy distribution. **Poor temperature control of the hot tip/high temperature.** Check thermocouple placement in hot tip.

Continued on next page

Problem	Possible causes and recommendations
	Gate location. The gate should not be in the center of the nominal wall. Plastic must have something to stick to or impinge upon as it enters the cavity. **Gate type.** Break any sharp corners of the gate. **No flow front formation.** Provide an impingement pin 40 to 70% of the nominal wall thickness just opposite the gate. Withdraw pin during second stage or pack and hold by a spring or hydraulic mechanism. **Fast injection rate.** Slow injection rate, but consistency and overall part quality will suffer.
Knitliness	See *Weldlines*.
Mold build-up or deposits	**Additive package; i.e. mold release, stabilizers, colorant(s), antioxidants, antistatic agents and flame retardant packages.** Try natural resin with no colorant. Use lower additive content. **Additive package temperature stability.** Reduce temperature to lowest recommended by supplier. Minimize residence time in the barrel by shorter cycle times or using a small barrel. Shot size should be 25 to 65% of the barrel capacity. **Inadequate venting** Check vents for proper depth and number. Provide a vacuum to the cavity during fill. **Moisture.** Check moisture level in resin. If high humidity environment, run with dehumidifier over mold surface. **Mold temperature.** Check for high mold steel temperatures in the problem area. **Contamination.** Check for contamination. **Infrequent mold cleaning.** Clean molds. **Injection rate too high.** Fast injection rates can lead to shear separation of low viscosity additives which bloom to surface. **Mold steel-resin incompatibility.** Certain metals and resins are incompatible. **White powder on surface of mold or part.** Flame retardant additives sometimes bloom. Dry acetals before processing.
Nonfill	See **Shorts**.
Nonreturn valve	**Performance test.** Turn first-stage pressure limit down to the normal second stage pack and hold pressure for this mold. Bring the screw back to 90% of the full shot and inject for 10 seconds with the previous part still in the mold, runner and sprue. Note that screw should not drift forward. Repeat at the 50% shot size and again at the 10% shot position. Screw drift indicate barrel or nonreturn valve wear. **Proper design.** The seat and the sliding ring often incorrectly have mating angles. *Few valves are built correctly.* Install step angle valves.
Nozzle drool or stringing	See *Hot tip stringing*. **Volatiles.** Moisture and low molecular weight additives form gas. This provides enough pressure to push molten plastic out the nozzle. Trapped air in the screw flights will do the same.

Continued on next page

Problem	Possible causes and recommendations
	No decompression after screw rotate. Plastic is compressible. During screw rotation with 1000 to 1500 psi plastic pressure the plastic is compressed. Upon sprue break or mold open, plastic will be pushed out of the nozzle. Use screw decompress to relieve this pressure. **Incorrect or poor temperature control on the nozzle.** Nozzle temperature is often poorly controlled. Placement of the thermocouple should be one third the distance of the nozzle from its tip if he nozzle is more than 75 mm long. For 75 mm lengths and shorter it is OK to have the thermocouple imbedded into the hex on the nozzle body. Do not allow the thermocouple to be attached via the screws on the clamp for the heater band. **Wrong nozzle length.** Try a shorter nozzle. **Wrong nozzle tip.** Avoid standard tips, try a reverse taper, nylon or a full taper, ABS tip.
Nozzle free or cold slug	**Nozzle tip too cold.** The sprue bushing is a heat sink for the nozzle tip and draws heat from the tip while in contact. Use sprue break if possible. Insulate the tip from the sprue bushing with high temperature insulation. **Poor temperature control of nozzle.** Placement of the thermocouple should be one third the distance of the nozzle from its tip if he nozzle is more than 75 mm long. For 75 mm lengths and shorter it is OK to have the thermocouple imbedded into the hex on the nozzle body. Do not allow the thermocouple to be attached via the screws on the clamp for the heater band.
Nozzle stringing	See *Nozzle drool*.
Odor	**Polymer degradation.** When overheated some polymers break down to acid gases such as PVC, Acetal (POM), and certain flame-retardant resins. Check for excessive temperatures, high back pressure, worn or scored screw or barrel. **Contamination.** Check resin. Contaminates may cause the odor themselves or cause the plastic to degrade.
Orange peel.	**Mold build up or deposits** Check for residue or deposits on the mold/cavity surface. If there are mold deposits, see *Mold build up* **Mold surface finish.** Check surface polish, finish and dirt. Repair and clean. **Slow filling.** Increase injection rate, this decreases resin viscosity and allows more pressure to be transferred to the cavity if first- to second-stage switchover is <0.1 seconds. Ensure velocity control is not pressure limited. **Low cavity pressure.** Increase second-stage pressure. Increase second-stage time, maintaining cycle time constant. **Mold temperature.** Increase mold temperature. Decrease mold temperature. **Melt temperature.** Check if melt temperature is within recommendations. **Uneven filling of a single cavity.** Balance flow path with flow leaders if possible. Increase injection rate. **Unbalanced filling in multicavity molds.** Adjust runner size to balance filling.
Pinking of the part	**Carbon monoxide.** Carbon monoxide discolors certain resins. Remove parts to open area. Expose part(s) to sunlight. If discoloration reverses, remove all gas fueled lifts, etc., from storage area. Improve storage area ventilation. Go to battery-operated fork lifts.

Continued on next page

4.4 Injection Molding

Problem	Possible causes and recommendations
Pitting	**Trapped gases dieseling.** See *Burns*. Dieseling will damage mold. **Corrosion or chemical attack by the resin or additive on the steel.** Check for resin compatibility with the steel of the mold. If acid gases are possible, a more chemically resistant surface may be required. A different steel or coating the existing surface should be specified. **Abrasive wear, erosion.** Highly filled resins erode a mold. Change gate location, coat cavity with a wear resistant finish. Rebuild tool with appropriate hardened steel.
Poor color mixing	See **Color mixing**
Racetracking, framing, or nonuniform flow front	**Nonuniform wall thickness.** Preferential flow occurs in thicker sections. Round the edge or taper the junction between the nominal wall change. Use uniform nominal wall. **Gate location.** Gate into the thick area and provide flow leaders to the thin areas to provide uniform filling. **Hot surface or section in the mold.** Allow the mold to sit idle until mold is at uniform temperature. Make and save first shot for 99% full and compare to later shots.
Record groves, ripples, wave marks	**Pressure limited first stage or lack of velocity control.** Check that the pressure at transfer is 200 to 400 hydraulic psi lower than the set first stage limit. **Incorrect position transfer.** Take second-stage pressure to 1000 psi (70 bar) plastic pressure or, if the machine does not allow this, take second-stage time to zero. The part should be 95 to 99% full. **Melt temperature too low.** Check melt temperature via the hot probe or IR sensor. **Poor first- to second-stage switchover response.** Note response of hydraulic pressure at switchover. It should rise to the transfer point then drop rapidly to the set second-stage pressure. If hydraulic pressure drops much below set second-stage pressure then the flow front may be hesitating and building a high viscosity. Repair machine or valve. **Low pack rate or volume.** Increase pack rate or volume of oil available for second-stage. **Low mold temperature.** Increase mold temperature 20 to 30°F. Decrease cycle time to increase mold temperature.
Screw design	**Standard general purpose screw.** These may produce unmelt. Replaced with melt uniformity screws of $L/D > 20$. **Screw and barrel metals.** Use bimetallic or hardened barrels and soft screws like stainless steel. Chemically resistant screw material is especially critical for clear resins. Screw should be polished with large radii to prevent dead spots. A modified barrier should transition the transition zone to the metering zone. **Barrier flights.** Generally not recommended unless short and at the end of the transition zone or beginning of the metering section. **Vented barrels.** Vented barrels provide excellent melt uniformity and process resins that are not subject to hydrolysis more uniformly. The two-stage screw must be designed with a continuous flight through the decompression section. The first stage should be cut such that it cannot overpump the second stage. Vented barrels require near zero back-pressure to prevent vent flooding.

Continued on next page

Problem	Possible causes and recommendations
Screw recovery, slow recovery, screw slips, or does not feed	**Feed throat temperature.** Run throat temperature at 110 to 120°F for most resins. For high-end engineering resins you may want to go higher. Do not run feed throat at 60 to 80F. Feed throat should be PID temperature controlled. **Feed problems.** Check size of granules and flow through hopper and feed throat. Ensure that material gravity feeds correctly when resin is being loaded into the hopper. Vacuum loading may interrupt normal gravity feeding. **Heavily carbonized or blocked flights.** Check for dead spots behind flights of general purpose screws. **Worn screw and/or barrel.** Worn screws and barrels will provide better mixing, but slow recovery rates as plastic flushes back over. **Moisture.** Check moisture content of plastic; check feed throat for cracks, leaking water. **Granule size.** Plastic granules should be uniform in size and shape. A wide range in granule size, fines, and small granules, along with large chunks of regrind, will cause feeding problems. **High back pressure.** Target 1000 to 1500 psi melt back pressure. Try lower back pressure. **High RPM.** Try lower screw rotate speeds, better melt uniformity, and mixing is obtained with slow screw speeds. Use all but ~ 2 seconds of the cooling time for plasticating. Do not lengthen the cycle. **Incorrect barrel temperature settings.** Check recommended range. Set rear zone at the minimum of the range. Back pressure set at 1000 melt psi (70 bar). Average recovery time for 10 cycles. Repeat with rear zone 10°F higher until you have reached the rear zone setting at the maximum recommended by the resin supplier. Pick the temperature that gives you the minimum recovery time. **Poor screw design.** See *Color mixing* and *Screw design*.
Screw slip.	See *Screw recovery*.
Shorts or short shots or nonfill. (Consistent short shots)	**Incorrect shot size.** Take second-stage pressure to 1000 psi (70 bar) plastic pressure or, if the machine does not allow this, take second-stage time to zero. The part should be 95 to 99% full. Unless this is a thin-walled part, part should be full with only slight underpack near the gate. Adjust position transfer to provide appropriate fill volume. Do not use first-stage pressure limit to do this test or make appropriate volume on first stage. **Pressure limited first-stage or lack of velocity control.** Double check that the pressure at transfer is 200 to 400 hydraulic psi lower than the set first stage limit. **Injection rate.** Increase injection rate to decrease viscosity. **Nonreturn valve or barrel worn or broken.** Check nonreturn valve and barrel. If the nonreturn valve is OK, check barrel for wear and ovality. **Large pressure drop.** Find if there are blockages. **Trapped gas or air.** Short shoot the mold to determine racetrack, jetting and ribs covered before filling. Check for moisture. **Insufficient second-stage pressure.** Make sure first stage is at the right shot volume. If OK, raise second stage pressure. **No cushion.** Ensure adequate cushion to allow for packing or second-stage pressure.

Continued on next page

Problem	Possible causes and recommendations
	Melt temperature. Verify melt is within recommended range. **Mold temperature.** Try higher mold temperatures and or faster cycle times. **Resin viscosity.** Change resin with higher MFI. **Long flow length.** Add gates or flow leaders. **Thin nominal wall.** Add flow leaders.
Shorts or short shots or non-fill. (Intermittent short shots)	**Nonreturn valve or barrel worn or broken.** Check nonreturn valve and barrel. **Cushion not holding.** Check cushion repeatability. If variability by more than 0.200 in (5 mm), check nonreturn valve. Increase decompression stroke. **Contaminated material.** Check gates and parts for foreign material. Check regrind. **Melt temperature.** Verify melt is within the recommended range. **Mold temperature.** Try higher mold temperatures and or faster cycle times. **Unmelt.** Look for unmelted granules in the part, color streaks. See *Screw design and introduction to Color mixing*. **Cold slug.** Check nozzle for cold slug formation. See *Nozzle drool*. **Insufficient second stage pressure.** Raise second stage pressure. **Trapped gas.** See earlier; see *Bubbles*.
Shrinkage. (Too much shrinkage)	**Insufficient plastic.** Make sure cut-off position provides the right volume of plastic on first stage. Do a gate seal analysis. Add more second-stage time to achieve gate seal. Add more second-stage pressure. Check cushion. **Degree of crystallinity.** Lower mold temperature to increase cooling rate and decrease crystallinity. Add nucleating agent. **Low second-stage pressure.** Raise second stage pressure. See *Insufficient plastic*. See *Short shots*
Shrinkage. (Too little shrinkage)	**Too much plastic.** Make sure cut-off position provides the right volume of plastic on first stage. Do a gate seal analysis. Add more second-stage time to achieve gate seal. Add more second-stage pressure. Check cushion. **Degree of crystallinity.** Raise mold temperature to decrease cooling rate and increase crystallinity. **High second-stage pressure.** Lower second-stage pressure. Lower second-stage time to allow for gate unseal. See *Flash*.
Silvery streaks	See *Splay*
Sinks or sink marks	**Insufficient plastic.** Pack more plastic into the cavity with consistent cushion, higher second-stage pressures, longer second-stage time, slower fill rates, counter pressure, opening gate to increase gate seal time, or increase the runner diameter. Lower the mold temperature. Reduce melt temperature. Change gate location; fill thick to thin. **Postmold cooling.** Cool in water or between aluminum sheets rather than air. **Thick nominal walls.** Core out to thin the nominal wall.

Continued on next page

Problem	Possible causes and recommendations
Splay and silver streaks	**Moisture in the plastic granules.** Check moisture content. Check dryer.
	Dryer not functioning properly. Check for the correct drying temperature set point. Remove material from the dead space the cone of the hopper. Check for fines clogging the filters. Check that the regeneration heaters are functioning. Check the temperature of the air going into the desiccants. Check dew point of air before and after desiccants if possible.
	Condensed moisture on pellets. Check material.
	Moisture due to high humidity. Check relative humidity.
	Moisture in the feed throat. Check feed throat temperature and look for cracks.
	Moisture from mold. Check mold, particularly cores, for condensation. Raise mold temperature if condensation is present. Check for water leaks.
	Poor nozzle temperature control. Placement of the thermocouple should be one third the distance of the nozzle from its tip if the nozzle is more than 75 mm long. For 75 mm lengths and shorter it is OK to have the thermocouple imbedded into the hex on the nozzle body. Do not allow the thermocouple to be attached via the screws on the clamp for the heater band.
	Air trapped in melt from screw or decompression. Assure an L/D >18. Raise backpressure, slow screw rpm, and extend cooling time. Raise rear zone temperature. Use melt uniformity screw. See *Screw design*.
	Chips or flakes from three-plate molds. Open mold and carefully inspect for plastic chips, flakes, or particles. These can upset the flow front and cause marks similar to splay.
	Contaminants, dirt, or cold slugs. Solid particles can upset the flow front and cause marks similar to splay. Inspect pellets. Check for degradation on the screw.
	Unmelted granules and gels. Solid particles can upset the flow front and cause marks similar to splay. Unmelt can be a cause due to poor screw design or short L/D. Unmelt often occurs with general-purpose screws. Gels are high molecular weight or cross-linked resin that does not melt. Change resin.
	Volatiles due to degradation of polymer or additives. Gases can be generated by high shear, hot spots in the barrel, flame retardants, degraded regrind.
Sprue sticking	**Nozzle tip radius mismatch with sprue bushing.** Check if nozzle radius matches sprue bushing.
	Nozzle tip orifice too large. Check orifice of sprue and tip. Tip orifice should be at least 0.030 in (0.75 mm) smaller in diameter.
	Scratches or incorrect polish on sprue. Sprue should be draw polished to a ♯2 finish. Vapor honing is better for soft touch material. Circular polishing provides minute undercuts that get filled with plastic and "stick" the sprue. Look for scratches that form undercuts.
	Sprue puller problems. Check if sprue puller is large enough. Add undercuts, Z-puller, or more reverse taper.
	Sprue too soft, not frozen. If sprue is still soft at mold open, down size sprue or cool the sprue bushing.
	Improper taper on sprue. Taper should be 1/2 in (13 mm) per foot (305 mm).
	Overpacked sprue. Do a gate seal study. If possible take off some second-stage time and add to cooling or mold closed timer.

Continued on next page

4.4 Injection Molding

Problem	Possible causes and recommendations
Sticking in mold	**Mold build up or deposits.** Check for residue or deposits on the mold/cavity surface. See *Mold build up*. **Mold surface finish.** Check surface of cavity has proper finish and is clean. **Slow filling.** Increase injection rate to decrease resin viscosity and allow more pressure to be transferred to the cavity. Ensure velocity control is not pressure limited. **Low cavity pressure.** Increase second-stage pressure. Increase second-stage time, but, keep cycle time constant. **Mold temperature.** Increase mold temperature. Decrease mold temperature. **Melt temperature.** Assure that temperatures are within recommended ranges. **Uneven filling of a single or multicavities.** Balance flow path with flow leaders if possible. Increase injection rate. **Unbalanced filling in multicavity molds.** Adjust runner size to balance filling. **White powder build up on acetal parts.** Dry the polyacetal before processing. Improper polish or undercuts causing parts to stick to the "A" side of the mold upon mold opening. In manual mode, open the tool slowly and note any noises or cocking of the part. Before ejection, inspect part for deformation. High polish forms a vacuum. **Improper mold polish causing part to stick to the "B" side of the mold upon ejection.** Check if part sticks during ejection. Many soft-touch polymers need a vapor-honed surface to aid release. **Incorrect second-stage time or improper gate seal time.** Perform a gate seal analysis. Determine if gate seal or unseal changes the problem. **Overpacking, not enough shrinkage.** Reduce pack and hold second-stage pressure. Make a few parts with shorter pack and hold time to achieve gate unseal, maintaining constant cycle time. Try longer cycle time. **Underpacking causing excessive shrinkage.** Increase pack and hold pressure, increase second-stage time with constant cycle time. Try a shorter cycle. **Vacuum, high polish can cause a vacuum.** Provide a vacuum break before mold opening or ejection. **Excessive ejection speed.** Slow ejection velocity. **Parts handling or robot removal.** Check end of arm tooling and movement of the part on the robot arm. **Contamination.** Check regrind. **Plate out on the mold surface.** Inspect mold surface for build up. **Incorrect mold finish.** High a polish can cause vacuum. **Crazing.** Check for stress in the area near ejector pins or places where the part sticks. Check for chemical attack. **Mold temperature too cold or too hot.** Raise and or lower mold temperature significantly as long as you do not cause mold damage. Assure a Reynold's number of 5000 or greater to achieve turbulent flow. Make sure the difference in temperature between in and out of water lines is less than $4°F$ ($2°C$). **Ejector plate cocking.** Check for even length of knock out bars. Should be within 0.003 in (76μm) of each other.

Continued on next page

Problem	Possible causes and recommendations
	Mold deflection. Take dial gauge measurements on top and bottom of both sides of the mold. Repeat for both sides. Note mold deflection looking for deflections opposite each other. **Inadequate mold release in resin.** Add mold release.
Streaks	See *Color mixing* and *Black specks*
Surface finish	See *Orange peel* and *Color mixing*
Unmelted particles	See *Screw design* and *Color mixing*
Vent clogging	See *Mold build up*
Warpage(Warp for amorphous resins, low shrink.)	**Molecular orientation.** Try fast to slow injection rates. Raise mold temperature 20 to 30°F (10 to 15°C) may also help. Nonuniform nominal wall. **Pressure or pressure gradient.** Run a gate seal study. Change gate location. **Nonuniform cooling/nonuniform shrink.** Mold steel cooling problems or nonuniform wall sections. Adding cooling lines. Cored out thick sections. **Nonuniform wall thickness.** Cored out thick sections. **Warp** (Warp for semi-crystalline resins, high shrink.) **Molecular orientation.** Same as above. **Pressure or pressure gradient.** Same as above. **Nonuniform cooling/nonuniform shrink.** Same as earlier. Try a filled or nucleated version of the resin. **Nonuniform wall thickness.** Same as above. **Crystallinity.** Try an amorphous polymer. If the amorphous parts do not shrink, you have a crystallinity issue. Try a nucleated or filled version of this semicrystalline resin.
Weldlines	**No molecular chain entanglement.** Increase injection rate. Raise the mold temperature. Try a higher MFI, check properties and performance. Improve venting. Provide a flow channel at the weldline to aid venting and provide some movement at the weldline. This will help with some chain entanglement. Try resin without filler. Change gate location. If multigated part, try blocking some gates.
	Trapped air. Improve venting. Provide a flow channel to aid venting. Vent core pins. Add flow leaders to direct flow path. **Low pressure at weldline.** Increase second-stage pressure. Increase injection rate. Raise melt temperature as last resort.

4.4.4 Troubleshooting Injection Molding

Correct process control is critical for making identical parts to tight tolerances and meeting numerous quality standards. The troubleshooting table (Table 4.6[5]) presented above lists possible injection molding or process defects, with possible causes and remedies.

---

[5]This table is based on the table developed by J. Bozzelli and presented in T.A. Osswald, L.S. Turng and P.J. Gramann, *Injection Molding Handboook*, Hanser Publishers, Munich (2002).

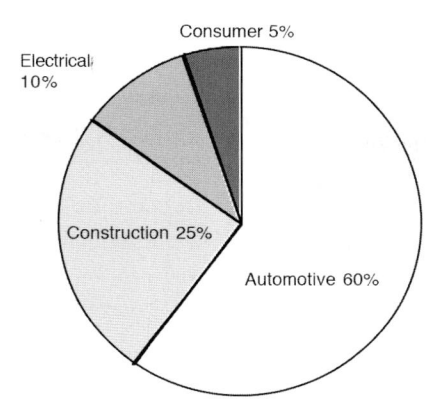

Figure 4.92: Application areas of SMC/BMC in North America.

4.5 COMPRESSION MOLDING

There are numerous types of compression molded parts and processes. The parts distinguish themselves from each other by the type of material used, the length of the reinforcing fibers, their size, and their surface quality. By default every type of material and part has its own unique variation of a compression molding or injection-compression molding process. However, here we will concentrate on four widely used processes: the compression molding of sheet molding compound (SMC) parts, the injection-compression molding of bulk molding compound (BMC) parts, and the compression molding of glass matt reinforced thermoplastics (GMT) and long fiber reinforced thermoplastics (LFT) . Both, SMC and BMC are fiber reinforced thermosetting materials, while GMT and LFT are fiber reinforced thermoplastic matrices. The common matrix in SMC and BMC materials is unsaturated polyester, cross-linked with styrene. Figure 4.92 presents a break-down of the industries that use SMC and BMC compression molded materials [4]. The automotive industry is the largest user of SMC/BMC materials. The total market of SMC/BMC was about 650 million pounds in North America for 2002. Of this, 60% was used in the automotive industry. Phenolic, the oldest synthetic polymer, is used with parts that require excellent part stability at elevated temperatures and high creep resistance. Furthermore, phenolic delivers those qualities at prices that are compatible with ABS.

The heart of the compression molding process is mold filling. Today, most compression molds are positive compression molds, such as the one presented in Fig. 4.93. With this type of mold the charge weight is measured exactly before it is placed inside the cavity. Since there is some variation in the weight of sheet

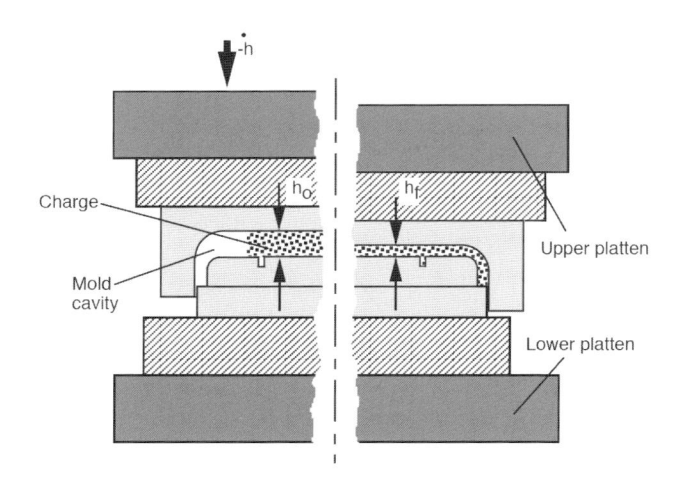

Figure 4.93: Schematic of the compression molding process.

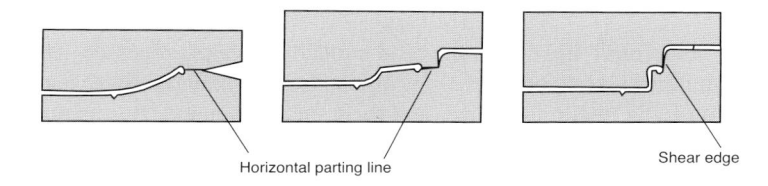

Figure 4.94: Schematic diagram of compression molds with their respective parting lines.

molding compound, it is often necessary to place small pieces of SMC on the center on the charge to complete the full weight of the part. Figure 4.94 presents three different types of compression molds with horizontal and vertical parting lines.

A schematic diagram of an SMC production line is depicted in Fig. 4.95.

The automotive industry has moved beyond just using compression molding to manufacture fiber reinforced components to reduce weight and increase fuel efficiency. Its versatility to produce high strength parts that are cosmetically superior to parts manufactured by other processes has led to its use for critical structural members and body panels on the most showcased sports cars in the world.

4.5 Compression Molding 371

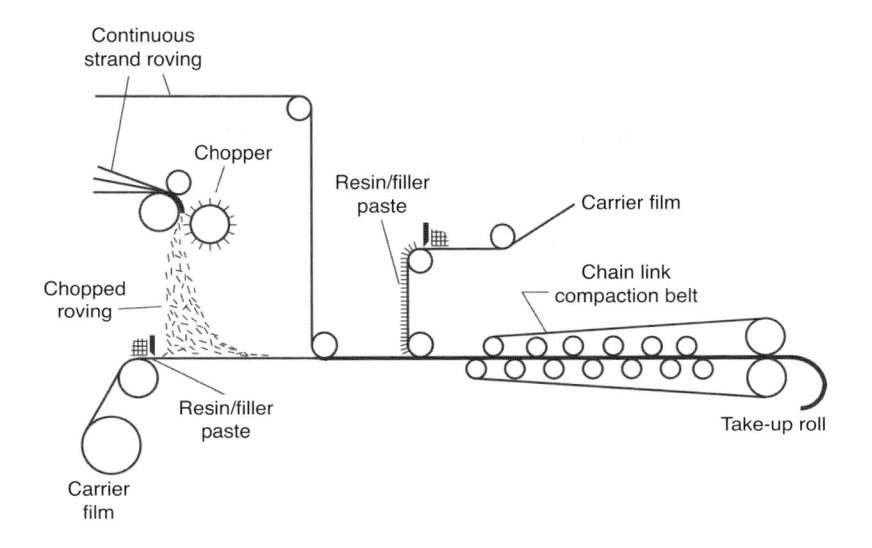

Figure 4.95: SMC production line.

4.5.1 Compression Molding of SMC and BMC

The compression molding process of sheet molding compound is schematically depicted in Fig. 4.96. As shown in the figure, a charge or charges are cut from a roll of SMC in a cutting table or station. Due to the somewhat inhomogeneous nature of the SMC roll, it is difficult to gage the weight of the charge from its area. Hence, often small pieces are either cut from the charge, or placed on top of the charge to achieve the desired weight of the final part. Once a finished part is removed from the mold cavity, a new charge is placed on the heated mold surface. At this stage, some one-sided heating may take place. Preheating of the charge will speed-up the curing process and is therefore sometimes desirable. The mold is closed rapidly until both mold halves touch the SMC charge, and then slows down, causing the material to flow and fill the mold cavity. During flow, the orientation of the fibers is changed, which has a profound impact on the properties of the finished part. In addition, if two or more fronts of material meet during mold filling, knitlines form which are the weakest area in a the part. Flow and deformation also forces air voids out of the material, eliminating some of the porosity inside the sheet. This will have a direct influence on the quality of the surface in the finished product.

It is generally desirable to have a random orientation in a final part. Due to the random loading a part may experience, it is necessary for it to be equally strong

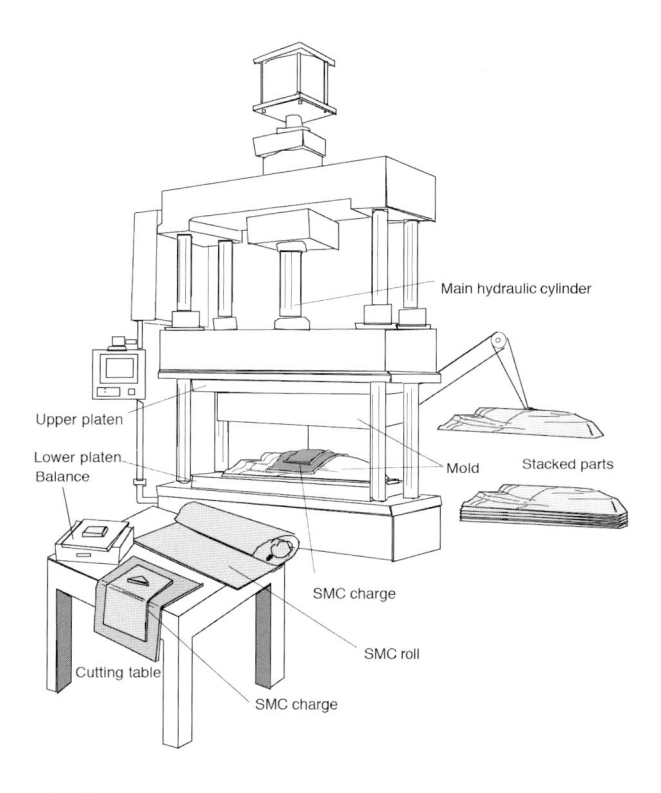

Figure 4.96: SMC compression molding process.

in all directions. A randon orientation can be attained when there is little to no flow inside the mold cavity. On the other hand, if little flow occurs during mold filling, a part may have poor surface finish. This is detrimental when molding class A automotive panels. Once the mold is full, a constant force is maintained on the material, as the curing of the thermosetting resin progresses. When the reaction is almost complete, the part is removed and is allowed to cool down. During the curing reaction and the cooling process, residual stresses form that sometimes, in conjunction with the anisotropy caused by the fiber orientation, cause the part to shrink and warp. The sequence of events during compression molding of an SMC part, as shown in Fig. 4.97, is called the SMC compression molding cycle. The cycle begins when the prepared charge is placed on the mold cavity and ends when the finished part is removed from the mold. As can be seen, the dominant event during the cycle is the curing process.

4.5 Compression Molding

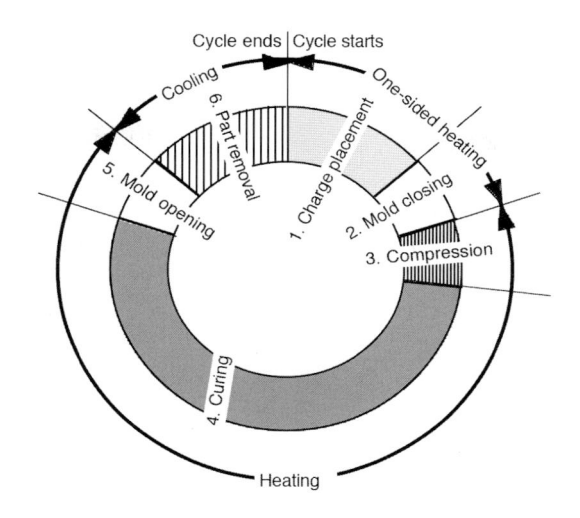

Figure 4.97: SMC compression molding cycle.

Depending on the thickness, type of matrix material, and processing conditions the curing time ranges from 20 s to over 1 min. Charge placement and part removal can take 10 s each, and the mold closing, compression and mold opening take approximatelly 5 s each. The compression molding of bulk molding compound (BMC) is very similar to the transfer molding process. The basic ingredients of BMC are very similar to SMC. The material is compounded in an internal batch mixer and subsequently extruded into the shape of a large sausage that is fed into a cylinder above the BMC press. From there, a charge is injected into the cavity and then compressed into the finished product.

4.5.2 Compression Molding of GMT and LFT

The compression molding process of glass matt reinforced thermoplastics is schematically depicted in Fig. 4.98. The process begins when the GMT preform is placed on the conveyor belt, that will guide it through the heating unit. The heating taking place inside the unit is often radiative heating; however, it can be contact heating, convective heating, or a combination of all three. Next the charge is lifted from the conveyor belt and placed on the cooler mold surface.

As in compression molding of SMC, here too, the mold cavity is closed rapidly first, until both mold halves touch the preheated charge. Next the mold is closed slowly as the charge deforms and fills the mold cavity as the part rapidly cools. Once the mold is full, the part is cooled further under pressure, allowing it to attain

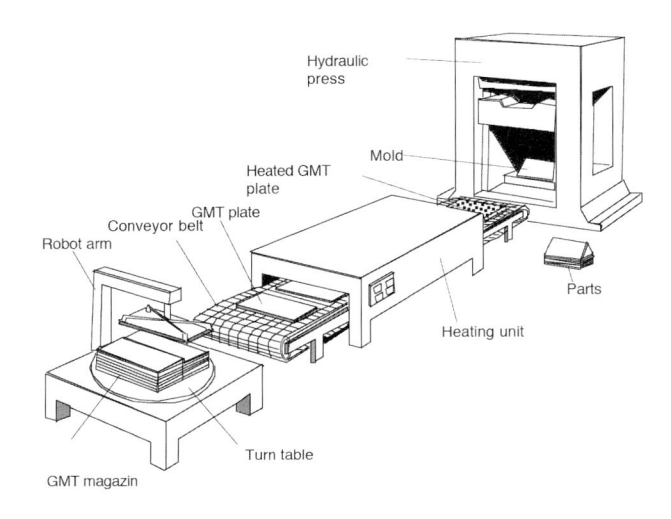

Figure 4.98: GMT compression molding process.

its structural integrity. Once cooled sufficiently, the finished part is removed from the mold cavity. Similar to the manufacturing of SMC parts, during molding of GMT, orientation as well as residual stresses play a significant role. The sequence of events during compression molding of a GMT part, as shown in Fig. 4.99, is called the GMT compression molding cycle. The figure clearly shows that by far the dominat event is the heating cycle of the GMT preform. This heating cycle can take between 60 and 90 s, and sometimes more.

Long fiber reinforced thermoplastics are a more recent development. Today there are various systems, ranging from in-line compounding systems to the extrusion of prefabricated thermoplastic resin coated glass. Basically, in all these systems a fiber filled sausage or charge is deposited in the mold cavity before mold closing. The LFT charge is deposited by an extruder or a reciprocating screw system. A schematic of an LFT molding process is presented in Fig. 4.100. The process starts with long fiber reinforced pellets being heated, plasticated, and deposited on the mold surface using an extruder or a reciprocating screw system. Long fiber reinforced pellets are prepared by wire coating, crosshead extrusion, or pultrusion.

Another variation of this process is combining the compounding of the fibers and the thermoplastic resins with the compression molding of the final part. This process completely eliminates the need of preparing or manufacturing the LFT pellets or GMT preforms. Here, the fiber rovings are continuously fed into the screw channel of the extruder in the region after plastication has occured. One

4.5 Compression Molding 375

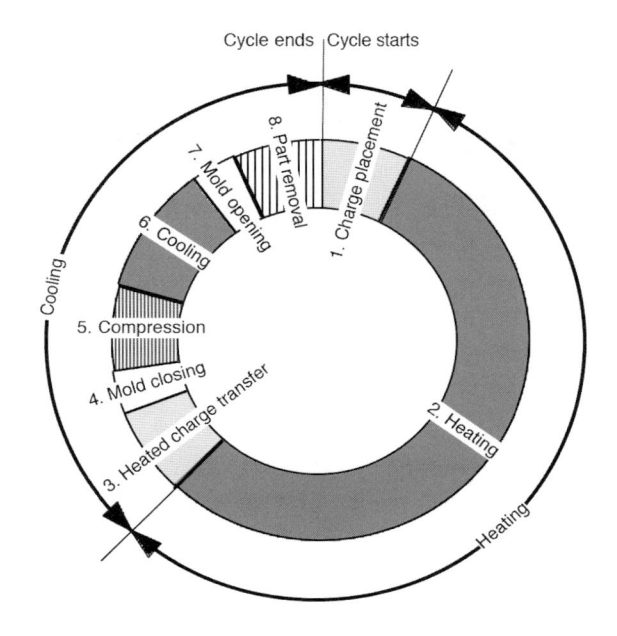

Figure 4.99: GMT compression molding cycle.

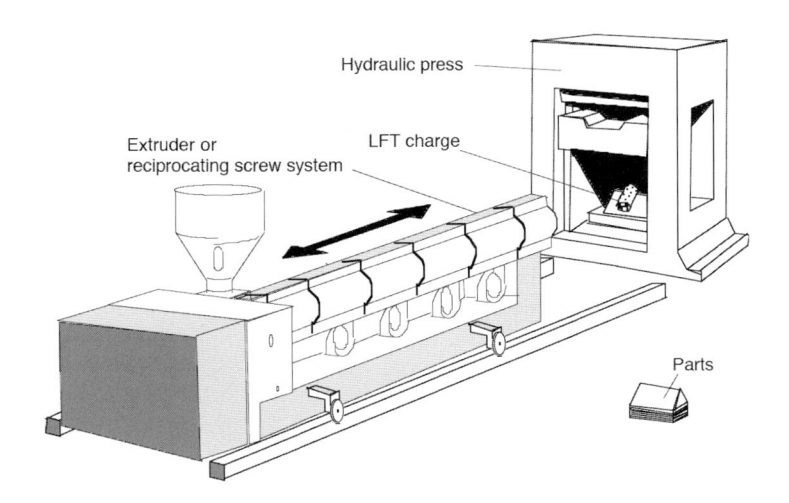

Figure 4.100: LFT molding process.

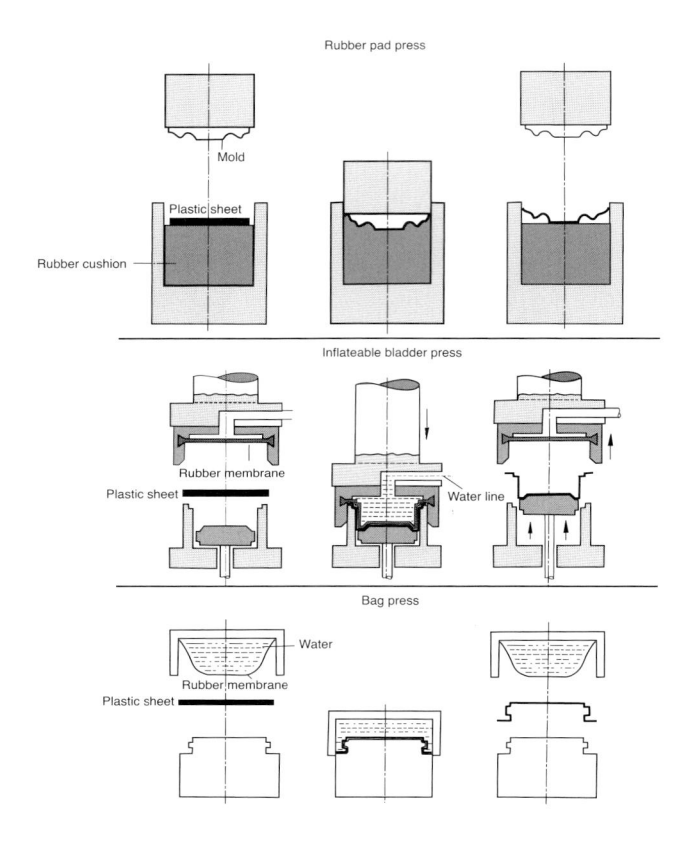

Figure 4.101: Schematic diagram of various fluid cell or flexform presses.

draw back of all LFT systems, whether in pellet form or in-line compounding, is that the fibers suffer significant damage during the extrusion process. The LFT molding cycle is very similar to the GMT molding cycle. The heating process of the GMT process is replaced by the plastication or the compounding of the LFT process.

4.5.3 Cold Press Forming

Semi-crystalline polymers have the ability of being shaped at slightly elevated temperatures, and in some cases even at room temperature, as is often done for metals. One technique, used for shaping plates or sheets, is the so-called *flexform press*, also referred to as *fluid cell press*. Here, a flexible rubber diaphragm, some-

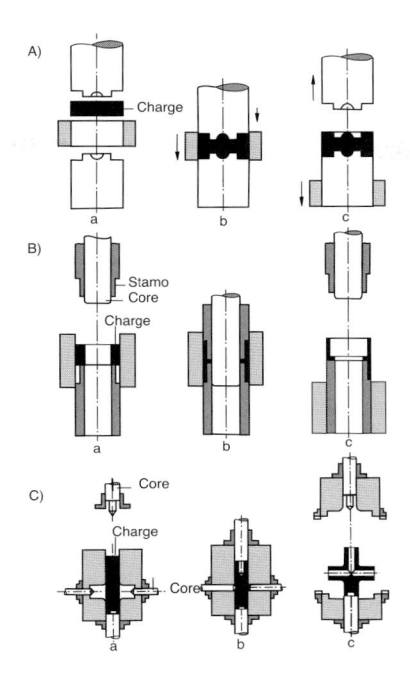

Figure 4.102: Schematic diagram of various drop- or die-forging applications.

times referred to as *bladder*, is used along with a single rigid tool, as schematically depicted in Fig. 4.101. The figure shows three different types of bladder presses.

To shape three-dimensional parts, of more complex geometries, it is common to use *drop-forging presses*, also referred to as *die-forging presses*. Figure 4.102 presents several examples for shaping wheels, tubular parts, and fittings.

4.5.4 Troubleshooting Compression Molding

This troubleshooting guide covers common problems encountered with compression molding of thermosetting parts. Table 4.7[6] presents several defects with possible causes and remedies.

---

[6]Parts of this table were contributed by J. Horner and presented in B.A. Davis, P.J. Gramann, T.A. Osswald and A.C. Rios, *Compression Molding*, Hanser Publishers, Munich (2003).

Table 4.7: Troubleshooting Guide for Compression Molding

Problem	Possible causes and recommendations
Air burns or dieseling	**Entrapped gases.** Clean shear edges and vents. Check for enough clearance at shear edge or vent ejector pins. Add vents or vent ejector pins as needed. Add a breath cycle. Vary initial charge layout.
Blisters	**Entrapped gases.** See above. Wrong combination of closing speed and mold temperature. Decrease mold temperature. Add or change breath cycle. Decrease clamping pressure. Increase material flow length. Vary initial charge layout. **Air within glass fiber** Fibers can sometimes be poorly impregnated with resin leading to voids and blisters. Lower glass content. Check for moisture and contamination. Check for dried out material **Under - cure.** Sometimes blisters can be confused with under-cured regions. Under-cured regions may leave a raised area and sometimes have material extruded from the opening.
Chipped edges	**Part sticking to the mold.** Decrease temperature differential between mold halves. Decrease mold opening speed. Decrease ejection speed. Decrease mold closing speed. Decrease clamping pressure. Vary initial charge layout and/or initial charge weight. Reduce undercuts. Clean and polish problem areas. Refit and chrome mold. Add draft angles. Add external mold release.
Contamination	**Mold/equipment contamination.** Check for dirty handling equipment. Clean or polish mold. Check compressed air lines for contaminants. Add external mold release. **Material contamination.** Check for moisture or impurities in the charge. In some cases it can lead to part discoloration and parts sticking in the mold. Contaminants can also be a cause of stress concentrations leading to part failure.
Cracking	**Part sticking to the mold (see also sticking part).** Vary mold halves temperature differential. Decrease ejection speed. Decrease mold opening speed. Decrease material flow length. Increase cure time. Vary initial charge layout and initial charge weight. Check and eliminate knit-lines at the problem area (see Knit Lines). Check for drag marks and undercuts (see Drag Marks). Clean, polish, or chrome mold. Add external mold release. Check parallelism during ejection. Check mold parallelism. Check for large clearance on ejector pin. Check or add ejector pins on the problem section. Choose tougher compound. Add proper draft angles. Eliminate abrupt changes in thickness.
Drag marks	**Excessive friction with mold (see also Sticking Part).** Increase curing time. Check for burrs at the shear edge, side walls, or ribs. Check for undercuts. Add proper draft angles. Polish or chrome mold.
Dull part surface	**Improper mold temperature (also see Heat Smear).** Examine the location where the initial charge was placed. If the part looks glossy here and diminishes in appearance further away from the point of flow origination (assuming uniform mold heat) then the mold is too hot. If the part looks dull at the initial charge location but gains in appearance as the material flows then the mold is too cold.

Continued on next page

4.5 Compression Molding

Problem	Possible causes and recommendations
	Charge placement. Irregular and isolated dull regions may be an indication of poor mold contact with the part. In this case, this defect is known as laking (see Laking). Increase initial charge weight. Decrease material flow length. Increase curing time. Increase mold clamping pressure. Clean or polish mold. Check mold parallelism. Check height of mold stops. Check for dried-out material.
Fiber pull	**Glass contacts mold surface** Chemical stick from the binder takes place. During part ejection the fiber is pulled out of the part which in some cases leads to chipped edges (see also Chipped Edges). Change initial charge location. Clean the problem area and add external mold release. Check the ejector pin for parallelism.
Flash	**Initial charge too large.** Decrease initial charge weight. Decrease clamp pressure. Decrease mold clamping pressure. Decrease mold closing speed. Decrease temperature differential of mold halves. Increase mold temperature. **Shear edge clearance too large.** Keep in mind that the optimal shear edge clearance depends on the material and process used (BMC materials require smaller clearance than SMC materials). Also, changes in mold temperature affect mold deflection, thus shear edge clearance.
Flow lines	**High fiber orientation.** In general, reducing fiber orientation reduces flow lines. Vary initial charge location. Decrease material flow length. **Components separated from the resin (See Streaking).** Increase mold closing speed. Add a breath cycle. Eliminate abrupt changes in thickness. Increase resin viscosity.
Heat smear	**Running the mold too hot.** Decrease mold temperature. Decrease curing time. Decrease mold clamping pressure. Vary initial charge location.
Knitlines	**Filling Pattern.** To eliminate knit lines the mold filling pattern must be changed. If the knit line cannot be eliminated then shift it to areas of the part that are not structurally vital. Another way to deal with knit lines is to drag them with material flow. When the knit line occurs early in the mold filling process it will be dragged by the flow and thus, increase the chance for fibers to bridge and strengthen the knit line. Vary initial charge location. Redesign obstructions such as core pins. Replace part holes with firing core pins or mash-offs. Eliminate abrupt changes in thickness.
Laking	**Poor contact between the mold and the area that appears dull.** Check the mold for areas where material might be leaking out. Sometimes the center of a part will exhibit laking while the outside of the part is leaking material. Increase mold temperature. Increase initial charge weight. Increase mold clamping pressure. Increase temperature differential of mold halves. Vary mold closing speed. Check height of mold stops. Check mold temperature uniformity. Check for excessive material leakage. Check shear edge clearance. Check for dried-out material. Change thickness of the problem area.
Mottling	**Separation of pigments (see also Streaking).** These darker areas show the start of gelling of the material. When mottling appears like paint brushed streaks directed towards a vent location, it is usually caused by a vacuum at that vent. Decrease mold temperature. Increase curing time. Increase mold clamping pressure. Vary temperature differential of the mold halves. Clean the mold and add external mold release.

Continued on next page

Problem	Possible causes and recommendations
Paint popping	**Entrapped gas.** The gas usually originates at cracks, voids, porosities, or exposed fibers. Improve handling of finished part. Use compound with tougher resin. Use a primer sealer designed to reduce paint pops.
Porosity	**Low molding pressure.** Increase clamping force. Increase initial charge. Vary mold closing speed. Decrease mold temperature. Increase material flow length. Increase temperature differential of mold halves. **Knitlies.** Porosity sometimes occurs around knit lines (see also Knit Lines). Vary initial charge location. **Abrupt thickness changes.** Porosity can also occur in areas where the part thickness changes abruptly. This abrupt change in thickness tends to trap air and create porosity. Eliminate abrupt changes in thickness **Material problems.** Check for dried-out material. Check for moisture and contamination. Change viscosity of the material.
Pre-gel	**Uneven cure.** Cure usually occurs first at the initial charge location or in areas where heat is excessive on the mold surface. Decrease mold temperature. Shorten the tool load time. Increase mold closing speed. Check for dried-out material. Use material with a less reactive catalyst (increase gel time).
Read-out	**Bonding parts.** A reinforcing part that is bonded to a class A part, shrinks and warps and deforms the class A surface. Use a stronger compound. Increase class A part thickness (2.5 mm or higher).
Resin rich areas or fiber-matrix separation or fiber jamming	**Resin squeeze out.** This defect is difficult to detect and sometimes is only revealed when the part cracks in the regions of high stress and low fiber content. Generally X-ray analysis or destructive testing, such as burn-out, have to be performed to show the fiber and resin concentrations. In most cases it occurs around edges (also see Chipped Edges), ribs, and bosses. Decrease material flow length. Increase mold closing speed. Decrease mold temperature. Vary initial charge location. Increase material viscosity. Check for initial material fiber-matrix homogeneity. Use material with smaller diameter fiber bundles. Eliminate abrupt changes in thickness.
Separation and phasing	**Low resin viscosity.** The low viscosity resin loses its ability to carry its components (usually fiber glass or thermoplastic additives). Phasing may be an indication that incompatibility between components leads to non-homogeneous pigmentation (see also Mottling). Shorten tool load time. Decrease mold clamping pressure. Decrease mold temperature. Increase material flow length. Check compatibility and level of thermoplastic additives. Increase material viscosity (See Resin rich areas).
Sink marks	**Ribs and bosses.** Use appropriate rib thickness relative to part thickness. Increase mold temperature. Increase temperature differential of mold halves. Increase initial charge weight. Increase mold closing speed. Vary initial charge location. Vary cure time. Reduce clamping pressure after mold fill. Use lower shrinkage material.
Sticking part	**Undercuts.** Inadequate ejector force. Decrease mold clamping pressure. Decrease ejection speed. Check for proper draft angle on side walls and ribs. **Low shrinkage material.** Increase cure time. Increase material flow length. Vary initial charge location. Vary level of material shrinkage. Try to open the mold or eject the part during the exothermic reaction of the material. **Ribs with EDM (electrical discharge machining) marks on them.** Polish or chrome mold. Clean problem area and add external mold release.

Continued on next page

Problem	Possible causes and recommendations
Streaking or abrasion	**Premature gelling** Decrease mold temperature. Decrease material flow length. Vary mold closing speed. Use material with a less reactive catalyst (increase gel time). Use higher viscosity material. **Contamination.** Clean, polish, or chrome mold. Check for contamination. Check for dried-out material. Check homogeneity of material.
Under-cure	**Cure time.** Increase mold temperature. Increase cure time. Increase mold clamping pressure. Preheat charges before molding. Check temperature uniformity. Check for appropriate catalyst and quantity. Check inhibitor quantity. Avoid excessively thick regions.
Voids or non-fills	**Mold is too cold.** Increase mold temperature. Check for dried-out material. **Not enough material.** Increase initial charge. Decrease material flow length. Increase mold closing speed. Shorten tool load time. Add a breath cycle. **Poor mold ventilation.** Check or add vents or vented ejector pins. Check height of mold stops. Clean vents and shear edge.
Warpage	**Uneven cooling.** Vary temperature differential of mold halves. **Uneven cure.** Increase cure time. Check mold and platen parallelism. Vary material formulation or shrinking control additives. Use a cooling fixture if needed. **Fiber orientation.** Decrease amount of deformation (reduce flow length). Change charge pattern. Increase or balance wall thickness or add ribs to stiffen part - this may cause other problems such as high internal stresses.
Surface waviness or ripples	**Low pressure at the flow front.** Ensure constant molding pressure throughout the part. Check height of mold stops. Decrease material flow length. Decrease mold closing speed. Increase mold clamping pressure. Vary initial charge location. **Viscosity too high.** *Process corrections:* Vary material formulation. Increase mold temperature. **heating lines.** Heating lines may be to far apart. Waves often reflect heating lines.

4.6 COMPOSITES PROCESSING

Fiber composites are broken down into continuous and discontinuous fibers that are randomly oriented and aligned. Additionally, some processes such as compression and injection molding result in a distinct fiber orientation distribution that depends on each process, mold geometry and processing conditions. In *fiber reinforced composite parts* the load and stresses are carried by the fiber. The matrix is in place to hold the system of fibers in place and to completely transfer the load onto the fibers. In previous sections injection, compression and injection-compression molding processes of fiber reinforced composites were presented, and this section presents some of the remaining main composites processing techniques.

4.6.1 Resin Transfer Molding and Structural RIM

Resin transfer molding (RTM) and SRIM are two similar liquid composite molding (LCM) processes that are well suited to the manufacture of medium-to-large, complex, lightweight, and high-performance composite components primarily for aerospace and automotive industries. In these processes, a reinforcement fiber mat (preform) is preplaced in a closed mold to be impregnated by a low-viscosity, reactive liquid resin in a transfer or injection process. These two processes differ in such areas as the resins used, mixing and injection set up, mold requirement, cycle time, fiber volume fraction, and suitable production volume. LCM processes such as resin transfer molding (RTM) and structural reaction injection molding (SRIM) are recognized as the most feasible and structurally efficient approach to mass produce lightweight, high-strength, low-cost structural composite components. In general, these processes consist of preparation of a reinforcing fiber mat (known as preform), preplacement of the dry preform in a closed mold, premixing and injection or transfer molding of reactive liquid resin, impregnation of the fiber mat by the curing resin, and removal of the cured, finished component from the mold (demolding). The preform is the assembly of dry (unimpregnated) reinforcement media that is preshaped (via, e.g., thermoforming) and assembled with urethane-formed cores, if necessary, into a three-dimensional skeleton of the actual part. Both of the RTM and SRIM processes make use of a wide variety of reinforcing media, such as woven and nonwoven fiber products, die-cut continuous strand mats (CSM), random mats consisting of continuous or chopped fibers laid randomly by a binder adhesive, knit braiding or two- and three-dimensionally braided products, or hybrid preforms made of layers of different types of media. The selection of the preform architecture depends on the desired structural performance, processability, long-term durability, and cost. The main difference between RTM and SRIM stems from the reaction activation mechanism of the resins used. This leads to different filling and cycle times, mixing and injection set-up, volume fraction and construction of the reinforcement, mold requirements, and suitable production volume. To be more specific, the chemical systems used in reaction molding processes can roughly be divided into two types: thermally activated and mixing activated. The resins used in RTM fall into the thermally activated category, whereas the resins used in SRIM are mixing activated. In the RTM process, a static mixer is used to produce the mixture at a typical mix ratio of 100 to 1 by volume. These thermally activated resin systems do not react appreciably at the initial resin storage temperature. They reply to the heated mold wall to accelerate the chemical reaction. As a result, the filling times for RTM can be as long as 15 minutes, and the cycle time is on the order of an hour or longer, depending on the resin and application. Given the slower injection and reaction rates with

RTM, high-volume fraction and a more complex preform can be used to improve the part strength and performance further. In addition, the low viscosity of the resin and the slow injection rate result in low injection-pressure and clamp-force requirements. As a result, RTM allows the use of so-called soft tools (e.g., wood-backing epoxy mold or aluminum molds). Because of the long cycle time, RTM is generally limited to a low-volume production (i.e., less than 10,000 parts). On the other hand, SRIM derives its name from the RIM process, from which its resin chemistry and injection techniques are adapted. That is, the chemical reaction is activated by impingement mixing of two highly reactive components in a special mixing head under high pressure. Upon mixing, the mixture is subsequently injected into the mold at a lower pressure. The resin starts to cure as it impregnates the preform and forms the matrix of the composite. Due to the fast reaction rate and rapid build up of viscosity from curing, the cavity has to be filled within a few seconds. In addition, the cycle time is as short as 1 minute. Flow distances for typical SRIM applications are therefore limited to 0.6 to 0.9 m from the inlet gate. Furthermore, the volume fraction and construction of the reinforcement have to be selected carefully to facilitate fast, complete filling before gelation occurs. Because of the high injection rate and short cycle time, SRIM generally uses steel tools and is suitable for medium- to high-volume production (10,000 to 100,000 parts). Prepreg/autoclave processes have traditionally been the major manufacturing technique for producing lightweight composite components in the aerospace industry. This process however, is slow, expensive, and labor intensive. In addition, common manufacturing processes in automotive industry, such as thermoplastic injection molding and compression molding of SMC, can only incorporate low-volume, short fiber reinforcement. Thus, they cannot produce high-strength products needed for structural applications. RTM and SRIM, therefore, offer viable options for producing lightweight, high-performance structural components for both aerospace and automotive industries. In addition, low injection pressure is another process advantage of both RTM and SRIM. Although the injection pressure varies with the permeability of the fiber mat, part geometry, and the injection rate, typical injection pressure varies from 70 to 140 kPa for low injection rate and reinforcement content (10 to 20%) to 700 to 1400 kPa for rapid mold filling and high reinforcement content (30 to 50%). Furthermore, the RTM and SRIM processes employ closed molds, which reduces or eliminates the emission of hazardous vapor. Other advantages include more repeatable part thickness and minimal trimming and de-flashing of the final part. Of the several resin systems employed by RTM, polyesters are the most commonly used resins because of the low cost. Other resins used in RTM include epoxies, vinyl esters, acrylic/polyester hybrids, acrylamate resin family, and methymethacrylate vinyl esters. On the other hand, common resins used for SRIM include urethane, acrylamate, and dicyclopentadiene. It is during the resin flow and fiber impregnation

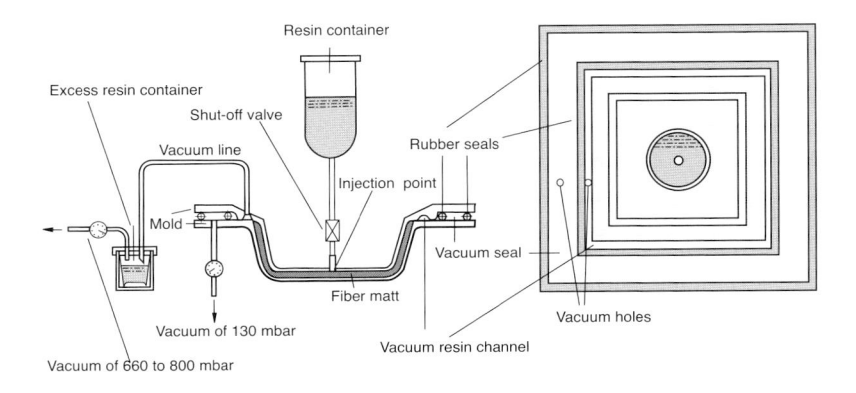

Figure 4.103: Schematic diagram of the resin infusion process (RIP).

step of the process where the main problem of RTM ans S-RIM lies. Poor fiber impregnation can lead to voids between fiber and matrix, leading to localized stress concentrations and possible crack initiation sites.

One way to overcome the shortcomings of RTM is to use a vacuum to force the resin through the bed of fibers. This technique, schematically depicted in Fig. 4.103, is often referred to as *vacuum injection process* or *resin infusion process*.

4.6.2 Filament Winding

Filament winding is used to manufacture high strength hollow structures as well as other strong components. In filament winding, a fiber or roving is passed through a resin bath where it is coated or impregnated with a polymer, often a thermoset such as liquid epoxy resin mixed with an initiator, pigments, UV protectors and other additives. The fiber tension is controlled by the guides located before the fibers enter the bath. After impregnation the fibers are usually pulled through a wiping device that removes the excess resin before gathering the fibers into a flat band. The band is wound onto the body with an orientation that aligns with the principal stresses and loads. Typical items that are manufactured by this technique are pressure vessels and pipes or tubular geometries.

Figure 4.104 presents a *polar filament winding* system of a pressure vessel. A polar winding system is where the filament passes tangent to the polar openings of the vessel. In other words, the fiber travels from pole to pole as the mandrel arm rotates about its longitudinal axis.

4.6 Composites Processing

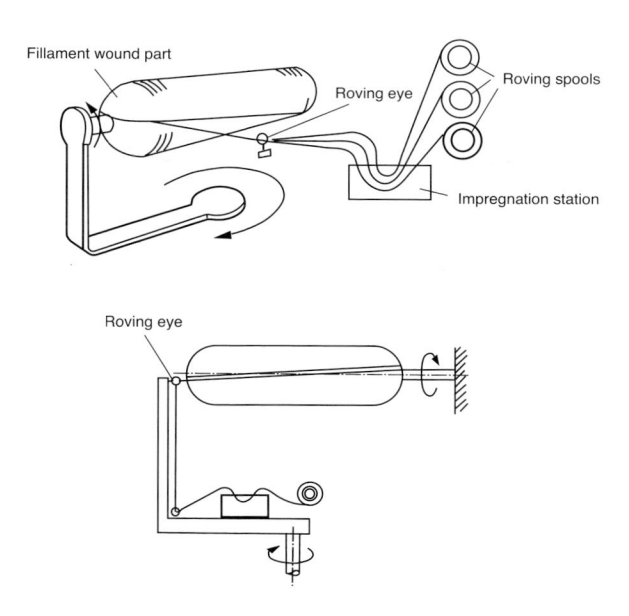

Figure 4.104: Schematic diagram of polar filament winding of a pressure vessel.

In contrast in *hoop winding* is where the fibers are wound circumferentially, and where the fibers advance only one band width per rotation. This pattern is often chosen for systems where the hoop stresses are dominant.

Helical winding is where the mandrel rotates at a constant speed as the fiber travels back and forth along the axis of the mandrel. The speed of the mandrel or of the fiber guide control the angle (helix) of the fiber.

More complex geometries are wound with the aid of a computer controlled robot arm, such as the one schematically depicted in Fig. 4.105.

4.6.3 Pultrusion

Pultrusion, schematicaly depicted in Fig. 4.106, is a continous, cost effective process to manufacture high strength profiles with controlled and uniform cross-sections. There are various types of reinforcing fibers that are used in the pultrusion process. These include fiber bundles, braided fibers, as well as woven and non-woven planar structures. During pultrusion the fibers and fiber structures are passed through a thermoseting resin bath containing additives, such as catalysts, initiators, pigments, and many more. Next, the excess resin is wiped from the fibers, before they are sent to a pre-forming vise which pre-shapes the

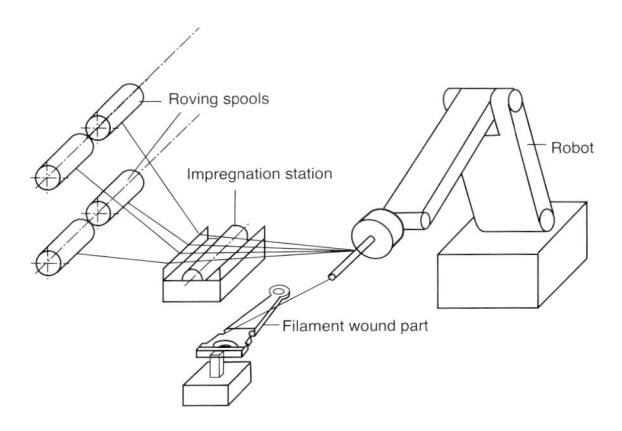

Figure 4.105: Schematic diagram of controlled robot filament winding system.

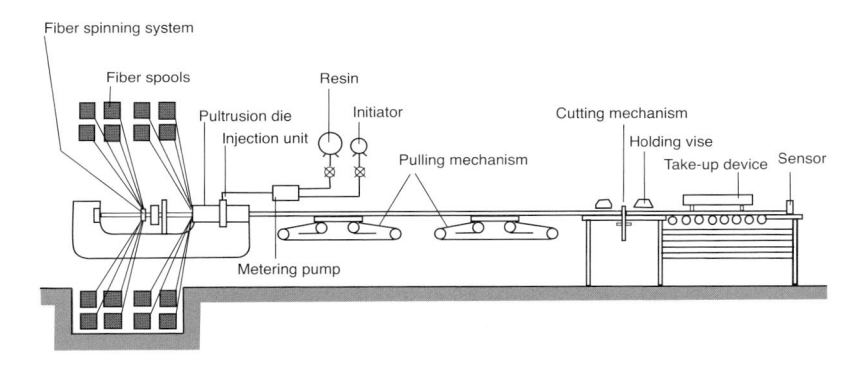

Figure 4.106: Schematic diagram of a pultrusion process.

profile. The pre-forming die often also removes the excess resin. The pre-formed profile is then pulled through a heated die that gives the part its final shape and surface finish. Typical pultruded parts are hollow tubes, solid profiled I-beams, flat structures, hand rails, and many more.

4.7 SECONDARY SHAPING

Secondary shaping operations, such as extrusion blow molding, film blowing, and fiber spinning occur immediately after the extrusion profile emerges from the die. The thermoforming process is performed on sheets or plates previously extruded

4.7 Secondary Shaping

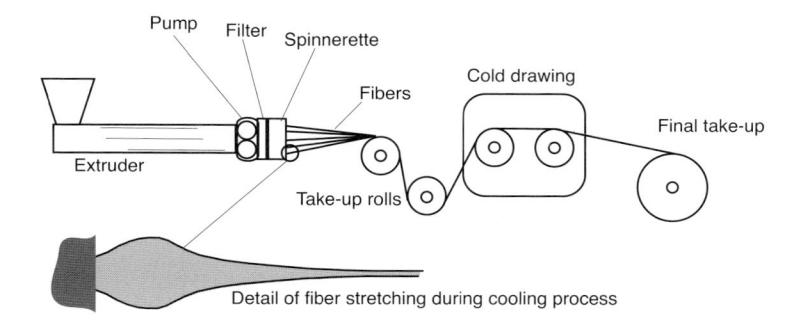

Figure 4.107: The fiber spinning process with detail of a stretching fiber during the cooling process.

and solidified. In general, secondary shaping operations consist of mechanical stretching or forming of a preformed cylinder, sheet, or membrane.

4.7.1 Fiber Spinning

Fiber spinning is used to manufacture synthetic fibers. During fiber spinning, a filament is continuously extruded through an orifice and stretched to diameters of $100 \mu m$ and smaller. The process is schematically depicted in Fig. 4.107 The molten polymer is first extruded through a filter or screen pack to eliminate small contaminants. The melt is then extruded through a spinneret, a die composed of multiple orifices. A spinneret can have between one and 10,000 holes. The fibers are then drawn to their final diameter, solidified, and wound onto a spool. The solidification takes place either in a water bath or by forced convection. When the fiber solidifies in a water bath, the extrudate undergoes an adiabatic stretch before cooling begins in the bath. The forced convection cooling, which is more commonly used, leads to a non-isothermal spinning process. The drawing and cooling processes determine the morphology and mechanical properties of the final fiber. For example, ultra high molecular weight PE-HD fibers with high degrees of orientation in the axial direction can have the stiffness of steel with today's fiber spinning technology. Of major concern during fiber spinning are the instabilities that arise during drawing, such as brittle fracture, Rayleigh disturbances, and draw resonance. Brittle fracture occurs when the elongational stress exceeds the melt strength of the drawn polymer melt. The instabilities caused by Rayleigh disturbances are like those causing filament break-up during dispersive

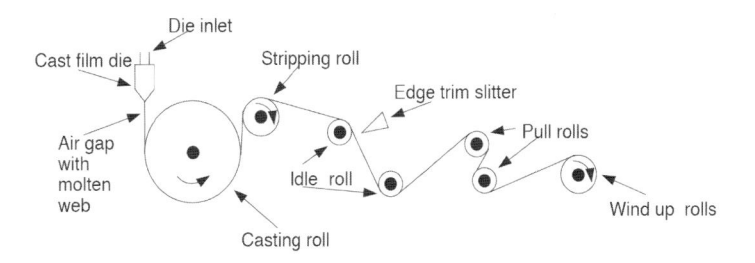

Figure 4.108: Schematic diagram of a film casting operation.

mixing. Draw resonance appears under certain conditions and manifests itself as periodic fluctuations that result in diameter oscillation.

4.7.2 Film Production

Cast Film Extrusion In a cast film extrusion process, a thin film is extruded through a slit onto a chilled, highly polished turning roll where it is quenched from one side. The speed of the roller controls the draw ratio and final film thickness. The film is then sent to a second roller for cooling of the other side. Finally, the film passes through a system of rollers and is wound onto a roll. A typical film casting process is depicted in Figs. 4.108 and 4.109 The cast film extrusion process exhibits stability problems similar to those encountered in fiber spinning.

Film Blowing In film blowing, a tubular cross-section is extruded through an annular die, normally a spiral die, and is drawn and inflated until the freezing line is reached. Beyond this point, the stretching is practically negligible. The process is schematically depicted in Fig. 4.110. The advantage of film blowing over casting is that the induced biaxial stretching renders a stronger and less permeable film. Film blowing is mainly used with less expensive materials such as polyolefins. Polymers with lower viscosities such as PA and PET are better manufactured using the cast film process. The extruded tubular profile passes through one or two air rings to cool the material. The tube's interior is maintained at a certain pressure by blowing air into the tube through a small orifice in the die mandrel. The air is retained in the tubular film, or bubble, by collapsing the film well above its freeze-off point and tightly pinching it between rollers. The size of the tubular film is calibrated between the air ring and the collapsing rolls.

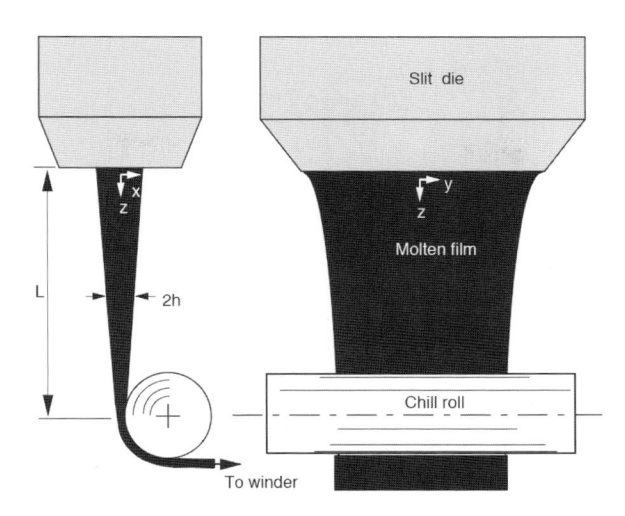

Figure 4.109: Film casting.

Blow molding produces hollow articles that do not require a homogeneous thickness distribution. Today, HDPE, LDPE, PP, PET, and PVC are the most common materials used for blow molding.

Extrusion Blow Molding In extrusion blow molding, a parison or tubular profile is extruded and inflated into a cavity with the specified geometry. The blown article is held inside the cavity until it is sufficiently cool. Figure 4.111 presents a schematic of the process steps in blow molding. During blow molding, the appropriate parison length must be generated such that the trim material is minimized. Another means of saving material is generating a parison of variable thickness, usually referred to as parison programming, such that an article with an evenly distributed wall thickness is achieved after stretching the material. An example of a programmed parison and finished bottle thickness distribution is presented in Fig. 4.112. A parison of variable thickness can be generated by moving the mandrel vertically during extrusion as shown in Fig. 4.113 A thinner wall not only results in material savings but also reduces the cycle time due to the shorter required cooling times. As expected, the largest portion of the cycle time is the cooling of the blow molded container in the mold cavity. Most machines work with multiple molds in order to increase production. Rotary molds are often used in conjunction with vertical or horizontal rotating tables (Fig. 4.114). Today, extrusion blow molding is used to manufacture multi-layered containers such as automotive fuel containers. Figure 4.115 presents a three-layer coextrusion die

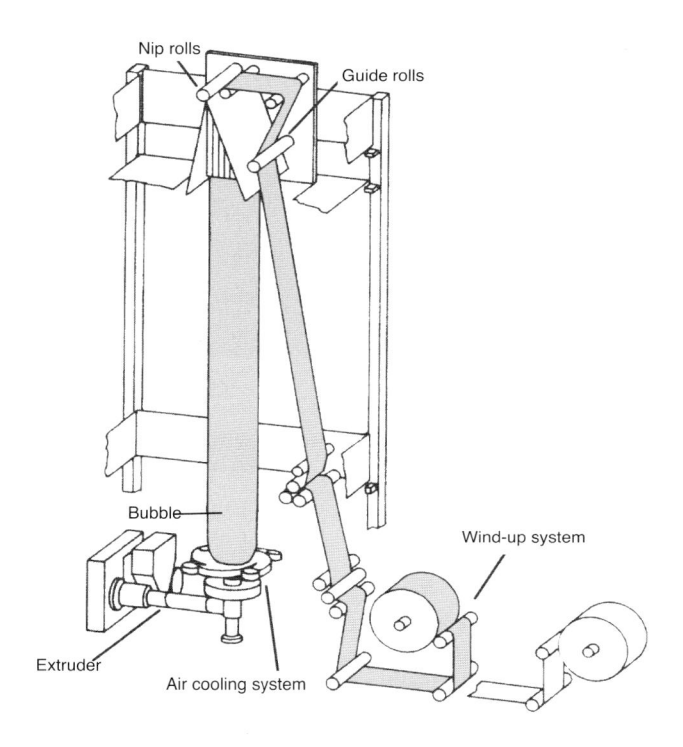

Figure 4.110: Schematic of a film blowing operation.

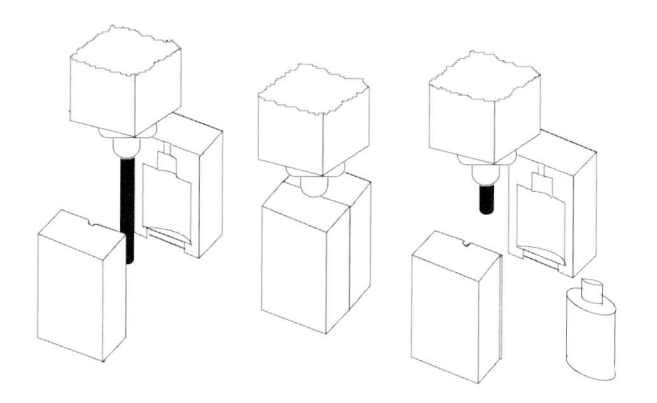

Figure 4.111: Schematic of the extrusion blow molding process.

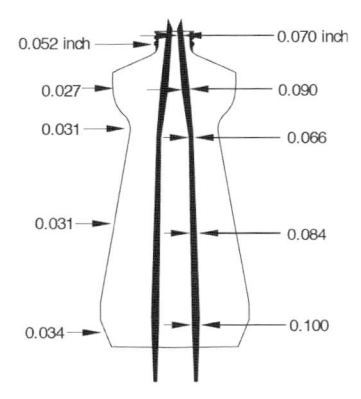

Figure 4.112: Wall thickness distribution in the parison and the bottle.

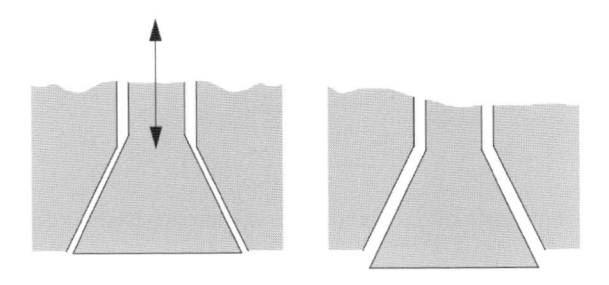

Figure 4.113: Moving mandrel used to generate a programmed parison.

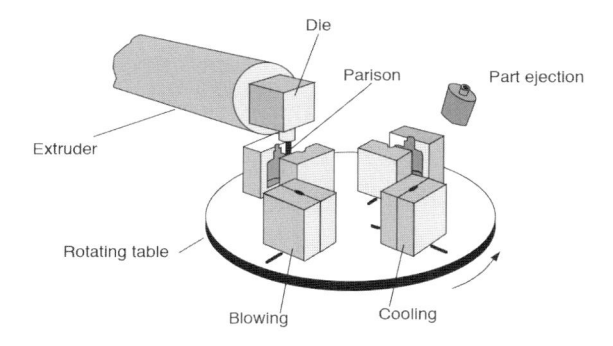

Figure 4.114: Schematic of an extrusion blow molder with a rotating table.

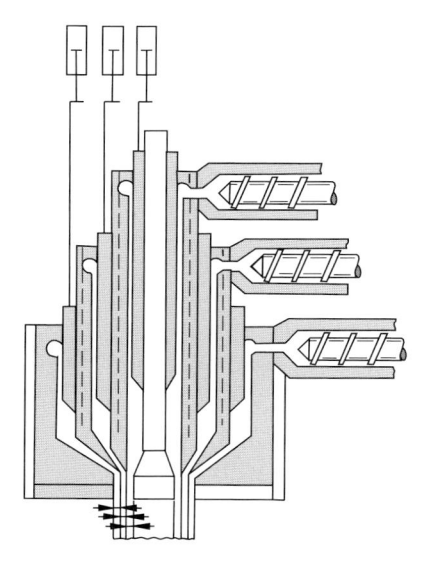

Figure 4.115: Three-layered coextrusion die for an extrusion blow molding system.

with a ring-piston accumulator for each individual extruder. In the shown system the layers are joined as they exit the die. Table 4.8 presents various problems that may arise during extrusion blow molding, their causes and suggestions for their solution.

4.7 Secondary Shaping

Table 4.8: Troubleshooting Guide for Blow Molding

Problem	Possible recommendations
Scratches and die lines	Increase stock resin temperature. Increase extrusion back pressure. Check for contamination in material (resin). Check tooling for damage. Remove burnt resin on tooling faces. Increase die temperature.
Melt fracture alligatoring	Change stock resin temperature. Change extrusion pressure/rate. Change extrusion back pressure. Decrease regrind level. Check for contamination in material (resin). Increase resin melt index. Increase container weight. Increase die temperature. Check head heater bands. Check design of flow path in die (angles). Check tooling for damage.
Streaks	Decrease stock resin temperature. Decrease extrusion pressure/rate. Decrease extrusion back pressure. Decrease regrind level. Check head heater bands. Check heat controllers. Check for contamination in material (resin). Check tooling for damage. Increase container weight. Check design of flow path in die (angles).
Rough/milky cold parison	Increase stock resin temperature. Increase extrusion pressure/rate. Increase extrusion back pressure. Increase regrind level. Check head heater bands. Check heat controllers. Increase resin melt index.
Shiny/clear	Decrease stock resin temperature. Decrease extrusion pressure/rate. Decrease extrusion back pressure. Decrease regrind level. Check head heater bands. Check heat controllers. Check for contamination in material (resin). Increase container weight.
Bubbles	Decrease stock resin temperature. Decrease extrusion pressure/rate. Check head heater bands. Check heat controllers. Check for contamination in material (resin). Increase extrusion back pressure. Check for moisture in resin. Check feedzone cooling for leakage. Check parison programmer slide plates for wear. Check screw/barrel for wear. Check screw gap adjustment. Check RPM vs. output (pounds per turn).
Smoking	Decrease stock resin temperature. Decrease extrusion pressure/rate. Decrease extrusion back pressure. Decrease regrind level. Check head heater bands. Check heat controllers. Check for contamination in material (resin). Check for moisture in resin.
Wrinkles, draping, curtaining	Decrease stock resin temperature. Decrease extrusion pressure/rate. Decrease extrusion back pressure. Increase container weight. Increase regrind level. Increase resin melt index. Decrease resin die swell. Check design of flow path in die (angles). Check parison programmer.
Drawdown, sag/stretch	Decrease stock resin temperature. Decrease extrusion back pressure. Decrease regrind level. Decrease resin die swell. Increase extrusion pressure/rate. Decrease resin melt index. Decrease container weight. Decrease mold open time. Increase accumulator pushout speed. Decrease material (melt) temperature.
Curls	Check head heater bands. Check heat controllers. Check for contamination in material (resin). Increase container weight. Check screw gap adjustment. Increase stock resin temperature. Check head air for leakage/drafts. Center mandrel in die. Check tooling for damage. Decrease support air. Increase die temperature. Check screw tip design.
Curves	Check head heater bands. Check heat controllers. Check for contamination in material (resin). Increase stock resin temperature. Check head air for leakage/drafts. Center mandrel in die. Check tooling for damage. Check/adjust choke valves. Check for damage in lower head parts. Check parison cutting.

Continued on next page

Problem	Possible recommendations
Intermittent feed	Decrease stock resin temperature. Decrease extrusion pressure/rate. Decrease regrind level. Check head heater bands. Check heat controllers. Check for contamination in material (resin). Increase extrusion back pressure. Decrease extruder mixing zone temperature. Check hopper feed for bridging. Decrease extruder feed zone temperature. Check head fill or backpressure. Check RPM vs. Amps. on extruder drive.
Length head to head	Check head heater bands. Check heat controllers. Check for contamination in material (resin). Check parison programmer. Center mandrel in die. Check/adjust choke valves. Check stock resin temperature.
Length shot to shot	Decrease extrusion pressure/rate. Decrease regrind level. Check heat controllers. Increase extrusion back pressure. Check parison programmer. Decrease resin melt index. Check hopper feed for bridging. Decrease extruder feed zone temperature. Check RPM vs. Amps. on extruder drive. Check machine cycle.
Excessive flash	Increase stock resin temperature. Decrease extrusion pressure/rate. Decrease extrusion back pressure. Decrease pre-blow air pressure. Increase regrind level. Increase container weight. Check mold alignment. Center mandrel in die. Decrease resin die swell. Decrease preblow time.
Excessive cycle time	Decrease material (melt) temperature. Decrease mold temperature. Decrease wall thickness. Increase blowing air pressure. Decrease stock temperature. Install reverse flush. Install exhaust hole in part. Install interval blow. Install cold blow air system. Install dry air curtain system.
Container hangs in mold	Center mandrel in die. Decrease mold temperature. Increase blowing air pressure. Decrease stock resin temperature. Increase air exhaust time. Check mold for damage. Increase cycle time. Check mold flash pocket volume. Check mold vents/surface. Check mold for undercuts. Check pinch-off for burrs/sharpness. Check pinch-offs for damage.
Thin mold parting line	Check mold alignment. Decrease mold temperature. Increase blowing air pressure. Increase air exhaust time. Check mold for damage. Check mold vents/surface. Check if mold is fully closed. Start blow air earlier. Check possibility for entrapped air to get out.
Thin bottom weld	Decrease extrusion pressure/rate. Decrease extrusion back pressure. Increase regrind level. Check mold alignment. Decrease mold temperature. Decrease stock resin temperature. Check mold vents/surface. Check pinch-off for burrs/sharpness. Increase mold closing speed on last 10% of way. Increase mold flash pocket volume. Increase parison wall thickness in this area. Decrease mold closing speed on last 10% of way. Check pinch-off design. Check pinch-offs for damage.
Thick bottom weld	Increase stock resin temperature. Decrease pre-blow air pressure. Increase blowing air pressure. Check pinch-offs for damage. Increase mold closing speed on last 10% of way. Increase mold flash pocket volume. Check pinch-off design. Increase extrusion pressure/rate. Increase extrusion back pressure. Decrease regrind level. Decrease container weight. Decrease parison wall thickness in this area.
Partial handle	Increase blowing air pressure. Check pinch-offs for damage. Decrease container weight. Increase preblow air pressure. Increase resin die swell. Increase resin melt index. Check handle blow needle sharpness. Check time for needle shooting in. Check time for start blowing. Check head tool design. Check head tool size. Check blow air start time. Check mold closing speed. Check pre-pinch device (if installed).

Continued on next page

4.7 Secondary Shaping

Problem	Possible recommendations
Orange peel	Increase stock resin temperature. Decrease pre-blow air pressure. Decrease preblow time. Increase blowing air pressure. Check mold vents/surface. Check possibility for entrapped air to get out. Increase extrusion pressure/rate. Increase extrusion back pressure. Check blow air start time. Increase mold temperature. Decrease mold open time. Decrease cycle time. Check molds for water leakage. Check for humidity of air (sweating of mold). Clean mold surface. Install dry air curtain system.
Warpage	Decrease extrusion back pressure. Decrease mold temperature. Increase blowing air pressure. Increase air exhaust time. Increase cycle time. Increase extrusion pressure/rate. Decrease container weight. Check head tool size. Decrease mold open time. Increase cooling cycle. Check cooling channels. Check parison programming. Ovalize tooling. Install PWDS/SFDR.
Excessive shrinkage	Increase blowing air pressure. Increase air exhaust time. Increase cycle time. Decrease container weight. Increase mold temperature. Decrease mold open time. Check parison programming. Ovalize tooling.
Neck underfills	Decrease pre-blow air pressure. Decrease resin die swell. Check mold for damage. Check mold vents/surface. Increase resin melt index. Increase mold temperature. Check parison programming. Increase blowpin temperature. Adjust parison programming. Check machine cycle time to parison cycle time. Check parison cutting delay.
Unenven wall thickness	Check mold alignment. Center mandrel in die. Decrease material (melt) temperature. Increase resin melt index. Check head tool design. Check head tool size. Check parison programming. Ovalize tooling. Check head heater bands. Decrease resin melt index. Decrease blow-up ratio. Check for contamination. Check die/mandrel for damage. Check for air leakage. Check/adjust choke valves. (Hay dos recomendaciones contrarias: increase resing melt index y decrease resin melt index!)
Poor pinch-off weld	Increase stock resin temperature. Check mold alignment. Increase blowing air pressure. Increase air exhaust time. Check pinch-off for burrs/sharpness. Check pinch-offs for damage. Increase extrusion pressure/rate. Check blow air start time. Check mold closing speed. Decrease mold open time. Check parison programming. Decrease mold flash pocket volume.
Poor part, multiple head	Check mold for damage. Check mold vents/surface. Check parison programming. Check for contamination. Check die/mandrel for damage. Check for air leakage. Check/adjust choke valves. Check each head temperature. Check mold temperature (in each cavity). Check air exhaust time. Check head tooling. Check programmer sync.-unit.
Blow outs and bubbles	Check mold for damage. Check pinch-offs for damage. Check if mold is fully closed. Increase extrusion back pressure. Check mold closing speed. Check cooling channels. Check parison programming. Check for contamination. Check die/mandrel for damage. Increase pinch-off land width. Check cooling system. Check mold hot spots. Check cycle time. Decrease blowing air pressure. Check for moisture in resin. Check for hopper fines. Check waterflow in mold. Check machine and head for dry air, oil leaks or spray onto parison.

Injection Blow Molding Injection blow molding depicted in Fig. 4.116 begins by injection molding the parison onto a core and into a mold with finished bottle threads. The formed parison has a thickness distribution that leads to reduced thickness variations throughout the container. Before blowing the pari-

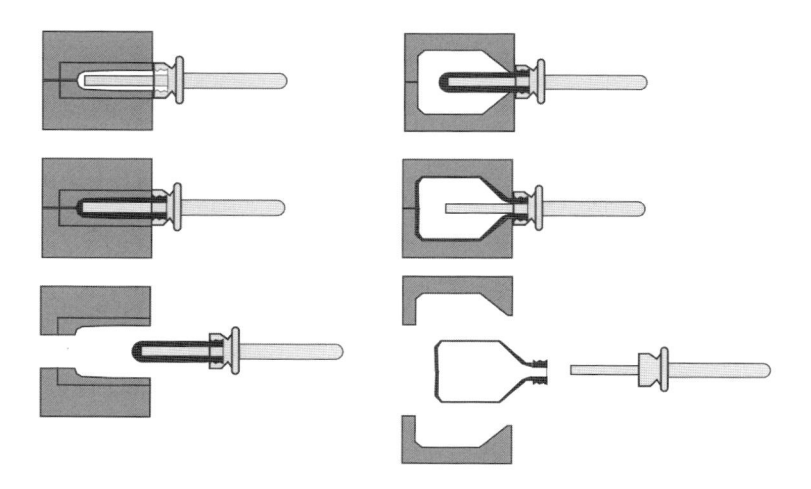

Figure 4.116: Injection blow molding.

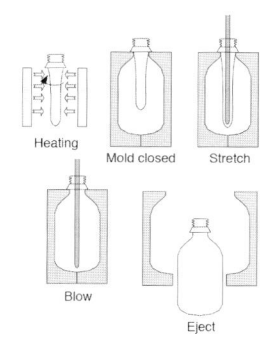

Figure 4.117: Stretch blow molding.

son into the cavity, it can be mechanically stretched to orient molecules axially, Fig. 4.117. The subsequent blowing operation introduces tangential orientation. A container with biaxial molecular orientation exhibits higher optical (clarity) and mechanical properties and lower permeability. In the injection blow molding process one can go directly from injection to blowing or one can have a re-heating stage in-between. The advantages of injection blow molding over extrusion blow molding are:

- Pinch-off and therefore post-mold trimming are eliminated

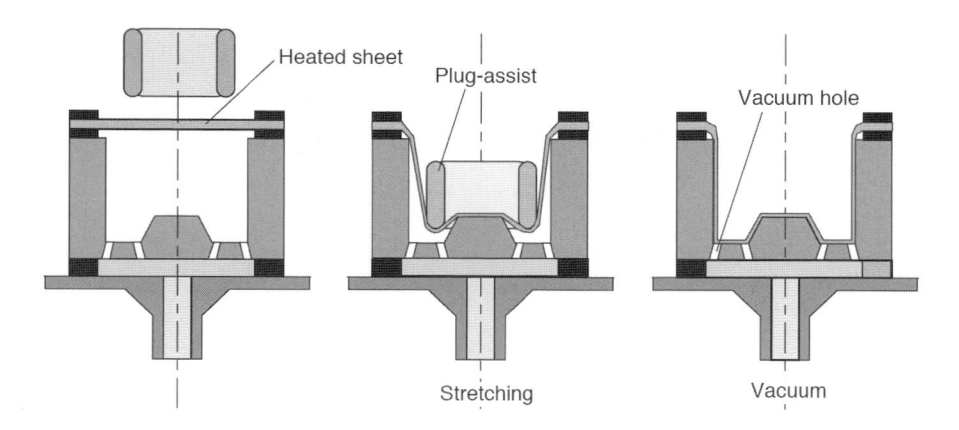

Figure 4.118: Plug-assist thermoforming using vacuum.

- Controlled container wall thickness
- Dimensional control of the neck and screw-top of bottles and containers

Disadvantages include higher initial mold cost, the need for both injection and blow molding units and lower volume production.

4.7.3 Thermoforming

Thermoforming is an important secondary shaping method of plastic film and sheet. Thermoforming consists of warming the plastic sheet and forming it into a cavity or over a tool using vacuum, air pressure, or mechanical means. In addition to packaging, thermoforming is used to manufacture refrigerator liners, pick-up truck cargo box liners, shower stalls, bathtubs, as well as automotive trunk liners, glove compartments, and door panels. A typical thermoforming process is presented in Fig. 4.118. The process begins by heating the plastic sheet slightly above the glass transition temperature for amorphous polymers, or slightly below the melting point for semi-crystalline materials. Although, both amorphous and semi-crystalline polymers are used for thermoforming, the process works best with amorphous polymers because they have a wide rubbery temperature range above the glass transition temperature. At these temperatures the polymer is easily shaped, but still has enough rigidity to hold the heated sheet without much sagging. Most semi-crystalline polymers lose their strength rapidly once the crystalline structure breaks up above the melting temperature. The heating is achieved using radiative heaters and the temperature reached during heating must be high enough for sheet shaping, but low enough so the sheets do not droop into

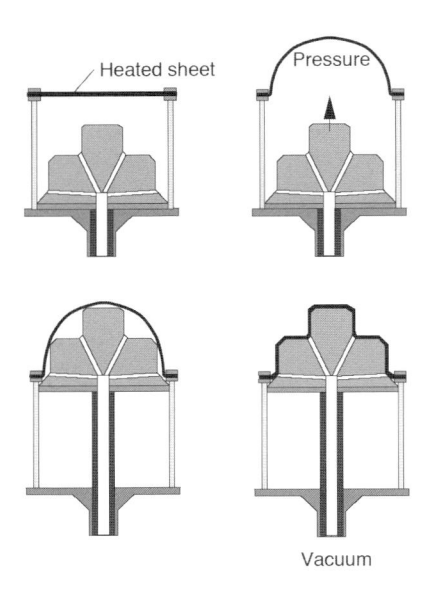

Figure 4.119: Reverse draw thermoforming with plug-assist and vacuum.

the heaters. One key requirement for successful thermoforming is to bring the sheet to a uniform forming temperature. The sheet is then shaped into the cavity over the tool. This can be accomplished in several ways. Most commonly, a vacuum sucks the sheet onto the tool, stretching the sheet until it contacts the tool surface. The main problem here is the irregular thickness distribution that arises throughout the part. Hence, the main concern of the process engineer is to optimize the system such that the differences in thickness throughout the part are minimized. This can be accomplished in many ways, but most commonly by plug-assist. Here, as the plug pushes the sheet into the cavity, only the parts of the sheet not touching the plug-assist stretch. Since the unstretched portions of the sheet must remain hot for subsequent stretching, the plug-assist is made of a low thermal conductivity material such as wood or hard rubber. The initial stretch is followed by a vacuum for final shaping. Once cooled, the product is removed. To reduce thickness variations in the product, the sheet can be pre stretched by forming a bubble at the beginning of the process. This is schematically depicted in Fig. 4.119. The mold is raised into the bubble, or a plug-assist pushes the bubble into the cavity, and a vacuum finishes the process.

One of the main reasons for the rapid growth and high volume of thermoformed products is that the tooling costs for a thermoforming mold are much lower than for injection molding. Some thermoforming processes are highly automated

4.7 Secondary Shaping

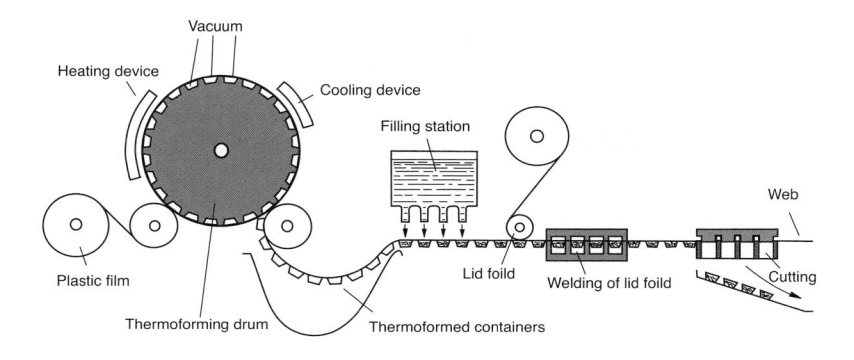

Figure 4.120: Rotational vacuum thermoforming process with integrated packaging and sealing stations.

and part of a larger assembly and packaging process such as the one shown in Fig. 4.120. Here, the thermoforming of the container is integrated with the filling station, the labeling, the sealing and the cutting of the filled container.

Table 4.9[7] presents several defects with possible causes and remedies when manufacturing thermoformed parts.

Table 4.9: Troubleshooting Guide for Thermoforming

Problem	Possible Causes and Recommendations
Incomplete forming part	**Sheet to cool.** Increase heat and heating time.
	Not enough vacuum/pressure Check lines.
	Sheet not flat Check sheets.
Blister or bubbles	**Sheet to hot.** Decrease heat and heating time.
	Excess moisture in sheet. Dry. Check relative humidity.
	Uneven heating. Check individual heaters.
Webbing or bridging	**Mold corners to sharp.** Round off corners
	Not enough vent holes. Add venting holes.
	Sheet too hot causing too much material in forming area. Decrease heat and heating time.
Mold release difficult.	**Draft insufficient.** Rework mold. Add appropriate draft angles.
	Undercuts. Reduce undercuts. Round off corners within undercut regions.
	Rough mold surface. Polish mold.
	Part has shrunk on mold. Change processing conditions and timings.
	Part temperture too high. Increase cooling time.
	Not cooled long enough. Increase cooling time.
	Forming pressure to high. Adjust process conditions.

Continued on next page

[7]This table is based on the table presented by B. Strong, *Plastics*, Prentice Hall, (2000).

Problem	Possible causes and Recommendations
Wraping	Sheet (all or part) too cool when formed. Poor design. Mold temperature too low. Cooling time too short.
Tearing	**Design exceeds maximum elongation.** Increase sheet thickness. **Plug speed too fast.** Adjust process conditions. **Sheet too hot.** Reduce heating time. **Not enough clearance between mold and plug or bubble.** Adjust plug path. **Sheet too cold(especially if very thin).** Increase heat and heating time.
Excessive shrinkage of part	**Stresses in sheet beacause it is too cool when molded.** Increase heat and heating time. **Part not cooled enough in mold.** Extend cooling time. **Sheet molecular orientation incorrect** Rotate sheet with respect to mold.
Cracking	**Angles in mold too sharp.** Round off corners. **Part to cool when molded.** Increase heat and heating time.
Pinhole or mold mark-off	**Vent holes too large.** Rework mold. **Sheet temperature too high.** Reduce heating time. **Vacuum or pressure to high.** Readjust process conditions.
Blushing	**Sheet too cool.** Increase heat and heating time. **Vacuum application too slow or not high enough vacuum.** Readjust process conditions.
Sheet scorched	**Outer surface of sheet too hot** Reduce heat and lengthen heating time. Heat both sides.
Mottled surface	**Entrapped air.** Check vent holes. Add vent holes. **Mold temperature.** Adjust process conditions. **Moisture in sheet.** Dry. Check relative humidity. **Mold surface too shiny.** Rework mold. **Oven too cold.** Increase heat.
Part corners to thin	**Improper heating.** Readjust process conditions. **Improper part or plug design.** Reevaluate plug/moold design. Increase plug speed. Increase sheet thickness.

4.8 CALENDERING

In a calender line, the polymer melt is transformed into films and sheets by squeezing it between pairs of co-rotating high precision rollers. Calenders are also used to produce certain surface textures which may be required for different applications. Today, calendering lines are used to manufacture PVC sheet, floor covering, rubber sheet, and rubber tires. They are also used to texture or emboss surfaces. When producing PVC sheet and film, calender lines have a great advantage over extrusion processes because of the shorter residence times, resulting in a lower requirement for stabilizer. This can be cost effective since stabilizers are a major part of the overall expense of processing these polymers.

4.8 Calendering

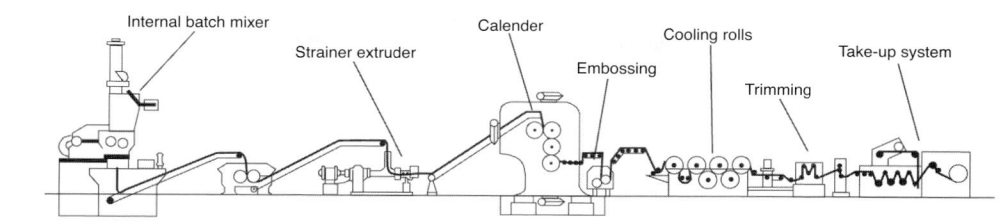

Figure 4.121: Schematic of a typical calendering process (Berstorff GmbH, Germany).

Figure 4.121 presents a typical calender line for manufacturing PVC sheet. A typical system is composed of:

- Plasticating unit
- Calender
- Cooling unit
- Accumulator
- Wind-up station

In the plasticating unit, which is represented by the internal batch mixer and the strainer extruder, the material is melted and mixed and is fed in a continuous stream between the nip of the first two rolls. In another variation of the process, the mixing may take place elsewhere, and the material is simply reheated on the roll mill. Once the material is fed to the mill, the first pair of rolls control the feeding rate, while subsequent rolls in the calender calibrate the sheet thickness. Most calender systems have four rolls as does the one in Fig. 4.121, which is an inverted $L-$ or $F-$type system. Other typical roll arrange-ments are shown in Figs. 4.122 After passing through the main calender, the sheet can be passed through a secondary calendering operation for embossing. The sheet is then passed through a series of chilling rolls where it is cooled from both sides in an alternating fashion. After cooling, the film or sheet is wound. One of the major concerns in a calendering system is generating a film or sheet with a uniform thickness distribution with tolerances as low as ± 0.005mm. To achieve this, the dimensions of the rolls must be precise. It is also necessary to compensate for roll bowing resulting from high pressures in the nip region. Roll bowing is a structural problem that can be mitigated by placing the rolls in a slightly crossed pattern, rather than completely parallel, or by applying moments to the roll ends to counteract the separating forces in the nip region.

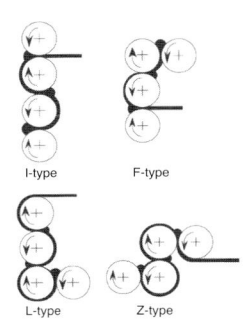

Figure 4.122: Calender arrangements.

4.9 COATING

In coating a liquid film is continuously deposited on a moving, flexible or rigid substrate. Coating is done on metal, paper, photographic films, audio and video tapes, and adhesive tapes. Typical coating processes include wire coating, dip coating, knife coating, roll coating, slide coating, and curtain coating. In wire coating, a wire is continuously coated with a polymer melt by pulling the wire through an extrusion die. The polymer resin is deposited onto the wire using the drag flow generated by the moving wire and sometimes a pressure flow generated by the back pressure of the extruder. The process is schematically depicted in Fig. 4.123[8]. The second normal stress differences, generated by the high shear deformation in the die, help keep the wire centered in the annulus. Dip coating is the simplest and oldest coating operation. Here, a substrate is continuously dipped into a fluid and withdrawn with one or both sides coated with the fluid. Dip coating can also be used to coat individual objects that are dipped and withdrawn from the fluid. The fluid viscosity and density and the speed and angle of the surface determine the coating thickness. Knife coating, depicted in Fig. 4.124, consists of metering the coating material onto the substrate from a pool of material, using a fixed rigid or flexible knife. The knife can be normal to the substrate or angled and the bottom edge can be flat or tapered. The thickness of the coating is nearly half the gap between the knife edge and the moving substrate or web. A major advantage of a knife edge coating system is its simplicity and relatively low maintenance. Roll coating consists of passing a substrate and the coating

---

[8]Other wire coating operations extrude a tubular sleeve which adheres to the wire via stretching and vacuum. This is called tube coating.

4.9 Coating

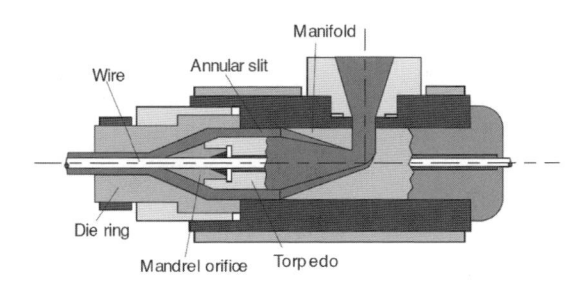

Figure 4.123: Schematic of the wire coating process.

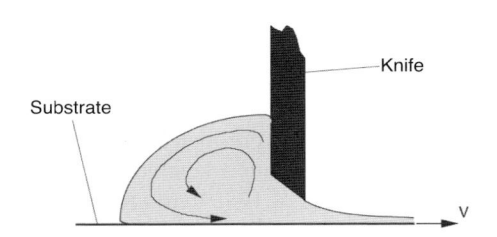

Figure 4.124: Schematic of the knife coating process.

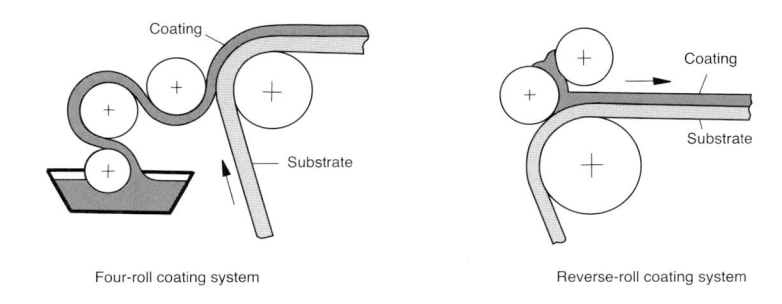

Figure 4.125: Schematic of a four-roll coating system with forward-coating rolls and a reverse-coating rolls system.

simultaneously through the nip region between two rollers such as shown in Fig. 4.125 for a four-roll with forward-roll system and reverse-roll coating system. As can be seen, the physics governing this process is similar to calendering, except that the fluid adheres to both the substrate and the opposing roll. The coating material is a low viscosity fluid, such as a polymer solution or paint and is picked up from a bath by the lower roll and applied to one side of the substrate. The thickness of the coating can be as low as a few μm and is controlled by the viscosity of the coating liquid and the nip dimension. As shown, roll coating systems can be configured as either forward-roll coating for co-rotating rolls or reverse-roll coating for counter-rotating rolls. The reverse roll coating process delivers the most accurate coating thicknesses.

In melt roll coating systems the polymer is delivered into the nip region in powder form, granules, or sometimes pre-plasticized from an extruder, such as depicted in Fig. 4.126.

Slide coating and curtain coating, schematically depicted in Fig. 4.127, are commonly used to apply multi-layered coatings. However, curtain coating has also been widely used to apply single layers of coatings to cardboard sheet. In both methods, the coating fluid is pre-metered by a single screw extruder. Another technique of manufacturing multi-layered film systems is the *hot melt* laminating process presented in Fig. 4.128. Here a molten polymer, delivered by an extrusion die, is sandwisched between two film substrates, cooled and spooled onto a roll.

It is also common to laminate finished films into multi-layered systems by heating the polymer films and pressing the laminate between roll systems as schematically depicted in Fig. 4.129.

Another common coating technique is to metallize films by evaporating aluminum onto a heated roll inside a vacuum chamber, such as shown in Fig. 4.130. For a successful process the required vacuum must be 10^{-4} mbar. This semi-

4.9 Coating

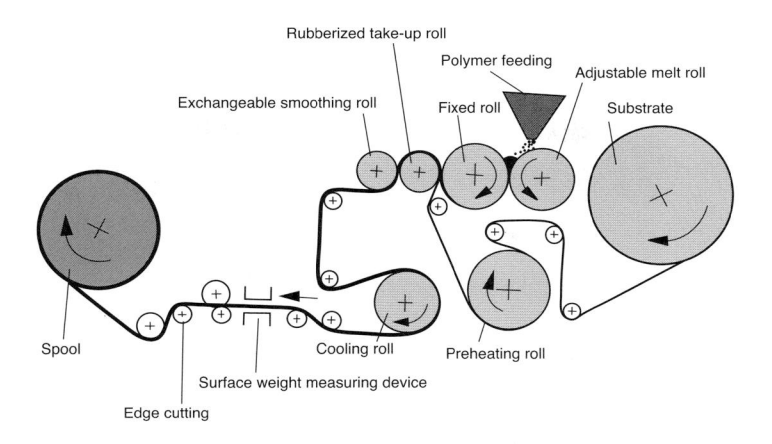

Figure 4.126: Schematic of a melt roll coating system.

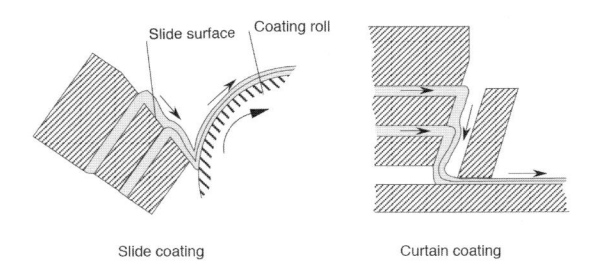

Figure 4.127: Schematic of slide and curtain coating.

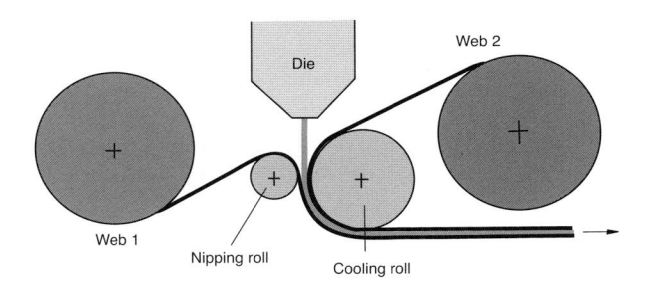

Figure 4.128: Schematic of a hot melt three-layer laminating system.

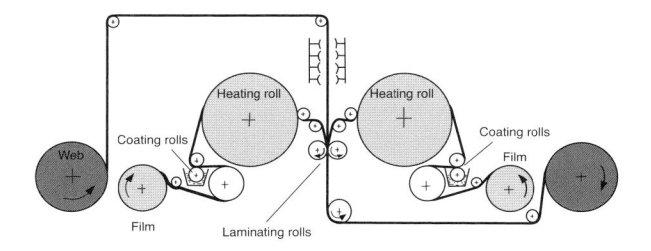

Figure 4.129: Schematic of a three-layer laminating system.

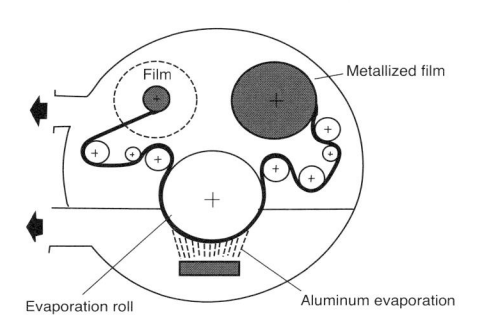

Figure 4.130: Schematic of a thermal evaporation film metallizing system.

continuous process is mostly used to metallize plastic film with aluminum, however, more recently it has been used to coat SiO_x onto films. The aluminum layers are between 0.03 and 0.04 μm thick, and the film thicknesses can be as low as 9 μm, for PET, and 15 μm, for PP and PC, with widths between 0.3 and 2.4 m. Such films are often implemented as gas and light barriers, decoration, conductive condensor films and light reflectors.

Finally, *plasma coating*, also referred to as *chemical vapor deposition* (CVD) is a widely used coating process to decrease the permeability of PET bottles, increase the scratch resistance of PC surfaces, make surfaces either hydrophilic or hydrophobic, add UV protection, modify the index of refraction, to name a few. During plasma coating, schematically depicted in Fig. 4.131, a monomer is added to a low pressure plasma (totally or partially ionized gas). The collisions between the plasma and the monomers cause the molecules to heat up spontaneously, leading to the opening of chemical bonds, which normally would only occur at high temperatures. Having chosen the appropriate monomer this will trigger a reaction that will lead to a thin coating on the substrate. For example, a 0.1

4.10 Foaming

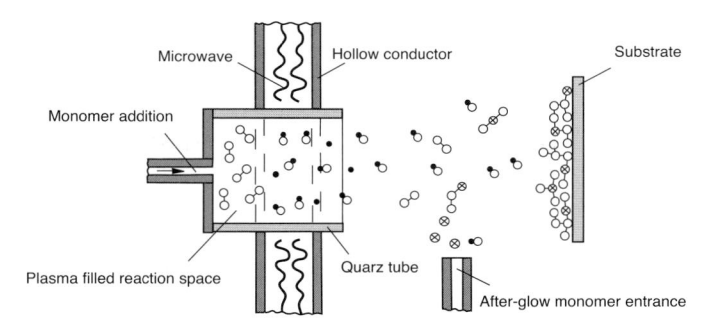

Figure 4.131: Schematic of plasma polymerization coating process.

to 0.8 μm thick layer of Si-O$_2$ can be deposited on PC, to make light lenses scratch-resistant for automotive applications.

4.10 FOAMING

In foam or a foamed polymer, a cellular or porous structure has been generated through the addition and reaction of physical or chemical blowing agents. The basic steps of foaming are cell nucleation, expansion or cell growth, and cell stabilization. Nucleation occurs when, at a given temperature and pressure, the solubility of a gas is reduced, leading to saturation, expelling the excess gas to form a bubble. Nucleating agents, such as powdered metal oxides, are used for initial bubble formation. The bubbles reach an equilibrium shape when their inside pressure balances their surface tension and surrounding pressures. The cells formed can be completely enclosed (closed cell) or can be interconnected (open cell). A physical foaming process is one where a gas such as nitrogen or carbon dioxide is introduced into the polymer melt. Physical foaming also occurs after heating a melt that contains a low boiling point fluid, causing it to vaporize. For example, the heat-induced volatilization of low-boiling-point liquids, such as pentane and heptane, is used to produce polystyrene foams. Also, foaming occurs during volatilization from the exothermic reaction of gases produced during polymerization such as the production of carbon dioxide during the reaction of isocyanate with water. Physical blowing agents are added to the plasticating zone of the extruder or molding machine. The most widely used physical blowing agent is nitrogen. Liquid blowing agents are often added to the polymer in the plasticating unit or the die. Chemical blowing agents are usually powders introduced in the hopper of the molding machine or extruder. Chemical foaming occurs when the blowing agent thermally decomposes, releasing large amounts

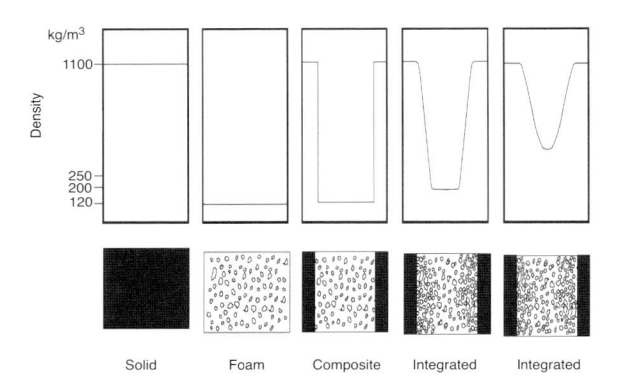

Figure 4.132: Schematic of various foam structures.

of gas. The most widely used chemical blowing agent for polyolefin is azodicarbonamide. In mechanical foaming, a gas dissolved in a polymer expands upon reduction of the processing pressure. The foamed structures commonly generated are either homogeneous foams or integral foams. Figure 4.132 presents the various types of foams and their corresponding characteristic density distributions. In integral foam, the unfoamed skin surrounds the foamed inner core. This type of foam can be achieved during injection molding and extrusion and it replaces the sandwiched structure shown in the center of Fig. 4.132 Today, foams are of great commercial importance and are primarily used in packaging and as heat and noise insulating materials. Examples of foamed materials are polyurethane foams, expanded polystyrene (EPS) and expanded polypropylene particle foam (EPP). Polyurethane foam is perhaps the most common foaming material and is a typical example of a chemical foaming technique. Here, two low viscosity components, a polyol and an isocyanate, are mixed with a blowing agent such as pentane. When manufacturing semi-finished products the mixture is deposited on a moving conveyor belt where it is allowed to rise, like a loaf of bread contained whithin shaped paper guides. The result is a continuous polyurethane block that can be used, among others, in the upholstery and matress industries. A schematic of a continuous process to manufacture a soft foam block is presented in Fig. 4.133. In contrast, Fig. 4.134 presents a double conveyor system used to manufacture hard polyurethane foam plates.

The basic material to produce expanded polystyrene products are small pearls produced by suspension styrene polymerization with 6-7 percent of pentane as a blowing agent. To process the pearls they are placed in pre-expanding machines heated with steam until their temperature reaches 80 to 100°C. To enhance their expansion, the pearls are allowed to cool in a vacuum and allowed to age and

4.10 Foaming

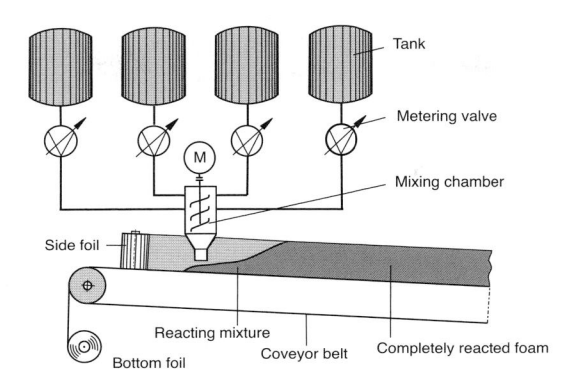

Figure 4.133: Schematic diagram of the manufacturing process of a soft polyurethane foam block.

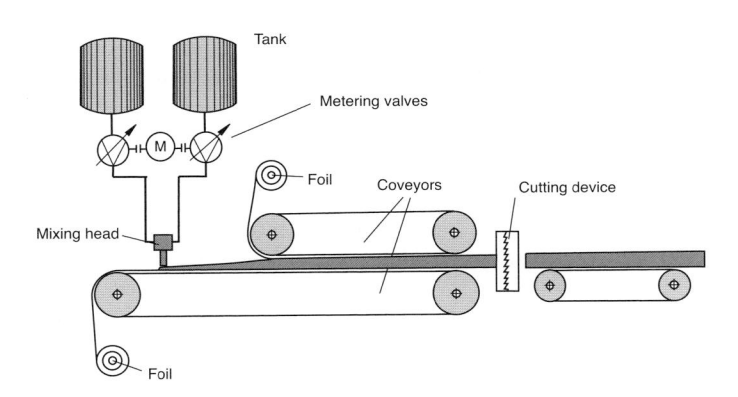

Figure 4.134: Schematic diagram of the manufacturing process of a hard polyurethane foam plate.

dry in ventilated storage silos before the shaping operation. Polystyrene foam is is used extensively in packaging, but its uses also extend to the construction industry as a thermal insulating material, as well as for shock absorption in children's safety seats and bicycle helmets. Expanded polypropylene particle foam is similar to EPS but is characterized by its excellent impact absorption and chemical resistance. Its applications are primarily in the automotive industry as bumper cores, sun visors and knee cushions, to name a few.

4.11 ROTATIONAL MOLDING

Rotational molding is used to make hollow objects. In rotational molding, a carefully measured amount of powdered polymer, typically polyethylene, is placed in a mold. The mold is then closed and placed in an oven where the mold turns about two axes as the polymer melts, as depicted in Fig. 4.135 During heating and melting, which occur at oven temperatures between 250 and 450°C, the polymer is deposited evenly on the mold's surface. To ensure uniform thickness, the axes of rotation should not coincide with the centroid of the molded product. The mold is then cooled and the solid part is removed from the mold cavity. The parts can be as thick as 1cm, and still be manufactured with relatively low residual stresses. The reduced residual stress and the controlled dimensional stability of the rotational molded product depend in great part on the cooling rate after the mold is removed from the oven. A mold that is cooled too fast yields warped parts. Usually, a mold is first cooled with air to start the cooling slowly, followed by a water spray for faster cooling. The main advantages of rotational molding over blow molding are the uniform part thickness and the low cost involved in manufacturing the mold. In addition, large parts such as play structures or kayaks can be manufactured more economically than with injection molding or blow molding. The main disadvantage of the process is the long cycle time for heating and cooling of the mold and polymer. Figure 4.136 presents the air temperature inside the mold in a typical rotational molding cycle for polyethylene powders. The process can be divided into six distinct phases:

- Induction or initial air temperature rise
- Melting and sintering
- Bubble removal and densification
- Pre-cooling
- Crystallization of the polymer melt
- Final cooling

The induction time can be significantly reduced by pre-heating the powder, and the bubble removal and cooling stage can be shortened by pressurizing the

4.11 Rotational Molding   411

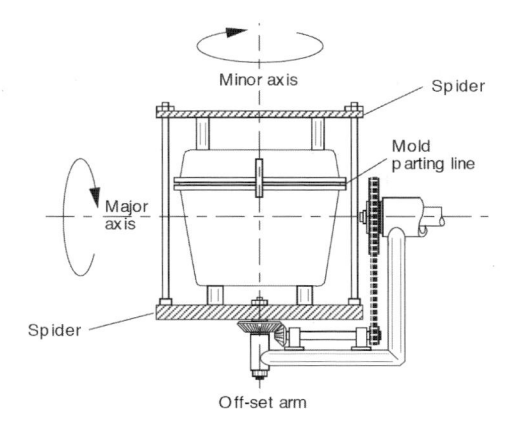

Figure 4.135: Schematic of the rotational molding process.

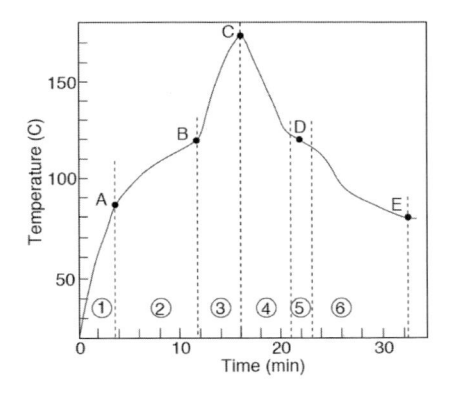

Figure 4.136: Typical air temperature in the mold while rotomolding polyethylene parts.

material inside the mold. The melting and sintering of the powder during rotational molding depends on the rheology and geometry of the particles.

4.12 WELDING

Joining or assembly of plastics and polymeric composites is an important step in manufacturing of parts from these materials. Joining methods for plastics can be divided into three major groups: mechanical joining, adhesive bonding, and welding (also called fusion bonding). In this section we will concentrate on welding. Welding of plastics is used to join large (e.g., vessels, tanks, and pipelines) and small (e.g. lighters, electronic dip switches, and cameras) parts. It is used to join simple and complex structures and in batch and mass production. It is used in numerous industries including automotive, aerospace, toy, medical, electronic, and infrastructure.

Examining most welding processes, one can identify five distinct steps that make up these processes. For some welding processes, these steps are sequential while for other processes some of the steps may occur simultaneously. The basic welding steps are:

- Surface Preparation
- Heating
- Pressing
- Intermolecular Diffusion
- Cooling

Welding processes are often categorized and identified by the heating method that is used. All the processes can be divided into two general categories: internal heating and external heating. Internal heating methods are further divided into two categories: internal mechanical heating and internal electromagnetic heating. Internal mechanical heating methods rely on the conversion of mechanical energy into heat through surface friction and intermolecular friction. These processes include: ultrasonic, vibration, and spin welding. Internal electromagnetic heating methods rely on the absorption and conversion of electromagnetic radiation into heat. These processes include: infrared/laser, radio frequency, and microwave welding. External heating methods rely on convection and/or conduction to heat the weld surface. These processes include: hot tool, hot gas, extrusion, implant induction, and implant resistance welding.

4.12.1 Hot Tool Butt Welding

Hot tool butt welding, which is sometimes also called hot plate welding is one of the most popular plastic welding methods because it is highly reliable, repeatable and economical. It is used in a wide range of industries including automotive, chemical, infrastructure, food, and housing. Although hot tool welding has long cycle times compared to internal heating methods, it is well suited for mass production applications that require high reliability like car batteries or fuel systems. In addition to high reliability it is highly flexible making it an attractive process for batch processing as well as manual and semi-automated welding (e.g. pipes, liners, and vessels). There are also a few variations of hot tool welding that were developed for specific applications like impulse heating for sealing of food packages or socket welding of pipes. Contact hot tool welding is the most common type of hot tool welding. In contact hot tool welding, the parts are heated by direct physical contact with the heated tool resulting in conduction heating of the surfaces. While contact hot tool welding equipment may be customized for specific applications, all welders have the following components:

- Heated tool. Usually a platen that is heated by either electrical cartridge heaters or hot gas. It is machined to conform to the weld surfaces of the parts. It is usually coated with a non-stick polytetrafluoroethylene (PTFE) coating to minimize sticking of the molten polymer to the tool; the PTFE coating is useable to temperatures of about 250°C. Dissimilar material welding is possible by using two heated tools that are set to different temperatures.
- Guided tracks. The parts and heated tool are usually mounted on guided tracks to insure proper alignment during heating and forging and to enable transmission on the necessary forces to the parts.
- Actuators. Pneumatic, hydraulic or electromechanical actuators are used to move the hot tool and the parts, and to apply the desired force or pressure.
- Clamping/holding devices. The parts are clamped or held to the welding fixtures that are mounted on the guided tracks by a variety of methods including vacuum, pneumatic, hydraulic, or manual clamps.
- Controller. A number of controllers are used to insure repeatability of the weld. One or more temperature controllers are used to control the heated tool temperature. Pressure controllers are used to control the pressure that is applied by the actuators. Frequently programmable controllers are used to control the cycle timing as well as incorporating safety features with limit switches. Mass production welding equipment may also include force and displacement sensors and capabilities for statistical process control.

The steps of heated tool butt welding, graphically depicted in Fig. 4.137, are:

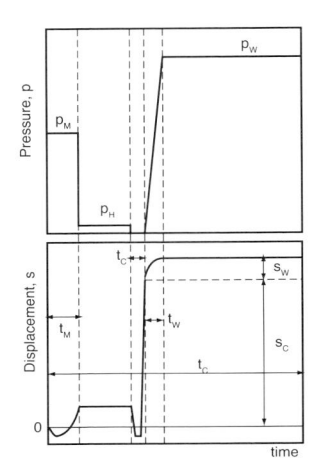

Figure 4.137: Process steps of heated tool butt welding. Subscripts: M=mating, C=change-over, W=welding and C=cooling.

- Surface Preparation. For manual or semiautomated systems, the parts are prepared for welding by either machining of the weld surfaces or by just cleaning with a solvent. For mass production surface preparation is rarely used.
- Matching. The parts are pressed against the heated tool for a predetermined time or displacement using fairly high pressure. In mass production, melt is squeezed out from warped regions in the parts to insure good matching of the weld interfaces.
- Heating. During the heating phase the pressure is either dropped to a very low level or mechanical stops are used to effectively drop the pressure to zero. Heating is done for a predetermined time to allow a melt layer of a desired thickness to form.
- Change-Over. After heating, the parts are retracted away from the hot tool, the tool is retracted and the parts are pressed together.
- Joining and Cooling. During the joining process, the parts are slowly brought in contact and pressed together using the preset joining pressure. For some machines, the joining pressure is maintained until the weld seam has cooled sufficiently. Other machines use mechanical stops allowing the pressure to be applied until the stops are reached.

Specific hot tool welding parameters depend on the material and application as well as the size of the parts. Therefore, design of experiments methodology

4.12 Welding

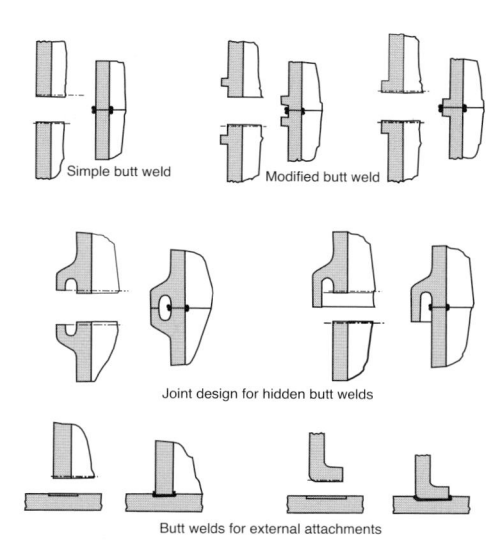

Figure 4.138: Variations of the hot tool butt welding process.

or other experimental methods can be used to optimize the welding conditions. A good starting point for the hot tool temperature is $T_H \approx T_g + 100°C$ for amorphous and $T_H \approx T_m + 50°C$ for semi-crystalline materials. Using hot tool welding in mass production requires greater emphasis on process optimization and cycle time minimization compared to manual or semi-automated systems. Luckily in most mass production applications the parts have thin walls (on the order of a few mm), so they require thin melt layers for optimum weld quality, which is possible even for short cycle times (on the order of tens of seconds to a few minutes). It is often acceptable to sacrifice weld quality in order to reduce the cycle time, so long as the joint is able to meet all the service requirements. For example, in cases where hermetic seals are needed and the force transfer through the joint is low, it is possible to reducing the melt layer thickness below the optimum thickness for maximum joint strength. Cycle time can also be reduced by using the highest possible hot tool temperatures, even if it may cause slight degradation of the polymer. In mass production it is usually economical to incorporate high levels of complexity into the parts, which often results in parts with complicated weld contours. One of the major advantages of hot tool welding is that it can be readily used with such parts. However, careful design of the hot tool is needed because in addition to matching the geometry, it must provide uniform heating throughout the weld area or the melt layer thickness will not be uniform, resulting in non-uniform weld quality.

In mass production applications the aesthetics of the joint are often just as important as the mechanical performance of the joint. Therefore, a wide variety of joint designs for hot tool welding have been developed. Fig. 4.138 shows examples of simple butt welds. The hidden butt welds that are shown in Fig. 4.138 are used when aesthetics are important or when reduced friction for flow in a pipe or a duct is important. When external attachments (e.g. mounts, nipples, or clamps) are needed near the weld area, the joint designs that are shown in Fig. 4.138 might be used.

During the heating phase, of *non-contact hot tool welding* the parts are brought very close to the heated tool without coming into contact with it and are heated by thermal radiation and convection. Here, the hot tool temperature can be raised up to 400-500 °C to speed up the heating phase. During the change-over phase, the hot tool is retracted and the parts are brought together. During the joining phase, they are pressed together to achieve intimate contact, and to cool and re-solidify the surface. Therefore, this process is identical to contact hot tool welding except that there is no matching phase. In mass production of moderate to large parts the matching phase is used to minimize molding effects such as warped surfaces. This is not possible with non-contact hot tool welding, which limits its use to relatively small parts with little warpage, or to parts with high quality molding. The noncontact hot tool welding process provides considerable advantages, such as eliminating the risk of contamination of the weld interface due to sticking to the hot plate, short cycle time, good weld quality and high process repeatability.

Hot tool socket welding is another variation of hot tool welding of pipes. As shown in Fig. 4.139 the pipe ends are heated at the end and on the outer surface. At the same time the socket is heated on the inside. Following heating, the pipe is inserted inside the socket to form the assembly. In contrast with traditional hot tool welding, the joining pressure develops through the thermal expansion on the outside of the pipe and inside of the socket. Therefore, dimensional accuracy of the outer diameter of the pipe and inner diameter of the socket is critical to achieve high weld quality repeatability.

Finally, *wedge welding* is used to weld or seal seams in films using a heated wedge (see Fig. 4.140). This is a continuous welding technique were the wedge is moved manually or automatically along the seam. Following the wedge are pressure rollers that keep the film under pressure while it cools.

4.12.2 Ultrasonic Welding

Ultrasonic welding is a very popular technique for fusion bonding or welding of thermoplastics and thermoplastic composites. Welding is accomplished by applying relatively low amplitude (1 µm - 250 µm) high frequency (10-70 kHz) mechanical vibrations to the parts to be joined. This results in cyclical deforma-

4.12 Welding

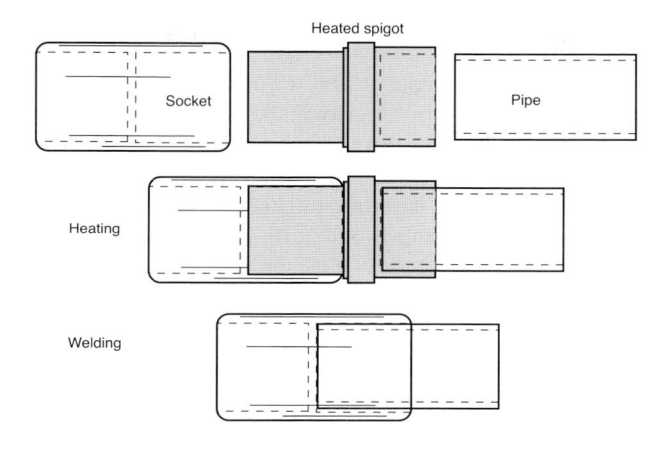

Figure 4.139: Configuration for hot socket welding of pipes.

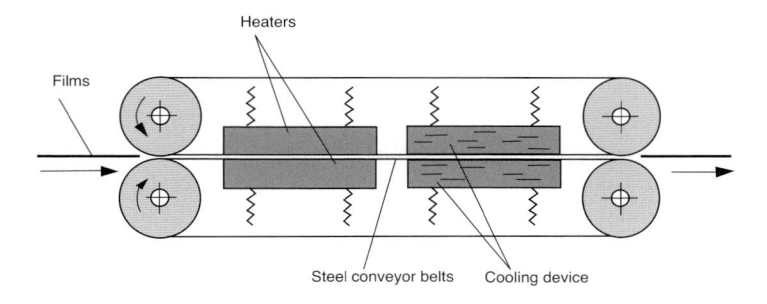

Figure 4.140: Schematic of a heated wedge laminating system.

tion of the parts and of any surface asperities. The cyclical energy is converted into heat - within the thermoplastic - through intermolecular friction. This is similar to heating that occurs in a metal wire that is bent back and forth repeatedly, or in general when materials are subjected to cyclical loading. The heat, which is highest at the surfaces (due to asperities straining more than the bulk), is sufficient to melt the thermoplastic and to fusionbond the parts. Usually, to improve consistency a man-made asperity in the form of triangular protrusion is molded into one of the parts. This protrusion, which is also called an energy director or concentrator, experiences the highest levels of cyclical strain producing the greatest level of heating. Therefore, the energy director melts and flows to join the parts. Ultrasonic welding is one of the most common methods used in industry to join plastics. There is no one reason for its popularity but some of its advantages include:

- Speed (typical cycle times less than 1 s)
- Ease of automation
- Relatively low capital costs
- Applicable to a wide range of thermoplastics

Ultrasonic welding is usually divided into two major groups: near-field and far-field welding. Current industry practice, which is based on the most extensively used 20 kHz welding system, considers cases where the distance between the horn/part interface and the weld interface is less than 6 mm to be near-field welding. Far-field ultrasonic welding is used for distances greater than 6 mm. At 20 kHz, the wavelength in the plastic component is between 6 and 13 cm depending on the specific polymer. Therefore, during near-field ultrasonic welding, the vibration amplitude at the weld interface is close to the amplitude at the horn face. For far-field welding, the amplitude of vibration at the weld interface depends on the ultrasonic wave propagation in the parts.

Ultrasonic welding of thermoplastics can be divided into two categories: plunge and continuous welding. In plunge welding, discrete welding of the parts is done. The ultrasonic tooling plunges against the parts to be welded/joined and retracts at the end of the process. In continuous welding, the ultrasonic tooling remains against the parts under a certain force or fixed gap and the parts (usually films or fabrics) are drawn through in a continuous fashion. This type of application is similar to sewing and a lot of continuous ultrasonic welding equipment is designed to look and function like sewing machines. Beyond standard welding techniques, it is possible to join parts that are not typically weldable, such as metals or wood products to thermoplastics by mechanical means with ultrasonic joining. One technique is ultrasonic insert. In this case, a metal insert, usually a threaded design, is driven into a thermoplastic substrate to accept a screw or

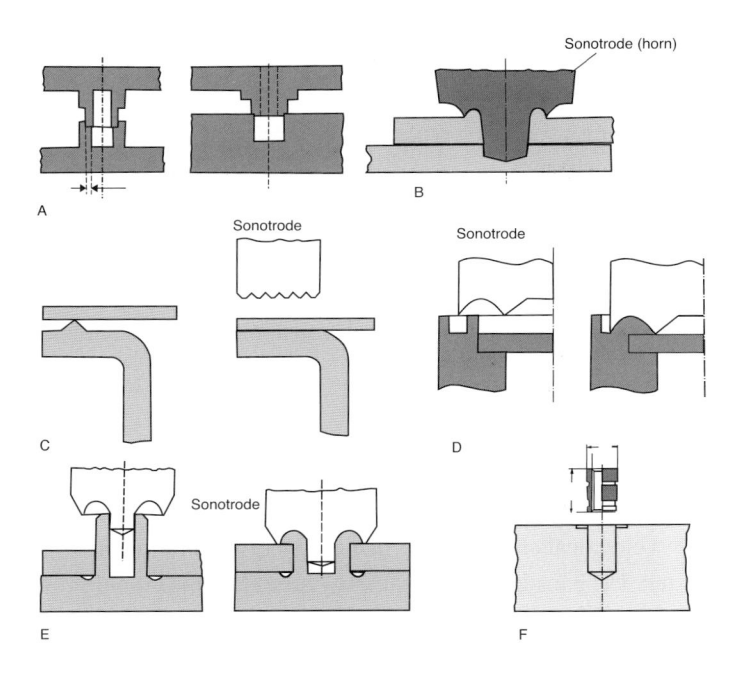

Figure 4.141: Schematic of various ultrasound welding joints.

similar device, see Fig. 4.141-F. Often, this is more economical than molded-in threads and can have superior strength.

Another design relying on mechanical fastening using ultrasonic energy, is ultrasonic staking, see Fig. 4.141-E. In staking, a thermoplastic protrusion is heated and reshaped by the ultrasonic tooling/ horn.

Yet another process variation that uses ultrasonic energy is swaging. In this case, there is a thermoplastic protrusion, similar to staking, however, in swaging the final shape of the reshaped protrusion is not limited to a "rivet-like" shape. Instead, the final shape can be straight and long or even cylindrical to capture complex geometries, such as capturing a lid on a can. Figure 4.141-D shows a typical example of ultrasonic swaging.

Most commercial equipment operates between 20 and 40 kHz. See Table 4.10 for general guidelines to the advantages and limitation to selected operating frequencies.

Table 4.10: Comparison of Operating Frequencies for Ultrasonic Welding of Plastics

Operating Frequency (kHz)	Typical power capabilities converter (W)	Comments
10 to 20	6000 to 3000	Relatively noisy, high power, possible part damage.
20 to 30	3000 to 1000	Moderate noise and power.
40 to 75	1000 to 400	Very quiet, low power, gentle process.

There are seven major components in a typical ultrasonic machine:

- Power Supply. This component converts line voltage to high frequency power.
- Controller. This component controls and monitors the system and interfaces with the user as well as with the PLC's, if applicable.
- Converter. In simple terms, the converter is a motor, which is used to convert electrical energy to mechanical energy (vibrations).
- Booster. The booster is used to amplify or re-duce the vibration amplitude.
- Sonotrode or horn. The horn has two main functions: (1) Further increase the amplitude of vibration (just like the booster) and (2) apply the ultrasonic energy to the work piece.
- Actuator. The actuator holds the stack assembly (converter, booster and horn) and brings the stack into contact with the part. In addition, the actuator provides the clamp force (weld force). Most actuators are based on pneumatics, how-ever, there are some on the market that are servo driven.
- Fixture. The fixture serves two functions: (1) Holding and securing the parts and (2) providing location of the parts.

Table 4.11 contains recommended amplitudes for ultrasonic assembly. These values are starting points only during equipment setup, and often fine-tuning of the values is needed.

It is possible to predict the weldability of two different materials using the following guidelines:

- Similar melt temperatures. The T_g or T_m, depending if the application is an amorphous or crystalline material, should be within 22 °C (40 °F) of each other.
- Similar melt flows. The melt flow index for the two materials should be within 10% of each other.

4.12 Welding

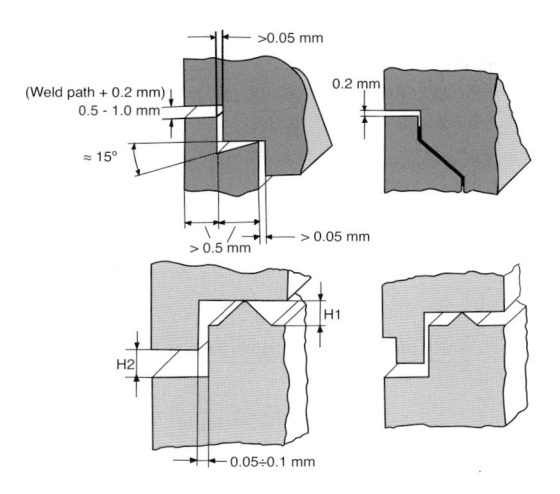

Figure 4.142: Examples of variation of the energy director.

- Similar surface energies. The relative surface energies should be within 10% of each other.

For any application, there are two major factors regarding design: part design and joint design. There are many variations on joint design, but most fall under two major categories: energy director and shear joint. Both promote stress at the bond line (faying surface) in order to assure that the energy is concentrated at the bond line. Several energy director ultrasound welds are presented in Fig. 4.142. Table 4.12 presents various problems that may arise during ultrasonic welding, their causes and suggestions for their solution.

Table 4.11: Range of Vibration Amplitudes at 20 kHz for Various Materials and Applications

Material	Welding (μm)	Insertion (μm)	Staking (μm)
PS	20-40	20-40	20-30
PC	50-70	50-70	40-60
ABS	40-60	40-60	30-50
PP	70-90	70-90	50-70
PE	70-90	70-90	50-70
PA	60-80	60-80	60-80
PPO	50-90	50-80	40-70
PEI	60-125	60-100	40-100
PVC	40-75	40-75	30-60

Continued on next page

Material	Welding (μm)	Insertion (μm)	Staking (μm)
SAN/NAS	30-65	30-65	30-60
POM	75-125	75-100	50-100
PEEK	60-125	60-100	60-100
PPS	80-125	75-100	75-100
PI	60-125	60-100	60-100
PBT	60-125	60-100	60-100

Table 4.12: Troubleshooting Guide for Ultrasonic Welding

Problem	Possible causes and recommendations
Welding problems	
Overwelding	**Too much energy input for welding.** Reduce weld pressure, weld time, energy, welding distance. Reduce amplitude of vibration by using lower power setup and/or lower gain booster. Reduce down speed.
Underwelding	**Insufficient energy input for welding.** Do the counter actions described for overwelding problem. **Energy loss to fixture (if fixture is made of Urethane).** Use a more rigid material.
Inconsistent welding from part-to-part	**Part dimension variation.** Run statistical study if a pattern develops with certain cavity combinations. Check part tolerances/dimensions, molding conditions, and cavity dimension variation. **Mold release.** Clean mating surface or replace mold release. **Variation in material characteristics.** Check regrind and/or filler content variation. Check molding condition variation. Check moisture content. **Variation in utility.** Check line voltage and air pressure fluctuations.
Marking	**Horn heats up.** Decrease amplitude. Check for cracked horn, booster, or converter. Check for loosened stud and horn. Utilize air nozzles for cooling. **Weld cycle is too long.** Increase amplitude and/or pressure. Adjust dynamic air pressure. **Improper fit of part to fixture.** Check for proper fixture support. Redesign fixture.
Flash	**Energy director is too large.** Reduce size of energy director, weld time, pressure. **Overwelding.** Refer to recommendation in overwelding. **Shear interference is too great.** Reduce amount of interference. **Insufficient flash trap.** Add flash trap. Increase flash trap.
Insertion problems	
Inserts pull out too easily	**Insufficient interference between hole and insert.** Reduce the size of hole. Increase amplitude and/or reduce the weld pressure. **Inserts get pushed in boss before plastic melts.** Use pre-trigger on. **Horn retracts before the plastic around the inserts is solidified.** Increase hold time.

Continued on next page

4.12 Welding

Problem	Possible causes and recommendations
Inconsistent insertion	**The plastic is not melting consistently around all inserts.** Increase the amplitude. **Inserts are seated at different heights.** If using a process controller, weld under specified distance (absolute or collapse). Use mechanical stop, if applicable.
The plastic around boss cracks	**The insert is pushed in before the plastic gets melted.** Reduce the down speed, weld pressure, and/or amplitude. Set pre-trigger on. Increase the thickness of the boss wall.
Staking problems	
Stake head is not uniform	**Staking cavity at horn is too large or insufficient volume in the stud.** Reduce cavity size or increase the stud height or diameter.
Excessive flash is formed around the stake head	**Staking cavity is too small or stud volume is too much.** Increase the staking cavity or reduce the stud height and/or diameter.
Parts are loose after staking	**The hole diameter is too large with respect to stud diameter.** Reduce hole diameter. Increase hold time. **The holding force was released before the molten stud is solidified.** Increase the stud diameter. During holding period, external clamps or nodal plunger can be used.
Premature staking head is formed at the end of cycle	**Insufficient weld time or energy input.** Increase weld time/amplitude, and energy.
Severe marking and distortion on opposite side of stake head	**Inappropriate fixture alignment.** Improve the fixture alignment. **Clamping pressure is too high.** Reduce the clamp pressure. If heat damages surface, placing a metal foundation at the fixture surface will serve as a heat sink to reduce marking. **Amplitude or energy input is too high.** Reduce amplitude and/or energy input.
The stud is broken at its base	**There is stress concentration at the sharp corner near the stud base.** Introduce radius at the stud corner. **Too much pressure is applied before melting.** Set up pre-trigger on. **The stud is aligned at the center of horn.** Align the stud at the center of horn cavity.
Equipment problems	
Heat develops at the interface between horn/booster or booster/transducer	**Loose connection at the interface.** Disassemble the stack and clean the interface before tightening the stack by the manufacturer's suggested torque. **A mounting stud is cracked.** Replace the stud. **A stud is loosened.** Remove the loose stud. Inspect the stud and retighten the stud. **There is a dirty interface between horn/booster or transducer/booster.** Clean the interface with clean cloth or paper towel and apply a thin film of grease before re-assembly.
Noise from the stack	**A loosened interface between components.** Check the stack assembly and retighten with the recommended torque. **The stack component gets cracked.** Replace any component as needed. **The stack frequency is out of the resonance range.** Retune the generator.

Continued on next page

Problem	Possible causes and recommendations
Welder will not tune in air	**Defective horn, converter, booster, stud, or junction.** Replace stack components and junctions as needed. **Defective power supplier.** Replace power supply. **Improper horn.** Check resonance frequency of horn and compare it with weld frequency range. If needed, replace it.
Welder tunes in air but overloads during welding	**Defective horn, converter, booster, stud, or junction.** Replace stack components and clean interfaces as needed. **Defective power supplier.** Replace power supplier. **Unstable horn.** Redesign the horn. **Horn requires high starting energy.** Reduce pre-trigger force. Use slower pressure buildup rate, and/or use lower amplitude gain horn.

4.12.3 Vibration Welding

Vibration welding is a well-established process that can join large, sealed, and mechanically strong weld seams with cycle times of only a few seconds. It is mainly used when short cycle times are demanded and the part is too large for the use of ultrasonic welding. However, because the process requires relatively flat (planner) joining surfaces, some applications are not suited for vibration welding. A slight incline in the joint of about 10° can be tolerated in the linear vibration welding components, although it is not recommended. In some cases the vibration welding process can be used to solve the warpage issues of glass-fiber-filled polyamide parts, because it is possible to clamp the parts so that they are forced in the correct geometry. Vibration welding can also be used as an alternative to adhering or mechanically joining different materials such as wood fiber reinforced materials to thermoplastics. In vibration welding, the two parts to be joined are clamped together under a relatively high force. At a preset force, one part is vibrated relative to the second, usually with a displacement between 0.5 and 1.5 mm (0.020 and 0.060 in.). The motion results in frictional heating. This heat results in the joining surface melting and fusing together.

There are several major components in a typical vibrational welding machine:

- Driving unit. The most common drive design consists of electro-magnetic drive systems in conjunction with springs.
- Fixtures. Fixtures must provide a support to parts to prevent movement or damage and must ensure the transmission of the clamping forces. Because the pressure may be as high as 8 N/mm (1160 psi), parts and fixture must be rigid.
- Control. Processsing programmable controllers (PLC) are used most often.

Linear and orbital welding processes can be divided into four discrete phases:

- Initial heating of the thermoplastic resin through the friction of solid-to-solid interfacial heating (Phase 1)

4.12 Welding

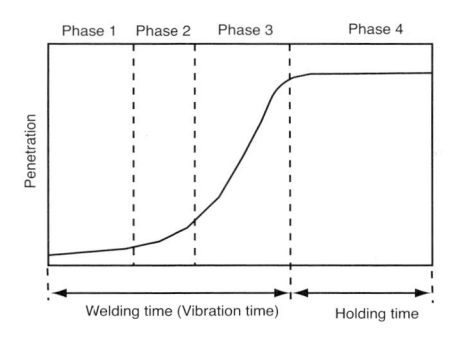

Figure 4.143: Schematic of the different phases in friction welding as a function of penetration (meltdown).

- Intermittent heating of the thermoplastic resin through shear within a thin layer of molten thermoplastic, transitional phase between Phase 1 and Phase 3 (Phase 2)
- Stationary thermoplastic temperature, steadystate melt displacement/penetration (Phase 3)
- Cooling (Phase 4)

The vibration amplitude is the horizontal peak-topeak movement of the vibrating welding head. In general, doubling of the vibration frequency will result in a reduction of the vibration amplitude by 1/2 and vice versa. Typically, a vibration frequency of 200 Hz has a vibrational amplitude of 1 to 1.8 mm (0.040 to 0.071 in) and a frequency of 100 Hz has an amplitude of 2 to 4 mm (0.078 to 0.158 in). The lower frequency/higher amplitude is used for components with flexible supporting walls and large joint areas, for example, car body components such as bumpers. The lower amplitude/ higher frequency is used for smaller applications, e.g., automotive air intake manifolds. The method for setting the vibrational amplitude varies from machine to machine. It is important that the vibrating system operates at or near the resonance frequency of the system to ensure optimum machine performance and longevity. Manufacturer guidelines should be observed when setting up new tooling. Generally, orbital vibration welding operates at lower amplitudes than linear vibration welding; typically it ranges from 0.25 to 1.5 mm (0.010 to 0.060 in). The values for amplitude, welding force, holding force, welding time, and holding time depend on the vibration machine, the type of plastic, and the part geometry. Therefore, the optimal parameters must be examined through an experimental design.

In both the orbital and linear vibration welding techniques, circular and non-circular components can be accommodated. Joint designs can range from the

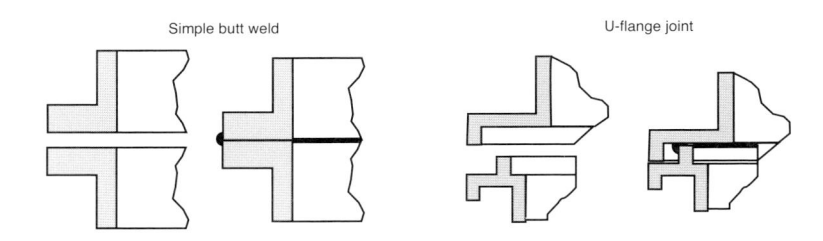

Figure 4.144: Examples of vibration welds.

simple butt joint to the U-flange joint with a flash trap (Fig. 4.144), each fulfilling a particular role depending on the complexity of the component or thick-ness of the component wall.

Table 4.13 presents various problems that may arise during vibration welding, their causes and suggestions for their solution.

Table 4.13: Troubleshooting Guide for Vibration Welding

Problem	Possible causes and recommendations
Operation problems	
Overwelding	**Too much weld time or weld displacement.** Reduce weld time or displacement.
	Poor flash trap design. Evaluate flash trap design.
Underwelding	**Too short weld time or insufficient displacement.** Increase weld time or displacement.
	Material difficult to weld due to low friction. Consider material change.
Non-uniform weld around component	**Warped parts/poor moldings.** Check part dimensions.
	Uneven weld interface. Check molding process conditions.
	Lack of parallelism between fixture and part. Shim fixture where necessary. Ensure tooling true to base. Check part dimensions.
	Wall flexure during welding. Design parts to incorporate strengthening ribs and U-flanges.
	Insufficient fixture support (urethane fixtures). Modify fixture to prevent outward flexure. Improve support in critical areas. Redesign fixture to improve rigidity. If large sections of urethane are deflecting, add rigid back-up. Check provisions for alignment in mating parts.
	Part tolerances. Improve part tolerances.
	Poor part alignment in fixture. Re-dimension parts.
	Mold release agent at weld. Check molding process conditions. Clean mating surfaces with suitable degreasing agent. Use a paintable/printable mold release if required.
	Fillers. Reduce amount of filler.

Continued on next page

Problem	Possible causes and recommendations
Inconsistent weld results part to part	**Mold release agent at weld interface.** Check molding process conditions. Clean mating surfaces with suitable degreasing agent. Use a paintable/printable mold release if required. **Part tolerance.** Improve part tolerances. Check part dimensions. Check molding process conditions. **Molding cavity to cavity variations on multi-cavity molds.** Run statistical study on molding process to see if pattern develops with certain cavity combinations. Check part tolerances and dimesions. Check for cavity wear in mold. Check molding process conditions. **Regrind/degraded plastic.** Reduce percentage of regrind. Improve quality of regrind. **Poor distribution of filler content.** Check processing conditions. **Incompatible materials or resin grades.** Check with resin supplier. **Moisture in molded part (usually nylon parts).** Weld directly after molding. Dry parts before welding. **Drop in air line pressure.** Raise compressor output pressure. Add surge tank with a check valve.
Flash	**Inadequate flash trap design.** Review flash trap design. **Overwelding.** See overwelding. **Non-uniform joint.** Check molding process conditions.
Misalignment of welded assembly	**Wall flexure during welding.** Add ribs to molded part. Use U-flange joint design. **Part tolerance/poor molding.** Tighten up part and tooling tolerances. Check molding process conditions. **Incorrectly aligned upper and lower tooling.** Realign tooling. **Very high weld pressure.** Reduce to recommended levels.
Component design problems	
Internal components damaged	**Overwelding.** See overwelding in operation problems section. **Internal components improperly mounted i.e. too close to joint area.** Make sure internal components are properly mounted. Isolate internal components from housing.
Melting/fracture of part sections outside of joint	**Overwelding.** See overwelding in operation problems section. **Internal stresses.** Check molding process conditions. Check part design. **Too high a weld force.** Lower pressure and re-evaluate component design.
Internal parts welding	**Internal parts same material as housing.** Change material of internal parts. Lubricate internal components. Consider component redesign.
Marking	**Incorrect fit of part to fixture.** Check for proper support. Redesign fixture. Check for cavity to cavity variations. **Movement during welding.** Redesign component to fit tooling correctly.

4.12.4 Spin Welding

Similar to vibration welding, spin welding of welding processes commonly used in industry to weld parts with round or cylindrical weld lines. Often the parts themselves are not round in shape. The process involves holding one part stationary and rotating the second part while the two parts are under a pre-load in order to assure proper alignment and heating. One of the benefits of spin welding is that the process is relatively fast, with typical cycle times between 1 and 5 s. In addition, capital equipment costs are relatively low compared to other frictional welding processes, such as linear vibration welding or orbital welding. One of the unique benefits of spin welding is that it can weld under water. One application that has benefited from this ability is welding of liquid-filled directional compasses. Spin welding also allows to weld a hermetic seal. There are various types of spin welding:

- Inertial Spin Welding Mode. Historically, spin welding has been used to weld circular parts where relative alignment between the two parts being joined is not critical.
- Drive Spin Welding Mode. In some applications, alignment between the two parts is critical and an additional level of control is required. Recently, machine designs have utilized servo motor technology so that once the faying surface is fully melted, the parts can be oriented within 1 degree relative to each other.
- Angular Spin Welding. In selected servo-driven machines, it is possible to program the head to rotate between two selected angles (orientations) without making an entire rotation. That means that the machine can be programmed to rotate back-and-forth between to angles, allowing applications, which cannot be fully rotated, to be welded with spin welding. Due to the accelerations of generating such a motion, this process is usually limited to parts less than 25 mm (1.0 in) in diameter.

There are several major components in a typical spin welding machine:

- Generator/Power Supply. This component sends line voltage to the circuits and servo motors if applicable.
- Controls. The controls establish an interface between the machine and the operator and monitor the system. They usually contain a PLC or similar type of logic circuitry.
- Actuator. This component is usually a pneumatic press that moves the upper fixture (spin head) to the lower fixture and applies force during the weld and hold cycle.
- Lower fixture. In many cases the lower fixture is a simple jig, which locates the parts being welded.
- Upper fixture. This component couples the drive mechanism to the part of the application being rotated. The upper fixture is often referred to as the drive head.

Most thermoplastic materials can be joined with a spin welding process. One of the more critical properties that a material must have is a coefficient of friction

4.12 Welding

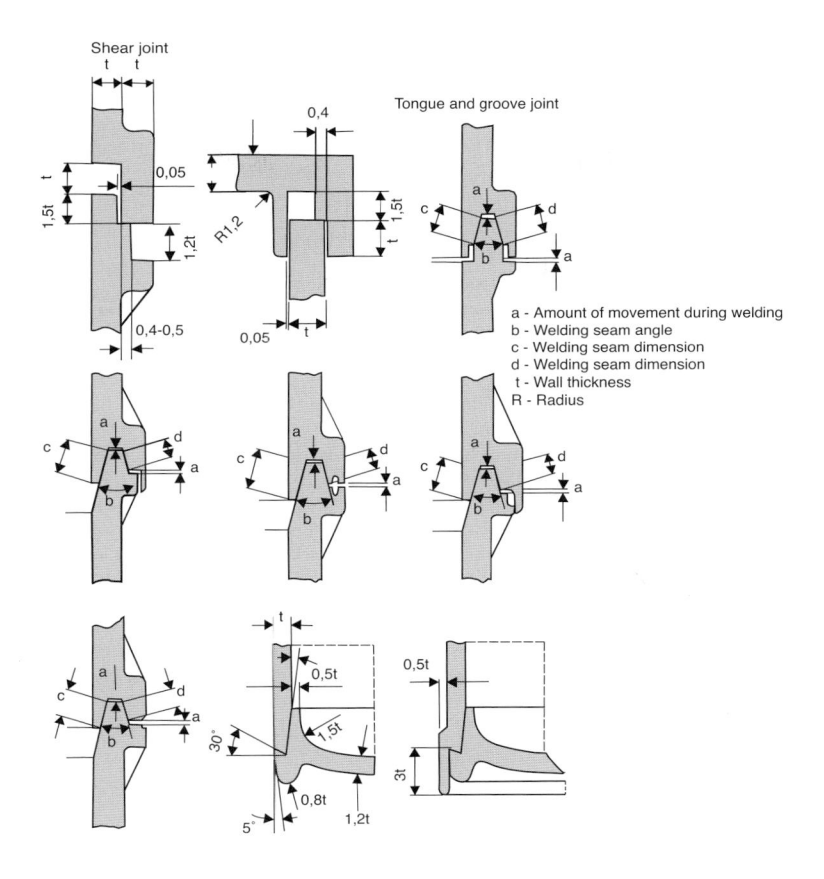

Figure 4.145: Typical joint design for spin welding.

sufficient to promote heating and melting. Thus, materials such as fluorpolymers that have a relatively low coefficient of friction are typically not joined with a spin welding process. Most parts that are joined with the spin welding process are circular. In some cases, only the bondline is circular while the part is not circular.

In addition, some designs incorporate appendages to the part design, which engage the fixture to allow mechanical coupling between the part and fixture. Once the faying surfaces are melted and the process is discontinued, the appendages are designed to fracture (break-away) as a result of resistance force when the material solidifies. This allows the spinning motion to be discontinued nearly instantaneously and enhances weld quality since the weld solidifies without any spinning motion. One of the most common joint designs is the tongue and groove, because it helps with flash containment and aids in part alignment. Figure 4.145 presents various spin welding joints.

Table 4.14 presents various problems that may arise during spin welding, their causes and suggestions for their solution.

Table 4.14: Troubleshooting Guide for Spin Welding

Problem	Possible causes and recommendations
Vibration while spinning	**Spin tool, or spin tool and part combination are out of balance.** Balance the spin tool. If the problem persists, balance the spin tool with the plastic part that is loaded into the spin tool installed.
Part marking	**Slippage in the fixture.** Friction drive tools will wear over time and start to slip while welding. The tool must be re-worked if made out of steel, or re-coated if silicone rubber is used.
Excessive flash	**The assembly is being overwelded.** Reduce the number of revolutions of weld or weld pressure.
Excessive/significant flash, but weak weld	**Lack of fusion at the joint line.** Make sure that the spin welder has rapid deceleration or active braking. If the spin tool coasts to a stop it is likely to tear bonds that are being formed as the tool slows down and the material cools. It is also possible that excesseive clamp pressure prevents proper fusion, thus a reduced clamp pressure may reduce this problem.
Excess particulate	**Lack of proper melting.** Increase spin RPM or weld force. If the spin RPM or clamp pressure is too low, the abrasion phase of the spin weld may be extended causing excess particulate generation. Increasing one or both can decrease the time to enter the melt phase.

4.12.5 IR and Laser Welding

Infrared (IR) and laser welding of plastics have been available for many years, but only recently, with the decrease in price for lasers and laser diodes and with a higher demands on part quality, IR/laser welding has become more popular. Sources of IR radiation include lasers, quartz lamps, and ceramic heaters. The emitted radiation for each source is specific and often a determining factor in selecting the source. The most basic form of laser welding is where two surfaces are heated simultaneously with a laser, as depicted in Fig. 4.146, and subsequently joined.

Advantages of IR/laser welding:

- Fast cycle time, typically 2 to 10s
- No part marking from tooling and fixture
- No particulate generated
- Flash is smooth and fully attached to part (not free to break away)
- Relatively low residual stresses

4.12 Welding

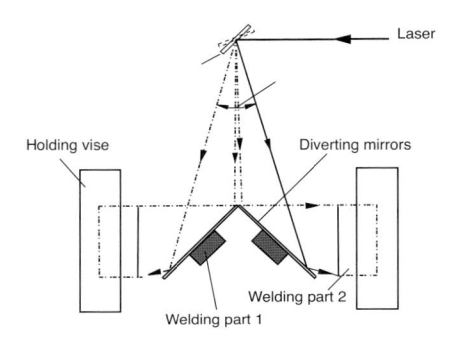

Figure 4.146: Schematic of laser heating.

- Heat effected zone can be well defined

Limitations of IR/laser welding:

- Capital costs can be relatively high
- Some materials are not well suited, such has highly filled crystalline materials
- Some applications' geometry is not well suited, such as applications with internal walls that cannot be exposed to the IR/laser radiation

There are several variations of IR and laser welding. These are:

- Through transmission IR welding (TTIr). The technique is based on passing IR/laser radiation through one component, which is IR transparent, and having the second compo-nent, absorb the radiation as schematically depicted in Fig. 4.147.
- Plunge/Simultaneous welding. The entire weld seam is simultaneously irradiated by a laser light with specially adapted optics, and welded within seconds. The advantages of the process are high process rate and the ability to tolerate part warpage.
- Surface heating. This technique is very similar to heated tool or hot plate welding. The surfaces of the components to be joined are heated by direct IR/laser exposure for a suffi-cient length of time to produce a molten layer, usually for 2-10 s. Once the surface is fully melted, the IR/laser tool is withdrawn from between the parts and the parts are forged together and allowed to solidify as shown schematically in Fig. 4.148.
- Scanning/Contour welding. A laser beam is translated along a weld seam. Either the laser is guided along the contour by a robot arm or a moving

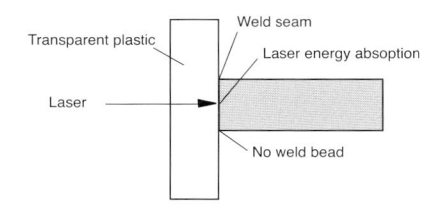

Figure 4.147: Schematic diagram of TTIr.

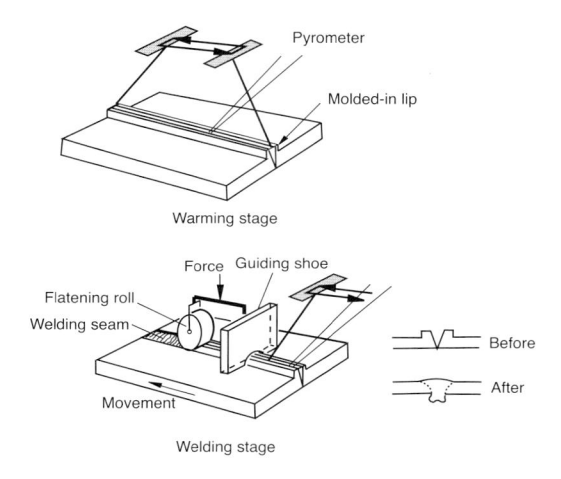

Figure 4.148: Schematic diagram of IR/Laser surface heating.

optical conductor transmits the beam to the weld seam. This has the advantage that a single machine can be programmed to weld a multitude of parts.

- Mask welding. A laser beam diverged out in the shape of a line is translated across the parts to be joined through a mask or shield. The laser beam heats the joint wherever a weld seam is not masked (shadowed). Mask welding can result in very high resolutions between welded and nonwelded areas. Today, it is possible to produce a weld width of less than 100 μm.
- Staking. This technique deforms a thermo-plastic stud into a button geometry to produce a mechanical fastener similar to a rivet. This allows dissimilar materials to be joined, such as metals to thermoplastics.

There are several components in IR-Laser welding equipment:

4.12 Welding

- Generator/Power supply. This component converts line voltage and frequency into the correct voltage, current, and frequency to power the IR source.
- Controls. The controls establish an interface between the machine and the operator and monitor the system.
- Actuator. This component is usually a pneumatic press that moves the upper fixture (platen) to the lower fixture and applies force during the weld and hold cycle.
- Lower fixture. In many cases the lower fixture is a simple jig, locating the parts being welded.
- Upper fixture. This component is often the most complex and most critical component of the entire machine. Typically, it is within this component that the IR/laser radiation is generated and delivered to the application being welded.
- IR/Laser enclosure. In nearly all equipment designs the operator needs to be protected from the radiation by some type of enclosure. If the machine contains laser(s), the enclosure needs to be FDA certified in order to assure that the operator is properly protected.

Since there are several modes of IR/laser welding, it is not possible to define weldable materials for the process in general. Therefore, any material considered for IR/laser welding should be tested first. As a general guideline, if a material is transparent in the visible spectrum it is also likely to be transparent in the near-IR spectrum. In addition, crystalline materials such as PE and PP are relatively transparent to near-IR radiation, however, because of their scattering light characteristics it is difficult to pass sufficient radiation through samples thicker than 5 mm to make a weld. A similar effect is seen when adding fillers such as glass fibers.

Because IR laser welding is a relatively new process, there is little guideline for part design. For TTIr, it is important to keep the thickness of the transparent component to a minimum, especially when welding crystalline materials.

4.12.6 RF/Dielectric Welding

Radio Frequency (RF) welding, which is also often referred to as "dielectric welding", is a process that relies on internal heat generation by dielectric hysteresis losses of thermoplastics. It is most commonly used to weld PVC bladders such as intra-vein drip bags for the medical industry. It is also used to weld book and binding covers. RF welding has the advantages that it is a relatively fast process with typical cycle times less than 2 to 5 s. It also does not require special joint

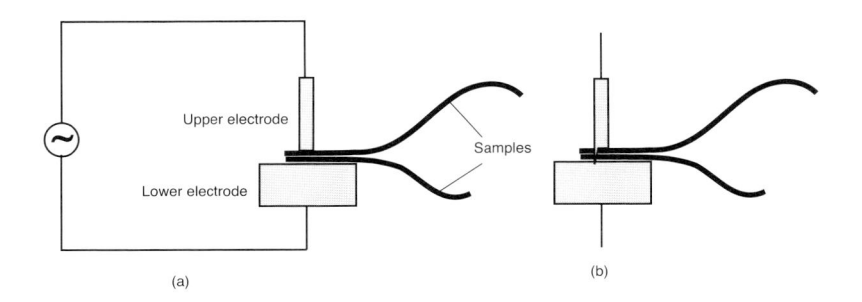

Figure 4.149: Typical electrode configuration for (a) welding and (b) cutting and sealing.

designs and produces cosmetically appealing welds. RF welding is almost exclusively used for welding relatively thin sheets or films. Thickness' usually range from 0.03 to 1.27 mm (0.001 to 0.050 in), depending on the material and application. The limitation of welding films is due to the fact that a strong electric field must be generated and this can only be achieved when the welding electrodes are brought together in close proximity (0.03 to 1.27 mm). If the welding electrodes are significantly further apart, the electric field density is too low to effectively heat and melt the plastic. It is important that the materials being joined have the proper electrical properties. One such property is a relatively high dielectric constant, typically > 2. This allows more current to flow through the material, promoting heating at a lower electrode voltage. For RF welding it is important that the material have significant dielectric loss.

There are two variations of RF/dielectric welding. These are:

- RF Welding/Sealing. In many applications the films or sheets are simply welded together. In this mode of operation, the films/sheets are usually ready in the final shape and size. Figure 4.149(a) shows the typical electrode configuration.
- RF Cut and Welding/Sealing. In many applications, the welding process trims or cuts the application to the final shape while at the same time sealing the films/sheets, see Fig. 4.149(b). Because the electrodes, even at the cutting edge, cannot be allowed to make contact while the electric field is applied (to prevent machine damage), there is a small amount of material remaining at the toe of the weld. The operator must tear the parts along this section of the weld.

There are several components in RF/dielectric welding equipment:

- Generator/Power Supply. This component converts line frequency into high frequency (27.12 MHz) high voltage.

- Controls. The controls interface the machine with the operator and monitor the system. It usually contains a PLC or similar type of logic circuitry.
- Actuator. This component is usually a pneumatic press that moves the upper fixture (platen) to the lower fixture and applies force during the weld and hold cycle.
- Lower fixture. In many cases the lower fixture is simply a large flat plate. It acts as a lower support for the parts being welded and the lower electrode to apply the electric field.
- Upper fixture. The upper fixture is often designed and built by the end user. The fixture applies the electric field and the localized clamp force to assure proper welding.
- RF enclosure (Faraday Cage). In higher power equipment (+5 kW), a cage is placed around the tooling/electrodes to protect the processor from the high voltage of the electrodes as well as from other RF radiation hazards.

For a material to be effectively welded with the RF process it must have a high dielectric loss, a high dielectric constant as well as a high dielectric break down.

4.12.7 Hot Gas Welding

Hot gas welding is a process where heated gas or air is used to heat and melt a filler (weld) rod into the joint. It is a very flexible process, which makes it well suited for short-runs or prototype welding of small items or for welding of large structures or tanks. Hot gas welding can be performed manually, in speed welding mode and automated mode as schematically depicted in Fig. 4.150. As shown in the figure, in manual operation pressure is applied by pushing the weld rod into the joint area by hand. In speed welding, a tip with a pressure shoe or tongue is used to apply pressure enabling higher welding speeds (Fig. 4.150). Automated welding is performed using custom equipment and is designed specifically for the application. The filler rod should be made from the same material as the parts that are to be welded. Usually, the filler rod has a round cross-section, but it is also available in oval, triangular and rectangular cross sections. Like in metal welding, for large joints multiple passes are used to fully fill the cavity. In some cases it may be desirable to use a small filler rod for the first pass to assure complete penetration and then use larger rods for subsequent passes.

The components during hot gas welding are:

- Hot gas gun. The hot gas gun usually includes an electrical heating element, which heats the gas that passes over it. The gas temperature is controlled by adjusting the flow rate through the gun. Guns without external compressor or air source are available. Incorporated into the gun is

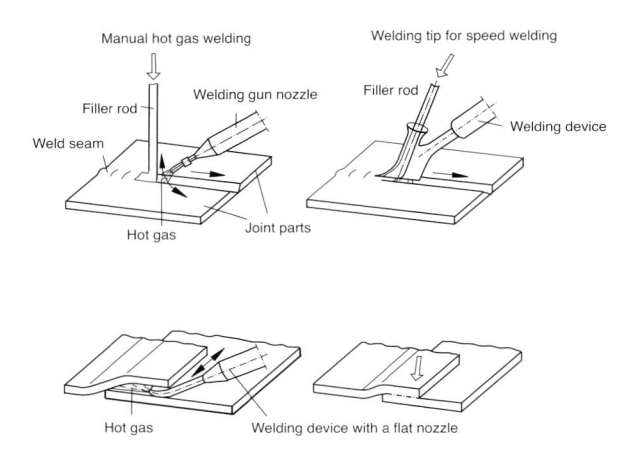

Figure 4.150: Various types of hot gas welding.

an electric fan, which blows the air over the heating element. These types of guns usually include temperature controllers to maintain the out-air at a constant temperature.
- Nozzle. The hot gas is directed to the filler rod and the weld cavity by a nozzle. For manual welding a round nozzle is used with a tack tip to allow tacking of the parts to each other prior to welding. Speed welding nozzles have a pressure shoe and a filler rod feed tube.

The steps during hot gas welding are:

- Surface Preparation. Bevels are machined into the parts followed by cleaning or degreasing of the weld surfaces and the filler rod.
- Fixturing. The parts are then fixed or held in place by clamps or other specialized fixtures. In some cases, the parts may be tack welded prior to clamping.
- Welding. Heated gas is directed to the end of the filler rod and to the cavity until melting of the filler rod begins. The filler rod is then pushed into the cavity as it melts. The nozzle is swept or fanned between the filler rod and the cavity to assure uniform melting of the polymer on the surface of the parts and the filler rod. For speed welding the nozzle includes holes to preheat the surface of the parts prior to deposition of the molten filler rod in the cavity.

Hot air or inert gas (usually nitrogen) that is filtered and clean of contaminants is used to heat and melt the weld rod and the base material. Inert gas is required

4.12 Welding

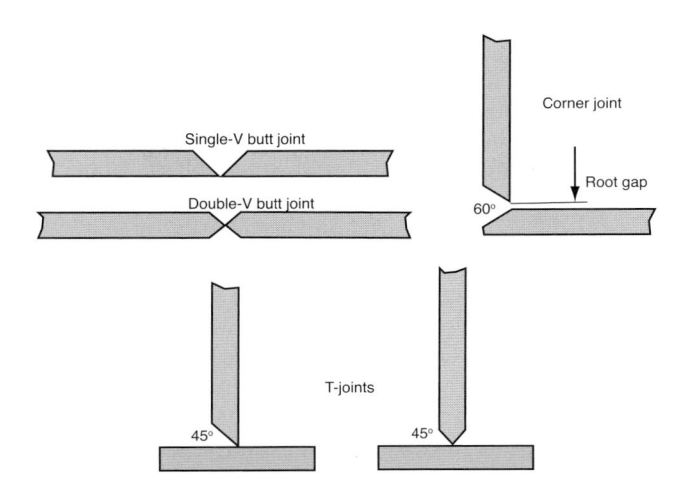

Figure 4.151: Various types of hot gas welding joints.

when the material can oxidize easily. Since heating is done by convection, the temperature of the gas and, to a lesser effect, the velocity of the gas will affect the rate of heating of the weld rod and base material surface. Depending on welder technique, a variety of gas temperatures and velocities can provide adequate heating conditions for welding. Optimal welding parameters are usually determined by trial and error and by operator experience. It should be noted that some variability in weld quality between welders is likely, and it should be considered in design of the parts and welds. In fact, even the same welder is likely to have variations in weld quality from one assembly to the next or from one day to the next. Figure 4.151 shows designs for hot gas joints joints. The single-V joints are normally used for thin plates. A root gap is used to insure complete penetration. For 6 mm or thicker plates a double-V butt weld, also shown in Fig. 4.151, is used. Passes should be alternated between sides to avoid excessive distortion.

T-joints, with single and double fillet welds, as well as corner welds are also presented in Fig. 4.151. Just like with butt joints, a root gap is recommended to assure complete penetration.

4.12.8 Extrusion Welding

Extrusion welding is very similar to hot gas welding except that the surface of the base material is heated by hot gas while the filler rod is extruded into the cavity. A welding shoe on the extruder is used to apply pressure. It is used in tank and pipe construction, welding of sheet liners, and sealing of geo-membranes. As shown in Fig. 4.152, an extrusion welder has three primary components: hot gas unit, extruder, and welding shoe. The hot gas unit is used to heat the surfaces of the cavity by convection just like in hot gas welding. The extruder operates just like a conventional extruding machine. Depending on the application, pellets or filler rod are fed into the extruder. The extrusion screw, which is turned by an electric motor, propels the pellets into the barrel, where they heat and melt. The melt is then extruded into the cavity. The welding shoe is attached to the nozzle of the extruder. It is used to apply pressure while keeping the extrudate from squeezing sideways out of the cavity. To minimize sticking and to reduce friction, the welding shoe is coated with PTFE. Two types of extruders are used. For larger extrusion rates and for welding of thicker sheets a stationary extruder on wheels with a movable welding shoe is generally used. For smaller extrudate volumes and thinner sheets, a hand held extrusion welder is used. Hand held extruders are generally fairly large and heavy and they can be cumbersome to move.

During extrusion welding the following steps are followed:

- Surface Preparation. Following machining of the bevels (if needed), the joint surfaces are cleaned to remove contaminants. The pellets or weld rod must be dry and clean.

- Fixturing. The parts are fixed or clamped prior to welding.

- Heating, joining and cooling. The extruder is placed over the joint interface. The weld surfaces of the parts are heated by the hotgas. The filler is melted and extruded by the extruder. The rate of extrusion determines the welding speed. The joining pressure is generated by the weight of the machine and the force applied by the operator. Cooling occurs under pressure as the weld shoe travels over the extrudate.

Except for very thick walls, extrusion welding is done in a single pass, where enough extrudate is used to completely fill the weld cavity. This results in a very uniform and aesthetically pleasing weld. During welding the operator must apply pressure and move the welder at a constant velocity. The operator must also make sure that the correct temperature settings for the hot gas unit and extruder are maintained.

A single-V butt weld, such as the one used in hot gas welding is typical of extrusion welding. Some excess extrudate material beyond the overlap can occur when it gets under the sealing faces of the weld shoe. Generally, little or no sideway excess occurs and no post welding operations are needed. In cases where significant excess of melt is observed it should be removed, which can easily be done by scraping.

4.12 Welding

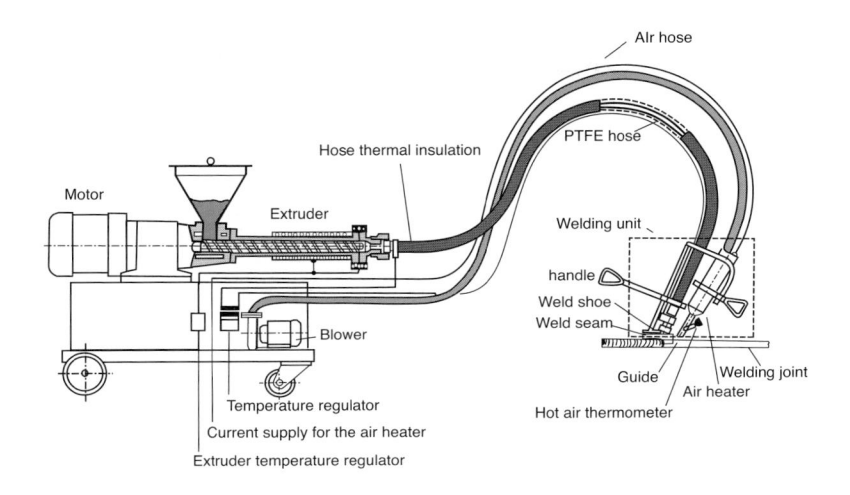

Figure 4.152: Schematic diagram of an extrusion welder.

Generally, extrusion welding produces one-layer seams. Sheets with difficult accessibility and thick walled sheets are welded with more layers. Joint design for a single-V butt joint should enable full penetration of the welds. This is accomplished by having a V groove of 90° for thin plates and 45° for thick plates. A root gap of 2 mm should also be used to insure complete heating and penetration all the way to the root of the weld. While single pass welding is preferred, for plates that are more than 30 mm thick multiple passes must be used because the required joining pressure cannot be applied by the operator.

Double-V butt welds such as used for hot gas welding are also used with extrusion welding. T-joints are also like the ones used for hot gas welding.

4.12.9 Implant Induction Welding

In implant induction welding, a magnetic field is applied to a gasket placed in the joint. This gasket is a composite of the polymer to be welded and conductive metal fibers or a ferromagnetic filler. In an alternating magnetic field, eddy currents are generated in conductive materials resulting in resistive heating. Ferromagnetic materials that are placed in an alternating magnetic field experience hysteresis heating. In either case, the gasket heats, resulting in melting of the polymer in the gasket and on the surface of the two parts. The polymer in the gasket must be the same or it must be compatible with the welded material to enable chain diffusion and entanglement. Then the electromagnetic field is turned off and the parts are allowed to cool under pressure. The gasket becomes a permanent part

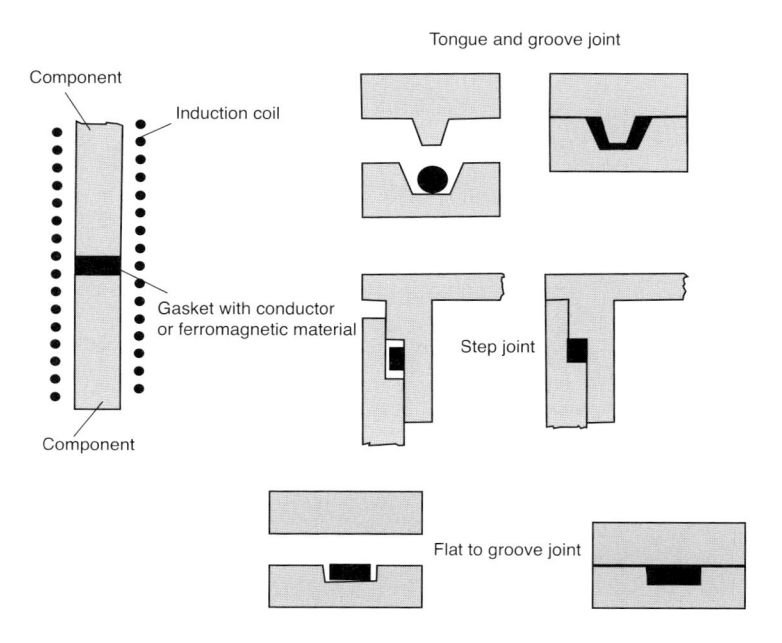

Figure 4.153: Setup for implant induction welding.

of the assembly as schematically depicted in Fig. 4.153. The components for an implant induction welding system include the generator, coil, fixtures and gasket. The induction generator converts line electrical power in the range of 20 kHz to 8 MHz high frequency power. The work coil, which is directly connected to the generator, produces an alternating magnetic field.

Since very high currents are generated in the coil it can get very hot. Therefore, hollow copper tubes are used to form the coil and cooling water is circulated through it. Coil design is very critical because the coil geometry and electrical impedance can greatly affect the efficiency and rate of heating. In addition, it is critical to maintain uniform proximity between the coil and the gasket or non-uniform heating would result. This is especially important when operating at high frequencies.

Fixtures provide the appropriate support for the parts to be welded, and when combined with a pneumatic system apply the desired pressure. Fixtures for implant induction welding have to be carefully designed to avoid being heated by the electromagnetic field while at the same time being able to fully support and apply pressure to the parts. The gasket is a composite of the polymer and either electrically conductive or ferromagnetic filler. The choice of filler material is very important for the operating frequency range and for the weld quality. For lower operating frequencies it is usually better to select conductive fillers, while at higher frequencies ferromagnetic fillers are more effective. In some cases both conductive and ferromagnetic fillers are used. As mentioned earlier, the polymer in the gasket must the same or at least be compatible with the polymer used in the parts. Finally, it is important to remember that the gasket will remain imbedded at the interface. Therefore, filler materials that may corrode over time or in some other way may affect the performance of the weld should be avoided. The following welding parameters are important to consider during implant induction welding:

- Power. Typically induction generators generate power in the range of 1 to 5 kW, although higher power generators are commercially available. Coil design and impedance matching are critical for efficient power transfer to the gasket.
- Welding Time. The power transmitted to the work coils determine the strength of the oscillating magnetic field produced. The more power generated the less time is needed to cause sufficient heating of the inductive polymer gasket. However, an appropriate balance between power and time must be achieved to allow sufficient heat flow to the parts without overheating the polymer in the gasket.
- Welding Pressure. Usually a pneumatic cylinder is used to apply a preset amount of pressure to an even distribution of the gasket inside the joint.
- Cooling Time and Pressure. Holding under pressure will allow the parts to cool and resolidify. Hold times vary according to each application, but are typically under 1s.

Implant induction welding is used in a wide range of applications including automotive, medical, food packaging, composite welding and much more. It can be used on small and large parts and on a variety of materials and even welding of some dissimilar materials is possible. Joint design is also important for achieving the desired short and long-term performance from the weld. Usually, the joint is designed to facilitate easy placement of a preformed gasket at the weld area, preventing it from falling out during the process. This is usually accomplished by having a groove on the bottom part to hold the gasket. During welding, the gasket flows so as to fill the available space in the groove. This assures complete contact between the gasket and the parts at the weld, thereby maximizing the weld strength. In most cases, the gasket and the weld perform better when loaded in shear. The flat to grove joint is quite effective when the plates are loaded in the plane or when the weld area is large. It does not perform well when loaded in tension. The tongue and grove joint is very strong because of the large contact area and loading in shear. It does require that the parts be thicker near the weld.

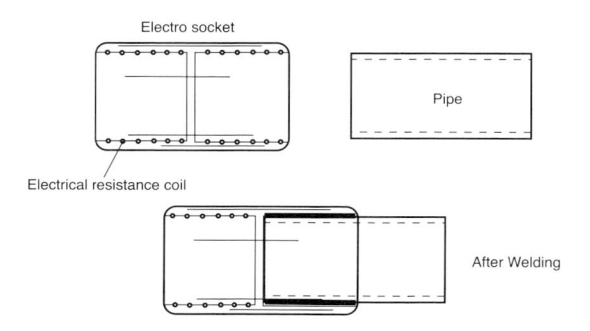

Figure 4.154: Electrofusion of a socket-pipe assembly.

While not as strong as the tongue and groove, the shear joint requires less space and it is sufficiently strong for many applications. Further sacrifice in strength for reduced space can be achieved by using the step joint. Both the shear and step joint are very effective for hermetic seals with good aesthetic finish.

4.12.10 Implant Resistance Welding

Implant resistance welding is a simple technique, which can be applied to any thermoplastic and almost any thermoplastic composite. The technique involves passing a direct or low frequency alternating current through an electrically conductive implant placed between the parts to be joined. Electrical resistance heating raises the temperature of the implant above the glass transition (T_g) or melting temperature (T_m) of the thermoplastic being joined, and fuses the parts. Depending on the joint and application configuration, pressure either is generated internally through thermal expansion or is applied by the welding system. In applications where the pressure is internally generated, part tolerance (fit-up) is usually critical to achieving a good weld. Implant resistance welding is often used when a high level of consistency is required and where applications can justify the cost of the implant (which in some cases can be significant). Examples of successful implant resistance welding include welding of polyethylene gas pipes (Fig. 4.154) and welding thermoplastic composites for aerospace applications.

After an implant resistance weld, the implant (which is usually metallic) remains in the welded component. This is an important issue if recycling has to be considered, because shredded components would contain a proportion of implant material. The implant can also act as a stress concentration point because it is a discontinuity in the base polymer. In addition, with metallic-type implants, corrosion can sometimes be an issue depending on the application and service environment.

There are two types of implant resistance welding. The most common one is *electrofusion* or pipe welding. This technique permits joining of pre-assembled pipes and fittings, to be carried out with minimum equipment and skill, thus reducing the necessity for extensive operator training and supervision. It also offers a number of practical advantages to the installer because it is easy to use for repairs where the available space and pipe movement is limited. The equipment is well suited for "in-the-field" use, and it allows fusion of a wide range of polyethylene resins. Thus, electrofusion is a popular joining technique for polyethylene pipes.

The second implabt resistance welding is *production implant resistance welding*. Although less popular than elctrofusion, implant resistance welding can be used to weld a wide range of components using a braided tape containing tinned copper wire strands interwoven with mono filaments of thermoplastic. This approach reduces the flow path required of the molten polymer during welding to make a fully consolidated joint. In other designs, a wire or wire mesh is simply placed in the bondline.

Electrofusion fitting must be properly designed for each application. For example, the clearance between the pipe's outside diameter (O.D.) and the fitting's inside diameter (I.D.) must provide clearance so the pipes and fitting can be assembled. However, the clearance must not be so great that the thermal expansion of the molten material does not produce sufficient bondline pressure to promote fusion. Out of roundness must not exceed 1.5% of the outer diameter of the pipe. The heating coils are generally monofilar in construction, with a continuous coil wound from one socket to another thus enabling fusion of both pipe ends in one operation. This arrangement minimizes the effect of any short-circuiting of wires, which may take place during joining. Coil terminations are electrically shrouded. The heating coil is typically embedded less than 0.25 mm from the fitting bore (I.D.). This provides protection from dislodgement during assembly, as well as placement of the heat source as close to the pipe surface as possible. A wide range of fittings, such as sockets, pipebranches, and tapping-tees, in numerous dimensions are commercially available. This is another reason why electrofusion is a popular welding technique for pipes. Another application utilizing implant resistance welding is the welding of advanced composites in the aerospace industry. In this application the implant is typically fabricated from graphite fiber that is insulated from the component with an additional layer of thermoplastic. The graphite, while only a semiconductor, acts as the conductive implant and also provides structural integrity in the final joint while maintaining a homogenous structure. Similar, two-part car bumpers have been welded using this technology, and in fact, a feature designed to accommodate the braided tape prior to welding was molded into one of the bumper components.

4.12.11 Microvawe Welding

The use of microwave radiation for welding thermoplastic materials is a relatively new technology. Therefore, it is not commonly used in industry. We only mention this technology and do not provide any details on the process. It is presented here to inform the reader of its existence. One of the unique benefits is that it is possible to weld internal walls that may not be accessable by other welding techniques, because most plastics are transparent to microwave radiation. However, microwave welding can produce non-uniform heating because of standing waves that can be generated within the welding cavity. It is possible to minimize the non-uniform heating by mixing the radiation with stirring devices or more commonly by rotating or translating the parts within the microwave oven. In nearly all microwave welding of thermoplastics, the process works by placing a foreign material into the bond line to enhance microwave absorption. This reduces the heating time, since most plastics are relatively transparent to the microwaves. During the process the gasket (insert) acts as a consumable and is squeezed out of the bond line. If the gasket/insert remains in the bond line at the end of the cycle, it can be reheated and disassemble the joint.

4.13 RAPID PROTOTYPING

The rapid prototyping revolution began in 1986 with the first stereo lithography patent. Today, there are many *rapid prototyping* technologies that are used to automatically manufacture near net-shape parts from existing CAD data. In rapid prototyping techniques, the three-dimensional parts are built layer by layer. For this reason rapid prototyping technology is often referred to as *layered manufacturing* or *freeform manufaturing*.

In addition to using rapid prototyping equipment to manufacture models or prototypes of a product, it is sometimes used for *rapid tooling* and *rapid manufacturing*. Rapid tooling (RT) is a way of using rapid prototyping equimpment to manufacture machine tools such as injection molds. Rapid tooling is divided into two categories: *soft tooling* and *hard tooling*.

In soft tooling, also referred to as *indirect tooling*, the rapid prototype is placed in a container filled with silicone rubber that is then allowed to vulcanize. After the rubber hardens the prototype is cut out of the silicone. The resulting mold is used to cast between 15 and 20 polyurethane replicas of the original rapid prototype.

Hard tooling, also referred to as *direct tooling* uses epoxy, aluminum reinforced resins and some metal alloys to make middle series production molds. In one example, polymer coated steel pellets are sintered using rapid prototyping equipment (described below) to produce the mold. The part is then place inside a furnace where the binder is burned off and replaced with a copper infiltration technique. This type of mold can produce tens of thousands of parts.

4.13.1 Stereo-Lithography (STL)

Stereo lithography, schematically depicted in Fig. 4.155, is the most widely used rapid prototyping thechnique in the United States. It is a technique where an ultra-violet laser beam is used to cure a liquid *photo-polymer* layer by layer. The method uses a platform that sits just below the surface of vat containing a liquid epoxy resin or an acrylate resin. An elevator is used to increm,entally lower the platform after each layer of the photo sensitive polymer has been exposed to the ultra violet light by a highly focused laser beam. After exposing every layer of the 3D object, the solid part is removed, rinsed of excess liquid resin and put in an ultra viuolet oven to complete the curing reaction. Stereo lithography's advantages are:

- Relatively accurate gemetries can be generated
- Applicable for complex geometries with fine features
- The process is continuously being improved

The main disadvantages of stereo lithography are:

- Difficult to predict shrinkage
- Occasional warpage
- Required support structures
- Material restrictions
- Involved postprocessing

4.13.2 Solid Ground Curing (SGC)

Solid ground curing, schematically depicted in Fig. 4.156, is a technique similar to stereo lithography, however, instead of using a laser to expose and harden the photo-polymer, it uses a mask to expose the whole layer at once with a single burst of ultra-violet light. The mask is generated using electrophotography or *xerography*. After the layer is cured, the excess polymer is vacuumed and the gaps are filled with wax. The surface is then prepared for the next step by milling it flat before spraying the next layer of photo sensitive polymer. Advantages of using SGC include:

- High productivity
- No support structure required

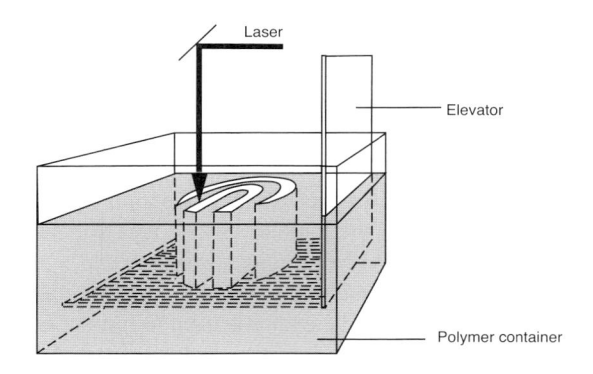

Figure 4.155: Schematic diagram of rapid prototyping by stereo lithography (STL).

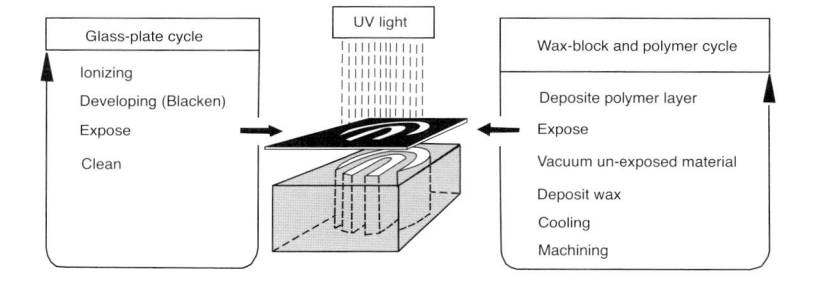

Figure 4.156: Schematic diagram of rapid prototyping by solid ground curing (SGC).

- No post curing cycle required
- No warpage

Disadvantages of using SGC include:

- Relatively complex equipment required
- Operator required
- Involved postprocessing
- Involved recycling operations

4.13.3 Selective Laser Sintering (SLS)

Selective laser sintering, schematically depicted in Fig. 4.157, uses a laser to locally heat-up and melt powder, which is then cooled and harden. Using a

4.13 Rapid Prototyping

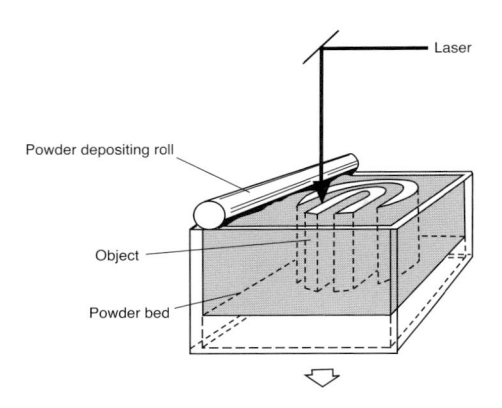

Figure 4.157: Schematic diagram of rapid prototyping by selective laser sintering (SLS).

powder feed roller a new layer of powder is deposited on over the sintered layer. This way the model is built-up layer by layer. The main advantages of SLS are:

- Various materials can be used
- No post-curing required
- No support structure needed

The main disadvantages of SLS are:

- Poor mechanical properties of the powders
- Limited surface quality
- Involved finishing operations

4.13.4 3D Printing or Selective Binding

3D printing or *selective binding*, schematically depicted in Fig. 4.158, is very similar to SLS, except that instead of using a laser, a liquid binder is applied to bond the particles. The printer first spreads a layer of powder from a feed box, followed by printing the binder using the ink-jet printing principle. The binder bonds the powder, leaving loose powder in the regions that are not printed on. The advantages of 3D printing are:

- Parts can be directly used for casting
- No support structure used
- Simple and reliable process

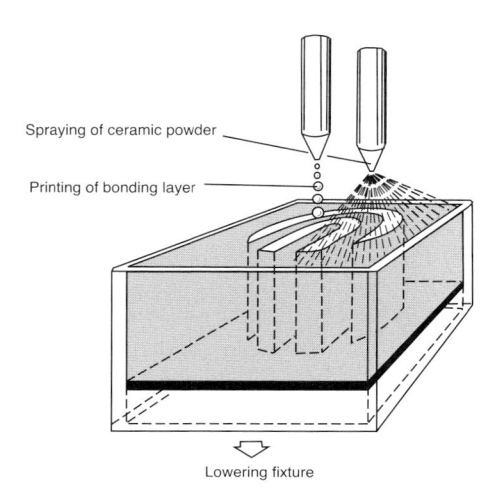

Figure 4.158: Schematic diagram of rapid prototyping by 3D-printing.

The main disadvantages of 3DS printing are:

- Two step process
- At the time only ceramic powders are used
- Surface and feature inaccuracies

4.13.5 Fused Deposition Modeling (FDM)

When manufacturing a model using the *fused deposition modeling*, schematically depicted in Fig. 4.159, a molten polymer filament is extruded through a heated nozzle and cooled on the object, layer by layer. The flow of the polymer through the nozzle can be turned on and off. The advantages of FDM are:

- Inexpensive process
- Can be implemented in an office environment
- No post-curing and post-processing required
- Can be used to verify CAD models

The main disadvantages of FDM are:

- Does not work well for thin parts and small features
- Not very accurate in the z direction
- Poor surface quality

4.13 Rapid Prototyping

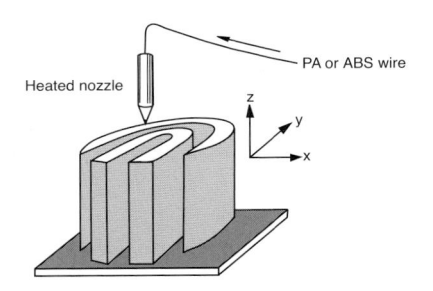

Figure 4.159: Schematic diagram of rapid prototyping by fused deposition modeling (FDM).

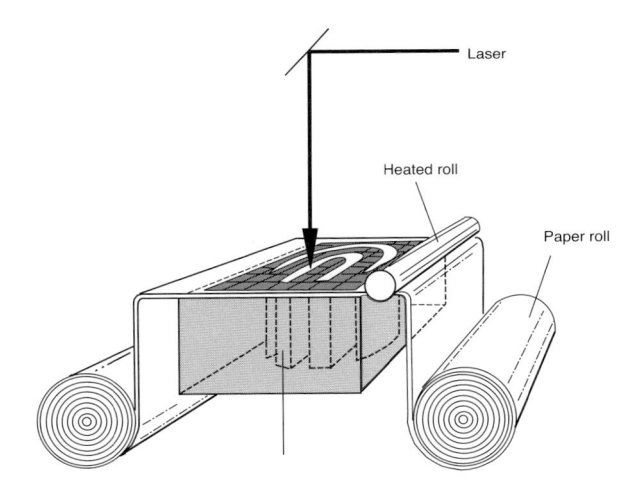

Figure 4.160: Schematic diagram of rapid prototyping by laminated object manufacturing (LOM).

4.13.6 Laminated Object Manufacturing (LOM)

In *laminated object manufacturing*, schematically depicted in Fig. 4.160, three dimensional objects are made by depositing layers of paper that are cut using a laser. A hot roll is used to melt the layer of polymer that bonds the new layer to the object. The residue paper is rolled-up and the part is finished. The advatges of using LOM are:

- No support structure required
- No post-curing required

- Low maintanance costs
- No residual stresses in the 3D model

The main disadvantages of LOM are

- Poor surface quality
- Warpage under humid conditions
- Part strength and stability depends on gemetry and size of features

CHAPTER 5

ENGINEERING DESIGN

When designing a product we must consider the interaction of three major components: material, design and manufacturing. Material and its selection is often the most involved, complex, and certainly most costly aspect of product design, which often requires development, testing and evaluation. Material - dependent design aspects must include non-linearities caused by temperature effects, time and aging. The design of the product requires drawings or CAD models, calculations, prototypes and experience. The manufacturing process, which is part - and material - dependent, requires process optimization and simulation to determine optimal conditions as well as process dependent properties such as fiber or molecular orientation, shrinkage and warpage. Furthermore, the functional and esthetic design requirements are often in conflict with manufacturing related issues. Hence, design for manufacturing becomes an important issue in plastics design.

5.1 DESIGN PHILOSOPHY

The goal of a systematic design process is to control the various stages that influence the finished product, and understand the interaction between those stages. Typically, a product design involves the following steps:

- Marketing
- Esthetic or industrial design
- Engineering design
- Tooling
- Manufacturing

Traditionally, these steps were performed sequentially. This often results in delays and additional costs if at one time of the design process it is decided that changes must be done in previous steps to allow completion. For example, a tooling problem may result in cosmetic changes which affect the industrial design

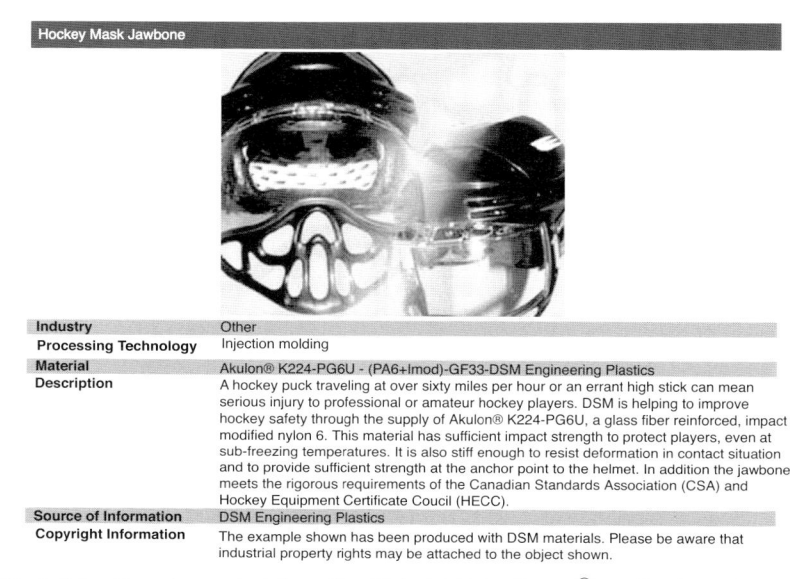

Figure 5.1: Sample product from the Material Data Center® applications data bank, (www.materialdatacenter.com), (Courtesy M-Base GmbH, Aachen, Germany).

aspects of the product, and consequently the engineering design performance. Therefore, today it is common to perform the above tasks concurrently, often referred to as *concurrent engineering*. One of the main advantages of concurrent engineering is that the whole project is done in a team effort with improved (and constant) communication between all players; from marketing to manufacturing. With concurrent engineering the various groups work in parallel. Although the functional and esthetic design of the part must be finalized before the engineering design calculations begin, other design aspects such as material choice and manufacturing process can be done concurrent with the industrial design aspects.

When developing a new product, as a starting point, an engineer can often draw from the experience of others. For example, today product design data banks are available which contain many applications, describing the design requirements, materials selected, and other design-dependent characteristics. Figure 5.1 presents an example from such a data bank.

The *engineering design* aspect of product design is broken down into many smaller stages. These are:

- Define product requirements
- Preliminary CAD model (concurrent with industrial design team)
- Material selection (concurrent with stress-strain calculations)

- Stress-strain calculations (concurrent with material selection)
- Design for manufacturing modifications (concurrent with industrial design team)
- Make a prototype (concurrent with tooling and manufacturing team)

These steps are discussed individually in the subsequent sections.

5.1.1 Defining Product Requirements

During design, the end use requirements of the product are decided in this stage of the development. At this point the preferred procedure is to specify quantitative requirements such as maximum loads and operating temperatures. In addition to the product dimensions, industrial design aspects, and economic issues, a designer must specify the loading and environmental conditions, as well as regulatory and standards restrictions.

Loading Conditions At this stage the type, magnitude, and duration of loading must be determined and foreseen. These loadings can occur during processing, such as stresses and deformations that may occur during demolding, assembly, packing, shipping and life of the product. If a product is subjected to loads for extended periods of time while being exposed to high temperatures, the designer should only consider materials for which appropriate data is available for the subsequent stress-strain calculations.

Environmental Conditions Here, all possible environmental scenarios must be specified. These can include exposure to sunlight, and consequently UV rays, humidity, as well as chemical environments. Chemical environments may include lubricants, cleaners, and other materials that can lead to chemical degradation or *environmental stress cracking* of the product. Environmental stress cracking is a result of loading and environment, where a product has the potential of failing at relatively low stresses. Here, special attention must be paid when the product comes in contact with dangerous materials such as concentrated acids. Although, the chemical resistance of many plastics has been tested and the respective values are available in the literature, for many applications the data is not available. At this stage of the design process, the chemical resistance of the material choices must be determined, often requiring actual testing.

Dimensional Requirements The load, functionality, tolerances, as well as industrial design aspects determine the size, thicknesses, and surface finish of a part. However, cost, manufacturability, material choice and polymer physics will bring several constraints at this stage of design. Manufacturability is related to design for manufacturing, such as undercuts, assembly, etc., discussed later in this

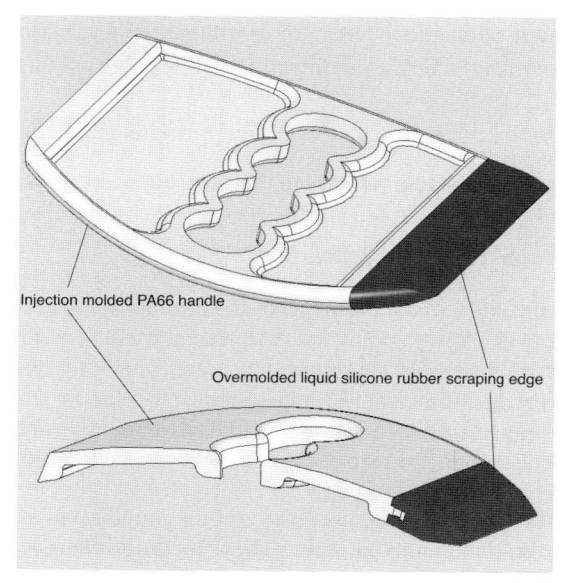

Figure 5.2: CAD model of a PA66-liquid silicone rubber spatula (Courtesy Simtec Silicone Parts, LLC, Madison, WI, USA).

chapter. Material choice and polymer physics will always have a profound effect on part design, manufacturing, and performance. For example, variability on thickness with semi-crystalline polymers will result in variabilities of the degree of crystallinity, and consequently in warpage.

Regulatory and Standards Requirements A certain application may be implemented in an area that is regulated by an agency. For example, construction, electronic and household applications must fulfill certain flammability requirements. Food packaging, cooking utensils and containers must fulfill Food and Drug Administration (FDA) standards. The fulfillment of certain standards or regulations often requires a prototype that can be used for testing.

5.1.2 Preliminary CAD Model

Once the load, strength and functional requirements are agreed upon, and preliminary decisions have been made, together with the industrial design team, the designer will generate a 3D CAD models that will help visualize the design and its functionality, and point to possible problems. During the design process this model will be modified and subsequently used to generate finite element models for stress-strain calculations, or to import into a rapid prototyping ma-

5.1 Design Philosophy

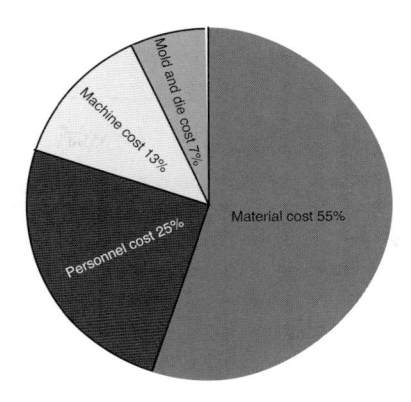

Figure 5.3: Cost break-down for a typical injection molded part. Part made of polyamide at weight of 190 g and a yearly production rate of 180,000 parts.

chine to make a prototype of the product. As an example of a preliminary CAD model of a part, Fig. 5.2 presents the kitchen spatula with tentative material and manufacturing choices.

5.1.3 Material Selection

For most plastic parts, material cost accounts for over half of the manufacturing expense. To illustrate the important of material selection, Fig. 5.3 presents a typical cost break-down for a 190 g part injection molded out of polyamide at a rate of 180,000 parts per year. Once all end functions and requirements for the part have been specified, the designer must begin searching for the best available material. This can be an overwhelming task since there are tens of thousands different grades of plastics available worldwide. Furthermore, applicable resins may exist in the thermoplastic, thermoset or elastomer categories. Today, such selection process is made easier with the use of material data banks such as CAMPUS, which will not only eliminate tedious searches through catalogs and material data sheets printed by the resin suppliers, but will also facilitate a fair comparison between the materials since all their properties are measured using the same standardized testing techniques. CAMPUS is available in six languages; Chinese, English, French, German, Japanese and Spanish. Typical screen captures of CAMPUS are presented in Figs. 5.4 and 5.5. Figure 5.4 shows the mechanical properties data for a POM material, and Fig. 5.5 presents the secant modulus for the same POM as a function of strain for several test temperatures as an illustration of a properties diagram.

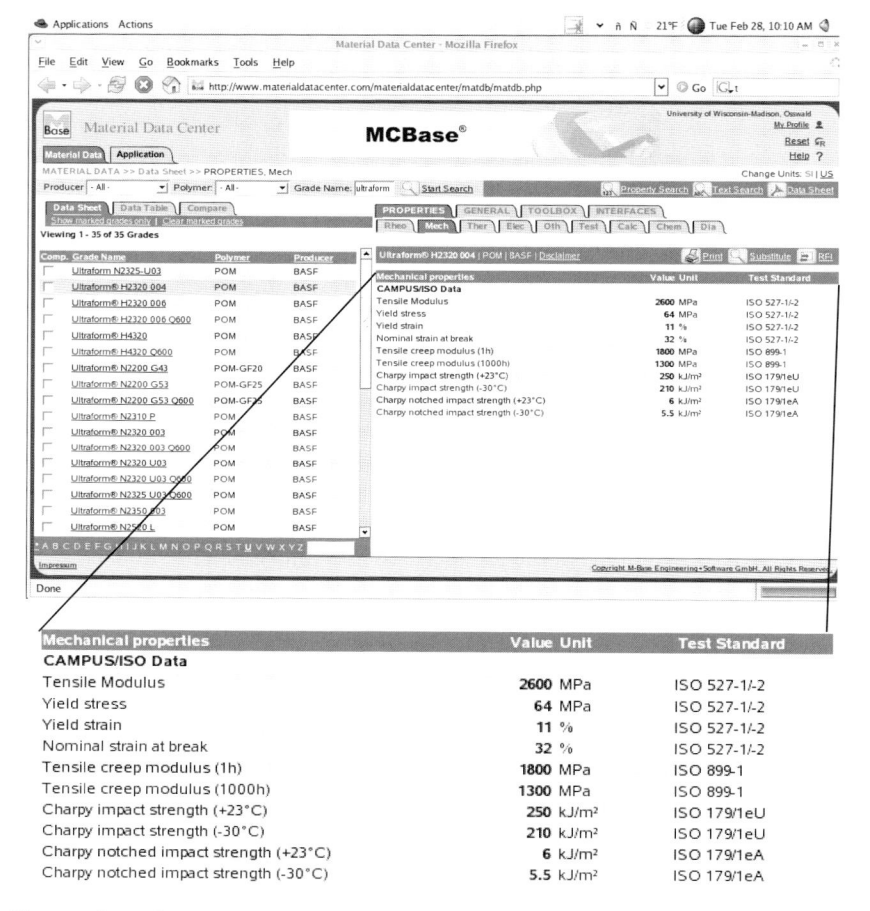

Mechanical properties	Value	Unit	Test Standard
CAMPUS/ISO Data			
Tensile Modulus	2600	MPa	ISO 527-1/-2
Yield stress	64	MPa	ISO 527-1/-2
Yield strain	11	%	ISO 527-1/-2
Nominal strain at break	32	%	ISO 527-1/-2
Tensile creep modulus (1h)	1800	MPa	ISO 899-1
Tensile creep modulus (1000h)	1300	MPa	ISO 899-1
Charpy impact strength (+23°C)	250	kJ/m²	ISO 179/1eU
Charpy impact strength (-30°C)	210	kJ/m²	ISO 179/1eU
Charpy notched impact strength (+23°C)	6	kJ/m²	ISO 179/1eA
Charpy notched impact strength (-30°C)	5.5	kJ/m²	ISO 179/1eA

Figure 5.4: Screen capture of mechanical properties from the data bank CAMPUS. An Ultraform® POM material from BASF is presented as an example (www.materialdatacenter.com). (Courtesy M-Base GmbH, Aachen, Germany)

During the material selection process, the designer will not only search for the suitable candidate, but also for a material whose properties, such as creep, heat deflection temperature or impact strength, to name a few, have been measured using the standard tests.

During a preliminary selection process, the designer will be able to choose several material grades and formulations suitable for the specific application. However, since materials contribute to over half the cost of the final product,

5.1 Design Philosophy   457

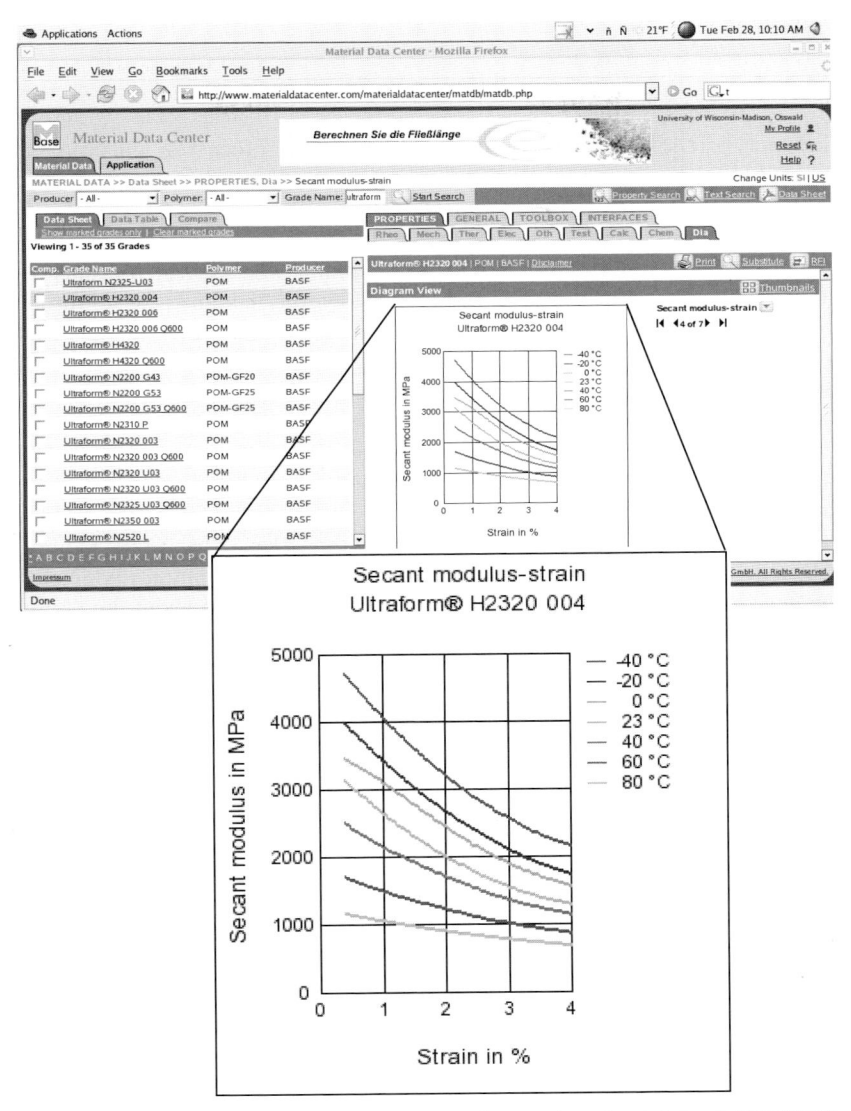

Figure 5.5: Screen capture of a secant modulus diagram from the data bank CAMPUS for an Ultraform® POM material from BASF (www.materialdatacenter.com). (Courtesy M-Base GmbH, Aachen, Germany)

many of the appropriate materials materials may be too expensive for a given application.

5.1.4 Process Selection

Ultimately part quantity and geometry determine which process should be used to manufacture a specific product. When designing a product and deciding which process should be selected to manufacture that product, the geometry and requirements for that part are often modified. For example, a given container can easily be manufactured using the injection molding process, which results in a product with an even thickness distribution. However, the thickness uniformity can be compromised by using the thermoforming process. Although this leads to non-uniform product, the process cost is significantly reduced. On the other hand, since the part thickness will vary, the part may be over-designed in some areas to achieve the minimum thickness requirements in the thinner regions.

The material type will also control the process choice. For example, long fiber reinforced composite parts require processes such as compression molding, resin transfer molding, pultrusion, etc. Thermosets require special injection molding machines or compression and transfer molding processes.

Details for existing and emerging plastic processing techniques is given in Chapter 4 of this handbook, and will aid the reader to learn about the various processing techniques available to the industry.

5.2 PROCESS INFLUENCES ON PRODUCT PERFORMANCE

The mechanical properties and dimensional stability of a molded polymer part are strongly dependent upon the anisotropy of the finished part. The structure of the final part, in turn, is influenced by the design of the mold cavity, e.g., the type and position of the gate, and by the various processing conditions, such as injection speed, melt or compound temperatures, mold cooling or heating rates, and others. The amount and type of filler or reinforcing material also has a great influence on the quality of the final part. After molding or shaping of the product, the part must set, either by cooling or cross-linking. As discussed in previous chapters, a thermoplastic polymer hardens as the temperature of the material is lowered below either the melting temperature for a semi-crystalline polymer or the glass transition temperature for an amorphous thermoplastic. A thermoplastic has the ability to soften again as the temperature of the material is raised above the solidification temperature. On the other hand, the solidification of a leads to cross-linking of molecules. The effects of cross-linkage are irreversible and lead to a network that hinders the free movement of the polymer chains independent of the material temperature. The solidification process will lead to residual stresses

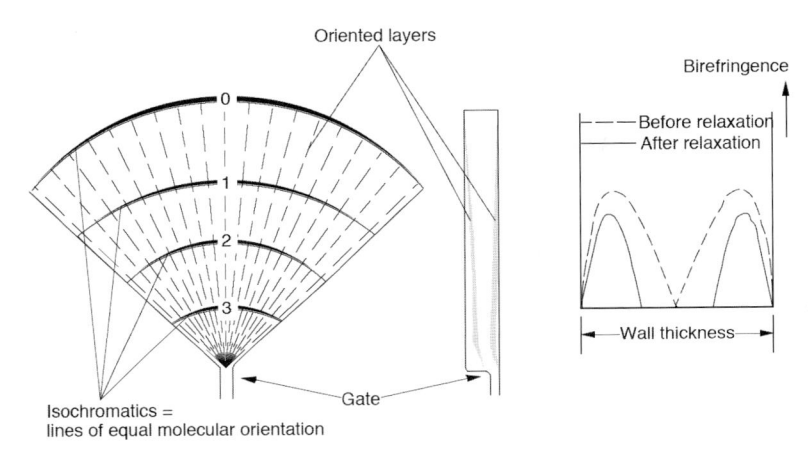

Figure 5.6: Orientation birefringence in a quarter disc (After Woebken).

and consequently to warpage, perhaps one of the biggest headaches of the design engineer

5.2.1 Orientation in the Final Part

During processing, the molecules, fillers, and fibers are oriented in the flow and greatly affect the properties of the final part.

When thermoplastic components are manufactured, the polymer molecules become oriented. The molecular orientation is induced by the deformation of the polymer melt during processing. The flexible molecular chains get stretched, and because of their entanglement they cannot relax fast enough before the part cools and solidifies. At lower processing temperatures this phenomenon is multiplied, leading to even higher degrees of molecular orientation. This orientation is felt in the stiffness and strength properties of the polymer component. Orientation also gives rise to *birefringence*, or *double refraction*, a phenomenon already discussed in Chapter 3. As polarized light travels through a part, a series of colored lines called isochromatics become visible, as shown in Fig. 5.6. The isochromatics are lines of equal molecular orientation and numbered from zero, at the region of no orientation, up with increasing degrees of orientation. A zero degree of orientation is usually the place in the mold that fills last and the degree of orientation increases towards the gate. Figure 5.6 shows schematically how molecular orientation is related to birefringence. The layers of highest orientation are near the outer surfaces of the part with orientation increasing towards the gate.

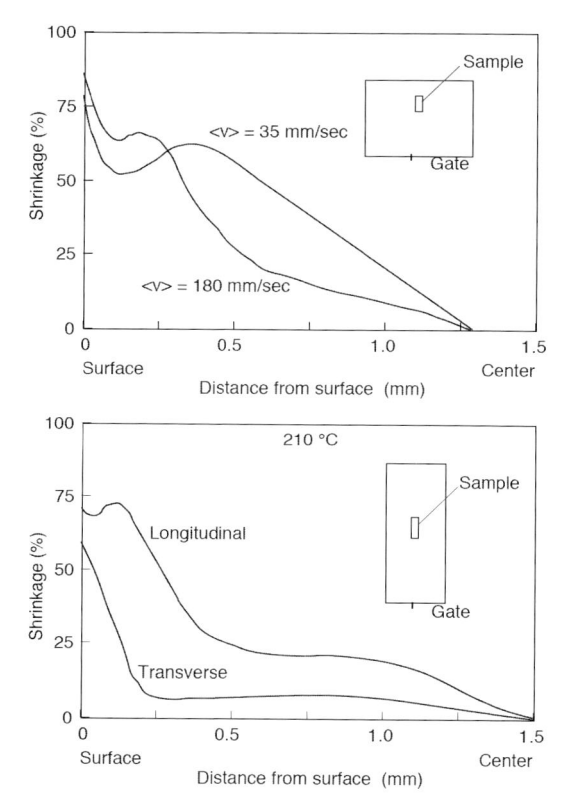

Figure 5.7: Shrinkage distribution of injection molded polystyrene plates (After Menges and Wübken).

Early studies have shown that a molecular orientation distribution exists across the thickness of thin injection molded parts. Figure 5.7 shows the shrinkage distribution in longitudinal and transverse flow directions of two different plates. The curves demonstrate the degree of anisotropy that develops during injection molding, and the influence of the geometry of the part on this anisotropy.

The degree of orientation increases and decreases depending on the various processing conditions and materials. For most materials the degree of orientation increases with decreasing wall thickness. An explanation for this is that the velocity gradients increase when wall thickness decreases, and the cooling rate increases, leaving less time for molecular relaxation. Orientation is also related to the process used to manufacture the part. For example, an injection molded disc will have a lower degree of orientation as one that is injection-compression

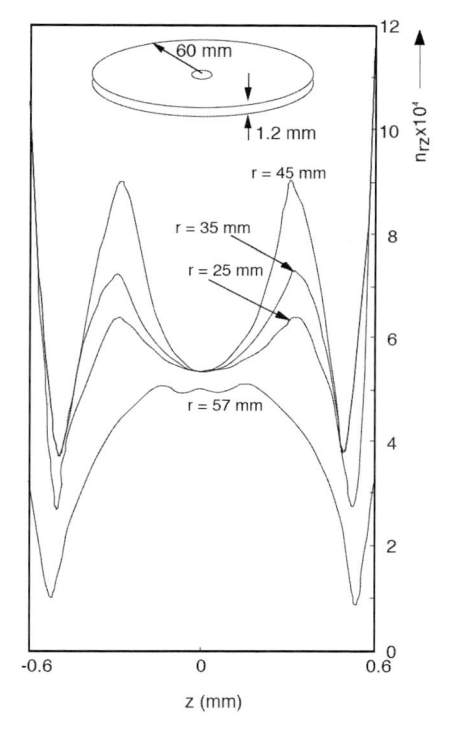

5.2 Process Influences on Product Performance   461

Figure 5.8: Birefringence distribution in the rz-plane at various radius positions. Numbers indicate radial position (After Wimberger-Friedl).

molded. An example of how to use the birefringence pattern of polymer parts to detect severe problems is in the manufacture of polycarbonate compact discs. Figure 5.8 shows the birefringence distribution in the rz-plane of a 1.2 mm thick disk molded with polycarbonate. The figure shows how the birefringence is highest at the surface of the disk and lowest just below the surface. Towards the inside of the disk the birefringence rises again and drops somewhat toward the central core of the disk. A similar phenomenon is observed in glass fiber reinforced and liquid crystalline polymer injection molded parts which show large variations in fiber and molecular orientation through the thickness.

It can be said that molecular or filler orientation in injection molded parts can be divided into seven layers schematically represented in Fig. 5.10. The seven layers may be described as follows:

- Two thin outer layers with a biaxial orientation, random in the plane of the disk

- Two thick layers next to the outer layers with a main orientation in the flow direction
- Two thin randomly oriented transition layers next to the center core
- One thick center layer with a main orientation in the circumferential direction.

There are three mechanisms that lead to high degrees of orientation in injection molded parts: *fountain flow effect*, radial flow, and holding pressure induced flow. The fountain flow effect is caused by the no-slip condition on the mold walls, which forces material from the center of the part to flow outward to the mold surfaces as shown in Fig. 5.9. As the figure schematically represents, the melt that flows inside the cavity freezes upon contact with the cooler mold walls. The melt that subsequently enters the cavity flows between the frozen layers, forcing the melt skin at the front to stretch and unroll onto the cool wall where it freezes instantly. The molecules which move past the free flow front are oriented in the flow direction and laid on the cooled mold surface which freezes them into place, though allowing some relaxation of the molecules after solidification.

Radial flow is the second mechanism that often leads to orientation perpendicular to the flow direction in the central layer of an injection molded part. This mechanism is schematically represented in Fig. 5.11. As the figure suggests, the material that enters through the gate is transversely stretched while it radially expands as it flows away from the gate. This flow is well represented in today's commonly used commercial injection mold filling software. Finally, the flow induced by the holding pressure as the part cools leads to additional orientation in the final part. This flow is responsible for the spikes in the curves shown in Figs. 5.7 and 5.8.

During the manufacture of thermoset parts there is no molecular orientation because of the cross-linking that occurs during the solidification or curing reaction. A thermoset polymer solidifies as it undergoes an exothermic reaction and forms a tight network of inter-connected molecules. However, many thermoset polymers are reinforced with filler materials such as glass fiber, wood flour, etc. These composites are molded via transfer molding, compression molding, or injection-compression molding. The properties of the final part are dependent on the filler orientation. In addition, the thermal expansion coefficients and the shrinkage of these polymers are highly dependent on the type and volume fraction of filler being used. Different forms of orientation may lead to varying strain fields, which may cause warpage in the final part. This topic will be discussed in the next chapter. Since the cool thermoset material flows in a heated mold, a lubricating layer of resin forms between the bulk of the resin and mold surfaces. Hence, the material deforms uniformly through the thickness with slip occurring at the mold surface as shown schematically in Fig. 5.12. For example, during

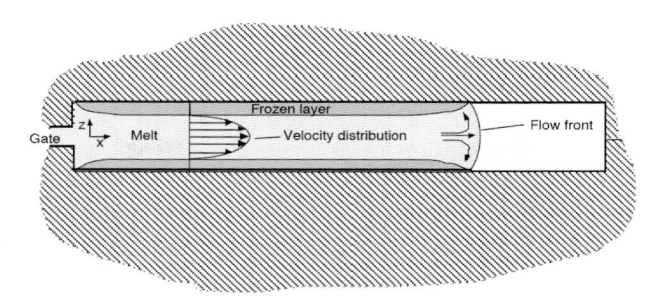

Figure 5.9: Flow and solidification mechanisms through the thickness during injection molding.

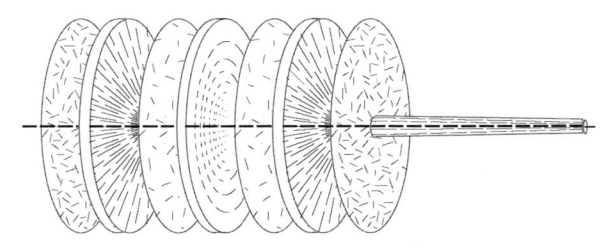

Figure 5.10: Filler orientation in seven layers of a centrally injected disc.

compression molding, an SMC charge is placed in a heated mold cavity and squeezed until the charge covers the entire mold surface.

Significant degrees of fiber orientation are reached during molding of fiber reinforced thermoset parts. Figure 5.13 shows a histogram of fiber orientation measured from a plate where the initial charge coverage was 33%. Such distribution functions are very common in fiber reinforced compression or transfer molded parts and lead to high degrees of anisotropy throughout a part.

To illustrate the effect that orientation or anisotropy has on mechanical properties of fiber reinforced parts, Fig. 5.14 presents the stress strain behavior of plates compression molded with different initial mold coverage, consequently with different degrees of deformation. For example, in the 33% mold coverage example, the initial charge thickness is 3 times the thickness of the final part.

During injection or compression mold filling, weldlines (known as knitlines with fiber reinforced parts) form. Weldlines form when fronts of material meet inside the mold cavity during mold filling. They will either form because a mold has multiple gates, because multiple charges are placed in the cavity, or because of geometric constraints in the mold cavities. Weldlines may also form when there

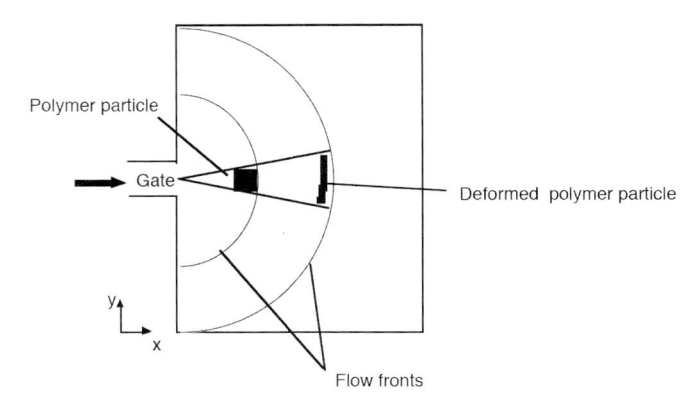

Figure 5.11: Deformation of the polymer melt during injection molding.

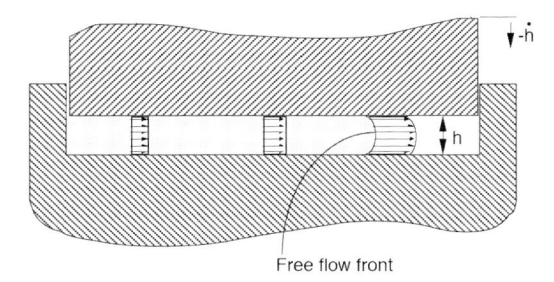

Figure 5.12: Velocity distribution during compression molding with slip between material and mold surface.

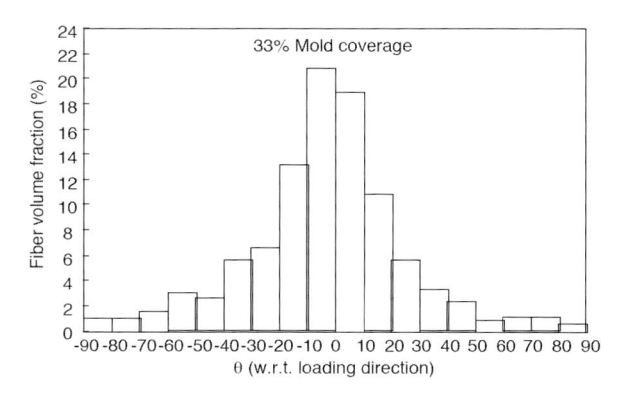

Figure 5.13: Measured fiber orientation distribution histogram in a plate with 33% initial mold coverage and extensional flow during mold filling (After Lee, Folgar and Tucker).

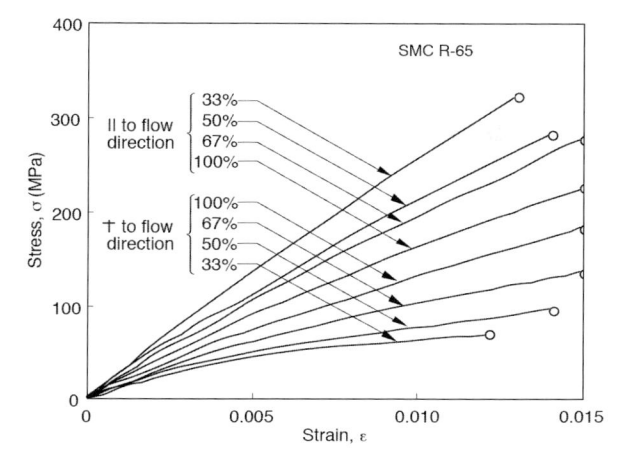

Figure 5.14: Stress-strain curves of 65% glass by volume SMC for various degrees of charge deformation and consequently fiber orientation (after Chen and Tucker).

are large differences in part thickness, sometimes referred to as race tracking, and when the material flows around thin regions.

5.2.2 Fiber Damage

One important aspect when processing fiber reinforced polymers is fiber damage or *fiber attrition*. This is especially true during injection molding where high shear stresses are present. As the polymer is melted and pumped inside the screw section of the injection molding machine and as it is forced through the narrow gate, most fibers shorten in length, reducing the properties of the final part (e.g., stiffness and strength). Figure 5.15 helps explain the mechanism responsible for fiber breakage. The figure shows two fibers rotating in a simple shear flow. Fiber **a**, which is moving out of its 0° position, is under compressive loading while fiber **b**, which is moving into its 0° position, is under tensile loading. It is clear that the tensile loading is not large enough to cause any fiber damage, but the compressive loading is potentially large enough to buckle and break the fiber. A common equation exists that relates a critical shear stress, τ_{crit}, to elastic modulus, E_f, and to the L/D ratio of the fibers

$$\tau_{crit} = \frac{ln(2L/D) - 1.75}{2(L/D)^4} E_f \tag{5.1}$$

where τ_{crit} is the stress required to buckle the fiber. When the stresses are above τ_{crit}, the fiber L/D ratio is reduced. Figure 5.16 shows a dimensionless plot of

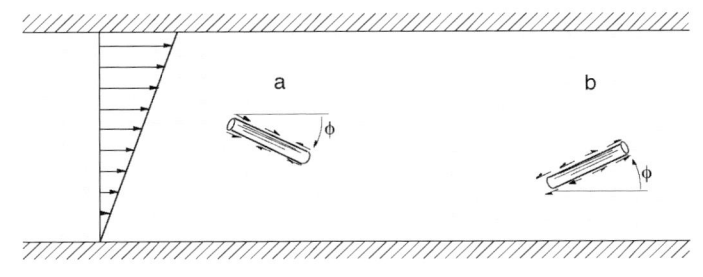

Figure 5.15: Fiber in compression (a) and tension (b) as it rotates during simple shear flow.

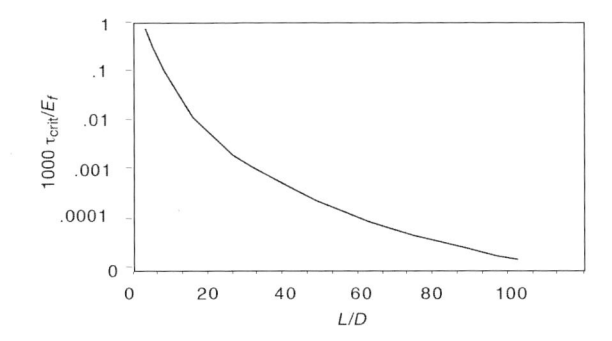

Figure 5.16: Critical stress, τ_{crit}, versus fiber L/D ratio.

critical stress versus L/D ratio of a fiber as computed using the above equation. It is worthwhile to point out that although the equation predicts L/D ratios for certain stress levels, it does not include the uncertainty which leads to fiber L/D ratio distributions - very common in fiber filled systems.

Figure 5.17 demonstrates that during injection molding most of the fiber damage occurs in the transition section of the plasticating screw. Lesser effects of fiber damage were measured in the metering section of the screw and in the throttle valve of the plasticating machine. The damage observed inside the mold cavity was marginal. However, the small damage observed inside the mold cavity is of great importance since the fibers flowing inside the cavity underwent the highest stresses, further reducing their L/D ratios. Another mechanism responsible for fiber damage is when the fibers that stick out of partially molten pellets are bent, buckled, and sheared-off during plastication.

5.2 Process Influences on Product Performance

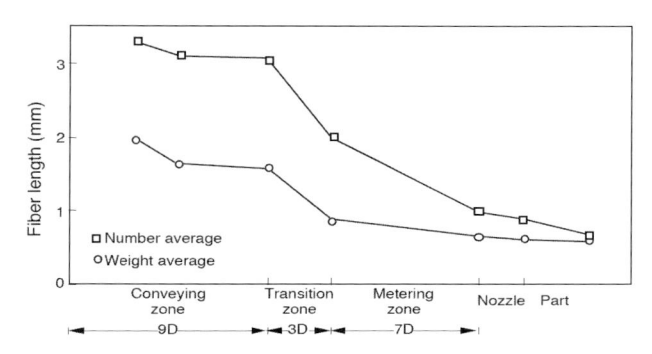

Figure 5.17: Fiber damage measured in the plasticating screw, throttle valve and mold during injection molding of a polypropylene plate with 40% fiber content by weight (after Thieltges).

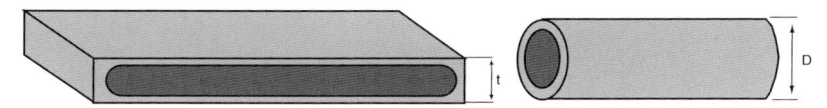

Figure 5.18: Notation used to predict cooling times in polymer processing.

5.2.3 Cooling and Solidification

Since polymer parts are generally thin, the energy equation can be simplified to a one-dimensional problem. Thus, using the notation shown in Fig. 5.18 the cooling time for a plate-like geometry of thickness h can be estimated using

$$t_{cooling} = \frac{t^2}{\pi \alpha} ln \left(\frac{8}{\pi^2} \frac{T_M - T_W}{T_D - T_W} \right) \qquad (5.2)$$

and for a cylindrical geometry of diameter D using

$$t_{cooling} = \frac{D^2}{23.14\alpha} Ln \left(0.692 \frac{T_M - T_W}{T_D - T_W} \right) \qquad (5.3)$$

where α represents the thermal diffusivity, T_M the injection melt temperature, T_W the mold wall temperature and T_D the average part temperature at ejection.

The solidification process of thermosets, such as phenolics, unsaturated polyesters, epoxy resins, and polyurethanes is dominated by an exothermic chemical reaction called curing reaction. A curing reaction is an irreversible process that results in a structure of molecules that are more or less cross-linked. Some thermosets cure under heat and others cure at room temperature.

In a cured thermoset, the molecules are rigid, formed by short groups that are connected by randomly distributed links. The fully reacted or solidified

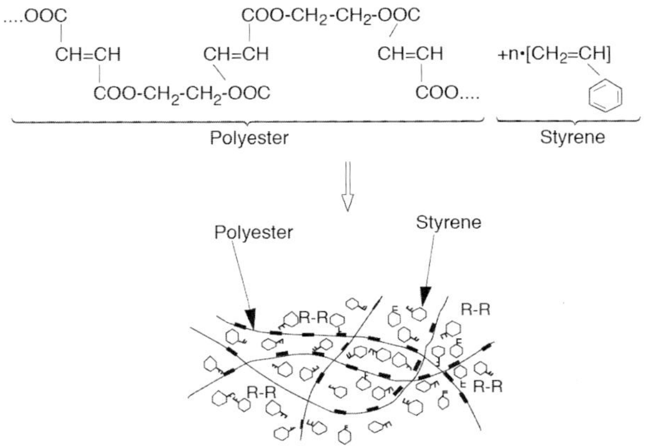

Figure 5.19: Symbolic and schematic representations of uncured unsaturated polyester.

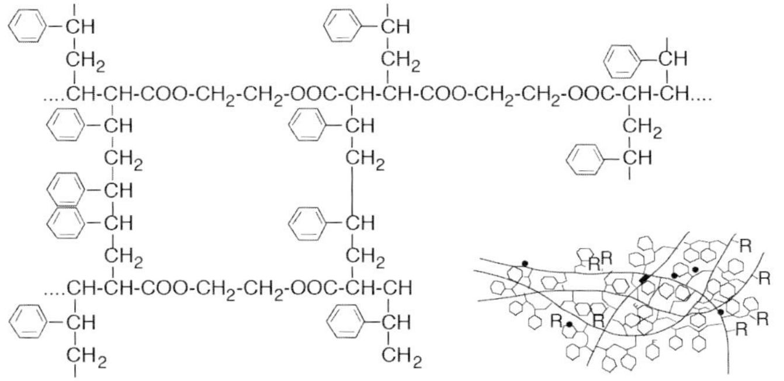

Figure 5.20: Symbolic and schematic representations of cured unsaturated polyester.

thermosetting polymer does not react to heat as observed with thermoplastic polymers. A thermoset may soften somewhat upon heating and but then it degrades at high temperatures. Due to the high cross-link density, a thermoset component behaves as an elastic material over a large range of temperatures. However, it is brittle with breaking strains of usually 1 to 3%. An example of a cross-linking reaction of a thermoset by *free radical reaction polymerization* is the co-polymerization of unsaturated polyester with styrene molecules, shown in Fig. 5.19. The molecules contain several carbon-carbon double bonds which act as cross-linking sites during curing. An example of the resulting network after the chemical reaction is shown in Fig. 5.20. Curing reactions are discussed in more detail in Chapter 3 of this handbook.

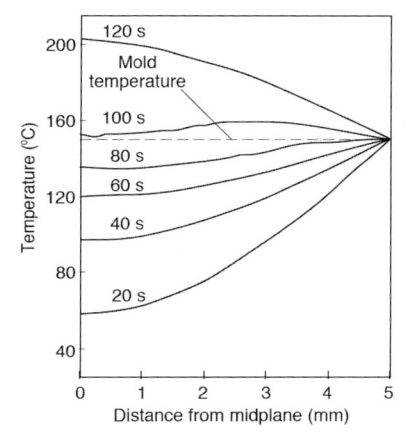

5.2 Process Influences on Product Performance 469

Figure 5.21: Temperature profile history of a 10 mm thick SMC plate (after Barone and Caulk).

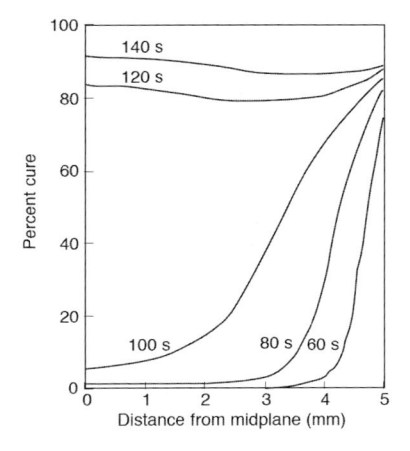

Figure 5.22: Curing profile history of a 10 mm thick SMC plate (after Barone and Caulk).

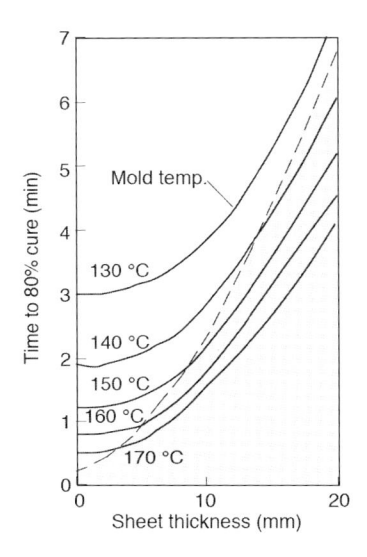

Figure 5.23: Cure times versus plate thickness for various mold temperatures. Shaded region represents the conditions at which thermal degradation may occur (after Barone and Caulk).

Figures 5.21 and 5.22 show typical temperature and degree of cure distributions, respectively, during the solidification of a 10 mm thick part. In Fig. 5.21, the temperature rise resulting from exothermic reaction is obvious. This temperature rise increases in thicker parts and with increasing mold temperatures. Figure 5.23 is a plot of the time to reach 80% cure versus thickness of the part for various mold temperatures. The shaded area represents the conditions at which the internal temperature within the part exceeds 200°C because of the exothermic reaction. Temperatures above 200°C can lead to material degradation and high residual stresses in the final part. Improper processing conditions can result in a non-uniform curing distribution which may lead to voids, cracks, or imperfections inside the part. It is of great importance to know the appropriate processing conditions which will avoid both over-heating problems and speed up the manufacturing process.

5.2.4 Shrinkage, Residual Stresses and Warpage

Some major problems encountered when molding polymeric parts are the control and prediction of the component's shape at room temperature. For example, the resulting sink marks in the final product are caused by the shrinkage of the material during cooling or curing. Warpage in the final product is often caused by processing conditions that cause unsymmetric residual stress distributions through the thickness of the part. Thermoplastic parts most affected by residual stresses are those that are manufactured by the injection molding process.

Shrinkage Prediction Using pvT Behavior The pvT diagram of a polymer can be used to trace the process and predict melt shrinkage. Since the pvT behavior of the plastic relates the thickness of the final part to the mold cavity thickness, the diagram can be used to estimate the effect of process condition changes during part and process design or optimization. For example, the ABS pvT diagram presented in Fig. 5.24 presents a trace of the injection molding process for a given application. The material is injected at 220°C at the point marked **1** in the diagram to a pressure of 350 bar, marked **2**. Initially, since cooling data is not available, the injection and mold filling is represented as an isothermal process. However, the diagram presents an average material temperature of 210°C after filling. At this point, the pressure is maintained constant until the gate freezes shut at location **3**. The packing process, which is isobaric (constant pressure) compensates for most of the 9% shrinkage this ABS material undergoes from injection to finished product. Once atmospheric pressure (1 bar) is reached at point **4**, the part starts to shrink away from the mold cavity until room temperature is reached at point **5**. The amount of shrinkage between points **4** and **5** is primarily compensated by through-the-thickness shrinkage. In the case illustrated in Fig. 5.24 this amounts to a 3% shrinkage. Hence, a mold must be designed according to this shrinkage. This shrinkage can be reduced by increasing the packing pressure.

Sink Marks A common geometry that usually leads to a sink mark is a ribbed structure. The size of the sink mark, which is often only a cosmetic problem, is not only related to the material and processing conditions but also to the geometry of the part. A rib that is thick in relation to the flange thickness will result in significant sinking on the flat side of the part. The last place to solidify in a ribbed structure is the center of the juncture between the rib an the flange. As the melt pool shrinks, it generates negative hydrostatic pressures, causing the flange surface to collapse inward. Although a thick rib will result in high reinforcement, it can render a cosmetically unacceptable part. Hence, it is better to have several thin ribs that will add to the same reinforcement as a thick one, without compromising the part's esthetics. As rule of thumb, a sink mark that measures less than $2\mu m$ is not discernible by the naked eye. Therefore it is advisable to use rib thicknesses that are the same thickness or less than the flange. However, with semi-crystalline materials a rib size that is smaller than the flange will lead to lower degrees of crystallinity in the rib due to its faster cooling. This will cause the part to shrink more in the flange region, causing the part to bow upward on the side of the flange. Figure 5.25 presents measured sink

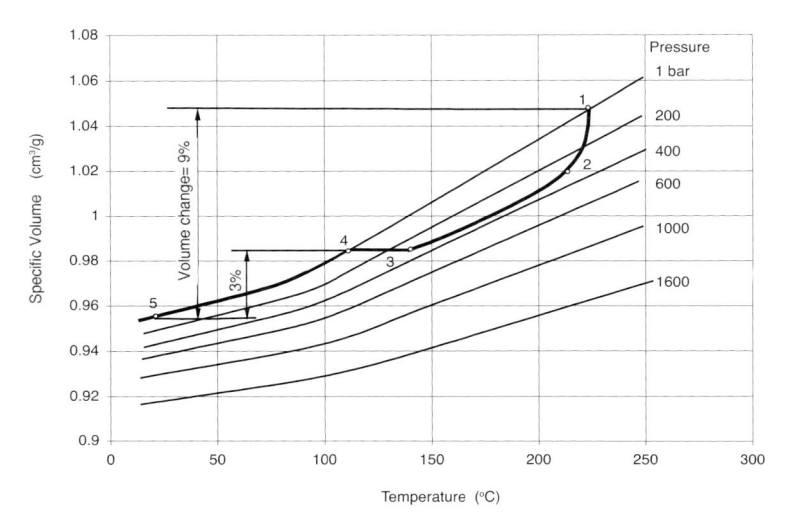

Figure 5.24: Injection molding process traced on a pvT diagram for ABS.

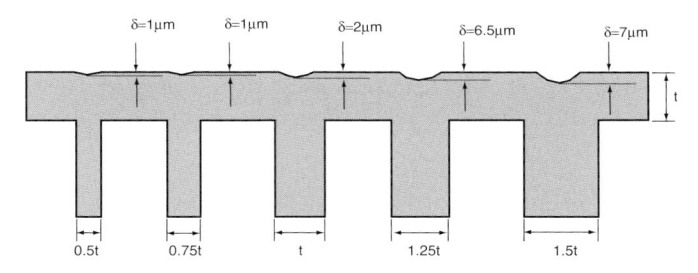

Figure 5.25: Measured sink mark depth as a function of rib relative to flange thickness for a PBT filled with glass beads.

mark depths as a function of rib-thickness for injection molded PBT parts filled with with glass beds.

Residual Stresses The formation of residual stresses in injection molded parts is attributed to two major coupled factors: cooling and flow stresses. The first and most important factor is the residual stress that is formed because of the rapid cooling or quenching of the part inside the mold cavity. This dominant factor is the reason why most thermoplastic parts have residual stresses that are tensile in the central core of the part and compressive on the surface. Typical residual stress distributions are shown in Fig. 5.26 [24], which presents experimental results for PMMA and PS plates cooled at different conditions.

5.2 Process Influences on Product Performance

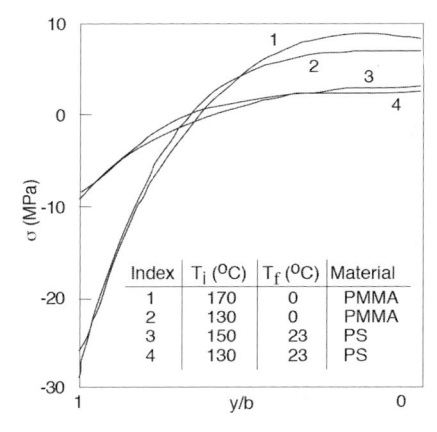

Figure 5.26: Residual stress distribution for 3 mm thick PMMA plates cooled from 170°C and 130°C to 0°C, and for 2.6 mm thick PS plates cooled from 150°C and 130°C to 23°C (after Isayev).

Residual stresses in injection molded parts are also formed by the shear and normal stresses that exist during flow of the polymer melt inside the mold cavity during the filling and packing stage. These tensile flow induced stresses are often very small compared to the stresses that build up during cooling. However, at low injection temperatures, these stresses can be significant in size, possibly leading to parts with tensile residual stresses on the surface. The resulting tensile residual stresses are of particular concern since they may lead to environmental stress cracking of the polymer component.

Shrinkage and warpage result from material inhomogeneities and anisotropy caused by mold filling, molecular or fiber orientation, curing or solidification behavior, poor thermal mold lay-out, and improper processing conditions. Shrinkage and warpage are directly related to residual stresses. Transient thermal or solidification behavior as well as material anisotropies can lead to the build-up of residual stresses during manufacturing. Such process-induced residual stresses can significantly affect the mechanical performance of a component by inducing warpage or initiating cracks and delamination in composite parts. It is hoped that an accurate prediction of the molding process and the generation of residual stresses will allow for the design of better molds with appropriate processing conditions. This section presents basic concepts of the thermomechanical behavior during the manufacturing process of polymeric parts. The formation of residual stresses during the fabrication of plastic parts is introduced first, followed by a review of simple models used to compute residual stresses and associated warpage of plates and beams under different thermal loadings. Several mod-

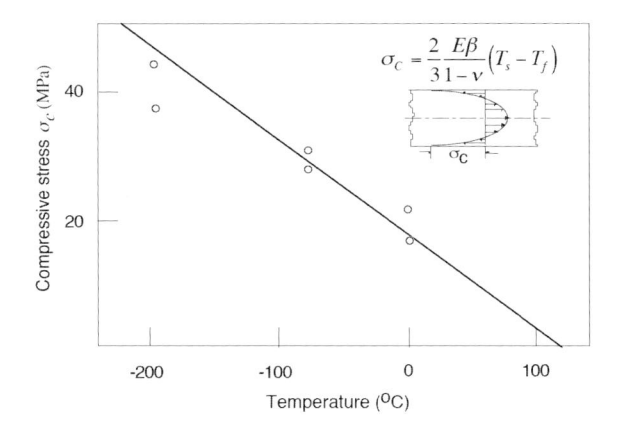

Figure 5.27: Comparison between computed and measured compressive stresses on the surface of injection molded PMMA plates (after Ehrenstein).

els, which characterize the transient mechanical and thermomechanical behavior of thermoplastic polymers will be reviewed and discussed next. Using these existing models, residual stresses, shrinkage, and warpage of injection molded thermoplastic parts can be predicted. Furthermore, results from the literature are presented. Since thermoset polymers behave quite differently from thermoplastic polymers during molding, other models need to be introduced to compute the thermomechanical behavior of thermoset polymers. Based on these models, results for predicting residual stresses and the resulting shrinkage and warpage for both thin and thick thermoset parts are also discussed.

The parabolic temperature distribution which is present once the part has solidified will lead to a parabolic residual stress distribution that is compressive in the outer surfaces of the component and tensile in the inner core. Assuming no residual stress build-up during phase change, a simple function based on the parabolic temperature distribution, can be used to approximate the residual stress distribution in thin sections

$$\sigma = \frac{E\beta}{1-\nu}(T_s - T_f)\left(\frac{4z^2}{t^2} - \frac{1}{3}\right) \qquad (5.4)$$

Here, T_f denotes the final temperature and T_s the solidification temperature; the glass transition temperature for amorphous thermoplastics, or the melting temperature for semi-crystalline polymers. Figure 5.27 compares the compressive stresses measured on the surface of PMMA samples to the above equation.

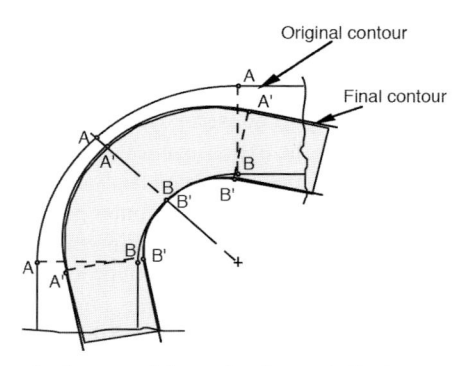

Figure 5.28: Schematic diagram of the spring-forward effect.

Warpage Caused by Mold Thermal Imbalance During molding, the mold wall surface temperatures may vary due to improper thermal mold layout, with variations typically in the order of 10°C. In addition, the temperatures on the mold surface may vary depending on where the heating or cooling lines are positioned, however, one often neglects this effect. Hence, the amount of warpage caused by temperature variations between the two mold halves can easily be computed using

$$\delta = \frac{\beta L^2}{2t}\Delta T \qquad (5.5)$$

where δ is the deflection caused by warpage, L the characteristic planar dimension of the part, t the thickness, β the thermal expansion coefficient, and ΔT the mold thermal imbalance.

Anisotropy Induced Curvature Change In the manufacturing of large and thin laminate structures or fiber reinforced composite parts with a large fiber-length/part-thickness ratio, the final part exhibits a higher thermal expansion coefficient in the thickness direction than in the surface direction. If the part is curved, it will undergo an angular distortion, as shown in Fig. 5.28, which is a consequence of the anisotropy of the composites. This phenomenon is usually called the spring-forward effect or anisotropy induced curvature change. Through-thickness thermal strains, which are caused by different thermal expansion coefficients, can lead to an angle distortion of a cylindrical shell experiencing a temperature change. As demonstrated in Fig. 5.28, when a curved part undergoes a temperature change of ΔT, the curved angle, θ, will change by $\Delta\theta$. The resulting $\Delta\theta$, therefore, is dependent on the angle θ, the temperature change ΔT, and the difference of the thermal expansion coefficients in the r (thickness) and θ (planar) directions

$$\Delta\theta = \Delta\beta\theta\Delta T \qquad (5.6)$$

5.2.5 Process Simulation as Integral Part of the Design Process

Computer simulation of polymer processes offer the tremendous advantage of enabling designers and engineers to consider virtually any geometric and processing option without incurring the expense associated with prototype mold or die making or material waste of time consuming trial-and-error procedures. The ability to try new designs or concepts on the computer gives the engineer the opportunity to detect and fix problems before beginning production. Additionally, the process engineer can determine the sensitivity of processing parameters on the quality and properties of the final part. For example, computer aided engineering (CAE) offers the designer the flexibility to determine the effect of different gating scenarios, runner designs or cooling line locations when designing an injection mold.

However, process simulation is not a panacea. As with any modeling technique, there are limitations caused by assumptions in the constitutive material models, or geometric simplifications of the model cavity. For example, the tendency of the industry is to continuously decrease the part thickness of injection molded parts. Thickness reductions increase the pressure requirements during mold filling, with typical pressures reaching 2,000 bar. Such pressures have a profound effect on the viscosity and thermal properties of the melt; effects that in great part are not accounted for in commercially available software.

The first step of CAE in process design and optimization is to transform a solid model, such as the PA6 housing presented in Fig. 5.29 into a finite element mesh that can be used by the simulation software package. Typically, a fairly three dimensional geometric model is transformed into a mid-plane model that essentially represents a two-dimensional geometry oriented in three dimensional space. A finite element model is then generated on the mid-plane surface. Basically, the most common injection molding models use this approach to represent the geometry of the part. While most injection molded parts are thin and planar, and would be well represented with such a model[1], some injection molded parts are of smaller aspect ratio, or have three dimensional features, making these models invalid.

Mold Filling Predictions Using a finite element mesh, such as the one presented in Fig. 5.29, in conjunction with the control volume approach a simulation package solves a coupled energy and momentum balance, bringing as a result a mold filling pattern that not only includes the non-Newtonian effects present in the flow of polymer melts, but also the effect that the cooling has on the melt flow inside the mold cavity.

---

[1] In the literature, this model is often referred to as the *Hele-Shaw flow model*. This model was first developed for injection molding by Prof. K.K. Wang's Cornell Injection Molding Program, and implemented into a commercial software called C-Mold in the late 1980's.

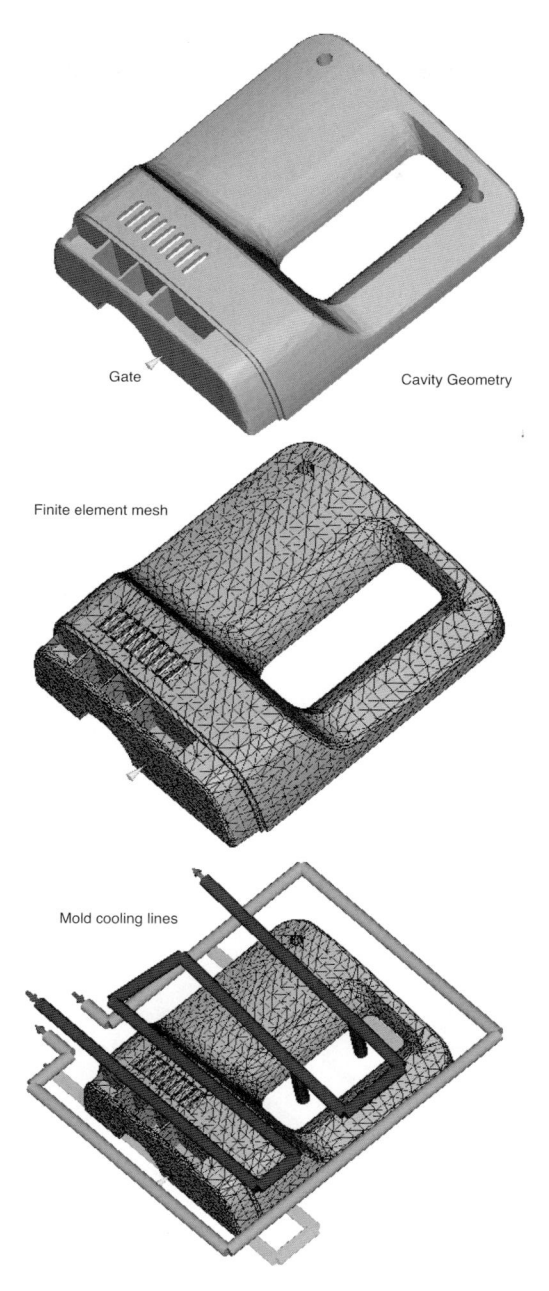

5.2 Process Influences on Product Performance 477

Figure 5.29: Geometric representation of the part, finite element mesh of the mid-plane surface and mold cooling line locations (Courtesy SIMCON Kunststofftechnische Software GmbH).

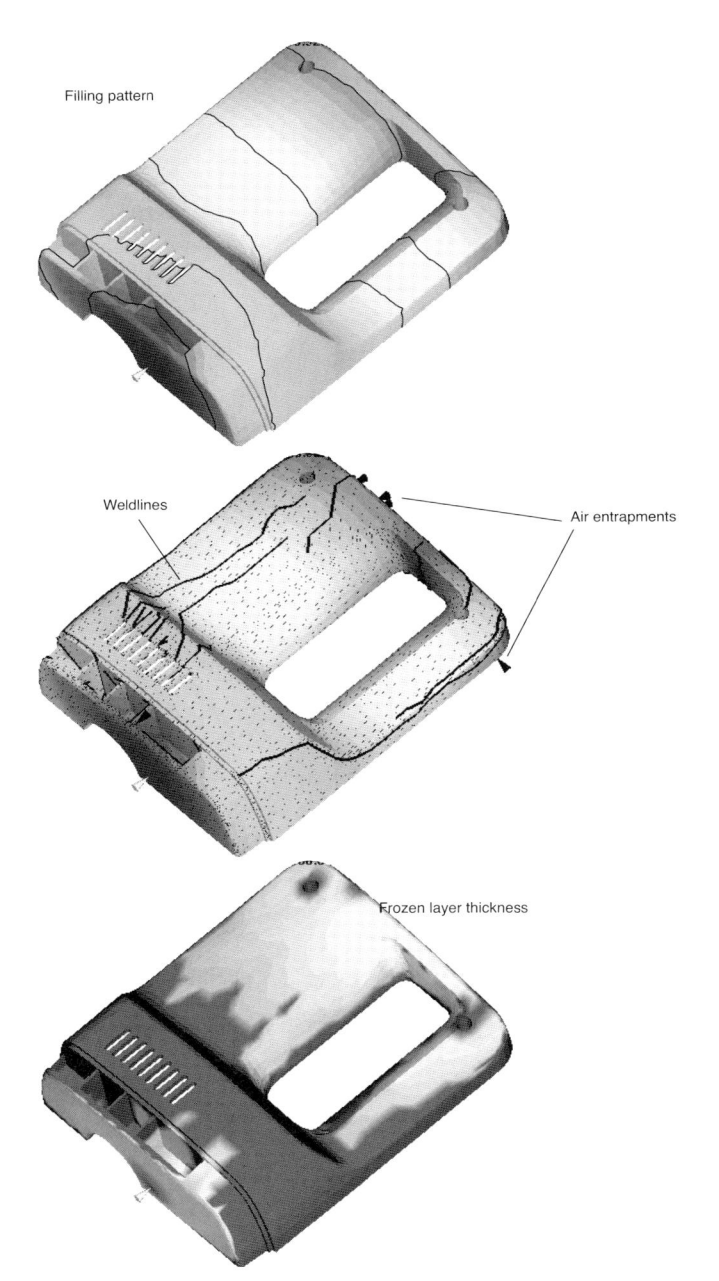

Figure 5.30: Mold filling pattern, weldline and possible air entrapment location as well as frozen layer diagram 13 seconds into the cycle (Courtesy SIMCON Kunststofftechnische Software GmbH).

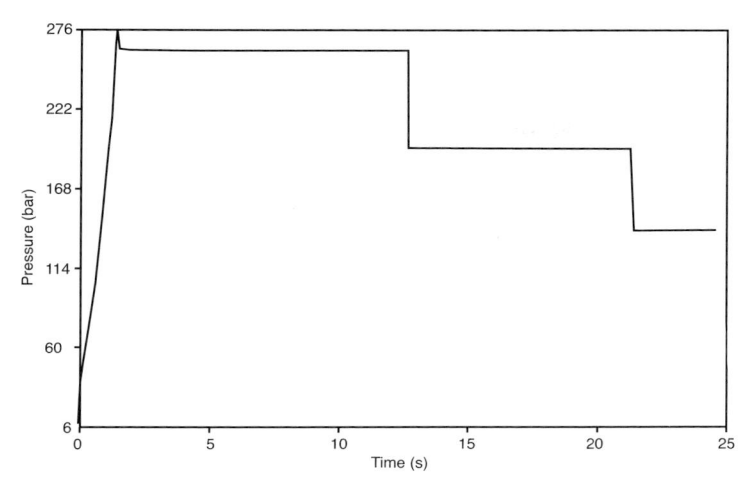

Figure 5.31: Predicted cavity pressure during injection molding of the housing presented in Fig. 5.29 (Courtesy SIMCON Kunststofftechnische Software GmbH).

The mold filling analysis and the resulting filling pattern can be used to predict the formation of weldlines (knitlines when dealing with fiber reinforced composite parts) and gas entrapment. These can cause weak spots and surface finish problems that can lead to cracks and failure of the final part, as well as esthetic problems in the finished product. Figure 5.30 presents the mold filling pattern as well as the weldline and last filling points of the part and gate presented in Fig. 5.29. The last filling points are necessary information for mold venting, and to avoid air entrapment and consequently dieseling and other issues associated with it.

The pressure and clamping force requirements are also needed information during part and process design. Both are computed by commercial injection molding software. For the part presented in Fig. 5.29 the pressure inside the cavity is given in Fig. 5.31 and the corresponding mold clamping force is presented in Fig. 5.32.

To illustrate the mold filling pattern of a compression molding process, Fig. 5.33 presents the initial charge location and filling pattern during compression molding of an automotive fender.

Orientation and Anisotropy Predictions Molecular and filler orientation have a profound effect on the properties of the finished part. Molecular orientation will not only influence the mechanical properties of the polymer but also its optical quality. For example, birefringence is controlled by molecular orientation, which must be kept low for products that require certain optical properties,

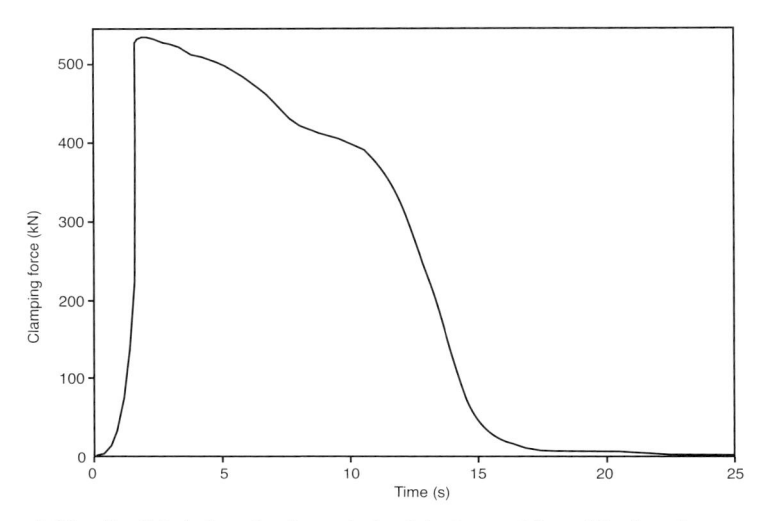

Figure 5.32: Predicted clamping force during injection molding of the housing presented in Fig. 5.29 (Courtesy SIMCON Kunststofftechnische Software GmbH).

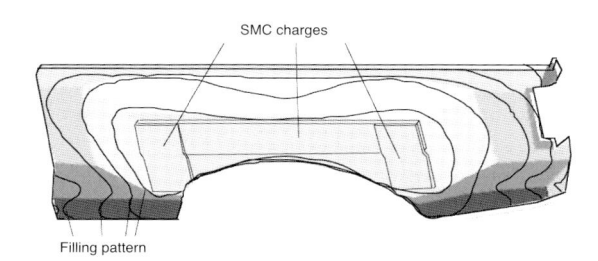

Figure 5.33: Initial charge and filling pattern during compression molding of an automotive fender.

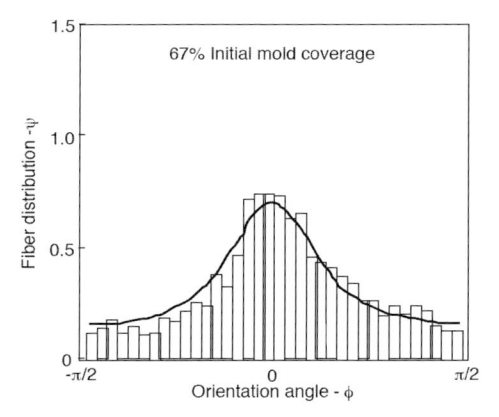

Figure 5.34: Comparison of predicted and experimental fiber orientation distributions for SMC experiments with a 67% initial mold coverage.

such as lenses. The Folgar-Tucker model has been implemented into various, commercially available injection and compression mold filling simulation programs. The model has proven to work well compared to experiments done with extensional flows. Figures 5.34, 5.35 and 5.36 compare the measured fiber orientation distributions to the calculated distributions using the Folgar-Tucker model for cases with 67%, 50%, and 33% initial charge mold coverage, respectively. Again, to illustrate the effect that fiber orientation has on material properties of the final part, Fig. 5.14 shown above, demonstrates how the fiber orientation presented in Figures 5.34, 5.35 and 5.36 affects the stiffness of the plates.

The fiber orientation distribution in a realistic part is solved coupled with the flow from the mold filling simulation.

Shrinkage and Warpage Predictions Shrinkage and warpage are directly related to residual stresses which result from locally varying strain fields that occur during the curing or solidification stage of a manufacturing process. Such strain gradients are caused by nonuniform thermomechanical properties and temperature variations inside the mold cavity. Shrinkage due to cure can also play a dominant role in the residual stress development in thermosetting polymers and becomes important for fiber reinforced thermosets, and are a concern when sink marks appear in thick sections or ribbed parts. When processing thermoplastic materials, shrinkage and warpage in a final product depend on the molecular orientation and residual stresses that form during processing. The molecular or fiber orientation and the residual stresses inside the part in turn depend on the flow and heat transfer during the mold filling, packing, and cooling stage of the injection molding process. To predict the residual stress in the finished part, modern soft-

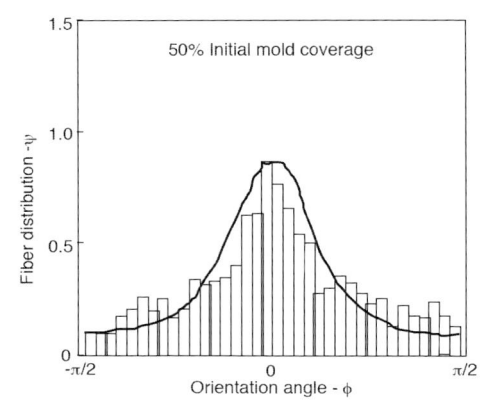

Figure 5.35: Comparison of predicted and experimental fiber orientation distributions for SMC experiments with a 50% initial mold coverage.

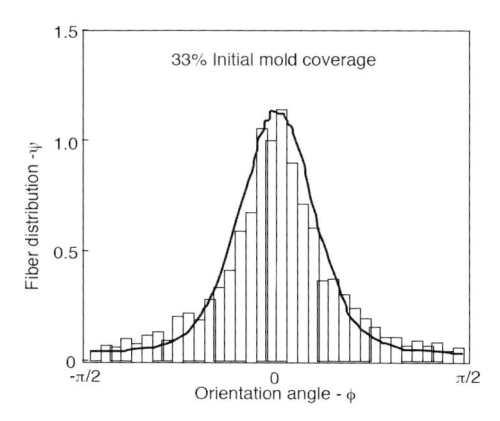

Figure 5.36: Comparison of predicted and experimental fiber orientation distributions for SMC experiments with a 33% initial mold coverage.

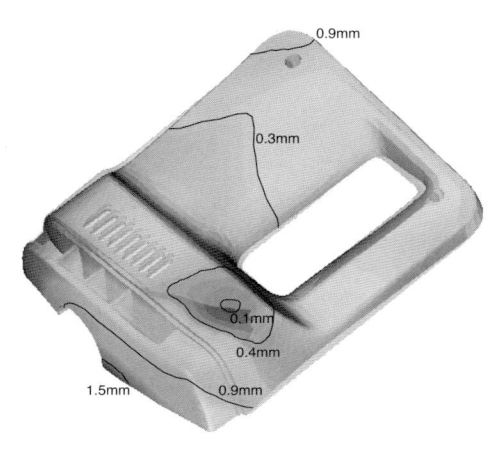

5.2 Process Influences on Product Performance 483

Figure 5.37: Predicted warped geometry after mold removal and cooling (Courtesy SIMCON Kunststofftechnische Software GmbH).

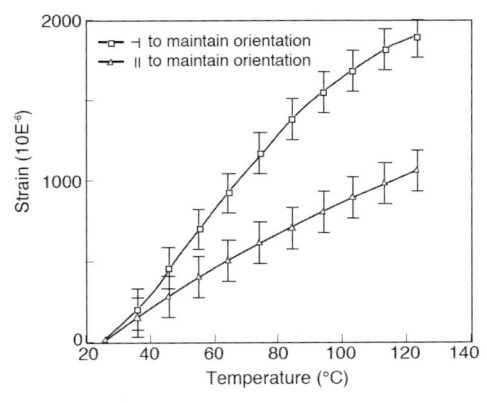

Figure 5.38: Experimentally measured thermal strains in an SMC plate with a fiber orientation distribution that resulted from a 25% initial mold coverage charge.

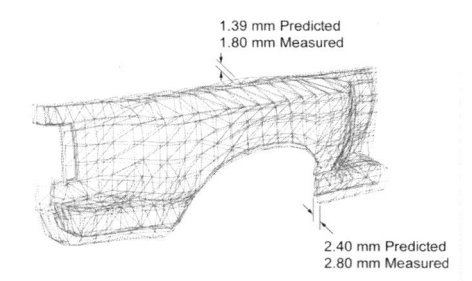

Figure 5.39: Simulated displacements of an automotive body panel. Displacements were magnified by a factor of 20.

ware packages characterize the thermomechanical response of the polymer from melt to room temperature, or from the pvT behavior to stress-strain behavior. Figure 5.37 presents the warped geometry of the part depicted in Fig. 5.29 after mold removal and cooling. The warpage is usually depicted graphically as total amount of deflection as well as superposing deflected part geometry and mold geometry. For clarity, the figure presented here was enhanced with iso-deflection lines. However, commercial software presents these in color graphs and scales that clearly show warpage in the part.

The shrinkage and warpage in thin compression molded fiber reinforced thermoset plates are predicted using non-planar finite element plate models. The anisotropic thermal expansion coefficient, caused by fiber orientation, is perhaps the largest cause of warpage in fiber reinforced parts. Figure 5.38 demonstrates how, for typical composite parts, the thermal shrinkage parallel to the main orientation direction is about half of that normal to the main orientation direction. The thermal shrinkage was measured from a rectangular plate molded with a charge that covered 25% of the mold surface and that was allowed to flow only in one direction.

To calculate the residual stress development during the manufacturing process, the heat transfer equation is coupled to the stress-strain analysis through constitutive equations. Figure 5.39 compares the mold geometry with part geometry for the truck fender shown earlier, after mold removal and cooling. The fiber content by volume in thepart was 21% and the material properties for the glass fiber and the unsaturated polyester resin are listed in Table 5.1.

Minimizing warpage is one of the biggest concerns for the design engineer. This is sometimes achieved by changing the formulation of the resin. Further reduction in warpage can also be achieved by changing the number and location of gates, when the part is injection molded, or the size and location of the initial charge, when the part is compression molded. Although trial-and-error solutions, which is still the most feasible with today's technology, are commonly done, computer optimization often reduces cost.

Table 5.1: Mechanical and Thermomechanical Properties for Various Materials

Property	Fiberglass	Polyester	Epoxy
E (MPa)	7.3×10^4	2.75×10^3	4.1×10^3
ν	0.25	0.34	0.37
β (mm/mm/K)	5.0×10^{-6}	3.7×10^{-5}	5.76×10^{-5}

5.3 STRENGTH OF MATERIALS CONSIDERATIONS

The mechanical behavior of plastics, especially thermoplastics, depend on temperature, time and structure developed during the process. In addition, environmental effects such as UV degradation, chemical attack, relative humidity, to name a few, play a significant role in the structural integrity and properties of a finished product. However, many basic concepts of stress and strain used with metals also apply for plastics, provided that the temperature and time dependent behavior of the material are included.

5.3.1 Basic Concepts of Stress and Strain

Strictly speaking, polymers cannot be modeled using linear theory of elasticity. However, if small deformations are used along with time dependence, for example from creep data, the stress-strain response of a linear elastic model for the polymer component can suffice in the evaluation of a design and the prediction of the behavior of the component during loading. For a full three-dimensional model, as shown for a small material element in Fig. 5.40, there are six components of stress and strain.

The stress-strain relation for a linear elastic material is defined by the following equations,

$$\sigma_{xx} = \mathcal{E}I_\epsilon + 2G\epsilon_{xx} \tag{5.7}$$

$$\sigma_{yy} = \mathcal{E}I_\epsilon + 2G\epsilon_{yy} \tag{5.8}$$

$$\sigma_{zz} = \mathcal{E}I_\epsilon + 2G\epsilon_{zz} \tag{5.9}$$

$$\tau_{xy} = G\gamma_{xy} \tag{5.10}$$

$$\tau_{yz} = G\gamma_{yz} \tag{5.11}$$

$$\tau_{zx} = G\gamma_{zx} \tag{5.12}$$

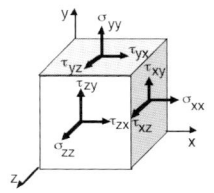

Figure 5.40: Differential material element.

where,
$$\mathcal{E} = \frac{\nu E}{(1+\nu)(1-2\nu)} \tag{5.13}$$

and I_ϵ is the first invariant of the strain tensor and represents the volumetric expansion of the material which is defined by

$$I_\epsilon = \epsilon_{xx} + \epsilon_{yy} + \epsilon_{zz} \tag{5.14}$$

The elastic constants E, ν and G represent the modulus of elasticity, Poisson's ratio and shear modulus, respectively. The shear modulus, or modulus of rigidity, can be written in terms of E and ν as

$$G = \frac{E}{2(1+\nu)} \tag{5.15}$$

The above equations can be simplified for different geometries and load cases. Two of the most important simplified models, the plane stress and plane strain models, are discussed below.

Plane Stress A common model describing the geometry and loading of many components is the plane stress model. The model reduces the problem to two dimensions by assuming that the geometry of the part can be described on the $x-y$ plane with a relatively small thickness in the z-direction. In this case $\sigma_{zz} = \tau_{zx} = \tau_{yz} = 0$ and the above equations reduce to

$$\sigma_{xx} = \frac{E}{1-\nu^2}(\epsilon_{xx} + \nu\epsilon_{yy}) \tag{5.16}$$

$$\sigma_{yy} = \frac{E}{1-\nu^2}(\nu\epsilon_{xx} + \epsilon_{yy}) \tag{5.17}$$

and

$$\tau_{xy} = G\gamma_{xy} \tag{5.18}$$

5.3 Strength of Materials Considerations

Plane Strain Another common model used to describe components is the plane strain model. Similar to the plane stress model, the geometry can be described on an $x - y$ plane with an infinite thickness in the z-direction. This problem is also two-dimensional, with negligible strain in the z-direction but with a resultant σ_{zz}. For this case, the above equations reduce to

$$\sigma_{xx} = \frac{E(1-\nu)}{(1+\nu)(1-2\nu)} \left(\epsilon_{xx} + \frac{\nu}{1-\nu} \epsilon_{yy} \right) \quad (5.19)$$

$$\sigma_{xx} = \frac{E(1-\nu)}{(1+\nu)(1-2\nu)} \left(\frac{\nu}{1-\nu} \epsilon_{xx} + \epsilon_{yy} \right) \quad (5.20)$$

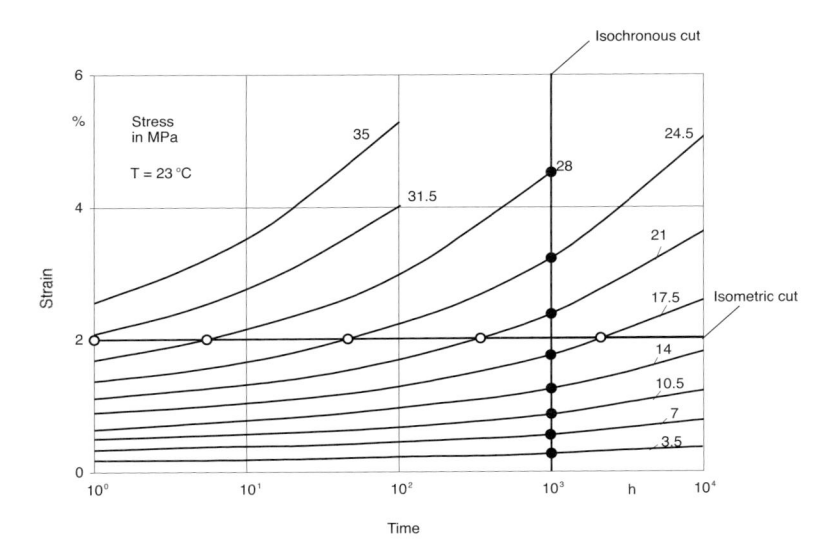

$$\tau_{xy} = G\gamma_{xy} \quad (5.21)$$

Figure 5.41: Creep response of a PBT at 23°C with a 1000 h isochronous cut and a 2% isometric cut.

Material Properties The most difficult aspect of carrying through predictions of stress and strain of a given design using the above models is to acquire the necessary material properties, namely, the modulus, E, and Poisson's ratio, ν. Most of the time these properties are needed for a specific time scale, associated with the time or speed of loading. In addition, many engineering applications are at temperatures significantly higher than room temperature. As presented in Chapter 3 of this book, all these factors significantly influence the the material

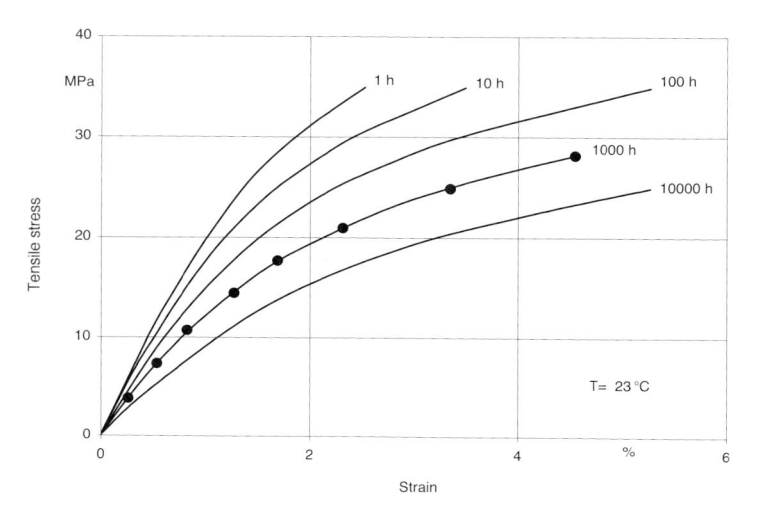

Figure 5.42: Isochronous stress-strain curves for the PBT at 23°C creep responses with the corresponding 1000 h cut points.

behavior of a plastic. The time dependent modulus can be extracted from standard creep tests as presented in Fig. 5.41 for a PBT at room temperature. As discussed in Chapter 3 this graph can be transformed into isochronous or isometric curves. The figure illustrates two cuts, one to generate a 1000 h isochronous curve, and one to generate a 2% isometric curve. The corresponding isochronous and isometric curves, with the transfered data points, are presented in Figs. 5.42 and 5.43, respectively.

■ **EXAMPLE 5.1**

Sample Calculation Using Isochronous Curves. Using creep data for PBT we are to determine the height, h, of the cantilevered bracket shown in Fig. 5.44. Use the geometry and load shown in the figure and a maximum strain of 2% after 6 weeks (1000h) of loading.

The first step in the solution of this example is to determine the maximum allowable stress. For a maximum strain of 2% we can read the stress from the 1000h isochronous curve presented in Fig. 5.42 to be 19 MPa. For cantilever beams the maximum stress is located in upper and lower surfaces of the beam and is given by

$$\sigma_{max} = \frac{Mc}{I} \tag{5.22}$$

5.3 Strength of Materials Considerations

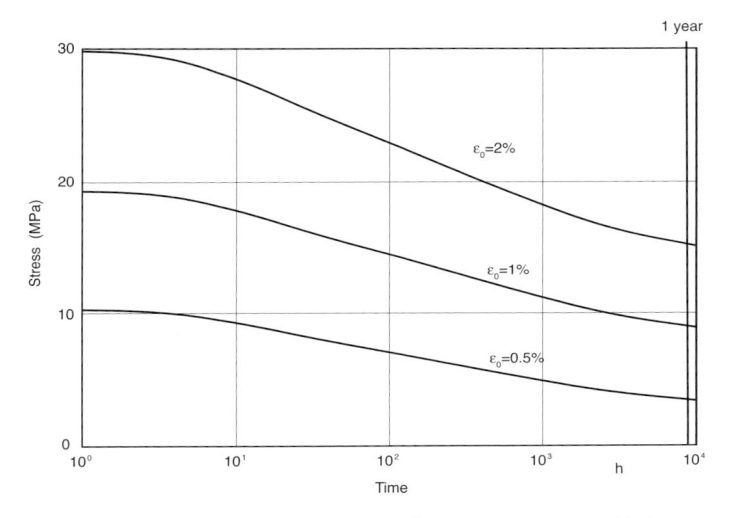

Figure 5.43: Isometric curves for the PBT at 23°C creep responses with the corresponding 2% cut points.

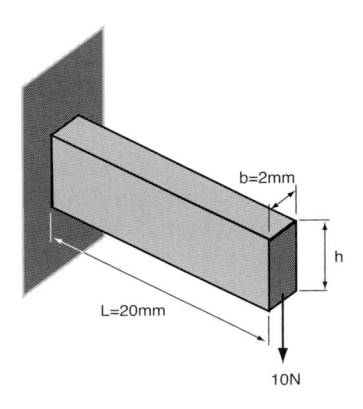

Figure 5.44: Cantilever beam geometry.

where, $M = FL$, $c = h/2$ and $I = bh^3/12$. Here, we can solve for h as

$$h = \sqrt{\frac{6FL}{\sigma_{max}b}} = \sqrt{\frac{6(10\text{N})(20\text{mm})}{19\text{MPa}(4\text{mm})}} = 3.97\text{mm} \qquad (5.23)$$

■ EXAMPLE 5.2

Sample Calculation Using Isometric Curves. In the assembly shown in Fig. 5.45, a tubular PBT feature is pressed on a 15 mm long steel stud. The inner diameter of the 1 mm thick PP tubular element is 10 mm. The metal stud is slightly oversized with a diameter of 10.22 mm. With a coefficient of friction $\mu = 0.3$ estimate the force required to disassemble the parts shortly after assembly, and after one year.

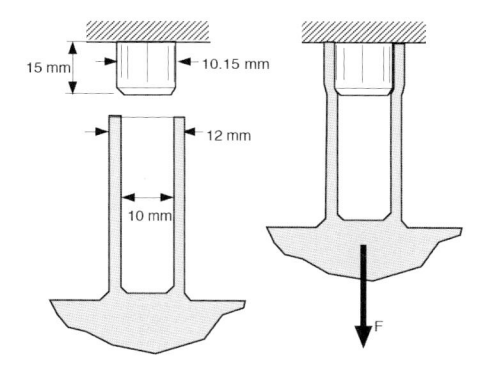

Figure 5.45: Press fit assembly.

This is a classic constant strain, ϵ_0, stress relaxation problem. The initial hoop stress that holds the assembly together can be quite high. However, as time passes the hoop stress relaxes and it becomes easier to disassemble the two components. Using thin pressure vessel theory, and neglecting the deformation of the steel stud, the strain in the system after assembly is computed using

$$\epsilon_0 = \Delta D/\bar{D} = 0.22\text{mm}/11\text{mm} = 0.02 \rightarrow 2\% \qquad (5.24)$$

In order to follow the hoop stress history after assembly we use the 2% isometric curve presented in Fig. 5.45. From the curve we can see a hoop stress, σ_H, of 29.5 MPa right after assembly and of 16 MPa after one year.

5.3 Strength of Materials Considerations

The pressure acting on the metal stud due to the hoop stress is computed using

$$p = 2h\sigma_H/\bar{D} = 2(1\text{mm})(29.5\text{MPa})/(11\text{mm} =) = 5.36\text{MPa} \quad (5.25)$$

right after assembly. The disassembly force is computed using

$$F = \mu p(\pi D_i L) = 0.3(5.36\text{MPa})(\pi \times 10\text{mm} \times 15\text{mm}) = 757\text{N}(170\text{lb}) \quad (5.26)$$

The required force after one year is 410N or 92lb.

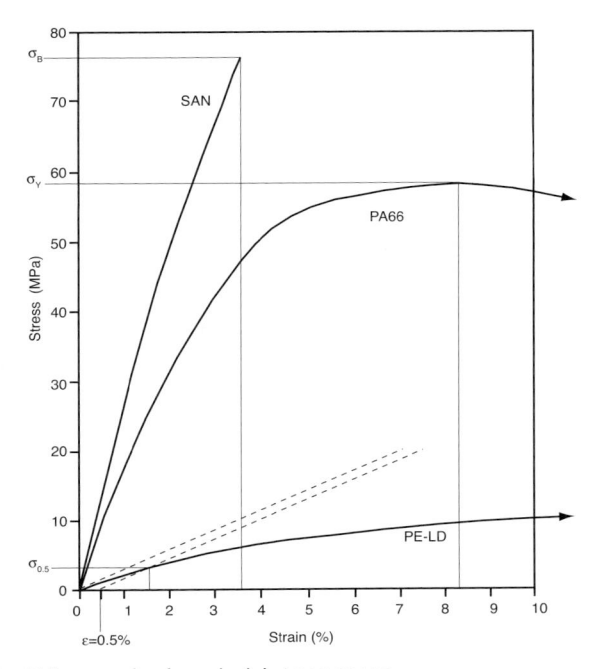

Figure 5.46: Failure modes for uniaxial stress cases.

Failure Criteria Depending on the type of material, plastic parts fail differently. Some plastics are brittle and fail at low strains, others undergo a distinct yield stress. However, in many cases there is no distinct mode of failure, or data is not available. For those situations, the 0.5% offset strain is a conservative value, that has proven to be quite useful to the design engineer. Figure 5.46 presents these three failure criteria. The stress-strain curve for the SAN material presents a typical brittle failure behavior with a maximum brittle stress, σ_B of 76 MPa and strain at failure of 3.7%. The PA66 material undergoes a clear yield stress,

σ_Y, of 58 MPa at a strain of 8.3%. On the other hand, the PE-LD continuously deforms as the crystalline structure orients and stretches. For such a case one can choose the stress, $\sigma_{0.5}$ that corresponds to the 0.5% offset strain, ϵ=0.5%. However, it must be stressed that failure modes are influenced by time scale. For example, creep rupture associates the loading stress to the time to fail by static fatigue. Cyclic fatigue, as with metals, is presented in form of Wöhler diagrams. As discussed in Chapter 3, a standard stress-strain test will yield different results if tested at different rates of deformation, and the failure stress must be chosen according to the corresponding loading conditions.

In any case, the appropriate failure criteria can be chosen from the three cases presented in Fig. 5.46, and compared to an equivalent stress, σ_{eq} obtained from a chosen strength hypothesis. For the cases where the compressive failure strength, σ_{CB}, equals the tensile strength, σ_{TB}, we can use the equivalent stress as the simple Huber-van Mises-Henky, $\sigma_{eq} = \sigma_{HMH}$, defined by

$$\sigma_{HMH} = \frac{1}{\sqrt{2}}\sqrt{(\sigma_1 - \sigma_2)^2 + (\sigma_2 - \sigma_3)^2 + (\sigma_3 - \sigma_1)^2} \tag{5.27}$$

However, in many polymers the compressive strength is higher than the tensile strength, and the Huber-van Mises-Henky criterion overestimates the tensile behavior of the material. For such cases we can use either the *conical fracture criterion* given by

$$\sigma_C = \frac{m-1}{2m}(\sigma_1 + \sigma_2 + \sigma_3) \pm \frac{m+1}{2\sqrt{2}}\sqrt{(\sigma_1 - \sigma_2)^2 + (\sigma_2 - \sigma_3)^2 + (\sigma_3 - \sigma_1)^2} \tag{5.28}$$

or the *parabolic fracture criterion* given by

$$\sigma_P = \frac{m-1}{2m}(\sigma_1 + \sigma_2 + \sigma_3) \pm$$
$$\sqrt{\left(\frac{m+1}{2m}\right)^2 (\sigma_1 + \sigma_2 + \sigma_3)^2 + \frac{1}{2m}((\sigma_1 - \sigma_2)^2 + (\sigma_2 - \sigma_3)^2 + (\sigma_3 - \sigma_1)^2)} \tag{5.29}$$

where m is the ratio of the compressive to the tensile strength of the material. This is given by $m = \sigma_{CB}/\sigma_{TB}$ for a brittle failure, and by $m = \sigma_{CY}/\sigma_{TY}$, for a ductile material with a distinct yield strength. For combined shear, τ_B, tensile stress, σ_{TB}, load cases one used the ratio $t = \tau_Y/\sigma TY$, for yielding and

5.3 Strength of Materials Considerations

$t = \tau_B/\sigma_{TB}$ for fracture, with

$$\sigma_C = \frac{\sqrt{3}t - 1}{\sqrt{3}t}(\sigma_1 + \sigma_2 + \sigma_3) \pm \frac{1}{\sqrt{6}t}\sqrt{(\sigma_1 - \sigma_2)^2 + (\sigma_2 - \sigma_3)^2 + (\sigma_3 - \sigma_1)^2} \quad (5.30)$$

for a conical criterion, and

$$\sigma_P = \frac{3t^2 - 1}{6t^2}(\sigma_1 + \sigma_2 + \sigma_3) \pm \sqrt{\left(\frac{6t^2 - 1}{6t^2}\right)^2 (\sigma_1 + \sigma_2 + \sigma_3)^2 + \frac{1}{2m}((\sigma_1 - \sigma_2)^2 + (\sigma_2 - \sigma_3)^2 + (\sigma_3 - \sigma_1)^2)} \quad (5.31)$$

for a parabolic criterion.

Figure 5.47 compares the HMH, conical and parabolic failure criteria with data from materials that have a yield strength. Figure 5.48 presents the three failure criteria for materials that undergo a brittle fracture.

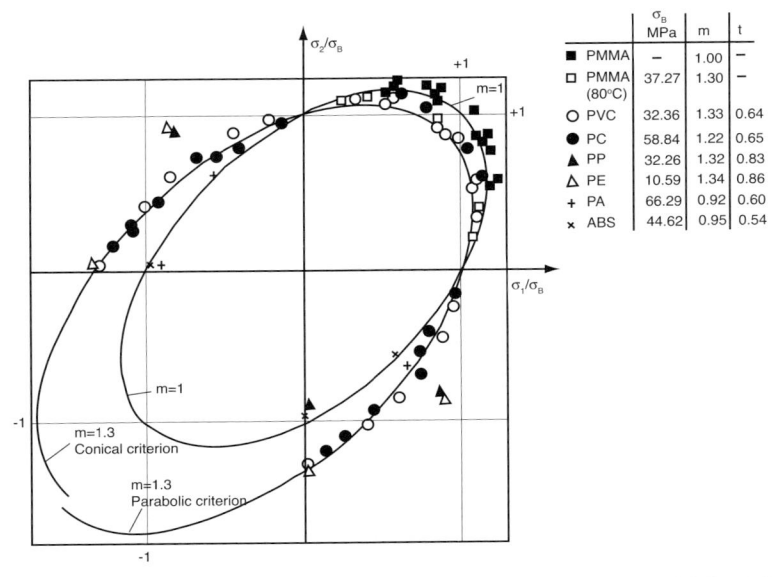

	σ_B MPa	m	t
■ PMMA	–	1.00	–
□ PMMA (80°C)	37.27	1.30	–
○ PVC	32.36	1.33	0.64
● PC	58.84	1.22	0.65
▲ PP	32.26	1.32	0.83
△ PE	10.59	1.34	0.86
+ PA	66.29	0.92	0.60
× ABS	44.62	0.95	0.54

Figure 5.47: Comparison between failure criterion and experiments of failure due to yielding under biaxial stress.

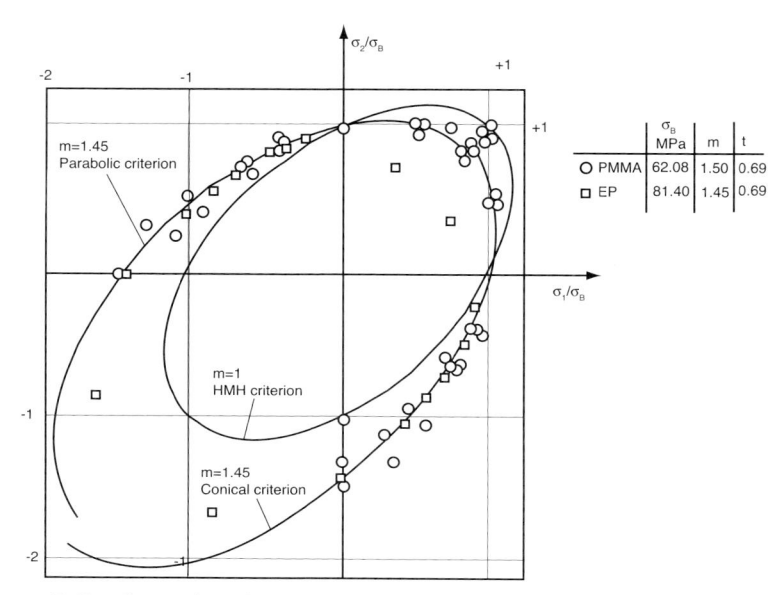

Figure 5.48: Comparison between failure criterion and experiments of failure due to fracture under biaxial stress.

5.3.2 Anisotropic Strain-Stress Relation

Filled polymers are often anisotropic, and the relations presented in the above equations are not valid. The three-dimensional anisotropic strain-stress relation is often written as

$$\epsilon_{xx} = \frac{1}{E_{xx}}\sigma_{xx} - \frac{\nu_{yx}}{E_{yy}}\sigma_{yy} - \frac{\nu_{zx}}{E_{zz}}\sigma_{zz} \tag{5.32}$$

$$\epsilon_{yy} = -\frac{\nu_{xy}}{E_{xx}}\sigma_{xx} + \frac{1}{E_{yy}}\sigma_{yy} - \frac{\nu_{zy}}{E_{zz}}\sigma_{zz} \tag{5.33}$$

$$\epsilon_{zz} = -\frac{\nu_{xz}}{E_{xx}}\sigma_{xx} - \frac{\nu_{yz}}{E_{yy}}\sigma_{yy} + \frac{1}{E_{zz}}\sigma_{zz} \tag{5.34}$$

$$\gamma_{xy} = \frac{1}{G_{xy}}\tau_{xy} \tag{5.35}$$

$$\gamma_{yz} = \frac{1}{G_{yz}}\tau_{yz} \tag{5.36}$$

$$\gamma_{zx} = \frac{1}{G_{zx}}\tau_{zx} \tag{5.37}$$

5.3 Strength of Materials Considerations

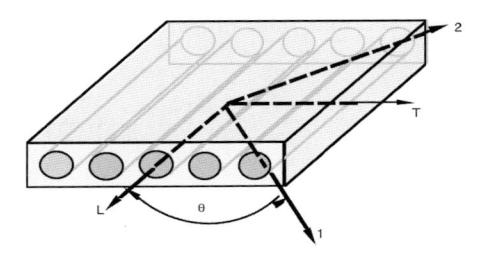

Figure 5.49: Schematic diagram of unidirectional continuous fiber reinforced laminated structure.

and in matrix form for the more general case,

$$\begin{pmatrix} \epsilon_{xx} \\ \epsilon_{yy} \\ \epsilon_{zz} \\ \gamma_{xy} \\ \gamma_{yz} \\ \gamma_{zx} \end{pmatrix} = \begin{bmatrix} S_{11} & S_{12} & S_{13} & S_{14} & S_{15} & S_{16} \\ S_{21} & S_{22} & S_{23} & S_{24} & S_{25} & S_{26} \\ S_{31} & S_{32} & S_{33} & S_{34} & S_{35} & S_{36} \\ S_{41} & S_{42} & S_{43} & S_{44} & S_{45} & S_{46} \\ S_{51} & S_{52} & S_{53} & S_{54} & S_{55} & S_{56} \\ S_{61} & S_{62} & S_{63} & S_{64} & S_{65} & S_{66} \end{bmatrix} \begin{pmatrix} \sigma_{xx} \\ \sigma_{yy} \\ \sigma_{zz} \\ \tau_{xy} \\ \tau_{yz} \\ \tau_{zx} \end{pmatrix} \quad (5.38)$$

where coupling between the shear terms and the elongational terms can be introduced.

Aligned Fiber Reinforced Composite Laminates The most often applied form of the above equations is the two-dimensional model used to analyze the behavior of aligned fiber reinforced laminates, such as that shown schematically in Fig. 5.49. For this simplified case, anisotropic stress strain equations reduce to

$$\epsilon_L = \frac{1}{E_L}\sigma_L - \frac{\nu_{TL}}{E_T}\sigma_T \quad (5.39)$$

$$\epsilon_T = -\frac{\nu_{LT}}{E_L}\sigma_L + \frac{1}{E_T}\sigma_T \quad (5.40)$$

$$\gamma_{LT} = \frac{1}{G_{LT}}\tau_{LT} \quad (5.41)$$

which can also be written as

$$(\epsilon_{LT}) = [S_{LT}](\sigma_{LT}) \quad (5.42)$$

where the subscripts L and T define the longitudinal and transverse directions, respectively, as described in Fig. 5.49, and $[S_{LT}]$ is referred to as the compliance matrix. The longitudinal and transverse properties can be calculated using the widely used Halpin-Tsai model as

$$E_L = E_m \left(\frac{1 + \xi\eta\phi}{1 - \eta\phi} \right) \tag{5.43}$$

$$E_T = E_m \left(\frac{1 + \xi\eta\phi}{1 - \eta\phi} \right) \tag{5.44}$$

$$G_{LT} = G_m \left(\frac{1 + \lambda\phi}{1 - \lambda\phi} \right) \tag{5.45}$$

where

$$\eta = \frac{\left(\frac{E_f}{E_m} - 1 \right)}{\left(\frac{E_f}{E_m} + \xi \right)} \tag{5.46}$$

$$\lambda = \frac{\left(\frac{G_f}{G_m} - 1 \right)}{\left(\frac{G_f}{G_m} + 1 \right)} \tag{5.47}$$

$$\xi = 2 \left(\frac{L}{D} \right) \tag{5.48}$$

Here, the subscripts f and m represent the fiber and matrix, respectively; L the fiber length; D the fiber diameter; ϕ the volume fiber fraction which can be expressed in terms of weight fraction, ψ, as

$$\phi = \frac{\psi}{\psi + (1 + \psi)(\rho_f/\rho_m)} \tag{5.49}$$

It should be pointed out that, in addition to the Halpin-Tsai model, there are several other models in use today to predict the elastic properties of aligned fiber reinforced laminates. Most models predict the longitudinal modulus quite accurately as shown in Fig. 5.50, which compares measured values to computed values using the *mixing rule*[2], where ϕ is the volume fraction of fibers. This comes as no surprise, since experimental evidence clearly shows that longitudinal modulus is directly proportional to the fiber content for composites with unidirectional reinforcement. However, differences do exist between the models when predicting the transverse modulus, as shown in Fig. 5.51.

---

[2]The mixing rule is the simplest form of calculating elongational modulus and is given by $E_L = E_m(1 - \phi) + E_f\phi$

5.3 Strength of Materials Considerations

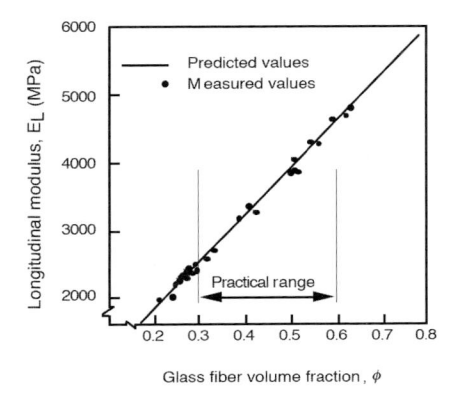

Figure 5.50: Measured and predicted longitudinal modulus for an unsaturated polyester/aligned glass fiber composite laminate as a function of volume fraction of glass content.

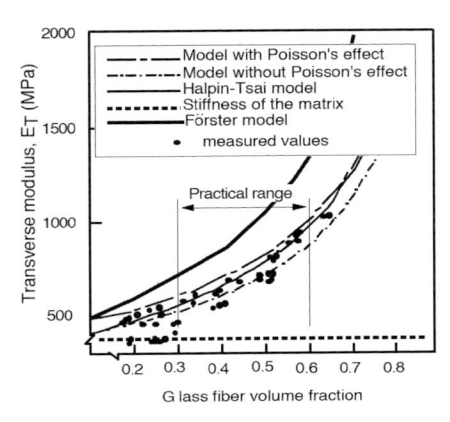

Figure 5.51: Measured and predicted transverse modulus for an unsaturated polyester/aligned glass fiber composite laminate as a function of volume fraction of glass content.

5.4 FUNCTIONAL ELEMENTS

Functional elements such as snap fits and living hinges must be designed such that they withstand mold ejection and assembly forces, as well as the daily loading and unloading, or opening and closing of a plastic assembly.

5.4.1 Press Fit Assemblies

Press fits are the simplest method of assembling plastic parts, where one part is force fit on another and held together by a combination of hoop stresses and friction. During press fitting an oversized shaft or boss is fitted into a hole in the mating part. One of the common examples of such an assembly are LEGO® toys. When designing this form of assembly for a long-term one must consider stress relaxation effects, deduced from creep data. As expected, the mating forces will reduce with time, rendering a less effective fit. Gear, bearings and pulleys are commonly attached to plastic or metal shafts using press fit principles. A schematic of a press fit assembly with common notation is shown in Fig. 5.52.

The interference between the shaft and hole, δ, leads to a hoop stress and strain. When calculating hoop stresses, sliding forces and torques, one must first compute the hub's geometric factor given by,

$$\beta = \frac{D_o^2 + D_i^2}{D_o^2 - D_i^2} \tag{5.50}$$

The pressure that holds the assembly together is a function of the interference between the hub and shaft as well as the properties of both the hub and the shaft

$$p = \frac{2\delta}{\frac{D_i}{E_h}(\beta + \nu_h) + \frac{D_i}{E_s}(\beta - \nu_s)} \tag{5.51}$$

Using the pressure, one can easily compute the force required to cause the shaft to slip axially using

$$F_{slip} = 2\pi L D p \mu \tag{5.52}$$

where μ is the coefficient of friction between the mating surfaces. Similarly, the maximum torque allowed before the shaft slips is computed using

$$T_{slip} = \pi L D^2 p \mu / 2 \tag{5.53}$$

Due to stress relaxation, the modulus of the plastic parts will decay with time. The modulus can be determined from creep data transformed to isochronous curves, as shown for a PBT at 23°C in Fig. 5.43. Figure 5.53 presents the shrinkage hoop stresses within a polyamide hub overmolded on a steel shaft under various conditions.

5.4 Functional Elements

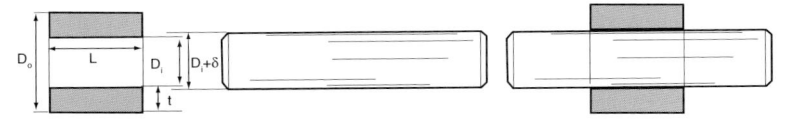

Figure 5.52: Schematic of a press fit assembly.

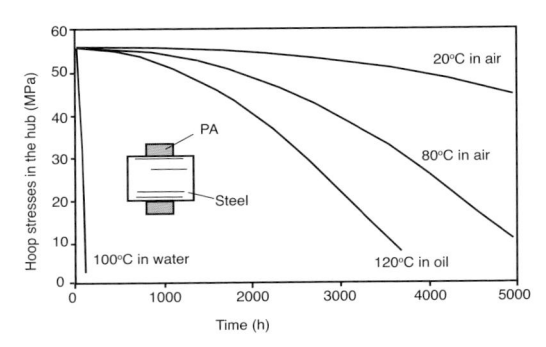

Figure 5.53: Shrinkage induce hoop stress as a function of time within hubs overmolded on a steel shaft, under various environmental conditions.

5.4.2 Living Hinges

A living hinge or integral joint is a flexible element that connects two sections of a component, such as the body and cover of DVD's. Living hinges are found in thermoplastic parts as an integral component of the plastic molding in injection molded parts, as well as in thermoformed and blow molded articles. These friction-free elements of the plastic parts suffer from very little wear, and will therefore have a long life, provided they are designed without sharp notches and stress concentrators. Figure 5.54 presents a typical injection molded living hinge geometry, with generous radii and smooth section transitions, alongside a poor living hinge design. Injection molded living hinge designs are found primarily in polypropylene parts, and in some polyethylene parts. Polymers with high molecular weights with a narrow distribution usually perform best in living hinge geometries. However, adding fillers or reinforcing fibers usually reduce the life and properties of living hinges.

5.4.3 Snap Fit Assemblies

The *snap fit* is one of the most versatile and economic joining elements. They are composed of a hook, at the end of cantilever beam, and a groove. The beam deflects as the mating parts are pushed together until the hook falls into the groove,

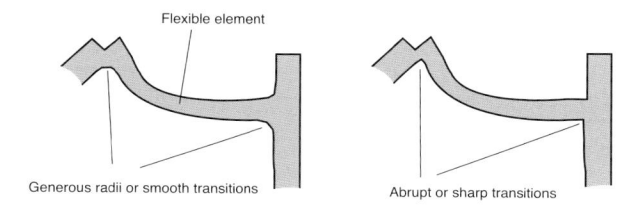

Figure 5.54: Schematic diagram of good (left) and bad (right) injection molded living hinge designs.

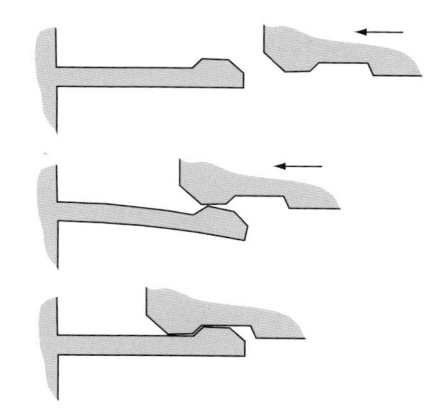

Figure 5.55: Schematic diagram of the insertion, deflection and recovery action of a snap-fit during assembly.

holding the parts in place. The basic snap fit assembly with the insertion, deflection and recovery action during assembly is schematically depicted in Fig. 5.55. The snap fit geometry shown in the figure shows as assembly as well as disassembly angle. Figure 5.56 presents two examples of snap fit assemblies with common cantilever beam geometries and Fig. 5.57 presents somewhat different snap fit type on a housing cover assembly. This type presents a thumb release snap fit.

On the other hand, snap fits can also be annular. The most common example is the lid of a pen or a marker. Other types of annular snap fit elements include segmented annular snap fit joints as the one presented in Fig. 5.58.

One important aspect of snap fit design is to predict and control the assembly and disassembly forces. Figure 5.59 presents the notation commonly used to predict these forces. From geometry and force balances, which includes the

5.4 Functional Elements   501

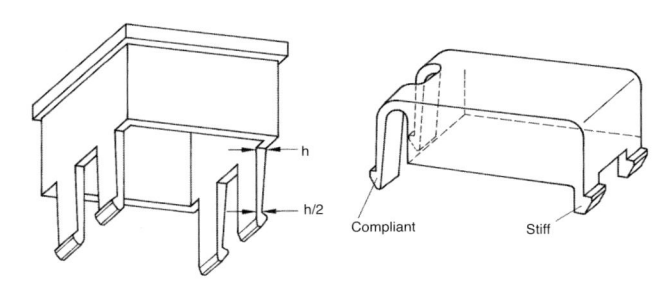

Figure 5.56: Common snap fit assemblies and geometries.

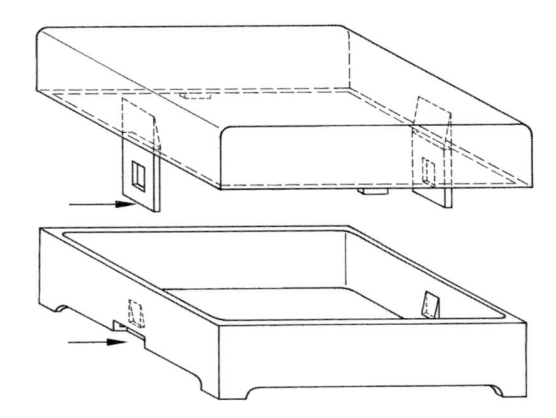

Figure 5.57: A thumb release snap fit on a housing cover assembly.

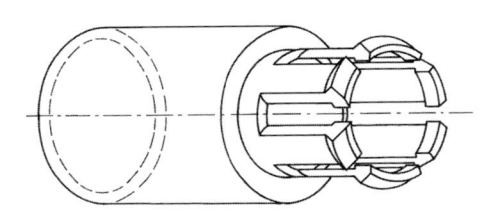

Figure 5.58: Segmented annular snap fit joint.

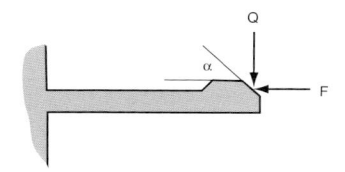

Figure 5.59: Notation used for snap fit assembly calculations.

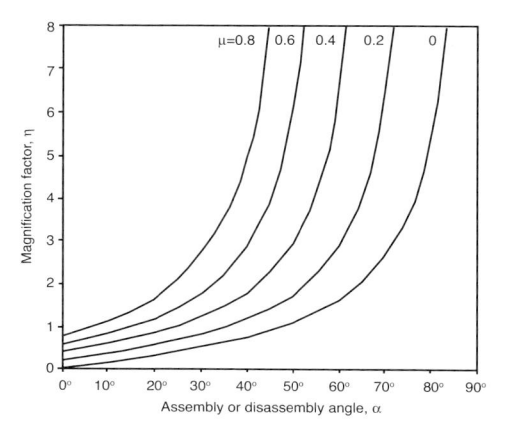

Figure 5.60: Force magnification factor as a function of assembly or disassembly angle for various coefficients of friction.

friction between the mating parts, the ratio between the bending forces, Q, and the assembly or disassembly forces, F, can be computed using,

$$\eta = \frac{F}{Q} = \frac{\mu + \tan\alpha}{1 - \mu\tan\alpha} \qquad (5.54)$$

The bending force, Q depends on the geometry of the cantilever beam that holds the hook as well as the amount of the deflection. If the maximum allowable strain is given, the deflection can be computed using

$$y_{max} = C\frac{\epsilon L^2}{h} \qquad (5.55)$$

where C is a geometric factor defined in Fig. 5.61 and h is the height of the beam at the root. Once the strain is known we can use isochronous or short term stress-strain curves to determine the stress. The maximum stress in the cantilevered beam, which occurs at the upper (tensile) and lower (compressive)

5.4 Functional Elements

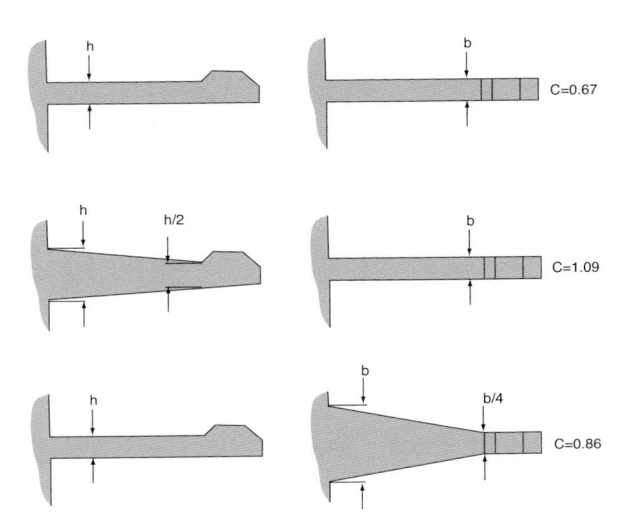

Figure 5.61: Different snap fit geometries and their corresponding geometric factor C.

surfaces at the root of the beam is given by

$$\sigma = \frac{QLc}{I} \tag{5.56}$$

where c is the distance from the mid-plane to the outer surfaces. For a rectangular cross-section

$$I = \frac{bh^3}{12} \tag{5.57}$$

The above equations can be used to solve for the force, Q, required to deflect the beam during assembly and disassembly.

■ EXAMPLE 5.3

Sample Snap Fit Calculation Using an Annular Assembly. Using the geometry given in Fig. 5.62 and a coefficient of friction (μ) of 0.3 find the mating force for the annular snap fit assembly.

The maximum strain is computed using

$$\epsilon = \frac{\Delta D}{D} = \frac{0.5mm}{12mm} = 0.042 \tag{5.58}$$

We can use this strain to determine the secant modulus for the material from the graph presented in Fig. 5.5. Using the curve that corresponds to 23°C we get E_{POM}=1700MPa.

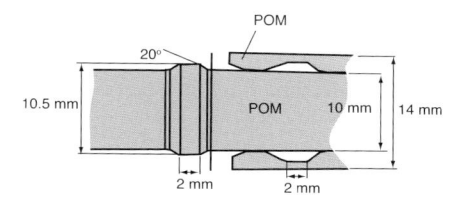

Figure 5.62: Geometry of an annular snap fit assembly.

We first need to compute a geometric factor given by

$$k = \frac{\pi(D_o/D_i - 1)\sqrt{(D_o/D_i)^2 - 1}}{5(D_o/D_i)^2(1-\nu) + 5 + 5\nu}$$
$$= \frac{\pi(1.4 - 1)\sqrt{1.4^2 - 1}}{5(1.4)^2(1-0.333) + 5 + 5(0.3333)} = 0.0932 \quad (5.59)$$

to calculate the force needed to deflect the annular fitting

$$Q = kE_{POM}D_i^2\epsilon = 0.0932(1700\text{MPa})(0.01\text{m})^2(0.042) = 665\text{N} \quad (5.60)$$

For an angle of 20° and a coefficient of friction of 0.3 we can read the force magnification factor, η=0.75, resulting in a mating force of

$$F = Q\eta = 665\text{N}(0.75) = 499\text{N} \quad (5.61)$$

5.4.4 Mechanical Fasteners

The traditional mechanical fastening techniques such as screws, bolts and nuts should be comparably unimportant in the plastics product design field. Such fasteners are an awkward inheritance from metals and wood and should be replaced by snap fits, press fits, living hinges and other integrated plastics oriented design elements. However, they are still very prevalent in the plastics industry. Today, we still find extensive use of these types of fasteners in plastic parts. These include:

- Threaded inserts
- Self-threading screws
- Rivets

5.5 SOFTWARE

As was evident in this chapter, software plays a significant role in modern engineering design with polymers. From first concept designs using CAD systems, to material data banks that will help the engineer thread through the vast sea of information, today's design is streamlined by the use of software.

Traditional design, which finds its roots in the metals industry, makes use of CAD systems in conjunction with FEA software to properly design a product. Although this approach is often used by plastics products designers, it not appropriate for the material. With plastics, the process plays a significant role in the performance of the finished product. Thus, a plastics product designer must have a full grasp of what is referred to as the 5 P's: polymer, processing, product, performance and post-consumer life. All but the last of these P's can be completely integrated through software.

Hence, material choice is linked to mechanical performance, which depends on the process. The process, in turn, depends on the choice of material. Therefore, a material data bank not only presents properties used in design, such as creep data, but also process relevant data such as shear and temperature dependent viscosity.

As a reference, Table 5.2 presents a list of common software used in the plastics industry.

Table 5.2: Commercial Software for Plastics Technology

Type	Name	Producer	Web address
CAD- Systems	CATIA	IBM, Dassault Aviation	www.ibm.com
	Pro/Engineer	Parametric Technology Corporation	www.ptc.com/
	Solid Works	SolidWorks Deutschland GmbH	www.solidworks.de/
	AutoCAD	Autodesk	www.autodesk.de
	I-Deas,	Unigraphics	www.eds.com
FEM software	Abaqus	Abaqus	www.abacus.com/
	Ansys	Ansys Inc.	www.ansys.com/
	Fidap, Fluent	Fluent Inc.	www.fluent.com
	Marc	MSC. Software	www.marc.com/
	Nastran	Macro Industries	www.macroindustries.com
	LS-Dyna	Livermore Software Technology Corporation	www.ls-dyna.com
	PAMCASH	ESI-Group	www.esi-group.com
Pre/post- Processors	ANSA	Beta-CAE System S.A.	www.beta-cae.gr/
	HyperWorks	Altair	www.altair.com/
	Animator	GNS mbh	www.gns-mbh.com/

Continued on next page

Type	Name	Producer	Web address
Material Data Banks	CAMPUS	CWFG mbH	www.campusplastics.com/
	MCBase	M-Base	www.m-base.de
Process Simulation			
Injection Molding	CADMOULD	Simcon GmbH	www.simcon-world-wide.com
	MOLDEX	SimpaTec GmbH	www.simpatec.com
	MOLDFLOW	Moldflow Corp.	www.moldflow.com
	SIGMASOFT	Sigma Engineerin GmbH	www.sigmasoft.de
Compression Molding	CADPRESS	The Madison Group	www.madisongroup.com
	EXPRESS	M-Base	www.m-base.de
Blow Molding	B-SIM	ACCUFORM	www.t-sim.com
Thermoforming	T-SIM	ACCUFORM	www.t-sim.com
3D Flow Simulation	FIDAP	Fluent	www.fluent.com
	FLUENT	Fluent	www.fluent.com
	POLYFLOW	Fluent	www.fluent.com

CHAPTER 6

MATERIALS

In this chapter, in order to compare the different types of plastics material, material properties according to standards adopted by material data banks such as CAMPUS will be used (see Table 6.1). The number of properties used will be limited to enhance the clarity and comparativeness of the information.

Table 6.1: Material Properties -Symbols

Property	Unit	Symbol
Density	g/cm^3	ρ
Tensile modulus of elasticity	MPa	E_t
Tensile stress at yield	MPa	σ_y
Elongation at yield	%	ϵ_y
Elongation at break	%	ϵ_B
Nominal Elongation at break	%	ϵ_{tB}
Stress at 50% strain	MPa	σ_{50}
Failure stress	MPa	σ_B
Failure strain	%	ϵ_B
Processing temperature	°C	T_P
Heat distortion temperature HDT/A 1.8 MPa	°C	HDT
Coeff. of linear expansion, parallel (23-55°C)	10^{-5}/K	α_p
Coeff. of linear expansion, perpendicular (23-55°C)	10^{-5}/K	α_n
Flammability UL 94 at 1.6 mm thickness	Class	UL94
Dielectric constant at 100 Hz	-	$\epsilon_r 100$
Dielectric loss factor at 100 Hz	$\cdot 10^{-4}$	$\tan\delta 100$
Specific volume resistivity	Ohm·m	ρ_e
Specific surface resistivity	Ohm	σ_e
Dielectric Strength	KV/mm	$E_B I$
Water absorption at 23°C, saturation	%	W_w
Moisture absorption at 23°C/50% r.h., satu	%	W_H
Glass content (volume)	%	ϕ
Glass content (weight)	%	ψ
Melting temperature	°C	T_m
Glass transition temperature	°C	T_g

For example, in general, no values for impact strength will be provided, because these values typically do not contribute significant information to either part design or a part's load bearing capacity.

As a general rule of thumb it can be assumed that for resin types with a lower modulus of elasticity and a higher elongation at yield (elongation at break), parts made from these resins will be more ductile (tougher) under high loads.

Unless particularly noted, specific property values are given for generic unmodified material classes. Because in general there will be no data provided for specific trade names, the property values given should be considered an indicating range. Specific material types may vary widely from these ranges, particularly regarding classification such as 'resistant' or 'non-resistant' to environmental conditions.

A breakdown of different plastics with their most common types, properties, applications and chemical composition are presented. The chapter serves a resource to the practicing engineer to help identify which material suits best a certain application, or which material was used for an existing product. This type of identification is a very important task of a plastics engineer. Tables 6.2 to 6.5 present a summarized guide to aid in the identification of plastics. These tables must be used in conjunction with common sense and engineering insight. Further, as already discussed in Chapter 3 of this handbook, analytical techniques such as TGA, IR-spectroscopy and Raman technology are great tools to identify materials and their composition.

6.1 POLYOLEFINS (PO), POLYOLEFIN DERIVATES, AND COPOPLYMERS

Polyolefins are polymers built from hydrocarbons with a double bond and the general chemical structure C_nH_{2n} (ethene, propene, butene-1, isobutene). They include polyethylene, polypropylene, polybutylene-1, polyisobutylene, poly-4-methylpentene as well as their respective copolymers. Today, the available homo- and copolymeric resins based on ethylene and propylene provide an extraordinary broad range of properties, as shown in Fig. 6.1 for the modulus of elasticity of PE.

The polyolefin architecture is determined in particular by the catalysts used, with metallocene catalysts gaining increasing importance. Polymeric resins made with metallocene catalysts provide a narrow molecular weight distribution and their polymerization allows for particular sequence and order of their monomer building blocks. In addition, it is possible to incorporate building blocks into the polyolefin structure that could not be incorporated previously by copolymerization. Figure 6.2 provides an overview of applications for metallocene polyolefin resins.

Table 6.2: Plastics Identification Table

Material Acronym	Density non-filled g/cm³	Density filled (up to) g/cm³	Transparent thin film	Crystal clear	Cloudy to opaque	Generally filled	Leathery or rubbery	Compliant	Stiff	Gasoline	Benzene	Methylene chloride	Diethylether	Acetone	Ethylacetate	Ethylalcohol	Water
PE soft	0.92		+		+			+		i/sw	sw	i	i/sw	i/sw	i/sw	i	i
PE rigid	0.96		+						+	i	i/sw	i/sw	i	i	i	i	i
PP	0.905	1.3*	+		+			+	+	i/sw	i/sw	i/sw	i	i	i	i	i
PB	0.915				+			+	+	sw	i/sw	i	i/sw	i	i/sw	i	i
PIB	0.93	1.7				+	+			s	s	s	sw	i	i	i	i
PMP	0.83			+					+	sw	sw	i	i	i	sw	i	i
PS	1.05		+	+					+	sw/s	s	s	i/sw	s	s	i	i
SB	1.05		+	+				+	+	sw/l	s	s	s	s	s	i	i
SAN	1.08	1.4*	+						+	i	s	s	s	s	s	i	i
										Dissolution rate dependent on type							
ABS	1.06				+			+	+	sw	s	s	s	s	s	sw	i
ASA	1.07				+				+	sw	s	s	sw	s	s	sw	i
PVC	1.39 / 1.35		+	+					+	i	i/sw	sw/s	i/s	i/sw	i/sw	i	i
										Copolymer easier sw/s than PVC							
PVC-C	ca. 1.5		+	(+)					+	i	i/sw	i	i	i/sw	i	i	i
PVC-HI																	
PVCEVA	1.2–1.35		+		+		+	+		i/sw	i/sw	sw/s	i/sw	sw	i/sw	i	i
PVCPEC	1.3–1.35		+		+		+	+		i	sw	sw	i	i	i	i	i
PE-C	1.1–1.3			+			+	+		sw Between PE and PVC depending on Cl-content							
PVC-P	1.2–1.35	1.6	+	+	+		+			i	sw	sw	sw	sw	sw	sw	i
										Plasticizer removed with diethylether							
PTFE	2-2.3																
FEP	2-2.3				+			+		i	i	i	i	i	i	i	i
PFA	2-2.3									PFA, ETFE, sw in hot CCl4 or similar							
ETFE	1.7																
PCTFE	2.1		+	+			+	+		i	i	i	i	i	i	i	i
PVDF	1.7–1.8		+	(+)	+				+	i	i	sw	i/sw	sw	i	i	i
PVAC	1.18		Dispersed solids				+	+		i	i	s	s	sw	s	s	i
PVAL	1.2-1.3		+				+	+		i	i	i	i	i	i	i	s
										Free of acetylene groups (also hot)							
PVB	1.1–1.2		Safety film				+			i	sw	sw/s	i	sw/s	sw/s	s	i
PAE	1.1–1.2		Dispersed solids				+	+		i/s	i	sw	i	sw	i	i	i
										Soluble in polyacrylic acid							
PMMA	1.18			+					+	i	s	s	i	s	s	i	i
AMMA	1.17		+ Yelow						+	i	i	sw	i	sw	i	i	i

Table 6.3: Plastics Identification Table

Material Acronym	Slow heating in test tube m = melts d = decomposes b = basic n = neutral a = acidic sa = strong acidic	Smoke smell	Ignition with small flame 0 = ignites with difficulty I = self extinguishing II = continues burning without flame III = burns violently, fulminates Color and type of flame	Odor of smoke during ignition in test tube or after ignition and extinguishing	Other comments, residue of other elements (N, Cl, F, S, Si)
PE soft	Turns clear, m, d	n	II			Different melting points 105 – 120 °C
PE rigid	Little visible smoke	n	II	Yellow with blue core. Drips burning	Yellow with blue core. Drips burning	125 – 130 °C
PP		n	II			165 – 170 °C
PB	m, gasifies, gases are flammable					130 – 140 °C
PIB		n	II	Yellow, soft flame	Yellow, soft flame	
PMP	m, d, gasifies, some white smoke	n	II	Yellow, blue core, drips	Yellow, blue core, drips	245 °C
PS	Melts and gasifies	n	II		Typical coal gas-like	When breaking: brittle fracture
SB	m, yellowish, d	n	II		Similar to PS+rubber	
SAN	m, yellow, d	b	II	Intense flickering yellow flame. Very sooty	Similar to PS, scratchy, cinnamon	Stress whitening N, brittle fracture
ABS	d, turns black	n	II			N, Stress whitening
ASA	m,d,black residue	a	II		Like PS, pepper	
PVC		sa	I	Yellow, sooty, lower flame edge	Salt acid, irritating smell	Cl, distinctive feature Cl-contents and softening temperature
PVC-C	softens, turns brown	sa	I	slightly green		
PVC-HI	Black		I/II			
PVCEVA		sa,a	I			
PVCPEC		sa				
PE-C	m, turns brown	sa	I/II	Intense yellow, sooty	HCl and paraffin	
PVC-P	Like PVC	sa	I/II	Luminous from plasticizers	HCl and plasticizer	Stiffens when plasticizer is eluted
PTFE	Turns clear, does not melt, d at red heat	sa	0	Blue/green, does not burn	At red heat, pungent (HF)	F
FEP						PFEP, PFA melt at 300 °C
PFA						
ETFE						ETFE at 270 °C
PCTFE	m, d at red heat	sa	0	Similar to PTFE	Hydrochloric+ hydrofluoric acid	F, Cl
PVDF	m, d at high temperat.	sa	0/I	Difficult to ignite	Pungent (HF)	F
PVAC	m, braun, gasifies	a	II	Luminous, sooty	Acetic acid smell	
PVAL	m, d, brown residue	n	I/II	Luminous	Irritating smell	
PVB	m, d, foams	a	II	Blue with yellow edges	Rancid butter	
PAE	m, gasifies	n	II	Luminous, slightly sooty	Penetrating smell	
PMMA	softens, m, expands, crackles, residue	n	II	Burns, crackles, drips, luminous	Fruity smell	Cast acrylic soften little
AMMA	Brown, then m, d black	a	II	Sooty, sputters	Irritating then penetrating	N

6.1 Polyolefins (PO), Polyolefin Derivates, and Copoplymers

Table 6.4: Plastics Identification Table

Material Acronym	Density non-filled g/cm³	Density filled (up to) g/cm³	Transparent thin film	Crystal clear	Cloudy to opaque	Generally filled	Leathery or rubbery	Elastic (Large strain)	Stiff	Gasoline	Benzene	Methylene chloride	Diethylether	Acetone	Ethylacetate	Ethylalcohol	Water
POM	1.41	1.6			+				+	i	i	i	i	i	i	i	i
PA	1.02-1.14	1.4	+		+			+	+	i	i	i	i	i	i	i	i
PC	1.2	1.4	+	+					+	i	i	sw	i	sw	i	i	i
PBT	1.35	1.5	+	+	+			+	+	i	sw	s	sw	sw	sw	i	i
PET	1.41	1.5	+	+	+			+	+	i	i	sw	i	i/sw	sw	i	i
PPS					+				+	i	i	i	i	i	i	i	i
PSU	1.24	1.5	(+)	+				+	+	i	s	s	i	sw	i/sw	i	i
PPE		1.3			+				+	i	s	s	i	i	i	i	i
PI	1.4		+ Yellow		+				+	i	i	i	i	i	i	i	i
CA	1.3		+	+				+	+	i	i	sw/s	i	sw/s	sw/s	i	i
											Depending on degree of acetylization							
CAB	1.2		+	+					+	i	sw	s	i	s	s	sw	i
CP	1.2			+					+	i	i	sw	i	s	s	sw	i
CN	1.35-1.4		+	+				+	+	i	i	i	sw	s	s	i	i
CH	1.45		+					+		i	i	i	i	i	i	i	i
													Softens					
VF	1.2-1.3				+		+	+		i	i	i	i	i	i	i	i
PF[1]	1.25-1.3	Resins	(+)						+	i	i	i	i	s	i	s	(s)
PF[2]		1.8-2				+			+	i	i	i	i	i	i	i	i
PF[3]		1.4				+			+	i	i	i	i	i	i	i	i
UF/MF[4]		1.5				+			+	i	i	i	i	i	i	i	i
MF[5]		2.0				+			+	i	i	i	i	i	i	i	i
MF+PF[6]		1.5				+			+	i	i	i	i	i	i	i	i
UP[7]	1.2-1.3			+					+	i	i	sw	i	sw	sw	i	i
UP[8]		1.4-2			+	+			+	i	i	sw	i	sw	sw	i	i
EP	1.2				+				+	i	i	sw	i	sw	sw	i	i
EP[9]		1.7-2				+			+	i	i	sw	i	sw	sw	i	i
PUR[10]	1.26				+		+	+	+	i	i	sw	i	sw	sw	i	i
PUR[11]	1.17-1.22		+		+		+			i	sw	sw	i	sw	sw	i	i
PUR[12]	<1				+		+	+	+	i	sw	sw	i	sw	sw	i	i
SI[13]	1.25					+	+	+		sw	sw	sw	i	i	i	i	i

[1] Uncured resin, [2] Mineral filled type 11-16, [3] Natural fiber filled types 30-84, [4] Natural fiber filled types 131, 152-154,
[5] Mineral filled types 155-157, [6] Natual fiber filled types 180-182, [7] Self extinguishing, [8] Mineral filled laminates,
[9] Glass fiber filled laminates, [16] Cured, [17] Rubber elastic, [18] Foam, [19] Silicone rubber

Table 6.5: Plastics Identification Table

Material Acronym	Slow heating in test tube m = melts d = decomposes b = basic n = neutral a = acidic sa = strong acidic	Smoke smell	Pyrolysis (Burning tests) Ignition with small flame 0 = ignites with difficulty I = self extinguishing II = continues burning without flame III = burns violently, fulminates	Color and type of flame	Odor of smoke during ignition in test tube or after ignition and extinguishing	Other comments, residue of other elements (N, Cl, F, S, Si)
POM	m, d, gasifies	n (a)	II	Blue, almost colorless	Formaldehyde	
PA	Turns clear, m	(b)	I/II	Difficult to ignite, blue-yellow, crackles	Typical burnt horn	N
PC	d, brown	(a)				
PBT	m, tough, colorless, d. brown	a	I	Luminous, sooty, blasig, charred	First faint, then phenolic	Detection of phenolic
PET	m, d, dark brown, white tarnish on top	a	I/II	Luminous, crackling, dripping, sooty	Sweetly, scratchy	
PPS		sa	I	Strong sooty	Like H_2S	
PSU	Turns clear, m then brown	b	II	Difficult to ignite, yellow, sootychars	First faint, finally in test tube H2S	
PPE	m, vapors not visible, brown	b	II	Difficult to ignite, then luminous, sooty	First faint, then phenolic	
PI	Turns black, m, d, brown vapors, does not melt, turns brown, incandescent	sa	0	Incandescent	At high temperatures phenolic	
CA	m, d, black	a	II	Melts and drips, yellow-green, spark	Acetic acid + burnt paper	
CAB	m, d, black	a	II	Yellow luminous, burning drips	Acetic acid, butyric acid, burnt paper	
CP	m, d, black	a	II	Like CAB	Propionic acid, burnt paper	
CN	d fiercly	sa	II	Bright, violent, brown vapors	Nitric oxide (camphor)	N
CH	d, chars	n	II	Like paper	Burnt paper	
VF	d, chars	n	I/II	Burns slowly	Burnt paper	
PF[1]	m, d	n	I	Difficult to ignite, luminous, sooty	Phenolic, formaldehyde or ammonia	
PF[2]	d, fractures		0/I	Luminous, sooty		
PF[3]	d, fractures	n(a)	I/II	Chars		
UF/MF[4]	d, fractures, color turns dark	b	0/I	Barely ignites, flame slightly yellow,	Phenolic, formaldehyde	
MF[5]				material chars with white edges,	Ammonia, amines, fishy smell	
MF+PF[6]				structure stays		
UP[7]	Turns dark, m,d,	n/b	I/II	Luminous yellow,	Styrene, irritating smell	Ignitibility also depends on filler and pigments
UP[8]	fractures sometimes white covering	(a)		sooty, crackles, chars, glass fiber residue		
EP	Turns dark from edges,			Difficult to ignite,	Depending on curing agent, ester-like or amines, later phenolic resins	N with amine curing agents or special
EP[9]	d, fractures, sometimes white covering			burns with small yellow flame, sooty		
PUR[10]	m when hot, then d	n	II	Difficult to ignite,	Typical unpleasant stinging (isocyanate)	N
PUR[11]	m	a		yellow luminous,		
PUR[12]	d	b	II	foams, drips		
SI[13]	d hot, white powder	n	0	Glows in flame	White smoke	Si - SiO_2 residue

[1] Uncured resin, [2] Mineral filled type 11-16, [3] Natural fiber filled types 30-84, [4] Natural fiber filled types 131, 152-154,
[5] Mineral filled types 155-157, [6] Natual fiber filled types 180-182, [7] Self extinguishing, [8] Mineral filled laminates,
[9] Glass fiber filled laminates, [16] Cured, [17] Rubber elastic, [18] Foam, [19] Silicone rubber

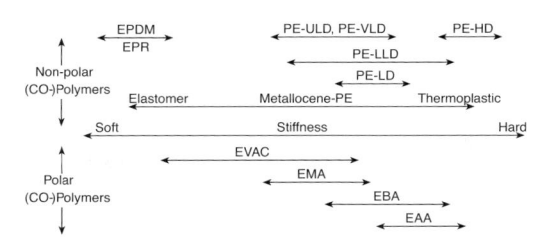

6.1 Polyolefins (PO), Polyolefin Derivates, and Copoplymers

Figure 6.1: Polyolefin family with the bandwidths of their respective moduli of elasticity.

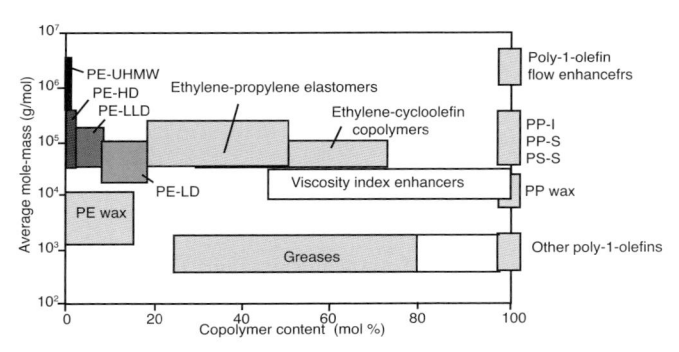

Figure 6.2: Applications for metallocene polyolefin resins.

6.1.1 Standard Polyethylene Homo- and Copolymers (PE-LD, PE-HD, PE-HD-HMW, PE-HD-UHMW, PE-LLD)

Polyethylenes (PE) are semicrystalline thermoplastic materials. Their structure, molecular weight, crystallinity, and thus their properties depend to a high degree on the polymerization method used. Equation 6.1 shows the basic structure of PE with possible side chain configurations, while Fig. 6.3 depicts the side chains for three different types of PE.

$$\text{Branched PE} \quad \left[-CH_2-\underset{\underset{(CH_2)_X}{|}}{\overset{\overset{CH_3}{|}}{C}}-(CH_2)_n-\underset{\underset{(CH_2)_Y}{|}}{\overset{\overset{H}{|}}{C}}-CH_2- \right]$$

$$\underset{CH_3}{}$$

(6.1)

The polymerization method determines the type and rate of the side chains. The relative molecular weight and its distribution can be influenced by high ther-

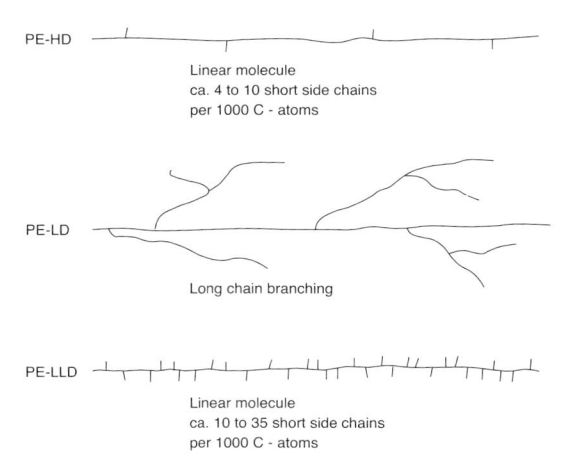

Figure 6.3: Branched PE configurations.

mal or mechanical stresses, which is typically avoided. The degree of crystallinity is determined by the structure of the polymer and the processing conditions. For industrial applications, ISO 1133 categorizes polyethylenes by their different densities, which depend on the respective degree of crystallinity of each specific PE grade. DIN EN ISO 1872 is used internationally and specifies PE resins mainly by density and melt flow rate (MFR); in addition, it uses a system of abbreviations indicating application, processing method, additives, fillers, and reinforcements. However, the characterization by these designations is inadequate in describing the properties and application ranges for the numerous PE grades available and is therefore rarely used in practice.

Polymerization, Chemical Constitution In general, polyethelenes are produced either by high-pressure processes in the presence of radicals (radical polymerization) or by medium- and low-pressure processes with the help of catalyst systems (anionic polymerization). In addition, polyethylenes are classified as suspension-, solution-, gas phase-, or mass polymerization grades depending on their state. High-pressure processes are used for highly branched homopolymers (LDPE), while medium- and low-pressure processes are used to synthesize linear homo- and copolymers (PE-HD, PE-MD, PE-LLD).

PE polymerization processes can be broken down as

- High-pressure process.polymerization,high pressure) PE-LD (LD = low density) is synthesized by a high pressure process (ICI 1939) from ethylene ($CH_2=CH_2$) under a pressure of 1000 to 3000 bar at 150 to 275°C with 0.05 to 0.1% oxygen or peroxides as catalysts; synthesis occurs either discontinuously in stirrer vessels or continuously in pipe reactors. The result is a highly branched PE with side chains of different lengths. Its crystallinity ranges from 40 to 50%, its density from 0.915 to 0.935 g/cm^3, and it has a molecular weight average up to 600,000 g/mol. With the help of high-performance catalyst systems, PE-LD synthesis equipment can also produce linear low density PE (PE-LLD).

- Medium- and low-pressure process.

PE-HD (HD = high density) is synthesized by either a medium-pressure (Phillips) or a low-pressure process (Ziegler), both of which are suspension processes. The Phillips method uses pressures from 30 to 40 bar, temperatures from 85 to 180°C, and chromium oxide or aluminum oxide as catalysts. The molecular weight obtained is $\approx$ 50,000 g/mol. The Ziegler method uses pressures from 1 to 50 bar, temperatures from 20 to 150°C, and titanium halides, titanium esters, or aluminum alkyls as catalysts, obtaining molecular weights of 200,000 - 400,00 mol/g. PE-HD is almost unbranched and therefore has a higher degree of crystallinity (60 - 80%) and higher densities (0.94 - 0.97 g/cm^3) than PE. PE-HD-HMW (HMW = high molecular weight) with a density of 0.942 - 0.954 g/cm^3 and PE-HD-UHMW (UHMW = ultra high molecular weight) with a density of 0.93 - 0.94 g/cm^3 offer high molecular weights together with their high densities. These resins are produced with special catalysts in a low-pressure process. The average molecular weight ranges from 200 - 500 kg/mol for PE-HD, from 500 - 1,000 kg/mol for PE-HD-HMW, and from 3,000 - 6,000 kg/mol for PE-HD-UHMW. PE-LLD (LLD = linear low density) is polymerized with high-efficiency catalysts (metal complexes) in four different processes: a low-pressure process in the gas phase, in solution, in suspension, or in a modified high-pressure process. Copolymerization of ethylene with 1-olefins, such as butylene-1 or hexane-1, creates short side chains. Compared to linear PE-HD, PE-LLD contains a higher ratio of co-monomer. The higher molecular weight and the low number of side chains lead to improved properties of these resins. PE-VLD (VLD = very low density) with a density of 0.905 - 0.915 g/cm^3 and PE-ULD (ULD - ultra low density) with a density of 0.890 - 0.905 g/cm^3 exhibit such a high degree of branching caused by their increased co-monomer content that their densities decrease below 0.915 g/cm^3. PE-(M) (polyethylene produced with metallocene catalysts) exhibits a narrow molecular weight distribution and can be produced in a broad range of densities. Although PE-MLLD is of the same composition as 'regular' PE-LLD, it shows a different sequence statistic. Transitional metal compounds activated by methyl aluminoxanes are used as catalysts, typically with Ti- or Zr-centers linked with cyclopentadienyl residuals.

The densities of PE-(M) grades range from

- PE-MLLD: 0.915 - 0.930 g/cm^3
- PE-MMD: 0.930 - 0.940 g/cm^3 (medium density)
- PE-MHD: 0.940 - 0.995 g/cm^3 (under development)
- In addition, PE-MVLD with 0.863 - 0.885 g/cm^3 (polyolefin-elastomer) and with 0.866 - 0.915 g/cm^3 (polyolefin-thermoplastics) can be produced.

Processing Processing of PE is non-critical. The broad variation of PE grades covers a wide range of processing conditions. Special grades with unique processing characteristics are used for particular applications and processes. For injection molding, the resin temperatures for PE-LD range from 160 - 260°C, for PE-HD from 260 - 300°C, and the mold temperatures range from 50 - 70°C and 30 - 70°C, respectively. Easy flowing PE grades are used for mass production. The density and the shrinkage of parts made of these semicrystalline resins are determined by the temperature history until demolding. Parts that were cooled quickly exhibit low crystallinity and minimal mold shrinkage but also a high degree of post-mold shrinkage, caused by post-crystallization at elevated temperatures. The result can be warpage and stress cracking caused by frozen-in stresses. This problem can be avoided with PE grades with lower melt flow rates. Gating brittleness, resulting from strong molecular orientation, can be alleviated by an increase in melt temperature or grades with the highest applicable melt flow rate. Only special grades of PE-HD-UHMW can be injection molded (parts up to 1 kg). Because of their low flow rates they require machines with high injection pressures ($\approx$ 1,100 bar), no backflow valves, grooved feed zones if possible, and short flow distances. Processing temperatures range from 240 - 300°C, mold temperatures range from 70 - 80°C. PE-LLD is more difficult to process than PE-LD. As a rule, screw driving horsepower should be increased, output decreased; shrinkage properties are less favorable. PE grades with higher melt viscosities and sufficient melt strength must be selected for blow molding so that the parisons do not tear under their own weight. Melt and mold temperatures range from 140°C for PE-LD to 160 - 190°C for PE-HD, respectively. PE-LLD is not well suited for blow molding because of its narrow molecular weight distribution and therefore its higher melt viscosity. It is well suited though for rotational molding. PE-LD is extruded at melt temperatures of 140 - 210°C (films and pipes), 230°C (cable sheathing), and 350°C (coating). PE-HD requires a temperature increase by 20 to 40 K. These grades are also used to produce boards and monofilaments. High-pressure plasticization (2,000 - 3,000 bar) in twin screw extruders and ram extruders is used for PE-HD-UDMW profiles. Co-rotating twin screw extruders, running at approximately 10 RPM, are also suited for profile extrusion of PE-HD-UHMW profiles (melt temperature 180 - 200°C). The extrusion of PE-LLD on equipment designed for PE-LD will result in a 20 - 30% reduction in output. This disadvantage results from the necessary reduction in screw length from 30 D to 25 - 20 D and from the rpm reduction by 50%. Measures to compensate for the loss in throughput include: using screws of larger diameter, increase of screw pitch, or increase in die gap. Optimum melt temperatures range from 210 - 235°C, for film extrusion from 250 - 280°C. By progressive, controlled orientation (30-fold) of fiber staple under conditions that lead to almost single-crystalline orientation of the crystallites, extremely tough reinforcement fibers are obtained with a strength of 1 - 5 GPa, modulus of elasticity from 50 - 150 GPa, and elongation at break of approximately 5%. Separation of PE from solution under shear results in cellulose-like fibers, so-called fibrides. The Neopolen process uses pre-foamed particles, obtained by quenching from a melt containing foaming agent, to form pellets or particles by steam sintering PE-E (E = expanded. Simple parts are formed by compression molding with 2 - 5 bar pressure. Conductive resins can easily be heated by a current flow. PE-HD and PE-LD can be pressed at 105 - 140°C. Powder techniques (rotational molding, fluidized bed sintering) use PE powders with particle diameters of 30 - 800 μm, molded densities of 0.92 - 0.95 g/cm^3, and low volume flow rates. PE grades with higher flow rates are used for carpet backing and in-mold decoration fabrics. Precipitated PE powders with uniform particle sizes of $\approx$ 50 μm are suitable for electrostatic coating of metals and fabrics. Still finer powders (8 - 30 μm) are dispersed in the beater in papermaking or in printing inks.

Post-Processing Treatment PE is easily weldable with standard welding techniques: heated element-, friction-, hot gas-, ultrasonic-, and extrusion welding. Induction welding of conductive grades is possible; however, high-frequency welding cannot be used because of the non-polar character of polyolefins. Adhesive joining and decorating of PE is difficult because of its non-polar character and low solubility. By grafting PE-LLD with maleic anhydride (see Table 6.9, No. 31), excellent bonding strength and high heat distortion stability can be achieved. Parts made from plasma-treated PE-MLLD powder can be coated with water-based lacquer or foamed with PU foam without pre-treatment when it is processed by rotational molding. During machining of PE parts it is important to make sure that the material does not get too hot, because it tends to become tacky due to frictional heating. Parts made of soft grades or with thin walls are easy to stamp.

Properties Numerous grades of PE with widely differing properties can be produced by homo- and copolymerization and by the creation of low, medium, and high densities, low, medium, high, and ultra-high molecular weights, or narrow and broad molecular weight distributions, respectively. Low-molecular weight PE is used as an aid in plastics processing. Compared to other polymeric materials, all high-molecular weight polyolefins share low densities, relatively

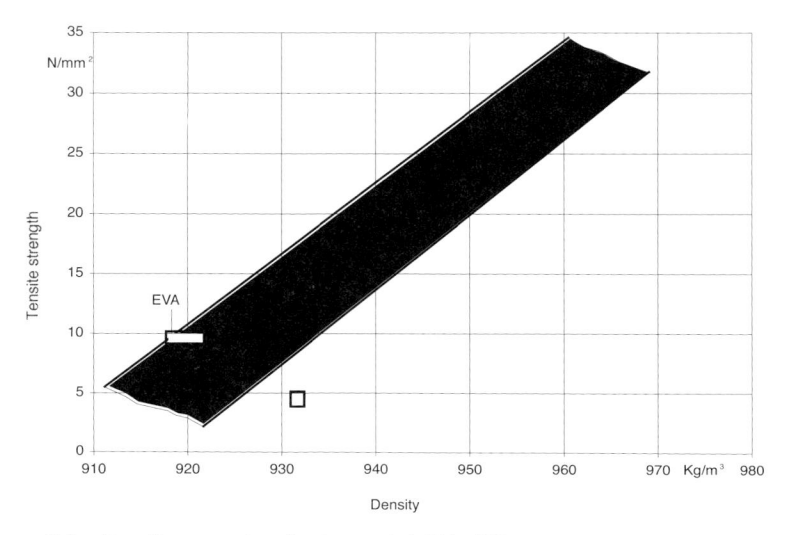

Figure 6.4: Density versus tensile stress at yield for PE.

low strength and stiffness, high toughness and elongation at break, good friction and wear behavior, and very good electrical and dielectrical properties (PE is non-polar). The tensile stress at yield increases ≈ linearly with density (see Fig. 6.4). Water absorption and water vapor permeability are low. Permeation of oxygen, carbonic acid, and many aromatic and odiferous agents is substantial, but it decreases with increasing density. The maximum allowable temperatures for short-time exposure range from 80 - 120°C (PE-LD-UHMW can be used from -268°C to 150°C, short-term) depending on grade; continuous service temperatures range from 60 - 95°C (PE-LD-UHMW: 100°C).

PE is resistant to water, saline solution, acids, alkalis, alcohols, and gasoline. Below 60°C, PE is insoluble in all organic solvents, but it increasingly swells in aliphatic and aromatic hydrocarbons with decreasing density. Individual grades of high-density PE have been approved as containers for heating oil, gasoline, and vehicle fuel tanks. If the internal surfaces have been fluorinated (by fluorine-nitrogen blends) or sulfonated (by SO_3), these containers are impermeable to all kinds of fuels and hydrocarbons. PE is not resistant to strong oxidizing agents, such as fuming sulfuric acid, concentrated nitric acid, nitration acid, chromium-sulfuric acid, and halogens as well as to some cleaning agents. Surfactants (washing and wetting agents) can induce environmental stress cracking in PE parts. PE-LLD and all PE grades with densities of ≈ 0.90 g/cm^3 provide the highest stress cracking resistance. Addition of carbon black protects

PE against photo-oxidation. PE can be crosslinked by high-energy radiation. PE burns like wax, flame-retardant types, e.g., for applications in construction, are available. PE is odorless, tasteless, and physiologically inert. Most PE grades meet the current FDA guidelines for use in contact with food. The property dependence on structural parameters (density, molecular weight, and molecular weight distribution) is summarized in Table 6.6.

Table 6.6: Structural Parameters and Properties of PE

Structural parameter	Density g/cm³		Molecular configuration	
Limiting values	0.915	0.97	Branched	Linear
Degree of crystallization	-/+	++	−	++
MFI	0	0	0	0
Processability	+	-	+	-
Tensile and flexural strength	→	→	→	→
Elongation at break	←	←	←	←
Stiffness and hardness	→	→	↔	↔
Impact strength	→	→	→	→
Stress cracking resistance	←	←	→	→
T_m and HDT	→	→	→	→
Low temperature range	→	→	→	→
Chemical and resistance	→	→	→	→
Permeability	→	→	→	→
Transparency	←	←	←	←

Limiting values	Low 20000 to 60000	High 20000	Narrow	Broad
Degree of crystallization	-	+	+	-
MFI	++	−	0	0
Processability	+	-	-	+
Tensile and flexural strength	→	→	←	←
Elongation at break	→	→	←	←
Stiffness and hardness	0	0	←	←
Impact strength	→	→	←	←
Stress cracking resistance	→	→	←	←
T_P and HDT	→	→	←	←
Low temperature range	→	→	←	←
Chemical resistance	→	→	0	0
Permeability	0	0	0	0
Transparency	0	0	0	0

+,-: high and low values, respectively
→ : arrow indicates directionof increasing positive effect
0: no significant influence

Metallocene PE grades (PE-M) distinguish themselves from conventional PE grades by a combination of properties: Polyolefin-Thermoplastics: a new polymer class with high comonomer content offers improved penetration resistance, gloss, sealing capability, oxygen- and hydrogen permeability (packaging of fresh foods). PE-MD/HD-(M): Modification of type and length of comonomer allows for tailoring of properties, such as toughness, stiffness, strength, and optical and organoleptic properties to meet particular requirements. Polyolefin-Elastomers (also as comonomers in TPO and PP) offer the following advantages compared to conventional polymers (in parantheses): thermoplastic processing (crosslinked elastomers), shelf life (TPSA), transparency, flexibility, and crack resistance (EVA, EMA), processability, cost effectiveness, and environmental compatibility (PVC-P), and after crosslinking: long-term and temperature resitance (EPDM, IIR, SBR, NR).

PE properties can be customized for specific fields of application by the use of additives: glass fibers are used to improve rigidity and strength, anti-oxidants and UV-stabilizers for outdoor use, flame-retardants, foaming agents, anti-static agents, carbon black and other means to enhance electric conductivity (EMI, electromagnetic interference), slip agents, small amounts of fluoroelastomers to enhance melt strength and flowability, and pigments for coloring. A comparison of properties is provided in Table 6.7.

Table 6.7: Comparison of Properties for Polyethylenes

Properties	Unit	Polyethylene		
		PE-LD	PE-MD	PE-HD
ρ	g/cm^3	0.915-0.92	0.925-0.93	0.94-0.96
E_t	MPa	200-400	400-800	600-1400
σ_y	MPa	8-10	11-18	18-30
ϵ_y	%	≈ 20	10-15	8-12
ϵ_{tB}	%	>50	>50	>50
σ_{50}	MPa	-	-	-
σ_B	MPa	-	-	-
ϵ_B	%	-	-	-
T_P	°C	105-118	120-125	126-135
HDT	°C	-	30-37	38-50
α_p	10^{-5}/K	23-25	18-23	14-18
α_n	10^{-5}/K	-	-	-
UL94	Class	HB*	HB*	HB*
ϵ_r 100	-	2.3	2.3	≈ 2.4
tan δ 100	$\cdot 10^{-3}$	2-2.4	2	1-2
ρ_e	Ohm $\cdot$ m	>10^{15}	>10^{15}	>10^{15}
σ_e	Ohm	>10^{13}	>10^{13}	>10^{13}
E_BI	kV/mm	30-40	30-40	30-40
W$_w$	%	<0.05	<0.05	<0.05
W$_H$	%	<0.05	<0.05	<0.05
Properties	Unit	Polyethylene		
		PE-UHMW	PE-LLD	PE-(M)**
ρ	g/cm^3	0.93-0.94	≈ 0.935	0.904
E_t	MPa	700-800	300-700	75
σ_y	MPa	≈ 22	20-30	7
ϵ_y	%	≈ 15	≈ 15	-

Continued on next page

6.1 Polyolefins (PO), Polyolefin Derivates, and Copoplymers

Properties	Unit	Polyethylene		
		PE-UHMW	PE-LLD	PE-(M)**
ϵ_{tB}	%	>50	>50	>50
σ_{50}	MPa	-	-	-
σ_B	MPa	-	-	-
ϵ_B	%	-	-	-
T_P	°C	130-135	126	100
HDT	°C	42-49	≈ 40	-
α_p	10^{-5}/K	15-20	18-20	-
α_n	10^{-5}/K	-	-	-
UL94	Class	HB*	HB*	-
ϵ_r 100	-	2-2.4	2.3	2.3
tan δ 100	$\cdot 10^{-3}$	≈ 2	2	-
ρ_e	Ohm · m	$>10^{15}$	$>10^{15}$	$2 \cdot 10^4$
σ_e	Ohm	$>10^{13}$	$>10^{13}$	-
E_BI	kV/mm	30-40	30-40	-
W_w	%	<0.05	<0.05	-
W_H	%	<0.05	<0.05	-

* also available from V-2 to V-0 ** produced with metallocene catalysts

Applications Table 6.8 gives an overview of PE use in injection molding.

Table 6.8: Processing Properties and Applications for PE Injection Molding Resins

Density	0.92 g/cm³	0.93 g/cm³	0.94 g/cm³ [2]
MFR 190/2,16:>25 to 15[1]	Easy flow; mass products, not exposed to particular stress	Easy flow; large surface parts, little warpage, good gloss	Easy flow; impact resistant parts, no particular requirements on stiffness
15 to 5	Parts with higher strength, little surface gloss	Low stress parts with good surface gloss	Good impact resistance, little susceptible to stress corrosion, high performance technical parts
≈ 1.	Very good mechanical properties and resistance to stress corrosion		Good creep resistance, little susceptible to stress corrosion, particularly stressed closures
Density	0.95 g/cm³	0.96 g/cm³	
MFR 190/2,16:>25 to 15[1]	Easy flow; little warpage, difficult to injection mold household	Easy flow; hard and stiff; bowls, sieves, dishes, bottle cases, hard hats	
15 to 5	Easily processed, good impact strength, screw caps, closures	Impact resistant, dimensionally stable, mechanically highly stressed parts	

Continued on next page

Density	0.95 g/cm^3	0.96 g/cm^3
≈ 1.	Resistant to stress corrosion, good surface quality, high performance technical parts	
<1	Highly molecular, usually highly stabilized; pressure valves, pipe fittings	

[1] extremely free-flowing PE-LD and -HD grades are aviable with MFR > 100
[2] often PE-LD/HD blends; similar applications as for PE-LLD of low density

PE-LD: Main application area are films for packaging, films for heavy-duty bags and sacks, shrink film, carrying bags, agricultural films, water vapor barriers in composite films, in which PE-LD copolymers are used as coupling agents with, e.g., EVA, EAA, or EEA; pipes; boards for thermoforming; sheathing for wires (also foamed and crosslinked PE); coating of steel pipes; flexible containers and bottles; canisters up the 60lcapacity; and tanks up to 200lcapacity. Blends with PE-LLD produce stretched films with higher extensibility. PE-LLD: Films (extendable to 5 μm) with good optical properties, improved strength at low temperatures, resistance to tearing and penetration, and lower tendency for stress cracking than films made from PE-LD. Blown film made from blends with PE-LD (also as composite films) are increasingly substituting PE-LLD in rotational molding of canisters, canoes, and surfboards. PE-HD: Housewares, storage and transportation containers, trash cans and trash containers up to 1,100lcapacity, bottle cases, gasoline canisters, automotive fuel tanks. Special grades are used for high-pressure pipes, fittings for drinking water supply and waste water disposal (extruded diameters up to 1,600 mm, even larger diameters for rolled pipes), boards (also glass fiber reinforced) for devices for the chemical industry and in the automotive industry (lower cover for engine compartment). Fibrides are used as hydrophobic but hydrocarbon-binding additives to patching compounds for the absorption of oils, and as reinforcement for paper, fibers, and highly strengthened reinforcement fibers. PE-HD-HMW: Surfboards up to 5 m in length; monofilaments for nets, cables, cords, ropes, and woven fabric; packaging film with a reduced thickness of 10 - 7 μm (from 20 μm) for carrying bags and with a thickness from 80 - 120 μm for inner linings of paper bags; sealing liners for landfills. PE-HD-UHMW: Pressed blocks and filter-pressed boards. Because of its excellent wear behavior it is used to line hoppers and chutes for abrasive materials, machine elements such as screw conveyors, pump and slide elements, pulleys, gears, bushings, and rolls; sintered spacers with a porous core for lead batteries, surgical implants, prostheses, anti friction coatings for skis. PE-(M): Packaging films, heat sealing coatings, surface protection films, highly flexible gaskets and covers.

6.1.2 Polyethylene Derivates (PE-X, PE + PSAC)

Cross-linked PE (PE-X)   By 3-dimensional cross-linking the linear PE macromolecules (PE-X), creep resistance, low temperature impact resistance, and stress cracking resistance are considerably improved, while hardness and stiffness are slightly reduced. Because PE-X behaves like an elastomer and will not melt, it can withstand higher thermal loads: short-term without additional mechanical loads up to 250°C, long-term up to 120°C. With increasing crosslinking density, the shear modulus increases even at elevated temperatures. There are four different types of crosslinking, which are all used for extrusion: peroxide-(A)-, silane-(B)-, electron beam-(C)-, and azo-(D)-crosslinking. In injection molding, PE-HD is processed with peroxide crosslinking agents in the barrel within a precisely defined temperature range of 130 - 160°C and then crosslinked in the mold at 200 - 230°C. Injection molded and blow molded parts can also be crosslinked by irradiation.

Peroxide cross-linking (PE-XA)   With the Engel process, a twin high-pressure plunger, continuously conveying machine sinters a pipe from a grit-like PE-HD/peroxide mixture (see Eq. 6.39), subsequently melts it in a heating barrel, molds it into its final shape, and crosslinks it at temperatures ranging from 200 - 250°C, above the crystallization temperature. This creates a uniform network of macromolecules with low stresses and high flexibility. The material is softer and tougher than electron crosslinked PE-X. In a modified process, the PAM (Pont-a-Mousson) process, the extruded pipe is crosslinked in the sizing zone in a hot salt bath. The two-step Daopex process uses normal extruded PE-LD pipes and crosslinks them from the outside by exposing them to an epoxide emulsion under pressure and at temperatures above the crystallite melting temperature of PE. DIN 163 892 specifies a degree of crosslinking of more than 75% for pipes made of PE-XA. The Engel process achieves crosslinking degrees up to 99.5%.

Silane cross-linking (PE-XB)   With the Sioplas-, Hydro-Cure-, Monosil-, and Spherisil processes, silane-grafted PE compounds are processed with the addition of a silane crosslinking catalyst. The crosslinking process is started by a hot water-pressure treatment with the formation od Si-O-Si bridges.

Electron beam cross-linking (PE-XC)   PE parts made without crosslinking aids are crosslinked in special irradiation equipment at temperatures below the crystallization point under separation of hydrogen. Radiation sources are electron beam accelerators or isotopes (β- or γ-rays). The penetration depth can be adjusted (β-rays up to 10 mm, γ-rays up to 100 mm) so that the core of thick walls or specified areas of a part will remain uncrosslinked. The crosslinking of PE-HD compounded with an inhibited version of butyl perbenzoate is initiated immediately after extrusion or molding, respectively, by UHF radiation.

Azo cross-linked (PE-XD) Azo compounds are added to PE, which will crosslink in a subsequent hot salt bath with nitrogen cross linkages (Lubonyl process). Applications fields for PE-X are medium- and high-voltage cable coatings, pipes for hot water and radiant heating installation, parts for electrical engineering, chemical engineering, and automotive applications. PE-X can be joined by hot plate welding; application: pipes.

Degradable PE (PE+PSAC) Compounds of biologically degradable polymers (polysaccharides, starch, PSAC) with conventional, non-degradable polymers (PE, up to 94%) are bio-degradable. Higher ratios of starch will cause processing problems. Generally, a starch-PE-pigment masterbatch is extruded with PE; applications in particular in packaging. The PE-enclosed starch particles are bio-degraded by the diffusion of moisture. However, the PE matrix is not bio-degradable. Photo degradation of polymers can be controlled fairly accurately by the integration of UV-sensitive molecules, such as keto-groups (ECO-copolymer), or by blending with photo-sensitizers (iron dialkylthiocarbamate). These products are not bio-degradable. Today's applications for photo-degradable polymers are focused on agro-films, carrier bags, and trash bags.

6.1.3 Chlorinated and Chloro-Sulfonated PE (PE-C, CSM)

Chlorination of polyolefins is achieved in solution, dispersion, or by direct exposure to chlorine gas. PE-C with a chlorine content of 25 - 30% is pliant to rubber-like soft and low temperature resistant. Applications: Because it is miscible with many polymers, PE-C is blended with polyolefins to reduce flammability, with PVC to increase impact resistance, with PS for recycled commingled compounds. PE-C+PVC compounds with 70 - 90% PE-C content are used as unplasticized and bitumen resistant roofing and water-proofing material, films for liners, and profiles. There are several methods to crosslink PE-C to become PE-C elastomers. Typically, PE-C is vulcanized by peroxides, because they provide higher valence forces than sulfur- or radiation crosslinking. The favorable price/property ratio has made PE-C elastomers competitive for cable applications and in the rubber industry with CSM (chloro-sulfonated PE), CR (polychloroprene), EPDM (ethylene propylene diene rubber), and NBR (acrylonitrile butadiene copolymer). Its outstanding properties include: resistance to aging, weather and ozone resistance, flame resistance, oil resistance (even at elevated temperatures), wear resistance, ductile-brittle transition temperature, and good processability. Treatment of PE in chlorinated hydrocarbon solution (PE-usually LD) with SO_2 and chlorine gas and simultaneous radiation with UV or other high energy rays (e.g., a Co^{60} source, see Eq. 6.2) yields chloro-sulfonated PE (CSM). Vulcanization is carried out with MgO or PbO and activators. Applications: liners for containers in transportation and in the chemical industry, cable sheathing (good oxidation- and ozone resistance), and in blends with other rubbers for coatings.

$$\left[-|(CH_2)_6-CH|_{12}-\underset{\underset{Cl}{\overset{\overset{|}{O=S=O}}{|}}}{\overset{|}{C}}- \right]_n$$

Chlorosulfonated PE, CSM

(6.2)

6.1.4 Ethylene Copolymers (ULDPE, EVAC, EVAL, EEAK, EB, EBA, EMA, EAA, E/P, EIM, COC, ECB, ETFE

Copolymerization of ethylene with propene, butene-1, vinyl acetate, acrylic acid ester, carbon monoxide, among others, interrupts the linear structure of the methylene chain and thus reduces the degree of crystallinity (for the chemical structure of the monomers, see Table 6.9). This results in a reduced melting temperature.

The intra-molecular forces and the glass transition temperature depend on the type and polarity of the comonomer. Table 6.10 provides a comparison of properties. Polar copolymers are exclusively produced with high-pressure processes. Copolymerization of PE with non-polar monomers, such as butene, hexene, and other higher α-olefins to produce linear PE of lower densities is carried out in low-pressure processes in gas phase-fluidized beds or in solution.

Ultra-light Polyethylene (PE-ULD, PE-VLD) PE-ULD and PE-VLD are co- and ter-poplymers of ethylene with up to 10% octene-, 4-methylpentene-1, and sometimes propylene with densities ranging from 0.91 - 0.89 g/cm^3. They have a low degree of crystallinity, are transparent, and, with an elongation at break of > 900%, they are flexible in a wide temperature range. Applications: penetration resistant stretch film for heavy loads, films for heavy-cargo bags, (multi-layer) barrier films and medical packaging, and parts manufactured by high-speed injection molding. These easily processable materials are also used to improve elastic properties and stress cracking of other PE grades. Butyl acrylate modified PE is similar to PE-LLD.

Table 6.9: Basic Structures of Olefin- and Vinyl Polymers and Their Copolymers (Terpolymers with PS, see Table 6.12)

$$-[\underset{\underset{H}{|}}{\overset{\overset{H}{|}}{C}} - \underset{\underset{R_2}{|}}{\overset{\overset{R_1}{|}}{C}}]-$$ General Structure of olefins and vinyls

	Chemical	Acronym	R_1	R_2	Copolymerized with
(1)	Ethylene	(PE)	-H	-H	Vinyl chloride:VCE
(2)	Propylene	(PP)	-H	-CH$_3$	Ethylene: E/P
(3)	Butene-1	(PB)	-H	-CH$_2$-CH$_3$	
(4)	Octene		-H	-(CH$_2$)$_5$-CH$_3$	
(5)	Styrene	(PS)	-H	–⟨phenyl⟩	
(6)	α-Methylstyrene	(PMS)	-CH$_3$	–⟨phenyl⟩	Styrene: SMS
(7)	Paramethylstyrene		-H	–⟨phenyl⟩–CH$_3$	
(8)	Vinyl chloride	(PVC)	-H	-Cl	
(9)	Chlorinated ethylene	(PE-C)	-H	-Cl	Vinyl chloride: VCPE-C
(10)	Vinylidene chloride	(PVDC)	-Cl	-Cl	Vinyl chloride: VCVDC
(11)	Acrylate (Acryl acid)	(PAA)	-H	-C=O, O-H	Ethylene: EAA
(12)	Methacryl acid	(PMA)	-CH$_3$	-C=O, O-H	Vinyl chloride: VCMAK
(13)	Methacrylate	(EMA)	-H	-C=O, O-CH$_3$	Ethylene: EMA
(14)	Methyl-methacrylate	(PMMA)	-CH$_3$	-C=O, O-CH$_3$	Styrene: SMMA, Vinyl chloride: VCMMA
(15)	Ethyl acrylate	(EA)	-H	-C=O, O-CH$_2$-CH$_3$	Ethylene: EEAK
(16)	Ethyl methacrylate		-CH$_3$	-C=O, O-CH$_2$-CH$_3$	
(17)	Butyl acrylate	(PBA)	-H	-C=O, O-(CH$_2$)$_3$-CH$_3$	Ethylene: EBA
(18)	Butyl methacrylate		-CH$_3$	-C=O, O-(CH$_2$)$_3$-CH$_3$	
(19)	Octyl acrylate		-H	-C=O, O-(CH$_2$)$_7$-CH$_3$	Vinyl chloride: VCOA
(20)	Vinyl acetate	(PVAC)	-H	-O-C=O, CH$_3$	Ethylene: EVAC, Vinyl chloride: VCVAC
(21)	Vinyl alcohol	(PVAL)	-H	-OH	Ethylene: EVAL
(22)	Vinyl ether		-H	-O-R, R= -CH$_3$, -CH$_2$-CH$_3$, etc.	
(23)	Vinyl methyl ether	(PVME)	-H	-O-CH$_3$	

Continued on next page

# 6.1 Polyolefins (PO), Polyolefin Derivates, and Copoplymers

Table 6.9: Basic Structures of Olefin- and Vinyl Polymers and Their Copolymers (Terpolymers with PS, see Table 6.12)

$$-[\overset{H}{\underset{H}{C}}-\overset{R_1}{\underset{R_2}{C}}]-$$ General Structure of olefins and vinyls

	Chemical	Acronym	R_1	R_2	Copolymerized with
(24)	Vinyl pyrrolidone	(PVP)	-H	$\begin{array}{c}\text{N}\\ \text{CH}_2\quad\text{C}=\text{O}\\ \text{CH}_2-\text{CH}_2\end{array}$	
(25)	Vinyl carbonzole	(PVK)	-H	N-(C$_6$H$_4$)$_2$	Ethylene: E/P
(26)	Acrylonitrile	(PAN)	-H	$-C\equiv N$	Styrene: SAN
(27)	4-Methyl pentene-1	(PMP)	-H	$-CH_2-CH\begin{array}{c}CH_3\\CH_3\end{array}$	
(28)	Vinyl butyral*	(PVB)	$-[CH_2-CH-CH_2-CH]_m-[CH_2-CH]_n-$ $\quad\quad\quad\backslash\quad\quad\quad\quad/\quad\quad\quad\quad\backslash$ $\quad\quad\quad O-CH-O\quad\quad\quad\quad OH$ $\quad\quad\quad\quad\quad\mid$ $\quad\quad\quad\quad(CH_2)_3$			
(29)	Vinyl formal*	(PVFM)	$-[CH_2-CH-CH_2-CH]_m-[CH_2-CH]_n-$ $\quad\quad\quad\backslash\quad\quad\quad\quad/\quad\quad\quad\quad\backslash$ $\quad\quad\quad O-CH_2-O\quad\quad\quad OH$			
(30)	Itaconic acid ester		$\begin{array}{c}-O-C=O\\ \mid\\ CH_3\end{array}$	$\begin{array}{c}-C=O\\ \mid\\ O-CH_2-CH_3\end{array}$	
(31)	Maleic acid anhydride*		$-[CH-CH]-$ $\quad/\quad\quad\backslash$ $CO-O-CO$		Styrene: SMAH, Vinyl chloride: VCMAH
(32)	Maleic(acid)imide*		$-[CH-CH]-\quad R=-CH_3$ $\quad/\quad\quad\backslash\quad\quad-[CH_2]_3-CH_3$ $CO-N-CO\quad\quad\text{etc.}$ $\quad\quad\mid$ $\quad\quad R$		Vinyl chloride: VCPE-C
(33)	Acrylic acid ester-elastomer	(ACM)	-H	$\begin{array}{c}-C-O-R\\ \parallel\\ O\end{array}$	$R=-CH_3$ $-CH_2-CH_3$ $-[CH_2]_3-CH_3$ etc.
(34)	EPDM-rubber*	(EPDM)	$-[CH_2-CH_2]_m-[CH_2-CH]_n-[CH_2-CH]_m-$ $\quad\quad\quad\quad\quad\quad\quad\quad\mid\quad\quad\quad\quad\quad\quad\mid$ $\quad\quad\quad\quad\quad\quad\quad\quad CH_3\quad\quad\quad\quad CH_2-CH=CH-CH_3$			Styrene: SEPDM
(35)	Butadiene rubber*	(IR)	$-[CH_2-CH=CH-CH_2]-$			Styrene: SBR
(36)	Isoprene rubber*	(EMA)	-H	$-[CH_2-C=CH-CH_2]-$ $\quad\quad\quad\mid$ $\quad\quad\quad CH_3$	
(37)	Thermoplastic PUR elastomer*	(TPU)		-[R-NH-COOR]-	
(38)	Ethene butene			$CH_3-CH-CH=CH_2$ $\quad\quad\mid$ $-[HC-CH_2]-$	
(39)	Ethene propene			$H_3C-C=CH_2$ $\quad\mid$ $-[HC-CH_2]-$	

* Full structure shown here

Table 6.10: Comparison of Properties for Ethylene Copolymers and Other Polyolefins

Property	Unit	EVAC	EIM Ionomer	COC 52% Norbornene	PDCPD
ρ	g/cm^3	0.93-0.94	0.94-0.95	1.02	0.93-0.94
E_t	MPa	30-100	150-200	2600-3200	1800-2400
σ_y	MPa	-	7-8	-	
ϵ_y	%	-	>20	2-5	4
ϵ_{tB}	%	>50	>50	-	
σ_{50}	MPa	4-9	-	-	
σ_B	MPa	-	-	46-66	46
ϵ_B	%	-	-	2-10	25
T_P	°C	90-110	95-110	80-180	
HDT	°C	-	-	75-170	90-115
α_p	10^{-5}/K	≈25	10-15	≈6	8.2
α_n	10^{-5}/K	-	-	-	-
UL94	Class	HB	HB	HB	
$\epsilon_r 100$	-	2.5-3	≈2.4	≈2.4	
$\tan \delta 100$	$\cdot 10^{-3}$	20-40	≈30	-	
ρ_e	Ohm·m	>10^{14}	>10^{15}	>10^{14}	
σ_e	Ohm	>10^{13}	>10^{13}	-	
$E_B I$	kV/mm	30-35	40	-	
W_w	%	<0.4	≈0.5	<0.01	
W_H	%	<0.2	≈0.3	-	

Property	Unit	EA	PB	PMP
ρ	g/cm^3	≈0.935	0.90-0.915-0	0.83-0.84
E_t	MPa	40-130	210-260 (420	1200-2000
σ_y	MPa	4-7	15-25	10-15
ϵ_y	%	>20	≈10	>10
ϵ_{tB}	%	>50	>50	>10
σ_{50}	MPa	-	-	-
σ_B	MPa	-	-	-
ϵ_B	%	-	-	-
T_P	°C	92-103	125-130	230-240
HDT	°C	-	55-60	40
α_p	10^{-5}/K	≈20	13	12
α_n	10^{-5}/K	-	-	-
UL94	Class	HB	HB	HB
$\epsilon_r 100$	-	2.5-3	2.5	2,1
$\tan \delta 100$	$\cdot 10^{-3}$	30-130	2-5	≈2
ρ_e	Ohm·m	>10^{14}	>10^{14}	>10^{14}
σ_e	Ohm	>10^{13}	>10^{13}	>10^{13}
$E_B I$	kV/mm	30-40	20-40	
W_w	%	<0.4	<0.1	<0.01
W_H	%	<0.2	<0.05	<0.05

6.1 Polyolefins (PO), Polyolefin Derivates, and Copoplymers

Ethylene Vinyl Acetate Copolymers (EVAC) Gas permeability of PE increases with increasing vinyl acetate content and the materials become softer because the crystallinity decreases, see Figs. 6.5. and 6.6. Table 6.11 shows the properties and applications of copoplymers with varying VAC content. Resins with more than 10% VA are more transparent, tougher, and easier to heat-seal than PE-LD; they are approved for use in food packaging. EVAC copolymers are grafted with vinyl chloride (VC), which makes them easier to process as plasticizers and impact modifiers. EVAC is processed similar to PE-LD; however, the melt temperature depends on the VA content: 175 - 220°C and mold temperatures of 20 - 40°C. Extrusion temperatures range from 140 - 180°C (for flat film up to 225°C). To avoid separation of acetic acid, the melt temperature should not exceed 230°C and residence time should be kept short. Flushing with PE-LD is recommended after production stops. Printing and coating of EVAC is easier than of PE because of its polar structure. Pretreatment, usually with corona discharge, improves bonding strength.

Table 6.11: Properties and Applications of EVAC Copolymers with Various VAC Content

VAC-content of copolymers (wt.-%)	Properties and applications
1 to 10	More transparent, flexible and tough compared to PE-LD (heavy-duty films, freezer packaging), easier to seal (bags, composite films), little susceptible to stress cracking (cable sheathing), increased shrinkage at lower temperatures (shrink films), less relaxation of pre-stretched films (stretch films)
15 to 30	Still processable like a thermoplastic, very flexible and soft, rubber-like (applications similar to PVC-P, in particular for closures, seals, carbon black filled compounds for cable industry)
30 to 40	High elastic elongation, softness, able to carry filler load, wide softening range, polymers with good rigidity and good adhesion for coatings and adhesives
40 to 50	Products still exhibiting strong rubber properties (peroxidic and radiation crosslinked), e.g., for cables; for grafting, e.g., for high impact modified PVC with excellent weatherability; polymers resulting from hydrolysis for coating of woven materials, melt adhesives, thermoplastic processing for parts and films with high rigidity and toughness
70 to 95	Used as latices for emulsion paints, paper coating, adhesives, and saponification products for films and special plastics

Applications: Flexible pipe, profiles, wire- and cable coating, bags, gaskets, dust- and anesthesia masks, packaging and green house films, foldable and deformable toys, material for compounds. EVAC (and ethylene/acrylic acid acrylate, EAA) are processed with paraffins, waxes, and synthetic resins and sometimes crosslinked with peroxides to manufacture polishing compounds, hot melt

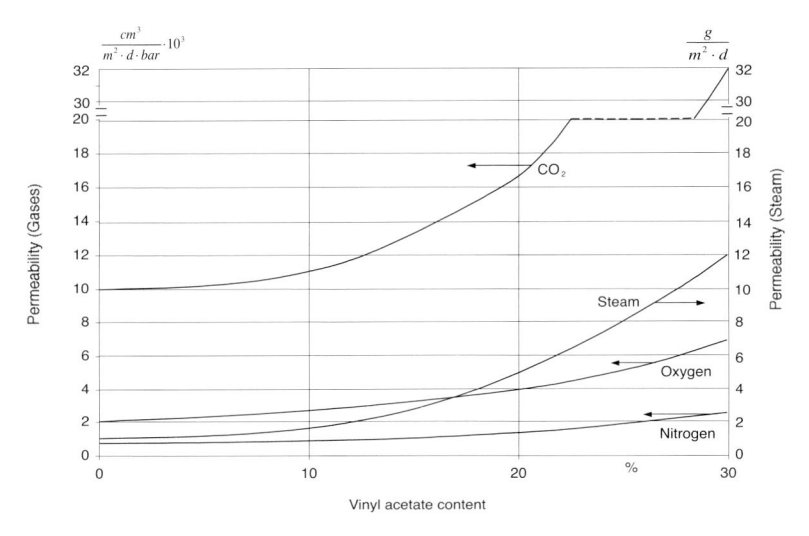

Figure 6.5: Water vapor and gas permeability of EVAC versus VAC content.

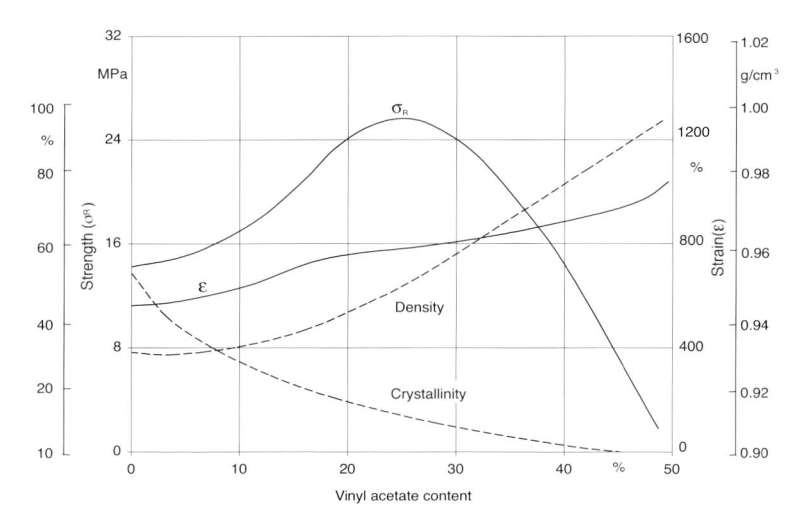

Figure 6.6: Properties of EVAC copoplymers versus VAC content.

adhesives, and coatings. Modified and partially saponified EVAC is used as interlayer for compound glass.

Ethylene/Vinyl Alcohol Copolymers (EVAL) Copolymers containing vinyl alcohol are produced by partial saponification of EVAC, just like poly vinyl alcohols (PVAL). EVAL containing between 24 - 30% VAL are available as powders for impact resistant electrostatic coatings (also adhering to metals) for corrosion and chip protection. Grades with VAL contents of 53 - 68% provide excellent barrier properties against N_2, O_2, CO_2, and fragrances but show considerable water vapor permeability and -absorption. EVAL with water contents of 3 - 8% loses its barrier properties and is therefore coextruded in multiple-layer films between PE, PP, and also together with PA or polyterephthalates. Glass transition temperature: 66°C, optimum processing temperatures range from 160 - 180°C. Short-term exposure to temperatures up to 200°C can be tolerated; however, the barrier effect against O_2 will be reversibly reduced. The reduction in barrier effect caused by boiling composite film packages is also reversed after cooling. PVALs are semi crystalline, water soluble polymers. Saponified grades dissolve at acceptable rates only in hot water. Partially saponified grades dissolve easier in cold than in hot water. Both PVAL types are used for the manufacturing of films: monofilms for watersolubable bags and as coextruded film because of their extremely low O_2 permeability.

Ethylene-Acrylic Copoplymers (EEAK, EBA, EAA, EAMA, EMA) Ethyl acrylate- (EA), butylacrylate-(BA), and methacrylate-(MA) copolymers with PE are used at low temperatures for high-elasticity, stress crack resistant, highly filled packaging films and hot sealing layers. EMA with up to 20% and EEAK with up to 8% comonomer content are approved for food contact applications. Carbon black-filled, semi conductive films and tubes are used for microchip packaging, explosive packaging, and medical and other applications where static electricity is a hazard. EAA, EAMA, and EMA, ethylene-meth(acrylic acid) copolymers, and terpolymers modified by acrylamide are used as intermediate bonding layers in multi-layer film, e.g., between PE and PA, and coextruded for metal coating. They are basic components of ionomers and useful as compatibilizers in blends, for example, of PE+PET. Applications: EAA see EVAC.

PE-α-Olefin Copolymers (PEα-PO-(M)) Copolymers of PE and α-olefins produced with metallocene catalysts (PE-mLLD) are polymers with a linear structure and narrow molecular weight distribution. They exhibit low densities (0.903 - 0.917 g/cm^3), favorable optical properties, and high toughness. Applications: Film in medical technology and food industry.

Cycloolefin Copolymers (COC, COP) COCs are copolymers of linear (ethene) and cyclic olefins (technically usually norbornene). They can be produced with metallocene catalysts in two categories: random (amorphous) and alternating (semicrystalline) polymers, the latter exhibiting better solvent and chemical resistance. The property range of COC can be varied in a broad range during polymerization. For example, the glass transition temperature of COC can be adjusted from 0°C for 12 mol-% to 230°C for 80 mol-% norbornene content. COC polymers are highly transparent, blood compatible, sterilizable, metalizable, easy processable, and hydrolysis resistant. Injection molded parts show very little optical anisotropy, the index of refraction is 1.53, the density 1.02 g/cm^3. Applications: CDs, other optical data storage media, optical, pre-filled transparent syringes, biaxially stretched condenser films, H_2O barrier films, blister films, and cosmetics. Table 6.10 provides a comparison of properties. Polymerization and subsequent hydrogenation of dicyclopentadiene produces an amorphous COC copolymer (COP), with high transparency, low birefringence, good chemical resistance, low moisture absorption, and temperature resistance from 120 - 130°C (new developments: T_g = 100 - 165°C). Applications: Medical devices, such as syringes, and blister film, LCD, optical projection plates, condenser films.

Ionomer Copolymers (EIM) Other than with conventional plastics, there are ionic bonds (electrostatic bonds) between the molecular chains, in addition to the secondary valence bonds. Ion contents up to 10% in otherwise non-polar materials define ionomers; materials with higher ion contents are called polyelectrolytes. Inonomers are typically synthesized by copolymerization of a functionalized monomer (e.g., acrylic acid, methacrylic acid, or p-styrene sulfonic acid) with an olefinic monomer and subsequent salt formation. An important group are thermoplastic copolymers of ethylene with monomers containing carboxylic groups, such as acrylic acid, some of which are free carboxylic groups, while the rest is linked with metal cations of the 1. and 2. group in the periodic table. This provides a certain physical crosslinking, see Eq. 6.3.

$$(CH_2-CH_2)_x-(CH_2-\underset{\underset{COO^-}{|}}{\overset{\overset{CH_3}{|}}{C}})_y-(CH_2-CH_2)_v-(CH_2-\underset{\underset{COOH}{|}}{\overset{\overset{CH_3}{|}}{C}})_z$$

$$Me^+ \quad Me^+$$

$$(CH_2-CH_2)_x-(CH_2-\underset{\underset{CH_3}{|}}{\overset{\overset{COO^-}{|}}{C}})_y-(CH_2-CH_2)_v-(CH_2-\underset{\underset{CH_3}{|}}{\overset{\overset{COOH}{|}}{C}})_z$$

ME = Metal cations (Na, Mg, Zn)

(6.3)

At elevated temperatures, these crosslinks are dissolved so that ionomers can be processed by all standard thermoplastic processes at melt temperatures between 150 - 260 (330)°C. The melt has a high degree of elasticity, thus facilitating the production of sheet and film. Because of their polar groups, ionomers bond well to various carrier materials and non-porous films of only 12 μm thickness can be extruded. These films are easily warm drawn with a favorable draw ratio.

General properties: Ionomers are non-crystalline. Over the range of their service temperature (-40 to +40°C) and above these materials are tough and glass-clear. They are resistant to alkalis, weak acids, fats, and oils. Organic solvents will merely swell them. They are not resistant to oxidizing acids, alcohols, ketones, aromatic and chlorinated hydrocarbons and their susceptibility to stress cracking is slight. Their permeability to water vapor, O_2, and N_2 is comparable to that of PE; they are less permeable to CO_2. Ionomers burn with a bright flame. There are grades available approved for food contact. Applications: Glass-clear pipes for drinking water, wine, and fruit juice; transparent films for fat-containing foods; parts for laboratory and medical applications; skin- and blister packaging; bottles for vegetable oil, liquid fats, and shampoos; shoe soles, ski shoe shells, tool grips, transparent coating, corona- and stress cracking resistant insulation. Ionomers act as bonding agents between polymers of varying polarity in alloys and composites. Low-molecular polymers are useful additives for the homogenization of dispersions and to reinforce adhesive bonds.

Ethylene Copolymer / Bitumen Blends (ECB, ECB/TPO) ECBs are blends of ethylene copolymers (ethylene/butyl acrylate copoplymer) and special bitumens. In ECB the polymer forms the continuous phase. ECB under the trade name Lucobit is plasticizer-free, weather- and aging resistant. It is available as black pellets (density = 0.97 g/cm^3) and typically extruded primarily into sheets at 140 - 190°C or injection molded at 160 - 220°C. At temperatures between 250 - 280°C it can be processed into thick-walled parts by nonpressure casting. Colored roofing and sealing sheets are available from ECB/TPO (thermoplastic polyolefins) blends.

Applications: Sealing films (also with glass- or polyester reinforcement) for flat roofs, tunnels, and in civil engineering, as additives to bitumen- and asphalt materials to improve structural stability. Copolymers with fluoropolymers (ETFE) are presented later in this chapter.

6.1.5 Polypropylene Homopolymers (PP, H-PP)

PP is polymerized from propylene (H_3C-CH=CH$_2$). Like PE it is a semicrystalline thermoplastic material; however, it exhibits higher strength, stiffness, and crystalline melting temperature at lower densities (0.905 - 0.915 g/cm^3). A wide variety of PP grades is available (see Table 6.12). The product palette and area of applications for these materials is continuously expanding. Worldwide, 34 million tons of PP were consumed in 2002. New PP grades with tailored property profiles are produced with special catalyst systems. High molecular homo- and copolymers, as well as block copolymers and elastomer-modified grades are of interest. Transparent grades, grades with low stiffness, grades with high stiffness and good heat distortion stability, and scratch resistant grades are available. In certain application areas, PP grades will replace higher priced materials, such as ABS, PS, PET, PC, and TPE.

Table 6.12: Comparison of PP Properties

Properties	Unit	Polyethylene				
		PP-H Homo-polymer	PP-R Random-polymer	PP-B Block-copolymer	(PP + EPDM)	PP-T 20 Talc
ρ	g/cm^3	0.90-0.915	0.895-0.90	0.895-0.90	0.89-0.92	1.04-1.06
E_t	MPa	1300-1800	600-1200	800-1300	500-1200	2200-2800
σ_y	MPa	25-40	18-30	20-30	10-25	32-38
ϵ_y	%	8-18	10-18	10-20	10-35	5-7
ϵ_{tB}	%	>50	>50	>50	>50	>20
σ_{50}	MPa	-	-	-	-	-
σ_B	MPa	-	-	-	-	28-30
ϵ_B	%	-	-	-	-	15-20
T_P	°C	162-168	135-155	160-168	160-168	162-168
HDT	°C	55-65	45-55	45-55	40-55	60-80
α_p	10^{-5}/K	12-15	12-15	12-15	15-18	10-11
α_n	10^{-5}/K	-	-	-	-	10-11
UL94	Class	HB*	HB*	HB*	HB*	HB*
$\epsilon_r 100$	-	2.3	2.3	2.3	2.3	2.4-2.8
tan $\delta 100$	$\cdot 10^{-5}$	2.5	2.5	2.5	2.5	7-10
ρ_e	Ohm·m	>10^{14}	>10^{14}	>10^{14}	>10^{14}	>10^{14}
σ_e	Ohm	>10^{13}	>10^{13}	>10^{13}	>10^{13}	>10^{13}
$E_B I$	kV/mm	35-40	35-40	35-40	35-40	45
W_w	%	<0.2	<0.2	<0.2	<0.2	<0.2
W_H	%	<0.1	<0.1	<0.1	<0.1	<0.1
Properties	Unit	Polyethylene				
		PP-T40 Talc	PP-Gf30 Glass fiber	PP-GFC30 Glass fiber chem. coup.	PP-B25 Barium	
ρ	g/cm^3	1.21-1.24	1.21-1.14	1.12-1.14	1.13	
E_t	MPa	3500-4500	5200-6000	5500-6000	1850	
σ_y	MPa	30-35	-	-	-	
ϵ_y	%	3	-	-	-	
ϵ_{tB}	%	4-10	-	-	-	
σ_{50}	MPa	-	-	-	-	

Continued on next page

6.1 Polyolefins (PO), Polyolefin Derivates, and Copoplymers

Properties	Unit	Polyethylene			
		PP-T40 Talc	PP-Gf30 Glass fiber	PP-GFC30 Glass fiber chem. coup.	PP-B25 Barium
σ_B	MPa	30	40-45	70-80	-
ϵ_B	%	3-15	3-5	3-5	-
T_P	°C	162-168	162-168	162-168	-
HDT	°C	70-90	90-115	120-140	53
α_p	10^{-5}/K	8-9	6	6	0,7
α_n	10^{-5}/K	8-9	7	7	-
UL94	Class	HB	HB	HB	HB
$\epsilon_r 100$	-	2.4-3	2.4-3	2.4-3	≈2.6
$\tan \delta 100$	$\cdot 10^{-5}$	12-15	10-15	10-15	≈20
ρ_e	Ohm·m	>10^{14}	>10^{13}	>10^{14}	<10^{15}
σ_e	Ohm	>10^{13}	>10^{13}	>10^{13}	<10^{13}
$E_B I$	kV/mm	45	45	45	≈28
W_w	%	<0.2	<0.2	<0.2	-
W_H	%	<0.1	<0.1	<0.1	<0.01

Chemical Constitution, Polymerization During polymerization, the CH$_3$ groups of PP can be sterically arranged in different configurations and thus create different properties, see Eq. 6.4.

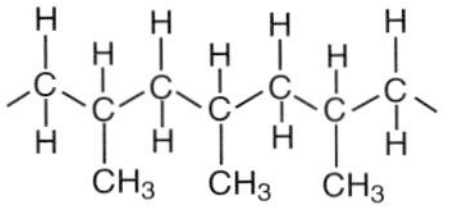

Isotactic PP (PP-I)

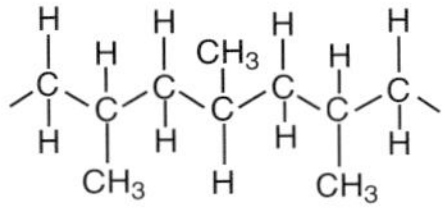

Syndiotactic PP (PP-S)

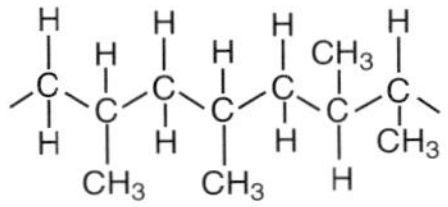

Atactic PP (PP-R)

(6.4)

The majority of CH$_3$ groups in isotactic PP (PP-I) is arranged either on the same side of the C-chain or helically turned outward. In syndiotactic PP (PP-S), the CH$_3$ groups alternate on the opposing sides of the main chain, while in atactic PP (PP-A or PP-R for random) the CH$_3$ groups are arranged randomly on both sides of the chain. Atactic PP has the consistency of non-vulcanized rubber. Quantitatively, isotactic PP is the most important grade; it is technically characterized by its index of isotacticity, which indicates the ratio of polymer non soluble in boiling xylol. The basic synthesis is low-pressure precipitation polymerization of propene gas at the surface of Natta (1955) organometallic catalysts. These are stereospecifically effective Ziegler catalysts dispersed in hydrocarbons. A certain ratio of atactic PP is generated as a by-product, creating a softer and less temperature-resistant material. This by-product is heptane-soluble and can be separated. The newer gas-phase polymerization processes produce high yields of pure products (97% isotactic PP) using minimum amounts of selectively adjustable high-efficiency catalysts. The Spheripol process with its 'Catalloy'-catalyst systems that can be tailored to produce different homo- and copolymer grades produces particles with 0.5 - 4 mm diameters processable without additional granulation. These gas-phase processes also allow the production of PP blends with normally incompatible amorphous thermoplastic materials, creating a continuous PP matrix with an evenly distributed amorphous phase (reactive blending). The properties of PP can be specifically adjusted by polymerization with metallocene catalysts. With this technology, tacticity, molecular weight, molecular weight distribution, and comonomer content can be tailored to meet customer requirements. In addition, PPs with narrow molecular weight distribution and lowered melt viscosities (CRPP, controlled rheology) are produced, e.g., by adding organic peroxides during mixing or processing.

Processing PP grades for injection molding cover a broad range of requirements ranging from high-temperature resistant, rigid to elastic, to low-temperature impact resistant grades. The melt temperature ranges from 250 - 270°C, the mold temperature from 40 - 100°C. Moisture can condensate on the pellets' surface in humid climates and should be removed by drying or with the help of vented extruders before processing. Blown films, flat films, sheet, pipe, blow molded parts, and monofilaments are extruded at melt temperatures from 220 - 270°C. Because of the high cooling demands during film production, chill-roll flat sheet die processes are preferred over tubular film blowing, which requires intensive water-cooling of the film tube. To achieve transparency (brilliant transparency only for PP-H) films must be shock-cooled below the crystalline melting temperature. Other than films for biaxial orientation, most films contain slip- and anti-blocking agents. Highly viscous PP-H grades are used for extrusion blow molding at temperatures from 190 - 220°C. Melt spin processes are used to produce staple fibers from low-viscosity, highly isotactic PP-H; orientation of blown and flat films produces woven film tape and fibrillated fibers; precipitation of PP under shear from solution produces fibrids.

PP Foams (EPP) PP foams are produced by the following processes (Also discussed in Chapter 4 of this handbook):

- Extrusion for flexible foams with small closed cells and extremely low densities (10 kg/m^3)
- High-pressure processes for rigid foams with densities from 50 - 120 kg/m^3
- Injection molding with gas or chemical foaming agents for structural foams with densities from 400 - 700 kg/m^3
- Extrusion of rigid profiles in the same density range

Post Processing Treatment The narrow thermo-elastic temperature range of PP makes hot forming of flat parts difficult. However, PP can be melted by solid phase pressure processes (SPPF), at temperatures just below the crystalline melting temperature (150 - 160°C) and molded, pressed, or rolled at room temperature. PP must be pre-treated prior to any surface finishing (coating, printing). PP can be vacuum-metallized; specific grades can be galvanized after surface activation by precious metal salts. In general, a nickel layer is applied first, which is subsequently plated with nickel or chromium. Similar to PE, PP can be joined by welding and adhesives. Contact adhesives on the basis of natural rubber or chlorobutadiene rubber, as well as adhesives based on silicone, epoxy resins, or polyurethanes are generally used. Surfaces are pre-treated with chromium sulfuric acid. Diffusion bonding is also possible. Machining of PP is easier than of PE because of its greater hardness; however, stamping is usually not possible.

Properties The variety of available PP grades is wider than that of most other plastic materials. Molecular structure, average molecular weight (200,000 - 600,000 g/mol), molecular weight distribution, crystallinity, and spherulite structure can be varied over a broad range, thus determining the properties. Stiffness and hardness range between those of PE and engineering plastics, such as ABS, PA, and others. The dynamic load capacity is relatively high. With a glass transition temperature of 0°C, all PP-H grades embrittle at low temperatures. The crystalline melting point ranges from 160 - 165°C, higher than that of PE. Therefore, maximum service temperatures are also higher: short-term 140°C, long-term 100°C. The electrical properties compare to those of PE and are not affected by exposure to water. The dielectric constant and the dielectric loss factor are largely independent of temperature and frequency (see Table 6.12). PP exhibits only minimal water absorption and permeability. Grades approved for food contact can be hot-filled and hot sterilized. Gasses, in particular CO_2, low-boiling hydrocarbons, and chloro-hydrocarbons diffuse through PP. PP swells in contact with chloro-hydrocarbons. Because of its non-polar structure, PP is chemically highly resistant: up to 120°C resistant to aqueous solutions of salts, strong acids and alkalis, as well as brines. High-crystalline PP grades provide particularly good resistance to polar organic solvents, alcohols, esters, ketones, fats, and oils. Only special grades are resistant to fuels at elevated temperatures. Strong oxidants, such as chlorosulfonic acids, oleum (disulfuric acid), concentrated nitric acid, or halogens attack PP even at room temperature. While PE is cross-linked in the presence of oxygen by radiation, PP is decomposed by oxygen. For outdoor exposure PP needs weatherproofing; yet, it still cannot compete with PE with regard to weather resistance. PP continues to burn with a light luminescent flame after removal of the ignition source; flame resistant grades are available. Parts made from PP are translucent. Orientation below the crystalline melt temperature increases crystallinity, as does the addition of nucleating agents to create a fine grain crystalline structure. Compared to conventional PP, the properties if sPP(M) can be modified to: lower the melting point to $\approx$ 150°C, increase gloss and transparency while maintaining stiffness (thus competing with PS because of the higher impact resistance), decrease of melt viscosity, and improve barrier properties. rPP(M) parts are very soft products with high elongation at break and a glass transition temperature of only 0°C (see Table 6.13).

Table 6.13: Comparison of Properties for Special PP Grades

Property	Unit	PP-R-LMW	PP-R-HMW	PP-R-HMW	PP-S-(M)	PP-I-(M)
ρ	g/cm^3	0.836	0.861	0.855	0.9	0.903
E_t	MPa	10	8	5	61	1000-1700
σ_B	MPa	1	1	2	2.4	20-35
ϵ_B	%	110	1400	2000	-	100-300
MFI	g/10min	670	7	0.1	3	1.8
Opacity	%	58	20	18	1.7	85
T_m	°C	-	-	-	168	163
Crystallinity	-	Low	amorphous	amorphous	30-40%	40-60%
HDT/A	°C	-	-	-	-	55

Applications Their special properties provide an exceptionally wide range of applications of PP, a fact that the resin suppliers pay tribute to in their formulations. Amorphous atactic PP and other atactic α-olefinic polymers are flexible to rigid plastics at temperatures as low as -30°C. They are used as (melt) coatings for paper packaging and carpet tile backing, automobile insulation, corrosion protection, paint for road markings, hot-melt adhesives, sealing compounds, bitumen blends, and aging-resistant roofing and weather strips. Injection molding of isotactic PP: sound absorbing automobile interiors (if they pass the head impact test), ventilation systems, dashboards, center consoles, front- and rear light housings, mass produced articles such as cups and food containers, tool boxes, suitcases, transport and stacking boxes, housings and functional parts, such as living hinges, for household appliances, such as coffee makers, toasters, fan heaters, dish washers, dryers, and washing machines; filled and reinforced grades for automotive body parts, automotive expansion tanks for coolants (for 2.3 bar at 125°C), lawn mowers, electric tools, parts for submerged pumps, electric installations, lawn furniture.

Blow molding: highly viscous grades (PP-Q, polypropylene with high impact resistance) for bottles for cosmetic and medical powders; somewhat less viscous grades for mass production of containers up to 5 l, also biaxially stretched; hot water containers, automotive air ducts, cases with living hinges for tools, sewing machines, electronic parts (antistatic), motorcycles, and surfboards. Extrusion: high-pressure warm water and waste water pipes, profiles, sheet, cable sheathing, unstretched and biaxially stretched (oriented) film (PP-BO, transparent) for packaging, insulation, and as composite film, foamed sheet, packaging tape, woven film tape, non-woven fabric for geo textiles and filters, cords for agricultural applications (degradable), sacks, glass mat reinforced sheet (GMT) for hot-press molding. Preferred applications for PP(M): transparent packaging (injection molded, flat sheet), thin-wall applications, carpet yarns, staple fiber; for PP-CR: packaging (thin-walled cups), coatings, and fibers for filter- and hygiene applications.

6.1.6 Polypropylene Copolymers and -Derivates, Blends (PP-C, PP-B, EPDM, PP + EPDM)

Copolymerization and blending with other polymers can also modify PP properties.

Chlorinated PP (PP-C) PP-C is of less importance than PE-C. It is used for chemical- and corrosion resistant coating.

PP Copolymers (PP-B) Ethylene, butene-1, and higher α-olefins are used as copolymers (see Table 6.9, Nr. 1, 3, and 27). Sometimes, PP-B grades are also called block-copolymers, although they are heterophasic blends of homo- and copolymers. The insertion of PE interrupts the molecular chain; however, the crystallinity of PP is maintained up to a PE content of 20%. PE lowers the glass transition temperature by 5°C. The melt temperature of a random PE copolymer is substantially lowered, even at low PE contents. If PE is inserted as a block, the melt temperature is not substantially lowered; 10% PE content significantly improves the impact resistance at low temperatures (-30 to $-40°C$); 20% PE results in flexible products, such as PE-LD with a melting point above 160°C. It is used for hot-sealing layers, flexible pipes, and transparent injection molded and blow molded parts.

Ethylene-Propylene (Diene) Copolymers (EPDM) Copolymerization of PE, PP, and ethylene/norbornene (for terpolymers) in hexane with Ziegler catalysts yields EPDM. Norbornene is obtained by synthesis of ethylene and cyclopentadiene; it is a raw material for synthetic rubbers and is pressed in bale form, while PP-EPDM compound is provided as pellets. Random copolymers (amorphous EPDM): The mobility of linear PE macro molecules diminishes only at temperatures below -100°C. However, this property does not result in the expected flexibility because PE-HD is semi-crystalline. Hence, in order to achieve an amorphous product, some H-atoms in the PE molecule must be replaced by random distributed polar groups, as a result hindering crystallization. This can be achieved by copolymerization with propylene. Copolymers with α-olefins such as PP or butene-1 with up to 70% PE are amorphous, 3-dimensionally cross-linked, exhibit extremely low densities of 0.86 - 0.87 g/cm^3, and a glass transition temperature sufficiently below room temperature. These products can be processed like rubbers. Prerequisite for being vulcanizable with sulfur and for predetermining the degree of cross-linking is the insertion of dienes. E/P colpolymers (EPM) can only be chemically cross-linked with the help of peroxides; however, this makes optimization of their cross-linking degree and -density impossible. EPDM with a PE content of more than 50% cannot be processed like a thermoplastic; it is used in automotive and construction applications and in the cable industry. Sequential copolymers (semi-crystalline EPDM) are physically cross-linked. The cross-linking bonds are formed between crystalline or glass-like solid parts of the polymer chain. In addition, they can be vulcanized. With a PE content of at least 70%, the PE sequences are long enough to form such domains. These materials can be processed like thermoplastic; however, the physical bonds have the disadvantage that, with rising temperatures and depending on block structure, they will start to break, thus losing the elastomeric character of the material. Both vulcanized and unvulcanized materials exhibit good weather resistance and weldability and are therefore used as sealing films for roofs and floors. Because they are easily filled with heavy fillers, they are used as sound insulating layers, e.g., in automotive applications.

PP + EPDM Elastomer Blends The ideal structure of a polyolefin elastomer displays blocks of amorphous ethylene and propylene sequences in a random distribution together with fixed PP blocks. These structures do not necessarily have to be connected to each other in a chain, but can be achieved by blending PP with EPDM. These products exhibit high stiffness and softening temperatures, are easily modified by copolymerization of PP, and compatible with EPDM. UV stabilization is achieved by addition of either carbon black or sterically hindered amines, if light-colored products are required. There is also the possibility of coating with flexible PUR to achieve UV resistance. Stiffness can be enhanced by chalk or glass fibers. The properties of PP + EPDM elastomers depend on blend ratio. 90% PP results in the properties of conventional PP with slightly lower stiffness and softening temperature, but also with increased impact resistance at $-40°C$. Blends with 40% PP exhibit the typical properties of thermoplastic rubbers. Other determining factors are crystallinity, molecular weight, and molecular weight distribution of PP. In addition, it is important whether a homo- or a copolymer, random or sequential PP is used. There is also the possibility to blend with PE. All processing methods used for PP can be used with PP-elastomer blends. Highly viscous grades are extruded, blown, or pressed at $\approx 250°C$. Less viscous grades are injection molded at melt temperatures from 220 - 260°C and mold temperatures of $\approx 60°C$. Main application area is the automotive industry: bumpers, spoilers, coverings for wheel houses and trunks, mud flaps, dashboards, consoles and other interior parts, steering wheel covers; flexible tubes and pipes in construction; shoe-, sport-, and toy industry.

Polypropylene Blends Blends of iPP and, e.g., methylmethacrylate or styrene, are produced by reactive blending. They have the following advantages: low density (0.91 - 0.96 g/cm^3), weather resistance, scratch resistance, little warpage and moisture absorption. Compounds with 3 - 6% hydrocarbon resins, for example, hydrogenated dicyclopentadiene (DCPD), increase the glass transition temperature of PP films by up to 25 K and thus the modulus of elasticity by up to 50%, while reducing water vapor permeability by up to 30%.

6.1.7 Polypropylene, Special Grades

Additives such as nucleating agents cause a finer spherulite structure and thus higher transparency and flexibility, but they also reduce stiffness and lower the heat distortion point. The addition of powdered peroxides during compounding or processing creates radicals that split hydrogen from the molecular chain, thus leading to a narrower molecular weight distribution polypropylene random copolymer (PP-CR). This reduces the melt viscosity and facilitates processing. Because copper ions catalyze the thermooxidation, PP should be appropriately stabilized for applications such as insulating copper wires. Applications in washing machines require stabilization against alkaline solutions and heat aging. Antimony trioxide in combination with halogene compounds and phosphoric acid esters are flame retardant (B1 according to DIN 4102 and HB to V0 according to UL 94). Grades suitable for plating contain pigments that cause finely cracked surfaces, which in turn increase the adhesion of the first metal layer during plating. Carbon black is used for weather stabilization; for colored applications, amines are used. Modified grades with higher melt stability are used for extrusion and blow molding. Talc is the most common filler used with PP. It improves stiffness, dimensional stability, heat resistance, and creep behavior and it also serves as a nucleating agent (see Table 6.12). Disadvantages are the lowering of the low temperature impact resistance, the reduction in weldability and oxidation resistance at elevated temperatures, and the formation of matte surfaces. Calcium carbonate has the same effect as talc but has additional advantages: easier to disperse, better flowing melt, higher UV- and oxidation stability, higher surface quality, less tool wear, and reduced cycle times for injection molding; 40% of mica increases stiffness to the same extent as 30% glass fibers at lower cost. Wood flour increases acoustic insulation. Calcium silicate increases impact resistance and electrical and thermal properties. Zinc oxide protects against microorganisms and increases UV-resistance. Scratch resistant surfaces similar to those of ABS can be achieved with special coated fillers in combination with additives and mineral fillers. Glass fiber reinforced PP is available with ground and cut glass fibers. Short glass fibers increase the stiffness and lower impact resistance, while longer glass fibers increase structural integrity, strength and creep resistance. Chemically linked glass fibers enhance this effect. Particularly long glass fibers cause anisotropic shrinkage because of their fiber orientation (thus causing warpage), matte surfaces, and increased wear. The addition of glass beads, also in combination with glass fibers, creates reinforced PP with higher stiffness and compressive strength as well as fewer tendencies for warpage (see Table 6.12). glass mat reinforced PP sheet (GMT) and long fiber reinforced thermoplastics (LFT) for compression molding (GMT and LFT are discussed in more detail in Chapter 4 of this Handbook) are used for high-load applications, particularly in automotive applications. Properties: Processing temperature 215°C, heat distortion point $\approx$ 150°C, short-term service temperature 140°C, long-term 100°C (see Table 6.14).

Table 6.14: Comparison of Characteristic Properties for GMT and Long Glass Fiber Reinforced PP.

Property	Unit	GMT		Direct process	
		PP-GM 30	PP-GM 40	PP-GLF 30	PP-GLF 40
		Glass mat reinforced		Reinforced w/ LGF	
ρ	g/cm^3	1.13	1.22	1.13	1.22
ψ	wt.-%	30	40	30	38
σ_B	MPa	70	90	70	80
E_t	MPa	4500	5500	5000	6000
ϵ_{tB}	%	1.8	1.6	2	1.8
α_p	10^{-6}K^{-1}	$\approx$30	25 to 30	30 to 40	20 to 30

6.1.8 Polybutene (PB, PIB)

There are two forms of isomeric butene of technical interest: butene-1 and isobutene. The linear monomer butene-1 is used for the thermoplastic polybutene-

1 (PB), while the branched isobutene is used for the rubber-like polyisobutene (PIB) (see Eq. 6.5)

$$\left[-CH_2-\underset{\underset{CH_3}{|}}{\overset{\overset{CH}{|}}{CH_2}}-\right] \quad \text{a) Polybutene (PB)}$$

$$\left[-CH_2-\underset{\underset{CH_3}{|}}{\overset{\overset{CH_3}{|}}{C}}-\right] \quad \text{b) Polyisobutylene (PIB)}$$

(6.5)

Polybutene-1 (PB)

Chemical Constitution. Polybutene-1 is created by stereospecific polymerization of butene-1 (see Eq. 6.5a) with specific Ziegler-Natta catalysts. PB is a mostly isotactic, semi-crystalline polymer with a high molecular weight of 700,000 - 3,000,000 g/mol and a low density of 0.910 - 0.930 g/cm^3. During cooling, it first crystallizes to $\approx$ 50% in a metastable, tetragonal modification (density $\approx$ 0.89 g/cm^3), forming a soft, rubber-like material. With corresponding shrinkage, this material transforms at room temperature within $\approx$ one week into a stable, double hexagonal modification. Under higher pressures the transformation completes faster; at higher or lower than room temperature the process is slower. During the transformation, density, elongation at break, and hardness increase. Atactic polybutene-1 is used as melt adhesive; syndiotactic polybutene-1 has no technical significance.

Processing. Hopper drying of pigmented grades is necessary for the major processing technologies (injection molding and extrusion). Melt temperatures for injection molding range from 240 - 280°C, for extrusion from 190 - 290°C. Mold temperatures range from 40 - 80°C. Note that PB recrystallizes after thermoplastic processing. This fact can be utilized when producing pipe bends: extruded pipes are wound over a drum and left there to transform into a stable modification.

Properties. The mechanical properties of PB in its stable modification at room temperature range between PE and PP. Even at elevated temperatures it exhibits high creep rupture strength, low creep, and high stress cracking resistance because of its high molecular weight and the strong bonds between the crystalline blocks. These favorable properties are sustained even at carbon black loads of 20%. Particularly notable are low creep in combination with chemical resistance and excellent abrasion resistance, for example, for the transportation of slurry or diluted solid particles. PB pipes meet international standards for stress cracking resistance. PB is resistant to non-oxidizing acids, oils, fats, alcohols, ketones, aliphatic hydrocarbons, and cleaning agents. It is non-resistant to aromatic or chlorinated hydrocarbons. Like all other polyolefins, PB burns easily and must be stabilized for out door use. Food contact is permitted, PB is physiologically harmless. Contact with and inhalation of butene-1 vapors should be avoided because they have anesthetizing effects. (Property comparison, see Table 6.10)

Applications. Hot water pipes, pipes for floor heating, fittings, extrusion blow molded hollow parts, containers for the chemical industry, telephone cable, two-layer blow molded or flat sheet film with PA/PE, PE, or PP as carrier and a peelable sealing layer made of PE/PB blends for hot- or cold-filled food or meat packaging. Adding 1 - 5% PB increases the extrusion speed of PE, PP, and PS. Blends of 25 - 30% PB with aliphatic adhesive components (tackifiers) und micro-crystalline waxes are used in special adhesive formulations, which sustain long exposure times and high temperature resistance; 1 - 2% PB as processing aid for PE-LLD reduces melt fracture during film blow molding.

Polyisobutene (PIB)

Chemical Constitution. PIB (raw density 0.91 - 0.93 g/cm^3) is a rubber-like thermoplastic material produced by cationic polymerization of isobutene (see Eq. 6.5(b)). Homo-polyisobutene is not vulcanizable because of its lack of functional properties (see also butyl rubber, IIR).

Processing. Solid rubber-like products are processed like rubber with kneaders, rolling mills, calenders, presses, extruders, and injection molding machines. Melt temperatures range from 150 - 200°C. If the temperatures are too low, PIB will mechanically degrade. Dispersions for coatings and foams are available.

Properties, Applications. Depending on molecular weight (MW), PIBs are viscous oils, tacky to soft/flexible, or rubber-like materials. Some applications are due to the high low-temperature compliance (glass transition temperature -73°C), others to the low gas permeability (D = 0.081 $\times 10^{-6}$ cm^2/s). PIBs of all MW are used as adhesives and sealers, because their properties do not change significantly between the glass transition temperature (-73°C) and almost 100°C. They are used for copolymerization with other polyolefins (to improve processability), styrenes, and other monomers. In rubber blends, PIB improves weather- and age-resistance and adhesive strength of tire treads, while reducing gas permeability. Oily fluids, MW 300 - 3,000 g/mol; electro insulation oils, viscosity improving additives to mineral oils, adhesion promoting additives for oriented films. Highly viscous materials, MW 40,000 - 120,000 g/mol: blends for laminating waxes, sealing resins, chewing gum. Elastic materials, MW 300,000 - 3,000,000 g/mol: polymer additives, electro insulation films, lining and sealing webs against acids and water pressure, for flat and conventional roofs (with light-stabilizing and mineral fillers). Most important properties for these applications: extremely high ductility, even at high filler loads, although at the cost of very low strength, low water vapor and gas permeability, excellent dielectric properties (δ_D = 10^{15} Ohm cm, ϵ_R = 2.2, and tan δ = 0.0004), short-term service temperatures -140 to 80°C, long-term -30 to 65°C. PIB is resistant to acids, alkalis, and salts, exhibits limited resistance to nitric and nitrided acids; it is not resistant to chlorine, bromine, chloride sulfuric acid; it is soluble in aromatic, aliphatic, and chlorinated hydrocarbons; it swells in butyl acetate, oils, and fats; it is insoluble in esters, ketones, and low alcohols. PIB must be stabilized against sun light and UV radiation. It burns like rubber.

6.1.9 Higher Poly-α-Olefins (PMP, PDCPD)

Poly-4-Methylpentene-1 (PMP)

Chemical Constitution, Properties. After polymerization, 4-methylpentene-1 is a highly branched glass clear (90% light transmission), hard, semi-crystalline thermoplastic material with a low density in the raw state (0.83 g/cm^3), see Table 6.9, Nr. 27). Its has a degree of crystallinity of 65%; it is highly transparent because both amorphous and crystalline domains have the same refraction index. Micro cracks forming between the amorphous and the crystalline domains because of their different expansion coefficients make PMP homopolymers slightly cloudy in appearance. The melting point is at 245°C and can be lowered by copolymerization; however, this will also lower the tendency for micro-crack formation. Short-term service temperatures are at 180°C, long-term at 120°C. Electrical properties resemble those of PE-LD. PMP is resistant to mineral acids, alkaline solutions, alcohols, cleaning agents, oils, fats, and boiling water; it is not resistant to ketones, aromatic and chlorinated hydrocarbons; it is also susceptible to stress cracking. Weather resistance, even of stabilized grades, is low. PMP will yellow and lose its good mechanical properties. It burns with a luminescent flame, is approved for food contact, and physiologically inert. For property comparison, see Table 6.10.

Processing. PMP is injection molded at melt temperatures ranging from 280 - 310°C and a mold temperature of ≈ 70°C. Because of the narrow melt temperature range, extrusion is difficult. Blow molding of hollow parts is possible at temperatures from 275 - 290°C; however, because of the low melt strength the size of the hollow parts is limited. Thermoforming is also possible. Like all other polyolefins, PMP is weldable and easier to join by adhesive bonds, as long as the surfaces are roughened. Treatment with chromium sulfuric acid enhances bond strength; however, it is recommended to use plasma- or jet blast corona treatment instead to avoid waste disposal issues.

Applications. Sight glass or observation windows, interior light fixtures, sterilizable parts and films for medical applications and packaging, packing for instant meals, coloring spools, Raschig rings, transparent pipes and fittings, cable insulation flexible at low temperatures (-10°C).

Polydicyclopentadiene (PDCPD)

Chemical Constitution, Processing. Dicyclopentadiene is a by-product of the crack process used to produce gasoline. At 99% purity in a mixture with norbornene it is liquid above -2 to 0°C. In the presence of catalysts, such as alkyl aluminum, it is polymerized to a cross-linked PDCPD by ring opening and splitting the double bonds. Mixed with additives, DCPD in form of a two-component system can be processed on conventional RIM equipment with reaction times of 15 s to 2 min.

Properties, Applications. Properties and applications compare to those of rigid PUR-RIM compounds. PDCPD exhibits higher strength, stiffness, and dimensional stability at elevated temperatures. Service temperatures range from -40 to 110°C. Typical applications are body parts for agricultural- and construction machinery and short runs for medical applications.

6.2 STYRENE POLYMERS

6.2.1 Polystyrene, Homopolymers (PS, PMS)

Polystyrene (PS), Poly-p-Methylstyrene (PPMS), Poly-α-Methylstyrene (PMS)

Chemical Constitution. From a chemical standpoint, polystyrenes are actually polyvinylbenzenes, in which the phenyl groups are normally randomly distributed along the chain. Therefore, crystalline domains cannot form, so that conventional PS or PS-R (random) are amorphous, transparent thermoplastic materials. Suspension-, bead-, or mass- polymerization produces glass-clear products. PS belongs to the most important materials for consumer goods that are mass produced and relatively inexpensive. The use of stereospecific Natta catalysts results in isotactic polymers, PS-I, in which all phenyl groups are in the same dimensional arrangement. These products are up to 50% crystalline and opaque when cooled slowly from the molten state or after exposure to temperatures of 150°C. PS polymerized with metallocene catalysts is syndiotactic (sPP, sPP(M)). It is crystalline and has a melt temperature of 270°C; therefore it is considered a high-temperature resistant plastic. Applications: injection molding and extrusion, electronics, heat resistant house hold products. Besides styrene, substituted styrenes such as p-methylstyrene and α-methylstyrene are polymerized (PS, PPMA, PMS, see Table 6.9, Nr. 5, 6, and 7). p-Methylstyrene is radical polymerized and exhibits higher heat distortion points and hardness than PS. α-Methylstyrene is difficult to radical polymerize into highly molecular products. Lower molecular poly-α-methylstyrenes are used as modifiers and processing aids, e.g., for thermoplastic elastomers, PVC, and ABS. Processing, properties, and applications, are discussed later.

6.2.2 Polystyrene, Copoplymers, Blends

Chemical Constitution A wide variety of copoplymers (also in combination) and blends were developed in order to modify specific properties, such as heat distortion resistance, stiffness, impact resistance, chemical resistance, stress cracking resistance, and to meet specific requirements. Table 6.9 listed the basic structures of these copolymers. Table 6.15 lists selected co- and terpolymers as well as common blends.

Table 6.15: Polystyrene Ter-and -Blockcopolymers, Blends

Terpolymers	Acronym	see Nr. in Table 6.9
Acrylonitrile-butadiene-styrene	ABS	5+26+35
Methacrylate-butadiene-styrene	MBS	5+12+35
Methylmethacrylat-acrylonitril-butadiene	ASA	5+17+26+35
Acryl ester-styrene-acrylonitrile	ASA	5+26+33
Styrene-butadiene-methylmethacrylate	SBMMA	5+35+17
Acrylonitrile-chlorinated PE-styrene	ACS	5+26+9
Acrylonitrile-EPDM rubber-styrene	AES (AEPDMS)	5+26+34
Styrene-maleic acid anhydride-butadiene	SMAB	5+31+35
Styrene-isoprene-maleic acid anhydride	SIMA	5+26+31
Blockcopolymers, TPE-S		
Styrene-acrylonitrile	SAN	5+26
Styrene-butadiene	SB	5+35
Styrene-butadiene-styrene	SBS	5+35+5
Styrene-isoprene	SIR	5+36
Styrene-isoprene-styrene	SIS	5+36+5
Styrene-ethene-propene	SEP	5+1+2
Styrene-ethene-butene-styrene	SEBS	5+38+5
Styrene-ethene-propene-styrene	SEPS	5+39+5
Blends of	**with**	
PS	PPE	
ABS	PC	
	PC-Blend	
	PA	
	TPU	
	PVC	
	SMA	
	PSU	
ASA	ABS	
	AES	
	PC	

An important representative of this material class is SAN, which is used to produce ABS and ASA by addition of rubbers. These are further processed into blends, such as ABS + PC and ABS + PA. MABS and MBS are also members of the styrene-copolymer family.

Processing Most styrene polymers are injection molded. Melt temperatures range from 180 - 280°C, mold temperatures from 5°C (short-lived mass produced articles) to 80°C. Low mold temperatures and high injection speeds cause frozen-in stresses and orientation and will reduce the long-term quality of the produced parts. Low shrinkage enables precision molding. Impact modifiers increase melt viscosity, thus lowering the melt flow rate. Grades with added blowing agents can be injection molded into structural parts. In general, all styrene polymers can be extruded into profiles, films, and sheets. ABS in particular is well suited for plating after a special surface pre-treatment. Painting, printing, and laser-labeling are possible. Adhesion can be achieved with glues that fit the polymer type, or with two-component adhesives. Common welding processes used for thermoplastic materials can be utilized.

Properties PS homopolymers are crystal clear and glossy. Easily flowing grades are stiff and hard, but also brittle and susceptible to fracture and therefore their suitability for use under dynamic load is limited. Wear- and abrasion resistance and water absorption are minimal, dimensional stability is high. Short-term maximum service temperature ranges from 75 - 90°C, long-term from 60 - 80°C. PS exhibits very good electrical and dielectrical properties. It is insensitive to moisture and resistant to brine, lye, and non-oxidizing acids. Esters, ketones, aromatic and chlorinated hydrocarbons act as solvents and the solutions are used as glues. Gasoline, essential oils, and flavors (spices, lemon peel oil) cause stress cracking. Stress-cracking tendencies are high, in particular when stresses are frozen-in because of rapid cooling of the melt or because of disadvantageously positioned gates. While PS retains its gloss and transparency in doors, it is not suitable for outdoor use. PS easily burns after ignition. Syndiotactic PS (PS-S, sPS(M)) behaves similar to standard random PS of equal density; however, it is stiffer and exhibits a much higher heat distortion point - in particular glass fiber reinforced grades - and better melt flowability. PS, as well as its copolymers and blends, are stabilized against degradation at high processing temperatures and yellowing caused by UV-exposure. There are easy flowing, glossy, antistatic, antiblocking, internally and externally lubricated, colored or pigmented (organic or inorganic), and galvanizable grades available. Fillers and reinforcements are rarely used with PS (particularly with pure PS), because they have little effect on hardness, brittleness, and stress crack resistance. Chalk, talc, and glass fibers/beads are more commonly used with ABS and its blends. The general effects of copolymer blocks are detailed in Table 6.16.

6.2 Styrene Polymers

Table 6.16: Effect of Comonomers on the Properties of Polystyrene

Monomer	Line Nr.in Table 6.9	Influence on PS
α-Methylstyrene, MS	6	Higher heat distortion point
Chlorinated ethylene, PE-C	9	Impact modifier
Methylmethacrylate, MMA	17	Creates transparency in PS and ABS
Vinyl carbazole	25	Higher heat distortion point, but toxic
Acrylonitrile, AN	26	Higher stiffness, toughness, resistance, water absorption, decreased electrical properties, yellowish
Acrylic acid ester-elastome	33	Impact modifier, decreased heat deflection temperature, stiffness, rigidity, increased weatherability
EPDM-rubber, EPDM	34	Same as acrylic acid ester-elastomer
Butadiene rubber, BR	35	Same as EPDM, but decreased heat deflection temperature
Isoprene rubber, IR	36	Same as BR
Thermoplatic PUR-elastomer,	37	Increased impact and wear resistance

To enhance mechanical properties and heat distortion temperature, styrene groups can be replaced by methyls styrene. At 50% content, the glass transition temperature increases from 95 to 115°C. AN and MA have the same effect. Examples: SAN, ASA, AES, or SMAB. AN increases toughness, resistance to oils and fats, but also water absorption and yellowing. SAN-modifications with polyvinyl carbazole (PVK is toxic!) results in extremely heat distortion resistant but toxic products. Modifications with carbonic acid dimethyl ester additionally results in higher chemical and UV resistance. In addition to increasing heat distortion resistance, PVK is also used as homopolymer molding material: density 1.19 g/cm^3, modulus of elasticity 3,500 MPa, short-time service temperature 170°C, long-term 150°C, brittle. Applications: insulation for high-frequency and television applications under high mechanical or thermal loads. High impact resistant PS (PS-HI) is produced by graft copolymerization with, e.g., 5 - 15% butadiene styrene rubber (BR, SBR). Copolymers with 15 - 40% BR are used as processing aids in rubber manufacturing. There are numerous modifications of multi-phase engineering plastics (graft polymers and/or thermoplastic-blends) with ACM, EPDM, PE-C, or thermoplastic PUR-elastomers, exhibiting well balanced thermal, mechanical, and toughness properties (up to a leathery toughness). Stiffness and strength of these products are reduced, in particular impact resistance at low temperatures, while stress crack resistance is improved. There are opaque and even transparent grades. Butadiene containing grades are less UV-resistant than SAN copolymers and butadiene-free impact resistant grades. ABS polymers have gained particular importance because of the broad range of variation in their properties. The two most significant manufacturing methods are:

- Graft polymerization of styrene and acrylo nitrile on butadiene latex. The resulting graft polymer is blended with SAN, coagulated, and dried
- The graft polymer and SAN are produced separately, dried, blended, and then pelletized. By appropriate selection of the rubber component and substitution of styrene by α-methylstyrene it is possible to specifically control glass transition temperature and thus heat distortion point and impact resistance.

In addition to the standard opaque ABS, transparent plastics based on ABS with suitable rubber components are available. With increasing PC contents, ABS + PC blends exhibit increasing heat distortion resistance; their stiffness and hardness compare to those of PC. The unfavorable properties of PA, such as shrinkage and little impact resistance at low temperatures, are alleviated in ABS + PA (PA 6) blends. The blends offer good toughness, processability, chemical resistance, and a lower density compared to ABS + PC. Because of their excellent heat distortion resistance (180 - 200°C, heat sag test) ABS + PA blends can be used for cathodic dip primer with subsequent drying at temperatures slightly above 200°C. ASA is distinguished by its weatherability and heat distortion resistance, caused by its acryl ester rubber content. ASA in outdoor applications does not require a lacquer or paint coat. Like ASA, AES is used in similar applications as ABS; however, it meets higher requirements regarding light fastness and weatherability. ASA + PC blends exhibit higher heat distortion resistance compared to ABS; in blends with PPE + PS, this resistance can be increased up to 160°C. Polystyrene copolymers are used extensively as blend components for other thermoplastic materials. In the case of ABS, this is the result of the polar nature of the CN bond and the relatively low melt viscosity of the SAN group. Special modifying resins are available for PVC. Coupling agents based on SB blends are used as bonding layers for composite sheet and coextruded film made of styrene copolymers with other polyolefins, PC, PMMA, or PA, and as coupling agents in mixtures of recycled plastic blends. Tables 6.17 and 6.18 compare the properties of some selected grades.

Applications. Standard PS is typically used for short-lived, non-technical applications, such as disposable packaging for food, pharmaceutical products, and cosmetics. sPS is supposed to substitute higher temperature resistant plastics such as PC, PT, LCP: head light reflectors, electronic, medical, codenser film.

6.2 Styrene Polymers

PS copolymers and blends, in particular ABS and similar products, are used in higher value products: insulation films and sheet for thermoforming, pipes, profiles; parts and housings for radios, TVs, notebooks, telephones, computers, printers, tools, housings for optical devices, household and gardening tools; plotters, luminescent screens, toys; inner linings for refrigerators and freezers; automotive: interior panels, dashboards, interior modules (e.g., from SMA-GF10), ventilation systems, roof liners, battery housings. Weather- and UV resistant grades: spoilers, radiators, fenders, and hubcaps for cars, exterior parts for tractors, trucks, motorcycles, RVs, boat hulls, and lawn furniture.

Table 6.17: Property Comparison of Styrene Homopolymers and -Copolymers

Properties	Unit	Styrene-Homopolymers and Copolymers					
		PS(PS-R)	PS-s(M)	SMS	SB-I	SB-T
ρ	g/cm^3	1.04-1.05	1.05	1.05-1.06	1.03-1.05	1.0-1.03
E_t	MPa	3100-3300	4500	3300-3500	2000-2800	1100-2000
σ_y	MPa	-	-	-	25-45	20-40
ϵ_y	%	1.5	1.4	-	1.1-2.5	26
ϵ_{tB}	%	-	-	-	10-45	20->50
σ_{50}	MPa	-	-	-	-	-
σ_B	MPa	30-55	-	50-60	-	-
ϵ_B	%	1.5-3	-	24	-	-
T_P	°C	-	270	-	-	-
HDT	°C	65-85	95[1)]	80-95	72-87	60-75
α_p	10^{-5}/K	6-8	-	68	8-10	7-14
α_n	10^{-5}/K	-	-	-	-	-
UL94	Class	HB*	-	HB*	HB*	HB
ϵ_r 100	-	2.4-2.5	3	2.4-2.5	2.4-2.6	2.5-2.6
tan δ 100	·10^{-5}	12	-	1-2	1-3	2-4
ρ_e	Ohm	>10^{14}	-	>10^{14}	>10^{14}	>10^{14}
σ_e	Ohm	>10^{14}	-	>10^{14}	>10^{13}	>10^{14}
E_BI	kV/mm	55-65	-	55-65	456-5	40-50
W$_w$	%	<0.1	-	<0.1	<0.2	<0.1
W$_H$	%	<0.1	-	<0.1	<0.1	<0.1

Properties	Unit	Styrene-Homopolymers and Copolymers			
		SB-HI	SAN	SAN-GF35	SMAHB
ρ	g/cm^3	1.03-1.04	1.07-1.08	1.35-1.36	1.05-1.13
E_t	MPa	1400-2100	3500-3900	10000-1200	2100-2500
σ_y	MPa	15-30	-	-	37
ϵ_y	%	1.5-3	-	-	-
σ_{50}	MPa	-	-	-	-
σ_B	MPa	-	65-85	110-120	-

Continued on next page

Properties	Unit	Styrene-Homopolymers and Copolymers			
		SB-HI	SAN	SAN-GF35	SMAHB
ϵ_B	%	-	2.5-5	2-3	11-26
T_P	°C	-	-	-	-
HDT	°C	60-80	95-100	100-105	104-115
α_p	10^{-5}/K	8-11	7-8	2.5-3	6-9
α_n	10^{-5}/K	-	2	6-9	-
UL94	Class	HB*	HB*	HB*	HB
$\epsilon_r 100$	-	2.4-2.6	2.,8-3	3.5	-
$\tan \delta 100$	$\cdot 10^{-5}$	1-3	40-50	70-80	-
ρ_e	Ohm	>10^{14}	10^{14}	10^{14}	10^{13}
σ_e	Ohm	>10^{13}	10^{14}	10^{14}	>10^{15}
$E_B I$	kV/mm	45-65	30	40	25
W_w	%	<0.2	0.2-0.4	0.2-0.3	-
W_H	%	<0.1	0.1-0.2	0.1	0.2

* also available from V-2 to V-0

Table 6.18: Property Comparison of Acrylonitrile-Styrene-Copolymers and Blends

Properties	Unit	Acrylonitrile-styrene-copolymers and blends			
		ABS	ABS-HI	ABS-GF20	ABS+PC
ρ	g/cm^3	1.03-1.07	1.03-1.07	1.18-1.19	1.08-1.17
E_t	MPa	2200-3000	1900-2500	6000	2000-2600
σ_y	MPa	45-65	30-45	-	40-60
ϵ_y	%	2.5-3	2.5-3.5	-	3-3.5
ϵ_{tB}	%	15-20	20-30	-	>50
σ_{50}	MPa	-	-	-	-
σ_B	MPa	-	-	65-80	-
ϵ_B	%	-	-	2	-
T_P	°C	-	-	-	-
HDT	°C	95-105	90-100	100-110	90-110
α_p	10^{-5}/K	8.5-10	8-11	3-5	7-8.5
α_n	10^{-5}/K	-	-	-	5-6
UL94	Class	HB*	HB*	HB*	HB*
$\epsilon_r 100$	-	2.8-3.1	2.8-3.1	2.9-3.6	3
$\tan \delta 100$	$\cdot 10^{-3}$	90-160	90-160	50-90	30-60
ρ_e	Ohm·m	10^{12}-10^{13}	10^{12}-10^{13}	10^{12}-10^{13}	>10^{14}
σ_e	Ohm	>10^{13}	>10^{13}	>10^{13}	>10^{14}
$E_B I$	kV/mm	30-40	30-40	35-45	24
W_w	%	0.8-1.6	0.8-1.6	0.6	0.6-0.7
W_H	%	0.3-0.5	0.3-0.5	0.3	0.2

Continued on next page

| Properties | Unit | Acrylonitrile-styrene-copolymers and blends |||
		ASA	ASA+PC	ABS+PA**
ρ	g/cm^3	1.07	1.15	1.07-1.09
E_t	MPa	2300-2900	2300-2600	1200-1300
σ_y	MPa	40-55	53-63	30-32
ϵ_y	%	3.1-4.3	4.6-5	
ϵ_{tB}	%	10-30	>50	>50
σ_{50}	MPa	-	-	
σ_B	MPa	-	-	
ϵ_B	%	-	-	
T_P	°C	-	-	
HDT	°C	95-105	105-115	75-80
α_p	10^{-5}/K	9.5	7-9	9
α_n	10^{-5}/K	-	-	
UL94	Class	HB*	HB*	HB
ϵ_r 100	-	3.4-4	3-3.5	
tan δ 100	$\cdot 10^{-3}$	90-100	20-160	
ρ_e	Ohm·m	10^{12}-10^{14}	10^{11}-10^{13}	2-10^{12}
σ_e	Ohm	>10^{13}	10^{13}-10^{14}	3-10^{14}
E_BI	kV/mm	-		30
W$_w$	%	1.65	1	
W$_H$	%	0.35	0.3	1.3-1.4

*also available as V-2 to V-0   ** conditioned

6.2.3 Polystyrene Foams (PS-E, XPS)

The market offers PS-, SB-, and ABS grades with chemical blowing agents, blowing agent containing concentrates, and ready-for-use blends for the production of structural foams (some with flame-retardant treatment). The blowing agents, together with $\approx 0.2\%$ butylstearate of paraffin oil as coupling agent, are blended by the processor. Structural parts are typically injection molded (thermoplastic foam injection, TSG) from impact resistant modified grades with densities ranging from 0.7 - 0.9 g/cm^3; they usually have a wall thickness of $\approx$ 5 mm and can weigh up to 30 kg. The skin of these parts is too rough for some applications, so that post-treatments such as sanding and coating are necessary. Profiles for construction and furniture are extruded. PS grades with incorporated physical blowing agents such as CO_2 are used for the extrusion of lighter foamed products, such as packing films with densities of 0.1 g/cm^3 or insulating boards with densities of 0.025 g/cm^3; physical blowing agents can also be metered directly into the extruder. Extruded rigid polystyrene foam (XPS) is a closed cell foam with a dense skin. It exhibits good compressive strength, minimal water absorption, and low flammability. Compressive strengths from 0.20 - 0.70 MPa can be achieved with raw densities ranging from 20 - 50 kg/m^3. PSE (PS particle foam, PS-expanded) are bead polymers with diameters ranging from 0.2 - 3 mm, which are polymerized in a low-boiling hydrocarbon, preferably pentane, as blowing agent. These grades are foamed in a 3-step process into rigid parts, e.g., for packaging, or into blocks measuring up to 1.25 m ×1.0 m×8.0 m. Foams of styrene copolymers with maleic acid are temperature resistant up to 120°C.

Applications for PSE Foams. General insulation, shock- and sound insulation, packaging, transport palettes, safety gear (bicycle- and other protective helmets, safety seats), surf board cores, and model airplanes. PSE parts are used in lost-core processes, e.g., in the manufacturing of intake manifolds. Concrete filled with PS-E results in light-weight concrete with a density from 700 - 900 kg/m^3. Shredded PS-E parts or production waste (chips with mesh sizes from 4 - 25 mm) are used as insulating material for thermal insulation or for soil improvement (styrene mulch). Incorporation of infrared absorbers or -reflectors into the foam structure reduces the radiation part in heat transmission and thus the thermal conductivity by 12 - 18%, depending on raw density. The pentane containing, expandable, black PS pellets are processed into silvery-grey foam blocks or parts. Incorporation of expandable hollow beads into TPS-SEBS pellets facilitates injection molding and extrusion of fine-pored foams with $\rho = 0.5$ g/cm^3 for the production of, e.g., handles, wheels, and profiles.

6.3 VINYL POLYMERS

6.3.1 Rigid Polyvinylchloride Homopolymers (PVC-U)

Chemical Constitution The major differentiating criteria for vinyl polymers are their different polymerization processes (emulsion polymerization, PVC-E, suspension polymerization, PVC-S, and mass polymerization, PVC-M) and their different properties (rigid grades without plasticizers, PVC-U), plasticized PVC (PVC-P), and PVC pastes. Table 6.9 provides an overview of the basic structures of vinyl chloride (VC) and its copolymers. Today, VC is typically produced by combining chlorine and ethylene to produce ethylrnene dichloride that is then cracked to produce vinyl chloride and subsequently polymerized. The high content of available chlorine resulting from NaCl electrolysis (homopolymeric PVC contains 56.7% chlorine) determines both the economic importance and the special properties of PVC polymers. PVC-M is extremely pure and exhibits a narrow distribution of grain sizes ($\approx$ 150 μm). Processability and thermal stability are better than those of PVC-S. Receptivity for plasticizers and liquid additives is very high.

Delivery Form, Processing While other thermoplastic materials are typically available as ready-to-use compounds, VC polymers are often mixed during processing from powdered raw materials and the necessary additives. Unless granulation of the raw materials by extrusion is technically required, free-flowing powder mixtures (grain sizes should not be too small), agglomerates, or dry-blends are used. These processing techniques are very economical and limit the exposure of the thermally sensitive PVC raw materials. PVE-E develops from primary particles with diameters of 0.1 - 2 mm. The final size and form of the powder grains is determined by the processing conditions (drying). This allows tailoring specific resin types for their respective applications: micro-sized resins for plastisol processing; fine-grained resins that are easiliy processed during calendering; coarse-grained, free-flowing resins with high bulk density for extrusion (PVC-U); and coarse-grained, porous resins for soft PVC applications. PVC-E contains up to 2.5% emulsifier and sometimes other inorganic additives. Depending on the type and level of these additives, transparency, water absorption, and electrical insulation properties are typically inferior to those of PVC-S and PVC-M. PVC-E possesses better processability, leading to finished products with smooth, non-porous surfaces, higher toughness, and low electrostatic chargeability. Both, PVC-S with a particle size distribution between 0.06 and 0.25 mm and PVC-M with a particularly uniform particle size distribution, are very pure products, due to their manufacturing process. With little stabilization, both are suitable for crystal-clear, high-quality products with high mechanical and electrical load-carrying capacity and favorable corrosion and weathering resistance. Resins with porous grains (bulk density ranging from 0.4 to 0.5 g/ml) are suitable for PVC-P; those with compact grains (bulk density ranging from 0.5 to 0.65 g/ml) are suitable for PVC-U. Micro-PVC-S polymers can be encapsulated (grain size 10 - 1 μm). Chlorinated PVC (PVC-C), with a chlorine content increased by up to 60% by post-chlorination, is more difficult to process than PVC, but thermally stable to above 100°C and shows even higher resistance to chlorine. VC polymers cannot be processed without stabilizers minimizing discoloration and further damage both by oxidation and HCl separation during processing at high temperatures and by elevated temperatures and UV exposure during end use. Also used are flow enhancers, lubricants, and UV-absorbers. Finely ground calcium carbonate at a filler load of 5 - 15% facilitates extrusion and increases notched bar impact toughness. PVC melts are highly viscous, shear- and temperature sensitive, but adhere less to the contact surfaces of the conveying equipment than other polymeric materials. Extrusion of pipes, profiles, and sheet and calendering and pressing of sheet are the major industrial processing methods. Injection molding and blow molding are also used. Calendering of rigid PVC sheet on an industrial scale is performed in several steps: metering and mixing of the compound ingredients, plasticization, typically with continuous screw kneaders, high-temperature calendering with temperatures increasing from 160 - 210°C, until sheet thicknesses ranging from 0.02 - 1 mm (predominantly 0.1 - 0.2 mm for PVC-S and PVC-M) are obtained. The low-temperature process (Luvitherm) is used for PVC-E films with a high K-value. Here, temperatures decrease from 175 - 145°C, the film is calendered, and subsequently stretched in one or both directions over a roller at 240°C. Calendered film with a thickness of $\approx$ 0.5 mm is used to laminate stress-free panels and blocks in multi-platen presses. Powdered or pelletized resin is used for extrusion of PVC-U at temperatures from 170 - 200°C. Powdered mixtures require twin screw extruders or single screw extruders with a length of 20D. Only PVC-S or PVC-M resins with low K values and appropriate processing aids are suitable for injection molding (melt temperatures from 180 - 210°C, mold temperatures from 30 - 50°C) or blow molding of bottles and other hollow articles. Processing has to be more careful than with other thermoplastics because over-heating must be avoided. It is particularly important to design the screws and molds for optimum flow conditions (no dead zones), to avoid degradation and subsequent separation of hydrochloric acid. In addition, corrosion resistant surface treatment of screws and molds is recommended. Welding of PVC-U can be achieved by all standard processes used for thermoplastic materials: hot gas-, hot plate-, ultrasonic-, and high-frequency-welding. Solvent adhesives based on tetrahydrofuran (THF) or two-component adhesives based on EP, PUR, or PMMA are used for bonding. Adhesive bonding is achieved with PUR, nitrile-, or chlorobutadiene rubbers. Thermoforming of panels, pipes, etc., is used on a large scale in the construction of chemical apparatus.

Identification VC polymers are identified according to their molecular weight by K-values or viscosity numbers. Both numbers are based on the relative viscosity in solution and are proportional to each other. There is a correlation between the following characteristic values: K-value (ISO 1628-2 and DIN 53726), specific viscosity, viscosity number J (DIN 53726-8), inherent viscosity (ASTM D 1234-T), melt flow index (ISO 1133 and ASTM D1238) and number and weight average molecular weight (see also resin standards ISO 1163 (PVC-U) and ISO 2898 (PVC-P)). PVC resins used for thermoplastic processing exhibit K-values between 50 and 80, according to DIN EN ISO 1628-2. Higher K-values indicate better mechanical and electrical properties of the molded parts; however, higher K-values also indicate more difficult processing conditions for PVC-U. Table 6.19 shows typical K-values for standard processing methods. DIN EN ISO 1060/1 contains a classification and identification for VC polymers.

6.3 Vinyl Polymers

Table 6.19: Areas of Application for Various Grades of PVC

PVC grades	PVC-U			PVC-P		
	E	S	M	E	S	M
Processing	K-values*			K-values*		
Calender	(60-65)	57-65	57-65	70-80	65-70	70
Thermal films	78	-	-	-	-	-
Flooring	-	-	-	65-80	65-80	
Extrusion	-		-		-	
PVC-U	-	-			-	
-Pipes	-	67-68	67-68	-		
-Window profiles	-	68-70	-	-		
-Construction profiles	60-70	60-68	60-68	-		
	-		-		-	
-Sheets	60-65	60	60	-		
-Blown films	60	57-60	57-60	-		
Extrusion	-	-	-		-	
PVC-P						
Preferred	-	-	-	65-70	65-70	65-70
Cable coating	-	-	-	-	70-90	
Preferred	-	-	-	-	70	70
Blow molding	-	57-60	57-60	-	65-80	60-65
Injection molding	-	50-60	56-60	-	65-70	55-60
Paste technology	-	-	-	65-80	(70-80)	

* according to DIN 53726: 0.25g PVC I 50 ml cyclohexanone

Properties and Applications PVC-U is a thermoplastic material with high rigidity and a high modulus of elasticity, but with low abrasion resistance, impact resistance at low temperatures, and long-term alternating fatigue strength. Service temperatures are also relatively low: short-term up to 75°C, long-term up to 65°C. Appropriate selection of stabilizers and additives ensures excellent electrical properties, particularly in the low voltage, low frequency range. Higher frequencies cause heating because of the high electric loss factor. N_2-, O_2-, CO_2-, and air permeability are lower than for polyolefins; water vapor permeability is higher. Up to temperatures of 60°C, PVC-U is resistant to most diluted and concentrated acids, except for oleum-containing sulfuric acid, lactid acid and concentrated nitric acid, diluted and concentrated base and saline solutions. Gaseous chlorine forms a protective layer of chlorinated material; liquid halogens corrode PVC. PVC is resistant to alcohols, gasoline, mineral oils, other oils and fats; esters, ketones, chlorinated hydrocarbons, and aromatic hydrocarbons swell or dissolve PVC to various degrees. PVC is largely stress crack resistant. Appropriately stabilized PVC-U is suitable for outdoor applications (window profiles, etc.). PVC-U is physiologically indifferent. Most PVC-U products are flame retardant[1], even without special flame inhibiting additives (B1 according to DIN 4102, up to V0 according to UL 94). The flame will extinguish after the ignition source has been removed. Depending on the type of polymerization, PVC is translucent to transparent and can be easily colored. With PVC, fillers are typically only used as extenders; glass fibers are rarely used as reinforcement (see Table 6.20).

Table 6.20: Properties of Filled and Reinforced PVC

PVC-U additive	ψ (%)	σ_B (MPa)	ϵ_B (%)	E_t (MPa)	Vicat-T (°C)	ρ (g/cm^3)
none	-	60	6-10	2700	85	1.36
CaCO$_3$	30	46	8	3200	94	1.53
	100	-	-	-	116	1.78
Precipitated CaCO$_3$	15	30-47	6	3100	87	1.45
Chalk	20	34	6	3500	-	1.48
Silicaflour	20	38	-	3100	-	-
Glass fibers	40	25	3	8000	85	-
Wollastonite	20	25	5.4	-	-	1.47

With PVC-U in pipe and profile production, 1 - 2% fine grain chalk (5 - 10 μm) is used as processing aid. A filler load of 5 - 15% can double notched impact toughness and more than 40% is used for pipes and profiles (pressure pipe, waste pipe, drainage pipe, sun shade profiles), depending on the mechanical stress to be sustained. Chalk will absorb separated acids, thus improving thermal stability. Chalk also counteracts any plating effect (deposit of exuded particles on calenders and extruders). Because kaolin increases the volume resistance it is used for PVC-P cable resins. Silicates reduce the plate-out effect, increase the thixotropy, and result in products with a matte surface. Aluminum hydroxide increases the flame retardance, e.g., for carpet backing. Barium ferrite is used as a filler for magnetic sealing profiles for refrigerators, among others. PVC- HI with 5 - 12% impact modifier content are 2-phase resins. Polyacryl acid ester - elastomers (ACM), PE-C, and EVAC as a dispersed soft phase are typical impact modifiers. ACM is used either as graft copolymer with VC with 6 - 50% AN content or as copolymer with methylmethacrylate (MA) with 60 - 90% AN content. High-grade ACMs are mixed with PVC-S, -M, or -E to a common impact modifier concentration of 5 - 7% or above. In contrast to ACM, PE-C and EVAC, to a certain degree, are shear sensitive impact modifiers; that is, the achievable impact resistance depends on the processing conditions. Polyacrylates modified with polystryrene are used to manufacture transparent parts with improved impact resistance.

---

[1] While PVC is self extinguishing, it does generate hydrochloric acid and dioxins when burned.

Table 6.21: Comparison of Properties of Vinyl Chloride Polymers and Blends

Properties	Unit	\multicolumn{4}{c}{Vinyl chloride polymers and blends}
		PVC-C	VCA acrylate mod.	PVC+VCA	PVC-PE-C
ρ	g/cm^3	1.55	1.34-1.37	1.42-1.44	1.36-1.43
E_t	MPa	3400-3600	2200-2600	2500-2700	2600
σ_y	MPa	70-80	45-55	45	40-50
ϵ_y	%	3-5	4-5	4-5	3
ϵ_{tB}	%	10-15	35->50	>50	10->50
σ_{50}	MPa	-	-	-	-
σ_B	MPa	-	-	-	-
ϵ_B	%	-	-	-	-
T_P	°C	-	-	-	-
HDT	°C	≈10	72	74	69
α_p	10^{-5}/K	6	8-9	7-7.5	8
α_n	10^{-5}/K	-	-	-	-
UL94	Class	V-0	V-0	V-0	V-0
ϵ_r 100	-	3.5	3.5	3.5	3.1
tan δ 100	·10^{-3}	140	120-140	120	140
ρ_e	Ohm·m	>10^{13}	>10^{13}	10^{13}	10^{12}->10^{14}
σ_e	Ohm	10^{14}	>10^{13}	>10^{13}	>10^{13}
E_BI	kV/mm	15	30	-	-
W$_w$	%	0.1	<0.25	0.5	0.1
W$_H$	%	0.01	<0.01	<0.1	0.03

Properties	Unit	\multicolumn{4}{c}{Vinyl chloride polymers and blends}
		PVC+ASA	PVC-P with 75/22	DOP 60/40	PVC-U
ρ	g/cm^3	1.28-1.33	1.24-1.28	1.15-1.20	1.38-1.4
E_t	MPa	2600-2800	-	-	2700-3000
σ_y	MPa	45-55	-	-	50-60
ϵ_y	%	3-3,5	-	-	4-6
ϵ_{tB}	%	≈8	>50	>50	10-50
σ_{50}	MPa	-	-	-	-
σ_B	MPa	-	-	-	-
ϵ_B	%	-	-	-	-
T_P	°C	-	-	-	-
HDT	°C	65-85	-	-	65-75
α_p	10^{-5}/K	7.5-10	18-22	23-25	7-8
α_n	10^{-5}/K	-	-	-	-
UL94	Class	V-0	-	-	V-0
ϵ_r 100	-	3.7-4.3	4-5	6-7	3.5
tan δ 100	·10^{-3}	100-120	0,05-0,07	0,08-0,1	110-140
ρ_e	Ohm·m	10^{12}->101	10^{12}	10^{11}	10^{13}
σ_e	Ohm	10^{12}-10^{14}	10^{11}	10^{10}	10^{14}
E_BI	kV/mm	-	30-35	≈25	20-40
W$_w$	%	0.4-0.8	-	-	0.1
W$_H$	%	0.1-0.3	-	-	0.01

All PVC-HI are long-term weather resistant and therefore suitable for outdoor applications. MABS and, particularly for hollow parts, MBS are used as impact modifiers for crystal clear packaging materials. Because of their butadiene contents they are less suitable for outdoor applications.

Applications, see also Table 6.19. Comparison of properties, see Table 6.21.

6.3.2 Plasticized (soft) Polyvinylchloride (PVC-P)

Chemical Constitution Hardness and brittleness of PVC can be influenced to a high degree by addition of plasticizers. They increase the distance or free volume between the PVC chain molecules, thus reducing the bonding forces. This effect leads to a deduction in glass transition temperature. When a low glass transition temperature polymer is blended into the PVC molecule, such as VCEVAC, the process is referred to as "internal plasticization." "External plasticization" is achieved by a low-molecular plasticizer penetrating the PVC and causing "swelling". This comes with the danger of exudation or seeping out, particularly at high temperatures and over longer time periods. Therefore, a distinction is made between primary plasticizers that are well gelled and do not exude, and secondary plasticizers that show little dipole effect and do not gel well. The latter are used in combination with primary plasticizers, which reduces migration tendencies and improves low-temperature toughness and extraction resistance. To reduce cost, primary plasticizers are often partially replaced by extenders, which are liquid fillers with low volatility and medium polarity. They do not act as gelatinizing agents; however, they can improve the rheological behavior of plastisols (fatty acid esters, alkylated aromates, naphtenes) or improve the flame retardance (liquid chloroparaffins). Polymeric plasticizers are resistant to solvents and migration.

Delivery Form, Processing PVC-P resins, already equipped with stabilizers, lubricants, etc., are available as pellets for injection molding and blow molding of hollow parts and as free flowing powders for extrusion. Powder mixtures are used for dip coating and electrostatic coating. PVC-P melts flow under comparatively low pressures; however, they should be injection molded at the maximum permissible melt temperature (170 - 200°C). If lower processing temperatures are used, the parts will not display optimum mechanical and electrical properties; post-molding shrinkage will be irregular and they will exhibit uneven, matte surfaces. Rapid cooling, i.e., a cold mold (mold temperatures from 15 - 50°C) will show similar results. As with PVC-U, molds and machines should be corrosion resistant. Resins with a hardness of Shore A 60 - 80 are extruded at melt temperatures of 120 - 165°C; harder resins require melt temperatures from 150 - 190°C. Blow molding of hollow parts is common practice. All adhesion and welding methods used for thermoplastic materials can be applied; high-frequency welding is used in particular for purses and bags.

Plasticizers Table 6.22 provides a overview of commonly used types of PVC primary plasticizers and their characteristic properties. DIN 53400 covers testing of characteristic values for uniformity and purity of plasticizers. Phthalate plasticizers (Group 1 in Table 6.22), primarily the all-purpose plasticizer DOP, account for 65 - 70% of all plasticizers consumed. For specific applications, phthalates of short-chained alcohols are used. Esters of aliphatic dicarbon acid (Group 2) are predominantly used in blends with phthalates to improve low temperature impact resistance of PVC parts. Phosphoric acid esters (Group 3) are favored for the manufacturing of technical parts with high flame-resistance. Alkylsulfonic acid esters of carbolic acid (Group 4), similar to DOP in their plasticizing behavior, exhibit minimal volatility. These esters provide good high-frequency weldability and weatherability, despite yellowing. Citric acid esters (Group 5) are special plasticizers for products subject to FDA regulation for food contact. Trimellitates (Group 6) are used for products subjected to elevated service temperatures for an extended time. Epoxidized products (Group 7) are added to PVC-C primarily because of their costabilizing properties. Their exclusive use in significant quantities can lead to exudation. Polyester plasticizers (Group 8) available as oligomers or polymer plasticizers offer a broad range of products, because they allow for the choice of esterification components and molecular weights ranging from 600 to 2000 g/mol and beyond. In addition to their low volatility, they are distinguished by their good extraction resistance against fats, oils, and fuels. Migration problems in contact with other substances can be resolved with these plasticizers, as long as the products are compatible.

Table 6.22: PVC Primary Plasticizers

Group	Name	Acronym	Characterization
1	Phthalate-plasticizers		
	Dioctylphthalate	DOP	Special plasticizer for plastisols
	Di-iso-heptylphthalate	DHP	Standard plasticizer for PVC, high gelling capacity, little volatile
	Di-2-ethylhexylphthalate	DEHP	Balanced heat-, low temperature-, water resistance and electrical properties
	Di-iso-octylphthalate (Phthalic acid mixture)	DIOP	
	Di-iso-nonylphthalate	DINP	Plasticizing effect, volatility, low temperature resistance decrease from DINP to DITDP (compared to DOP); heat resistant
	Di-iso-decylphthalate	DIDP	
	Di-iso-tridecylphthalate	DITDP	
	C_7-C_9 Phthalates	-	Mixed alcohol-esters;

Continued on next page

Group	Name	Acronym[1]	Characterization
	C_9-C_{11} linear alcohols	-	compared to DOP: low viscosity (for pastes),
	C_6-C_{10} n-Alkylphthalates	-	better low- temperature and water resistance,
	C_8-C_{10} Alkylphthalates	-	less volatile (important for artificial leather, floor coverings)
	Dicyclohexylphthalat	DCHP	Limited applications, resistant to
	Benzylbutylphthalate	BBP	fuel extraction; good gelling, for foam pasted, floor coverings
2	Adipic-, azelaic-and Sebacic acid		
	Di-2-ethylhexyl adipate	DOA, DEHA	DOA: outstanding low-temperature-resistant plasticizer, light fast; more volatile and water sensitive than DOP
	Diiso-nonyl adipate	DINA	DINA-DIDA: less low-temperature resistant and volatile than DOA
	Diiso-decyl adipate	DIDA	
	Di-2-ethylhexyl azelate	DOZ	Less water sensitive than adipate, similar to
	Di-2-ethylhexyl sebacate	DOS	DOS, best low-temperature resistance, little volatile
3	Phosphate Plasticizer		
	Tricresyl phosphate	TCF	Flame resistant, for heavy-duty mechanical and elctrical parts; not suitable for food contact;
	Tri-2-ethylhexylphosphat	TOF	flame retarded, light fast, less heat resistant than TCF; low-viscous (for pastes)
	Aryl-alkyl mixed phospates	-	similar to TOF, fuel resistant
4	Alcyl-sulfonic acid-phenyl-ester	ASE	Similar to DOP, less volatile than phthalates, tendency to discoloration, weather resistant
5	Acetyl-tributyl citrate	-	Similar to DOP, suitable for food contact
6	Tri-2-ethyl hexyl trimellitate	TOTM	Little volatile, thermally highly resistant
	Tri-iso-octyl trimellitate	TIOTM	High price (cable compounds)
7	Epoxidized fatty acid esters		Butyl-, Octyl-epoxy stearate; low-temperature resistant, little volatile, synergistically stabilizing with Ca
	Epoxidized linseed oil	ELO	ELO and ESO primarily for improving heat stability,
	Epoxidized soy oil	ESO	resistant to extraction
8	Polyester plasticizer	-	Polyester from (propane-, butane-, pentane- and hexane-) diols with dicarboxylic acids of group 1 and 2. Not volatile, little temperature dependent, largely resistant to extraction and migration.
	Oligomer plasticizer	-	Viscosity < 1000 mPa · s, also mixed with monomer plasticzers for pastes.
	Polymer plasticizer	-	Viscosity up to 300,000 mPa · s, for extrusion and calendering

6.3 Vinyl Polymers

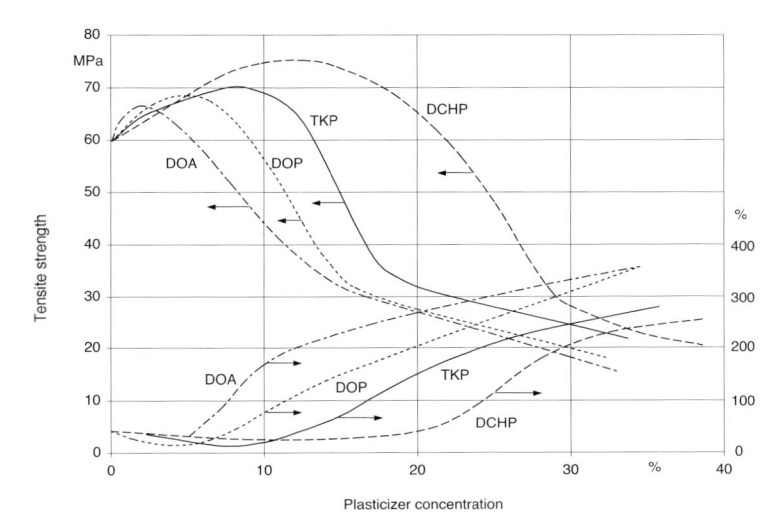

Figure 6.7: Tensile strength and elongation at break for PVC-P at 23°C, depending on type and amount of plasticizer. DOP: dioctylphthalate, DOA: dioctyladipate, TCP: tricresyl phosphate, DCHP: dicyclohexyl phthalate.

Properties and Applications PVC-P resins contain 20 - 50% plasticizers. Low plasticizer content that does not allow sufficient homogeneous gelation will cause embrittlement; rigidity will increase and elongation at break may decrease, see Fig. 6.7. Beyond this limit, PVC-P behaves more rubber-like with increasing plasticizer content. It is therefore common to characterize PVC-P grades by their Shore A hardness. Shore A hardness values of 96 to 60 are common for extrusion, 85 to 65 for blow molding, and up to 50 for injection molding. Highly filled resins with Shore A values from 85 to 70 for floor coverings and cable sheathing exhibit low elongation at break and are less flexible at low temperatures than unfilled resins. Low-temperature brittleness depends on the type and amount of plasticizer, see Tables 6.21 and 6.23. At 40 - 60°C, PVC-P exhibits distinct permanent deformation after higher mechanical loads for prolonged periods of time, which distinguishes it from rubber. The upper temperature limit for long-term use is at 80°C, because above this limit plasticizer loss increases markedly while rigidity decreases. For more temperature resistant resins, for example, for cable sheathing, trimellitate- and polyester plasticizers should be employed, despite their unfavorable low-temperature behavior. In general, the specific volume resistance will decrease with increasing plasticizer content. Addition of >13% carbon black will lower the volume- and surface resistivity by ≈ 6 to 8 orders of magnitude. Monomer plasticizers can migrate into other resins, such as thermoplastics, rubber, and lacquers on contact. This will cause embrittlement of the PVC-P and potentially of the absorbing material, and may initiating stress cracks. Although most plasticizers are physiologically indifferent, the use of plasticizer-containing plastics is undesirable in food contact applications, when there is a danger of plasticizer migration.

Blends of PVC-P with PVCEVAC-plasticizers, TPU, or NBR/AN copolymers exhibit excellent flexibility at low temperatures together with a general stability under continuous use in a wide temperature range. TPU- and NBR grades are also resistant to fats, oils, and fuels. Comparison of properties, see Table 6.21.

Table 6.23: Shore A and D Hardness of PVC-P

Shore-Hardness		General characteristics	Low-temperature brittleness range * (°C)
A	D		
98-91	60-40	Semi-rigid	0 to -20
90-81	39-31	Bend leather-like	-10 to -30
80-71	Not used	Taut rubbery	-10 to -45
70-61	Not used	Medium rubbery	-30 to -50
60	Not used	Very soft injection molded parts	-40 to -50

* measured by drop-hammer test

Applications

Injection molding: seals, protective caps, suction cups, bicycle and motorcycle handles, boots, sandals, shoe soles, shock absorbers.

Extrusion: pipes, tubes, hand rail profiles, sealing profiles, cable- and wire sheathing, insulating tapes and tubes (limited to low frequencies because of the high dielectric constant); also by calendering: sheet, film for shower curtains, coatings for weather gear, conveyor belts; floor coverings.

Blow molding: armrests in automotive applications that are PUR foam resistant, dolls, balls, tubes[2].

VCVAV copolymers are used as resins with up to 50% VAC content. Its properties compare to those of PVC; they have to be processed at lower melt temperatures and exhibit lower heat distortion resistance. Applications are special films for thermoforming. Copolymers with vinyl ethers and PP are also used for this purpose. VCVDC copolymers with high VDC contents are more temperature resistant than PVC; however, because of their tendency to separate HCl at elevated temperatures they are more difficult to process. Homopolymer

---

[2] While bottles are often blow molded out of PVC, they are often confused with PET when introduced into the recycling stream, contaminating a batch containing PET bottles. Due to higher processing temperatures of PET, a PVC bottle can destroy the PET batch due to production of hydrochloric acid.

PVDC decomposes below melting temperature and is therefore not commercially available.

6.3.3 Polyvinylchloride: Copolymers and Blends

Constitution Table 6.9 show a list of vinyl chloride copolymers, Table 6.15 lists terpolymers, and Table 6.24 shows the influence of comonomers on the properties of polyvinyl chloride.

Table 6.24: Influence of Comonomers on the Properties of Polyvinyl Chloride

Monomer	Line-Nr.in Tabl	Influence on PVC
Ethylene	1	Tougher at low temperatures, easier to process
α-Methylstyrene	6	Higher heat distortion point
Chlorinated polyethylene	9	Cl statistically distributed, increased impact resistance, more
Vinylidene chloride	10	Lower heat distortion point, easier to process
Methacrylate	16	Easier to process, easier thermoforming
Vinyl acetate Ethylene vinyl a	20	Lower heat distortion point, easier to process
Vinylether	22	Higher rigidity, more difficult to process
Acrylonitrile	26	Higher rigidity and impact resistance
Maleic(acid)imide	32	Higher heat distortion point, more difficult to process
Acrylic acid ester-elastomer	33	Increased impact resistance

Modifier resins, mostly based on high-molecular methacrylate-(multi-)-polymers are blendable in single-phase with PVC and result in crystal clear products. They improve the flow behavior of the melt and the Erichsen index in the thermoelastic range so that impact resistant and weather resistant products can be manufactured under gentle processing conditions. Modifier resins based on α-methylstyrene/acrylonitrile and styrene/maleic acid anhydride improve temperature resistance by 10 - 15 K. Other blends are manufactured with ABS, NBR, PE-C, PMMA. Comparison of properties, see Table 6.21.

Applications PVC-U, copolymers, and blends: semi-finished products such as films, sheet, corrugated sheet, profiles, pressure pipes for water and chemical industry, sewage and drainage pipes, fittings, chemical apparatus, shutter elements, eaves, window and door profiles, siding, reflecting road markers, housings, packaging films, bottles and other blow molded parts for packaging, foam, barrier layers against O_2, H_2, and flavors in food contact, plasticizer-free soft films (VCEVAC), fibers, filaments, nonwoven fabric, nets.

6.3.4 Polyvinylchloride: Pastes, Plastisols, Organosols

Pastes, Plastisols, Organosols Dispersion of fine grain PVC-U particles (K-values between 56 and 80) in plasticizers is used to manufacture PVC pastes. No reaction takes place between the PVC and the plasticizers at temperatures up to $\approx 35°C$. Gelation of the pastes will occur on heating to 150 - 220°C, which will then exhibit typical properties of PVC-P. VCVAC copolymers are used to lower the gelation temperature. The composition of the pastes defines plastisols, which consist primarily of PVC and plasticizers and sometimes small amounts of extenders, stabilizers, pigments, and fillers. Organosols are plastisols with a larger content of volatile diluents, such as gasoline or glycols, to reduce viscosity, thus making them suitable for lacquers. Plastigels are made kneadable and gel-like by addition of high amounts of colloidal silica or metal soaps. For selection of plasticizers, see above. Monomer glycol methacrylate is added as polymerizable plasticizer to lower the viscosity of the pastes. With additionally added catalysts, they polymerize during gelation to form materials with increased hardness. Pastes can be manufactured in PVC/plasticizer ratios ranging from 50:50 to 80:20; they result in rubber-like to PVC-U-like materials. At higher plasticizer contents, the materials will turn jelly-like. Fast turning mixers with subsequent filtration and venting, particularly required for dipping- and casting pastes, are used for manufacturing. Temperature-sensitive resins are produced in slow moving mixers with subsequent passage over a water-cooled calender. PVC can be further processed in a variety of methods: spread coating, dipping methods, cast and spray processes as well as rotary screen printing. Laminar carriers (textile, paper, glass- and mineral non-woven materials, sheeting) are typically converted into artificial leather, textiles for rain gear, tarps, floor and wall coverings, and sidings by several subsequent spread coatings with either compact or foamed material. The dipping method is used to manufacture protective gloves and coatings on metallic carrier materials. Casting methods, in particular rotational casting, are used for hollow parts, such as balls, dolls, bellows, and so forth. Sprayable plastisols are typically used with airless-methods for metal coatings, e.g., for underbody coatings in automotive applications. With rotary screen printing, PVC pastes are (partially) applied to paper or textiles (among others) to manufacture tarps, artificial leather, and floor and wall coverings (wall paper).

Foams Ready-to-use PVC-U resins containing blowing agents (PVC-E, expanded) are available for extrusion of structural foamed pipes and other profiles and for thermoplastic foam casting. They are used for door frames, window sills, and furniture parts with overall densities from 0.7 - 0.9 g/cm^2. PVC-P compounds (e.g., 100/70 PVC/DOP) with 0.5 part foaming agent and blended with 20 - 30 parts nitrile butadiene rubber (NBR) or EVA are used for sealing profiles, shoe soles, bumpers, and shock absorbers. At pressures from 200 - 600 bar, closed-cell blocks of PVC-U or -P are produced with the Airex technique. Cross-linking with diisocyanate increases the heat deflection temperature. Pastes for foaming on carrier materials are formulated to allow for gelling and foaming in one step, e.g., at 180 - 200°C in the heating channel of a spread coating machine. With this method, single- and multi-layer foamed artificial leathers and carrier-free foamed film are produced. Chemically embossed cushioned vinyl floor and wall coverings are produced by printing an inhibitor (TMSA, benzotriazol) on the pre-gelled foam layer, which will locally prevent foaming during the subsequent foaming step at 190 - 220°C.

6.3.5 Vinyl Polymers, other Homo- and Copolymers (PVDC, PVAC, PVAL, PVME, PVB, PVK, PVP)

The general chemical structure of the homopolymers is shown in Table 6.9.

Polyvinylidene Chloride PVDC decomposes below melting temperature and therefore it is only used as comonomer with PVC.

Polyvinyl Acetate (PVAC) PVACs (density = 1.17 g/cm^3, MM = 35,000 - 2,000,000 g/mol) are crystal clear, soft to rigid resins. Because of their insufficient temperature resistance they are not suitable as molding materials; however, they are easily soluble in most solvents (except for aliphatic hydrocarbons and water-free alcohols). They form lightfast, gasoline-, oil-, and water resistant films that slightly swell in water (water absorption up to 3%) and exhibit high pigment-binding properties and limited resistance to nitrocellulose and plasticizers. PVAC is susceptible to saponification.
Applications: Paints and coating compounds, also in the form of dispersions, adhesives, and finishes.

Polyvinyl Alcohol (PVAL) PVAL is produced by saponification of polyvinyl acetate; it is a white-yellowish powder that does not dissolve in organic solvents. Partially saponified grades with ≈ 13% PVC content are easily water soluble, better than fully saponified PVAL. Applications: clear films; with hydrophilic alcohols such as glycerin as plasticizer, PVAL can be processed like a thermoplastic material to age-resisting, leather-like products (fuel-, oil-, and solvent-resistant tubes, membranes, gaskets, and release sheets, e.g., for processing of UP resins); UV radiation renders PVAL with a bichromate addition impermeable by water; it is used in graphic art.

Applications for solutions: protective colloids for dispersion and stabilization, thickeners for the plastics, cosmetics, and pharmaceutical industry, textile dressings and glazes, paper glues, binders for color pencil leads, paints and printing inks, adhesives, coatings.

Polyvinyl Methyl Ether (PVME) PVMEs are soluble in cold water, and almost all organic solvents, except gasoline-like hydrocarbons, ethyl ether, isobutyl ether, and decalyl ether. Depending on the degree of polymerization, they are oils, or soft or rigid, non-saponifiable resins.

Applications: electric insulations, self-adhesive resins, blended resins for chewing gum and dental resins, octadecyl ether for high-gloss floor polish (V-wax).

Applications for solutions, see PVAL.

Polyvinyl Butyral, Polyvinyl Formal (PVB, PVFM) Polyvinyl acetals are created by conversion of polyvinyl alcohols with aldehydes. PVB and PVFM are solid resins soluble in organic solvents.

Applications: lacquers, (including hot-sealable lacquers, enamels, gold lacquers, printing inks, primers), impregnating resins, adhesives, shrink crown caps, peelable packaging, oil- and gasoline-resistant tubes and gaskets. High-molecular PVB is used for interlayers in safety glass; together with PVFM it is used for gasoline-resistant coatings for gasoline tanks; together with phenolic resins it is used for wire coating and heat-setting metal adhesive films.

Polyvinyl Carbazole (PVK) PVK is a crystal clear thermoplastic material like PS; however, because of the voluminous side groups (see Table 6.9), it exhibits significantly higher thermal resistance: short-term 170°C, long-term 150°C. Copolymers with styrene are known. The electrical properties are very good. PVK is resistant to alkaline and saline solutions, acids (except concentrated chromate, nitric and sulphuric acid), alcohols, esters, ethers, ketones, carbon tetrachloride, aliphatic hydrocarbons, mineral oils with low aromatic contents, caster oil, water and water vapor up to 180°C. PVK is non resistant to dimethylformamide, fuels, aromatic hydrocarbons, chlorinated hydrocarbons, such as chloroform, methylene chloride, and tetrahydofuran. Resistance against stress cracking and weathering are excellent. PVK is self-extinguishing; however; it must not be used in food contact.

Applications: PVK is used in the chemical industry because of its good thermal and chemical resistance; it is used in high-frequency applications with additional mechanical loads.

Polyvinyl Pyrrolidone and Copolymers (PVP) PVP is available in water-soluble and in highly water-swelling modifications. PVP and its copolymers are used as harmless non-skin irritating binders in the cosmetic and medical industry and for numerous other industrial applications. The blood substitute Periston is a solution of polyvinyl pyrrolidone.

6.4 FLUOROPOLYMERS

6.4.1 Fluoro Homopolymers (PTFE, PVDF, PVF, PCTFE)

Chemical Constitution. In fluoropolymers, the hydrogen atoms (H) in the main carbon chain have been completely or partially substituted by fluorine atoms (F), see Table 6.25. The F-atoms have a higher volume than the C-atoms and form a close, protective layer around the carbon chain. In addition, the F-C bond is very stable; therefore, these polymers exhibit excellent chemical resistance, even at elevated temperatures. They are weather-resistant without further stabilization, physiologically indifferent, non-flammable, and embrittle only at low temperatures. PTFE exhibits the highest heat deflection temperature of all engineering plastics: the use-temperature's upper limit short-term is 300°C, long-term 250°C. Stiffness (fracture strength below 50 MPa) and rigidity (modulus of elasticity 350 - 1800 MPa) are rather low, elongation at break is above 100%. Depending on molecular structure and processing, the degree of crystallinity can be up to 94%. Depending on fluorine content, the gross density can be up to 2.2 g/cm^{3}cm^3; therefore, PTFE is one of the plastic materials with the highest densities. Mold shrinkage is high, depending on degree of crystallization.

The following sections will describe the general properties of fluoropolymers and their areas of application. Table 6.25 provides information on service temperatures, and processing methods. Table 6.26 provides an overview of characteristic values.

Polytetrafluoroethylene (PTFE) PTFE undergoes a phase transition at 19°C, associated with a 1.2% increase in volume, which has to be taken into consideration when dimensioning and machining parts (advisable at 23°C). The volume increases by 30% when heated from 20°C to the range of crystal melting temperature of $\approx$ 327°C, at which PTFE transforms into a clear gel-like material. This causes a corresponding, process-dependent anisotropic shrinkage when the melt is cooled.

Table 6.25: Fluorine-Containing Polymers

	Chemical	Acronym	Chemical formula	Service temperatures (°C) Short term	Long term	Processing
(1)	Polytetrafluorpethylene	(PTFE)	$-[CF_2-CF_2]-$	300	-270 to 260	Compression Sintering Ram-extrusion
(2)	Polyvinyledene flouride	(PVDF)	$-[CH_2-CF_2]-$	-	-60 to 150	Injection Extrusion Coating
(3)	Polyvinyl fluoride	(PVF)	$-[CH_2-CHF]-$	-	-70 to 120	Semifinished product
(4)	Polychlorotrifluoro-ethylene-copolymer	(PCTFE)	$-[CFCL-CF_2]-$	180	-40 to 150	Injection Rod extrusion
(5)	Ethylene-chlorotri-	(ECTFE)	$-[CFCL-CF_2]-[CH_2-CH_2]-$	60	-75 to 140	Injection Extrusion Coating
(6)	Ethylene-tetra-fluoroethylene-copolymer	(ETFE)	$-[CF_2-CF_2]-[CH_2-CH_2]-$	200	-155 to 190	Injection Extrusion
(7)	Polyfluoroethylene-propylene	(FEP)	$-[CF_2-CF_2]-[CF_2-CF]-$ $\mid$ CF_3	250	-200 to 205	Injection Extrusion Compression Sintering
(8)	Perfluoropropylvinylether	(PFA)	$-[CF_2-CF_2]-[CF_2-CF]-$ $\mid$ $R=C_nF_{2n}$ $O-R$	$_nF_{2n}$	PFA: -270 to 260 TFB: -100 to 130	Injection Extrusion Coating
(9)	2,2 bis(trifluoromethyl)-4,5-difluoro-1,3-dioxolan	(AF)	$-[CF_2-CF_2]-[CF-CF]-$ $O\ \ O$ $\backslash\ /$ C $/\ \backslash$ $CF_3\ CF_3$	300	-270 to 260	Injection Extrusion Coating
(10)	Tetrafluoroethylene-hexafluoropropylene-vinylidenefluoride copolymer	(THV)	$-[CF_2-CF_2]-[CF_2-CF]-[CH_2-CF_2]-$ $\mid$ CF_3		-50 to 130	Injection Extrusion Coating

Table 6.26: Property Comparison of Fluoropolymers

Properties	Unit	Vinyl chloride polymers and blends			
		PTFE	PCTFE	PVDF	FEP
ρ	g/cm³	2.13-2.23	2.07-2.12	1.76-1.78	2.12-2.18
E_t	MPa	400-750	1300-1500	2000-2900	400-700
σ_y	MPa	-	-	50-60	≈ 10
ϵ_y	%	-	-	7-10	-
ϵ_{tB}	%	>50	>50	20->50	>50
σ_{50}	MPa	20-40	30-40	-	15-25
σ_B	MPa	-	-	-	-
ϵ_B	%	-	-	-	-
T_P	°C	325-335	210-215	170-180	255-285
HDT	°C	50-60	65-75	95-110	-

Continued on next page

6.4 Fluoropolymers

Properties	Unit	Vinyl chloride polymers and blends			
		PTFE	PCTFE	PVDF	FEP
α_p	10^{-5}/K	15-20	6-7	10-13	8-12
α_n	10^{-5}/K	-	-	-	-
UL94	Class	V-0	V-0	V-0	V-0
$\epsilon_r 100$	-	2.1	2.5-2.7	8-9	2.1
tan $\delta 100$	$\cdot 10^{-3}$	0.5-0.7	90-140	300-400	0.5-0.7
ρ_e	Ohm·m	>10^{16}	>10^{16}	>10^{13}	>10^{16}
σ_e	Ohm	>10^{16}	>10^{16}	>10^{13}	>10^{16}
E_BI	kV/mm	40	40	40	40
W_w	%	<0.05	<0.05	<0.05	<0.05
W_H	%	<0.05	<0.05	<0.05	<0.05

Properties	Unit	Vinyl chloride polymers and blends			
		PFA	ETFE	ETFE GF25	ECTFE
ρ	g/cm^3	2.12-2.17	1.67-1.75	1.86	1.86-1.70
E_t	MPa	600-700	800-110	8200-8400	1400-1700
σ_y	MPa	≈50	25-35	-	≈50
ϵ_y	%	-	15-20	-	-
ϵ_{tB}	%	>50	>50	-	>50
σ_{50}	MPa	20-35	40-50	-	40-50
σ_B	MPa	-	-	80-85	-
ϵ_B	%	-	-	8-9	-
T_P	°C	305	265-270	270	240
HDT	°C	45-50	70	210	≈75
α_p	10^{-5}/K	10-12	7-10	2-3	7-8
α_n	10^{-5}/K	-	-	>3	-
UL94	Class	V-0	V-0	V-0	V-0
$\epsilon_r 100$	-	2.1	2.6	2.8-3.4	2.3-2.6
tan $\delta 100$	$\cdot 10^{-3}$	0.5-0.7	5-6	30-50	10-15
ρ_e	Ohm·m	>10^{16}	>10^{14}	>10^{14}	>10^{13}
σ_e	Ohm	>10^{16}	>10^{14}	>10^{15}	10^{12}
E_BI	kV/mm	40	40	40	40
W_w	%	<0.05	<0.05	<0.05	<0.05
W_H	%	<0.05	<0.05	<0.05	<0.05

Powdered PTFE suspension polymers are used to compression mold preforms at 20 - 30°C, which are then processed above 327°C by the following techniques:

- Mold-free sintering: At 20 - 30°C and pressures of 20 - 100 MPa, simple preforms are compression molded in automatic presses; preforms with undercuts or hollow sections are compression molded under isostatic pressure with flexible tools. Subsequently, a pre-determined temperature profile is used to heat the free-standing molded parts in an oven, sintering them at 370 - 380°C, and slowly cooling them. Mold-free sintered parts (density ≈ 2.1 g/cm^3) are porous.

- Pressure sintering or sintering with post-molding pressure: Sintering is achieved by pressure applied in the mold on parts made of conductively modified PTFE (easily heated by conduction), or by post-forming pressure on the heated parts formed in the mold without pressure and subsequent cooling under pressure, or by blow forging of pre-sintered preforms. These processes result in non-deformable, non-porous parts of highest density and rigidity. Parts hot-formed slightly below the melting point have a tendency to revert, which is exploited for lip seals that adhere to worming.
- Ram extrusion (powder extrusion) of rods and thick-walled pipes: The entering molten resin is discontinuously pressed into tablets by a reciprocating ram in the first section of the long cylindrical barrel. The following section of the barrel is heated to 380°C, where the tablets are sintered under counter pressure by heat expansion and barrel friction to form continuously discharging (if necessary additionally decelerated at the end of the sintering pipe) sintered rods or pipes.
- Films are peeled from sintered cylindrical block; they can be treated by rolling.

Long, thin-walled products, mainly tubes up to 250 mm diameter with wall thicknesses from 0.1 - 4 mm and cable sheathing, are produced by paste-extrusion of emulsion polymers - made kneadable by the addition of 18 - 15% naphtha solvent - in a ram extruder. In a downstream continuous oven, first the lubricant volatilizes before the parts are sintered at 380°C. Thread sealing tapes, which should remain porous, are not sintered after extrusion, but simply rolled and dried. PTFE dispersions are used to cast thin films or to impregnate glass fiber products or parts made from graphite or porous metals, which are subsequently sintered and compression molded at 380°C. Anti-friction coatings made of PTFE dispersions on metal or ceramic surfaces (if necessary with bonding agents) and subsequently annealed, are porous and therefore not suitable as anti-corrosion coatings. Slight etching with solutions of alkali metals makes PTFE adhesive; there are special adhesives for temperatures up to 130°C. Machining requires keen edge tools.

Form of Delivery, Properties. PTFE is the most important fluoropolymer. Because of its high melt viscosity (melting point > 327°C), it cannot be processed by standard methods for thermoplastic processing. It is available as suspension polymerized powder for compression molding, sintering, and ram extrusion. Emulsion polymer powder is intended for paste extrusion, in dispersion for coatings and impregnations, or as an additive to other plastics to reduce dynamic friction. PTFE is a plastic exhibiting low rigidity and stiffness. Its advantages are the broad range of use-temperature (-270 to 300°C, PTFE embrittles only below -270°C), universal chemical resistance, insolubility in all known solvents below

300°C, weather resistance without stabilization, flammability rating of SE-O according to UL 94, excellent electrical and dielectrical properties, and the best slip and antistick properties of all plastics. Its wear properties are less outstanding; however, wear properties as well as stiffness and rigidity can be improved by modification (5 - 40 vol.%) with graphite, bronze, steel, MoS_2, or glass fibers.

Applications. Statically and dynamically loaded gaskets, bellows, pistons, and other machine elements, crucibles, coatings (for glass fabrics) and sheathings, carriers for printed circuit boards, films and other semi-finished products for subsequent machining.

Polyvinylidene Fluoride (PVDF)

Processing. Processing melt temperatures (injection molding and extrusion) range from 230 - 270°C; mold temperatures range from 60 - 90°C. Temperatures for coating with PVDF by spraying, dipping, or casting range from 190 - 215°C. PVDF can be thermoformed at 180°C. PVDF melts must not contact boron- or MoS_2-containing materials (screws, cylinders, certain glass fibers) to avoid spontaneous melt decomposition. In addition, manufacturers' recommendation for colorants, fillers, and reinforcing agents should always be followed. All standard welding processes can be used; also adhesion with glues or 2-component adhesives. More flexible modifications have a melting point at $\approx 165°C$.

Properties. PVDF contains 57% fluorine. Crystallinity depends on the thermal preconditioning of the parts: fast cooling of thin films results in transparent products, while annealing at 135°C results in highly crystalline and rigid products. Service temperatures range from -60 to +150°C. The polar structure excludes the use of high-frequency techniques. Weather exposure (radiation wave lengths from 200 - 400 nm) causes gradual decomposition. PVDF is highly resistant to chlorine and bromine, is UV resistant, and surpasses all other fluoropolymers in high-energy radiation resistance. It meets the highest requirements for purity and is therefore used for high-purity water conduits, packaging of chemicals (gas- and flavor impermeable bottles), for chemical apparatus, and semi-conductor manufacturing.

Applications. Gaskets, membranes, pumps- and valve parts, pipes, shrink-tubes, fittings, slide rails, linings, laminates for outdoor use, packaging film.

Polyvinyl Fluoride (PVF)

Properties. PVF is only available as film (crystal clear) and sheet. Lightly oriented or biaxially stretched films are on the market. Compared to PVDF, PVF exhibits higher rigidity and stiffness, lower density, ductility, and long-term service temperature. Because of its good UV-, IR-, weather-, and corrosion resistance, PVF is used for coatings of other materials in outdoor use. Thermoplastic processing is carried out at 260 - 300°C. Semi-finished products are weldable and can be adhesively bonded with EP resins.

Applications. Roofing membranes, roof coverings, sun collectors, laminates for out door use, road signs, packaging films, shrinktubes.

Polychlorotrifluoroethylene (PCTFE)

Processing. PCTFE is generally processed by compression molding, injection molding, and extrusion. Melt temperatures range from 270 - 300°C, mold temperatures range from 80 - 130°C. Dispersions can be used to coat metal parts. Because copper- and iron-containing metals catalyze the decomposition of PCTFE, preparatory electroless nickel plating is necessary when sheathing wires. High-frequency- and ultrasonic welding are possible; adhesive bonding requires pretreatment.

Properties. The symmetry of the macromolecules is disturbed by the integration of chlorine atoms, because they are bigger than fluorine atoms. Therefore, the crystallinity is lower and the distance between chains is larger. Nonetheless, the higher polarity of the chlorine atoms results in higher stiffness and rigidity. However, high-frequency applications are only limited. The chemical resistance is lower than that of PTFE. PCTFE has the lowest water vapor permeability of all transparent films; it is weather resistant; however, high-energy radiation causes the separation of chlorine; it does not burn, is physiologically inert, is used in medical applications, and can be used in food contact applications.

Applications. Fittings, tubes, membranes, crucibles, printed circuit boards, coil cores, insulating film, packaging.

6.4.2 Fluoro Copolymers and Elastomers (ECTFE, ETFE, FEP, TFEP, PFA, PTFEAF, TFEHFPVDF (THV), [FKM, FPM, FFKM])

Numerous copolymers with modified properties are available, in particular with improved thermoplastic processability, see Table 6.25. The nomenclature for these copolymers is not standardized; however, the names often include the abbreviation for the comonomer. Table 6.26 provides a comparison of properties.

Ethylene-Chlorotrifluoroethylene Colopymer (ECTFE) This block copolymer belongs to the fluorine-containing polymers with the highest modulus of elasticity and highest stiffness, while also exhibiting high impact strength between -4 to +150°C. It is available in pellet- and powder form for injection molding, extrusion, rotational molding, vortex sintering, and for foamed cable insulation. The melting temperature ranges from 260 - 300°C. Welding and adhesive bonding are possible. The barrier properties against oxygen, carbon dioxide, chlorine gas, and hydrochloric acid are 10 to 100 times better than those of PTFE, which allows for many applications in the chemical industry.

Applications. Cable sheathing in the electronics-, aero-, and oil-industry, radio chemistry, linings and coatings for containers in the chemical industry, packaging for pharmaceuticals, flexible printed circuits, films for lamination, fibers for filter cloths and fire-resistant upholstery, parts for clean applications.

Ethylene-Tetrafluoroethylene Copolymer (ETFE) Approximately 25% ethylene content in PTFE greatly improves thermoplastic processability: melt temperatures from 300 - 340°C for injection molding (the most important processing method). However, the maximum service temperature is lowered by $\approx$ 100°C. Stiffness and strength are improved, resistance is comparable. Stabilization against thermal and photochemical decomposition is necessary. Glass fiber reinforcement results in significant increase in stiffness and strength.

Applications. Gear wheels, parts for pumps, packaging, laboratory equipment, linings, cable insulation, blown film for transparent very durable roofing.

Polyfluoroethylenepropylene (FEP); Tetrafluoroethylene-Hexafluoropropylene-Copolymers (TFEP)
Processing. PTFE is copolymerized with 50 - 90% hexafluoropropylene to achieve thermoplastic processability. Injection molding and extrusion are carried at out high temperatures: melt temperatures from 315 - 360°C and mold temperatures from 200 - 230°C. Extrusion blow molding is possible. The barrels of the processing machines must be made of iron-free alloys, such as Hastelloy, Xaloy, Reiloy, or Monel. FEP is susceptible to melt fracture. Crystallinity can be increased up to 40 - 67% by annealing at 210°C. FEP powders are used for vortex sintering.

Properties. Compared to PTFE, FEP has a lower melt viscosity, higher impact resistance, lower stiffness and strength, lower long-term service temperature, but comparable chemical and weather resistance, flammability, and electrical properties. Graphite and milled glass fibers are typically used as reinforcing agents to increase stiffness and wear resistance. The relatively high melt viscosity limits the filler content.

Applications. Cable insulation and coatings, lining of containers, flexible printed circuit boards, injection molded parts for the electric, electronic, and chemical industries, packaging films, impregnations, heat-sealing adhesives.

Perfluoropropylvinylether Copolymer; Perfluoroalkoxy (PFA) Copolymers of perfluoropropylvinylether (PFA) with tetrafluoroethylene (TFE) reach the long-term service temperatures of PTFE while being thermoplastically processable because of their lower melt viscosity. Other properties are also comparable to those of PTFE. Melt temperatures reach 330 - 425°C for processing, mold temperatures 90 - 200°C. All melt conveying parts of the machine must be corrosion resistant; molds should be hard-chromium plated or made from nickel alloy.

Applications. Linings, technical products, cable insulation, heat-sealing adhesives for service temperatures from -200 to +260°C, chimney linings energy efficient heating systems. At 160 - 185°C, PFA is an easy flowing terpolymer, which is used for coatings and as adhesive film, CF-, AF-, GF-prepregs.

PTFE Copolymers with AF (PTFEAF) PTFEAF is a copolymer of PTFE and 2,2 bis(trifluoromethyl)-4,5-difluoro-1,3-dioxolan (AF). All fluorine-containing polymers described so far are semi-crystalline and therefore opaque to translucent and insoluble. PTFEAF is amorphous, transparent, and soluble, rendering it suitable for coating of substrates (anti-stick coatings). It is stiff and does not creep; the melt exhibits good flow properties and the lowest dielectric constant in the GHz range of all plastics.

Applications. Anti-corrosion and anti-stick coatings, fiber optics, parts for electro-and electronic industries.

Tetrafluoroethylene-Hexafluoropropylene-Vinylidenefluoride Copolymer (TFEHFPVDF [THV]) This terpolymer has a melting temperature of 160 - 185°C, low strength, a very low modulus of elasticity, and a very high elongation at break (even at low temperatures). Transparency in the visible spectrum is at 97%. The chemical resistance compares to that of PVDF, its flame resistance is better, and it holds up to weathering for 10 to 15 years. It can be processed like a thermoplastic material and is weldable by heat or high-frequency methods.

Applications. Fabric coatings (also with dispersions), fiber optics, solar cells; flexible tubes and pipes for the medical and chemical industries, cable insulation.

Other Fluoropolymers Tetrafluoroethylene-perfluoromethylene-vinylether copolymer (MFA) and vinylidenechloride-hexafluoropropylene (VDFHFP). Fluorinated Rubber: FKM, EPM, EFKM (covered under rubber in this chapter).

6.5 POLYACRYL- AND METHACRYL COPOLYMERS

All polymers based on acrylic acid and methacrylic acid belong to the group of acrylate polymers, see Table 6.9, No. 11 to 16. They exhibit particularly good transparency and weatherability.

6.5.1 Polyacrylate, Homo- and Copolymers (PAA, PAN, PMA, ANBA, ANMA)

Chemical designation, see Chapter 1 of this handbook.

Polyacrylonitrile (PAN) PAN is produced by polymerization of acrylonitrile, general structure see Table 6.9, No. 26. Major area of application is the production of fibers and use as copolymer with styrene (ABS, SAN) and butadiene

(NBR; nitrile rubber). PAN has good barrier properties against most gasses. Because pure PAN cannot be processed like a thermoplastic material, acrylonitrile copolymers are used for barrier layers in packaging films.

Polyacrylate, Special Products Homopolymer acrylic acid esters (PAA), see Table 6.9, No. 11, are soft resins, exhibiting good resistance against light, oxidation, and heat. They are used for co- and terpolymerization with PS, PVC, VA, MA, AN, and acrylic esters because of their plasticizing effects and are available as solid resins, in solution, but mainly in dispersion. Oxalidine-modified acrylic paint resins can be cross-linked with isocyanates. Elasto-plastic copolymers are used as base compounds for seals; those with > 20% acrylic acid are water soluble. Polyhydroxyethylene-methacrylate saturated with ≈ 40% water is used for contacts and coatings (e.g., for glasses) and encapsulation with controlled water absorption in medical and other technical applications. Acrylic resins soluble in the digestive tract are used for encapsulation of drugs. Mono-polymerization or polymerization of cross-linked two-component resins is used to produce heat- or radiation-curable paint resins. Rigid modified methacrylate copolymers are used as finishes for artificial leather and other fuel-resistant varnishes. MMA-VC copolymers are plasticizers for PVC. Unsaturated aliphatic polyurethane acrylate resins can be cross-linked with H_2O_2 to form tough, flexible glass fiber laminating resins.

6.5.2 Polymethacrylates, Homo- and Copolymers (PMMA, AMMA, MABS, MBS)

Polymethylmethacrylate (PMMA)

Chemical Constitution. PMMA is best known acrylate grade. Block-, emulsion-, or suspension polymerization of methacrylic acid yields a thermoplastic material with high molecular weight; for basic structure see Table 6.9, No. 17. Block polymerization between glass plates or in molds produces semi-finished products such as sheet, solid profiles, or pipes of high strength, high modulus of elasticity, and good surface finish. Peroxides are used as polymerization initiators for hot-setting processes; the same peroxides with the addition of amino-activators or other Redox-polymerization-starter systems are used for cold-setting processes. Addition of multi-functional components produces semi-cross-linked polymeric materials. Their further processing is only feasible by thermoforming, adhesion or machining because of the extremely high molecular weight. In order to produce resins for injection molding, blocks are polymerized in flat bags and subsequently milled. Uniform pellets are produced by extrusion and subsequent pelletizing.

Processing. PMMA resins can be injection molded and extruded; melt temperatures range from 200 - 230°C, mold temperatures range from 50 - 70°C.

Pre-drying of the resin or use of vented screws is required. When printing or painting on PMMA surfaces it is important to ensure that the parts do not retain any residual stresses and that the appropriate kind of color pigment is selected in order to avoid stress cracking. PMMA surfaces can be metallized, both with and without pre-treatment. Hot air-, high-frequency-, and ultrasonic welding are possible, as is adhesive bonding with single- or two-component methylmethacrylate adhesives or with glues based on methylene chloride or chlorinated carbon hydroxides. However, glues containing solvents should not be used because of their environmental impact; particularly, since today adhesives based on epoxy resins, PUR, or cyanoacrylate are available that are adapted for polymeric materials.

Properties. PMMA is brittle with high strength, high modulus of elasticity, high surface hardness (scratch resistance), and, if polymerized between polished glass plates, with high surface gloss. It can be polished, is weather resistant without further stabilization, and resistant to weak acids and alkalis, non-polar solvents, fats, oils, and water. PMMA is preferably used in the lighting industry because of its excellent light transmission and the fact that it can be colored to a high degree of saturation, see Figs. 6.8 and 6.9. For a comparison of properties see Table 6.27.

Applications. Molding resins: Piping/tubing elements, sanitation, glasses, lenses, housings, covers, viewing glasses, fiber optics, solar technology, cockpit and stadium window shields, light domes, filter for specific wavelengths, jewelry, optical data storage, "electronic paper" (60 μm thin film with finely distributed liquid crystalline drops: polymer dispersed LC, PDLC). Casting resins: gap fillers, safety glass, wood impregnation, integral casting of displays, glass fiber reinforced light displays, artificial stone tiles with, e.g., marble inlays, two-component systems for highly resistant pavement coatings, concrete repair, road markings, special resins in rigid or rubber-elastic modifications for surgical, orthopedic, and dental prosthetics. Modified, thermosetting, free-flowing, and highly reinforceable reaction resins based on MMA are used as bonding agents for BMC- and SMC molding resins, semi-finished products and pultrusion, resulting in highly moisture-, chemical-, and corrosion-resistant parts, some of which with increased shape stability. Dispersions with fine-grain mineral fillers are used to cast and heat cured kitchen sinks, water basins, sanitary wares, and marble-like sheets that can be thermoformed or finished like wood.

6.5 Polyacryl- and Methacryl Copolymers

Table 6.27: Comparison of Properties of Methylmethacrylate Polymers

Properties	Unit	PMMA	PMMA-HMW	PMMA-HI	MBS-HI
ρ	g/cm³	1.17-1.19	1.18-1.19	1.12-1.17	1.05-1.16
E_t	MPa	3100-3300	3300	600-2400	2000-2800
σ_y	MPa	-	-	20-60	30-55
ϵ_y	%	-	-	4.5-5	2-6
ϵ_{tB}	%	-	-	20->50	25-30
σ_{50}	MPa	-	-	-	-
σ_B	MPa	60-75	70-80	-	-
ϵ_B	%	2-6	4.5-5.5	-	-
T_P	°C	-	-	-	-
HDT	°C	75-105	90-105	65-95	85
α_p	10^{-5}/K	7-8	7	8-11	9
α_n	10^{-5}/K	-	-	-	-
UL94	Class	HB	HB	HB	HB
$\epsilon_r 100$	-	3.5-3.8	3.5-3.8	3.6-4.0	3.0-3.2
$\tan \delta 100$	$\cdot 10^{-3}$	500-600	600	400-600	270-290
ρ_e	Ohm·m	$>10^{13}$	$>10^{13}$	$>10^{13}$	$>10^{13}$
σ_e	Ohm	$>10^{13}$	$>10^{13}$	$>10^{13}$	$>10^{13}$
$E_B I$	kV/mm	30	≈ 30	30	25-30
W_w	%	1.7-2.0	1.7-2.0	1.9-2.0	0.4-0.6
W_H	%	0.6	0.6	0.5-0.6	0.2-0.3

Properties	Unit	MABS	AMMA sheet	PMMI
ρ	g/cm³	1.07-1.09	1.17	1.22
E_t	MPa	2000-2200	4500-4800	4000
σ_y	MPa	40-50	90-100	-
ϵ_y	%	3-5.5	10	-
ϵ_{tB}	%	20-30	40->50	-
σ_{50}	MPa	-	-	-
σ_B	MPa	-	-	90
ϵ_B	%	-	-	3
T_P	°C	-	-	-
HDT	°C	90	75	1.6
α_p	10^{-5}/K	9	6.5-7	4.5
α_n	10^{-5}/K	-	-	-
UL94	Class	HB	HB	HB
$\epsilon_r 100$	-	2.9-3.2	4.5	-
$\tan \delta 100$	$\cdot 10^{-3}$	160-200	600	-
ρ_e	Ohm·m	$>10^{13}$	$>10^{13}$	-
σ_e	Ohm	$>10^{13}$	$>10^{13}$	-
$E_B I$	kV/mm	35-40	≈ 30	-
W_w	%	0.7-1.0	2-2.25	5
W_H	%	0.35	0.7-0.8	2.5

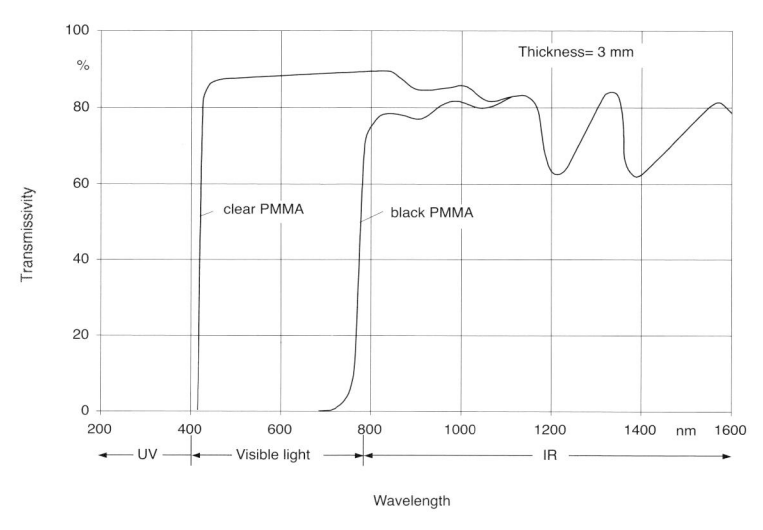

Figure 6.8: Spectral degree of transmission for crystal clear PMMA (a) and IR-transmitting black colored PMMA (b) (specimen thickness 3 mm).

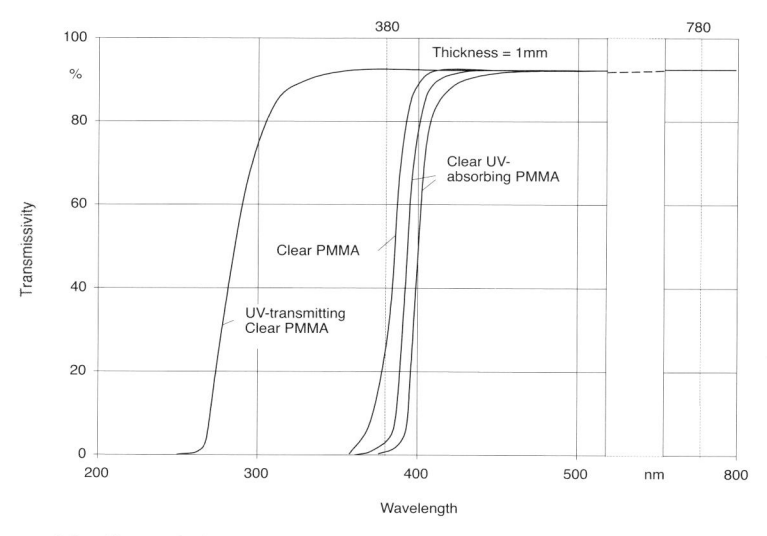

Figure 6.9: Spectral degree of transmission of crystal clear (a), crystal clear, UV-transmitting (b), and two crystal clear, UV-absorbing (c) grades of PMMA (specimen thickness 1 mm).

Methylmethacrylate Copolymers (AMMA) Copolymers and blends with methylmethacrylates are used to increase heat distortion temperature, impact resistance, and stress crack resistance against alcohols. Basic structure, see Table 6.9, No. 14 and 26.

Processing. High melt viscosity and a wide softening range render AMMA particularly suitable for blow molding and thermoforming. The pellets must be pre-dried for injection molding at melt temperatures not to exceed 230°C.

Properties. Solvent resistance and heat distortion point, and with an AN content of more than 50% also impact resistance, are increased. Because AN tends to form ring structures, AMMA has a slight yellowish tinge. Light stabilization is not required. Cast sheet can be produced by in-situ copolymerization. Biaxial stretching up to 70% increases impact, burst, and tear resistance; AMMA is therefore suitable for heavy duty glazing applications, e.g., in stadiums. The low gas permeability of polyacrylonitrile determines the performance of (graft-) copolymers with methacrylate or styrene. These "barrier" plastics contain up to 70% nitrile and sometimes additional butadiene-conating, plasticizing components. Permeability decreases with increasing biaxial stretching; fracture strength and impact resistance increase. AMMA is resistant to moderately concentrated acids and alkalis and most solvents. It swells in methanol and ketones; it is soluble in dimethylformamide and acetonitrile. Comparison of properties, see Table 6.27.

Applications. Packaging for liquid and solid foods such as spices, vitamin supplements, instant soups, and meat products; car care products, and cosmetics. Beverages must not come in contact with nitrile-containing plastics; therefore multi-layer packaging is used here.

Methylmethacrylate Acrylonitrile Butadiene Styrene Copolymers (MABS) and Methacrylate Butadiene Styrene Copolymers (MBS)
Basic structure, see Table 6.9, No. 17, 26, 35, and 5 MABS and MBS are impact resistant, clear, and translucent even at low temperatures. Because of their butadiene content they are not weather resistant; however, they are resistant to oils, fats, and fuels. They are suitable for sterilization by gamma radiation.

Processing. Injection molding, extrusion, blow molding, heat sealing, adhesive bonding, ultrasonic-, hot plate-, and friction welding.

Applications. Bottles for cosmetic, spray, and cleaning products; hollow technical parts, disposable medical devices, controls, and packaging.

6.5.3 Polymethacrylate, Modifications and Blends (PMMI, PMMA-HI, MMA-EML Copolymers, PMMA + ABS blends

Polymethacrylmethylimide (PMMI) Technically, PMMI is a copolymer of methylmethacrylate (MMA) and glutaramide; however, it is produced by reacting

PMMA with methylamine (MA) at high temperatures and pressure, see Eq. 6.6.

$$\text{PMMA} \xrightarrow[-2\,CH_3OH]{+\,CH_3NH_2} \text{PMMI}$$

(6.6)

Colorless and with high transparency, crystal clear and free of cloudiness its properties closely resemble those of PMMA. The ring termination provides increased chain stiffness and dimensional thermal stability (heat distortion temperature). Depending on the degree of imidization, it is possible to obtain all intermediate property values compared to PMMA. PMMI exhibits low oxygen permeability and is less susceptible to stress cracking caused by ethanol, ethanol/water, and isooctane toluol mixtures than PMMA. These properties depend on the imide content, see Fig. 6.10.

PMMI can be injection molded (after pre-drying) with melt temperatures ranging from 200 - 310°C and mold temperatures ranging from 120 - 150°C. Comparison of properties, see Table 6.27.

Applications. Headlamp diffusers for automotives, street light covers, fiber optics, blend components for fiber reinforced engineering materials.

Impact Resistant PMMA, PMMA-HI Impact resistant resins are two-phase systems manufactured by suspension or emulsion polymerization. The toughening disperse phase is formed in the PMMA matrix by, for example, styrene-modified acrylic elastomers (< 30%, with matched refractive index). This composition results in impact resistant molding compounds which can be blended as required with PMMA and which are as weather resistant and crystal clear as pure PMMA. Their stress cracking sensitivity is lower, their resistance to hot water is higher than PMMA's. The compounds are either pre-dried or processed with vented screws by injection molding or extrusion at melt temperatures of 210 - 230°C and mold temperatures of 60 - 80°C. Comparison of properties, see Table 6.27.

Applications. Anti-weathering coatings for construction profiles, for example PVC window profiles, household appliances, drawing and writing utensils; sanitary installations, light covers.

Methylmethacrylate exo-Methylene Lactone Copolymer(MMA-EML Copol, MMAEML) Compared to PMMA, copolymers of methylmethacrylate with exo-methylenelactones (EML), for example with methylene-methylbutyrolactone, see Eq. 6.7, exhibit improved properties: at EML contents of 15 - 40% glass transition temperatures range from 140 - 180°C, tensile strength from

6.5 Polyacryl- and Methacryl Copolymers

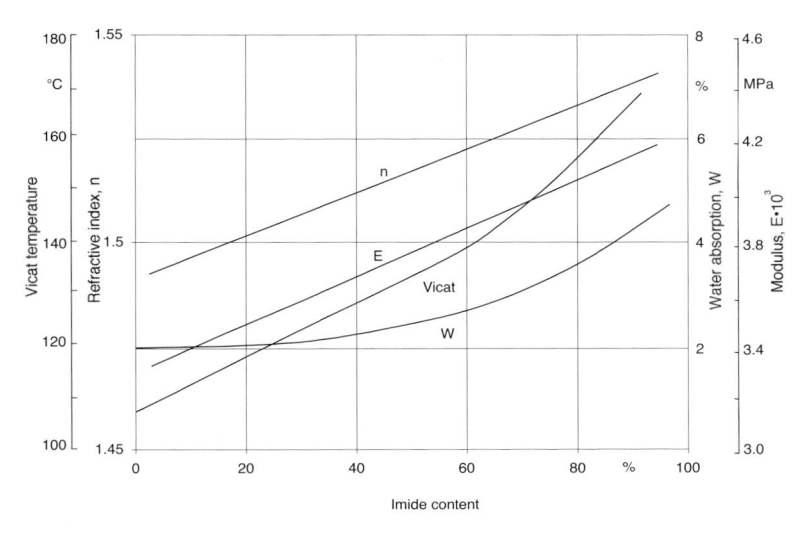

Figure 6.10: Dependency on imide content (I^c) for properties such as Vicat softening temperature ($VSTa$), modulus of elasticity (E), refractive index (n), and water absorption (W).

82 - 46 MPa, modulus of elasticity from 3600 - 4000 MPa, elongation at break from 4.7 - 1.2%, and water absorption from 2.5 - 4.3%. The refractive index of 1.54 is higher than that for PMMA (1.49). Melt temperature ranges from 230 - 280°C.

(6.7)

Applications. High hardness, optical brilliance, weather- and UV resistance, and paintability render these resins suitable for automotive head lights (diffusers and reflectors); fiber optics.

PMMA + ABS blends This type of blend is used for automotive parts (housings, reflectors), sanitary installations (surface coatings for tubs and shower bases), and in electric applications. The material can be metallized, has good welding properties, and superior weathering resistance and stiffness to ABS.

6.6 POLYOXYMETHYLENE, POLYACETAL RESINS, POLYFORMALDEHYDE (POM)

6.6.1 Polyoxymethylene Homo- and Copolymers (POM-H, POM-Cop.)

Chemical Constitution. POM (polyoxymethylene, polyformaldehyde, polyacetal) are semi-crystalline thermoplastic materials, formed by homo- (POM-H) or copolymerization (POM-Cop.) of formaldehyde, see Eq. 6.8.

$$\pm CH_2-O\pm \qquad \pm (CH_2-O)_n-(CH_2-CH_2)_{\overline{m}}\pm$$

Acetal homopolymer POM-H   Acetal copolymer POM-CO

(6.8)

In general, all manufacturing and post-manufacturing processes used for thermoplastics can be used with POM. The most important process is injection molding. Higher molecular grades can be extruded; weakly cross-linked grades can be blow molded. To achieve good crystallinity and surface textures, injection molds and polishing stacks for extruded products should be heated to 60 - 130°C. With decreasing mold temperature mold shrinkage decreases from 3 to $\approx$ 1%, while post molding-shrinkage increases accordingly. Process temperatures of 220°C cause decomposition and the formation of gaseous formaldehyde and are therefore dangerous. Because of its low tanδ value, POM cannot be high-frequency welded.

Properties. Non-reinforced POM is among the stiffest and toughest thermoplastic materials and exhibits good dimensional stability. It embrittles only below $-40°C$ and its short-term service temperature is at 150°C, its long-term service temperature at 110°C. Because of their surface hardness and low coefficients of friction, POMs show good slip and wear resistance. Their good insulation and dielectrical properties are little temperature- and frequency-dependent. Permeability to gasses and vapors, also to organic substances, is low. Both types of POM are attacked by strong acids (pH 4) and oxidants. They are not soluble in any of the usual solvents, including fuels and mineral oils, and barely swell.

Specially stabilized grades are resistant in up to 100°C hot diesel fuels and in aggressive gasoline. Non-stabilized grades are attacked by UV radiation; therefore, UV-stabilized or carbon black filled grades are recommended. Stabilization for colored grades is available; also slip-modified grades that prevent creaking. POM films are translucent. POM burns with a weak blue flame and drips. They are physiologically inert and grades for food contact are available. Comparison of properties, see Table 6.28.

Table 6.28: Property Comparison for Polyacetals

Properties	Unit	Polyacetals				
		POM-H	POM-H-HI	POM-CO	POM-CO-HI	POM-CO-GF30
ρ	g/cm^3	1.40-1.42	1.34-1.39	1.39-1.41	1.27-1.39	1.59-1.61
E_t	MPa	3000-3200	1400-2500	2600-3000	1000-2200	9000-1000
σ_y	MPa	60-75	35-55	65-73	20-55	-
ϵ_y	%	8-25	20-25	8-12	8-15	-
ϵ_{tB}	%	20->50	>50	15-40	>50	-
σ_{50}	MPa	-	-	-	-	-
σ_B	MPa	-	-	-	-	125-130
ϵ_B	%	-	-	-	-	3
T_P	°C	175	175	164-172	164-172	164-172
HDT	°C	105-115	65-85	95-110	50-90	155-160
α_p	10^{-5}/K	11-12	12-13	8-11	13-14	2.5-4
α_n	10^{-5}/K	-	-	-	-	6
UL94	Class	HB	HB	HB	HB	HB
ϵ_r 100	-	3.5-3.8	3.8-4.7	3.6-4	3.7-4.5	4.0-4.8
tan δ100	·10^{-3}	30-50	70-160	30-50	50-200	40-100
ρ_e	Ohm·m	>10^{13}	>10^{12}-10^{13}	>10^{13}	>10^{11}	>10^{13}
σ_e	Ohm	>10^{14}	>10^{14}	>10^{13}	>10^{11}-10^{12}	>10^{13}
E_BI	kV/mm	25-35	30-40	35	30-35	40
W_w	%	0.9-1.4	1.6-2.0	0.7-0.8	0.8-1.2	0.8-0.9
W_H	%	0.2-0.3	0.9	0.2-0.3	0.2-0.3	0.15

Applications: POM injection molded parts have largely substituted precision parts traditionally made from metals, such as gear wheels, levers, bearings, screws, coils; parts for textile machinery, fittings, hot water and fuel bearing pumps; tanks for diesel (stabilized grades); with 'outsert' injection molding, circuit boards with up to 120 functional POM elements are molded; POM-PUR combinations are used for chain wheels under impact loads, housings, living hinges, ski bindings, and zippers.

6.6.2 Polyoxymethylene, Modifications and Blends (POM + PUR)

To increase stiffness and strength, 10 - 40% glass fibers, glass beads or other mineral fillers (orientation-independent reinforcements) are incorporated into these

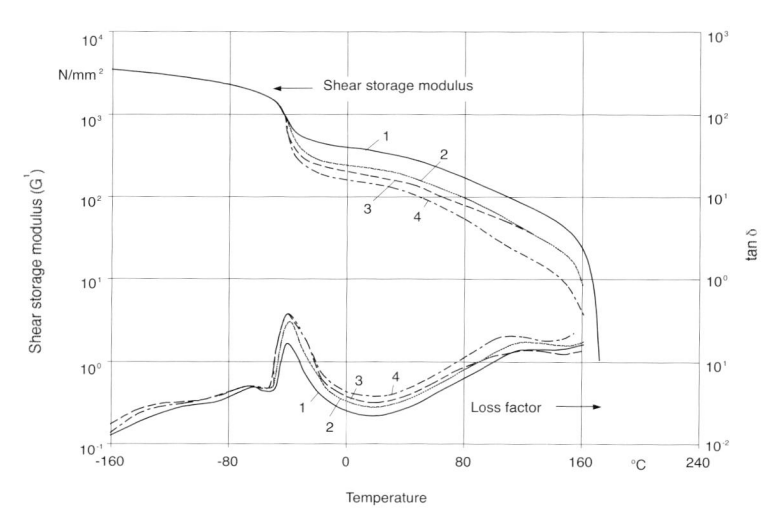

Figure 6.11: Shear modulus of acetal copolymer/PUR blends, ratio 1:1 (graph 1 to 4 with increasing mixing).

resins. Blends with cross-linked rubber and with up to 50% PUR elastomers and particularly free-flowing grades for thin-walled parts are produced; they exhibit increased toughness and simultaneously decreased stiffness and strength, see Fig. 6.11. Friction properties are improved by addition of MoS2, PTFE, PE silicone oils or special chalks. Powdered aluminum or bronze is used to increase heat distortion point and electric conductivity.

6.7 POLYAMIDES (PA)

6.7.1 Polyamide Homopolymers (AB and AA/BB Polymers) (PA6, 11, 12, 46, 66, 69, 610, 612, PA 7, 8, 9, 1313, 613)

Chemical Constitution The amide group shown in Eq. 6.9 is characteristic for all polyamides.

$$-\underset{\|}{C}-\underset{|}{N}-$$
$$OH$$

Imide group

(6.9)

The macromolecules of AB polymers (PA 6, PA 11, PA 12) consist of a basic unit (see Table 6.29). The numbers contained in the acronyms indicate the number

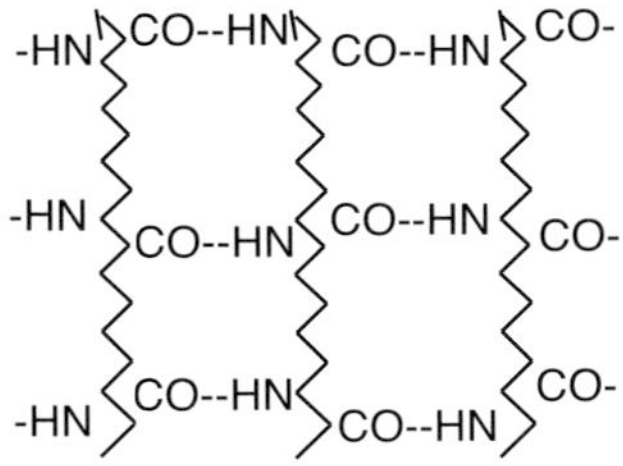

6.7 Polyamides (PA)

Figure 6.12: Formation of hydrogen bridges in PA 6.

of C atoms in the basic unit. AA/BB grades of polyamide (PA 46, PA 66, PA 69, PA 610, PA 612) are characterized by two basic units; the numbers in their acronyms indicate the number of C atoms in each of the two units. A number of additional letters also indicate the utilized monomer units in the polyamide (see Table 6.29). For example, I: isophthalic acid, N: 2,6-naphthalene-dicarbon acid, T: terephthalic acid, ND: 1,6-diamino-2,2,4-trimethylhexane, MC: 1,3-bis(aminomethyl)cyclohexane, IND: 1,6-diamino-2,4,4-trimethylhexane, IPD: isophoronediamine (1-amino-3-aminomethyl-3,5,5-trimethylcyclohexane), MTD: m-toluylenediamine, MXD: m-xylilenediamine, MACM: 3,3'-dimethyl-4,4'-diaminodicyclohexylmethane, PTD: p-toluylenediamine, PAPC: 2,2-bis(p-amino-cyclohexyl) propane, PPGD: polypropyleneglycoldiamine, PBGD: polybutylene-glycoldiamine, PAPM: diphenyl-methane-4,4-diamine, PACM: bis(p-aminocyclo-hexyl) methane, TMD: 6-3: 2,2,4-trimethylhexamethylenediamine and 2,4,4-trimethylhexame-thylenediamine.

Other polyamide grades based on these building blocks can be produced but have not gained significant industrial importance: PA 4 (production of fibers, high water absorption), PA 1313, PA 613. The high polarity of the CONH group causes the formation of hydrogen bridges between neighboring molecular chains, see Fig. 6.12. These bonds determine toughness, heat distortion point, and the high modulus of elasticity. Grades with smooth aliphatic segments between the CONH groups are highly crystalline.

Processing All standard processing methods for thermoplastic materials can be used. Pre-drying of molding compounds is required. Semi-crystalline PAs exhibit very low melt viscosities (shut-off nozzles are required for injection molding!) with well defined melting and freezing temperatures. Therefore, special processing measures or special grades are necessary for extrusion and blow molding. Melt freezing is associated with volume contraction of 4 - 7%. Accordingly, shrinkage is relatively high (up to 3%); voids are often encountered in thick-walled parts. Films are preferably manufactured by chill-roll processes. Table 6.30 provides an overview of processing parameters.

Table 6.29: Structure and Building Blocks of Polyamides

Acronym	General Structure	CH₂/CONH ratio	Raw Materials	Service temperatures (°C) Short term	Long term
	AB-Polymers —[NH—(CH₂)ₓ—CO]ₙ—				
PA6	x=5	5	e-Caprolactam	140-180	80-100
PA11	x=10	10	Amino undecanoic acid	140-150	70-80
PA12	x=11	11	Dodecanoic acid	140-150	70-80
	AA/BB-Polymers —[NH—(CH₂)ₓ—NH—CO—(CH₂)ᵧ—CO]ₙ—				
PA46	x=4, y=4	4	1,4 Diamino butene and adipic acid	280	140
PA66	x=6, y=4	5	Hexamethylene diamine and adipic acid	170-200	80-120
PA69	x=6, y=7	6.5	Hexamethylene diamine and azelaic acid		
PA610	x=6, y=8	7	Hexamethylene diamine and sebacic acid	140-180	80-100
PA612	x=6, y=10	8	Hexamethylene diamine and decanedicarboxylic acid	130-150	80-100
PA6T	—[NH—(CH₂)₆—NH—CO—⌬—CO]ₙ—		Hexamethylene diamine and terephtalic acid	120-130	70-90
PA NDT/INDT (PA6-3T)	—NH—CH₂—C(CH₃)(CH₃)—CH₂—CH—(CH₂)₂—NH—CO—⌬—CO—		Trimethylhexamethylene-diamine and terephthalic acid	130-140	80-100
PA MXD6 PARA	—[NH—CH₂—⌬—CH₂—NH—CO—(CH₂)₄—CO]ₙ—		m-Xylylenediamine and adipic acid	190-230	110-145
PA6I	—[NH—(CH₂)₆—NH—CO—⌬—CO]ₙ—		Hexamethylene diamine and isophthalic acid		
PA6-6T	see PA6 and PA6T		Caprolactam/hexame-thylene diamine and terephthalic acid		90
PA-elastomer			PA 12 and tetrahydrofuran- or polyether blocks	130	100
TPA-ET			Polyamide and dihydroxipolyether	130	100
PA-RIM			PE-; PP-glycol or polybutadiene with caprolactam		
PMPI	—[CO—⌬—CO—NH—⌬—NH]ₙ—		Isophthaloylchloride and m-phenylene diamine	260	
PPTA	—[CO—⌬—CO—NH—⌬—NH]ₙ—		Terephthaloylchloride and phenylene diamine	>250	>200

6.7 Polyamides (PA)

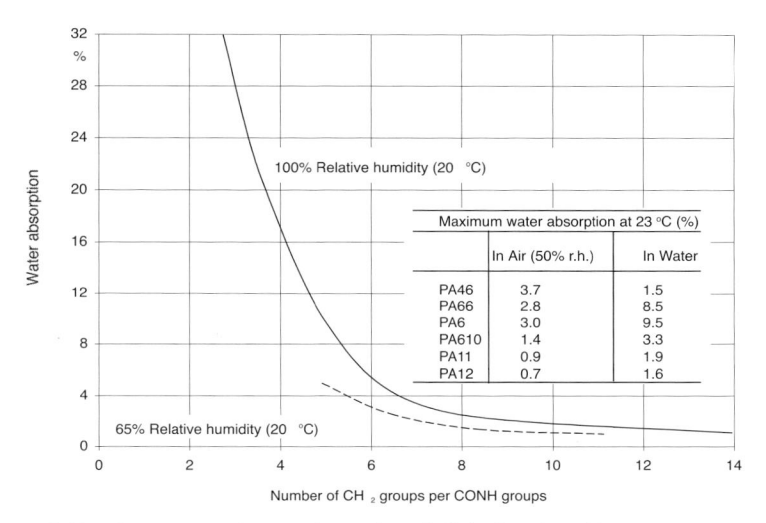

Figure 6.13: Structure and water absorption of aliphatic polyamides.

Table 6.30: Processing Parameters for Polyamides

PA-grade	Injection molding			Extrusion	Extrusion blow molding	
	Melt temp.°C	Melt temp.°C	Shrinkage %	Melt temp.°C	Melt temp.°C	Melt temp.°C
6	230-280	80-90 (120)	0.5-2.2	240-300	250-260	80
46	295-330	-	-	-	-	-
66	260-320	80-90 (120)	0.5-2.5	250-300	270-290	90
610	230-280	80-90 (120)	0.5-2.8	230-290	230-250	80
11-12	210-250	40-80	0.5-1.5	230-290	200-230	70
NDT/INDT	260-290	70-90	0.4-0.6	250-280	240-255	40
MXD6	250-280	100-140	-	-	-	-

With increasing ratio of CH_2 groups to $CONH$ groups the water absorption rate characteristic for PAs decreases (see Fig. 6.13). Figure 6.14 shows the dependency of equilibrium moisture content on relative humidity.

With increasing distance between the amide groups (increasing number of CH_2 groups), the strength of the intermolecular bonds decreases. Therefore, PA

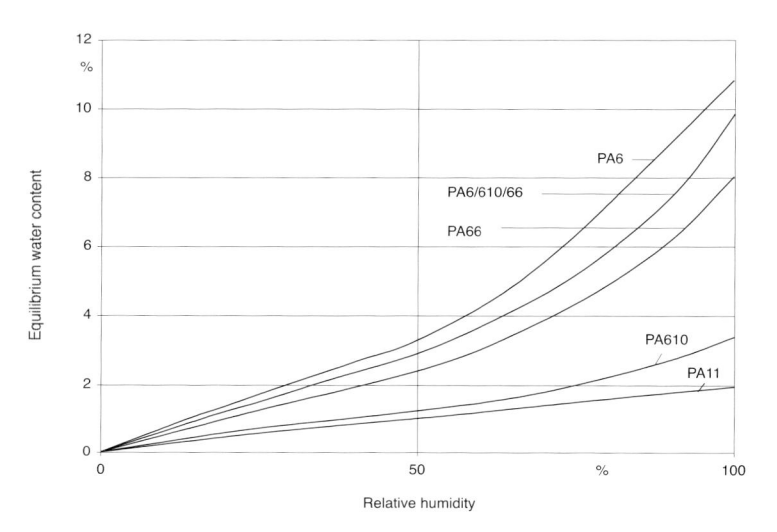

Figure 6.14: Storage conditions and water absorption of different PA grades.

11 is softer and has a lower melting temperature than PA 6. The crystallinity of PA varies, depending on the cooling rate, between 10% (high cooling rate: fine grained structure, high toughness) and 50 - 60% (slow cooling rate: large spherulites, high strength, high modulus of elasticity, high abrasion resistance, low water absorption).

Caprolactam and laurinlactam are polymerized on an industrial scale at 250 - 300°C to form molten PA 6 and PA 12. Co-catalysts (Acrylic agents, particularly isocyanates) facilitate fast, pressure-less, activated anionic polymerization of high-molecular cast PA 6 and PA 12 at 100 - 200°C. This way, large, thick-walled parts with weights up to 1000 kg can be pressure-less cast in simple molds such as those used, e.g., in metal casting. Hollow objects are produced by rotary casting or rotational molding. PAs can be welded by melting the interfaces; they can be adhesively bonded by special adhesives based on cyanacrylates or two-component EP resins and are easy to machine. Although adhesive bonding by resorcinol or cresols is possible, these methods should not be employed because of environmental and health concerns.

Properties The properties of the various PA grades differ only slightly. In a dried state, immediately after thermoplastic processing, they are hard and more or less brittle. After water has been absorbed, either from the atmosphere or from storage in water, they become tougher and more abrasion resistant and their modulus of elasticity decreases. Water absorption causes an increase in

6.7 Polyamides (PA)

volume and thus in dimension, a fact that has to be considered when designing with PAs. Slip- and wear properties of PA are very good: they exhibit good dry run properties; they are not sensitive to contamination and chemical resistant. Because the glass transition temperature T_g is in or only slightly above room temperature range, PAs soften at relatively low temperatures and should then not be submitted to constant loads, although they can be used at temperatures close to their melting point (the creep modulus is extremely time-dependent).

Reinforced or filled grades exhibit significantly improved load bearing capacity, even above T_g. Both electrical and mechanical properties deteriorate with increasing temperatures and water content. Electric surface resistivity decreases to such low values that dust or electrostatic attraction do not have to be considered. With decreasing water absorption, water vapor permeability of PA decreases; however, the normally low permeability to gasses (O_2, flavors) slightly increases.

The latter renders PAs suitable for packaging film, often as composite film, e.g., with polyolefins. PAs are resistant to solvents, oils, fats, fuels, weak alkalis, ketones, and boiling water (they can be sterilized); they are nonresistant to strong alkalis and acids. Natural colorants (tea, coffee, fruit juices, etc.) can stain PA. PA should be stabilized for long-term use above 100°C or out doors (e.g., by addition of $\approx$ 2% carbon black). The surface of glass fiver reinforced PA is less weather resistant and long-term exposure (several years) can lead to surface erosion. Non-modified PAs will continue to burn after the ignition source is removed. Comparison of properties, see Table 6.31.

Table 6.31: Comparison of Properties for Aliphatic Unfilled Homopolyamides

Property	Unit	PA 6 dry	PA 6 cond.*	PA 12 dry	PA 12 cond.*	PA 66 dry	PA 66 cond.*
ρ	g/cm^3	1.12-1.14	-	1.01-1.03	-	1.13-1.15	-
E_t	MPa	2600-3200	750-1500	1300-1600	900-1200	2700-3300	1300-2000
σ_y	MPa	70-90-	30-60	45-60	35-40	75-100	50-70
ϵ_y	%	4-5	20-30	4-5	10-15	4.5-5	15-25
ϵ_{tB}	%	20->50	>50	>50	>50	10-40	>50
σ_{50}	MPa	-	-	-	-	-	-
σ_B	MPa	-	-	-	-	-	-
ϵ_B	%	-	-	-	-	-	-
T_P	°C	220-225	220-225	175-180	175-180	255-260	255-260
HDT	°C	55-80	-	40-50	-	70-100	-
α_p	10^{-5}/K	7-10	7-10	10-12	10-12	7-10	7-10
α_n	10^{-5}/K	-	-	-	-	-	-
UL94	Class	HB-V-2*	HB-V-2*	HB*	HB*	V-2*	V-2*

Continued on next page

Property	Unit	Aliphatic homopolyamides					
		PA 6 dry	PA 6 cond.**	PA 12 dry	PA 12 cond.**	PA 66 dry	PA 66 cond.**
$\epsilon_r 10$	-	3.5-4.2	12-20	3.7-4	5-6	3.2-4	5-11
$\tan \delta 10$	$\cdot 10^{-3}$	60-150	2100-3500	300-700	800-1000	50-150	1000-2400
ρ_e	Ohm·m	$>10^{13}$	$>10^{10}$	$>10^{13}$	$>10^{12}$	$>10^{12}$	10^{10}
σ_e	Ohm	$>10^{12}$	$>10^{10}$	$>10^{13}$	$>10^{12}$	$>10^{10}$	$>10^{12}$
$E_B I$	kV/mm	30	25-30	27-29	28-32	25-35	25-35
W_w	%	9-10	9-10	1.3-1.7	1.3-1.7	8-9	8-9
W_H	%	2.5-3.4	2.5-3.4	0.7-1.1	0.7-1.1	2.6-3	2.6-3

Property	Unit	Aliphatic homopolyamides			
		PA 610 dry	PA 610 cond.**	PA 46 dry	PA 46 cond.	PA-MXD6-GF 30
ρ	g/cm^3	1.06-1.09	-	1.18	-	1.43
E_t	MPa	2000-2400	1300-1600	3300	1000	11500
σ_y	MPa	60-70	45-50	100	55	-
ϵ_y	%	4	15	-	-	-
ϵ_{tB}	%	30->50	>50	-	-	-
σ_{50}	MPa	-	-	-	-	-
σ_B	MPa	-	-	-	-	-
ϵ_B	%	-	-	-	-	190
T_P	°C	210-220	210-220	295	-	-
HDT	°C	60	-	160	-	-
α_p	10^{-5}/K	8-10	8-10	0.,8	-	1.8
α_n	10^{-5}/K	-	-	1	-	-
UL94	Class	V-2*	V-2*	V-2 (0.75)	V-2 (0.75)	HB
$\epsilon_r 10$	-	3.5	4	-	-	3.9
$\tan \delta 10$	$\cdot 10^{-3}$	70-150	1000-1800	-	-	100
ρ_e	Ohm·m	$>10^{13}$	10^{10}	10^{13}	10^6-10^9	$2 \cdot 10^{13}$
σ_e	Ohm	$>10^{12}$	$>10^{10}$	$>10^{15}$	10^{13}-10^{14}	-
$E_B I$	kV/mm	-	-	>25	15-20	30
W_w	%	2.9-3.5	2.9-3.5	-	-	3.2
W_H	%	1.2-1.6	1.2-1.6	3.7	-	1

** exposure of specimens at 23 °C/50 % r.h. until saturation
* also available as V-1 and V-0

Applications Technical parts such as bearings, gear wheels, rolls, screws, gaskets, fittings, linings, housings; parts of pumps, coils, carburetors, automotive pedals, intake manifolds for combustion engines, ventilators, appliances, consumer goods; extruded semi-finished products such as rods, pipes, tubes, sheet, cable sheathing; ski boots, shoe soles, membranes, seals; blown and extruded film, packaging, blow molded parts; fibers, bristles, fishing lines, glass-mat reinforced sheet (GMT), CF-, AF-, GF-prepregs. Cast polyamides: thick-walled

6.7.2 Modifications

Additives PAs are typically stabilized against the degrading effects of oxygen during processing, high temperatures, and UV radiation. Inorganic pigments, which are stable up to 300°C, are used for coloring. Cadmium pigments and organic colorants come with a risk of degradation. Flame resistant modified grades are available. Free-flowing injection molding resins are modified with nucleating agents for faster and finer crystallization; they exhibit better mechanical properties, lower water absorption and toughness. Lubricated grades are easier to demold.

Reinforcements, Fillers Because PA softens at elevated temperatures, many grades with up to 50% reinforcement with carbon and other fibers are available on the market. They exhibit increased strength, modulus of elasticity, and heat distortion point. Silica, talcum, chalk, and glass beads increase stiffness while reducing warpage and shrinkage. These grades have gained industrial importance and are described in Table 6.32.

Reinforcing PA 6 with 1 - 2 vol.% PA clay nanocomposites results in a 25 - 50% reduction in O_2 permeability while only slightly reducing puncture resistance. Metal powders, such as aluminum, copper, bronze, steel, lead, zinc, or nickel increase the heat distortion point and create electrically conductive materials. Magnets are produced with 80% barium ferrite filler. Slip- and abrasion properties are improved by addition of MoS_2, PTFE, HDPE, and graphite.

Table 6.32: Properties of Reinforced and Filled PAs, Conditioned at 23 °C, 50% r.h.

PA grade	ϕ (%)	ρ (g/cm^3)	σ_B (MPa)	ϵ_B (%)	E_t (MPa)	HDT/A (°C)
PA6	-	1.13	64	220	1200	80
Short glass fibers	30	1.37	148	3.5	5500	-
Glass spheres	30	1.35	65	20	3000	208
Carbon fibers	20	1.23	100	-	8000	-
Silicon dioxide	10	1.19	57	140	1000	-
Chalk	30	1.35	50	30	3000	60
PA66	-	1.14	63	60-300	1500	66-85
Short glass fibers	30	1.37	153	3	7200	204-249
Glass spheres	30	1.35	81	5	3700	74
Carbon fibers	20	1.23	197	4	16900	257
Mica	30	-	39		6900	
PA610	-	1.19	60	85-300	1900	60
Short glass fibers	30	1.3	128	3	7800	204

Continued on next page

PA grade	ϕ (%)	ρ (g/cm³)	σ_B (MPa)	ϵ_B (%)	E_t (MPa)	HDT/A (°C)
Talc	20	1.25	60	5	4000	
PA11	-	1.04	58	325	1200	58
Short glass fibers	30	1.26	93	4	6200	173
Bronze powder	90	4	34	4	5500	100
PA12	-	1.02	60	270	1200	40-50
Short glass fibers	30	1.23	83	6	5700	155
Glass spheres	30	1.23	45	25	2500	120
PA46	-	1.18	100	40	3300	160
Short glass fibers	30	1.41	175	2.5	10000	290

6.7.3 Polyamide Copolymers, PA 66/6, PA 6/12, PA 66/6/610 Blends (PA +: ABS, EPDM, EVA, PPS, PPE, Rubber)

Mixed PAs such as PA6/66 are soluble in alcohols. The solutions are used to produce fuel-resistant electro-insulating varnishes, coatings that adhere well to metals, wood, cardboard, and glass, and to cast thin films. Ready-to-use solutions for textile adhesion and coating are available. PA 12 is used as hot melt adhesive. Residual monomers behave as plasticizers (just like absobed moisture) with PA 6 and PA 66; 10 - 20% aliphatic glycol or aromatic sulfonamides (benzosulfonic-acid-n-butylamide) are excellent plasticizers for low-crystalline PAs such as PA 11 and PA 12. Most highly dispersed two-component blends or grafts with HDPE, ABS, PPS, PPE, elastomers such as EPDM or EVAC, ABR, BR, SBR, acrylate- or other synthetic rubbers also exhibit higher toughness (PA-HI). Grafting of 17% acrylate elastomer (see Table 6.9, No. 33) on PA 6 chains results in a high-impact resistant polymer. Ready-to-inject, dry, impact resistant molding compounds contain 10 - 20% PE, which is coupled either with the help of coupling agents (ionomers) or chemically (carboxylization, grafting with maleic anhydride or acrylic acid), see Table 6.33 and 6.34.

Table 6.33: Comparison of Properties of Aliphatic Homopolyamides, Filled and Modified

Property	Unit	Aliphatic homopolyamides, filled and modified			
		PA 6-GF 30		PA66-GF 30		PPA-GF33.
		Dry	Cond.*	Dry	Cond.*	Dry
ρ	g/cm³	1.35-1.37	-	1.36	-	1.46
E_t	MPa	9000-10⁸00	5600-8200	9100-10000	6500-7500	11700
σ_y	MPa	-	-	-	-	-
ϵ_y	%	-	-	-	-	-
ϵ_{tB}	%	-	-	-	-	-
σ_{50}	MPa	-	-	-	-	-

Continued on next page

6.7 Polyamides (PA)

Property	Unit	Aliphatic homopolyamides, filled and modified				
		PA 6-GF 30		PA66-GF 30		PPA-GF33.
		Dry	Cond.*	Dry	Cond.*	Dry
σ_B	MPa	170-200	100-135	175-190	115-14	220
ϵ_B	%	3-3.5	4.5-6	2.5-3	3.5-5	2.5
T_P	°C	220-225	-	255-260	-	
HDT	°C	190-215	-	235-250	-	285
α_p	10^{-5}/K	2-3	-	2-3	-	
α_n	10^{-5}/K	6-8	-	6-8	-	
UL94	Class	HB**	-	HB**	-	
$\epsilon_r 100$	-	3.8-4.4	7-15	4	8	-
$\tan \delta 100$	$\cdot 10^{-3}$	100-150	2000-3000	140	1300-2300	-
ρ_e	Ohm·m	10^{13}	10^{11}	10^{13}	10^{11}	
σ_e	Ohm	$>10^{13}$	$>10^{11}$	$>10^{13}$	$>10^{11}$	-
$E_B I$	kV/mm	35-40	25-35	40	35	-
W_w	%	6.0-6.7	-	5.0-5.5	-	
W_H	%	1.4-2.0	-	1.0-1.7	-	

Property	Unit	Aliphatic homopolyamides, filled and modified				
		PA 6-HI		PA 66-HI		PA 12-P
		Dry	Cond.*	Dry	Cond.*	
ρ	g/cm³	1.01-1.13	-	1.04-1.13	-	1.0-1.05
E_t	MPa	1100-2800	450-1200	1800-3000	900-2000	220-750
σ_y	MPa	25-80	20-45	50-80	40-55	15-35
ϵ_y	%	4-5	15-30	5-7	15-30	20-45
ϵ_{tB}	%	>50	>50	20->50	>50	>50
σ_{50}	MPa	-	-	-	-	-
σ_B	MPa	-	-	-	-	-
ϵ_B	%	-	-	-	-	-
T_P	°C	220	-	255	-	160-175
HDT	°C	45-70	-	60-75	-	40-50
α_p	10^{-5}/K	8.5-15	-	7-8.5	-	12-17
α_n	10^{-5}/K	-	-	-	-	-
UL94	Class	HB**	-	HB**	-	HB**
$\epsilon_r 100$	-	3-4	5-14	3.5-4	7-9	4-24
$\tan \delta 100$	$\cdot 10^{-3}$	100-140	500-3000	70-240	900-1800	900-3500
ρ_e	Ohm·m	$>10^{13}$	$>10^{10}$	$>10^{12}$	10^{10}-10^{12}	10^9-10^{11}
σ_e	Ohm	10^{10}-10^{12}	10^8-10^{10}	$>10^{13}$	$>10^{13}$	10^{11}-10^{15}-
$E_B I$	kV/mm	30-35	25-30	30-35	30-35	20-35
W_w	%	6.5-9.0	-	6.5-8.0	-	0.8-1.5
W_H	%	1.8-2.7	-	2.2-2.5	-	0.4-0.7

* exposure of specimens at 23 °C/50 % r.h. until saturation
** also available V-1 and V-0

Table 6.34: Comparison of Properties of Aromatic Polyamides, Copolyamides, and Polyether Block Amides

Property	Unit	\multicolumn{6}{c}{Aromatic polyamides, copolyamides and polyether block amides}
		PA 6/6T Dry	PA 6/6T Cond.*	PA 6-3-T Cond.*	PAMXD6 -GF 30	PA 61 Dry	PA 61 Cond.*
ρ	g/cm^3	1.18	1.18	1.12	1.43	1.18	1.18
E_t	MPa	3500	3000	2800-3000	11800	3300	3000
σ_y	MPa	110	100	80-90	-	110	90
ϵ_y	%	5	6	7-8	-	5	6
ϵ_{tB}	%	10-20	10-20	>50	-	>50	>50
σ_{50}	MPa	-	-	-	-	-	-
σ_B	MPa	-	-	-	185	-	-
ϵ_B	%	-	-	-	2.5	-	-
T_P	°C	295-300	295-300	-	225-240	175-180	175-180
HDT	°C	110	-	120	228	105	-
α_p	10^{-5}/K	6-8	6-8	5-6	1.5-2	6	6
α_n	10^{-5}/K	-	-	-	-	-	-
UL94	Class	V-2***	V-2***	V-2***	HB***	V-2***	V-2***
$\epsilon_r 1$	-	4	4.5	4-4.2	3.9	4.3	4.6
tan $\delta 1$	·10^{-3}	300	400	170-210	100	400	480
ρ_e	Ohm·m	10^{13}	10^{13}	>10^{13}	>10^{13}	>10^{13}	>10^{13}
σ_e	Ohm	10^{14}	10^{13}	>10^{14}	>10^{14}	>10^{15}	>10^{15}
E_BI	kV/mm	50	80	25	30	25	28
W$_w$	%	6.5-7.5	6.5-7.5	6.5-7.5	3.5-4	6	6
W$_H$	%	1.8-2.0	1.8-2.0	2.8-3	1.6	2	2

Property	Unit	\multicolumn{6}{c}{Aromatic polyamides, copolyamides and polyether block amides}
		PA 6/66 PA 66/6	PEBA 12 Shore D40-	PEBA 12 Shore D55-	PEBA 6 Dry	PEBA 6 Cond.*	PPA
ρ	g/cm^3	1.13-1.14	0.99-1.02	1.02-1.03	1.03	1.03	$\approx$1.5
E_t	MPa	2200-3000	70-250	270-450	90-250	60-140	
σ_y	MPa	80	-	20-25	-	-	
ϵ_y	%	-	-	30-35	-	-	
ϵ_{tB}	%	>50	>50	>50	>50	>50	
σ_{50}	MPa	-	10-20**	-	10-15**	8-12**	
σ_B	MPa	-	-	-	-	-	
ϵ_B	%	-	-	-	-	-	
T_P	°C	200-245	140-155	160-170			
HDT	°C	50-60	>50	50-55	34	-	
α_p	10^{-5}/K	-	18-23	14-18	15-20	15-20	
α_n	10^{-5}/K	-	-	-			

Continued on next page

Property	Unit	Aromatic polyamides, copolyamides and polyether block amides					
		PA 6/66 PA 66/6	PEBA 12 Shore D40-55	PEBA 12 Shore D55-65	PEBA 6 Dry	PEBA 6 Cond.*	PPA
UL94	Class	V-2***	HB	HB**	HB	HB	
$\epsilon_r 1$	-	3.7	6-11	6-11	4	6	
$\tan \delta 1$	$\cdot 10^{-3}$	300	400-1300	400-1300	300-500	950-1100	
ρ_e	Ohm·m	10^{13}	10^{10}-10^{12}	10^{10}-10^{12}	10^{11}	10^{10}	
σ_e	Ohm	10^{12}-10^{13}	10^{12}-10^{13}	10^{12}-10^{13}	10^{12}	10^{11}	
E_BI	kV/mm	-	30-40	30-40	35-40	30-35	-
W_w	%	9-10	0.6-1.5	0.6-1.5	3.5-5.0	3.5-5.0	-
W_H	%	3-3.2	0.3-0.7	0.3-0.7	1.0-1.5	1.0-1.5	-

* exposure of specimens at 23 °C/50 % r.h. until saturation
** breaking
*** available up to V-0

6.7.4 Polyamides, Special Polymers (PA NDT/INDT [PA 6-3-t], PAPACM 12, PA 6-I, PA MXD6 [PARA], PA 6-T, PA PDA-T, PA 6-6-T, PA 6-G, PA 12-G, TPA-EE)

Basic structures and further information, see Table 6.29. Incorporating bulky segments, such as aromatic dicarbon acid or branched or acyclic diamines in PA molecules instead of the smooth CH_2 segments, results in semi-crystalline or amorphous crystal clear PAs. For example, PA NDT/INDT (acronym according to ISO 1874, old acronym: PA 6-3-T) and PA 6I. Other partially aromatic, amorphous PA are listed in Table 6.35. PA 12, based on dodeca-n-diacid (DDS) and a cyclic aliphatic diamine (PA CM), acronym according to ISO 1874: PA-PACM, is micro-crystalline and therefore long-term transparent, exhibiting color fastness on weathering comparable to PMMA.

Table 6.35: Partially Aromatic, Amorphous Polyamides

PA grade	Monomers
PA 6-I-6-T [1]	Hexamethylenediamine, isophthalic acid, terephthalic acid
PA 6I/6T/PACM/PACMT	As PA 6-I-6-T + diaminodicyclohexylmethane
PA 12/MACMI	Laurinlactam, dimethyl- diaminodicyclohexylmethane, isophthalic acid
PA 12/MACTM	Laurinlactam, dimethyl- diaminodicyclohexylmethane

[1] Polyphthalamide, acronym also PPA, $T_g = 130$ °C, $T_m = 330$ °C

Increasing the number of amide groups and implementation of aromatic monomers into the molecular units increases the melt temperature: PA 6-T: 371°C and

PA-PDA-T (PDA = phenylenediamine): 500°C (no longer processable in the molten state). Major applications are high-strength, temperature resistant fibers. PA 6/6T copolymers (melt temperature 295°C) are more heat resistant than PA 66 and only absorb 6% water, which makes them suitable for applications that require PA with high long-term service temperatures and low water absorption, particularly when reinforced with glass fibers. Amorphous or low-crystalline polyamides, such as PA 6-T and PA MXD6 stand out from other PAs. As a rule, water absorption is significantly lower than for PA 6 and PA 66. They are more difficult to color, exhibit higher dimensional stability, less warpage, and less shrinkage. Total shrinkage is at $\approx$ 0.5%. Because the temperature range for softening is wider and the melt viscosity is higher, injection molding, extrusion, and particularly blow molding are made easier. Resins such as terephthalate-containing copolymers (PA 6-6-T, modulus of elasticity 3.5 MPa) or poly-m-xylylene adipamide (polyarylamide PA MXD6, PARA) with glass fiber reinforcements of 30 - 60% (also with increased impact resistance for electronic applications) exhibit the highest strength and stiffness of all polyamides, while maintaining good heat distortion stability and chemical resistance comparable to other PA grades. Pure PA MXD6 is also used as barrier layer against O_2, CO_2, hydrocarbons, solvents, and flavors with other PAs or with PET in bottles. Copolymers of PA 6-T and PA 6-I (PA 6-T-6-I) can be injection molded at melt temperatures of 340 - 350°C and mold temperatures from 140 - 160°C, despite their high crystalline melting point (330°C). Usually it is reinforced with 50% short strand glass fibers; it is resistant to all media typically encountered in automotive applications, hot water, and decalcification solutions; approved for food contact. Comparison of properties, see Table 6.34.

6.7.5 Cast Polyamides (PA 6-C, PA 12-C).

Properties compare to those of PA 6 and PA 12. The higher molecular weight increases stiffness and can complicate or even impede processing like a thermoplastic material. Volume shrinkage for cast parts is $\approx$ 15% and has to be considered when designing the casting mold.

6.7.6 Polyamide for Reaction Injection Molding (PA-RIM)

PA-RIM is in fact an NBC-RIM (nylon block copolymer) that is processed by two-component RIM techniques. Long polyether blocks (polymers with oxygen bridges in their chains) implemented in the molecular chain increase the impact resistance and decrease stiffness; therefore, properties may vary from those of PA 6 to those of a PA elastomer. The low viscosity of the formulation facilitates high filler and reinforcement loads. These resins are used, for example, for prototypes in vacuum casting systems.

6.7.7 Aromatic Polyamides, Aramides (PMPI, PPTA)

Poly-m-Phenylene Isophthalamide (PMPI) PMPI has a high heat distortion point and even at 260°C still maintains ≈ half of its mechanical properties. The crystalline melting point is at 368 - 390°C; for structure, see Table 6.29.

Applications. Fibers for heat protection gear and electrical insulation at elevated temperatures.

Poly-p-Phenyleneterephthalate (PPTA) PPTA is used for the manufacture of fibers. These fibers exhibit the highest ratio of strength to density of all commercially available fibers. After exposure to 250°C for 100 hours, they still exhibit 50% or their initial strength. PPTA does not melt and begins to char at 425°C. Density ≈ 1.45 g/cm^3, tensile strength ≈ 3.5 GPa. Modulus of elasticity ≈ 15 GPa, elongation at break 2.0 - 2.5%, for structure see Table 6.29.

6.8 AROMATIC (SATURATED) POLYESTERS

Chemical Constitution. The chains of thermoplastic (saturated linear) polyesters contain ester groups in regular intervals, see Eq. 6.10. In most cases, they are produced as condensation polymers of dicarbonylic acids and diols or their derivatives.

$$\overset{O}{\underset{}{\overset{\|}{-C}}}-O-R$$
Ester group

(6.10)

Polyesters containing aromatic groups (benzene rings) are basic polymers for engineering materials. The aromatic rings stiffen the molecular chains, thereby raising heat distortion and melting temperatures. The more frequently these groups occur in the chain, the higher are these temperatures. Fully aromatic polyesters (PAR, polyarylates) have outstanding thermal stability. Semi-aromatic polyesters are not attacked by either aliphatic hydrocarbons nor by ethanol and higher alcohols. They absorb only minimal amounts of water and are physiologically inert. Because they contain saponifiable ester groups, they are degraded by alkalis. Their resistance to oxidizing acids and to long-term exposure to water and water vapor above 70°C is limited. The resin has to be completely dried to avoid damage by hydrolysis during thermoplastic processing.

6.8.1 Polycarbonate (PC)

Polycarbonate Based on Bisphenol A (PC)

Chemical Constitution. Polycondensation and general structure, see Eq. 6.11.

$$\text{Bisphenol A (BPA)} + \text{Phosgene} \longrightarrow \text{Polycarbonate (PC)}$$

(6.11)

The most important PC grade is produced by metathesis of bisphenol A (generated from phenol and acetone) with phosgene. The melt condensation process from bisphenol A and diphenylene carbonate has lost industrial importance. The molecular weight generally ranges below 30,000 g/mol, because otherwise the melt viscosity would be too high. PC is an amorphous thermoplastic material.

Processing. PC can be processed and post-treated by all standard thermoplastic processing methods. Its high melt viscosity requires high injection pressures or relatively low ratios of flow distance to wall thickness. Melt temperatures for injection molding range from 280 - 320°C (mold temperatures form 80 - 120°C), for extrusion from 240 - 280°C. Drying for 4 - 24 h at 120°C is required to reduce residual moisture to below 0.01 - 0.02%. Vented screws may be of advantage. Shrinkage during processing ranges from 0.6 - 0.8%, mold shrinkage is negligible. PC is well suited for precision injection molded parts for optical and electrical applications. Extremely thin films can be cast with solutions of PC in methylene chloride. Very large integral foamed parts can be manufactured from pellets containing foaming agents. PC can be adhesively bonded with glues (e.g., from PC and methylene chloride) or with reaction resins; it can be welded by ultrasound or high-frequency.

Properties. Unreinforced PC is crystal clear, has high surface gloss, and is available in any color and color intensity. It is impact resistant and has a high strength and stiffness in a temperature range from -150 to +135°C. Maximal service temperature short-term is at 150°C, long-term at 130°C. PC is notch-sensitive, a fact also reflected in its low long-term alternating fatigue strength. Its abrasion resistance is limited; its good electrical properties are not affected by moisture. PC shows good resistance to high-energy radiation; with thicker walls or, in the case of sheets, UV stabilization on the exposed surfaces it shows good weather resistance. Its chemical resistance is limited. PC is susceptible

6.8 Aromatic (Saturated) Polyesters

to stress cracking. Permeability of CO_2 is relatively high; therefore bottles for CO_2-containing liquids should have a barrier layer, e.g., of PET or PBT. PC is sterilizable and self-extinguishing once the source of ignition is removed. Free-flowing grades are available for the production of large-sized parts (for example, covers for light fixtures) or compact disks. Grades with 10 - 40% short-strand glass fiber reinforcement are available to increase stiffness, in particular at elevated temperatures, and to reduce stress cracking tendencies. Grades reinforced with MoS_2, graphite, or PTFE exhibit improved slip and wear properties; reinforcement with aluminum powder improves the electrostatic shield of housings. Comparison of properties, see Table 6.36.

Table 6.36: Property Comparison of Polycarbonate and PC Blends

Property	Unit	Polycarbonate and PC Blends				
		PC (BPA)	PC-GF30	PC-GF 30 PEC var. ester cont.	(PC+ABS)**	(PC+ABS)-GF20
ρ	g/cm³	1.2	1.42-1.44	1.19-1.21	1.08-1.17	1.25
E_t	MPa	2300-2400	5500-5800	2000-2400	2000-2600	6000
σ_y	MPa	55-65	-	65-70	40-60	-
ϵ_y	%	6-7	-	7-9	3-3.5	-
ϵ_{tB}	%	>50	-	>50	>50	-
σ_{50}	MPa	-	-	-	-	-
σ_B	MPa	-	70	-	-	75
ϵ_B	%	-	3.5	-	-	2
T_P	°C	-	-	-	-	-
HDT	°C	125-135	135-140	135-165	90-110	115
α_p	10^{-5}/K	6.5-7	2.5-3	7-8	7-8.5	3-3.5
α_n	10^{-5}/K	-	-	-	5-6	-
UL94	Class	V-2*	V-1*	HB*	HB*	HB*
ϵ_r 100	-	2.8-3.2	3.3	2.8-3.3	3	3.2
tan δ 100	$\cdot 10^{-3}$	7-20	9-10	10-20	30-60	20-30
ρ_e	Ohm·m	>10^{14}	>10^{14}	>10^{14}	>10^{14}	>10^{14}
σ_e	Ohm	>10^{14}	>10^{14}	>10^{14}	>10^{14}	>10^{14}
$E_B I$	kV/mm	30-75	30-75	35-45	24	30
W_w	%	0.35	0.28-0.30	0.32	0.6-0.7	0.4-0.5
W_H	%	0.2	0.11-0.15	0.15	0.2	0.15-0.2

Property	Unit	Polycarbonate and PC Blends				
		(PC+PET)	(PC+PBT)	(PC+PBT)-GF30	PC + LPC	PC (TMC)
ρ	g/cm³	1.22	1.2-1.26	1.43-1.45	-	1.18-1.14
E_t	MPa	2100-2300	2300	7000	2600-4000	22509
σ_y	MPa	50-55	50-60	-	66	65
ϵ_y	%	5	4-5	-	5.6-2.9	7

Continued on next page

Property	Unit	Polycarbonate and PC Blends				
		(PC+PET)	(PC+PBT)	(PC+PBT)-GF30	PC + LPC	PC (TMC)
ϵ_{tB}	%	>50	25->50	-	-	>50
σ_{50}	MPa	-	-	-	-	-
σ_B	MPa	-	-	90	74-82	-
ϵ_B	%	-	-	3	-	-
T_P	°C	-	-	-	-	-
HDT	°C	105	70-95	150	120-135	140-180
α_p	10^{-5}/K	9-10	8-9	3	-	7.5
α_n	10^{-5}/K	-	-	-		
UL94	Class	HB*	HB*	HB*	-	HB*
ϵ_r 100	-	3.3	3.3	4	-	3.0-2.8
tan δ 100	$\cdot 10^{-3}$	200	20-40	30-40	-	16-13
ρ_e	Ohm·m	>10^{13}	>10^{14}	>10^{14}	-	>10^{14}
σ_e	Ohm	>10^{15}	>10^{14}	>10^{14}	-	>10^{14}
$E_B I$	kV/mm	30	35	35	-	35
W_w	%	0.35	0.35	0.25	-	
W_H	%	0.15	0.15	0.1	-	0.15

* also available as V-0
** with small PC content available as ABS+PC

Polycarbonate Copolymers Polyethercarbonate is a block copolymer of "rigid" BPA carbonate units, see Eq. 6.11, and "soft" polyethylene glycol units, see Eq. 6.12. They are used in a coagulation process to produce blood compatible dialysis membranes with better barrier properties than cellulose membranes.

$$-[O-(CH_2-CH_2-O)-CO]_n-$$
Polyethylene glycol

(6.12)

Co-condensates of BPA with fluorenon bisphenol exhibit increased heat distortion properties (up to 220°C). Co-condensation of BPA with long-chain aliphatic dicarbonylic acids results in very tough, free-flowing resins; however, they exhibit a lower heat distortion point.

Polycaronate Based on Trimethylcyclohexane Bisphenol (PC-TMC). PC copolymers based on bisphenol A and bisphenol TMC (trimethylcyclohexane, see Eq. 6.13) are also crystal clear and widen the service temperature range, depending on TMC content, to 160 - 220°C (Vicat softening temperature). They are suitable for automotive applications (head lights) and medical applications (sterilizable at 143°C). With rising TMC content toughness will decrease.

Trimethylcyclohexane-Bisphenol (TMC)

(6.13)

6.8 Aromatic (Saturated) Polyesters

Polyphthalate Carbonate (PPC). PPC is a transparent PC copolymer, see Eq. 6.14, with increased heat distortion point (> 10 K) and impact resistance.

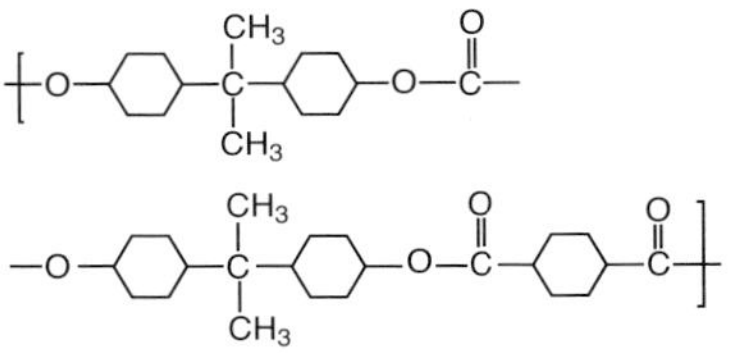

Polyphthalate-carbonate (PPC)

(6.14)

Copolymers with halogenated bisphenols, particularly with tetrabromobisphenol (Eq. 6.15), exhibit increased flame retardancy; however, alternative flame retardants are gaining importance.

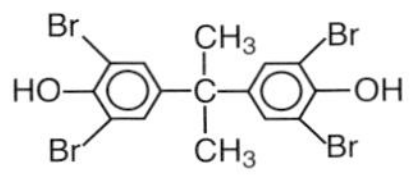

Tetrabromobisphenol A (TBBPA)

(6.15)

Increasing bisphenol S (dihydroxy diphenylene sulfide, Eq. 6.16) content increases impact resistance.

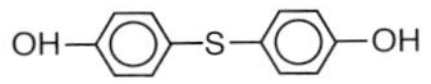

Dihydroxydiphenylsulfide (Bisphenol S)

(6.16)

Polycarbonates Based on Aliphatic Dicarbonylic Acids. These PC grades are based on a random copolymerization technology that can be used to electively increase flow behavior or mechanical properties, without affecting impact resistance in the first or processability in the latter case.

Diphenylene Polycarbonate (DPC). DPC is producded by a recently developed phosgene free process by conversion of bisphenol A and diphenylene carbonate.

Polycarbonate Blends (PC +: ABS, ASA, AES, PMMA+PS, PBT, PET, PPE+HISB, HIPS, PPE, PP Copolymers, SMA, TPU) Approximately 15% of all PCs are used for the production of blends, particularly for those with 10 - 50% SAN or ABS. Many applications do not require the high heat distortion temperature offered by PC, while that of polystyrene is often not sufficient. This gap can be closed by competitive blends, whose service temperature range can be extrapolated from those of its components by a linear ratio rule. The following polystyrenes or polystyrene-similar materials are used in blends

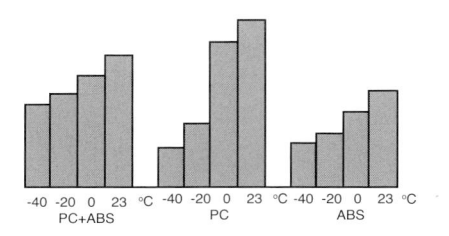

Figure 6.15: Comparison of Izod impact resistance for ABS, PC, and an ABS-PC blend.

in varying ratios: ABS, ASA, SMA, AES. As shown in Fig. 6.15, the impact resistance of an ABS blend may be higher in certain temperature ranges than the one of its respective components. ASA and AES blends result in more weather resistant resins, while the heat distortion point can be increased by the use of SMA, methylstyrene-containing ABS, or special PC grades. Flame retardant (also chlorine- and bromine-free), 10 -30% short-strand glass fiber reinforced, and foamable grades are available.

PC blends with PMMA increase UV resistance. Blends with modified PPE or with PP copolymers are also produced. Blends of PC with LCP (LCP content well below 50%) exhibit excellent flow behavior (it is possible to reduce minimal wall thickness by 50% for injection molding compared to PC+ABS). In addition, its strength and stiffness (depending on orientation) without glass fiber reinforcement is increased compared to PC. The reinforcing effect is caused by orientation of the LCP into fibrils, as they are sheared within the incompatible PC melt. To increase impact resistance, PC is reinforced with TPU, PBT, or PET. The heat distortion resistance of PC remains largely unchanged in blends with PBT and PET; fuel resistance is increased. Comparison of properties, see Table 6.36.

Applications. Utilizing transparency, heat distortion resistance, and impact resistance: parts and housings in electrical engineering and electronics applications, appliances, measuring equipment, binoculars, chronometers, projectors, power strips, CDs and DVDs, precision injection molded parts made from free-flowing grades, street lights, traffic lights, indicator lights, cover boxes and plates for switch and measuring equipment, glazing for stadiums and green houses made from extruded sheet, laminated safety glass, lenses; protective helmets, premium tableware, reusable drink bottles, milk bottles, containers for drinking water up to 20 l. Large parts, such as community mailboxes, junction boxes, mounting plates, street light posts made of structural PC foam with or without glass fiber reinforcement. PC copolymers are used when high heat distortion resistance is required. Coated (surface modified) PC is used in the automotive industry (for window glazing, side and rear panels, and roof modules). PC-TMC/BPA: light fixtures and cast film with increased temperature resistance, scratch resistant diffuser lenses with 4 - 8 μm Siloxane coating. PC and PC+PBT blends laminated with electro-luminescent film: 3-dimensional glowing plastic films.

6.8.2 Polyesters of Therephthalic Acids, Blends, Block Copolymers

Polyethylene Terephthalate (PET)
Chemical Constitution. The raw materials for condensation of PET are terephthalic acid and ethylene glycol. Equation 6.17 shows the general structure.

COOH—⟨◯⟩—COOH   OH—CH$_2$—CH$_2$—OH
Terephthalic acid   Ethyleneglycol

$$\left[-\overset{O}{\overset{\|}{C}}-\langle\bigcirc\rangle-\overset{O}{\overset{\|}{C}}-O-CH_2-O- \right]$$

Polyethylene terephthalate (PET)

(6.17)

PET is a semi-crystalline thermoplastic material. Initially it was only used for the manufacture of fibers; eventually is was used for films and bottles. Today, PET ranks among the most important packaging materials. Higher molecular grades, modified with nucleating agents to accelerate crystallization, can be used for injection molding. Three genral PET grades are distinguished: crystalline PET (PET-C), amorphous PET (PET-A), and PET copolymers with increased impact resistance (glycol-modified PET). Implementation of voluminous comonomers such as isophthalic acid or 1,4 cyclohexane dimethylol (CHDM), see Eq. 6.18, is used to lower crystallinity to facilitate the manufacture of transparent parts (e.g., bottles).

COOH COOH
   ⟨◯⟩
Isophthalic acid

OH—CH$_2$—⟨◯⟩—CH$_2$—OH
Cyclohexane-1,4-dimethylol

(6.18)

Processing. The major processing methods for PET are injection molding and one- and two-stage stretch blow molding of PET bottles and containers. Films, sheet, and solid profiles are extruded. Prior to thermoplastic processing, moist pellets must be dried for $\approx$ 10 h at 130°C. For injection molding of amorphous parts, the melt temperature should range between 260 and 290°C; the mold temperature should be above 60°C for amorphous and $\approx$ 140°C for semi-crystalline parts (wall thickness > 4 mm). Despite its high shrinkage of 1.2 - 2.5% for semi-crystalline parts, PET is well suited for overmolding of metal inserts, as long as the wall thickness is adequately high. Joints can be established by the following methods: ultrasonic, friction, hot plate, and hot gas welding and adhesive bonding with cyanoacrylate-, EP-, or PUR adhesives.

Properties. Mechanical properties depend on the degree of crystallinity, which in turn depends on the processing conditions during injection molding. Mold temperatures of 140°C, long residence time, and annealing result in a degreeof crystallinity of 30 - 40%. These parts exhibit high stiffness and strength below 80°C and low creep under constant static load. However, their impact resistance is low, slip- and wear properties are good. Amorphous parts are aimed for when, in addition to high transparency, high toughness, excellent slip- and wear properties, low shrinkage, and high dimensional stability are required. At $\approx 80°C$ amorphous PET components have a glass transition region, in which the modulus of elasticity, particularly of non-reinforced grades, decreases sharply. Highly reinforced grades have a structural integrity up to $\approx 250°C$. Long-term service temperature ranges from 100 - 120°C. The good electrical properties are little frequency- and temperature-dependent. Because PET has low permeability for O_2 and CO_2, it is well suited for bottling alcohol-containing or carbonated beverages. It is resistant to weak acids and alkaline solutions, oils, fats, aliphatic and aromatic hydrocarbons, and carbon tetrachloride. It is not resistant to strong acids and alkaline solutions, phenol, and long-term use in hot water above 70°C. PET is not prone to stress cracking. It exhibits good weatherability, particularly when carbon black stabilized against UV radiation. PET's general property spectrum explains its fast-paced application in packaging. Compared to glass, it is nonbreakable and significantly lighter (30 - 40 g/l for PET compared to 500 - 900 g/l for glass). Without added flame retardants, PET burns with an orange-yellow flame. It meets requirements for food contact; however, it can only be sterilized by radiation or in ethylene oxide atmosphere because of its low hydrolysis resistance. Addition of carbon- or glass fiber reinforcement increases stiffness and strength and reduces shrinkage to 0.4 - 0.8%. At the same time, anisotropy caused by shrinkage increases so that for applications in which strength is required, e.g., electrical or electronic, micro glass spheres are generally used. Comparison of properties, see Table 6.37.

Applications. Staple fibers and filaments (72% of total consumption), PET bottles (disposable, reusable, hot filling), wear resistant parts (also glass fiber-reinforced PET) such as bearings, gear wheels, shafts, guides, couplings, locks, knobs. Insulating, magnetic, and anti-stick films for cast resin processing, ink ribbons for printers, carrier films for photographic film, shrink film, fibers, PET-A films for blister packaging, foamed PET-C deep drawing film for fast-food trays.

6.8 Aromatic (Saturated) Polyesters

Table 6.37: Property Comparison for Polyalkylene Terephthalates and Polyester Elastomers

Property	Unit	\multicolumn{4}{c}{Polyalkylene Terephthalates and Polyester Elastomers}			
		PET-A	PET-C	PBT -GF 30	PBT	PBT elast.mod.
ρ	g/cm^3	1.33-1.35	1.38-1.40	1.56-1.59	1.30-1.32	1.2-1.28
E_t	MPa	2100-2400	2800-3100	9000-11000	2500-2800	1100-2000
σ_y	MPa	55	60-80	-	50-60	30-45
ϵ_y	%	4	5-7	-	3.57	6-20
ϵ_{tB}	%	>50	>50	-	20->50	>50
σ_{50}	MPa	-	-	-	-	-
σ_B	MPa	-	-	160-175	-	-
ϵ_B	%	-	-	2-3	-	-
T_P	°C	-	250-260	250-260	220-225	200-225
HDT	°C	60-65	65-75	220-230	50-65	50-60
α_p	10^{-5}/K	8	7	2-3	8-10	10-15
α_n	10^{-5}/K	-	-	7-9	-	-
UL94	Class	HB*	HB*	HB*	HB*	HB*
ϵ_r100	-	3.4-3.6	3.4-3.6	3.8-4.8	3.3-4.0	3.2-4.4
tan δ100	·10^{-3}	20	20	30-60	15-20	20-130
ρ_e	Ohm·m	>10^{13}	>10^{13}	>10^{13}	>10^{13}	>10^{13}
σ_e	Ohm	>10^{14}	>10^{14}	>10^{14}	>10^{14}	>10^{14}
E_BI	kV/mm	250	30	30-35	25-30	25
W$_w$	%	0.6-0.7	0.4-0.5	0.4-0.5	0.5	0.4-0.7
W$_H$	%	0.3-0.35	0.2-0.3	0.2	0.25	0.15-0.2

Property	Unit	PBT GF 30	(PBT+ASA)	TPC-EE Shore D35-50	Shore D55-75
ρ	g/cm^3	1.52-1.55	1.21-1.22	1.11-1.20	1.22-1.28
E_t	MPa	9500-11000	2500	30-150	200-1100
σ_y	MPa	-	53	-	-
ϵ_y	%	-	3.6	-	-
ϵ_{tB}	%	-	>50	>50	>50
σ_{50}	MPa	-	-	10-30	30-50
σ_B	MPa	130-150	-	-	-
ϵ_B	%	2.53	-	-	-
T_P	°C	220-225	225	155-210	215-225
HDT	°C	200-210	80	-	50-55
α_p	10^{-5}/K	3-4.5	10	15-22	10-18
α_n	10^{-5}/K	7-9	-	-	-
UL94	Class	HB*	HB*	HB*	HB*
ϵ_r100	-	3.5-4.0	3.3	4.4-5	-

Continued on next page

Property	Unit	Polyalkylene Terephthalates and Polyester Elastomers			
		PBT	(PBT+ASA)	TPC-EE	
		GF 30		Shore D35-50	Shore D55-75
$\tan \delta 100$	$\cdot 10^{-3}$	20-30	10	100-200	-
ρ_e	Ohm·m	$>10^{13}$	$>10^{14}$	≈ 10	-	-
σ_e	Ohm	$>10^{14}$	$>10^{15}$	$>10^{13}$	-
$E_B I$	kV/mm	30-35	30	20-25	-	-
W_w	%	0.35-0.4	0.5	0.6-1.2	0.4-0.8	-
W_H	%	0.1-0.15	0.2	0.3-0.6	0.2-0.4	
* also available as V-0					
PET-A (amorphous), PET-C (semi-crystalline)					

Polybutylene Terephthalate (PBT)

Chemical Constitution. The chemical constitutions and properties of PBT and PET are very similar. During condensation, 1,4-butanediol is used instead of ethylene glycol, see Eq. 6.19.

COOH—⟨◯⟩—COOH   OH—[CH₂]₄—OH

Terephthalic acid   1,4-butanediol

[−O−C(=O)−⟨◯⟩−C(=O)−O−[CH₂]₄−O−]

Polybutylene terephthalate (PBT)

(6.19)

PBT is semi-crystalline; however, it crystallizes faster and is therefore better suited for injection molding than PET.

Processing. PBT is mostly injection molded at melt temperatures of 230 - 270°C. Mold temperatures below 60°C are common; however, optimum surface quality is only reached at 110°C. As with PET, resin drying is necessary. PBT can be joined by ultrasonic-, friction, hot plate-, and hot gas welding as well as by two-component adhesives.

Properties. Stiffness and strength are slightly less than those of PET, toughness at low temperatures is slightly better. Friction and wear properties are excellent. Glass transition of the amorphous phase is at 60°C. Maximum form stability temperature is at 180 - 200°C, the long-term heat distortion temperature at 100 - 120°C. PBT is a good electrical insulator and its properties are little influenced by water absorption, temperature, or frequency. Permeability for CO_2 is significantly higher than that of PET, chemical and weather resistance and burn rates are comparable. Resistance to hot water is better. PBT grades meet the requirements for food contact. PBT grades with the following modifications are available: free-flowing, flame retarded, higher impact resistance by elastomer

6.8 Aromatic (Saturated) Polyesters

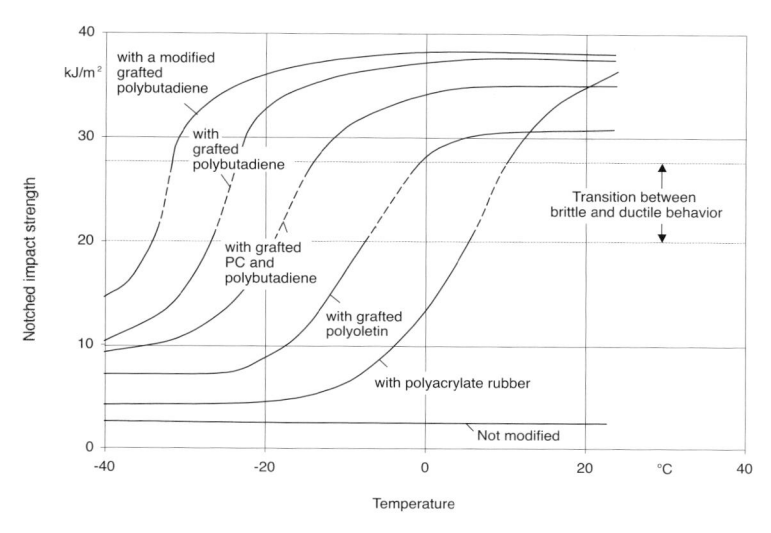

Figure 6.16: Notched impact toughness of unreinforced modified polybutylene terephthalate depending on temperature.

modification, see Fig. 6.16, reinforced or filled to increase stiffness, strength, wear resistance, or to lower the coefficient of dynamic friction.

Comparison of properties, see Table 6.37.

Applications. Friction bearings, roller bearings, valve parts, screws, multipoint connectors, pump housings, parts for small appliances such as coffee makers, egg cookers, toasters, blow dryers, vacuums.

Polytrimethylene Terephthalate (PTT) PTT's properties are comparable to those of PBT. The glass transition temperature T_g is slightly higher, so that with increasing temperature the decrease in modulus of elasticity is less pronounced.

Thermoplastic Polyester-Elastomers (TPC) See thermoplastic elastomers later in this chapter.

Polyterephthalate Blends (PET + :PBT, MBS, PMMA, PSU, Elastomer)
Blends with PBT, MBS, PMMA, and PSU exhibit improved processability. Blends of MBS with any of the three PET grades are used to improve impact resistance while retaining transparency. MBS (methacrylate-butadiene-styrene-copolymers) are manufactured by polymerizing a rigid shell on a soft rubber core.

Applications: packaging suitable for both freezer and microwave. Blends with PEN have a higher heat distortion point. PBT + PC is tougher and is increasingly used for large exterior parts in the automotive industry. PBT + PET-GF50 is extremely rigid, UV resistant, with a smooth surface. It is used for automotive windscreen wiper arms, exterior mirror fixtures. PBT + ASA-(GF) exhibits less shrinkage at a higher melt flow rate and improved surface quality. Automotive applications: headlight frames, electronics housings. Blends with EPDM, LCP, and SMA are available.

6.8.3 Polyesters of Aromatic Diols and Carboxylic Acids (PAR, PBN, PEN)

Polyarylate (PAR)

Chemical Constitution, Properties. PAR are thermoplastic condensation products of pure aromatic polyesters and polyester carbonate (PEC). Equation 6.20 exemplifies a structure.

$$\tag{6.20}$$

PAR are transparent and in their mechanical and electrical properties and chemical resistance comparable to PC. Grades for food contact are available. PAR can be sterilized. The glass transition temperature ranges from above 180 to 325°C, the long-term service temperature is at 150°C. By their nature, PAR are difficult to ignite, very UV resistant, and suitable for outdoor use even without stabilization. However, without UV absorbers they tend to yellow. Today, two PAR grades(PAR 15 and PAR 25) with very high temperature resistance are available, see Eq. 6.21.

Their glass transition temperatures are at 250°C (PAR 15) and 325°C (PAR 25), respectively. Both grades can be used to sinter rods and sheet. PAR 15 can also be injection molded. Both PAR 15 and PAR 25 are distinguished by the uniformity of their mechanical and electrical properties in a temperature range up to 200°C and 300°C, respectively. Because of their amorphous structure, molded parts and cast film are crystal clear. Applications: UV filters, membranes, Lichttechnik, welding equipment, electrical, and electronic applications. Comparison of properties, see Table 6.38.

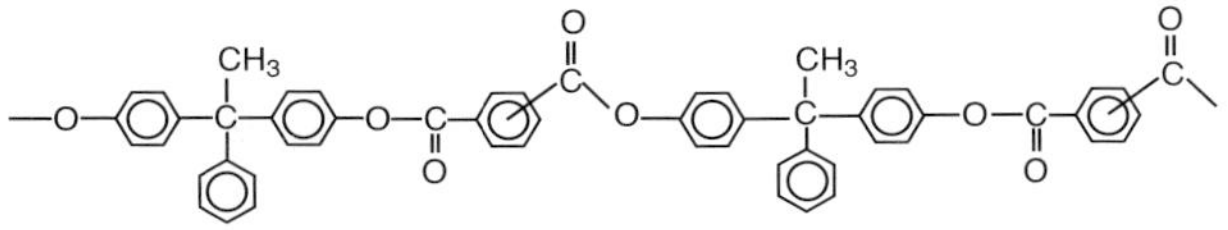

6.8 Aromatic (Saturated) Polyesters

PAR 15

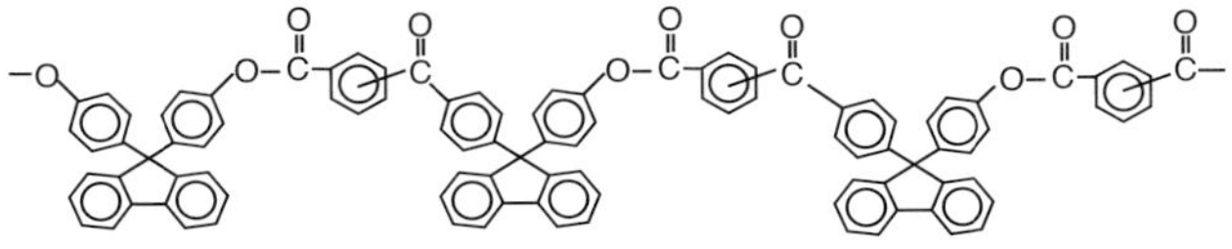

PAR25

(6.21)

Table 6.38: Comparison of Properties for PAR

Property	Unit	PAR 15/25	PBN-GF-30	PEN
ρ	g/cm^3	1.21/1.22	1.63	-
E_t	MPa	2350/2800	-	2400
σ_y	MPa	-	-	81
ϵ_y	%	-	-	7.2
σ_{50}	MPa	-	-	-
σ_B	MPa	76/100	153	-
ϵ_B	%	9-115	5	-
HDT	°C	235/305	228	87
UL94	Class	V-0	-	-
ρ_e	Ohm m	1 to 20·10^{14}	-	-
$E_B I$	kV/mm	-	66	-

Processing. PAR has to be dried to a moisture content of less than 0.02% at a temperature between 120 and 130°C before it can be processed (primarily injection molding and extrusion at melt temperatures of 340 - 400°C). The mold temperature may reach up to 150°C, shrinkage in flow direction is $\approx$ 0.2%, perpendicular to flow it is $\approx$ 0.7 - 0.9%. All other processing and post-processing operations as with PC.

Modifications. Addition of fibers and fillers, in particular glass and potassium titanate fibers, increases strength. Talc is also used as reinforcing filler. Flame retardant additives improve the already good burn rate (V2 according to UL 94). Blends with other polymers yield certain advantages, in particular with regard to processability, resistance to fuels, risk of hydrolysis, and price reduction. The lower impact resistance at low temperatures is compensated by grafted silicon-rubbers, see Table 6.39.

Applications. PAR are used when the heat distortion temperature of PC is not sufficient and PSU is too expensive; for example, for control panels of kitchen appliances, parts for blow dryers and microwaves, lamp housings and reflectors, tools, elements for office- and other machines. PAR 15/25: sintered sheet, pipes, tubes, rods; transparent cast film for insulation or multi-layer laminates in electrical or electronic devices, liquid crystal monitors, adhesive tapes, UV filters; functional elements in all technical fields that are exposed to elevated temperatures.

Table 6.39: Advantages of Blending Polyarylates with other Polymers

Polyarylate mixed with	Fuel resistance	Hydrolysis-resistance	Processability	Price	Other
Polyalkylene terephthalate	x		x	x	
Aliphatic polyamide	x		x	x	
Aromatic polycarbonate			x		
Polyolefin			x	x	Higher impact resistance
ABS polymers		x	x		Higher impact resistance
Polyphenylene sulfide		x			Higher flame resistance
Polyether imide	x				
Polycarbonate-siloxane block copolymers	x		x		Higher impact resistance
Grafted silicone rubber	x				Higher impact resistance

(x) signifies improvement

Polybutylene Naphthalate (PBN) PBN is a polymer based on dimethyl-2,6-naphthalene dicarboxylate (NDC) and butanediol. Compared to PBT it exhibits a higher heat distortion point (solder resistant), and it shows less permeability for methane and methanol than PA 11 or PE-HD. Property comparison, see Table 6.38.

Applications. Electronic devices, fuel carrying parts (copolymers (TPE) for hoses and cables).

Polyethylene Naphthalate (PEN) PEN is a polymer based on dimethyl-2,6-naphthalene dicarboxylate (NDC) and ethylene glycol. It is amorphous and transparent and is used as an alternative to or blend partner with PET for bottles (hot cleaning for multiple use), films, and fibers. For the production of polyester bottles the use of PEN is still cost prohibitive. Only for special applications PEN could substitute PET because of its better barrier properties and heat stability.

Processing. Injection molding, extrusion, thermoforming, blow molding.

Properties. See Table 6.38; low oxygen permeability, good UV resistance and resistance to chemicals and hydrolysis.

Applications. In the medical field it implemented in blood sampling devices, packaging, containers.

6.9 AROMATIC POLYSULFIDES AND POLYSULFONES (PPS, PSU, PES, PPSU, PSU + ABS)

6.9.1 Polyphenylene Sulfide (PPS)

Chemical Constitution PPS are semi-crystalline polymers in which aromatic monomer units are connected by sulfur atoms. They are distinguished by their extremely high heat stability (melt temperature up to 445°C), high chemical resistance and strength, see Eq. 6.22.

$$\text{—} \bigcirc \text{—S—}$$

Polyphenylene sulfide (PPS)

(6.22)

They can be produced as either cross-linkable thermosets or as thermoplastics; the latter being by far the more relevant.

Processing The processing techniques for PPS are:

- *Injection molding.* Injection molding is the most important processing method; melt temperatures ranging from 315 - 370°C, mold temperatures from 25 - 200°C. Above a mold temperature of 120°C part surfaces turn smooth and glossy, at 40°C the highest degree of toughness is achieved. Because glass fiber reinforced resins contain less than 0.05% water, pre-drying is only necessary in special cases, e.g., when hydrophilic fillers are used. In any case, storage at 150°C for 6 h has proven advisable. As PPS has a very low melt viscosity, even the filled grades are easily flowing and thin-walled parts can be molded.
- *Pressing, sintering.* PPS can be processed by pressing or sintering (e.g., for surface coating). Oxidative cross-linking prior to processing is used to

achieve the necessary melt viscosity of the powders. The next processing steps for molded parts are: compressing in the cold mold at 70 MPa, heating of the mold to 360°C within 1 h, recompressing at 70 MPa (compressing time 2 min per 2.5 mm wall thickness), cooling at a rate of less than 10 K/h, deforming at ≈ 150°C.

- *Post processing operations.* Carbide tools are recommended for machining of highly filled resins. Ultrasonic and hot plate welding are possible, while high-frequency welding is not suitable. Adhesion with solvent adhesives is impossible because of PPS' high chemical resistance; however, two-component adhesives based on epoxy or polyurethane resins yield good bonds. Before lacquering, the surfaces have to be pre-treated either by flame or plasma treatment or a primer based on PUR has to be used.

Properties Thermoplastic PPS is only little branched and therefore highly crystalline. The degree of crystallinity depends on the temperature-time history during processing. The long-term service temperature ranges from 200 - 240°C, the maximum service temperature is at 300°C. There are no known solvents below 200°C. At elevated temperatures PPS can be swelled. PPS is resistant to alkalis and non-oxidizing acid, with the exception of hydrochloric acid. It is attacked by oxidizing agents such as nitric acid. Light attacks the surface of non-stabilized and non-pigmented grades. Gas permeability is higher than for other semi-crystalline thermoplastics. PPS is flame retardant by nature and with a wall thickness of 0.4 mm it is rated V0 by UL 94. PPS is a brittle thermoplastic and therefore rarely used as an unfilled resin for injection molding. Applications include films produced by flat sheet die extrusion and fibers. For injection molding, compounds with fibrous reinforcements, mineral fillers, or combinations thereof are preferred. Reinforcing agents used are mainly glass- but also carbon- and aramid fibers; fillers used are calcium carbonate, calcium sulfate, kaolin, mica, talc, and silica. Filler loads up to 70 weight % are possible. The mechanical properties are strongly influenced by the type and load of filler/reinforcement, see Table 6.40. Blends with fluoropolymers improve the tribological properties. Conductive compounds are available. Comparison of properties, see Table 6.40.

Table 6.40: Property Comparison of Polyarylsufides and -Sulfones

Property	Unit	Polyarylsufides and -Sulfones					
		PPS GF40	$\phi=0\%$	PES GF30	$\phi=0\%$	PSU GF30	(PSU+ABS)
ρ	g/cm^3	1.60-1.67	1.36-1.37	1.58-1.6	1.24-1.25	1.44-1.45	1.13
E_t	MPa	13000-1900	2600-2800	9000-11000	2500-2700	7500-9500	2100

Continued on next page

6.9 Aromatic Polysulfides and Polysulfones (PPS, PSU, PES, PPSU, PSU + ABS)

Property	Unit	PPS GF40	PES ϕ=0%	PES GF30	PSU ϕ=0%	PSU GF30	(PSU+ABS)
σ_y	MPa	-	75-80	-	90	-	50
ϵ_y	%	-	5-6	-	6-7	-	4
ϵ_{tB}	%	-	20-50	-	20->50	-	>50
σ_{50}	MPa	-	-	-	-	-	-
σ_B	MPa	165-200	-	125-150	-	110-125	-
ϵ_B	%	0.9-1.8	-	1.9-3	-	2-3	-
T_P	°C	275-290	-	-	-	-	-
HDT	°C	≈26	200-205	210-225	170-175	185	150
α_p	10^{-5}/K	1.5-2.5	5-5.5	2-3	5.5-6	2	6.5
α_n	10^{-5}/K	3.5-5	-	4-4.5	-	-	-
UL94	Class	V-0	V-0	V-0	V-2/HB*	V-0/V-1	HB*
ϵ_r 100	-	3.9-4.8	3.5-3.7	3.9-4.2	3.2	3.5-3.7	3.1-3.3
tan δ 100	$\cdot 10^{-3}$	10-20	10-20	20-30	8-10	10-20	40-50
ρ_e	Ohm·m	>10^{13}	>10^{13}	>10^{13}	>10^{13}	>10^{13}	>10^{13}
σ_e	Ohm	>10^{14}	>10^{13}	>10^{13}	>10^{15}	>10^{15}	>10^{14}
E_BI	kV/mm	20-30	20-30	20-30	20-30	30-35	20-30
W_w	%	<0.1	1.9-2.3	1.5	0.6-0.8	0.4-0.5	0.3
W_H	%	<0.05	0.6-0.8	0.6	0.25-0.3	0.15-0.2	0.1

* also available in V-0

Applications Micro-precision injection molding, encapsulation of chips and other electronic devices., lamp- and headlight sockets, pump housings and other parts, structural foam parts, films, CF-, AF-, GF-prepregs. Their high heat distortion and chemical resistance makes PPS suitable for high-performance composite materials, with continuous reinforcing fibers (mainly glass-, carbon-, and aramid). Various coupling agents and impregnation techniques have been developed.

6.9.2 Polyarylsulfone (PSU, PSU + ABS, PES, PPSU)

Chemical Constitution Polysulfones are polycondensates, containing diarylsulfone groups as their characteristic molecule groups in the chain, see Eq. 6.23.

Diarylsulfone group

(6.23)

Well known products are polyarylates (aromatic rings, connected by O-, S-, SO_2-, or other bridges) that have additional ether (-O-) and sometimes also -C-$(CH_3)_2$- bonds. Chemical notations and symbols as well as basic structures, service temperatures, and processing temperatures are shown in Table 6.41.

Table 6.41: Polysulfones and Polysulfone Modifications

Acronym	General Structure	Service temperatures Short term/Long term	Melt Temperature	Mold Temperature
PES		220/200	340 to 390	Up to 170
PES so		- /200	345 to 390	140 to 165
PPSU		- /190	360 to 390	140 to 165
PSU		180/160	345 to 390	120 to 160
PSU/ABS		HDT-A=149	280 to 310	70 to 120
PES/PA-GF30			310 to 330	80 to 140

Processing PSU can be processed like any other thermoplastic material. Injection molding is the most common processing method. Pre-drying of the resin at 150 - 260°C for 3 - 6 h is necessary. Because the melt is highly viscous, melt and mold temperatures have to be high during processing. This reduces molecular orientation and internal stresses and thus the susceptibility for stress cracking. Heated tool and ultrasonic welding are used to join PSU parts; adhesion by glueing (PSU dissolved solvents) as well as two component adhesives.

Properties PSU are amorphous thermoplastics exhibiting a high degree of transparency when unfilled. Service temperatures (see Table 6.41) are higher than for common engineering plastics. PSU exhibits high elongation at break and high stiffness and strength; however, notch sensitivity is also high. Wear and slip properties are good and can be even improved by the addition of PTFE or graphite. PSU is susceptible to stress cracking. For outdoor use, PSU should be protected against UV by pigmentation, e.g., with carbon black, or by protective surface layers. PSU is resistant to high-energy radiation; depending on grade, it is classified V0 to V2 by UL 94. The following modifications of PSU are available: higher molecular grades with increased melt viscosity for extrusion, flame-retarded grades (also transparent), reinforced grades with glass fiber contents of 10 - 30% to increase stiffness and strength, and grades filled with minerals to increase heat distortion point and to decrease stress cracking. PSU blends with ABS, PAS and other plastics for technical applications meet requirements for increased chemical resistance, metal plating suitability, and low price. PES + PA is manufactured by reactive blending and exhibits high stiffness and strength together with good impact strength and high-temperature aging resistance. Applications: under the hood. Property comparison, see Tables 6.40 and 6.41.

Applications Multi-way connectors, coil bobbins, insulators, condensers, brush holder, alkali batteries, printed circuit boards, temperature resistant parts for appliances, medical devices, and dairy equipment, filtration membranes (micro- and ultra-filtration, dialysis), lenses, spotlights, reflectors, lamp sockets, protective layers for metals.

6.10 AROMATIC POLYETHER, POLYPHENYLENE ETHER, AND BLENDS (PPE)

6.10.1 Polyphenylene Ether (PPE)

Chemical Constitution. Modified polyphenylene ethers (or oxides, PPE, PPO), see Eq. 6.24, have gained significant technical and economical importance because of their unlimited compoundability with thermoplastics such as PS, PA, or PBT. The resulting resins are competitively priced, with good impact resistance at temperatures below 320°C and service temperatures > 100°C. Pure PPEs have no practical industrial applications.

(6.24)

6.10.2 Polyphenylene Ether Blends

Blends of polyphenylene ether and impact resistant PPS at a ratio of 1:1 have gained economical importance. These blends exhibit better oxidation resistance than pure PPE, which tends to rapid oxidative degradation at temperatures above 100°C. Processability is improved; the melt temperature ranges between 260 and 300°C. Service temperatures however range much lower: short-term from 120 - 130°C; long-term from 100 - 110°C. The styrene content causes higher stress cracking susceptibility. Laminating or coextrusion of highly viscous (also reinforced) PPE grades with sulfur cross-linkable rubber blends and subsequent hot vulcanization creates plastic-rubber composites with a composite strength exceeding the tear strength of the vulcanized rubber. Blends or alloys of amorphous PPE with semi-crystalline polycondensates offer optimized property profiles with better solvent- and stress cracking resistance than the amorphous and lower shrinkage and warpage than the semi-crystalline part. Heat distortion stability up to 210°C with excellent impact resistance at low temperatures can be reached with both reinforced and unreinforced blends based on PPE + PA 66. These grades are used for in-line paintable automotive parts, as well as for oil- and fuel resistant parts under the hood. PPE + PBT alloys absorb less water but are also less temperature resistant. The Radlite process is used to manufacture laminar glass-mat reinforced non-wovens from PPE/PA powder in aqueous foam emulsion for large surface parts. Foamable blends of PS-E (expandable PS) with PPE can be processed into foams with an overall density of 25 - 250 kg/m^3 and respective flexural strengths of 0.5 - 8.8 MPa (particle foams). Their heat distortion point ranges from 104 - 118°C (PS-E: 95°C). Low-molecular PPE is easy to process and has good adhesivity and compatibility with SB copolymers and aromatic and cycloaliphatic EP resins. It is used as an additive for SB and EP in adhesives and sealant compounds as well as for coatings and surface clays with increased heat distortion point; it also improves the melt flow rate of amorphous thermoplastics such as PEI at low loads of 5 wt%. Comparison of properties, see Table 6.42.

Applications. PPE + HIPS: Automotive instrument panels and interior lining, hub caps, radiator grill, housings, large parts for office machines, TV- and electric devices in structural foam, appliances parts, as masterbatch to improve thermal and mechanical properties of PS, halogen-free flame retardants. PPE + PA: Instrument panels and vehicle front ends, automotive body parts. PPE + PS-E: Energy absorbers in vehicles.

6.10 Aromatic Polyether, Polyphenylene Ether, and Blends (PPE)

Table 6.42: Comparison of Properties for Modified Polyphenylene Ether and Polyether Ketone

Property	Unit	Modified polyphenylene ether and polyether ketones				
		(PPE+SB) $\phi=0\%$	(PPE+SB) GF30	(PPE+PA66) $\phi=0\%$	(PPE+PA66) GF30	PEEKK $\phi=0\%$
ρ	g/cm^3	1.04-1.06	1.26-1.29	1.09-1.10	1.32-1.33	1.3
E_t	MPa	1900-2700	8000-9000	2000-2200	8300-9000	4000
σ_y	MPa	45-65	-	50-60	-	105-110
ϵ_y	%	6-7	-	5	-	6
ϵ_{tB}	%	20->50	-	>50	-	30-35
σ_{50}	MPa	-	-	-	-	-
σ_B	MPa	-	100-120	-	135-160	-
ϵ_B	%	-	2-3	-	2-3	-
T_P	°C	-	-	-	-	365
HDT	°C	100-130	135-140	100-110	200-220	165
α_p	10^{-5}/K	6.0-7.5	3	8-11	2-3	4.5-5
α_n	10^{-5}/K	-	-	-	-	-
UL94	Class	HB[1]	HB[1]	HB[1]	HB[1]	V-0
ϵ_r 10	-	2.6-2.8	2.8-3.2	3.1-3.4	3.6	3.6
tan δ 10	$\cdot 10^{-3}$	5-15	10-20	450	2-24	1-9
ρ_e	Ohm·m	>10^{14}	>10^{14}	>10^{11}	>10^{11}	>10^{13}
σ_e	Ohm	>10^{14}	>10^{14}	>10^{12}	>10^{12}	>10^{13}
E_BI	kV/mm	35-40	45	95	65	20
W_w	%	0.15-0.3	0.15	3.4-3.5	2.7-3.6	0.45
W_H	%	<0.1	<0.1	1.1-1.2	0.8-1.2	0.18

Property	Unit	Modified polyphenylene ether and polyether ketones				
		PEEKK GF 30	PEEKK CF 30	PEEKEKK Unreinf.	PEEKEKK GF 30	PEEK
ρ	g/cm^3	1.55	1.45	1.3	1.53	1.32
E_t	MPa	13500	22500-2300	4000	12000	3500
σ_y	MPa	-	-	105-115	-	-
ϵ_y	%	-	-	5-5.5	-	5
ϵ_{tB}	%	-	-	30->50	-	>60
σ_{50}	MPa	-	-	-	-	-
σ_B	MPa	170-180	220	-	190	100
ϵ_B	%	2.2-2.4	2	-	2.5-3.5	-
T_P	°C	365	365	375-380	375-380	343
HDT	°C	350	360	170	350	155
α_p	10^{-5}/K	1.5	1.2	4	2	4.7
α_n	10^{-5}/K	0.4	-	-	-	-
UL94	Class	V-0	V-0	V-0	V-0	V-0

Continued on next page

Property	Unit	Modified polyphenylene ether and polyether ketones				
		PEEKK GF 30	PEEKK CF 30	PEKEKK Unreinf.	PEKEKK GF 30	PEEK
$\epsilon_r 10$	-	3	-	3.4	3.9	3.2
$\tan \delta 10$	$\cdot 10^{-3}$	10	-	35-30	25-30	
ρ_e	Ohm·m	$>10^{13}$	10^3-10^4	$>10^{13}$	$>10^{13}$	51014
σ_e	Ohm	$>10^{13}$	-	$>10^{13}$	$>10^{13}$	
$E_B I$	kV/mm	20-30	-	20	20-30	
W_w	%	0.4	0.42	0.8	0.5	0.8
W_H	%	0.12	0.14-0.18	0.25	0.1	0.25

6.11 ALIPHATIC POLYESTER (POLYGLYCOLS) (PEOX, PPOX, PTHF)

Depending on molecular weight Polyethylene(propylene)oxide (PEOX, PPOX) are (PEOX: 200 - 20,000 g/mol; PPOX: 400 - 2000 g/mol), liquid, waxy to rigid polymers are created. Basic chemical structure, see Eq. 6.25

$$\underset{\text{Ethylene oxide}}{CH_2\overset{O}{\diagup\!\!\diagdown}CH_2} \quad \underset{\text{Propylene oxide}}{CH_3-C\overset{O}{\diagup\!\!\diagdown}C-CH_3}$$

(6.25)

PEOX is soluble in water and many other solvents, colorless, odorless, and non-toxic; it is used in pharmaceutical and cosmetic products, such as skin and hair tonics, lip sticks, ointments, lubricants, anti-blocking agents. PPOX is liquid even at higher molecular weights; in contrast to PEOX; it is used as lubricant additive and as break- and hydraulic fluid. PTHF (polytetrahydrofuran) is produced by cationic polymerization of tetrahydrofuran. PTHF are strictly linear polyetherdiols with molecular weights from 650 - 3000 g/mol. Poly(aryl Ether Ketones) (Aromatic Polyether Ketones: PAEK, PAE, PEK, PEEK, PEEEK, PEKK, PEEKK, PEEKEK, PEKEEK, PAEK +PEI).

Chemical Constitution In their polymer chain, PAEK contain ether(-oxygen) bridges and keto-groups (C=O) and all technically important grades also contain 1,4-phenylene units (alternative name: polyaryl ether, PAE). The simplest polyether ketone is depicted in Eq. 6.26. PAEK are semi-crystalline polymers, whose melt temperature depends on the ratio of keto-groups, see Table 6.43. Their acronyms reflect the number and configuration of the various groups, e.g., two ether- and one keto-group: polyether ether ketone, PEEK.

$$\left[\!\!\!-\!\!\!\bigcirc\!\!\!-\!\!\!O\!\!\!-\!\!\!\bigcirc\!\!\!-\!\!\!\overset{\overset{O}{\|}}{C}\!\!\!-\right]_n$$

PEK

(6.26)

Table 6.43: Thermal Properties of Various Polyaryl Ether Ketones

Material	Ketone group content %	Melt temperature °C	Glass temperature °C
PPE	0	285	110
PEEEK	25	324	129
PEEK	33	335	141
P(E)$_{0.625}$(K)$_{0.375}$	37.5	337	144
PEEKEK	40	345	148
PEK	50	365	152
PEEKK	50	365	150
P(E)$_{0.43}$(K)$_{0.57}$	57	374	157
PEKEKK	60	384	160
PEKK	67	391	165

Processing PAEK can be processed by injection molding or extrusion at melt temperatures of 350 - 420°C and mold temperatures of 150 - 190°C (for more complex geometries up to 250°C). If amorphous surface layers are created by fast cooling, the degree of crystallization can be increased by post-annealing. Pre-drying at 150 - 200°C for at least 3 h is necessary. Coating, e.g., by vortex sintering is possible. The melt flow rate is lower than for other high-temperature thermoplastics; therefore, special free-flowing grades are recommended for thin-walled injection molded parts.

Properties PAEK are highly rigid and stiff polymers over a wide temperature range. They are semi-crystalline thermoplastics and exhibit high fatigue strength. Impact resistance is good; however, unreinforced grades in particular are notch sensitive. Long-term service temperature is at 250°C. PAEK are largely chemical resistant to non-oxidizing acids, bases, oils, fats, and break fluids in automotive applications, even at elevated temperatures. They are not hydrolysable by hot water or steam and are not susceptible to oxidative degradation. They are conditionally resistant to oxidizing acids. Concentrated sulfuric acid is the only known solvent at room temperature. Appropriate pigmentation or coating is recommended for long-term outdoor use because of PAEK's low UV resistance. Even without flame-retardant modification, PAEK are classified V0 by UL 94 and exhibit extremely low flue gas densities. Various grades exhibit complete miscibility with isomorphous properties in a certain range. This range is within a difference in the keto-content of 25%. The crystallite melting point of such blends is linearly dependent on the keto-group content, see Table 6.43.

PAEK is completely miscible with polyether imide (PEI) in a range from below 30 to above 70 weight-%. These blends exhibit a higher impact resistance than their components. With a PAEK content lower than 80%, the blend can be used to injection mold amorphous, transparent parts with improved chemical resistance compared to PEI. Property comparison, see Table 6.42.

Applications Utilization of good flammability properties and heat- and chemical resistance: injection molded parts for automotive, aeronautical, electro, and medical industries; cable insulation, films, tapes, and sheet made from carbon fiber reinforced PEEK. Special, ultra-pure PEEK for implants, special PEEK for short-term use in medical applications such as catheter, endoscopes, etc. PEEK prepregs with carbon-, aramide-, steel-, and glass fibers compete with thermosets, see Table 6.44.

Table 6.44: Mechanical Properties of Carbon Fiber Reinforced PEEK

Property	Unit	Range
Unidirectional		
ϕ	%	40 - 65
ρ	g/cm^3	1,51 - 1,63
σ_B	MPa	1570 - 2600
E_t	GPa	78 - 155
ϵ_B	%	1,8 up to 2,0
Quasi-isotropic, Pressed mat 0°/90°/±45°, FVG: 60%		
σ_B at 23 - 200°C	MPa	1100 - 700

6.12 AROMATIC POLYIMIDE (PI)

Chemical Constitution. Polyimides exhibit the highest heat distortion point. They contain a characteristic imide-group, see Eq. 6.27.

$$-R\diagdown\begin{array}{c}C=O\\|\\C=O\end{array}\diagup N-R'$$

Imide group

(6.27)

6.12 Aromatic Polyimide (PI)

In an intermediate stage of polycondensation of aromatic diamines with aromatic dianhydride, fusible products are created, which are transformed into an insoluble and no longer fusible state by heating. They are used both as thermosets and as thermoplastics. To process the latter, special techniques have to be employed: pressing or sintering of a powdered intermediary product. Some grades can be injection molded or extruded at high temperatures ($\approx 350°C$). Polyaddition products are created from pre-polymers with unsaturated aliphatic end groups, which are saturated by thermal polymerizing groups. These grades exhibit slightly lower heat distortion points than the polycondensates. They are used to manufacture composites in heated presses or autoclaves or are processed by winding. During curing, volatile materials may form. Grades for spraying or coating, e.g., of sheet metal, as well as for adhesion and welding are available. Tables 6.45 and 6.46 provide basic structures and properties of selected thermosetting and thermoplastic polyimides.

6.12.1 Thermosetting Polyimide (PI, PBMI, PBI, PBO, and Others)

Polyimide (PI) The production of pure aromatic PI by polycondensation as insoluble and non-melting resins is possible only by special processes because of the need to vent the volatile by-products. Resin manufacturers offer solid semi-finished products, precision sintered parts, as well as 7.5 - 125 mm thick films for electric applications and cable insulation for aeronautical applications that are fatigue resistant at temperatures from -240 to 260°C in air and up to 315°C in vacuum or inert atmosphere. The same applies accordingly for very high-molecular polyimides based on carboxylic acid dianhydrides and diisocyanates and polybenzimidazoles (PBI) that are produced under secession of CO_2. PI shrink film, electrical and/or heat conductive and films with adhesive PFA layers are also available. Similar films made from urea derivates (parabanic acid) are heat resistant up to 155°C. Comparison of properties, see Tables 6.46 and 6.69.

Table 6.45: Polyimide and Selected Polyimide Modifications

Name	Acronym	General Structure	Service temperatures (°C) Short term/Long term
Thermosets			
Polyimide	PI		<400/260
Polybismaleinimide (Polyaminobis-maleinimide)	PBMI (PAMI)		250/190
Polybenzimidazole (Semiconductor)	PBI		<500/ - (no air) <300/ - (in air)
Polyoxadiazo-benzimidazole	PBO		<500/ -
Thermoplastics			
Polyimide	PAI		300/260
Polyetherimide	PEI		>200/170
Polyimidesulfon	PISO		>250/210
Polymethacrylimide (Hard foam)	PMI		- /180
Modified PMI	PMMI		- /120-150
Polyesterimide (semi-conductor)	PESI		- /200
Ladder polymers			
Polyimidazopyrrolone (pyrrone)			
Polycyclone			1000/ - (Fiber)

6.12 Aromatic Polyimide (PI)

Table 6.46: Comparison of Properties for Thermoplastic Polyimide

Property	Unit	Thermoset PI	Thermoplastics PAI ϕ=0%	PAI GF30	PAI CF30	PEI ϕ=0%	PEI GF30
ρ	g/cm^3	1.43	1.38-1.40	1.59-1.16	1.45-1.50	1.27	1.49-1.51
E_t	MPa	2300	4500-4700	12000-14000	24500	2900-3000	9000-11000
σ_y	MPa	210	-	-	-	85-100	-
ϵ_y	%	-	-	-	-	6-7	-
ϵ_{tB}	%	-	-	-	-	30->50	-
σ_{50}	MPa	-	-	-	-	-	-
σ_B	MPa	-	150-160	205-220	250	-	150-165
ϵ_B	%	88	7-8	2-3	1.2	-	≈2
T_P	°C	-	-	-	-	-	-
HDT	°C	>400	275	280	280	190	205-210
α_p	10^{-5}/K	20	3.0-3.5	1.6	0.9	5.5-6.0	2
α_n	10^{-5}/K	20	-	-	-	-	5
UL94	Class	-	V-0	V-0	V-0	V-0	V-0
ϵ_r 100	-	-	3.5-4.2	4.4	-	3.2-3.5	3.6
tan δ 100	·10^{-3}	-	10	-	-	10-15	15
ρ_e	Ohm·m	-	>10^{15}	>10^{15}	-	>10^{15}	>10^{14}
σ_e	Ohm	>10^{15}	>10^{16}	>10^{16}	-	>10^{16}	>10^{15}
E_BI	kV/mm	200	25	25-35	-	25	25
W$_w$	%	3.5	-	-	-	1.25	0.9
W$_H$	%	2	-	-	-	≈0.5	≈0.4

Polybismaleinimide (PBMI) PBMI are created by condensation with p-diamines and are either self-curing through terminal double bonds or they cure with reaction partners while gradually cross-linking. The formation of volatile products during polycondensation necessitates special measures during processing. Outstanding properties are high heat deflection temperature, radiation resistance, self-extinguishing after removal of the ignition source, chemical resistance, and efficient processability. Transfer molding resins contain 20 - 50% glass-, graphite-, carbon-, or synthetic fibers or PTFE to improve slip and wear behavior. Pressing requires mold temperatures from 190 - 260°C, transfer molding requires ≈ 190°C. Injection molding grades can be processed on thermoset injection molding machines with mold temperatures ranging from 220 - 240°C and cycle times from 10 - 20 s/mm wall thickness. Sintering powders are successively processed. Annealing for 24 h at 200 - 250°C is always required.

Applications. Substitution of metals, e.g., for vacuum pumps, integrated circuits, printed circuits, electric cable connectors, missile nosecones.

Polybenzimidazole (PBI) PBI belong to a group of polymers that do not adhere to the ladder principle of the double chain but rather contain additional single chains. They are formed by melt condensation. In vacuum they exhibit a small decrease in weight only above 500°C, while in regular atmosphere fast degradation starts above 200°C. Heat resistance increases, if hydrogen in the labile NH group is substituted by a phenyl group. During condensation, PBI adhere very well to metal surfaces and are therefore used as heat resistant metal adhesives. Because of the limited exposure of the glue line to environmental influence, oxidation of PBI at elevated temperatures can be neglected. PBI are resistant to all solvents, oils, acids, and bases; they are therefore used as coatings for surface protection in special applications. PBI are used in aerospace and military applications.

Polytriazine Equation 6.28 shows the triazine ring, which is resonance-stabilized similar to the benzene ring and therefore high temperature resistant. In industrial practice materials are often characterized as polytriazine although (with R = O) they are in fact polycyanurates. Successively cross-linkable polymers are formed with a R = bisphenol A remnant, which are mostly modified with bismaleinimide.

$$—R—C\overset{N}{\underset{N}{\diagup}}\overset{}{\underset{C}{\diagdown}}\overset{}{\underset{|}{C}}—R—$$

Triazine ring

(6.28)

As pre-polymers they can be melted at 30 - 130°C with low viscosity; they are easily proccessable and when cured reach a glass transition temperature of 200 - 300°C. Their excellent mechanical, electrical, and dielectrical properties lead to applications as copper laminated composites (carrier for printed circuits), other applications in the microelectronic field, insulation for large engines, insulation- and bearing material in the automotive and aeronautical industries.

Poly(oxadiazo)benzimidazole (PBO) The resin melts at 525°C and degrades at 550°C.

6.12.2 Thermoplastic Polyimides (PAI, PEI, PISO, PMI, PMMI, PESI, PARI)

Polyamidimide (PAI) PAI is an engineering thermoplastic material used primarily in the cooling industry and for aeronautical and aerospace applications exhibiting 220 N/mm^2 tensile strength and 6% elongation at break at -96°C, 66 N/mm^2 (reinforced up to 137 N/mm^2) tensile strength at 232°C, and high creep rupture strength. It is resistant to aliphatic, aromatic, chlorinated and fluorinated hydrocarbons, ketones, esters, ethers, high-energy radiation, weak acids and bases. PAI is attacked by water vapor and bases at elevated temperatures. Resins modified with PTFE or graphite are tried and tested for unlubricated bearings with minimal friction coefficients up to 250°C. For optimum application values, parts must be polymerized by curing for several days at 250°C. In contrast to polyimides, which are difficult to process, PAI can be injection molded into parts with intricate geometries; PAI can also be extruded. Melt temperatures range from 330 - 370°C, mold temperatures from 200 - 230°C. Pre-drying for 8 - 16 h at 150 - 180°C is required. To increase wear resistance it is recommended to cure parts at 245 - 260°C for 24 h to 5 days. Modification with graphite and PTFE are used as bearing materials, glass- and carbon fibers and minerals are used to increase stiffness and strength and to reduce cost. Processability can be improved by mixing with low-viscous engineering thermoplastics such as PSU, PEI, PA, PPS, and PC. Comparison of properties, see Table 6.46.

Applications. Constructional parts under mechanical or electrical loads up to 260°C such as impellers for hydraulic or pneumatic pumps, bearings and housings for fuel gages, bearings, solutions in polar solvents for wire coating and as glues, nano-filtration membranes with TiO$_2$ filler.

Polyetherimide (PEI) Because of its excellent melt flow rate, PEI can be injection molded (runner less), injection blow molded, extruded, and foamed. With good creep rupture strength under load and elevated temperatures, good electrical properties, the highest flame resistance of all thermoplastics other than fluoropolymers measured by oxygen index, and reasonably priced, it is suitable for a wide range of applications. PEI is amorphous, amber transparent when unpigmented, and soluble in methylene chloride and trichloroethylene. It is resistant to alcohols, automotive and aeronautical fuels, lubricants and cleaners, even when under load. It is also resistant to acids and weak bases (pH < 9), to hydrolysis by hot water and steam, UV and high-energy radiation. PEI is typically processed by injection molding at melt temperatures from 375 - 425°C and mold temperatures from 100 - 150°C. Pre-drying for 7 h at 120°C or for 4 h at 150°C is required. Because compounds with up to 40% glass fiber content with high melt viscosity are also used, the maximum injection pressure of the molding machine should range between 1500 and 2000 bar. PEI grades with release agents exhibit an increase in flow distance of 10%. Grades modified with mineral fillers and/or with PTFE exhibit improved slip and wear properties. Grades reinforced with up to 40% glass or carbon fibers have increased stiffness and strength. Comparison of properties, see Table 6.46.

Applications. High voltage breaker housings, multi-way connectors, parts for microwaves, solder bath resistant parts, parts for pistons and break cylinders, carburetor housings, bearings, gear wheels. Blends of PEI and PEC (polyestercarbonate) are used because of their thermal properties, hydrolytic stability, and stain resistance in applications such as microwave dishes and automotive headlight reflectors, Blends with free flowing PPE (≤ 5 wt%) improve processability (melt temperature for 30% glass fiber reinforced grades from 360 - 380°C to 310 - 340°C) without having major impact on mechanical properties.

Polyimidesulfone (PISO) PISO is transparent, with a glass transition temperature from 250 - 350°C and a service temperature of 208°C; it is resistant to all common solvents and easier processable than the PI system it is based on. PTFE- and graphite filled injection molding grades based on PISO are available (processing temperature < 370°C). Unmodified linear PISO with the structure depicted in Table 6.45 (T_g > 310°C) do no longer flow. Film is cast from dimethylformamide solution.

Polymethacrylimide (Rigid Foam) In a first processing step, the monomers methacrylonitrile and methacryl acid together with chemical (formamide) or physical blowing agents (isopropanol) are used to cast a polymer. In a second processing step, sheet with densities ranging from 30 - 300 kg/m^3 are foamed during a thermal process at temperatures between 170 and 220°C. The imide reaction takes place during the foaming process.

Delivery forms. Sheet of max. 2500 mm × 1250 mm, 1 - 65 mm thickness, with densities from 30 - 300 kg/m^3 (not available for all grades).

Properties. Closed-cell, vibration resistant rigid foam with high stiffness and high heat distortion point. Properties depending on grade and density:

- Modulus of elasticity (DIN 53 457) : 20 - 380 N/mm^2
- Compressive strength (DIN 53 421) : 0.2 - 15.7 N/mm^2
- Shear strength (DIN 53 294) : 0.4 -7.8 N/mm^2
- Heat distortion point (DIN 53 424) : up to 215°C
- Long-term service temperature : up to 180°C

All grades are resistant to most solvents and fuel components; they are not resistant to bases. Excellent creep behavior, even at elevated temperatures, and therefore suitable for autoclave manufacture of sandwich parts at 180°C and 0.7 N/mm^2. Also suitable for co-curing processes, which means curing of the complete sandwich structure (prepreg layers and core) in one step and as thermoelastically pre-formable sandwich core for in-mold pressing. The latter technique is based on the generation of pressure by the foam core in the closed mold. With standard PMI grades, this pressure is generated by post-foaming; with prepressed grades the pressure is created by the expansion of the core. Because the core expansion is constricted in the closed mold, pressure builds and presses the cover layer of the sandwich against the mold surface. PMI rigid foam can be easily adhesively bonded (both with itself and with common cover layer laminates) by two-component adhesive systems.

Applications. Core material for structural parts built in series for aircrafts (flaps, tabs, stringer), helicopters (rotor blades), in shipbuilding (cores for hulks), automotive (cores for body parts); sandwich cores for high performance bicycle frames and rims; core for cross-country and down hill skis; cores for tennis racks; parts for antennae and satellites, self-supporting, X-ray transmitting stretchers for X-ray labs.

Polymethacrylate Methylimide (PMMI) The insertion of methylmethacrylate in the PMI chain creates a transparent plastic material with certain advantages compared to PMMA: increased heat distortion point (120 - 150°C), higher stiffness, strength, chemical resistance, and index of refraction (1.53). The linear coefficient of expansion is lower, light transmission is at 90%. PMMI can be injection molded or extruded at 270 - 300°C and exhibits good compatibility when co-processed with PET, PA, PVC, S/AN, ABS, even without the use of compatibilizers.

Applications. Head- and rear lights, instrument covers, optical fibers, sun roofs and side windows in automotives, packaging for food, cosmetics, and medical supplies.

Polyesterimide (PESI) Polyesterimides can be injection molded. Branding will modify resins soluble in organic solvents and make them suitable for wire coating in the electronic industry. These wires are long-term temperature resistant up to 200°C.

6.13 LIQUID CRYSTALLINE POLYMERS (LCP)

Chemical Constitution Chemically LCPs are largely based on the following polymers:

- Polyterephthalates and -isophthalates; LCP-PET, LCP-PBT
- Poly(m-phenylene isophthalamides); LCP-PMPI
- Poly(p-phenylene phthalamides); LCP-PPTA

- Polyarylates; LCP-PAR
- Polyester carbonates; LCP-PEC

Research is also being conducted on base materials such as polyazomethines, polyesteramides, polyesterimides. Certain materials, polymers among them, form rigid crystalline regions in the liquid state or when in solution. This highly ordered state in solution or in the melt is in stark contrast to the disordered macromolecules of conventional polymers. LCP form a liquid crystalline state between the liquid and the solid state called 'mesomorphous.' The rigid crystalline areas in a molecular chain are called 'mesogenous,' see Figure 6.17. Only after LCP are heated above the so-called isotropic melting temperature, an isotropic liquid is obtained. An LCP consisting of mesogenous areas only in the main chain cannot be processed like a thermoplastic, because it melts at 400 - 600°C, which is above degradation temperature. Selective introduction of disturbances between the mesogenous areas of these molecules results in 'thermotropic' main chain LCPs, which melt at 220 - 400°C. Such disturbances may take the form of flexible -CH$_2$ sequences, built-in angles (e.g., by isophthalic acid) parallel shifts (2,6-hydroxynaphtholic acid, HNA, among others, 'crankshaft polymers') or voluminous substitutions. Monomer units creating rigid linear chain segments are suitable for main chain LCPs, see example in Eq. 6.29.

(6.29)

In side chain polymers, the mesogenous areas are located in side chains using flexible spacers to graft them to meltable macromolecules. In the melt, mesogenes can be oriented in the electric field (birefringence), which makes side chain LCPs suitable as 'functional polymers' in electro-optical information storage systems. Polyesters are almost exclusively used for injection molding and extrusion and for applications with fillers and reinforcements, such as glass and carbon fibers, mineral, graphite, or PTFE modification. In addition, conductive and metal-coatable grades are available. Liquid crystalline polyamides (so called aramides) are difficult to melt and are spun into fibers directly out of solution.

Processing LCPs are mostly injection molded. Extrusion, extrusion blow molding, and coextrusion are also commonly used processing methods. For injection molding, melt temperatures range from 280 - 330°C, mold temperatures range from 70 - 130°C. Perfectly closing molds are important to avoid flash-related problems during injection molding. Gating is critical because of the highly oriented state of the melt; weld lines are weak spots. LCP parts (both injection

6.13 Liquid Crystalline Polymers (LCP)

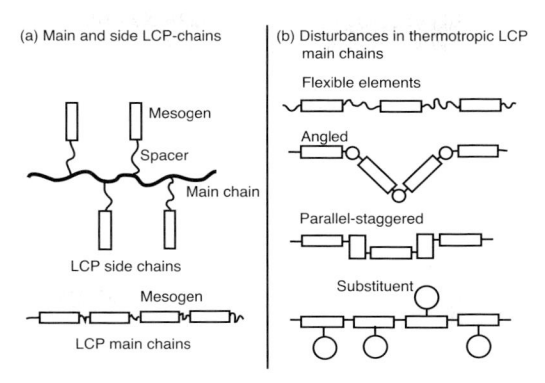

Figure 6.17: General configuration of liquid crystal polymers; (a) main chain- and side chain-LCP; (b) disturbances in thermotropic main chain LCP.

molded and extruded) can be machined by all standard methods; they can be joined by adhesion with adhesives suitable for polyester and by ultrasonic. LCP's suitability for metal coating makes them material of choice for MID (molded interconnect devices) parts and three-dimensional electronic circuit boards.

Properties LCPs for thermoplastic processing exhibit low viscosity melts so that filigree parts with thin walls can be manufactured. LCPs have a good melt flow rate, flame resistance, dimensional stability at elevated temperatures, chemical resistance, low thermal expansion, and good mechanical properties. During processing, the molecules are oriented in strain and shear flows, resulting in highly anisotropic properties and a so-called 'self-reinforcing' effect in the direction of orientation. Most data on toughness and strength (up to 240 MPa), the high modulus of elasticity (40 GPa), and the low coefficient of thermal expansion are only available in the direction of molecular orientation. The long-term heat distortion temperature without mechanical load ranges from 185 - 250°C, depending on grade. In a wide temperature range, LCPs are resistant to hydrolysis, weak acids and bases, alcohols, aromates, chlorinated hydrocarbons, esters, ketones, and all chemicals usually causing stress cracking, except strongly oxidizing acids and strong alkalis. Weather resistance and resistance to γ-radiation and short wavelengths are good. LCP are intrinsically flame retardant (V-0 according to UL 94) and other than for low tracking resistance they exhibit very good electrical properties. Glass- and carbon fiber reinforced resins exhibit slightly higher strength and stiffness. Mineral fillers reduce the anisotropy of properties in injection molded parts. LCPs are added to other thermoplastics to enhance the melt flow properties (5-30%). Vice versa, addition of low amounts of other thermoplastics is said to decrease the dependency on orientation for LCP properties. Newer resins (with fewer ions) are offered for electronic applications. They are used to minimize corrosion or short circuits caused by electrically charged particles migrated to the surface in metal coated conductor paths with narrow distances. LCPs have good barrier properties against gasses and water vapor, which is used in coextruded films for packaging applications. Comparison of properties, see Table 6.47.

Table 6.47: Comparison of Properties for Liquid Crystalline Copolyesters

Property	Unit	LCP-PC $\phi=0\%$	LCP-PET $\phi=0\%$	GF30	CF30
ρ	g/cm^3	-	1.4	1.6	1.5
E_t	MPa	2500-4300	10400	16100	23000
σ_y	MPa	-	-	-	-
ϵ_y	%	-	-	-	-
σ_{50}	MPa	-	-	-	-
σ_B	MPa	62-94	156	188	167
ϵ_B	%	5.6-3.7	2.6	2.1	1.6
T_P	°C	-	280	280	280
HDT	°C	125-135	168	232	240
α_p	10^{-5}/K	-	-0.3	-0.1	-0.1
α_n	10^{-5}/K	-	6.6	4.7	5.2
UL94	Class	-	V-0	V-0	V-0
ϵ_r 100	-	-	3.2	3.4	-
$\tan \delta$ 100	·10^{-3}	-	160	134	-
ρ_e	Ohm·m	-	>10^{14}	>10^{14}	50-100
σ_e	Ohm	-	>10^{13}	>10^{13}	104
E_BI	kV/mm	-	39	43	-
W$_w$	%	-	<0.1	<0.1	0.1
W$_H$	%	-	<0.05	<0.05	0.06

Applications Applications in electronics and electrical industries, connectors, electro-mechanical components, automotive applications, glow plugs, chip card readers, sun light sensors, couplers for optical fibers, bearings, gaskets, functional elements for pumps, measuring devices, parts in contact with fuels (under the hood), precision parts for aggressive environments, MID, multi-layer films with good barrier properties. Application of LCPs in PEM (polymer-electrolyte-membrane) fuel cells is currently under research.

6.14 LADDER POLYMERS: TWO-DIMENSIONAL POLYAROMATES AND -HETEROCYCLENES

Chemical Constitution Ladder polymers consist of linear chains, which are cross-linked at constant intervals (see examples in Table 6.45, pyrrone and polycyclone). This composition makes for their extremely high temperature resistance and stiffness. In principle, the synthesis is a connection of monomers with ring structures, which predominantly consist of aromatic (benzene-) rings or of nitrogenous heterocyclenes. During synthesis, bi-functional or multi-functional polymer blocks are formed, which will form ladder-like or three dimensional network structures with further cross-linking. Both multi-stage synthesis and processing are difficult and expensive. Ladder polymers can neither be processed like thermoplastics nor in solution. Shape forming has to take place from the soluble or fusible starting material during molecular or ladder crosslinking, respectively.

Linear Polyarylenes Linear or branched polyphenylenes in their final stage are insoluble and non-melting. Their long-term service temperatures - except for molten alkali metals - range from 200 - 300°C, under oxygen exclusion up to 400°C.

Poly-p-Xylylenes (Parylenes) They are dielectrics for long-term use at temperatures from -160 to 275°C in form of thin films or coatings. The polymers are polymerized at 600°C in vacuum from gasified xylol dimer radicals (Eq. 6.30) or from chlorinated xylols and precipitated on cold surfaces. Because of their -CH$_2$-group contents they are oxidation resistant only up to ≈ 80°C and resistant to all organic solvents up to 150°C. Polymonochloroparaylylene is a chlorinated variant of poly-p-xylylene; it can be solved in chloronaphthalene and is therefore used as coating on parts.

Poly-p-Hydroxybenzoate (Ekonol) Linear polyarylesters are polymerized from p-hydroxybenzoic acid (Eq. 6.30) monomers, which contain only aromatic rings besides the -O-CO- ester groups. The polymer melts at 550°C and can be used by flame-spraying or, rendered malleable by additives, as bearing journal material. Copolymers with ≈ 40% diphenol terephthalate are self-reinforcing injection molding materials. The long, relatively rigid macromolecules order themselves in the melt to form 'thermotropic' liquid crystalline phases. Parts injection molded from this anisotropic melt are by an order of magnitude more rigid and stiffer in flow direction than perpendicular to flow. Lyotropic liquid crystalline solutions of non-melting aramides are used to produce fibers.

HO—⟨◯⟩—COOH   (CH$_3$—⟨◯⟩—CH$_3$)

Xylol   p-Hydroxybenzoic acid

(6.30)

Polyimidazopyrrolone, Pyrone Polypyrrones are true ladder polymers and exhibit good resistance to high-energy radiation. They were developed for applications in space.

Polycyclone Polycyclobutadiene and polycycloacrylonitrile both belong to this group of ladder polymers. They can neither be processed like thermoplastics nor in solution and are used in particular for fibers that do exhibit little strength, but are short-term temperature resistant up to 1000°C. In air, beginning at 200°C, fast oxidation results in a rapid decrease in strength; therefore, the high heat distortion point can only be utilized if the fibers are either imbedded in a resin matrix or metallized.

Other Ladder Polymers The synthesis of long-chain ladder polymers with heterocyclic cores, similar to PI, is still an area of increased research. Their configurations and nomenclature is shown in Table 6.48.

Table 6.48: Heterocyclenes for the Synthesis of Ladder Polymers

Hydantoin	Urazol	Parabanic acid
Chinoline	Chinoxaline	Benzimidazole

They are used as impregnation or adhesive resins for printed circuit boards, in microelectronics, and as radiation-crosslinking photoresist coatings for printed circuits on semiconductor chips in the computer industry, which can resist temperatures up to 400°C and other stresses in subsequent operations. The syntheses also utilize the versatile reactivity of multi-functional isocyanates. The goal are 'open-chain' prepolymers that are liquid crystalline in concentrated solution or in the melt. They are then applied to the waver in a pre-ordered fashion, to form insoluble and non-melting end products. Poly-2,6-diphenyl-phenyleneoxide belongs to the PPE group, although it contains only aromatic rings. It has a T_g = 235°C and a T_m = 480°C and is spun out of organic solution to form fibers for filter fabrics and for high-frequency cable insulation papers.

6.15 POLYURETHANE (PUR)

6.15.1 Fundamentals

By their chemical constitution, polyurethanes are never "pure" polymers, such as PVC from vinyl chloride or polyethylene from ethylene, but rather chemically-structural mixed polymers. The urethane group, lending its name for this polymer class, is often contained in the macromolecule only to a limited degree. Even "PUR" products with no urethane group at all are known. However, all variations have the same "polyisocyanate-chemistry" in common. The property profile of PUR resins is mostly determined by other than the urethane structural characteristics. This fact constitutes their unique variety among polymers: polyurethanes are primarily known as foams, elastomers, and solid parts. In addition, they play a significant role in technical applications as PUR-coatings and paints, adhesives, plasticizers for other polymers, and as fibers.

Chemical Constitution Isocyanates are characterized by the energy-rich, high reactivity isocyanate group $-N = C = O$. This group reacts exothermally not only with hydrogen-active compounds, but under suitable conditions even with itself. How fast an NCO-group reacts with its partner depends on the partner as well as on the structure of the rest-molecule, to which it is bonded. The most important hydrogen-active compounds exhibit OH- or NH_2-functions, they are alcohols or amines. Water plays an important role as an OH-component in the production of foams. During the metathesis with isocyanates, gaseous CO_2 is created which acts as a foaming agent. CO_2 is also created when NCO-groups react with organic acids, which are characterized by a carboxyl group -COOH. In addition, unused or excessive isocyanate in the reaction mixture can lead to secondary subsequent reactions with already formed primary reaction products. This will create new chemical structures, which may strongly influence the properties of the manufactured polyurethane. Finally, isocyanates also react with each other, creating yet additional structural elements in the polymer with the respective effects on the polymer properties: isocyanurate structures (trimerisates of isocyanate) cause a drastic decrease in flammability in rigid foams for the construction industry, which is utilized in particular in rigid polyisocyanurate foams manufactured with pentane blowing agents. The $-N = C = O$ group can also react with compounds lacking active hydrogen such as epoxides. This results in products that have gained importance as cast resins. Table 6.49 summarizes the principles of polycyanate chemistry.

Table 6.49: Principles of Polycyanate Chemistry

R — NCO +			Isocyanate group
HO — R'	→	R — NH — C(=O) — OR'	Urethane
HOH + 2 OCN — R	→	R — NH — C(=O) — NH — R + CO_2	Symmetric urea
H_2N — R'	→	R — NH — C(=O) — NH — R'	Unsymmetric Urea
R — NH — C(=O) — OR'	→	R — N(C(=O)OR')—C(=O)—NH—R (allophanate structure)	Allophanate
R — NH — C(=O) — NH — R'	→	R — N(C(=O)NHR')—C(=O)—NH—R (biuret structure)	Biuret
HOOC — R'	→	R — NH — C(=O) — R' + CO_2	Amide
OCN — R	Cat. A →	R — N=C=N — R + CO_2	Carbodimide
R — N=C=N — R	Cat. B →	Uretonimine ring structure	Uretonimine
OCN — R	Cat. C →	Uretdione ring structure	Uretdione
3 OCN — R	→	Isocyanurate ring structure	Isocyanurate
R' — CH — CH_2 (epoxide)	→	Oxazolidone ring structure	Oxazolidone

Manufacture of the Polymer The generally liquid raw materials are mixed at room temperature, while adhering to exact stoichiometric volume ratios[3]. The respective calculation details are provided by the resin suppliers. The reference number or descriptive code is the most important parameter: it references the percentage of the effectively used isocyanate amount to the stoichiometric amount (that is the calculated amount); for example:

---
[3] See Chapter 4 of this handbook on polyurethane processing technology

6.15 Polyurethane (PUR)

- Reference number = 100 : used amount of isocyanate equals calculated amount
- Reference number = 110 : used amount of isocyanate is 10% higher than calculated amount
- Reference number = 90 : used amount of isocyanate is 10% lower than calculated amount

The reference number "play" allows influencing the processability of the reaction mixture and the properties of the PUR grade - within certain limits. The raw material suppliers often recommend an optimal reference number range. The PUR manufacturer is free to develop an individual recipe. Two-component systems, containing a polycyanate on the one hand and a polyol component on the other hand, are economically (or for logistic or ecological reasons) preferred. The polyol component typically also contains all further additives necessary for processing. For some foam processes, the blowing agents must be added separately. PUR raw materials are delivered in hobbocks, barrels, reusable containers, or tankers. There are two general processing techniques for PUR, based on the specifically adjustable difference in reactivity, both for isocyanates and polyols:

- *One-shot-process.* the complete amount of polycyanate and the complete amount of polyol and additives are mixed in one shot and reacted.
- *Pre-polymer process.* The complete amount of polycyanate is combined with a fraction of polyol in a first step. In a second step, this product, which still contains reactive NCO-groups, is reacted with the remaining polyol and the necessary additives.

Although the one-shot-process is faster, the slower pre-polymer process has the advantage of creating better ordered PUR polymers. Which process to use depends on the type of PUR to be produced and on the infrastructure of the manufacturer; in particular, whether he is capable of producing NCO pre-polymers. In general, processable pre-polymers or so-called modified isocyanates are offered by the resin suppliers together with the remnant polyol and additives as formulations and as 2-component systems. The PUR manufacturer only completes the second step of the pre-polymer process and can thus utilize the advantage of the one-shot process. In addition to OH- and NH_2-groups there are also special -CH_2-compounds with an active hydrogen group that can react with isocyanates to form so-called "disguised isocyanates". At room temperature, these products are inert towards polyols and decompose only at elevated temperatures into the (-CH_2-active) disguise medium and isocyanate, which can now react with the polyol. Disguised isocyanates are used for one-component systems (varnishes), among others. Because of the flexibility in reactivity of the polyisocyanate and polyol components together with an extensive range of additives, various processing techniques are available for the production of polyurethanes, such as:

casting, spraying, foaming, injection molding, extrusion, rolling (rubber industry), coating, and spinning to produce: blocks, sheet, molded parts, foaming of hollow parts, composites, coatings, impregnations, films, fibers, and others.

Flammability Polyurethanes are flammable like all organic materials; both, manufacturing and use therefore come with a certain fire hazard. During storage and handling of the raw materials the fire hazard is minimal because of the relatively high flashpoints of these products. However, during, for example, the production of flexible foam blocks, the high risk of self-ignition has to be taken into account. Incorrect metering can push the already exothermic PUR reaction through discoloration/burning in the core to self-ignition. To comply with insurance regulations, newly foamed blocks have to be stored in fire-proof rooms until they are completely cooled. Depending on their application, polyurethanes can be equipped with flame-retardant properties. They are subject to standard flame testing and classification. In addition, other fire-related hazards are classified: melting behavior (dripping, burning), smoke density and -toxicity, etc.). Not all fire testing methods are internationally standardized; in general, they depend on the country as well as on the application (furniture, flooring, building, automotive, etc.). An up-to date overview can be found in Troitzsch (Ed.) "Flammability Handbook", Hanser, 2004.

Quality Assurance While the raw material suppliers guarantee the data specified for the polyisocyanate- and polyol- components, the PUR producer is responsible for adherence to processing parameters.

Raw Material Handling, Safety Depending on the amount of material produced, the equipment for the production of PUR may be subject to local or federal regulations. The pertinent regulations and haz-mat declaration standards apply for the transport, loading and unloading operations and storage of the raw materials to minimize environmental impact. In addition, local and federal legislation regulates the workplace safety issues for employees handling raw materials ($\approx$ 500,000 worldwide). The raw material suppliers offer extensive information, including information on inertization of unused raw material and up-to date safety data for each product.

Environmental Protection, Safety, and Recycling PUR end products do not pose any hazardous risks for environment or health as long as they are used as intended. In general, PUR has a longer life span than its raw materials. PUR processing waste, such as cuttings, sprues, rejects, etc. cannot be reused. Table 6.50 summarizes the most important means of PUR waste recycling. While mechanical recycling is generally performed directly by the PUR producer, feedstock recycling requires detailed chemical knowledge, typically only available on an industrial scale. Depending on local conditions, thermal recycling (incineration) can be performed by PUR producers or through modern incineration plants.

Table 6.50: Methods of Recycling for PUR Waste

Mechanical recycling	Feedstock recycling	Thermal recycling (energy recovery)
Press-bonding	Glycolysis	Consumer waste incineration
Particle board	Hydrolysis	Fluidized bed furnace
Powdering	Pyrolysis	Revolving cylindrical furnace
Injection molding	Hydrogenation	Pyrolysis
Extrusion-forging	Glass production	Metallurgical recycling

Cost or logistic problems (collection, separation, identification, sorting) may prevent the utilization of all technical possibilities to provide recylates to the market. Therefore, landfilling of PUR waste is still inevitable. An example of mechanical recycling are flake composites from PUR flexible foam production waste: among others they are used for gym mats, special upholstery, sound absorbing elements. Also marketed are particle boards from reduced rigid foam waste. RIM-PUR pellets from automotive waste are used in certain building applications. Glycol production is a typical example of feedstock recycling.

6.15.2 Raw Materials and Additives

The raw material system of polyurethanes consists of three components: polyisocyanates, polyols, and additives.

Di- and Polyisocyanates Polyurethane chemistry relies on a few different grades of basic isocyanate (see Table 6.51), typically diisocyanates. Higher functional triisocyanates are special products, used for coatings and adhesives. Polymeric diphenylenemethane-4,4'-diisocyanate (PMDI) is another higher than difunctional isocyanate. Quantitatively it is the most important polyisocyanate, even surpassing TDI. The following special products find increasing applications: para-phenylene diisocyanate (PPDI) for PUR-TPE (TPU) and bis-(isocyanatomethylene)-norbornane (NBDI) for coatings.

Polyols, Polyamines These products are the most important reaction partners of isocyanates. The variety of grades available exceeds the number of isocyanate products by orders of magnitude. Skilled combination of different polyols, or polyamines and/or crosslinking agents where appropriate, facilitates the wide variety of available polyurethanes. In rigid foam formulations, polyols based on rape oil, castor oil or other renewable sources find increasing application.

Table 6.51: Di- and Polyisocyanates of Technical Importance in Polyurethane Chemistry

TDI	NDI
Toluylene-2,4-diisocyanate + toluylene-2,6-diisocyanate 80% + 20% = TDI 80/20 65% + 35% = TDI 65/35	Naphtylene-1,5-diisocyanate
MDI monomer Diphenylmethane-4,4′-diisocyanate	MDI polymer (PMDI)
IPDI Isophorondiisocyanate	HDI Hexane-1,6-diisocyanate

Polyetherpolyols are the most common backbone for polyurethanes. A wide palette of long- and short chained polyetherpolyols with 2 - 8 OH-groups per molecule (functionality) are available, due to technically easy access to alcolhols (di- and higher functionality) and epoxides (propylene- and/or ethylene oxide). In polyurethanes, this covers a property spectrum reaching from little crosslinked/linear-flexible to highly crosslinked/rigid. Polyetherpolyols are largely hydrolysis resistant; they require stabilizing additives to protect against photooxidation. Polyesterpolyols (PEUR) are used to a much lesser extent compared to polyetherpolyols. Their production is more expensive and at comparable chain lengths they exhibit higher viscosity. On the other hand, they are less susceptible to photooxidation, but more hydrolysis prone. Their undisputed advantage is their contribution to higher strength in the final PUR products. Numerous polyetherpolyesterpolyols based on long-chain fatty acids are used on an industrial scale. Polyols with organic fillers are special grades that find a surprisingly broad field of applications. They are known both as graft-, SAN-, or polymer polyols and as PHD-polyols. They are milky-white, stable dispersions of styrene-acrylonitrile polymers in the first and of polyureas in the second case. They provide flexible foams with relatively low densities, higher indentation hardness and elasticity. Poly-(amines) (e.g., polyoxypropylenediamine) are highly reactive reaction partners to isocyanates. They are either higher or lower molecular compounds with two or more amino- (NH_2-) groups. An example of the former is polyetheramine, the latter lead to crosslinking agents.

Crosslinking Agents, Chain Extenders The NH_2-groups help implementing urea segments into the PUR structure (see Table 6.49), which provide strength and improved temperature resistance to the polymer. In addition, amino compounds can be used to increase the reaction rate of polyaddition. Lower molecular bi- and higher functional alcohols, such as butanediol or glycerin can also be used as so-called OH-crosslinking agents to tailor the crosslinking density of the polymer and influence its properties, e.g., its density.

Additives Catalysts are used to accelerate the rate of reaction; for delayed acceleration (imidazo derivates) so-called latent catalysts are used. For this purpose, typically tertiary amines and/or organo-zinc compounds are used. Surfactants, such as emulsifiers effect better miscibility of the normally incompatible reaction partners polyisocyanate/polyol/water and together with catalysts contribute to reaction rate acceleration. Special organic silicon compounds are used as foam stabilizers and/or cell size controller in foam manufacturing. They stabilize the growing cells until the foam is cured. They also control open- or closed-cell development and cell size. Blowing agents are used to produce foam from liquid, viscous resins. There are chemical and physical blowing processes. The former are based on isocyanate-water reaction (Eq. 6.31) and create gaseous CO_2 blowing agents. During the physical blowing process, the exothermically reacting resins are foamed by the evaporation of low-boiling liquids that are added to the mixture. Environmental considerations have led to the use of HF (C) KW and/or hydrocarbons (pentane, cyclopentane), which have less ozone depletion potential than the formerly used FCKWs.

$$R-N=C=O + H_2O \longrightarrow \left[R-NH-\overset{O}{\underset{\|}{C}}-OH \right] \xrightarrow{R-NCO} R-NH-\overset{O}{\underset{\|}{C}}-NH-R + CO_2$$

(6.31)

The use of i-hexane results in pronounced skin formation in flexible integral foams and thus in good end use properties, see Table 6.52.

Processes such as Novaflex or CarDio use conventional multi-component metering devices to add 4 wt.% liquid CO_2 in addition to 3 - 5 wt.% water to the polyol mixture to produce extremely flexible block foams with uniform cell structure. Further developments such as the Novaform or the CannOxide process allow for the production of similar flexible molding foams. Flame retardants are added to decrease the flammability of PUR. Typical candidates are inorganic (aluminum hydroxide, magnesium hydroxide, aluminum oxide hydrates, ammonium polyphosphates) or organic (chlorine, bromine and/or phosphorus, sometimes also nitrogen) containing compounds: tetrabromide benzoeacidesters or aliphatic bromine compounds. Fillers (e.g., carbon black, chalk, silicates, barite, wollastonite) can be used to lower material cost. Glass fibers are most often used as reinforcing fillers to improve physical properties. Lately, hemp, jute, etc. have also gained importance. In many PUR applications, antiaging agents are indispensable to protect against photooxidation and hydrolysis. They are available in a wide variety of products, e.g., lactones and substituted phenols. Coloring agents are available as pastes or doughs. They are compounds of inorganic or organic colorants or pigments in polyols. Antistatic agents decrease the electrostatic charge; biozides protect PUR against microorganisms (bacteria, fungi). Release agents for fast and easy ejection of PUR molded parts from the mold are included in the polyopl formulation as so-called internal release agents. Typically materials used are Ca-, Mg-, Al-, and Zn-salts of fatty acids such as stearates, laureates, and others. In general, surface treatment of the mold is also required.

Table 6.52: Properties of Flexible Integral Foam Produced with *i-Hexane*

Property	Unit	Density (kg/m^3)			
		200	300	400	600
Skin density	kg/m	500	-	950	1000
Core density	kg/m	120	-	290	500
Hardness	Shore A	39	59	70	80
Shrinkage	%	2.78	1.52	1.01	0.25
Compression hardness at 10% deformation	kPa	20	66	115	306
Skin tensile strength	MPa	3.3	5.1	6.1	8.1
Core tensile strength	MPa	0.37	0.83	1.1	2.6
Skin elongation at break	%	1[1)]	120	125	150
Core elongation at break	%	125	135	120	130
Core compression hardness	kPa	17	60	[1)]0	270

6.15.3 PUR Polymers

Figure 6.18 gives an overview of the different forms of PUR products. It also exemplifies that there are no general "properties of polyurethanes". Rather, the question can be answered for one particular polyurethane, preferably in connection with its actual application. The information provided in Table 6.53 is therefore also only a first indication on the properties of different foam types.

Flexible Foams (PUR-F) Flexible foams arte open-celled and exhibit relatively little resistance to deformation under compression load (DIN 7726). Block foams are produced either continuously or discontinuously and subsequently cut

6.15 Polyurethane (PUR)

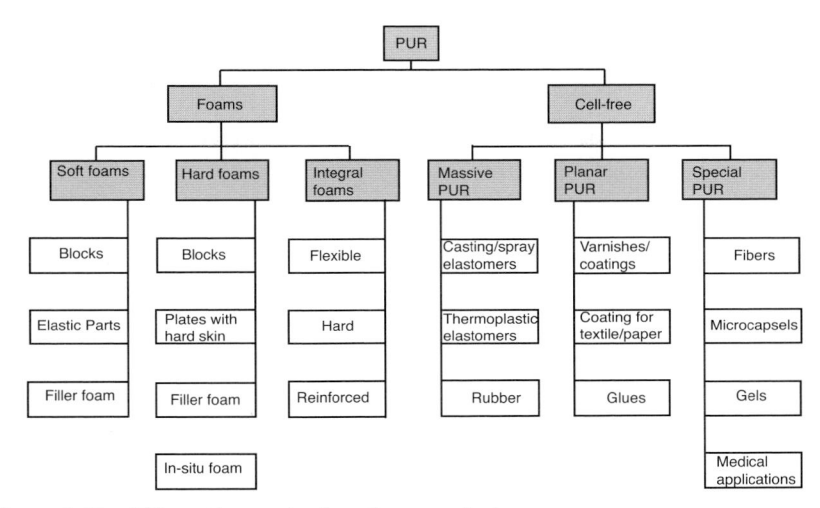

Figure 6.18: Different forms of polyurethane products.

to measure or to contour. Typical densities range from 20 - 40 kg/m³. Increased amounts of CO_2 reduce the densities by 10 - 20%. Figure 6.19 shows some exemplary pressure-deformation curves. Main areas of application are upholstered furniture (40%), mattresses (25%), and automotive interior (20%).

Molded foams are produced as either cold foam or hot foam (i.e., without or with external heat input until final curing), depending on the resin system. The densities range from 30 - 300 kg/m³ depending on application.

Table 6.53: Comparison of Properties for PUR Foams

Foam type			ρ (kg/m³)	σ_B (kPa)	ϵ_B(%)	σ_c[1] (kPa)	k[2] (W/m/K)
Soft	Polyester		35	160 - 120	200 - 450	6.5	
	Polyether	High load	36			6	
		Normal	36	100-180	100-400	4.5	
		Elastic	34	50-120	80-200	2.7	
Molded		Cold	38-44	110-152	125-150	3.3-5.6	
		Hot	33	90 - 112	190	4-4.7	
Semi-rigid		Damper	80-100	350-450	10-25	150-320	
Rigid			32	200	4	200	0.021
			50	270	5.2	350	0.021
			90	900	5	700	0.027

[1] compression stress at 40% compression
[2] thermal conductivity

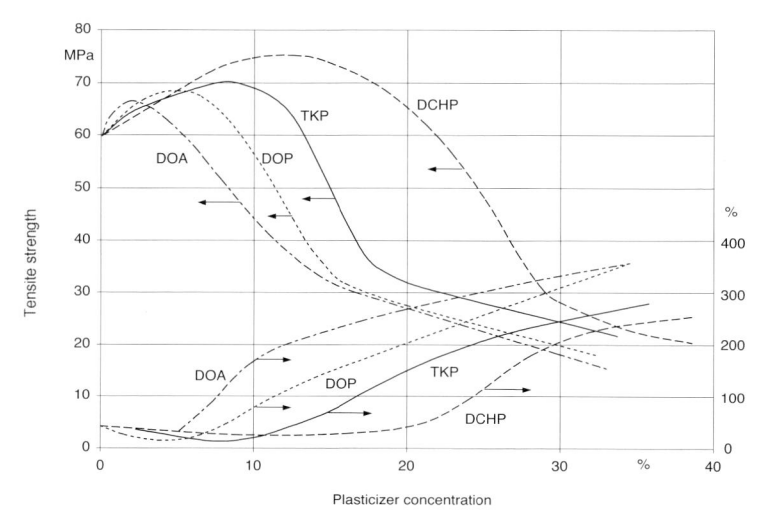

Figure 6.19: Hysteresis curves; (a) load curve, (b) release curve.

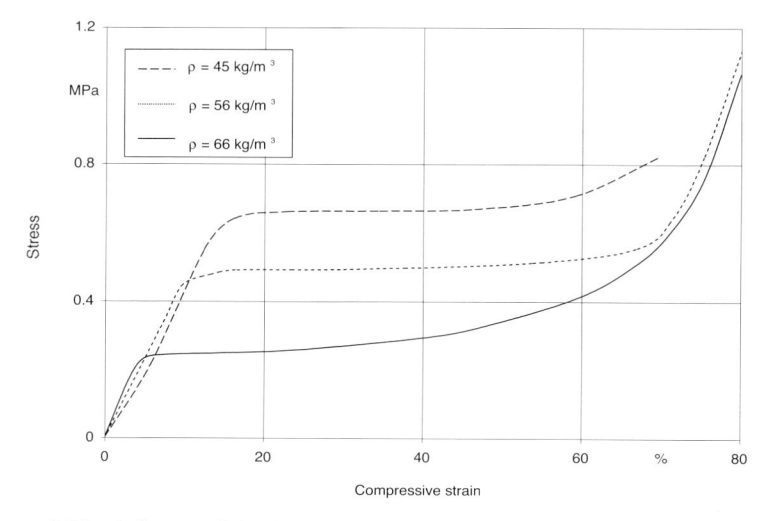

Figure 6.20: Influence of density on material properties of flexible PUR.

6.15 Polyurethane (PUR)

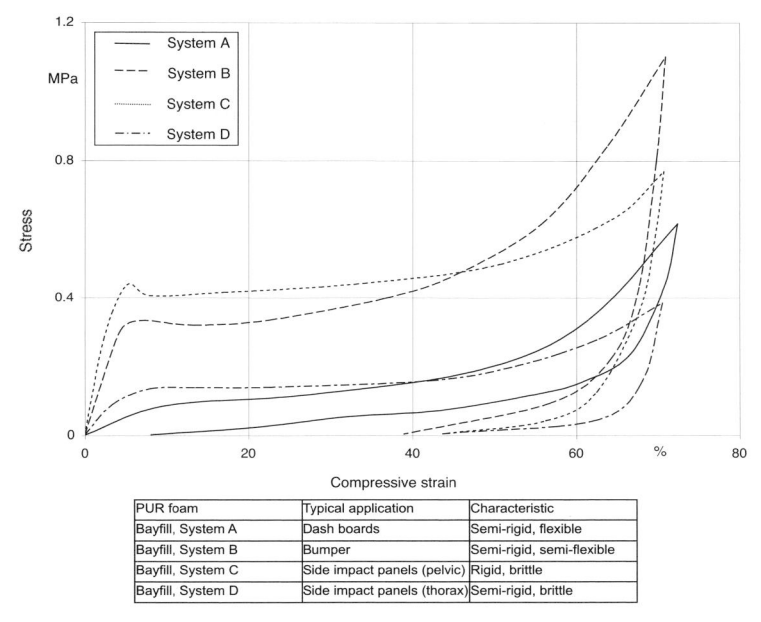

Figure 6.21: Compressive behavior of PUR foams with different energy absorption.

PUR foam	Typical application	Characteristic
Bayfill, System A	Dash boards	Semi-rigid, flexible
Bayfill, System B	Bumper	Semi-rigid, semi-flexible
Bayfill, System C	Side impact panels (pelvic)	Rigid, brittle
Bayfill, System D	Side impact panels (thorax)	Semi-rigid, brittle

Flexible molded PUR foams are mainly used for seats in automotive and aircraft construction; they are also used for upholstered furniture and for technical products (e.g., sound absorbing material). Flexible filling foams with densities ranging from 150 - 200 kg/m^3 are used as automotive interior and dashboard padding material. They are manufactured by back foaming of PVC- or ABS skins in closed molds. The compressive stress behavior of PUR foams can be influenced physically by the density (see Fig. 6.20) and chemically by the foam system (see Fig. 6.21, exemplified by energy-absorbing foams).

Compression properties of different energy-absorbing PUR foams Cellular PUR elastomers (TPU) represent an intermediate state compared to solid PUR. A water-containing reaction mixture is filled into pressureless steel molds and then foamed under separation of CO_2. These molded parts are particularly suitable for automotive cushioning because of their extremely progressive characteristics (see Fig. 6.22).

Rigid Foams (PUR-R) Rigid foams exhibit relatively high resistance to deformation under compression load. Their most important properties are their excellent thermal conduction because of the insulating gases in the closed cells

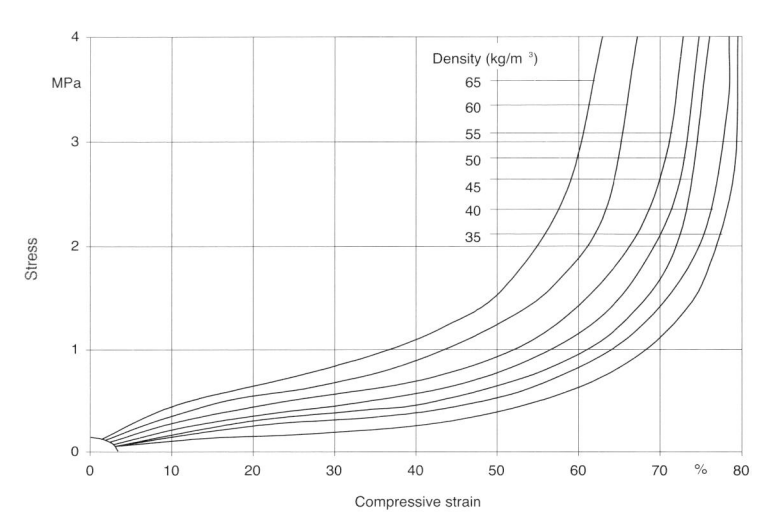

Figure 6.22: Stress-compression diagrams for cellular TPU of various densities.

and their ability to form solid composites with almost any flexible or rigid skin layer. Rigid foam blocks range in density from 30 - 90 kg/m^3 and are used for various insulation applications, automotive interior, and others. Comparison of properties, see Table 6.53.

Foamed panels with protective layers are produced either continuously or discontinuously. Aluminum, particle or plaster boards, coiled steel or beaded sheet metal, and coated glass fiber mats are only some carrier material examples. Areas of application are: building/construction as roofing insulation, sheet rock and plaster boards, sandwich elements for industrial buildings and cooling facilities. The core density of the foams ranges from 30 - 40 kg/m^3. Foams are introduced into hollow parts or hollow spaces as liquid reaction mixtures and fill these spaces after foaming. With densities from 30 - 60 kg/m^3 they serve as thermal insulators in cooling appliances of all sorts, for hot water boilers and district heating pipes. In situ foams for roofing and wall insulation, window- and door installation, and sealing is produced directly on site by mobile foaming machines or by pressure vessels from single- or two-component foam systems. Densities are $\approx$ 30 kg/m^3; spray foams for roof sealing purposes have densities up to 55 kg/m^3.

Integral Foams (PUR-I) While flexible and rigid foams are fundamentally different in their chemical structure (highly crosslinked/little crosslinked), integral foams present a variation in production technology: they are available both as rigid and as flexible molded foams only. The liquid, highly reactive mixture

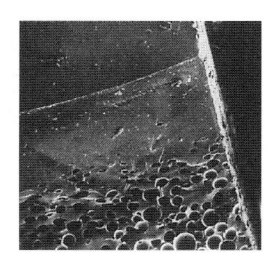

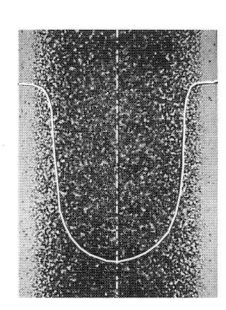

Figure 6.23: Integral foams; (left) foam structure, (right) density transition across specimen Flexible integral foams with densities from 200 - 1100 kg/m^3 are used for automotive body parts, bicycle saddles, shoe soles and other applications.

is injected into a closed mold, in which foam generation and mold temperature control are used to produce parts with cellular cores and solid skins. The transition from core to skin is not abrupt but rather continuous (see Fig. 6.23). This technique of foaming highly reactive PUR systems with particularly short holding times is generally called reaction injection molding (RIM); it is used to produce large parts as well as parts with minimum dimensions and with flow length-to-wall thickness ratios of up to 1:1000.

Rigid integral foams with comparable densities are used for technical parts, in construction, furniture, and sport- and leisure products of various designs. Glass fiber reinforced integral foams (R-RIM) with densities from 1.0 - 1.4 g/cm^3 are used for automotive body parts and housings. PUR -R-RIM is used to produce stiff, laminar parts, e.g., automotive interior side panels. Long fibers, fabrics, non-woven material, or fiber mats, for which bio renewable materials such as flax, sisal, or coir (coco fibers) are increasingly used, are sprayed or impregnated with PUR mixture and laid into the mold where they are pressed. Table 6.54 provides an overview of the properties of various integral foam systems, while Table 6.52 shows the dependency of certain properties on density.

Table 6.55 provides an example of flexible integral foam properties in automotive applications.

Applications. R-RIM parts (with 20 - 25% wollastonite reinforcement) can be coated in-line, e.g., for exterior automotive panels. The plastic parts are connected to the car body before electro immersion coating. Shrinkage is 0.8 or 0.57%, respectively, for temperature loads of 2 times 45 min at 200°C. These systems are also suitable for thin-wall technology: automotive door sills with 2 mm thickness and a length of 2 m.

Table 6.54: Property Comparison for PUR Flexible Integral Foam, R-RIM

System	Filler	ϕ (%)	ρ (g/cm^3)	Shore-H	E_f* (MPa)	σ_B (MPa)	ϵ_B (%)	A**
Flexible		0	1.05-1.1	D66	500-760	26-33	135-150	A
	Glass flakes	20	1.2	D75	1550	31	30	A
	Mica	22	1.25	D75	1720	30	25	A
	Glass fibers	15	1.2	D68	950	31	140	A
	length 180 μm	20	1.24	D70	1400	28	130	A
		25	1.3	D71	1700	27	120	A
For shoe soles		0	0.55-0.60	A55-60		8-10	550-460	B
		0	0.6	A60	-	25	600	C
Rigid	Glass mats 225 g/m^2	30	0.25	-	250	5.8***	2.3	D
	300 g/cm^2	20	0.48	-	900	25***	2.2	E
	600 g/cm^2	20	1	-	3500	110***	2.2	F
	1800 g/cm^2	32	1.4	-	650	170***	2.8	G
S-RIM		25	1.05	-	4200	70	1.9	H
LFI****		25-30	0.7-1.0	-	4500-6000	54-68	2.2-1.8	H

*flexural modulus
**application type
A:Car panels, B:Integrated shoe soles, C:Shoe soles, D:Car tops
E:Car roof-frame, column-,seat- and trunck-dressing
F:Car dashboard-frame, consoles, seat-frame, seat-back, spare-tire, motor-cover
G:Car seat-frame, bumpers, H:sun-roof dressing
***bending strength
****long fiber injection

Solid PUR (PUR-S) Solid PUR materials are available as molded parts, sheet, films, and as 'thick coatings (> 1 mm). For cast and spray elastomers, the liquid (or molten) raw materials are mixed and cast in open molds. Depending on raw material type, the part will cure at room temperature (cold cast systems) or at elevated temperatures (60 - 130°C: hot cast systems). PUR elastomers are available in Shore hardnesses ranging from 10A to 70D. They are used for solid tires, rotor bearings, sieve bottoms, sealings and casings, and as coatings in shipbuilding, automotive and in building/construction. Table 6.56 provides an overview of properties for selected systems.

Table 6.55: Property Comparison of Flexible PUR Integral Foam in Automotive Applications (RIM)

Property	Unit	Application, see below					
		A		B	C	D	
		Core	Skin			$\phi=0\%$	GF 22
ρ	g/cm^3	0.120-0.175	0.7	0.95-1.08	1.0-1.1	1.22	
σ_B	MPa	0.27-0.50	1.7-3.5	5.7	14-31	17-28	23
E_f^*	MPa	17	75-720	40-350	1300		
ϵ_B	%	220-125	220-120	230	220-140	>300-230	130
Shore	-	-	A 75	D 39-69	D 33-57	D67	

Type A: Motor cycle seats, automotive interior safety equipment
Type B: Skin for automotive bumpers
Type C: Bumpers, truck splash boards, exterior automotibe body parts
Type D: Automotive body parts

Table 6.56: Property Comparison of Cast PUR Resins; A: Polyether-Urethane, B: Polyester-Urethane

Formulation	Units	A, rigid	B, rigid	A/B, flexible	A,flexible	B,flexible
Shore D		86	85	61	20	21
Shore A		100	99	96	70	71
σ_B	MPa	86	60	20	4.5	-
ϵ_B	%	6.5	5.9	87	73	200
T_g	°C	138	90	47	15	-21

Thermoplastic PUR elastomers are discussed later. PUR rubber is produced by rubber processing technologies, e.g., on rollers (also crosslinked with sulfur), and molded, pressed or extruded to form parts with Shore hardnesses from 50A to 60D. Examples are shocks, grindings disks, and pump linings. Non-cellular, laminar PUR grades produced in thin layers (< 1mm) are typically coatings (DD-coatings) - solvent or water based, also UV curable -, for textiles, paper, leather, and other substrates, and PUR adhesives and special polycyanates for particle board production. A special class are PUR fibers, used for textiles and swim wear. Last but not least are special areas of applications such as PUR micro-capsules (e.g., for copy paper), PUR gels, medical applications in dialysis equipment, artificial blood vessels, and bandages replacing casts.

6.16 BIOPOLYMERS, NATURALLY OCCURRING POLYMERS AND DERIVATES

6.16.1 Cellulose- and Starch Derivates (CA, CTA, CAP, CAB, CN, EC, MC, CMC, CH, VF, PSAC)

Chemical Constitution Cellulose is a carbohydrate forming a frame in all organic matter. Cellulose is extracted from wood by eluating the lignin and also typically the pectin. The seed hairs of cotton provide cellulose in almost pure form. Cotton linters (short-haired fiber waste) are usually used as raw material in the plastics industry. Cellulose derivates are generated by etherification (with alcohols) or esterification (with acids) of cellulose. During these processes, up to three OH-groups of the ring system (glucose remnant) may react, see Table 6.57. The properties are commonly modified by addition of plasticizers. CA is created by esterification with glacial acetic acid, acetic acid anhydride, and sulfuric acid or zinc chloride as catalysts and usually methylene chloride as solvent. In further steps, acetic acid contents of $\approx$ 44 - 61% can be adjusted, thus creating low- to highly viscous products.

Table 6.57: General Structure of Cellulose Derivates

Chemical	Acronym	Basic Structure
		(ring structure with OR groups and ROH$_2$C)
		R
Cellulose (tri)acetate	CA (CTA)	-COCH$_3$
Cellulose propionate	CP (CAP)	-COCH$_2$CH$_3$
Cellulose acetobutyrate	CAB	-COCH$_3$/-CO(CH$_2$)$_2$CH$_3$
Cellulose nitrate	CN	-NO$_2$
Methylcellulose	MC	-CH$_3$
Ethylcellulose	EC	-CH$_2$CH$_3$
Carboxymethylcellulose	CMC	-CH$_2$COOH
Oxethylcellulose		-(CH$_2$)$_2$OH

For CTA, each of the three OH-groups of the glucose remnant is esterified with acetic acid. CAP is created by esterification of cellulose with propionic acid. Besides 55 - 62% propionic acid, it also contains 3 - 8% acetic acid (hence the A in CAP). Other than with CA, for CAB the cellulose is also esterified with butyric acid. The acetic acid content ranges from 19 - 23%, the butyric acid content from 43 - 47%. CN, often falsely called nitrocellulose, is created by conversion of cellulose with a mixture of sulfuric and nitric acid, by which a nitrogen content of 10.6 to max. 16% can be set. MC (EC) is created by a conversion of 1.3 - 1.5 (2.0 - 2.6) of the possible three OH-groups of the glucose remnant with methyl chloride (ethyl chloride). Starting material here is alkali cellulose, which is created by dipping of cellulose in 18% caustic soda. For CMC, alkali cellulose is converted with the sodium salt of monochloroacetic acid, whith up to 6 - 10 carboxymethyl groups for each 10 glucose units. Oxythylcellulose is created by conversion of cellulose with ethylene oxide.

CH (hydrate cellulose) is generated by intramicellular swelling of cellulose in caustic soda. This only extends the distance of the chain molecules, no chemical conversion takes place. PSAC (polysaccharides). By applying 10% humidity and temperatures from 100 - 150°C, starch is destructured to form a biological degradable material. Parts from thermoplastic starch are produced in compounding extruders at 120 - 220°C under addition of a swell- or plasticizing agent (e.g., glycerin or sorbitol). PHA (polyhydroxy alkaline) is based on saccharide and produced by fermentation with microorganisms.

Processing The common cellulose esters (CA, CAB, CAP) can be processed like any other thermoplastic material. Their flow behavior is good and provides for pin-point gating. Tools must be well vented to avoid plasticizer deposits. Pre-drying for ≈ 3 h at 60 - 90°C or for ≈ 1.5 h in a flash dryer is required. For injection molding, water content should be below 0.15%, for extrusion, it should be below 0.05%. Plasticizer vapors must be extracted. Melt temperatures for injection molding range from 180 - 230°C, mold temperatures from 40 - 80°C, where the higher temperatures apply for more rigid products with less plasticizer content. Films and sheet can be extruded with flat-sheet dies. The chill-roll process is used to cast films thinner than 0.8 mm. Automotive trim is produced by extrusion sheathing of 50 - 100 μm aluminum films with CAB (using a coupling agent). CAB- and CAP powders are used for rotational molding and dip coating of metals. Machining, polishing, welding (ultrasonics less suitable), adhesion with glues (CA and CAP: 50% methylglycol acetate and 50% ethyl acetate) or with 2-component adhesives based on PUR or EP resins, coating, printing, and metallizing are possible.

Properties, Applications
Cellulose(tri)acetate (CA, CAT). With increasing acetate contents, the viscosity

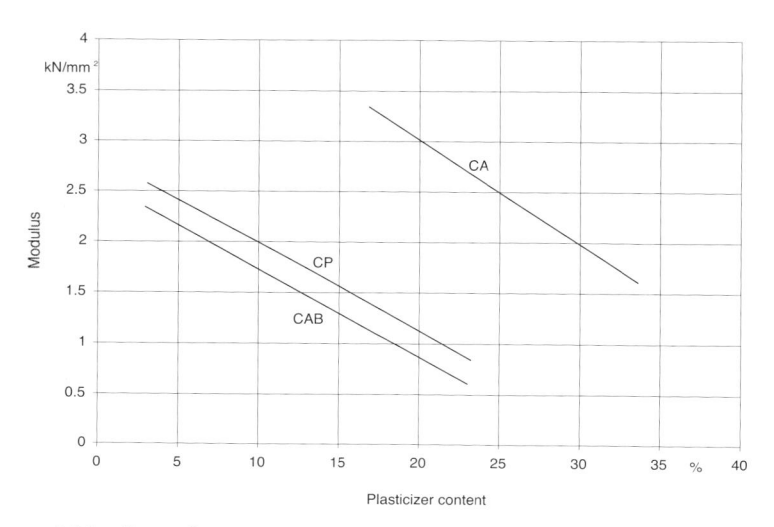

Figure 6.24: Dependency of modulus of elasticity of CA, CAB, and CAP on plasticizer content.

of CA increases: ≈ 44 - 48% low viscous, application: textile production, printing inks; ≈ 52 - 56% medium viscous, application: molding grades, coatings, adhesives; 56 - 61% high viscous, application: safety- and electro-insulation films. To enable thermoplastic processing, 15 - 20 (38)% mass plasticizers are added to the resin: dimethyl-; diethyl- and dimethylglycol phthalate, as well as trichloroethyl phosphate in combination with dimethyl- or diethyl phthalate and only in combination with these dibutyl-, diidopropyl-, and di-2-ethylhexyl phthalate. CA properties can vary in a wide range, depending on type and amount of plasticizer used, see Figs. 6.24 and 6.25. With increasing plasticizer content the heat distortion point decreases and the melt flow rate as well as the creep tendencies under mechanical load increase. These products are very suitable for tool handles, because they relieve the stresses created by shrinking the coating on a metal surface. CA is crystal clear and transparent and can be dyed in brilliant and deep colors. It is tracking resistant and because of the water content antistatic (not attracting dust). CA is resistant to fats, oils, mineral oils, and aliphatic hydrocarbons; the resistance to fuels, benzene, chlorinated hydrocarbons, and ethers depends on the composition. Alkali- and acid resistance is low. CA cannot be made permanently weather resistant. CA is little susceptible to stress cracking because stresses are relieved.

Comparison of properties, see Table 6.58.

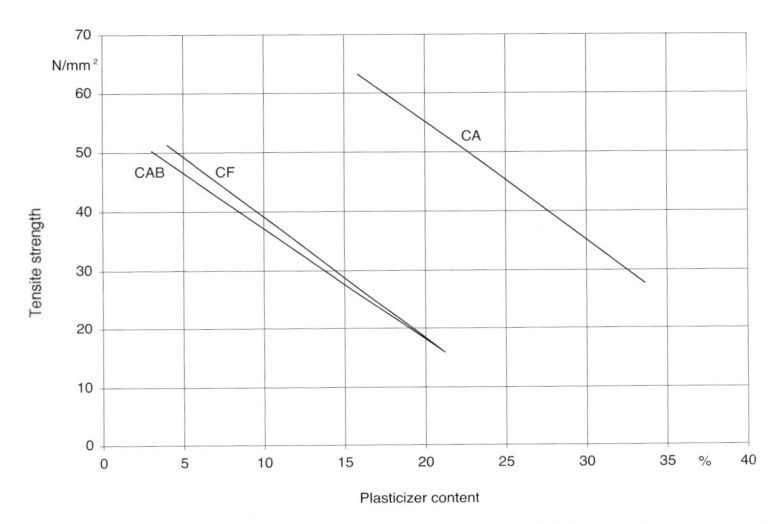

6.16 Biopolymers, Naturally Occurring Polymers and Derivates

Figure 6.25: Dependency of yield stress of CA, CAB, and CAP on plasticizer content.

Table 6.58: Comparison of Properties for Various Cellulose Derivates

Property	Unit	Cellulose derivates			
		CA	CAB	CP	EC
ρ	g/cm^3	1.26-1.32	1.16-1.22	1.17-1.24	1.12-1.15
E_t	MPa	1000-3000	800-2300	1000-2400	1200-1300
σ_y	MPa	25-55	20-55	20-50	35-40
ϵ_y	%	2.5-4	3.5-5	3.5-4.5	
HDT	°C	45-80	45-95	45-95	≈ 50
α_p	10^{-5}/K	10-12	10-15	11-15	10
UL94	Class	HB	HB	HB	HB
ϵ_r 100	-	5-6	3.7-4.2	4.0-4.2	≈ 4
tan δ 100	$\cdot 10^{-3}$	70-100	50-70	50	100
ρ_e	Ohm·m	10^{10}-10^{14}*	10^{10}-10^{14}*	10^{10}-10^{14}*	10^{11}-10^{13}*
σ_e	Ohm	10^{10}-10^{14}*	10^{12}-10^{14}*	10^{12}-10^{14}*	10^{11}-10^{13}*
E_BI	kV/mm	25-35	32-35	30-35	≈ 30
W$_w$	%	3.5-4.5	2.0-2.3	1.9-2.8	
W$_H$	%	≈ 1.2	0.6-0.8	0.6-0.9	

* conditioned to dry

Applications for molding resins: insulating tool grips and handles, writing utensils, combs, buckles, buttons, eyewear frames (cut from sheet), hollow fibers for dialysis. General applications, also for CAP and CAB: fibers, films, coatings, adhesives, low-profile additives for SMC and BMC.

Cellulose Propionate (CAP). CAP resins are also compounded with 2 - 5% mass plasticizer: dibutyl-, and di-2-ethylhexyl phthalate or -adipinate, dibutyl- and dioctylsebazate, dibutyl- and dioctylacetate. For applications with food contact, citric acid- and palmitic acit esters are used. The general properties resemble those of CA. Weather resistance of stabilized CAP is better than that of CA. Monomeric plasticizers tend to migrate at elevated temperatures. Coplymers of CAP with EVA graft polymers exhibit a higher property level with regard to stability, stiffness, creep resistance, heat distortion point, and migration resistance. Polymer modification content may reach 5 - 30% mass. Good solubility of UV- and IR absorbers is used for applications as optical fillers. Comparison of properties, see Table 6.58. Applications: Similar applications as for CAB, model railway components, powder boxes, high-quality frames for sunglasses and other glasses, stencils, toothbrush handles.

Celluloseacetobutyrate (CAB). Comparable types and amounts of plasticizers as for CAP are used to make CAB suitable for thermoplastic processing. The general properties compare to those of CA and CAP; however, stabilized CAB exhibits better weather resistance. Copolymerization with EVAC enhances the property levels similarly as in CAP. Comparison of properties, see Table 6.58. Applications: Automotive steering wheel sheathing, control knobs, handles, exterior lights, dome lights (thermoformed), advertising signs, ski goggles, trim, ear marks for livestock.

Cellulose Nitrate (CN), Celluloid. CN with nitrogen contents of 10.6 - 10.8% are alcohol soluble thermoplastics; with 12 - 12.2% nitrogen they are ester soluble. N-contents of >13 up to max. 16% results in gun cotton. CN with an N-content of $\approx 11\%$ is compounded with camphor to produce celluloid. Celluloid is the oldest thermoplastic material manufactured in large amounts. Today it is only used for the production of table tennis balls, toys, toothbrushes and combs because of its high flammability. Other areas of application for CN are as bonding agents for coatings and adhesives and coating material for artificial leather.

Methyl Cellulose (MC). MC with 1.3 - 1.5 OH-groups per glucose remnant is soluble in cold water; it coagulates at elevated temperatures depending on molecular weight, degree of etherification, and concentration. MC with more than 2 OH-groups is not soluble in water. Applications: Impregnation resins, surface films on paper (fat- and oil resistant), thickener, adhesives, paint binder.

Ethyl Cellulose (EC). EC is available in various viscosity grades. High- and medium viscosity grades are used for the manufacture of films and molded parts; medium- and low viscous grades are used for coatings and adhesives.

Carboxymethyl Cellulose (CMC). Other than MC, all viscosity grades of CMC are soluble in warm and cold water. Applications: Adhesives, paint binders, additives for detergents and drilling fluids, aid in the textile industry, emulsifying agent, and thickener.

Oxethyl Cellulose. Oxethyl cellulose exhibits the same properties in water as CMC and is used in the same areas of application.

Cellulose Hydrate (CH), Cellulose Film; Vulcanized Fiber (VF). CH, also called regenerated cellulose. Applications: Crystal clear films (cellulose films), which can be weather stabilized by double sided coating. Cellulose film exhibits low permeability to water vapor, gasses, oils, and others. To produce vulcanized fibers and vegetable parchment (Echtpergament), absorptive papers or cellulose are swelled in high-concentrated warm zinc chloride solution, stacked to form panels, washed out, and subsequently pressed. Applications: highly loaded parts such as gear wheels, grinding wheels, and suitcase reinforcements.

Polysaccharides (PSAC); Thermoplastic Starch. Today, almost 50% of starch produced is used in non-food applications. In the beginning, starch was used as filler in thermoplastic materials to obtain biodegradable grades: the starch content decayed causing the decomposition of the matrix resin into small particles (PE + starch for films used for plastic bags). Lately, plastics are produced from starch without using synthetic compounds. Only biodegradable plasticizers (water), oils, and fats are added. PSAC are biodegradable resins. They are used in combination with other materials or as copolymers. Pellets produced on a twin screw extruder from ground wood and corn can be injection molded at melt temperatures of ≈ 150°C and mold temperatures of 60°C. Properties: density 1.4 g/cm^3, tensile strength 25 MPa, elongation at break 1 - 1.5%, modulus of elasticity 13,000 MPa. Applications: wood substitute in the toy industry and for decorative applications, handles, knobs, ornaments. Thermoplastic starch is biodegradable and compostable. It is available as pelletss and can be injection molded or blow molded into films. Properties: density 1.1 - 1.2 g/cm^3, tensile strength 30 MPa, elongation at break 600 - 850% (films), modulus of elasticity 1000 - 2600 MPa, Vicat A/ 50 73 - 103°C. Applications: compostable packaging, perishable consumer goods (films and molded products).

Polyhydroxybutyric Acid (PHB); Polyhydroxyvalerate acid (PHV); Polyhydroxy Fatty Acid/Polyhydroxy Alkanoate (PHA). PHB and PHV belong to the group of polyhydroxy fatty acids (polyhydroxy alkanoates) that are created by many bacteria as storage or reserve medium. This fact is used today for the industrial production of PHB. Polyhydroxy butyric acid is a linear polymer; it is water resistant, mostly resistant to fats and oils, impermeable to oxygen, carbon dioxide, and water vapor, and also biodegradable. It is highly crystalline, exhibits similar properties to polypropylene (slightly more brittle and stiffer than PP), and can be injection molded or extruded. Properties of a compound based on starch: density 1.17 - 1.25 g/cm^3, tensile strength 15 - 27 MPa, modulus of elasticity 900 - 2000 MPa. Applications: Laboratory analyses, medical, injection molded parts.

6.16.2 Casein Polymers, Casein Formaldehyde, Artificial Horn (CS, CSF)

Casein is an important protein in milk. Caseins are used for glues and as binders in paints. Curing with formaldehyde turns casein extracted from skim milk into a polymer that is chemically related to natural horn and resembles it regarding appearance and machinability. CSF is mainly used for jewelry and for buttons and buckles.

6.16.3 Polylactide, Polylactic Acid (PLA)

Polylactide is a biodegradable polyester that does not occur naturally; it is synthesized from sugar in a multi-step process. During this process, sugar is fermented into lactid acid and then polymerized into PLA via dilactides. PLA is transparent, crystalline, rigid, exhibits high mechanical stability, and can be processed like a thermoplastic material (in particular injection molding, thermoforming).

Properties: density 1.25 g/cm^3, tensile strength 70 MPa, modulus of elasticity 3,600 MPa.

Applications: injection molding, surgical threads, implants, thermoforming films, packaging (e.g., yoghurt cups).

6.16.4 Polytriglyceride Resins (PTP®)

PTP[4] resins are of a family of triglycerides-based liquid epoxies made from renewable resources. The use of different triglycerides allows for the customization of different performance properties for specific customer requirements. PTP resins exhibit comparable properties to commercially available liquid epoxies.

Uncured Properties Density (20°C): 1.070 g/cm^3, viscosity (20°C): 450 mPa-s, viscosity (120°C): 20 mPa-s, cure temperature: 90-190° C.

Cured Properties Cure time: >1.0 min., T_g= 140-150°C, flexural strength: 80-90 N/mm^2, flexural modulus: 2055 N/mm^2, impact strength (Charpy): 8-15 kJ/m^2.

Applications PTP resins have been used to make natural fiber filled SMC, which was used to manufacture large panels for buses. Other applications may include advanced composites, aerospace, automotive/transportation, electrical/electronics, building/construction, adhesives and sporting goods.

---

[4]PTP® resins have recently been introduced in the United States from Europe, and licensed to IF Technologies, LLC, Westlake, OH. At the time of printing of this handbook, the material still did not have a tradename.

6.16.5 Natural Resins

Canada-balsam is extracted from American balsam firs and used to cement of optical lens systems. Amber is the cured resin of amber spruces and other coniferous woods. Its main application is in jewelry.

6.17 OTHER POLYMERS

6.17.1 Photodegradable, Biodegradable, and Water Soluble Polymers

Degradable polymers can be categorized in photodegradable and biodegradable polymers[5]. Water soluble polymers represent a special case; they are partially or completely (bio-) degradable in aqueous solution. Photochemical degradation of polymers is generally undesirable, because it comes with deterioration of physical properties, while polymers are generally optimized for longevity. Stabilization is used to avoid photodegradation. Degradable polymers have been developed to reduce the amount of plastic waste by decomposition of the polymer chains. Usually, the resulting polymer fragments cannot be further decomposed. In contrast, because of their molecular structure biodegradable polymers are completely decomposed into carbon dioxide, water and biomass by microorganisms, fungi, and bacteria. Biodegradability is verified by the test methods described in DIN V 54 900 and EN 13 432, respectively. Water soluble polymers (PVAL, PEOX, PPOX, MC, CMC, PSAC) include:

- Polyvinyl alcohols (PVAL)
- Polyethylene oxides (PEOX)
- Polypropylene oxides (PPOX)
- Methyl celluloses (MC)
- Carboxymethyl celluloses (CMC)
- Hydroxypropyl celluloses (HPC)
- Polysaccharides (PSAC, thermoplastic starch (TPS))

Photodegradable Polymers Selective implementation of UV-sensitive molecular structures, such as ketone groups (e.g., E-CO-copolymers), as well as compounding of photo-sensibilizers (e.g., ferric dialkylthiocarbamate or other metal-organic compounds) allows relatively precise control of photodecomposition of polymers. These products are not biodegradable. Applications of photodegradable polymers are concentrated on areas such as agricultural films, plastic bags, and trash bags.

[5]Contributed by Bettina Schmidt, www.carmen-ev.de

Biodegradable Plastics, Bioplastics Biological decomposition may be either aerobic or anaerobic; however; the presence of moisture (water) is always required. Therefore, biodegradable plastics have to exhibit hydrophilic surfaces. In some bioplastic grades, the absorption of moisture causes a decrease in mechanical properties. Bioplastics can be created from fossil as well as from natural raw materials and can be classified as follows:

- Compounds of plastics with water soluble polymers; they decompose into their components under the influence of moisture. However, these grades are only biodegradable if each component is biodegradable.
- Synthetic (co-) polyesters are made from crude oil; they are (semi-) crystalline polymers with mechanical and thermal properties comparable to those of PE-LD. As PE, they can be processed by all common processing methods, but other than PE they are biodegradable according to DIN V 54 900 and EN 13 432, respectively. Applications: pouches, bags, mulch film, catering articles, packaging.
- Poly lactic acid (PLA)
- Polyhydroxy alkanoate (PHA, PHB, PHV)
- Cellulose acetate (CA)
- Compounds of biodegradable polymers based on natural (polysaccharide/-derivates) or synthetic raw materials (some (co-) polyesters)

"Mater-bi", a bioplastic material composed of up to 90% starch and a biodegradable, synthetic, thermoplastic matrix takes a special position in this group of materials. It is processable by all common thermoplastic processing methods and is biodegradable - including the matrix polymer - according to DIN V 54 900/EN 13 432. The starch content provides good oxygen- and fat barrier properties and higher water vapor permeability. Applications: films, hygiene and horticultural applications, thermoforming, and injection molded articles. Properties of blown films based on starch: film thickness = 10 - 100 μm, density = 1.1 - 1.3 g/cm^3, tensile strength longitudinally/diagonally = 26 - 36 MPa, elongation at break longitudinally = 300 - 900%.

- New "thermoplastic" starch derivates are being developed, among them hydroxypropylienated and acetylenated starch.
- Potato and corn starch, sometimes compounded with plant fibers or recovered paper, is processed under high pressure, temperatures, and moisture to produce substitutes for EPS packaging chips (loose fill) or EPS molded parts for packaging (PSP paper foam, Zerzog, Cornpack, Storopack).
- Polymers only degradable in aqueous solution for water-soluble packaging applications based on polyvinylalcohol (PVAL) produced either by casting or by cheaper extrusion blow molding methods. Cold water soluble grades

are used in particular for toxic substances in powder form, e.g., pesticides; hot water soluble grades for packing applications impermeable by bacteria, e.g., for contaminated hospital laundry.

Water soluble temporary packagings always need a second outer packaging layer.

6.17.2 Conductive/Luminescent Polymers

For many electro-technical applications, the natural good insulation properties of polymeric materials are a prerequisite. However, polymeric materials with tailored conductivity are required in certain applications, such as to avoid electrostatic charging, to provide electromagnetic shielding for plastics housings, for the production of electrodes, light-emitting diodes, field effect transistors, and others. Conductive thermoplastic compounds are created by incorporation of conductive fillers or reinforcements. Here, we will cover conductive polymers:

Chemical Constitution Intrinsically conductive polymers (ICP) have been a research topic for many years and find increasing applications. These organic materials consist of conjugated double bonds (alternating single- and double bonds), as perfectly displayed in polyacetylene (see Table 6.59). Addition of electron donors (Na, K, Cs) or acceptors (J2, SbCl5, FeCl3, among others), atoms or molecules that release electrons (reduction) or accept electrons (oxidation), results in increased electron mobility and conductivities up to 10^5 S/cm. Similar to metals, single free electrons remain that are no longer bound to the atom residues but rather slide along the molecules and transport electric charges. This process is called doping in semi-conductor technology. Equation 6.32 exemplifies this with polyacetyls, PAC (polyenes). Figure 6.26 provides an overview on conductivity attainable by doping.

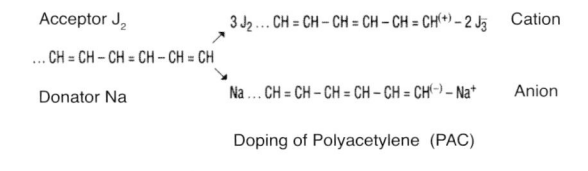

$$\begin{array}{c} \text{Acceptor J}_2 \qquad\qquad 3\,J_2 \ldots CH=CH-CH=CH-CH=CH^{(+)} - 2\,J_3^- \qquad \text{Cation} \\ \ldots CH=CH-CH=CH-CH=CH \\ \text{Donor Na} \qquad\qquad Na\ldots CH=CH-CH=CH-CH=CH^{(-)} - Na^+ \qquad \text{Anion} \end{array}$$

Doping of Polyacetylene (PAC)

(6.32)

Electro luminescence is defined as the ability of certain materials to emit light under application of voltage. This led to the development of OLEDs (organic light emitting diodes). They consist of a quartz- or glass wafer that, in the simplest case, carries three layers: an indium-tin oxide layer as anode, a thin film of luminescent polymer (poly-3,4-ethylene dioxythiopene), and finally a layer of calcium, magnesium, or aluminum as cathode. First generation conductive polymers were completely insoluble and non-melting. Soluble conductive polymers

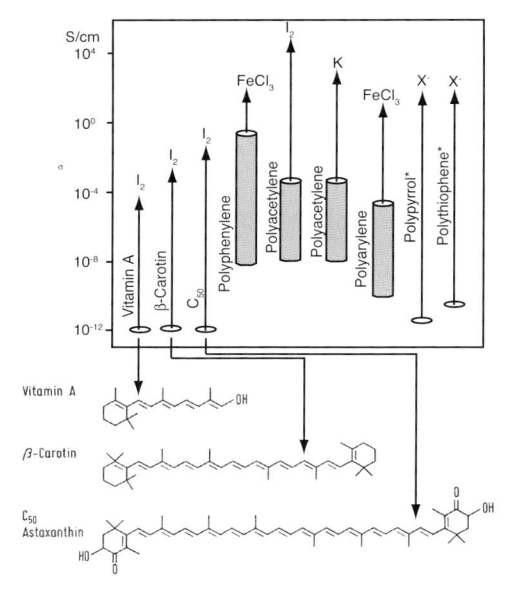

Figure 6.26: Increase in electric conductivity depending on the structure of the original polymer and the doping agent.

were developed to cast thin layers of conductive polymers: second generation polymers, such as those based on polyphenylene vinylene. Today, flexible, 3-dimensionally ductile and transparent electro-luminescence systems (TOLED: transparent organic light emitting devices) are available, consisting of a multilayer film. An electrically conductive polymer is spin coated, dipped, sprayed, or vacuum deposited on a carrier film of typically 0.15 - 0.8 mm thickness. When AC is connected to the electrode, the intermediate layer will glow. It is characteristic for this so-called 'smart surface technology' that it hardly consumes energy and does not emit heat. Light emitting plastic parts are suitable for automotive interior and console lighting, orientation- and safety signage, and for folding monitors. OLED can be applied at temperatures from -4 to +70°C; because of their chemical variability, they can be produced in any color, including white. Photoluminescence is created by the splitting of double bonds under the absorption of light. When the double bonds reconnect, the freed bond energy is emitted as light. Masterbatches based on PE, PP, PA, PS, and other engineering plastics are also available.

6.17 Other Polymers

Table 6.59: General Structures of Intrinsically Conductive Polymers (ICP)

Chemical	Acronym	Basic Structure	Spec. volume resistivity (Ωm)	Oxidation- or reduction agent, Counter-ions
Polyacetylene	PAC		10^{-3}–$6 \cdot 10^{-6}$	I_2, AsF_5, Li, Na
Polyparaphenylene	PPP		10^{-2}–10^{-3}	AsF_5, Li, K, Na
Polyphenylene sulfide	PPS		10^{-2}–$2 \cdot 10^{-3}$	AsF_5
Polyaniline	PANI		$3 \cdot 10^{-2}$–$5 \cdot 10^{-3}$	HCl, BF_4
Polyparaphenylvinylene	PPV		$3 \cdot 10^{-1}$–$2 \cdot 10^{-4}$	AsF_5
Polyphenylene butadiene	PPB			
Polyparapyridine	PPYR			
Polyparapyridinvinylene	PPYV			
Polypyrrol	PPY		10–2–$5 \cdot 10$–3	BF_4, ClO_4, Tos, FCl_4
Polythiophene	PT		10^{-1}–10^{-3}	BF_4, ClO_4, Tos, FCl_4
Polyfuran	PFU			CF_3SO_3
Polyethylene dioxythiophene	PEDT		ca. $3 \cdot 10^{-2}$	
Polyacene				

Properties Polyacetylenes (PAC) are available as powders, gels, or films (also transparent, orientation of up to 600% resulting in highest conductivity of $> 10^5$ S/cm). They are insoluble and exhibit conductivities from 10^{-5} to 10^5 S/cm, depending on polymerization process. Prior to doping, the densities range from 0.4 - 0.9 g/cm^3, after doping from 1.12 - 1.23 g/cm^3.

Polyphenylenes. Non-doped polymers are thermally stable up to ≈ 450°C. Positive doping with, e.g., FeCl$_3$ provides conductivities of ≈ 10^2 S/cm. The products are sensitive to hydrolysis; those with negative K-doping are also sensitive to oxygen.

Polyparaphenylene vinylene (PPV), polyparaphenylene (PPP). Applications: as planar light emitting diodes, electro luminescence.

Polyhetero aromates, polypyrroles (PPY), polyfurans (PFU), polythiophenes (PT). These materials are electro-chemically polymerizable and can be stripped from the anode as films. Self-supporting film thicknesses start at ≈ 30 μm. Coatings as thin as 0.01 μm, conductivities from 10^{-4} - 10^2 S/cm, tear strength 20 - 80 N/mm^2, elongation at break 10 - 20%. Examples of applications: antistatic equipment (PSU films), heating bands, fuses, sensors, batteries. Polyanilines (PANI) can achieve a conductivity of ≈ 10 S/cm by doping with various acids in aqueous solution. *Polyphenylene amines* are insoluble and non-melting polymers. As finely dispersed compounds in coatings (nanopowders) they shift the corrosion potential of iron, steel, aluminum, zinc, stainless steel and copper that are coated with polyphenylene towards the precious metals, thus slowing the corrosion process, in particular with marginal damages. Other applications: solar cells, membranes for gas separation, fuel cells, batteries, sensors. *Polyethylene dioxythiophene* (PEDT) is polymerized in solution as an insoluble, non-melting powder with powder conductivities of up to 30 S/cm. The surface resistivity σ_e of a layer with ≈ 0.1 μm thickness on PC or glass is ≈ 1000 Ω/square. Applications: solid electrolyte condensers, conductive coatings on plastics or glass by oxidative polymerization with, e.g., Fe-III-toluenesulfonate solution in n-butanol.

Polyethylene dioxythiopene-polystyrene sulfonate (PEDT/PSS). Oxidative polymerization of the monomeric PEDT in the presence of aqueous polysulfonic acid creates colloidal solutions of PEDT that can be processed. Potassium peroxide disulfate is used as oxidizing agent. Minimal surface resistivity σ_e of ≈ 150 Ω/square. Applications: transparent, antistatic coatings of photographic films, glass, light emitting diodes, OLED.

6.17.3 Aliphatic Polyketones (PK)

Chemical Constitution PK is synthesized by a special catalyst system from carbon monoxide, ethylene, and propylene, see Eq. 6.33. Ethylene- and propylene molecules are statistically distributed in the chain. PK is semi-crystalline with a degree of crystallization of ≈ 30 - 40%.

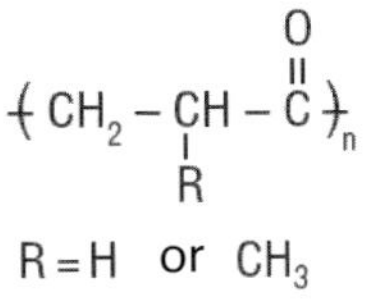

6.17 Other Polymers

$$\{CH_2 - \underset{R}{CH} - \underset{\|}{\overset{O}{C}}\}_n$$

R = H or CH$_3$

(6.33)

Processing PK absorbs ≈ 0.5% moisture in air and 2.3% in water and should therefore be pre-dried at 60°C for 4 h to avoid surface defects. Primary processing method: injection molding, melt temperatures 240 - 270°C, mold temperature 60 - 80°C (20 - 120°C). PK crystallizes quickly, allowing for short cycle times; no post-tempering necessary; processing shrinkage 1.8 - 2.2%; little or no post shrinkage. Film-, sheet-, and profile extrusion, blow molding, rotational casting and coatings are possible.

Properties PK exhibits high impact resistance and elongation at break (25%), short-term service temperatures of 180°C, and is resistant to aliphatic hydro carbons, saline solutions, weak acids and bases, as well as to media used in automotive applications. Its wear resistance is higher than that of PA or POM, in particular as friction partner of other plastics. Fire retarded grades reach V0 at thicknesses of 3.2 and 1.6 mm. Comparison of properties, see Table 6.60.

Table 6.60: Property Comparison of Aliphatic Polyketones

Property/Product	Unit	PK	PK-GF 30
Density	g/cm^3	1.24	1.46
Tensile modulus of elasticity	GPa	1.4	7.3
Tensile stress at yield	MPa	60	-
σ_B	MPa	-	1.2
ϵ_B	%	-	3
T_P	°C	220	220
HDT	°C	100	215
UL 94	Class	HB	HB
tan δ100	$\cdot 10^{-4}$	250	500
ρ_e	Ohm·m	10^{11}	10^{11}
σ_e	Ohm	10^{14}	10^{14}
$E_B I$	kV/mm	18 (d = 1.6 mm)	24 (d = 1.6 mm)
W_w	%	2.2	
W_H	%	0.5	0.3

Applications: fuel systems in automotives, pipe systems, electrical and electronic applications (flame retarded, halogen- and phosphorus-free grades).

6.17.4 Polymer Ceramics, Polysilicooxoaluminate (PSIOA)

Polymer ceramics represent a new class of materials. Further developments may be expected because of their inexpensive raw materials, easy and energy-saving processability compared to ceramics, and high temperature resistance. Polymer ceramics are polymer-clay mineral composites created by the reaction of flake silicates (phyllosilicates), in particular their reactive OH-groups, with the functional bonds of the silicium in alkali methylsiliconates. They are processed by dry pressing at temperatures up to 150°C. Parts with a high degree of spatial crosslinking are created without the need for post-sintering as with normal ceramics. Metallization and machining with any of the standard processes are possible. Polymer ceramics are resistant to corrosion, oxidation, and many chemicals. To produce foams, a solid mixture, a liquid aqueous curing agent, and a foaming agent are mixed and poured into a mold. An exothermal reaction triggers the polymerization to open- or closed-cell foams. Properties, see Table 6.61.

Table 6.61: Properties of Polysilicooxoaluminate Foam (trade name Trolit)

Property	Unit	
ρ	kg/m^3	100→800
Long-term T*	°C	1000
Short-term T*	°C	1200
k**	W/m-K	$\geq$0.037
Pore diameter	mm	0.5-3
Flammability		Not-flammable
σ_B compression	MPa	0.5→2.0
σ_B tension	MPa	0.11
E_t	MPa	$\approx$ 250
Post-shrinkage at 800 °C	%	<1.5
C_p between 20 and 60 °C	KJ/kg-K	$\approx$ 1.2
α_t	1/K	9·10^{-6}

* service temperature
** thermal conductivity

Applications: conducts, bearings, bushings, foams; elements for sound, heat, fire, and in-situ insulations, composites, catalyst carrier.

6.18 THERMOPLASTIC ELASTOMERS (TPE)

The acronyms for thermoplastic elastomers are presented in Table 6.62.

6.18 Thermoplastic Elastomers (TPE)

Table 6.62: Overview of Thermoplastic Elastomers

Acronym	Chemical description	Some trade names
TPE	**Thermoplastic elastomers**	
TPA	**Polyamide-TPE**	Bebax
TPA-EE	TPA, soft segments with ether- and ester bonds	Grilamid
TPA-ES	TPA with polyester soft segments	
TPA-ET	TPA with polyether soft segments	
TPC	**Copolyester-TPE**	Arnite, Bexloy, Ecdel, Hytrel, Lomod, Pibiflex, Riteflex
TPC-EE	TPC, soft segments with ether- and ester bonds	
TPC-ES	TPC with polyester soft segments	
TPC-ET	TPC with polyether soft segments	
TPO	**Olefin-TPE**	Dexflex, Engage
TPO-(EPDM+PP)	Ethylene/propylene/diene + polypropylene	Exact, Forflex, Hifax
TPO-(EVAC+PVDC)	Ethylene/vinylacetate + polyvinylidene chloride	Hybrar, Kelbuton, Keltan, Milastome
TPS	**Styrene TPE**	Hybrar, Multiflex
TPS-SBS	Styrene/butadiene block copolymer	Kraton D, Sofprene, Stereon, Styrof
TPS-SIS	Styrene/isoprene block copolymer	
TPS-SEBS	Styrene/ethenbutene/styrene block copolymer	Bergaflex, Europrene, Kraton G
TPS-SEPS	Styrene/ethenpropene/styrene block copolymer	Septon
TPU	**Urethane-TPE**	Desmopan, Esthane, Elastollan, Pell
TPU-ARES	Aromatic rigid segments, polyester soft segments	
TPU-ARET	Aromatic rigid segments, polyether soft segments	
TPU-AREE	Aromatic rigid segments, soft segments with ether and ester bonds	
TPU-ARCE	Aromatic rigid segments, polycarbonate soft segments	
TPU-ARCL	Aromatic rigid segments, polycaprolacton soft segments	
TPU-ALES	Aliphatic rigid segments, polyester soft segments	
TPU-ALET	Aliphatic rigid segments, polyether soft segments	
TPV	**TPE with crosslinked rubber**	Forprene, Santoprene, Sarlink
TPV-(EPDM-X+PP)	Highly crosslinked EPDM + PP	
TPV-(NBR-X+PP)	Highly crosslinked acrylonitrile/butadiene	
TPV-(NR-X+PP)	Highly crosslinked natural rubber + PP	
TPV-(ENR-X+PP)	Highly crosslinked epoxidized natural rubber + PP	

Continued on next page

Acronym	Chemical description	Some trade names
TPV-(PBA-X+PP)	Crosslinked polybutylacrylate + PP	
TPZ	**OtherTPE**	
TPZ-(NBR+PVC)	Acrylonitrile/butadiene rubber + polyvinyl	

6.18.1 Physical Constitution

TPEs combine the elastomeric properties of crosslinked elastomers with those of rubbers, with the advantage of thermoplastic processability. Their composition allows the classification in two groups: Polymer blends consist of a "rigid" thermoplastic polymer matrix, into which either non-crosslinked or crosslinked elastomer particles are incorporated as a "flexible phase". Examples are thermoplastic polyolefin elastomers (TPO or TPV) that consist of PP with up to 65% of incorporated ethylene-propylene-[diene] rubber (EP[D]M). Graft- or copolymers contain thermoplastic sequences A and elastomeric sequences B in their polymer molecule. Both components A and B are incompatible and demulsify locally so that the rigid A sequences form physical crosslinking points in the continuous matrix of flexible B sequences. An example is styrene block copolymer TPS, in which blocks of polystyrene (S) and butadiene (B) alternate: SSSSSSSS-BBBBBBBB-SSSSSSSS. At service temperature, the flexible B sequences are above their glass transition temperature (freezing point); however, the rigid A sequences are below their glass transition temperature (for amorphous polymers) or their melt temperature (for semi-crystalline polymers). Above their transformation temperature, the A sequences soften so that TPEs can be processed like thermoplastics.

6.18.2 Chemical Constitution, Properties, and Applications

The advantages of TPEs compared to rubber are: being suitable for thermoplastic processing and therefore also for recycling, weldability, transparency for some grades, colorability. An important area of application is therefore rigid-flexible combinations in 2-component injection molding and in coextrusion. Table 6.62 provides an overview of acronyms and chemical designations, while Fig. 6.27 provides a comparison of properties.

Table 6.63 offers an indication for which plastic and rubber grades may be substituted by TPEs in certain applications.

In the following sections, both TPE groups will be described in detail (see also the sections covereing the base polymers).

6.18 Thermoplastic Elastomers (TPE)

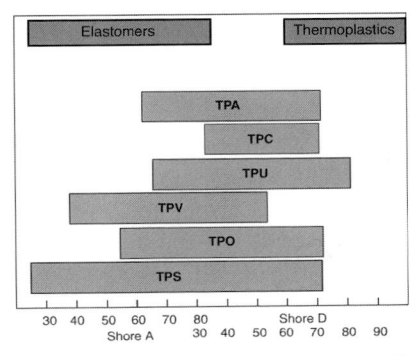

Figure 6.27: Classification of TPE in relation to elastomers and thermoplastics by range of hardness.

Table 6.63: Examples of Substitutions of Plastics and Rubbers by TPEs

Plastic material	Rubber	TPE
PE, PTFE	CR	TPA
PE, PTFE	CR, NBR, EPDM, ECO	TPC
PVC-P, PC+PBT	NR, SBR	TPO
PVC-P, PA+PPE, PC-Blends, PUR	NR, CR, SBR, NBR, EPDM, ECO	TPV
PVC-P, PUR	NR, CR, SBR, EPDM	TPS
PE, PP, PTEE, PVC-P, ABS, PUR	NR, CR, SBR, EPDM, NBR	TPU

Copolyamides (TPA) Polyamide elastomers are block copolymers with rigid polyamide blocks and flexible polyether- (TPA-ET) or polyester- (TPA-ES) blocks or with both flexible blocks (TPA-EE). Components for the production of TPA-ET are lactam, dihydroxypolytetrahydrofuran and dicarboxylic acid; for the production of TPA-EE: laurinlactam, adipic acid, and bishydromethylcyclohexane. TPEs are resistant to oils and fuels, sensitive to organic solvents and thermooxidative attack.

Applications: seals and gaskets, tubes in automotive and medical applications, sports goods, processing aids for TPU, antistatic agents for thermoplastics.

Copolyester (TPC) TPC are block copolymers consisting of flexible polyalkylene etherdiols and/or long-chain aliphatic dicarboxylic acid ester segments and semi-crystalline PBT segments. The properties range from rubber-like to highly flexible engineering plastics. Service temperature ranges from -4 to 100°C; at higher temperatures heat stabilization is required. TPCs are resistant to fuels, lubricants, and hydrolysis, and can be stabilized against UV and weather influences. Prior to processing the pellets should be dried; TPC should be injection molded or extruded at temperatures of $\approx 220°C$ with minimum residence times to avoid degradation.

Applications: membranes, hoses for compressed air and hydraulics, cable sheathing, bellows, caps, coupling and drive elements, seals, soles for ski boots and soccer shoes, rolls, bearings, and fasteners in automotive applications.

Polyolefin Elastomers (TPO) TPOs are blends based on isotactic PP and ethylene-propylene rubbers (EPDM).

Applications: cable, shoes, soles, tubes and hoses, seals and gaskets, conveyer belts, wire- and cable sheathing, and automotive applications, such as bumpers and mud flaps. TPO-(EVAC-PVDC) is a chlorine containing thermoplastic olefin-elastomer based on alloys of PVDC-rigid domains with a partially crosslinked flexible EVAC-copolymer matrix in a Shore hardness range from A 60 to 70. It is used to substitute oil-resistant vulcanized nitrile (NBR)- or chloroprene (NCR)-rubber in a wide range of service temperatures (-40 to 120°C) in static applications. These TPOs are age-, weather-, and ozone resistant to a higher degree than vulcanizates. They are not suitable for applications under high dynamic loads, such as tires, because of their high mechanical damping properties. TPO grades are available as pellets with 10% carbon black filler, even in pastel tones, and can be processed at 170°C on corrosion-resistant PVC processing machines as well as on rubber processing machines.

Polystyrene Thermoplastic Elastomers (TPS) Anionic polymerization with, e.g., lithium butyl, allows to successively create blocks of styrene and butadiene and then again styrene. This results in a three-block copolymer TPS-SBS. Polymers of the SIS type (I = isoprene) can be produced by a similar method, TPS-SIS. The elastic properties of these polymers are caused by the fact that the polystyrene chain segments aggregate to rigid domains, while the polybutadiene segments aggregate to flexible rubber domains (2-phase plastics). The rigid domains act as physical crosslinking points, which will dissolve in the temperature range of the PS melting point to form a melt that can be processed like any thermoplastic. Because of the susceptibility of the BR- and IR-chains to oxidation, these polymers are usually hydrogenated. This transforms SBS-three-block copolymers into styrene-ethene butene-styrene-three block copolymers (TPS-SEBS) and SIS into styrene-ethene propene-styrene-three block copolymers (TPS-SEPS). However, these products no longer exhibit the high elasticity of SBS and SIS grades, respectively, but are more rigid. TPS are substitutes for vulcanized rubber with the advantage of effective thermoplastic processability: tubes and profiles, medical devices, injection molded soles in the shoe industry, cable insulation and sheathing grades, sound absorbing elements under the hood, bellows. TPS are also used as flexile components in rigid-flexible-multi component injection molding. Good to excellent bonding is achieved with the following plastics: PE, PP, PS, ABS, PA, PPO, PBT. Incorporation of expandable hollow spheres into the TPS-SEBS pellets allows for the production of fine pored foams

6.18 Thermoplastic Elastomers (TPE)

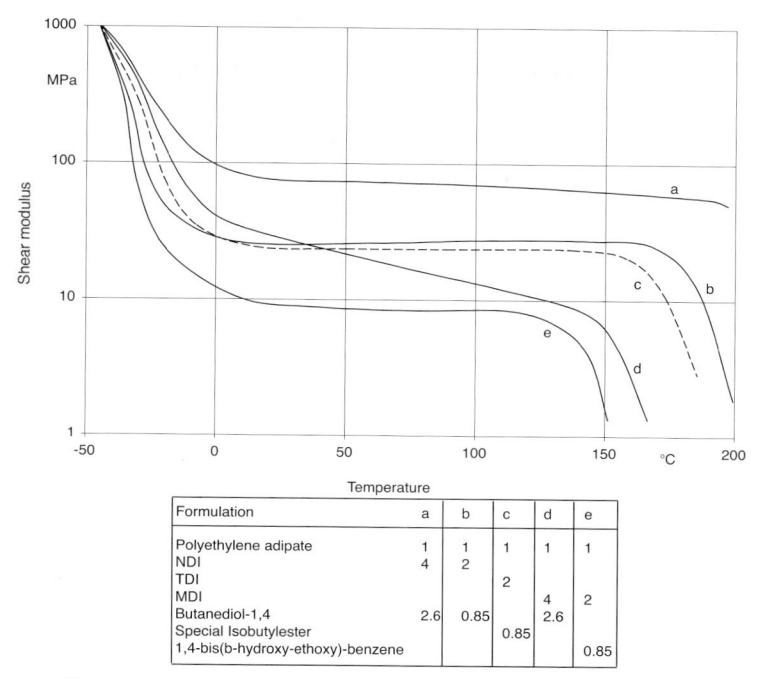

Formulation	a	b	c	d	e
Polyethylene adipate	1	1	1	1	1
NDI	4	2			
TDI			2		
MDI				4	2
Butanediol-1,4	2.6	0.85		2.6	
Special Isobutylester			0.85		
1,4-bis(b-hydroxy-ethoxy)-benzene					0.85

Figure 6.28: Influence of formulation on the shear modulus of TPU.

with densities of 0.5 g/cm^3 that can be injection molded or extruded for handles, wheels, or profiles.

Polyurethane Elastomers (TPU) TPUs are block copolymers and are created by polyaddition of long-chain diols (chain extenders:1,4-butanediol or 1,6-hexanediol; long-chain diols: polyetherdiols, polyesterdiols, polycarbonatediols) with aliphatic diisocyanates (IPDI, HDI, see Table 6.51). They are considered "bio-stable". Figure 6.28 shows the influence of the formulation on the shear modulus.

TPU exhibits high wear resistance and flexibility, even at low temperatures, damping, transparency, resistance to fats, oils, and solvents, and high long-term service temperatures.

Applications: tubes, hinge seals, attenuators, shoe industry, impact modification of polar thermoplastics, migration-resistant plasticizers for PVC.

Polyolefin Blends with Crosslinked Rubber (TPV) These TPEs are crosslinked with flexible rubber segments. They exhibit increased elasticity and resistance.

Oil-resistant and foamable grades are available. *Applications:* door- and window seals, air intake manifolds in automotive applications.

Other TPE, TPZ This category includes TPEs based on PVC with NBR [TPZ-(NBR + PVC)] or PBA [TPZ-(PBA + PVC)]. They are used for extruded seals. TPEs with crosslinked NBR are considered TPVs.

6.19 THERMOSETS

Thermosets is a general term describing polymeric materials consisting of close-meshed spatially crosslinked macromolecules. In general, these materials are rigid and behave as elastics up to their degradation temperature so that that they cannot be processed like thermoplastics. Molding and forming occurs either at the same time or prior of the chemical crosslinking (curing). Raw materials for thermosets are reaction resins (curable resins), which will react either at room temperature after the addition of curing agents or at elevated temperatures without curing agents. Curing can also be effected by electron- or UV radiation. Curable molding compounds are resins that can be processed by compression, transfer molding, extrusion, or (for special resins) by injection molding under simultaneous formation of macromolecules (crosslinking). They are filled and reinforced (some to a high degree). Deliverey forms: dust-free grounds, pellets, free-flowing rods and granules. Prepregs (preimpregnated) are laminar or tape-like materials, in which reinforcement can be incorporated as mats, fabrics, or rovings (oriented from anisotropic to uniaxial).

6.19.1 Chemical Constitution

Formaldehyde Resins (PF, RF, CF, XF, FF, MF, UF, MUF) The polycondensation of formaldehyde resins can be exemplified by phenol formaldehyde (PF). It begins with fundamental reactions (see Eq. 6.34 to 6.36), which involve the formation of various cross-linkages during curing: Addition to phenol alcohols:

Phenol + 3 CH₂O → Methylphenols (Resols)

(6.34)

Condensation with the elimination of water:

$$\begin{array}{c}\text{OH}\\ \text{CH}_2\text{OH}\\ n\end{array} + n\ \begin{array}{c}\text{OH}\\ \end{array} \longrightarrow \left(\begin{array}{c}\text{OH}\\ \end{array}-\text{CH}_2-\begin{array}{c}\text{OH}\\ \end{array}\right)_n - n\text{H}_2\text{O}$$

Methylolphenol   Phenol   Methylbridge

(6.35)

Condensation happens step-wise to facilitate the escape of volatile reaction products. In stage "A" (resol), the reaction product is still soluble and can be melted. In stage "B" (resistol), the product can only be swelled and softens only at elevated temperature, while in stage "C" (resite) complete crosslinking has occurred (Eq. 6.36). The product is insoluble and cannot be melted. The manufacturing of parts occurs sometimes in stage A, mostly though in stage B. Condensation with further elimination of water:

$$\text{Resole} \longrightarrow \text{Resite} - H_2O$$

(6.36)

Table 6.64 shows the fundamental building blocks of formaldehyde resins that are used either alone or in combination. The condensation process is similar to that of PF. Furan resins are not produced from furan directly, but from furan derivates, such as furfurol, furfuryl alcohol, and tetrahydrofurfuryl alcohol. PF, CF, RF, and XF are also called Phenolics; UF and MF are also called aminoplastics.

Unsaturated Polyester resins (UP) Unsaturated polyesters are soluble polyesters that can be melted (the building blocks of the polyester chains are connected by oxygen bridges), which contain at least one unsaturated component. They are typically copolymerized in a mixture with polymerizable compounds, such as styrene or other monomeric vinyl-, allyl-, or acrylic compounds, with the addition of peroxide compounds to form rigid, insoluble polymeric materials that can no longer be melted. The first step in the manufacture of a linear polyester (pre-condensate) is the pre-condensation of an acid, e.g., maleic acid anhydride, and an alcohol, e.g., ethylene glycol, see Eq. 6.37.

$$\text{HO}-\text{CH}_2-\text{CH}_2-\text{OH} + \text{CO}\underset{\text{O}}{\overset{\text{Ethyleneglycol}}{\diagup\diagdown}}\text{CO} \rightarrow -\text{O}-\text{CH}_2-\text{CH}_2-\text{O}-\text{CO}-\text{CH}=\text{CH}-\text{CO}-\text{O}$$

Maleic acid anhydride — Polyester

(6.37)

Table 6.64: Fundamental Building Blocks of Formaldehyde Resins

Chemical	Acronym	Basic Structure
Phenol-formaldehyde	PF	Phenol + Formaldehyde
Cresol-formaldehyde	CF	Cresol
Resorcin-formaldehyde	RF	Resorcin
Xylenol formaldehyde	XF	Xylenol
Urea formaldehyde	UF	Urea
Melamine formaldehyde	MF	Melamine
Furfuryl alcohol formaldehyde, furan resin	FF	Furfuryl alcohol

The use of high-order glycols, such as propylene glycol or butanediol, leads to more flexible and water resistant plastics. Linear polyesters can be crosslinked with the help of the double bond present in the maleic acid group, see Eq. 6.38:

$$HOOC - CH = CH - COOH$$
Maleic acid

(6.38)

Crosslinking typically occurs in a solution of monomeric compounds and continues until there is no solvent left. The transition (polymerization) from the unsaturated stage to the saturated, spatially crosslinked stage only occurs when the mixture is catalyzed, e.g., by peroxides, see Eq. 6.39.

Cross-linked polyester

$$R-O-O-R$$
Peroxide

(6.39)

UP resins based on dicyclopentadiene (DCPD) exhibit a higher heat distortion point, lower styrene content, less fiber see-through and surface tack than standard UP.

Vinylester Resins (VE), Phenacrylate Resins, Vinylester Urethanes (VU) VE and VU are reaction resins from, e.g., bisphenol A-glycidethers, see Table 6.65, or epoxidized novolacs, with terminal veresterter acrylic acid and/or methacrylic acid, see Table 6.9, Nr. 11 and 16. Dissolved in styrene (or a similar solvent), they cure during processing with special peroxide-cobalt-amine systems by crosslinking copolymerization with the solvent (similar to UP resins). The terminal crosslinking creates vibration-proof tough-rigid products. VE and VU exhibit better mechanical properties and resistance compared to UP.

Epoxy Resins (EP) The richness of epoxide chemistry (Table 6.65) relates to the capacity of the epoxy (or oxiran-) group to combine with "active" hydrogen (alcohols, acids, amides, amines) under suitable reaction conditions and with suitable catalysts. The hydrogen is displaced relative to the epoxy oxygen in such a way that it gives rise to an active HO group in the addition product. This can be used for further epoxide, but also for other addition reactions, e.g., with isocyanates.

Table 6.65: EP Chemistry

EP –chemistry, basic addition reaction
$-CH_2-CH-CH_2- + H-R \longrightarrow -CH_2-CH-CH_2-R$ with Epoxy/Glycide Group, product has OH

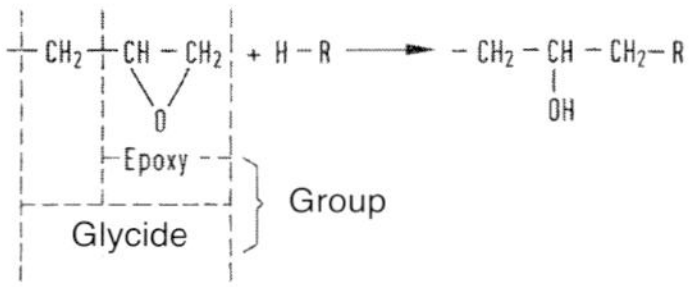

Examples of di-epoxy prepolymers

(1) Bisphenol glycidyl ether types

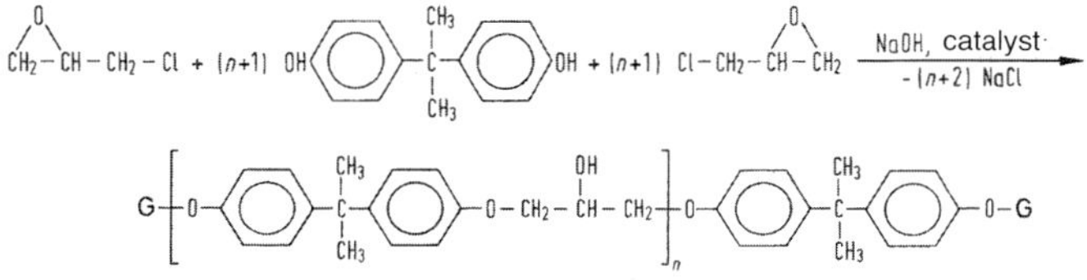

$CH_2-CH-CH_2-Cl + [n+1]$ Bisphenol A $+ [n+1]$ $Cl-CH_2-CH-CH_2$ $\xrightarrow{NaOH, catalyst}_{-[n+2] NaCl}$

„Dian"-resin:
n = 0: Bisphenol A diglycidyl ether (BADGE), 0< n < 10: liquid to solid resins
Variations: flame-retarded by halogenated dian or other bisphenoles

(2) Epoxy (cresol)

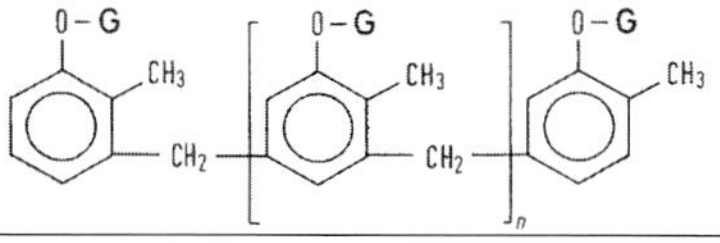

(3) Aliphatic diglycidyl ether type

$G-O-[CH_2]_n-O-G$ flexible resins with, e.g., n = 4

(4) Diglycidyl amine derivate

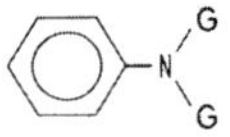

Diglycidyl aniline
(bifunctional diluent)

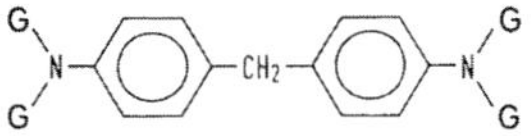

Multifunctional
for HT-coating resins

(5) Cycloaliphatic dialycidyl ester type

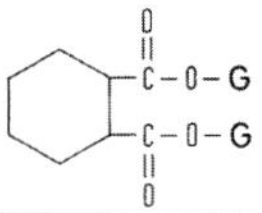

Hexahydrophthalic acid diglycidyl ester
for arc- und tracking resistant HT-resins

(6) Cycloaliphatic Type with directly bound EP-groups

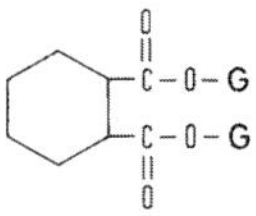

3,4-epoxycyclohexylmethyl-
3,4-epoxycyclohexancarboxylate
Materials for electrotechnical encapsulating
and casting resins

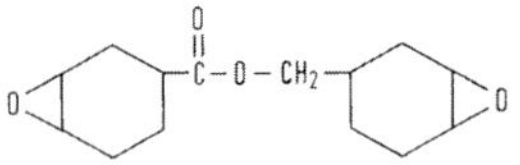

Vinyl cyclohexene dioxide

6.19 Thermosets

With either epichlorohydrin or bi- or multi-functional components, monomolecular or low polymeric "diglycydil" compounds (G in Eq. 1 - 5 in Table 6.65) with terminal epoxy groups can be produced that are also called "epoxide resins". Using different reactions, these products can also be directly incorporated in preproducts according to Eq. 6 in Table 6.65. Reactants for the synthesis of highly crosslinked EP resin products include bi- or multi-functional low molecular weight products containing active hydrogen. They do not simply act as catalysts in initiating crosslinking of pre-polymers, as is the case of curing agents and accelerators for UP resins. Rather they are chemically incorporated into the macromolecule by addition to the epoxy group.

It is therefore necessary to mix exactly measured quantities of the crosslinking components during epoxy resin processing. An additional possible variation is the use of "reactive diluents" with only one epoxy group to reduce the viscosity of liquid resins and to increase the flexibility of molded materials. Reactants for the synthesis of highly crosslinked EP resin products include bi- or multi-functional low molecular weight products containing active hydrogen. They do not simply act as catalysts in initiating crosslinking of pre-polymers, as is the case of curing agents and accelerators for UP resins. Rather they are chemically incorporated into the macromolecule by addition to the epoxy group. It is therefore necessary to mix exactly measured quantities of the crosslinking components during epoxy resin processing. An additional possible variation is the use of "reactive diluents" with only one epoxy group to reduce the viscosity of liquid resins and to increase the flexibility of molded materials. Liquid aliphatic polyamines and polyamidoamines are predominantly used to cold-cure liquid epoxy resins; tertiary amines are used as catalytic curing accelerators. For warm curing > 80°C, either aromatic amides or their derivates, or anhydride of phthalic acids or related acids (for flame retarded laminates) are used, sometimes together with accelerators. Systems cured with aliphatic amines exhibit highest resistance against chemicals; those cured with aromatic amines are solvent resistant, while those anhydritically cured show the highest resistance against weathering and acids. Many reactants, particularly amines, are corrosive and otherwise dangerous chemicals; their use is regulated and both manufacturers and the chemical industry offer specific guidelines for their use.

Diallyl Phthalate Resins, Allyl Esters (PDAP) PDAP is produced by the conversion of allyl groups (CH_2=CH- CH_2) with phthalates or isophthalates to two different grades of molding compounds, identified as ortho- or meta-grades. Because of their low viscosity, they are pre-polymerized during the production of molding compounds. Crosslinking occurs primarily peroxidically, sometimes with the addition of accelerators such as co-naphthenate or amines. They cure under pressure at 140 - 180°C.

Silicone Resins (Si) Silicones contain silicon and oxygen as parts of their chains (see Eq. 6.40)

$$-\underset{\underset{R}{|}}{\overset{\overset{R}{|}}{Si}}O-$$

(6.40)

Alkyl-, aryl-, or chlorine alkyl groups are used as organic remnants "R". When trifunctional organo(trichloro)silanes or silicon tetrachloride are used during polycondensation, crosslinked or branched polymers are created. Filled with inorganic fillers, the compounds are compression molded at 175°C.

6.19.2 Processing, Forms of Delivery

Curable molding compounds are generally used as compounds of hot-curing resins with organic or inorganic fillers, reinforcements, colorants or other additives. Special cold-curing molding compounds are compression molded (cold) and subsequently cured in an oven. Curable molding compounds are available as: dust-free grounds, pellets, free-flowing rods or granules. Aminoplasts are considered fast molding charges. The following acronyms and notations are established:

- GMC (granulated molding compounds) and PMC (pelletized molding compounds), respectively: dry, free-flowing granulated or pelletized (rods) molding compounds
- BMC (bulk molding compounds): wet, doughy-fibrous compounds, chemically thickened, ISO 8606
- DMC (dough molding compounds): wet, doughy-fibrous compounds with increased filler content, ISO 8606
- SMC (sheet molding compounds, resin mats) and SMC-R (R = random) are impregnated (prepreg) molding compounds with 2-dimensionally non-oriented reinforcing fibers (typically 25 - 50 mm long, glass fiber rovings 25 - 65 wt.%), along and perpendicular equally able to flow. Production, see Chapter 4
- SMC-D (D = direct) containing 75 - 200 mm long fibers, oriented longitudinally; longitudinally barely flowing
- SMC-C (C = continuous) containing continuous longitudinal fibers; longitudinally not able to flow
- Fabric prepregs are impregnated prepregs; glass fiber contents 35 - 60 wt.%
- Prepreg tapes: made from continuous fibers; unidirectionally reinforced tapes

6.19 Thermosets

UP molding compounds contain ≈ 40% fine mineral fillers together with 12 - 25 (65) wt.% glass fibers. For special compression molding compounds organic fillers and textile fibers are also added. GMC are available styrene-free, usually with diallyl phthalate as crosslinking agent. Shelf life until processing is ≈ one year. BMC and DMC contain resins dissolved in styrene. Though they are more difficult to meter, they do not reduce the length of the reinforcing fibers during processing; shelf life several months. EP molding compounds are produced from the melt as granular, flaked, or rod-shaped compounds, sometimes colored with (chemically linked) inorganic fillers. Depending on curing agent, they have a limited shelf life above 20°C. Because of their excellent flowing properties, these compounds are particularly suited for injection compression molding at low pressures, see Table 6.66.

Table 6.66: General Guidelines for Processing of Curable Molding Compounds

Molding resins	Barrel temperatures		Mold T	Stagnant pressure	Injection pressure	Holding pressure
	Conveying zone °C	Nozzle °C	°C	bar	bar	bar
PF, type 11-13	60-80	85-95	170-190	up to 250	600-1400	600-1000
type 31	70-80	90-100	170-190	300-400	600-1400	800-1200
type 51, 83, 85 type 15, 16, 57, 74, 77	70-80	95-110	170-190	up to 250	600-1700	800-1200
MF, type 131	70-80	95-120	150-165	300-400	1500-2500	1000-1400
type 150-152	70-80	95-105	160-180	up to 250	1500-2500	800-1200
type 156, 157	65-75	90-100	160-180	up to 150	1500-2500	800-1200
MF/PF, type 180, 182	60-80	90-110	160-180		1200-200	
UP, type 802	40-60	60-80	150-170	-	200-1000	600-800
EP	≈ 70	≈ 70	160-170	-	up to 1200	600-800
PE, crosslinking	135-140	135-140	180-230			

Continued on next page

	Compression molding	
Molding resins	Platten T °C	Pressure bar
PF, type 11-13	150-156	150-400
type 31	155-170	150-350
type 51, 83, 85	155-170	250-400
type 15, 16, 57, 74, 77	155-170	300-600
MF, type 131	135-160	250-500
type 150-152	145-170	250-500
type 156, 157	145-170	300-600
MF/PF, type 180, 182	160-165	250-400
UP, type 802	130-170	50-250
EP	160-170	100-200
PE, crosslinking	120°C melt, to 200°C crosslinking temperature	

[1] Temperatures not accurate because of exothermic reaction effects

Because the curing reaction is accelerated above T_g ($\approx 45°C$), the compounds must be quickly transferred to the mold at temperatures only slightly above the melting temperature of the resins at 70°C. EP-prepregs are produced by impregnating of glass, carbon, or synthetic fibers with resins solutions and subsequent evaporation of the solvent. The molding compounds are plastified under pressure and at elevated temperatures in the plasticating unit of the processing machine or in the mold, respectively; they are then molded and cured under pressure. Pre-heating of the compounds improves and accelerates processing. Table 6.66 provides data about processing conditions of the most common curable molding compounds. Condensation resins, such as PF, UF, MF, require mold venting and higher processing pressures and temperatures compared to easier flowing UP- and EP resins. Injection molding cycle times of these molding compounds are shorter than those of thermoplastics, because the molded parts can be ejected from the mold while still hot and because the exothermic reaction facilitates rapid heating with wall thicknesses > 4 mm.

6.19.3 Properties

General Properties Curable molding compounds contain not only resins but also combinations of fillers and reinforcements: wood flour, cellulose, cotton fibers, cotton fabric cuttings, mineral fibers, stone powder, mica, short and long fibers, glass mats, and others.

Cured thermoset parts exhibit combinations of properties that make them suitable for engineering applications:

- Appropriate selection of fillers (content 40 - 60 wt.%) allows to meet a wide range of end use requirements and enhances properties.

- Suitable for highly flame retarded electrical insulating applications.
- Stiffness and strength up to high temperatures barely decreasing, at low temperatures unchanged.
- Long-term service temperatures ranging from 100 to > 200°C, combined with short-term overload capacity of several 100 K result in resistance to damage.
- Dimensional accuracy, also very little shrinkage of parts - depending on resin grade and processing method - of < 0.1%, particularly if molding shrinkage is anticipated by tempering.

The brittle fracture behavior of highly crosslinked molding compounds can be compensated by appropriate part design. In addition, new PF molding compounds with up to 2% elongation at break and good thermal stability are now available. Molding compounds with inorganic fillers have a higher long-term service temperature and better moisture resistance than compounds with organic fillers, see Fig. 6.29. The molding compounds' notch sensitivity decreases with increasing length (or volume) of the fillers; however, at the same time processing becomes increasingly more difficult, particularly with respect to the free-flowing properties of the compound, the flow properties of the melt, and the surface quality of the finished part.

The properties of the molded compounds (cured molding compounds) are controlled by the type of fillers and reinforcements.

Phenoplastic Molding Compounds (PF, CF, RF, XF) Cured PF compounds darken after exposure to sunlight or extended exposure to elevated temperature and turn yellowish-brown. Therefore, they are offered only in dark colors. All PF moldings are resistant to organic solvents, fuels, fats and oils, even at elevated temperatures. They have only limited resistance to more concentrated acids and alkalis. Because of their increased water absorption, those PF compounds with organic fillers are more susceptible to attack on longer exposure than those filled with inorganic materials. The rate of water absorption (Fig. 6.29) is so small for all PF compounds that they are not damaged by short-term exposure to water. Adequately cured moldings should show no signs of damage after boiling in water or solvents for half an hour. Inorganic compounds should be used for insulating parts exposed to weather or damp conditions because the slow water absorption causes a decrease in electrical insulating properties in organic filled compounds.

Aminoplastic Molding Compounds (UF, MF) A resin content of $\approx 60\%$ is necessary to obtain a sufficient melt flow rate in UP molding compounds. The light colored compounds filled with bleached cellulose make for translucent molding compounds with good coloring properties and light resistance. The parts

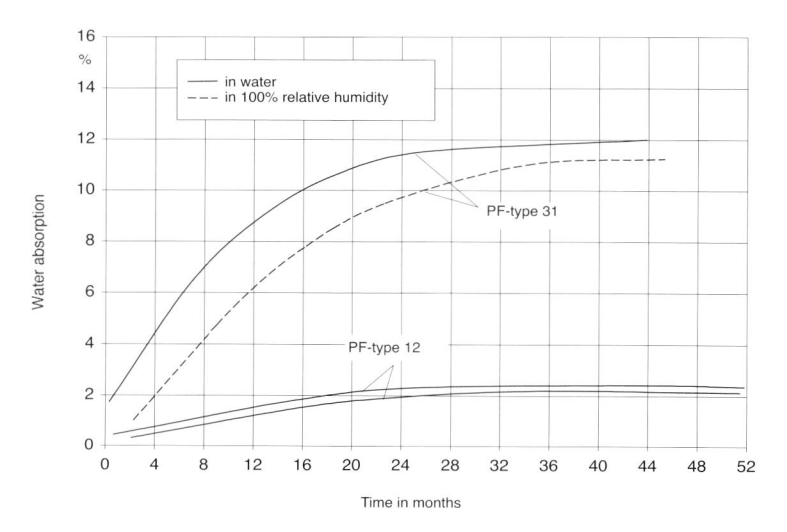

Figure 6.29: Moisture absorption of organic (type 31) and inorganic (type 12) filled PF molding compounds. Measured specimens 15 × 10 × 120 mm^3; linear change in dimension: type 31 ± 1.2%, type 12 ± 0.2%.

are taste- and odorless, resistant to food and oils, less resistant to acids and alkalis. Parts made form MF molding compounds filled with 60% a-cellulose are white and in all other colors light-, hot water-, and dish-washing detergent resistant. Their scratch resistance is higher than that of MPF (melamine phenolic resin compounds) or UF. Grades 152, 153, and 154 are insulation materials that are tracking resistant, moisture resistant, and resistant to high mechanical loads. The inorganic filled grades 155 and 156 exhibit arc and incandescence resistance and are almost inflammable. MF is attacked by acids and strong alkalis; it is largely resistant to fuels, oils, solvents, and alcohols. Only special MF compounds can be injection molded or compression injection molded. Because of their high shrinkage, MF parts are susceptible to stress cracking.

Melamine Phenolic Molding Compounds (MPF) MPF are mixed resin molding compounds that do not match MF with regard to tracking, distortion, temperature, and light resistance, but that come in light colors and meet the requirements for color fastness of many applications. Advantageous compared to MF is the significantly reduced shrinkage.

Melamine Polyester Resin Molding Compounds (MF + UP) Parts molded from MF + UP (filled with either cellulose or with a mixture of organic/inorganic materials) combine the brilliant color, tracking and arc resistance of MF with the minimal shrinkage, crack resistance, and increased dimensional stability under high temperature loads of UP.

Polyester Resin Molding Compounds (UP) Under appropriate processing temperatures and low pressures, UP resins for molding compounds have good melt flow properties. They quickly and completely cure by polymerization without separation of volatile by-products. The cured parts exhibit very little shrinkage and reliably hold size; they are also not susceptible to stress cracking. They are light resistant in any light colored pigmentation, resistant to alcohols, ethers, gasoline, lubricants, and fats; conditionally resistant to benzene, esters, weak acids, and boiling water; non resistant to alkalis and strong acids. They exhibit high glass transition temperatures and good electrical properties with high tracking resistance and low dielectrical losses. While GMC, PMC, BMC, DMC, and SMC-R exhibit largely isotropic properties in planar direction, SMC-C and SMC-D can be used to produce parts with optimized stability in the direction of load. Long-fiber strands inserted lengthwise into the mold result in molded parts with extremely high oriented resistance to loads. Processing shrinkages of 0.2 - 0.4% result in unacceptable rough surfaces, particularly in large automotive parts. Addition of 5 - 25% PVAC, PS, PMMA, or CAB dissolved in styrene results in low profile resins (LP resins). During curing of the UP resin, the styrene dissolved in the thermoplastic evaporates and its vapor pressure compensates for the shrinkage of the UP resin, thus creating smooth surfaces; recommended values, see Tables 6.67 and 6.68.

Vinyl Ester Molding Compounds (VE) VE exhibits higher elongation at break than UP (3.5 - 6%), and long-term service temperatures of 100 - 150°C. They are resistant to 37% HCl and 50% NaOH solution, liquid chlorine, chlorine dioxide, hypochlorite in all concentrations, hydrocarbons, and oxygen-containing organic media. The resins are temperature resistant to a higher degree and also resistant to chlorinated and aromatic solvents.

Epoxy Resin Molding Compounds (EP) EP molding compounds are distinguished by little processing shrinkage and basically no post-shrinkage even at elevated temperatures, together with high dimensional accuracy. Because of EP's good melt flow properties it is possible to embed very fine metal particles that will not undergo deformation during processing. Also, tight and well adhering coating of large metal parts can be achieved. EP is resistant to fuels and hydraulic oils. Characteristic values, see Tables 6.67 and 6.68.

Diallyl Phthalate Molding Compounds (PDAP) PDAP exhibits particularly good electrical properties, even at elevated temperatures, as well as good heat stability (to more than 200°C) and weather resistance. Most molding compounds are self-extinguishing. Characteristic values, see Tables 6.67 and 6.68.

Table 6.67: Charactristic Values for Reaction Resin Compounds

Property	Resin			UP		
	Reinforcement		GF-short fiber	GF<20 mm	GF	GF-cut matts
	Form of delivery		Pellets	Logs, BMC	Pellets	Cut matts
	Identification		Type** 802/804	Type** 801/803	Ceramic-like	SMC, Type 830-8
ρ	g/cm^3		1.9-2.1	1.8-2.0	≈ 2.1	1.7-2.4
σ_B	MPa		>30	>25	≈ 35	50-2304)
E_f	GPa		10-15	12-15	5-9	1-7
Charpy	kJ/m^2		4,5-6	22	5-6	50-70
Notched Charpy	kJ/m^2		2,5-4	22	3-4	40-60
HDT	°C		>200	200-260	≈ 265	180->200
Service T	°C		>160	150	200	150
α_t	10^5 · K-1		2-4	2-5	1.5-3.0	2-4
k	W/K · m		0.8	0.4	0.9	0.5
ρ_e	Ohm · cm		10^{12}	10^{12}	10^{14}	10^{12}-10^{15}
σ_e	Ohm		10^{12}	10^{10}-10^{12}	10^{13}	10^{10}-10^{11}
$E_B I$	kV/cm		120-180	130-150	-	130-150
$\epsilon_r 100$	50 Hz-1 MHz		4-6	4-6	4.5-7	4-6
$\tan \delta 100$	50 Hz-1 MHz		0.04-0.01	0.06-0.02	0.02	<0.1-<0.01
W_w	Mg		45	100-60	30-40	<100
W_H	%		0.1-0.5	<0.5	0.2	0.1-0.3

Table 6.68: Charactristic Values for Reaction Resin Compounds

Property	EP		DAP		SI
	Mineral	GF	Mineral	GF	Quartz
		Ground granulate or rod granulate			
ρ	1.6-2.0	1.6-2.1	1.65-1.85	1.7-2.0	1.9
σ_B	30-85	35-140	35-56	42-77	≈ 40
E_f	11-13	10-25	7-10	8.5-11	10-18
Charpy	6-7	9-100	-	-	4.3-4.6
Notched Charpy	2	3-100	-	-	-
HDT	107-260	107-260	160-288	166-288	>250
Service T	150->200	150->200	150-180	150-180	300
α_t	2-4	1-3	2-7	2-5	3.1-3.5
k	0.6	0.6	0.3-1.0	0.2-0.6	0.4-0.6

Continued on next page

Property	EP		DAP		SI
	Mineral	GF	Mineral	GF	Quartz
		Ground granulate or rod granulate			
ρ_e	10^{14}-10^{16}	10^{14}-10^{16}	10^{11}-10^{15}	10^{13}-10^{16}	10^{16}
σ_e	>10^{12}	>10^{12}	10^{10}-10^{14}	10^{10}-10^{14}	
$E_B I$	130-180	140-150	-	-	250
$\epsilon_r 100$	3.5-5	3.5-5	4-5	4-4.5	3.4-3.6
$\tan \delta 100$	0.07-0.01	0.04-0.01	0.007-0.015	0.004-0.009	0.001-0.002
W_w	30	30	-	-	-
W_H	0.03-0.2	0.04-0.2	0.2-0.5	0.1-0.3	0.1

Silicone Resin Molding Compounds (SI) They belong to the group of high temperature resistant plastics and exhibit a heat deflection temperature of 250 - 300°C after curing. Characteristic values, see Tables 6.67 and 6.68.

6.19.4 Applications

Phenoplastic Molding Compounds (PF, CF, RF, XF, FF) Special compounds filled with wood flour and rubber with increased impact resistance (but significantly reduced heat distortion temperature) are used for parts in electrical applications, such as wire plugs, housings and sockets. Resins filled with short glass fibers (heat distortion point up to 180°C) are used in telecommunication and electronics. Compounds filled with minerals and increasingly also filled with glass fibers that have short-term service temperatures up to 280°C, long-term up to 180°C, are used for temperature- and dishwasher resistant handles and fittings of small appliances, pots and pans, and in automotive applications (fuel resistant), e.g., for ignition electronics, carburetor heads, valve covers, complete cooling fluid pumps, multiple air intake manifolds (manufactured, e.g., by lost core techniques), brake- and clutch reinforcing pistons. Brake and clutch pads are short-term resistant to temperatures up to 600°C; rocket parts are exposed to similar high loads. High quality insulation resins (short-term 400°C) are also available as copper adhesive grades for collectors with high rotational speeds. Other special application areas include slip ring assempblies for ship generators, water pump impellers, (liquid-) gear drives. Molding compounds filled with Cu-powder are good heat conductors; they can be magnetized with Fe or barium ferrite; suitable for shielding of X-rays with Pb-powder; compounds with relatively high graphite filler content are used for semi-conductors, pump parts, or lubricating pin in automotive axels. Glass fiber reinforced PF-prepregs und SMC have gained importance in transportation-, in particular in airplane equipment, tunnel- and mine facing, because of their temperature resistance ($T_g = 300°C$), low smoke gas density and relative inflammability. The SMC webs are saturated with highly concentrated liquid resol and special curing agents; they require 6 - 8 days of curing time and then have a shelf life of ≈ two months.

Aminoplastic Molding Compounds (MF, UF, MPF) MF: Compounds filled with inorganic materials (types 155 and 156) are arc and incandescent resistant. They do not burn, and are therefore suitable for rooms with high fire hazards and for ship building. MF molded parts are attacked by acids and strong alkalis; they are largely chemically resistant to fuels, oils, solvents, and alcohols. Only special grades of MF molding compounds are suitable for injection and compression injection molding. Because of their high post-shrinkage, they are susceptible to stress cracking. UF: Preferably in white for closures in cosmetic applications, sanitary objects, small appliances, and for electrical insulation. UF resins compounds do not meet requirements for food contact (dishes, etc.) because of their post-processing release of small amounts of formaldehyde. MPF: Light colored parts for electrical applications, small appliances, screw connections. MF + UP: Electrical-, appliance-, control units applications; light sockets. Main area of application are parts under high loads (with complicated geometries) in electrical and electronics applications. Automotive electrics, headlight reflectors, and parts of appliances. Other than grades 801 and 802, grades 803 and 804 are flame retarded (UL 94 V-0, non-burning according to ASTM D 635), but otherwise similar.

Unsaturated Polyester Resin Molding Compounds (UP) UP-SMC (resin mats): The flow behavior of the resin mats allows the production of large double-curved parts, even with bosses and ribs, such as cabins, moon roofs, engine coverings, tailgates, bumpers, seat shells for trucks and cars, large parts for the interior of ships and aircraft, and others; grade 834 fulfills the fire protection requirements. To avoid bubble formation during subsequent branding of such parts, SMC/IMC (in mold coating) is used to apply an ≈ 0.1 mm thick skin of liquid PUR glue (under a pressure of 400 bar) to the part's surface. For automotive parts in the so-called crush zone, modified flexible LP-SMCs with low modulus of elasticity are available. Low-density, high impact resistant structural foam parts ($\rho \leq 1$ g/cm^3) are produced by the addition of ≈ 1% microencapsulated fluoroalkanes to the prepreg. Continuous mat prepregs (Unipreg) are particularly suitable for thin-walled complicated parts. With special UP- or phenacrylate resins, also carbon fiber reinforced, they almost reach the property profile of the much more expensive high-tech EP prepreg composites (see Table 6.69). Automotive bumpers and door frames are examples of applications for these HMC (= high modulus continuous) advanced SMC. Glass reinforced preglass surface mats are used as top coats for wood materials under high loads. Adhesive prepregs for the electronics industry are UP resin laminates on flexible carriers that will melt during coil winding at 100 - 120°C and then cure during annealing. The most highly mechanically stressed glass mat prepregs are useful only for flat or uni-directional curved moldings. Similar high mechanical values (750 N/mm^2 bending strength, 200 kJ/m^2 notched impact strength) are achieved by directional forming of cross-wound, filler-free XMC mats with 65 - 72% glass fiber content. Prepregs from light-curing UP resins - filled with aluminum hydroxide for opaque products - and granules or cut rovings with up to 35% glass fiber content are produced continuously (0.5 - 6 mm thick) between PVAL cover films in one process step. The leather-like soft material can be stored at room temperature for several weeks, as long as it is protected from light and UV radiation. These prepregs are used to thermoform (at 80 - 90°C) parts with limited degrees of deformation. Standard thermoforming machines can be used; however, larger parts (boat hulls, spoiler, covers) can be formed in simple box molds with edge clamps and vacuum connection. The molded parts are generally stable enough to be cured in an adjacent lamp area with 40 - 50 s exposure time for 4 mm thickness. After that, the cover films can be easily removed; lamination is also possible during thermoforming.

Table 6.69: Characteristics of Selected Fiber Reinforced Composite Parts

Resin/fiber group		Pultruded pipes and profiles			
		UP/GF	EP/GF	EP/AF	EP/CF
Property	Unit		Rovings		
ϕ_{resin}	%	≈ 30			
ρ	g/cm^3	1.9	2.1	1.4	1.6
W_w	%	≈ 1	0.2-0.3	0.5	0.2
σ_B	MPa	700	700	1300	1400
ϵ_B	%	2	2	1.8	0.6
E_t	GPa	35	35	75	130
Service T	°C	<150	<180	-	-
α_t	10^{-6} · K	10	10	0	0.2
k	$W/k \cdot m$	0.2	0.24	-	-

	Fiber reinforced plastic sheet						
	EP/GF	EP/GF	PI/GF	EP/GF	EP/AF		PI/CF
	Bisphenol-type	TGDA-typ					
Property	Fabric	Fabric	Fabric	Unidirectional Fabric	Fabric	Fabric	Unidirectional Fabric
ϕ_{resin}				32	-	50	40
ρ	1.9-2.0	1.9-2.0	1.9-2.0	2.1	-	1.5	1.6
W_w	0.2	0.2	0.2	0.2	-	0.8	0.8
σ_B	300	350	350	900	350-500	550-450	1400
ϵ_B	2	2	2	<2	-	-	-
E_t	23	22	21	35	29	-	-
Service T	130	155	200	130	-55 to 80	-	-

Epoxide Resin Molding Compounds (EP) EP, dry compounds: generally used for high-quality precision parts, down to very small dimensions. They are used particularly with metal inserts; in the electronics industry for collectors and coating of wrapped capacitors. Coating of electronic semi-conductor components (chips) by compression injection molding in multi-cavity molds with 240 - 360 cavities constitutes the largest consumption of this material. The grades used for this application have to be highly-pure, silica-filled, and cured with high T_g (e.g., by benzophenonetetracarboxylic acid anhydride). EP-woven prepregs: Laminates and composites with honeycombs or other core materials, particularly in the aerospace industry; copper laminated for printed circuits. EP-unidirectional prepregs, tapes: Applications as construction material for laminated (oriented) autoclave-cured tail units in airplanes, Compression elements in space applications, high-performance sports gear.

Diallylphthalate Molding Compounds (PDAP) PDAP molding compounds: for electronic components in military and space equipment; in these applications in particular, the high dimensional stability up the 200°C and the temperature- and weather stability as well as the excellent electrical properties under extreme conditions and climate changes are utilized.

Silicone Molding Compounds (SI) Electronic components for high temperature applications.

6.20 RUBBERS

6.20.1 General Description

Many macromolecular materials that can be crosslinked and have a low glass transition temperature are called rubbers. They are largely amorphous at room temperature and exhibit a low glass transition temperature. Uncrosslinked they exhibit thermoplastic properties, i.e., with increasing temperature they become softer and the rubbery elasticity caused by the tangled structure decreases gradually. By vulcanization with sulfur or a comparable chemical or physical process the macromolecules are crosslinked - wide-meshed to elastomers (soft rubber) or close-meshed to hard rubber (ebonite). The former exhibit high, reversible expansion properties. Elastomers take an intermediate position between the thermoplastics that are still able to flow (rubbers) and the rigid thermosets (hard rubber). They no longer soften at elevated temperatures and are therefore not processable like thermoplastics. The basis for classification of rubbers in DIN ISO 1629 is different than the one for plastics und therefore leads to different acronyms than the ones used by DIN EN ISO 1043. In the following short descriptions, the acronyms are matched with their related chemical descriptions and their basic chemical structure. Natural rubber and analogous synthetic rubbers, polymers of conjugated dienes (e.g., isoprene, butadiene, chlorobutadiene) as well as copolymers of conjugated dienes and vinyl derivates (e.g., styrene, acrylonitrile) are the most important rubber grades with more than 80% market share. They contain numerous unsaturated double bonds in their molecular chain, of which only a fraction is saturated during conventional vulcanization with sulfur and organic plasticizers to form soft rubber. The more double bonds are left in the vulcanizate, the lower is its oxidation- and weather resistance, although it can be significantly improved by the addition of antioxidants. Products with low diene content, e.g., those produced by polymerization of isobutylene with small amounts of isoprene, exhibit respective improved aging resistance; however, because of the low number of double bonds, their vulcanizationability is rather low. Products with low double bond contents can also be created by ring-opening polymerization, e.g., trans-1,5-polypenteneamer (from cyclopentene by ring-opening polymerization) and polyocteneamer (from cyclooctene). The latter plays a certain role as a processing aid rather than as a rubber. Copolymerization of ethylene and propylene creates ethylene-propylene rubbers that are completely saturated and therefore cannot be vulcanized with sulfur (EPM). Terpolymerization of ethylene and propylene with non-conjugated dienes (in particular ethylidene norbornene) produces saturated polymer chains with double bonds in the side chain, which can be vulcanized with sulfur (EPDM)r, or more advantageously with peroxides (R-O-O-R'). EP rubbers exhibit excellent aging resistance and play a significant commercial role. Other synthetic rubbers are: propylene oxide rubber, polyphosphazene, and polynorbornene. Polymers without double bonds require different vulcanization methods: some require alkaline crosslinking agents, others require oxidative crosslinking with peroxides. Both saturated and unsaturated polymers are crosslinked with electron beams. Most grades of synthetic rubber can be blended; however, a combined polymerization by sulfur or peroxides is not possible.

6.20.2 General Properties

A relevant criterion for the selection of rubbers for a specific application is their temperature resistance, in particular in automotive applications. Figure 6.30 shows the increased demands on temperature and oil resistance on high-performance products according to internationally recommended ASTM standards. This trend is caused largely by increasing temperatures under the hood.

The mastication of bales, the customary form of raw material delivery for rubber processing, requires much work. The effort to make synthetic rubber available as more easily processable crumbs or immediately extrudable or injec-

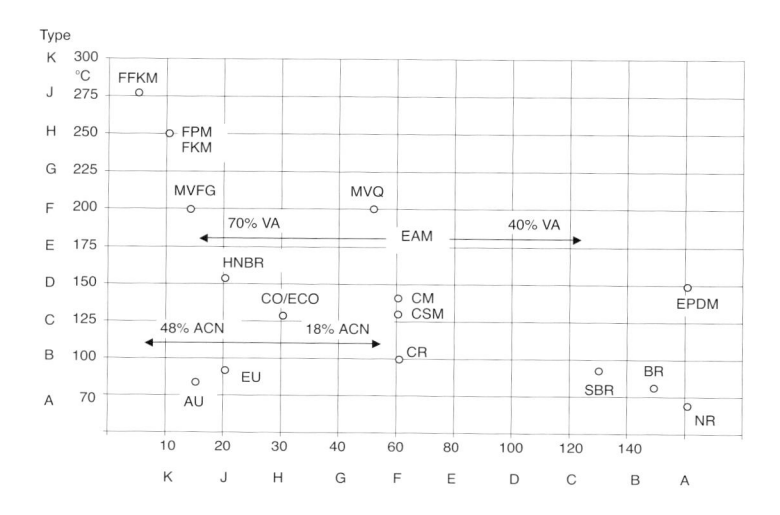

Figure 6.30: Classification of elastomers depending on temperature and oil resistance (in accordance to ASTM-D- 2000/SAE J200). Acronyms are explained in the next sections.

tion moldable powdered granulates with additives necessary for vulcanization already added has so far proven unsuccessful. Although relatively low molecular "liquid rubbers" with reactive end groups offer easier processing methods, such as those of liquid plastic pre-products, the rubber industry has yet to embrace these opportunities. The "reinforcement" of the formulation by active fillers is a very important issue in rubber technology. These fillers are typically carbon blacks for black and highly dispersed silica for light-colored products. Sometimes the fillers are already added to the latex. Conventional higher molecular synthetic rubber extended with oil results in easily processable formulations (oil extended rubber). Many rubber formulations also contain plasticizers. In the following, the most important elastomers are briefly described. Tables 6.70 and 6.71 provide an overview of selected properties and chemical resistance for the most important rubber grades.

6.20 Rubbers

Table 6.70: Property Comparison for Selected Vulcanized Rubber Grades (Scale: 1 = excellent, 2 = very good, 3 = good, 4 = sufficient, 5 = poor)

	NR	BR	CR	SBR	IIR	NBR	EPDM	CSM
σ_B	1	5	2	2	4	2	3	3
Elasticity	1	1	2	4	5	4	3	5
Wear resistance	3	1	2	2	4	3	4	3
Weather/ozone resistance	5	5	4	5	3	5	1	1
Heat resistance	5	5	4	4	3	4	2	3
Low T Flexibility	1	1	3	2	3	4	2	4
Gas permeability	4	4	2	4	1	2	3	2
Service T (°C)*	100	100	120	110	130	120	140	130

	EAM	ACM	ECO	AU	MVQ	FKM	FVMQ
σ_B	3	3	3	1	4	3	4
Elasticity	5	5	3	2	2	5	3
Wear resistance	4	4	4	1	5	4	5
Weather/ozone resistance	1	2	2	2	1	1	1
Heat resistance	2	2	3	3	1	1	2
Low T Flexibility	4	5	3	4	1	5	2
Gas permeability	2	3	2	2	5	2	5
Service T (°C)*	170	160	130	130	200	210	180

* for optimum compound ≈ 1000 h

6.20.3 R-Rubbers (NR, IR, BR, CR, SBR, NBR, NCR, IIR, PNR, SIR, TOR, HNBR)

Unsaturated chains, partially or completely made from diolefins

Natural Rubber (NR)

$$\left[-CH_2-\underset{\underset{CH_3}{|}}{C}=CH-CH_2-\right]$$

(6.41)

Even when unfilled, NR exhibits high strength and elasticity. Thermal range: long-term -50 to +70°C, special grades up to +90°C; short-term up to 120°C. NR exhibits little mechanical damping, is not susceptible to creep up to 50°C, is not resistant to oil, and must be ozone-stabilized.

Applications: rubber springs, truck tires

Isoprene Rubber (IR)

$$\left[-CH_2-\underset{\underset{CH_3}{|}}{C}=CH-CH_2-\right]$$

(6.42)

Table 6.71: Chemical Resistance of Selected Vulcanized Rubber Grades (up to 100 °C)

	NR	BR	CR	SBR	IIR	NBR	EPDM	CSM
Paraffin hydrocarbon	C	C	B	C	C	A	C	C
Fuels	C	C	C	C	C	B	C	C
Aromatics	C	C	C	C	C	C	C	C
Chlorinated hydrocarbon	C	C	C	C	C	C	C	C
Motor oils	C	C	B	C	C	A	C	C
Hypoid oils	C	C	C	C	C	B	C	C
Mineral lubricants	C	C	C	C	C	B	C	C
Alcohols	A	A	A	A	A	A	A	A
Ketones	A	A	B	A	A	C	A	B
Esters	B	B	C	B	C	C	C	C
Water	A	A	A	A	A	A	A	A
Acids (diluted)	A	A	A	A	A	A	A	A
Alkalis (diluted)	A	A	A	A	A	A	A	A
Brake fluids	A	A	B	A	A	C	A	C

	EAM	ACM	ECO	AU	MVQ	FKM	FVMQ
Paraffin hydrocarbon	A	A	B	A	C	A	A
Fuels	C	B	B	B	C	A	B
Aromatics	C	C	C	C	C	A	B
Chlorinated hydrocarbon	C	C	C	C	C	A	C
Motor oils	B	A	B	B	B	A	A
Hypoid oils	C	A	B	B	C	A	A
Mineral lubricants	B	A	B	B	B	A	A
Alcohols	B	B	B	B	B	C	C
Ketones	C	C	C	C	C	C	C
Esters	C	C	C	C	C	C	C
Water	A	B	B	C	B	A	C
Acids (diluted)	B	C	B	C	B	A	C
Alkalis (diluted)	B	C	B	C	B	A	C
Brake fluids	C	C	C	C	A	C	A

A: little or no attack (max. +10% change in volume)
B: weak to medium attack (max. +25% change in volume)
C: strong attack (more than 25% change in volume)

IR is a synthetic version of NR and their properties are comparable. IR exhibits slightly higher elasticity.

Applications: same as NR

Butadiene Rubber (BR)

$$-[CH_2-CH=CH-CH_2]-$$

(6.43)

BR is typically blended with other rubbers. It is available in wear- and low temperature resistant grades. XBR is a carboxylic group containing BR (-COOH).

Applications: Automotive stepping boards

Chloroprene Rubber (CR)

$$\left[CH_2 - \underset{Cl}{C} = CH - CH_2 \right]$$

(6.44)

CR exhibits improved aging resistance compared to NR, IR, and SBR; it is flame resistant and to some degree resistant to oils and fats. Thermal range: long-term -40 to +110°C, short-term up to +130°C. Long-term storage below 0°C results in reversible hardening by crystallization. Low gas permeability. Styrene-chloroprene rubber contains an additional -CH-C6H5 element; SCR and XCR contain a carboxylic group, see XBR.

Applications: bellows, grommets, cooling water hoses, sealing profiles in construction, roofing materials, conveyer belts, cable sheathing, protective gear.

Styrene-Butadiene Rubber (SBR)

$$\left[\underset{C_6H_5}{CH} - CH_2 - CH_2 - CH = CH - CH_2 \right]$$

(6.45)

SBR is produced by cold polymerization of 75% butadiene and 25% styrene. It is available in numerous grades with different stabilizers for both light-colored and dark products. It often substitutes NR as a general purpose rubber and represents the largest rubber class. Thermal range: long-term -40 to +100°C, short-term up to +120°C. SBR is not resistant to mineral oils. XSBR is an SBR containing carboxylic groups.

Applications: Tire formulations, cable sheathing, technical rubber parts, hoses and profiles, foam rubber.

Nitrile-Butadiene Rubber (NBR)

$$\left[\underset{CN}{CH} - CH_2 - CH_2 - CH = CH - CH_2 \right]$$

(6.46)

NBR (20 - 50% acrylonitrile content) gains resistance to oils, fats, fuels with increasing CAN content. At the same time, it becomes less flexible at low temperatures. Thermal range: long-term -30 to +100°C, short-term up to +130°C, special grades -40 to +150°C. Low gas permeability. NBR + PVC is ozone resistant. NBR + phenolic resin is tough-elastic and hot water resistant. XNBR is is an NBR containing carboxylic groups.

Applications: Most important sealing material in automotive and mechanical engineering; hoses, brake pads.

Nitrile-Chloroprene Rubber (NCR)

$$\left[\begin{array}{c} CH-CH_2-CH_2-C=CH-CH_2 \\ | | \\ CN Cl \end{array} \right]$$

(6.47)

NCR exhibits slightly better oil resistance than CR.

Butyl Rubber (IIR, CIIR, BIIR)

$$\left[\begin{array}{c} CH_3 \\ | \\ CH_2-C-CH_2-C=CH-CH_2 \\ | | \\ CH_3 CH_3 \end{array} \right]$$

(6.48)

IIR is a copolymer. The very high isobutene content is responsible for the good chemical and aging resistance and the low gas permeability (air); the low isoprene content for the vulcanizability. Thermal range: long-term -40 to +130°C. Chlorobutyl(CIIR)- and bromiumbutyl (BIIR) rubbers show good processability and aging resistance as well as low air permeability.

Applications: inner tubes, inner liners of tires, gas-proof membranes, pharmaceutical plugs, roofing material.

Isoprene-Styrene Rubber (SIR)

$$\left[\begin{array}{c} CH-CH_2-CH_2-C=CH-CH_2 \\ | | \\ C_6H_5 CH_3 \end{array} \right]$$

(6.49)

Special rubber with little commercial interest.

Polynorbornene Rubber (PNR)

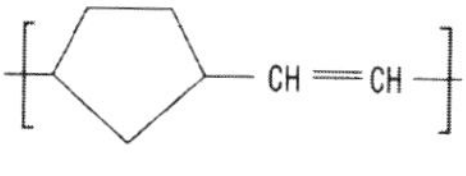

$$\left[\langle \rangle - CH=CH \right]$$

(6.50)

Plasticizers and fillers can easily be added to PNR, which results in very soft elastomers. Temperature resistance and susceptibility to creep are unfavorable compared to NR; ozone resistance is similar to NR.

Applications: soft roller coatings, foam rubber.

Trans-Polyoctenamer Rubber (TOR)

$$\left[CH_2-(CH_2)_5-CH=CH \right]$$

(6.51)

TOR is primarily used as a blending component with other rubbers with loads ranging from 5 - 30%. These blends provide better filler absorption and distribution, a significant reduction of viscosity of the blends at processing temperatures, and a considerable increase in melt flow rate.

Hydrated NBR Rubber (HNBR)

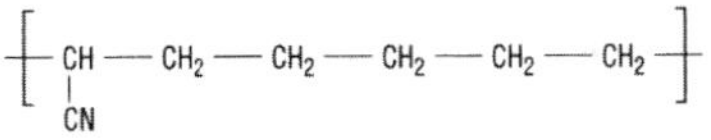

$$\left[\begin{array}{c} CH - CH_2 - CH_2 - CH_2 - CH_2 - CH_2 \\ | \\ CN \end{array} \right]$$

(6.52)

HNBR is periodically crosslinked, up to 150°C significantly more oil- oxidation-wear resistant than NBR.

Applications: Automotive, oil production equipment

6.20.4 M-Rubbers (EPM, EPDM, AECM, EAM, CSM, CM, ACM, ABM, ANM, FKM, FPM, FFKM)

Saturated chains of polymethylene type.

Ethylene-Propylene-(Diene) Rubber (EPM, EPDM)

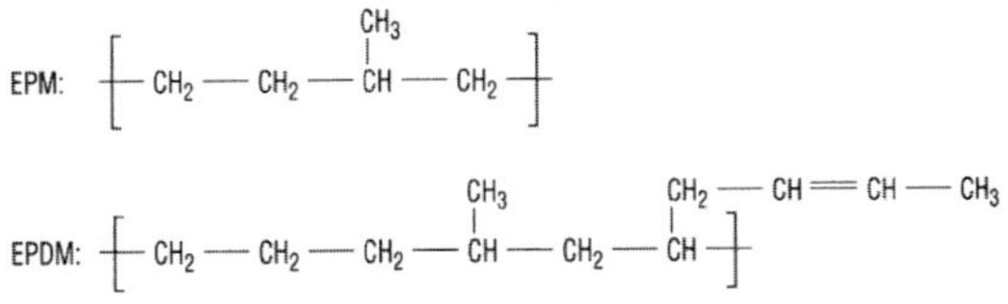

$$\text{EPM:} \quad \left[CH_2 - CH_2 - \underset{\underset{CH_3}{|}}{CH} - CH_2 \right]$$

$$\text{EPDM:} \quad \left[CH_2 - CH_2 - CH_2 - \underset{\underset{CH_3}{|}}{CH} - CH_2 - \underset{\underset{CH_2 - CH = CH - CH_3}{|}}{CH} \right]$$

(6.53)

EPDM are atactic copolymers of ethylene and propylene. EPM can only be crosslinked with peroxides; EPDM is also vulcanizable with sulfur. Because the unsaturated sites are not located on the main chain in terpolymers, they exhibit the chemical resistance and - with appropriate stabilization - the good weather-, ozone-, and aging resistance of saturated poyolefins. Thermal range: long-term -40 to +130°C, special grades up to +150°C. Highly pure grades are obtained by polymerization with metallocene catalysts.

Applications: automotive exterior, solid and foam sealing profiles, O-rings, hoses, cable sheathing.

Ethylene-Acrylic Ester Rubber (AECM)

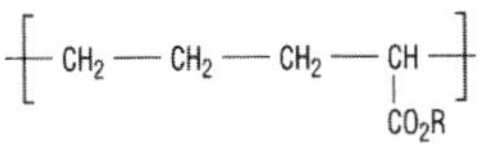

$$\left[CH_2 - CH_2 - CH_2 - \underset{\underset{CO_2R}{|}}{CH} \right]$$

(6.54)

AECM is a newly developed elastomer with excellent temperature resistance und medium oil resistance. Thermal range in air: long-term -25 to +170°C, short-term up to +200°C.

Applications: special seals in automotive, cooling water hoses, cables.

Ethylene-Vinylacetate Rubber (EAM)

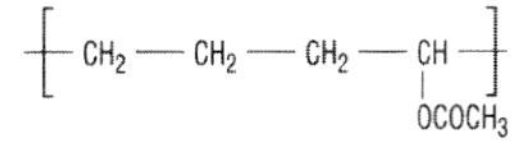

$$[CH_2-CH_2-CH_2-CH(OCOCH_3)]$$

(6.55)

EAM combines high temperature resistance with a limited oil resistance. Low temperature toughness is low. Thermal range: long-term -10 to +150°C.

Chlorosulfonated PE Rubber (CSM)

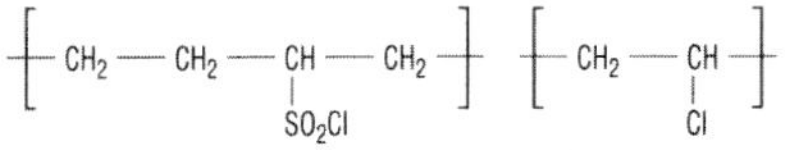

$$[CH_2-CH_2-CH(SO_2Cl)-CH_2][CH_2-CH(Cl)]$$

(6.56)

CSM is a light-colored special rubber with good weather- and chemical resistance. Thermal range: long-term -20 to +120°C. Self-vulcanizing, mineral filled compounds are used as spraying mixtures or weldable webs for roofing and linings.

Chlorinated PE Rubber (CM)

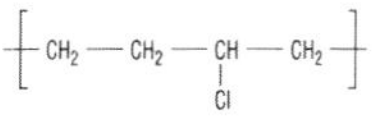

$$[CH_2-CH_2-CH(Cl)-CH_2]$$

(6.57)

CM exhibits slightly better hot air and hot oil resistance as well as a lower brittleness temperature than CSM; it is also less expensive.

Applications: technical rubber parts, cable sheathing, hoses for engines, impact modifier for PVC.

Acrylate Rubber (ACM, AEM, ANM)

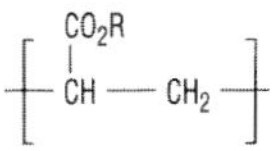

$$[CH(CO_2R)-CH_2]$$

(6.58)

ACM exhibits higher temperature resistance than NBR. Thermal range: long-term -25 to +150°C, short-term up to +170°C. Resistant to oils, ozone, UV radiation.

Applications: engine seals in automotive applications, self-vulcanizing acrylate latex for adhesion of non-woven materials.

Fluorine Rubber (FKM)

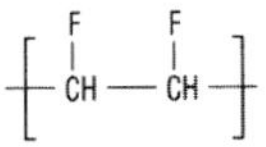

$$\left[\begin{array}{cc} \overset{F}{\underset{|}{CH}} - \overset{F}{\underset{|}{CH}} \end{array} \right]$$

(6.59)

FKM is resistant to most liquids. Thermal range: long-term -20 to +200°C, short-term up to +250°C.
Applications: Special seals.

Propylene-Tetrafluoroethylene Rubber (FPM)

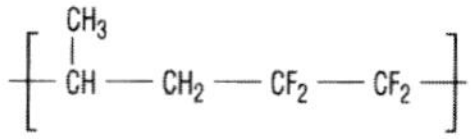

$$\left[\begin{array}{c} \overset{CH_3}{\underset{|}{CH}} - CH_2 - CF_2 - CF_2 \end{array} \right]$$

(6.60)

FPM is a copolymer of polypropylene and tetrafluoroethylene and exhibits extremely high chemical resistance.

Perfluoro Rubber (FFKM)

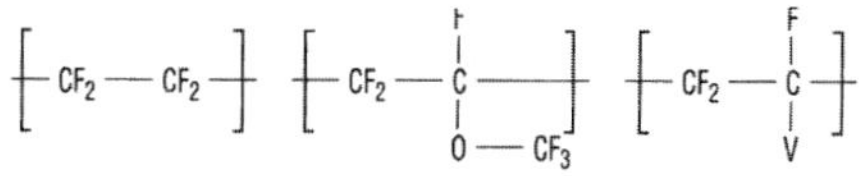

$$\left[CF_2 - CF_2 \right] \quad \left[CF_2 - \overset{F}{\underset{\underset{CF_3}{|}}{\overset{|}{C}}} \right] \quad \left[CF_2 - \overset{F}{\underset{\underset{V}{|}}{\overset{|}{C}}} \right]$$

V=cross-linking component

(6.61)

Because of its very high price, FFKM is only used for special seals in the chemical industry or in oil drilling equipment.

6.20.5 O-Rubbers (CO, ECO, ETER, PO)

Chains with oxygen.

Epichlorohydrine Homopolymer, Copolymer, and Terpolymer Rubbers (CO, ECO, ETER)

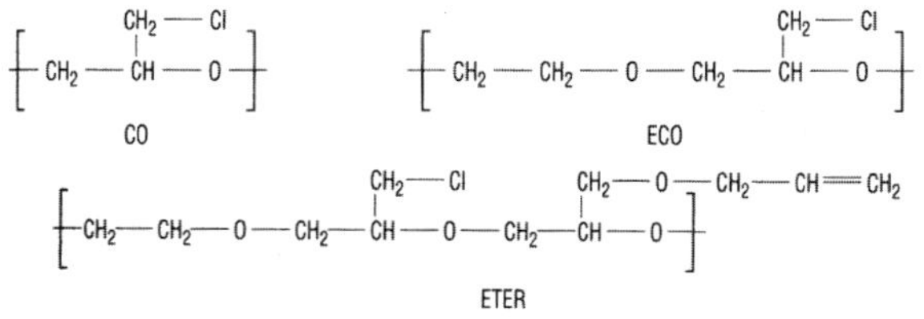

$$\left[CH_2 - \overset{CH_2-Cl}{\underset{|}{CH}} - O \right] \qquad \left[CH_2 - CH_2 - O - CH_2 - \overset{CH_2-Cl}{\underset{|}{CH}} - O \right]$$

CO ECO

$$\left[CH_2 - CH_2 - O - CH_2 - \overset{CH_2-Cl}{\underset{|}{CH}} - O - CH_2 - \overset{CH_2-O-CH_2-CH=CH_2}{\underset{|}{CH}} - O \right]$$

ETER

(6.62)

Epichlorohydrin elastomers are homopolymeric (CO) or with ethylene oxide copolymerized (ECO) polyether. They can be crosslinked with amines. Terpolymers containing vinyl groups (ETER) are also vulcanizable with sulfur and peroxide. Compared to NBR they exhibit similar oil- and slightly better temperature resistance. Low temperature impact strength and elasticity are significantly higher. Thermal range: long-term -40 to +120°C.

Applications: hoses and special seals in automotive applications.

Propylene Oxide Rubber (PO)

$$\left[CH_2 - \underset{\underset{CH_3}{|}}{CH} - O \right]$$

(6.63)

PO exhibits good heat and cold temperature resistance, but only low oil resistance.

6.20.6 Q-(Silicone) Rubber (MQ, MPQ, MVQ, PVMQ, MFQ, MVFQ)

Chains with siloxane groups.

General Properties Silicone rubbers are electrical high performance materials; they are physiologically benign and difficult to crosslink, as many other silicones. Most properties change only very little over a temperature range from -100°C for special grades, -60°C for standard grades 180°C long-term, 300°C short-term in dry conditions. Silicone rubbers are increasingly attacked in water vapor above 100°C.

Applications With peroxides at 200°C vulcanized rubber: seals (stationary and in motion), hoses, electro insulation, non-stick transport belts, foamed parts with smooth surfaces. Cold vulcanized, solvent-free, pasty one- or two component compounds: permanently elastic sealant material in construction, filling compounds for soft-elastic molds in electrical applications. Copolymers with thermoplastics to increase temperature resistance, processability, and flexibility. Boron containing rubbers that self adhere at room temperature: self adhering insulation tapes (Silicor). The major rubbers with siloxane groups are:

- Polydimethylene Siloxane Rubber (MQ)

$$\left[\underset{\underset{CH_3}{|}}{\overset{\overset{CH_3}{|}}{Si}} - O \right]$$

(6.64)

- Methylene-Phenylene-Siloxane Rubber (MPQ)

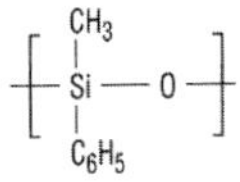

$$\left[\begin{array}{c} CH_3 \\ | \\ -Si-O- \\ | \\ C_6H_5 \end{array} \right]$$

(6.65)

- Methylene-Vinyl-Siloxane Rubber (VMQ)

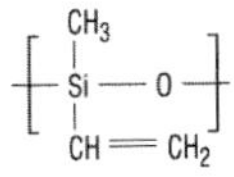

$$\left[\begin{array}{c} CH_3 \\ | \\ -Si-O- \\ | \\ CH=CH_2 \end{array} \right]$$

(6.66)

- Methylene-Phenylene-Vinyl-Siloxane Rubber (PVMQ)

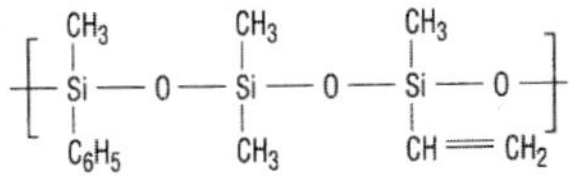

$$\left[\begin{array}{ccc} CH_3 & CH_3 & CH_3 \\ | & | & | \\ -Si-O- & Si-O- & Si-O- \\ | & | & | \\ C_6H_5 & CH_3 & CH=CH_2 \end{array} \right]$$

(6.67)

- Methylene-Fluorine-Siloxane Rubber (MFQ)

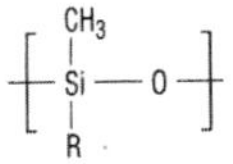

$$\left[\begin{array}{c} CH_3 \\ | \\ -Si-O- \\ | \\ R \end{array} \right]$$

R = Fluorinated alkyl remnant

(6.68)

- Fluorine-Silicone Rubber (MVFQ) General structure as MFQ, but contains additional unsaturated side groups. MVFQ swells less than MVQ and has better low temperature resistance than FKM. Thermal range: long-term -60 to +175°C.

  Applications: Seals against fuels, ATF oils, aerospace.

- Liquid Silicone Rubber (LSR) LSR has high resistance to hot air (180°C, stabilized up to 250°C, short-term up to 300°C)., high aging- and chemical resistance, transparency, and physiological compatibility. The addition of chemical bonding agents during crosslinking results in good adhesion on substrates such as thermoplastics, steel, or aluminum.

6.20.7 T-Rubber (TM, ET, TCF)

Chains with sulfur.

Polysulfide Rubber (TM, ET)

$$\left[S-CH_2-CH_2-S \right]$$

(6.69)

TM/ET exhibits excellent solvent resistance (esters, ketones, aromates) and aging resistance. Odor, low strength and heat resistance are disadvantages.

Applications: Container linings, sealants comounds (liquid rubber)

Thiocarbonyldifluoride Copolymer Rubber (TCF)

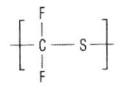

$$\left[\begin{array}{c} F \\ | \\ C-S \\ | \\ F \end{array} \right]$$

(6.70)

6.20.8 U-Rubbers (AFMU, EU, AU)

Chains with oxygen, nitrogen, and carbon.

Nitrose Rubber (AFMU)

Urethane Rubber, Polyester/Polyether (EU, AU)

$$\left[R_1-O-\overset{O}{\underset{}{C}}-\underset{H}{N}-R-\underset{H}{N}-\overset{O}{\underset{}{C}}-O-R_2-O \right]$$

R = diisocyanate remnant, R_1 = polyester chain, R_2 = polyether chain

(6.71)

Other than the rest of the elastomers, polyurethanes are produced by polyaddition of low molecular and thus low viscous pre-products. Processing takes the form of either thermoplastic molding of special pellets (thermoplastic PUR elastomers), casting of reactive mixtures, or the classic methods of elastomer processing: production of a pre-polymer, addition of fillers and crosslinking on rolls with subsequent vulcanization in heated molds. AU and EU exhibit exceptional high strength, flexibility, and elasticity; their properties range between those of the other elastomers and thermoplastics and thermosets, respectively. They are wear resistant, oil-, fuel-, and ozone resistant; they exhibit relatively high mechanical damping. AU exhibits better water resistance than EU.

Applications: hydraulic seals, chain wheels, rollers, solid wheels, bellows, drive belts, fenders.

6.20.9 Polyphosphazenes (PNF, FZ, PZ)

Chains alternating with phosphorus and nitrogen.

Fluorine-Phosphazene Rubber (PNF)

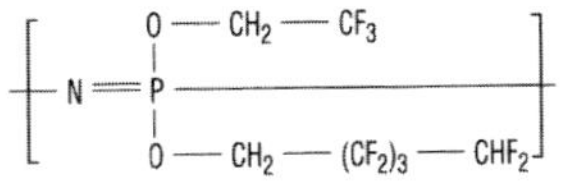

$$\left[-N=P \begin{matrix} O-CH_2-CF_3 \\ | \\ | \\ O-CH_2-(CF_2)_3-CHF_2 \end{matrix} \right]$$

(6.72)

PNF are produced by ring-opening polymerization of trimer chlorinated phosphazene with subsequent exchange of the chlorine atom by alkoxy side groups. PNF containing substituents with highly fluorinated unsaturated sites can be crosslinked with sulfur as well as with peroxides. The main chain containing only sulfur and nitrogen makes PNF absolutely resistant to oxygen and ozone. PNF is rubber-elastic from -50 to +150°C; in the case of fire, it does not melt, drip, or develop smoke; it is resistant to fuels, oils, and hydraulic liquids.

Applications: Seals for high-altitude airplanes and arctic oil- and fuel lines. Halogen-free grades with similar good fire behavior are used as open- or closed cell foams for ship insulation and aircraft upholstery.

Other polyphosphazenes: FZ, PZ Phosphazene rubber with fluoroalkyl- or fluoroxyalkyl groups. Phosphazene rubber with phenoxy groups.

6.20.10 Other Rubbers

Polyperfluorotrimethyltriazine Rubber (PFMT)

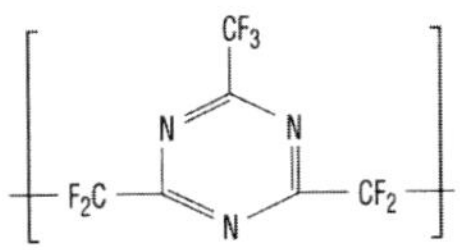

$$\left[-F_2C-\underset{N}{\overset{N=\!\!\!\!\!\underset{|}{C}-CF_3}{\underset{\|}{C}}}-CF_2- \right]$$

(6.73)

CHAPTER 7

POLYMER ADDITIVES

A polymer is rarely sold as a pure material. More often, a polymer contains several additives to aid during processing, add color, or enhance the mechanical properties. In fact, when we refer to a *plastic* we refer to a polymer with one or several additives.

7.1 ANTIBLOCKING AGENTS

Polyethylene and polypropelene films often stick together, making it diffficult to separate them, for example when we try to open a polyethylene produce bag. It is believed that this stick condition between layers of film is caused by the fraction of low molecular weight molecules that make the surfaces to adhere to each other. When adding antiblocking agents to the resin, the blocking force between the film layers can be reduced significantly. The mechanism that lowers the blocking forces is the surface roughness caused by the well dispersed particles.

The most commonly used antiblocking angents are:

- *Limestone* - This is a low cost antiblocking agent for the production of lower quality film. It is primarely calcium carbonate ($CaCO_3$), but may also contain magnesium carbonate ($MgCO_3$).
- *Natural silica*
- *Synthetic silica gel* - This antiblocking agent is formed by polymerized particles of 2-10 nm diameter, which results in a material with a very high surface area. Synthetic silica gel is widely used for high quality film applications.
- *Talc* - This is a widely mined, readily available antiblocking agent composed of magnesium hydrosilicate.
- *Zeolites* - These are crystalline hydrated alumosilicates with a highly uniform, three dimensional porous structure. They are not very common antiblocking agents.

7.2 SLIP ADDITIVES

Polyolefin films are very tacky, which causes them to adhere to themeselves and other surfaces during processing. The desired coefficient of friction when processing film should be around 0.2. Therefore slip additives must be blended with the material in order to modify the surface. Slip adddtitives should be incompatible with the polymeric material, causing them to *bloom* or migrate to the surface.

The two most commonly used slip additives are:

- *Erucamide* - This slip additive is derived from mono-unsaturated C_{22} erucic acid. Although erucamide migrates slowly to the surface, eventually it will reach a coefficient of friction lower than oleamide. The lower volatility of this additive means that it stays on the surface longer.
- *Oleamide* - This chemical is the amide from C_{18} mono-unsaturated acid, and is often referred to as the fast blooming slip additive, since it migrates quicker than erucamide. This is the additive of choice in large in-line plastic bag conversion operations.
- *Stearamide* - This is another slip additive that is used in conjunction with erucamide and oleamide when the fatty acid must also provide an antiblock effect.

7.3 PLASTICIZERS

Solvents, commonly called plasticizers, are sometimes mixed into a polymer to dramatically alter its rheological or mechanical properties. Plasticizers are used as processing aids since they have the same effect as raising the temperature of the polymer. The resulting lowered viscosities reduce the risk of thermal degradation during processing. For example, cellulose nitrite thermally degrades if it is processed without a plasticizer. Plasticizers are more commonly used to alter a polymer's mechanical properties such as stiffness, toughness, and strength. For example, adding a plasticizer such as dioctylphthalate (DOP) to PVC can reduce its stiffness by three orders of magnitude and can lower its glass transition temperature by 35°C. In fact, highly plasticized PVC is rubbery at room temperature.

7.4 STABILIZERS

The combination of heat and oxygen can result in thermal degradation in a polymer. Heat or energy produce free radicals which react with oxygen to form carbonyl compounds, giving rise to yellow or brown discolorations in the final product.

7.4.1 Antioxidants

Thermal degradation can be suppressed by adding stabilizers, such as antioxidants or peroxide decomposers. Today, the state-of-the-art of stabilizers is the use of stailizer blends to create a synergistic stabilizing effect. In any case, these additives do not eliminate thermal degradation but slow it down. Once the stabilizer or stabilizers have been consumed by reaction with oxygen, the polymer is no longer protected against thermal degradation.

This section covers the most common additives to inhibit oxidation with plastics.

Hydrogen Donors Hydrogen donors, more commonly referred to as *H-donors*, are used to offer the peroxy radical in a polymer backbone a more easily abstractable hydrogen to form the more stable hydrogen peroxide.

The most commonly used H-donor antioxidants are:

- *Aromatic amines* - Secondary amines and sterically hindered phenols are excellent H-donors. A great disatvantage is that they often discolor and stain the final plastic parts, and are therefore seldom used with thermoplastic products. However, they are often used with carbon black filled rubber products and sometimes with polyurethane applications
- *Phenols* - Phenolics that serve as H-donors are some of the most widely used antioxidants. The main reaction that governs this antioxidant is the formation of hydrogen peroxide by hydrogen extraction from the phenolic group, with formation of the phonoxyl radical.

Hydroperoxide Decomposers Hydroperoxide decomposers transform hydroperoxides, ROOH, into non-reactive thermally stable materials. During the decomposition, the hydrogen peroxide group in the polymer is reduced to an alcohol, ROH, as the hydroperoxide decomposer is oxidized in a stoichiometric reaction. These stabilizers are often used with H-donors such as phenols.

The most commonly used hydroperoxide decomposers are:

- *Phosphites and phosphonites* - Phosphorus compounds are the preferred hydroperoxide decomposers and are mostly used blended together with sterically hindered phenols.
- *Thiosynergists* - These sulfur based hydroperoxide decomposers react as a thermolysis of the initially formed sulfoxide to a sulfenic acid.

Alkyl Radical Scavengers Scavenging alkyl radicals, R*, immediately stop autooxidative processes, because they render a very high reaction rate with oxygen.

The most commonly used alkyl radical scavengers are:

- *Hindered amine stabilizers* - Sterically hindered amines based on tetramethyl piperidine derivatives are converted to nitroxyls by oxidation of the parent amine with the peroxide radicals, as well as to hydroxylamines by oxidation with peracids. Hindered amine stabilizers are also very effective UV stabilizers.
- *Hydroxyl amines* - During reaction, these stabilizers form intermediate nitrone which is capable of scavenging C-radicals.
- *Benzofuranone derivatives* - Even in small amounts, benzofuranone derivatives are powerful radical scavengers, and actually help control melt staibility during polymer processing operations.
- *Acryloyl modified phenols* - These stabilizers are very efficient C-radiacal scavengers, and are known to prevent styrenic copolymers from cross-linking and degrading during processing.

Metal Deactivators Metal deactivatiors are used to hinder decomposition of peroxides in the presence of metal ions. This is especially important for polymers that are in contact with copper, for example cable insulators.

7.4.2 Flame Retardants

Since polymers are organic materials, most of them are flammable. The flammability of polymers has always been a serious technical problem. However, some additives that contain halogens, such as bromine, chlorine or phosphorous, reduce the possibility of either ignition within a polymer component or once ignited, flame spread. Table 7.1 presents commonly used flame retardants and threir applicaton.

Many flame retardants are in the process of being reviewed or even banned around the world. The flame retardants in question are polybrominated biphenyls (PBB) and polybrominated diphenylether (PBDE). Currently, the application of tetrabromo-bisphenol-A is also being questioned. In the European Union, the flame retardants have been banned and regulated by the Waste Electrical and Electronic Equipment (WEEE) regulatory agency. Assessment of these flame retardants is also undergoing in the United States and other countries.

Many flame retardants are used with synergists. Antimony oxide has been widely used because of its low cost and its effectiveness in reducing the amounts of halogen-containing flame retardants. Other synergists are iron oxide, zinc borate, zinc phosphate, and zinc stannate.

Table 7.1: Commonly Used Flame Retardants for Specific Plastics

Flame retardant	Polymer
Aluminum hydroxide	UP
Aluminum oxalate	PA, PBT
Alumina trihydrate	PVC
Ammonium polyphosphate	UP, EP, gelcoats
Brominated pSSolystyrene	PBT
Dibromostyrene	ABS, PS-HI, Styrene-PAE elastomers, Thermosets
Magnesium hydroxide	PA
Octabromodiphenyloxide[a]	ABS
Octabromotrimethyl-phenylindane	PS-HI
Tetrabromo-bisphenol-A[b]	ABS, EP, PBT, PP
Zinc borate	PVC-P
Zinc chloride	PA, PP, PVC
Zinc sulfide	PA, PP, PVC

[a] Banned by the European Union
[b] Undergoing risk assessment by the European Union

7.4.3 UV Stabilizers

The photo-oxidation of plastics is a result of the combined action of light and oxygen. Most of the damage occurs because of the action of UV rays in sunlight, which leads to a deterioration of appearance, due to microcrack formation, and consequently a reduction in mechanical performance, as well as chemical resistance, to name a few. To reduce the action of photo-oxidation, specialty chemicals called light stabilizers or UV stabilizers were developed. These chemicals interfere with the physical and chemical processes of photo-oxidation.

The protection offered by UV stabilizers, sometimes also referred to as UV protectors, is basically blocking, absorbing and dissipating harmful UV radiation, so that it does not turn into heat.

The most important UV stabilizers used today are:

- *Sterically hindered amines (HALS)* - The hindered amines are the latest type of light stabilizers, and have demonstrated superior performance when compared to other UV protectors available.
- *2-hydroxybenzophenones*
- *2-hydroxyphenylbenzotriazoles*
- *Organic nickel compounds*

Carbon black and other pigments are also often used to protect plastics from UV rays.

7.4.4 PVC Stabilizers

Polyvinyl chloride is probably the polymer most vulnerable to thermal degradation. In polyvinyl chloride, scission of the C-Cl bond occurs in the weakest point of the molecule. The chlorine radicals react with their nearest CH group, forming HCl and creating new weak C-Cl bonds. During degradation, the macromolecules will also cross-link and the material will undergo discoloration. As expected, degradation will lead to significant reductions in physical performance of the material.

A stabilizer must therefore be used to neutralize HCl and stop the autocatalytic reaction, as well as preventing corrosion of processing equipment. Historically, lead cadmium containing stabilizers were used to stabilize thermal degradation of PVC-U. Some of these were barium-cadmium (BaCd) and lead-barium-cadmium (PbBa Cd). Today, these stabilizers have been replaced by lead-free systems. Some of the most widely used PVC stabilizers are:

- *Alkyltim stabilizers -*
- *Mixed metal stabilizers -* Some metal carboxylates derived from K, Ca or Ba act as HCl scavengers. Metals such as Zn and Cd serve as scavengers and substitute corboxylate for allytic chlorine atoms.
- *Alkyl phosphites stabilizers -* These types of stabilizers serve as scavengers of HCl through a reaction that forms dialkyl phosphites.

7.5 ANTISTATIC AGENTS

Since polymers have such low electrical conductivity, they can easily build up electric charges. The amount of charge build-up is controlled by the rate at which the charge is generated compared to the charge decay. The rate of charge generation at the surface of the component can be reduced by reducing the intimacy of contact, whereas the rate of charge decay is increased through surface conductivity. Hence, a good antistatic agent should be an ionizable additive that allows the charge to migrate to the surface. At the same time it should be creating bridges to the atmosphere through moisture in the surroundings.

Antistatic agents can be apppplied externally to the polymer surface by spraying or dipcoating, dissolved in an appropriate solvent carrier. The advantages of externally applied antistatic agents is that relatively small amounts are needed and that the effect of the additive is immediate. However, externally applied agents wear off due to friction, moisture run off, and other external factors. The internally applied antistatic agents are directly blended into the polymer matrix. These either work as lubricants that lower the surface friction, reducing the static charge generation, or as conductors, by creating a water absorbing layer on the surface of the polymer that creates a conductive path to the atmosphere.

The main categories of antistatic agents are non-ionic and anionic. The main non-ionic antistatic agents are:

- *Fatty acid esters* - Most commonly used fatty acid ester in polyolefin applications is glycerol monostearate (GMS). GMS primarely acts as a lubricant and meets most food and drug agencies requirements and is therefore often used in food packaging applications.
- *Ethoxylated alkylamines* - Since it is very compatible with polyolefins, it migrates much slower to the surface than GMS. It takes longer to take effect, but it provides excellent long term performance. Used with GMS it generates a synergistic effect.
- *Diethanolamides* - This agent does not meet all food and drug regulating agencies requirements for food packaging applications. A major area of application is in electronic packaging because they are non-corrosive and because they perform well even at low relative humidity.
- *Ethoxylated alcohol* - Primarily used in PVC-P applications. In addition to their antistatic attributes, they also lower the viscosity of the PVC resin.

Anionic antistatic agents are used mostly in conjuction with more polar polymers such as PVC and styrenics. The main anionic antistatic agents are *alkylsulfonates* and *alkylphosphates*.

7.6 ANTIMICROBIAL AGENTS

Microbial or bacterial growth is very high in wet and damp environments such as bathrooms or kitchens, and greenhouses, ponds and awnings. This growth leads to health hazzards, are unsightly and deteriorate the performance of the plastic. Antimicrobial agents, also referred to as *fungizides* are used to hinder the growth and propagation of bacteria such as E-Coli and salmonella in polyolefin films, and the formation and propagation of fungi, yeasts and molds in PVC-P.

Plasticized PVC, PE-LD, and polyesters are particularly susceptible to microbial attack. PVC and PU foams are by far the main plastics where fungizides are incorporated. While PVC itself is resistant to microbial attack, the plasticizers used often serve as nutrients to fungal growth. The platicizers most susceptible to microbial attack are: Sebacates, epoxidized oils, polyesters, and glycolates. The least susceptible ones are: phthalates, phosphates, and chlorinated hydrocarbons.

Today, because of their price and efficacy, over half of antimicrobial agents, also known as biostabilizers, are arsenic-based products, particularly *10,10'-oxybisphenoxarsine* (OBPA). However, due to ecological concerns and the toxicity of OBPA biotabilizers, non-arsenic-based products are gaining a share in the market. The fastest growing ones are isothiazolins, and somewhat less N-(trichloromethyl-thio)phthalamide. Research is under way on natural substances or enzymes, such as peroxidase, for plastics in food contact applications.

7.7 ANTIFOGGING AGENTS

Fogging or condensation will occur when a surface cools down below the wet bulb or dew point temperature of its surrounding air. Hence, this phenomenon depends on the environment's relative humidity. In plastics applications such as food packaging, agricultural film and greenhouses, to name a few, this effect is critical and must sometimes be avoided. The equilibrium position of each droplet is dominated by the difference between the surface tension of the polymer ($\sigma_s^{PP} = 0.030$N/m, $\sigma_s^{PE} = 0.035$N/m) and the surface tension of water ($\sigma_s^{Water} = 0.072$N/m) as schematicall depicted in Fig.3.67 (Chapter 3).

The antifogging agent is either externally applied to the polymer surface by spraying or dipcoating, or it is blended into the polymer and allowed to migrate to the surface. These agents make the surface of the polymer more polar, raising its surface tension. The antifogging agent in-turn disolves with the water, lowering the surface tension of water. As the surface tension difference between the polymer surface and the water decreases, so does the wetting angle. This angle becomes zero when the difference between the two surface tensions is zero. At that point, the water condensate spreads into a thin continuous film. In green houses and agricultural films, this causes the water to run off the surface into gutters. This means that the water will wash off the antifogging agent, making the internally blended fogging agents appropriate for these applications.

Fogging agents are chosen according to the application. Factors that are cosidered when choosing the agent are polymer type, thickness of film, whether the film will be in contact with food, as well as FDA (in the US) approval, or food and drug regulations in the intended country of application. The most commonly used fogging agents are:

- Glycerol esters
- Polyglycerol esters
- Sorbitan esters and their ethoxylates
- Nonyl phenol ethoxylates (not approved by European countries food and drug regulating agencies)
- Alcohol ethoxylates

7.8 BLOWING AGENTS

The task of blowing or foaming agents is to produce cellular polymers, also called expanded plastics. The cells can be completely enclosed (closed cell) or can be interconnected (open cell). Polymer foams are produced with densities between 1.6 kg/m^3 and 960 kg/m^3. There are many reasons for using polymer foams, such as their high strength to weight ratio, excellent insulating and acoustic properties, and high energy and vibration absorbing properties.

Polymer foams can be made by mechanically whipping gases into the polymer, or by either chemical or physical means. The basic steps of the foaming process are (1) cell nucleation, (2) expansion or growth of the cells and (3) stabilization of the cells. Cell nucleation occurs when, at a given temperature and pressure, the solubility of a gas is reduced, leading to saturation, and expelling the excess gas to form bubbles. Nucleating agents are used for initial formation of a bubble. The bubble reaches an equilibrium shape when the pressure in the bubble balances the surface tension.

When subjected to high temperatures, chemical blowing agents undergo a decomposing chemical reaction that liberates the blowing gas. Physical blowing agents do not undergo a chemical reaction and transformation. The gas is liberated by a physical process; either vaporization of a low boiling point liquid or by a pressure reduction of a compressed gas, previously introduced and diffused into the polymer melt.

In this chapter, we will concentrate on *chemical blowing agents*. The most commonly used chemical agents are:

- *Azodicarbonamide* - Depending on the grain size, this foaming agent is a pale yellow to orange powder with a density of 1.65 g/cm^3, a specific heat of 1087 J/kg/K, and a decomposition temperature of 205 to 215°C. A gas yield of 220 ml/g makes it one of the most economical blowing agents in the market.

- *Hydrazine derivatives* -
  p-Toluenesulfonylhydrazide. This foaming agent is a slightly yellow, crystalline powder with a density of 1.42 g/cm^3 and a decomposition temperature of 110 to 120°C. The gas yield is 115 ml/g with volatile products of nitrogen and water.
  Benzenesulfonylhydrazide. This foaming agent is a white, crystalline powder with a density of 1.53 g/cm^3, a specific heat of 878 J/kg/K and a decomposition temperature of 157 to 160°C. The gas yield is 125 ml/g.

- *Semicarbazides* - This foaming agent is a white crystalline powder with a density of 1.44 g/cm^3 and a decomposition temperature of 228 to 235°C in air, but only 213 to 225°C during processing. The gas yield is 140 ml/g.

- *Tetrazoles* - This foaming agent is a white crystalline powder with a density of 1.42 g/cm^3 and a decomposition temperature of 240 to 250°C. The gas yield is 210 ml/g at an experimental temperature of 260 to 265°C.
- *Nitroso compounds* - This foaming agent is a yellow crystalline powder with a density of 1.45 g/cm^3 and a decomposition temperature 190 to 205°C. The gas yield is of 190 to 210 ml/g.
- *Carbonates* - Sodium bicarbonate is one of the oldest blowing agents often used in household kitchens. It is a white powder with a decomposition temperature of 130 to 180°C and a gas yield of 125 ml/g.

7.9 COLORANTS

Colorants can be found in two general categories: pigments and dyes. Pigments are organic or inorganic materials that are practically incompatible with the polymeric material and that must be dispersed into the melt using relativelly intensive mixing operations. Pigments are found in several forms, such as powder pigments, pigment granulate, liquid or solid pigment concentrates and masterbatches tailored by compunders. On the other hand, dyes are compatible with the polymer melt. Because dyes dissolve in the resin, there are no visible particles, therefore they not affecting the transparency of the material.

Pigment suppliers assess their product using various criteria. The most important ones are:

- **Color Index (CI)** - The color index is a classification used for pigments and dyes. Within this classification, the generic names of *C.I. Pigment* and *C.I. Solvent* are used, e.g., C.I. Pigment 254 (for a red pigment) and C.I. Solvent Yellow 21.
- **Heat resistance** - High tempearture can affect the performance of a colorant. Hence a supplier will always give the highest processing temperature and the allowable time of exposure.
- **Light fastness** - Most of the time it is desired that the color of a part remains unchanged after an extended period of exposure to light. However, UV rays will ofen affect the color. The light fastness is rated between 8, for outstanding properties, and 1, for very poor performance.
- **Weather fastness** - As discussed in Chapter 3, the environment can play a significant role in the performance of a part. In addition to UV rays, IR rays, moisture, rain, and temperature fluctuations can affect the color quality of a part. Many of the weather fastness tests are performed in Florida, with samples facing south.
- **Migration** - Many pigments and dyes can migrate to the surface over time. The phenomenon is often referred to as *blooming* or *bleeding*.

7.9 Colorants

- **Abrasion** - Many inorganic pigments such as titanium dioxide are very abrasive, causing long-term damage to screws, mixing heads and other processing components.
- **Plate-out** - This is a phenomenon where the incompatible pigment or additive will deposit itself on the surface of the processing equipment, particularly on calendering rolls, eventually resulting in poor film quality.
- **Chalking** - Often, when too much pigment is used, the surface of a part rapidly degrades during a weathering test. This may be caused by a reaction between titanium oxide and moisture.

Some of the most commonly used inorganic pigments are presented in Table 7.2 while Table 7.3 presents a break-down of presently used inorganic and organic colorants.

Table 7.2: Common Inorganic Pigments

Pigment	Chemical family	Highest temperature (°C)	Chemical formula
White	Lithopone	300	$ZnS*BaSO_4$
	Titan dioxide	300	TiO_2
	Zinc oxide		ZnO
Black	Iron oxide black	240	$Fe_3O_4, (Fe,Mn)_2O_4$
	Spinnel Black	300	$Cu(Cr,Fe)_2O_4$
			$Cu(Cr,Mn)_2O_4$
			$(Fe,Co)Fe_2O_4$
Yellow-Orange	Cadmium yellow[a]	300	$CdS, (Cd,Zn)S$
	Chrome yellow[b]	260-290	$PbCrO_4$
			$Pb(CrS)O_4$
	Chromerutil yellow	300	$(Ti,Sb,Cr)O_2$,
			$(Ti,Nb,Cr)O_2$
			$(Ti,W,Cr)O_2$
	Iron oxide yellow	220-260	$\alpha\text{-FeO(OH)}$
			$\mathcal{X}\text{-FeO(OH)}$
	Nickelrutil yellow	300	$(Ti,Sb,Ni)O_2$
			$(Ti,Nb,Ni)O_2$
	Bismutvanadate/molybdate	280	$BiVO_4Bi_2MoO_6$
	Zinc ferrite		$ZnFe_2O_4$
Brown	Chrome iron brown	300	$(Fe,Cr)_2O_3$
	Iron oxide Mangan brown	260-300	$(Fe,Mn)_2O_3$
	Rutil brown		$(Ti,Mn,Sb)O_2$,
			$(Ti,Mn,Cr,Sb)O_2$
	Zinc ferrite brown	260	$ZnFe_2O_4$
Red	Cadmium red-orange[a,c]	300	$Cd(S,Se)(Cd,Hg)S$
	Iron oxide red	300	αFe_2O_4

Continued on next page

Pigment	Chemical family	Highest temperature (°C)	Chemical formula
	Molibdate red[b]	260-300	$Pb(Cr,Mo,S)O_4$
Green	Chrome oxide green	300	Cr_2O_3
	Cobalt spinell green	300	$(Co,Ni,Zn)_2(Ti,Al)O_4$
Blue	Cobalt blue	300	$CoAl_2O_4$
			$Co(Al,Cr)O_4$
	Ultramarine blue	300	$Na_8(Al_6,Si_6,O_{24})S_x$
Metallic	Aluminum	300	Al
	Copper	260	Cu-Zn alloy

[a] Should be avoided due to cadmium content
[b] Should be avoided due to lead content
[c] Should be avoided due to mercury content

Table 7.3: Break-Down of Colorants Used

Colorant	Share (%)
Titan dioxide	72
Carbon black	14
Inorganic color pigments	8
Organic color pigments	4.5
Dyes	1.5

7.10 FLUORESCENT WHITENING AGENTS

Fluorescent whitening agents (FWA) are used in order to mask the natural yellowish color of plastics, improve the whiteness, or prepare the polymers to better incorporate colorants. Many of the factors that affect colorants also apply when selecting FWA's. Hence, one must consider factors such as achievable whitening effect, campatibility and light fastness during the decision process.

The most commonly used FWA's are bis-benzoxazoles, phenylcoumarins and bis-(styril)biphenyls and are incorporated at fairly low concentrations, between 50 and 500 ppm.

7.11 FILLERS

Fillers can be classified three ways: those that reinforce the polymer and improve its mechanical performance; those used to take-up space and thus reduce the amount of resin to produce a part - sometimes referred to as extenders; and those, less common, that are dispersed through the polymer to improve its electric conductivity.

7.11 Fillers

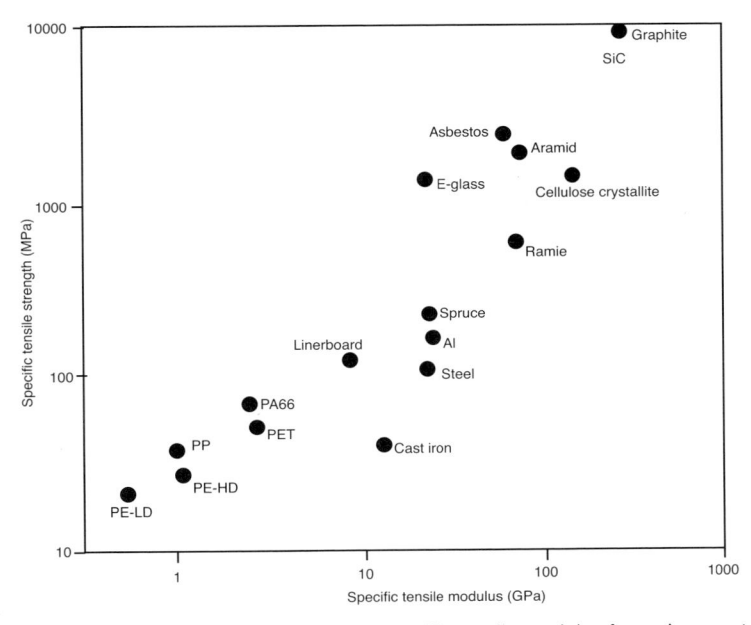

Figure 7.1: Specific tensile strength versus specific tensile modulus for various materials.

Polymers that contain fillers that improve their mechanical performance are often referred to as reinforced plastics or composites. Composites can be furthermore divided into composites with high performance reinforcements, and composites with low performance reinforcements. The high performance composites are those in which the reinforcement is placed inside the polymer so that optimal mechanical behavior is achieved, such as unidirectional glass fibers in an epoxy resin. High performance composites usually have 50 to 80% reinforcement by volume and usually have a laminated tubular shape containing braided reinforcements. The low performance composites are those in which the reinforcement is small enough that it can be well dispersed into the matrix. These materials can be processed the same way as their unreinforced counterparts. As an engineering guide, Fig. 7.1 presents several fillers and fibers on a specific tensile strength versus specific tensile modulus diagram.

The most common filler used to reinforce polymeric materials is glass fiber. However, wood fiber, which is commonly used as an extender, also increases the stiffness and mechanical performance of some thermoplastics. To improve the bonding between the polymer matrix and the reinforcement, coupling agents such as silanes and titanates are often added.

Extenders, used to reduce the cost of the component, are often particulate fillers. The most common of these are calcium carbonate, silica flour, clay, and wood flour or fiber. As mentioned earlier, some fillers also slightly reinforce the polymer matrix, such as clay, silica flour, and wood fiber. Polymers with extenders often have significantly lower toughness than when unfilled.

Table 7.4: Properties of Various Fillers

Filler	Shape	Diameter (μ m)	Density (g/cm^3)	L/D ratio	Specific Area (m^2/g)	Surface tension (mN/m)	E-moldulus (GPa)
Aerosil	Spheres	0.01	2.2	1	380	800	
Aluminiumnitrite	Powder	20-150					
Aluminiumsilicate	Spheres	<300	0.63	1			
Calciumcarbonate	Cubes	3-7,(0.7)<30	2.7	ca. 1		200	35
Calciumsulfate	Cubes	4	2.96				
Glass	Spheres	5	2.5	1	1.3	800	70
Glass	Hollow spheres	2-250	0.2-1.1				
Porcelain clay	Plateletss	1-10		<10	1040		
Ceramic	Hollow spheres	10-300	0.4-0.7				
Nano-glass particles	Spheres	0.25	2.5	1			
Nano-whisker	Needles	0.35	3.3	50		650	280
Talc	Platelets	10,<30	2.8	2-20	3.6	80	20
Wollastonite	Needles	<20	2.85	10-50			

Table 7.5: Chemical Formula for Various Fillers

Filler	Chemical Symbol
Aluminumhydroxide	Al(Oh)$_3$
Aluminumnitrite	AlN
Aluminumsilicate	AlSi
Bariumsulfate	BaSO$_4$
Calciumcarbonate	CaCO$_3$
Calciumsulfate	CaSO$_4$
Porcelain clay	Al$_2$(Si$_2$O$_5$)(OH)$_4$
Quarz	SiO$_2$
Talc	Mg$_6$(OH)$_4$(Si$_8$O$_{20}$)
Tonerde clay	Al$_2$O$_3$
Wollastonite	CaSiO$_3$

7.11 Fillers

Table 7.6: Effect of Various Fillers on the Polymer Matrix

Property	1	2	3	4	5	6	7	8	9	10	11	12	13	14	15	16	17
Tensile strength	++	+		+	+−			+	○					+			
Comprehensive strength	+								+		+	+		+	+		
E-modulus	++	++	++	++	+			++	+		+	+		+	+	+	+
Impact strength	−+	−	−	−	−	++	+	−+	−		−	−	−	−	−+	−	+
Reduced thermal expansion	+	+			+			+	+		+	+	+		+		
Reduced shrinkage	+	+	+	+				+	+	+	+	+	+	+	+	+	+
Thermal conduction		+	+	+				+	+	+	+			+		+	
HDT	++	+	+	++				+	+			+		+	+		
Electric conductance				+					+								+
Electric strength		+						++	+		+	++			+		
Thermal resistance		+						+	+		+	+	+			+	+
Chemical resistance		+	+					+	○	+		+	+				
Wear resistance				+				+	+	+		+					
Extrusion speed	−+	+						+				+		+			
Abrasion to equipment	+	○			○	○	○		○	○	−			○	○		○
Cost reduction	+	+	+				+	+	+	+	++	+	+	+	++		

(++) Significant improvement, (+) Positive effect, (○) No influence, (−) Adverse effect

1. Textile glass, 2. Asbestos, 3. Wollastonite 4. Carbon fibers, 5. Whiskers, 6. Synthetic fibers 7. Cellulose, 8. Mica, 9. Talc, 10. Graphite, 11. Sand -/quarz powder, 12. Silica, 13. Clay, 14. Glass beads, 15. Calcium carbonate, 16. Metal oxides, 17. Carbon Black.

APPENDIX A

MATERIAL PROPERTY TABLES

Table A.1: Processing Relevant Characteristic Values for Selected Plastics (Injection Molding)

Plastic	T_{P*} °C	Drying T/t °C/h	Mold temperature °C	Shrinkage %	Flow length**
PE-LD	160-220	-	20-60	1.5-5.0	550-600
PE-HD	180-250	-	10-60	1.5-3.0	200-600
EVAC	130-240	-	10-50	0.8-2.2	320
PP	200-270	-	20-90	1.3-2.5	250-700
PB	220-290	-	10-60	1.5-2.6	300-800
PIB	150-250	-	50-80	-	
PMP	280-310	-	≈ 70	1.5-3.0	
PS	170-280	-	10-60	0.4-0.7	200-500
SAN	200-260	85/2-4	50-80	0.4-0.7	-
SB	190-280	-	10-80	0.4-0.7	200-500
ABS	200-275	70-80/2[2)]	50-90	0.4-0.7	320
ASA	220-260	70-80/2-4	50-85	0.4-0.7	-
PVC-U	170-210	-	20-60	0.4-0.8	160-250
PVC-P	160-190	-	20-60	0.7-3.0	150-500
PMMA	190-270	70-100/2 to 6	40-90	0.3-0.8	200-500
POM	180-230	110/2[2)]	60-120	1.5-2.5	500
PA6	240-290	80/8-15[2)]	40-120	0.8-2.5	400-600
PA66	260-300	80/8-15[2)]	40-120	0.8-2.5	810
PA610	220-260	80/8-15	40-120	0.8-2.0	-
PA11	200-250	70-80/4-6	40-80	1.0-2.0	-
PA12	190-270	100/4[2)]	20-100	1.0-2.0	200-500
PA6-3-T	250-320	80-90/10	70-90	0.5-0.6	-
PC	270-320	110-120/4	80-120	0.6-0.8	150-220
PET-K	260-300	120/4[2)]	130-150 20[3)]	1.2-2.0 0.2[3)]	200-500
PBT	230-280	120/[2)]	40-80	1.0-2.2	250-600
PPE + PS	240-300	100/2	40-110	0.5-0.8	260
PSU	340-390	120/5	100-160	0.6-0.8	-
PPS	320-380	-	20-200	≈ 0.2	-
PES	320-390	160/5	100-190	0.2-0.5	-
PVDF	250-270	-	90-100	3-6	-
PTFE	320-360	-	200-230	3.5-6.0	-
Perfluoroalkoxy normal	380-400	-	95-230	3.5-5.5	80-120
PEEK	350-390	150/3	120-150	≈ 1	-
PA I	330-380	180/8	≈ 230	-	-
PE I	340-425	150/4	65-175	0.4-0.7	-
PEK	360-420	150/3	120-160	≈ 1	-
CA	180-220	80/2-4	40-80	0.4-0.7	350
CP	190-230	80/2-4	40-80	0.4-0.7	500
CAB	180-220	80/2-4	40-80	0.4-0.7	500
PE type 31	60-80	-	170-190	1.2	-

Continued on next page

Plastic	T_P* °C	Drying T/t °C/h	Mold temperature °C	Shrinkage %	Flow length**
MF type 131	70-80	-	150-165	1.2-2	-
MF/PF type 180/82	60-80	-	160-180	0.8-1.8	-
UP type 802	40-60	-	150-170	0.5-0.8	-
EP type 891	≈ 70	-	160-170	0.2	-
TPO	180-200	75/2 or 65/3	10-80	1.5-2.0	-
SBS	175-250	-	10-90	0.3-2.2	-
TPA	170-230	110/2-4 or 100/3-6	15-80	1.0-2.0	-
TPC	160-220				-
TPS	180-220				-
TPU	180-220	110/0.5 or 100/2	20-40	0.8-1.5	-
PP-GF20	-	-	-	1.2-2.0	-
PS-HI	-	-	-	0.4-0.7	-
SAN-GF30	-	-	-	0.2-0.3	-
ABS-GF30	-	-	-	0.1-0.3	-
POM-GF30	-	-	-	0.5-1.0	-
PA6-GF 30	-	-	-	0.2-1.2	-
PA66	-	-	-	0.2-1.2	-
PET-A	-	-	-	0.2	-
PET-K-GF-30	-	-	-	0.2-2.0	-
PBT-GF-30	-	-	-	0.5-1.5	-
PPE+PS-GF-30	-	-	-	0.2	-
PSU-GF40	-	-	-	0.2-0.4	-
PPS-GF-40	-	-	-	0.2	-
PES-GF-40	-	-	-	0.15	-
PEK	-	-	-	0.7-0.9	-
PEK-GF-30	-	-	-	0.3-0.8	-
PC-GF-30	-	-	-	0.2-0.4	-

* Processing temperature, ** flow length at average melt temperature, melt pressure, and mold wall temperature; ***sometimes not necessary because delivered in pre-dried form; [3] for amorphous grades

Table A.2: Shrinkage Characteristic Values and Tolerance Groups for Plastic Resins (according to DIN 16 901)

Shrinkage characteristic value	Dimensional tolerance groups (according to Table A.3)			
	No tolerance value provided	Tolerance value Row 1	Row 2	
				Amorphous thermoplastics
0-1	130	120	110	PS; SB; SAN; ABS*; ASA; PPE+PS*; PPE; PMMA; PVC-U; PA*; PC*; PET; PES; PSU*; PPS-GF
1-2	140	130	120	EVAC; PPEPDM; TPU 50 Shore D; CA, CAB, CAP, CP;
2-3	150	140	130	TPU 70-90 Shore A
3-4	160	150	140	
				Semi-crystalline thermoplastics
0-1	130	120	110	PA6/66/610/11/12-GF; PET-GF; PBT-GF: POM-GF
1-2	140	130	120	PP-GF >4 mm; PA 6/66/610/11/12; PET, PBT, POM >150 mm
2-3	150	140	130	PE, PP 4 mm; PP-GF >4 mm; fluorinated PEPP; POM >150 mm
3-4	160	150	140	PE, PP >4 mm
				Thermoset resins
0-1	130	120	110	EP; PF, MF, DAP with inorganic fillers; UP-GF
1-2	140	130	120	PF, UF, MF, DAP with organic fillers; UP-GF, MPF; PF-cold molding resins; UP-Prepreg
2-3	150	140	130	
3-4	160	150	140	

*Reinforced and non-reinforced

Table A.3: Shrinkage and Tolerances for Polymers

Tolerance group Table A.2	Code letter*	Nominal dimension (mm)					
		over 0 up to 1	1 3	3 6	6 10	10 15	15 22	22 30
		Allowable deviation					
160	A	±0.28	±0.30	±0.33	±0.37	±0.42	±0.49	±0.57
	B	±0.18	±0.20	±0.23	±0.27	±0.32	±0.39	±0.47
150	A	±0.23	±0.25	±0.27	±0.30	±0.34	±0.38	±0.43
	B	±0.13	±0.15	±0.17	±0.20	±0.24	±0.28	±0.33
140	A	±0.20	±0.21	±0.22	±0.24	±0.27	±0.30	±0.34
	B	±0.10	0.11	0.12	0.14	0.17	±0.20	±0.24
130	A	±0.18	±0.19	±0.20	±0.21	±0.23	±0.25	±0.27
	B	±0.08	±0.09	±0.10	±0.11	±0.13	±0.15	±0.17
		Tolerance					
160	A	0.56	0.6	0.66	0.74	0.84	0.98	1.14
	B	0.36	0.4	0.46	0.54	0.64	0.78	0.94
150	A	0.46	0.5	0.54	0.6	0.68	0.76	0.86
	B	0.26	0.3	0.34	0.4	0.48	0.56	0.66
140	A	0.4	0.42	0.44	0.48	0.54	0.6	0.68
	B	0.2	0.22	0.24	0.28	0.34	0.4	0.48
130	A	0.36	0.38	0.4	0.42	0.46	0.5	0.54
	B	0.16	0.18	0.2	0.22	0.26	0.3	0.34
120	A	0.32	0.34	0.36	0.38	0.4	0.42	0.46
	B	0.12	0.14	0.16	0.18	0.2	0.22	0.28
110	A	0.18	0.2	0.22	0.24	0.26	0.28	0.3
	B	0.08	0.1	0.12	0.14	0.16	0.18	0.2
Precision engineering	A	0.1	0.12	0.14	0.16	0.2	0.22	0.24
	B	0.05	0.06	0.07	0.08	0.1	0.12	0.14

Continued on next page

Group	Code	Nominal dimension (mm)						
		30–40	40–53	53–70	70–90	90–120	120–160	160–200
		Allowable deviation						
160	A	±0.66	±0.78	±0.94	±1.15	±1.40	±1.80	±2.20
	B	±0.58	±0.68	±0.84	±1.05	±1.30	±1.70	±2.10
150	A	±0.49	±0.57	±0.68	±0.81	±0.97	±1.20	±1.50
	B	±0.39	±0.47	±0.58	±0.71	±0.87	±1.10	±1.40
140	A	±0.38	±0.43	±0.50	±0.60	±0.70	±0.85	±1.05
	B	±0.28	±0.33	±0.40	±0.50	±0.60	±0.75	±0.95
130	A	±0.30	±0.34	±0.38	±0.44	±0.51	±0.60	±0.70
	B	±0.20	±0.24	±0.28	±0.34	±0.41	±0.50	±0.60
		Tolerance						
160	A	1.32	1.56	1.88	2.3	2.8	3.6	4.4
	B	1.12	1.36	1.68	2.1	2.6	3.4	4.2
150	A	0.98	1.14	1.36	1.62	1.94	2.4	3
	B	0.78	0.94	1.16	1.42	1.74	2.2	2.8
140	A	0.76	0.86	1	1.2	1.4	1.7	2.1
	B	0.56	0.68	0.8	1	1.2	1.5	1.9
130	A	0.6	0.88	0.76	0.88	1.02	1.2	1.5
	B	0.4	0.48	0.58	0.68	0.82	1	1.3
120	A	0.5	0.54	0.6	0.68	0.78	0.9	1.06
	B	0.3	0.34	0.4	0.48	0.58	0.7	0.86
110	A	0.32	0.36	0.4	0.44	0.5	0.58	0.68
	B	0.22	0.26	0.3	0.34	0.4	0.48	0.58
Precision engineering	A	0.26	0.28	0.31	0.35	0.4	0.5	
	B	0.16	0.18	0.21	0.25	0.3	0.4	

Continued on next page

Group	Code	Nominal dimension (mm)						
		200	250	315	400	500	630	800
		250	315	400	500	630	800	1000
		Allowable deviation						
160	A	±2.70	±3.30	±4.10	±5.10	±6.30	±7.90	±10.00
	B	±2.60	±3.20	±4.00	±5.00	±6.20	±7.80	±9.90
150	A	±1.80	±2.20	±2.80	±3.40	±4.30	±5.30	±6.60
	B	±1.70	±2.10	±2.70	±3.30	±4.20	±5.20	±6.50
140	A	±1.25	±1.55	±1.90	±2.30	±2.90	±3.60	±4.50
	B	±1.15	±1.45	±1.80	±2.20	±2.80	±3.50	±4.40
130	A	±0.90	±1.10	±1.30	±1.60	±2.00	±2.50	±3.00
	B	±0.80	±1.00	±1.20	±1.50	±1.90	±2.40	±2.90
		Tolerance						
160	A	5.4	6.6	8.2	10.2	12.5	15.8	20
	B	5.2	6.4	8	10	12.3	15.6	19.8
150	A	3.6	4.4	5.6	6.8	8.6	10.6	13.2
	B	3.4	4.2	5.4	6.6	8.4	10.4	13
140	A	2.5	3.1	3.8	4.6	5.8	7.2	9
	B	2.3	2.9	3.6	4.4	5.6	7	8.8
130	A	1.8	2.2	2.6	3.2	3.9	4.9	6
	B	1.3	1.8	2	3	3.7	4.7	5.8
120	A	1.24	1.5	1.8	2.2	2.6	3.2	4
	B	1.04	1.3	1.6	2	2.4	3	3.6
110	A	0.8	0.96	1.16	1.4	1.7	2.1	2.6
	B	0.7	0.86	1.06	1.3	1.6	2	2.5

* A for non mold-related measurements; B for mold-related measurements

Table A.4: Possible Post-Processing Operations for Selected Plastics

Plastic	Postprocessing[1]				
	Galvanizing	Coating and printing	Hot embossing	Vapor coating	Solvent adhesion
PE-LD	-	+	++	+	-
PE-HD	-	+	++	+	-
EAVC	-	+	++	+	-
PP	++[4]	+	++	+	-
PB	-	+	++	++	+
PS	-	++	++	+	++
SAN	-	++	++	+	++
SB	+[4]	++	++	+	++
ABS	++[4]	++	++	+	++
ASA	-	++	++	+	++
PVC-U	+[4]	++	++	+	++
PVC-U-E	-	+	++	+	++
PVC-P	-	+	++	-	++
PVC-P-E	-	+	++	-	++
PMMA	-	++	++	+	++
POM	-	+	++	+	++[5]
PA 6	-	++	++	+	++[5]
PA 66	-	++	++	+	++[5]
PA 610	-	++	+	+	++[5]
PA 11			++	++	++[5]
PA 12	-	+	++	++	++
PA 6-3-T	-	++	++	+	++
PC	+	++	++	+	++
PET	-	++	++	+	-
PBT	-	+	++	-	-
PPE+PS	+	++	++	+	++
PSU	+	++	++	+	++
PPS	-	+	++	+	-
PES	-	++	++	+	++
PVDF	-	++	++	+	++
PTFE	-	+	+		-
PFA	+	+	+	+	-
PEEK		++	++	+	
PAI		++	++	+	
PEI		++	++	+	
PEK		++	++	+	
CA	-	++	++	+	++
CP	-	++	++	+	++
CAB	-	++	++	+	++
PF type 31		++	-	+	-
MF type 131		++	-	+	-
MF/PF type 180/182		++	-	+	-

Continued on next page

Plastic		Postprocessing[1]			
	Galvanizing	Coating and printing	Hot embossing	Vapor coating	Solvent adhesion
UP type 802		++	-	+	-
EP type 891		++	-	+	-
TPO	-	+	++	+	-
TPA	-	++	++	+	++
TPE	-	++	++	+	-
TPS	-	++	++	+	+
TPU	-	++	++	+	++

	Ultrasonic welding	Hot plate welding	Vibration welding	High frequency welding
PE-LD	-	+	+	-
PE-HD	+[3]	+	+	-
EAVC	o[3]	+	+	o
PP	+	+	+	-
PB	+[3]	+	+	-
PS	+	+	+	-
SAN	+	+	+	o
SB	+	+	+	o
ABS	+	+	+	+
ASA	+	+	+	o
PVC-U	+	+	+	+
PVC-U-E	o	+	o	+
PVC-P	-	+	-	+
PVC-P-E	-	o	-	+
PMMA	++	++	++	
POM	++	++	++[3]	
PA 6	+	+	+	+
PA 66	+	+	+	+
PA 610	+	+	+	+
PA 11	+	+	+	+
PA 12	+	+	+	+
PA 6-3-T	+	+	+	+
PC	+	+	+	-
PET	+	+	+	-
PBT	+	+	+	-
PPE+PS	+	+	+	-
PSU	+	+	+	-
PPS	+	+	+	-
PES	+	+	+	-
PVDF	o[3]	+	o	-
PTFE	-	-	-	-
PFA	-	+	o	-
PEEK	+	+	+	-
PAI	+	+	+	-
PEI	+	+	+	-

Continued on next page

Plastic	Postprocessing[1]			
	Ultrasonic welding	Hot plate welding	Vibration welding	High frequency welding
PEK	+	+	+	-
CA	o	+		+
CP	o	+		+
CAB	o	+	-	o
PF type 31	-	-	-	-
MF type 13	-	-	-	-
MF/PF type	-	-	-	-
UP type 80	-	-	-	-
EP type 89	-	-	-	-
TPO	o[3]	+	+	
TPA	o[3]	+	+	
TPE	o[3]	+	+	
TPS	+	+	+	
TPU	o[3]	+	+	

[1] Adhesion ++ = possible without pre-treatment, + = possible with pre-treatment, - = not possible
[2] Welding + = possible, o = possible under certain condiotions, - = not possible
[3] for ultrasonic - welding only possible in near field applications
[4] for certain grades
[5] not recommended

Table A.5: Diode Laser Welding of Thermoplastic Elastomers and Thermoplastics

	Fastened to							
	PR	HF	TF	PUR	PUR/GF	HW	BF	Exotic woods
PS		O	O	O	O		—	
ABS	—	—	—	—	—	—	—	—
PMMA		O	O	O	O			
PVC		—	O	O	O			
PPE+PA	—	—	—	—	—	—	—	
PPE+PS	—	—	—	O	O			
PC	—	—	—	O	O	—	—	—
PC+ABS	—	O	O	O	O			
PE-HD, LD		O	O	O	O			
PP	—	—	—	O	O	—		
PA	—	—	—	—	—	—	—	—
PAA		O	O	O	O			
POM	—	—	—	O	O			
PET		O	O	O	O			
PBT		O	O	O	O			
PBT+PC		O	O	O	O			
SAN		O	O	O	O			
PSU		O	O	O	O			
PEI		O	O	O	O			

PR = Particle board   — = Experimentally proven
HF = Wood fiber   O = Mechanical bond possible
TF = Textile fiber
GF = Long glass fiber
HW = Wood flour
BF = Bast fiber

Table A.6: Possible Combinations of TPE and Thermoplastics for Laser Welding

Rigid component	TPV	TPS	TPU
PVC	-	x*	x
PA 6	-	x*	x
PA 6.6	-	x*	x
PP	x	x	-
PE	x	x	-
POM	-	-	x
ABS	-	x*	x
PSU	x	-	-
PBT	-	x*	x
PMMA	-	x*	-
SAN	-	x*	x

-: no bond, x: bond
* modification with bonding agent necessary

Table A.7: Adhesive Bonding of Selected Plastics

Adhesive bond	Plastic	Polarity + polar - non polar	Solubility + soluble; - insoluble / Hard to extinguish	Possibility of Diffusion - bonding	Possibility of Adhesion - bonding
Good	PS	+/-	+	+[1]	+
	PVC-U	+	+	+	+
	PVC-P	+	+	+[2]	+
	PMMA	+	+	+/-	+
	PC	+	+	+	+
	ABS	+	+	+	+
	CA	+	+	+	+
	PUR	+	-	-	+
	UP	+	-	-	+
	EP	+	-	-	+
	PF	+	-	-	
	UF-MF	+	-	-	+
Under certain conditions	PA	+	-	+/-	+
	POM	+	-	-	+
	PET	+	-	-	+[3]
	Rubber	+	-	+/-	+
Difficult	PE	-	-	-	+/-[4]
	PP	-	-	-	+/-[4]
	PTFE	-	-	-	+/-[4]
	SI	+/-	-	-	+

[1] not possible for foamed PS; [2] follow instructions of PVC-P supplier; [3] after pre-treatment with caustic soda (80°C, 5 min); [4] only after pre-treatment, PP + PE reach parent material strength, low peel strength

Table A.8: Risk Potential of Solvent-Based Cleaners

Solvent	MAC-value mg/m^3	Ignition point °C	Saturation concentration g/m^3	Risk quotient RQ[3)
Trichloroethylene	-	-	417	[1)
Cyclohexanone	-	43	18	[2)
Methylene chloride	360	-	1535	4263
Methanol	260	11	168	646
Toluol	190	6	110	578
Methylethylketone	590	-1	311	527
Acetone	1200	-17	555	462
Dimethylformamide	30	58	10.6	353
Ethylacetate	1400	-4	350	250
Resin-Clean A[4)	80	95	2.1	26
Resin-Clean VF970[4)	80	93	1.9	23
Resin-Clean VF610[4)	160	98	1.5	9
Resin-Clean 5A-DE[4)	400	85	2	5
Resin-Clean E-90[4)	350	>100	0.5	1.5
Elasto-Clean E[4)	400	>100	0.5	1.2

[1) No RQ-value, because carcinogenic
[2) No RQ-value, because possibly carcinogenic
[3) Risk quotient = Saturation concentration [g/m^3] : MAC-value [µg/m^3] · 1000
[4) Organic solvent and cleaner formulation based on propoxilated alcohols, amines and esters (Färber & Schmid AG, Oetwil, Switzerland)

Table A.9: Selection of Cleaners for Plastics and Elastomers

Product[2]	Plastic or elastomer				
	PUR	Epoxide	Polyester	Silicone	PVC
Resin-Clean 5A-DE	++	+	+	o	+
Resin-Clean VF 970 [1]	++	+	++	o	++
Resin-Clean A [1]	++	+	++	o	++
Resin-Clean VF 610	++	+	+	o	+
Elasto-Clean E	++	+	+	o	+
Resin-Clean E-90 [1]	+	++	++	+	o

	GFK	Polyamide	PET	PMMA	Adhesives
Resin-Clean 5A-DE	+	-	+	+	+
Resin-Clean VF 970 [1]	++	o	++	++	++
Resin-Clean A [1]	++	o	++	++	++
Resin-Clean VF 610	++	o	+	++	+
Elasto-Clean E	o	-	+	+	-
Resin-Clean E-90 [1]	+	+	+	++	+

++ = excellent releasing force, + = good releasing force, o = moderate releasing force, - = poor releasing force

[1] These products also release cured resins
[2] see [4] in Table A.8

Table A.10: Characteristic Mechanical and Thermal Properties of Selected Plastics

Plastic	ρ g/cm^3	σ_B MPa	ϵ_B %	E_t MPa
PE-LD	0.914-0.92	8-23	300-1000	200-500
PE-HD	0.94-0.96	18-35	100-1000	700-1400
EVAC	0.92-0.95	10-20	600-900	7-120
IM	0.94	21-35	250-500	180-210
PVK	1.19	20-30	-	3500
PP	0.90-0.907	21-37	20-800	1100-1300
PB	0.905-0.92	30-38	250-280	250-350
PIB	0.91-0.93	2-6	>1000	-
PMP	0.83	25-28	13-22	1100-1500
PS	1.05	45-65	3-4	3200-3250
SAN	1.08	75	5	3600
SB	1.05	26-38	25-60	1800-2500
ABS	1.04-1.06	32-45	15-30	1900-2700
ASA	1.04	32	40	1800
PVC-U	1.38-1.55	50-75	10-50	1000-3500
PVC-P	1.16-1.35	10-25	170-400	-
PTFE	2.15-2.20	25-36	350-550	410
PFEP	2.12-2.17	22-28	250-330	350
PCTFE	2.10-2.12	32-40	120-175	1050-2100
PETFE	1.7	35-54	400-500	1100
PMMA	1.17-1.20	50-77	2-10	2700-3200
POM	1.41-1.42	62-70	25-70	2800-3200
PA6	1.13	70-85	200-300	1400
PA66	1.14	77-84	150-300	2000
PA11	1.04	56	500	1000
PA12	1.02	56-65	300	1600
PA6-3-T	1.12	70-84	70-150	2000
PC	1.2	56-67	100-130	2100-2400
PET	1.37	47	50-300	3100
PBT	1.31	40	15	2000
PPE+PS	1.06	55-68	50-60	2500
PSU	1.24	50-100	25-30	2600-2750
PPS	1.34	75	3	3400
PES	1.37	85	30-80	2450
PAI	-	100	12	4600
PEI	1.27	105	60	3000
PI	1.43	75-100		3000-3200
PEEK	1.32	90	50	3600
CA	1.3	38	3	2200
CP	1.19-1.23	14-55	30-100	420-1500
CAB	1.18	26	4	1600
VF	1.1-1.45	85-100	-	-

Continued on next page

Plastic	Properties			
	ρ	Mechanical Properties		
	g/cm^3	σ_B MPa	ϵ_B %	E_t MPa
PUR	1.05	70-80	3-6	4000
TPU	1.2	30-40	400-450	700
PF	1.4	25	0.4-0.8	5600-1200
UF	1.5	30	0.5-1.0	7000-10500
MF	1.5	30	0.6-0.9	4900-9100
UP	2	30	0.6-1.2	14000-20000
DAP	1.51-1.78	40-75	-	9800-15500
SI	1.8-1.9	28-46	-	6000-12000
PI	1.43	75-100	4-9	23000-28000
EP	1.9	30-40	4	21500

Table A.11: Characteristic Thermal Properties of Selected Plastics

Plastic	Properties		
	Thermal properties		
	α_t $K 1\cdot10^6$	k W/mK	C_p kJ/kgK
PE-LD	230-250	0.32-0.40	2.1-2.5
PE-HD	120-200	0.38-0.51	2.1-2.7
EVAC	160-200	0.35	2.3
IM	120	0.24	2.2
PVK	-	0.29	-
PP	110-170	0.17-0.22	2
PB	150	0.2	1.8
PIB	120	0.12-0.20	-
PMP	117	0.17	2.18
PS	60-80	0.18	1.3
SAN	80	0.18	1.3
SB	70	0.18	1.3
ABS	60-110	0.18	1.3
ASA	80-110	0.18	1.3
PVC-U	70-80	0.14-0.17	0.85-0.9
PVC-P	150-210	0.15	0.9-1.8
PTFE	120-250	0.25	1
PFEP	80	0.25	1.12
PCTFE	60	0.22	0.9
PETFE	40	0.23	0.9
PMMA	70-90	0.18	1.47
POM	90-110	0.25-0.30	1.46
PA6	60-100	0.29	1.7
PA66	70-100	0.23	1.7
PA11	130	0.23	1.26
PA12	110	0.23	1.26
PA6-3-T	80	0.23	1.6
PC	60-70	0.21	1.17
PET	40-60	0.24	1.05
PBT	60	0.21	1.3
PPE+PS	60-70	0.23	1.4
PSU	54	0.28	1.3
PPS	55	0.25	-
PES	55	0.18	1.1
PAI	36	0.26	
PEI	62	0.22	
PI	50-60	0.29-0.35	
PEEK	47	0.25	
CA	120	0.22	1.6
CP	110-130	0.21	1.7
CAB	120	0.21	1.6
VF	-	-	-
PUR	10-20	0.58	1.76

Continued on next page

Plastic	Properties		
	Thermal properties		
	α_t K $1\cdot 10^6$	k W/mK	C_p kJ/kgK
TPU	110-210	1.7	0.5
PF	10-50	0.35	1.3
UF	50-60	0.4	1.2
MF	50-60	0.5	1.2
UP	20-40	0.7	1.2
DAP	10-35	0.6	-
SI	20-50	0.3-04	0.8-0.9
PI	50-63	0.6-0.65	0.8
EP	11-35	0.88	

Table A.12: Electrical Properties of Plastics

Plastic	Specific volume resistance Ωcm	Electric surface resistance Ω	Dielectric constant 50 Hz	Dielectric constant 10^6 Hz	Dielectric loss factor tanδ 50 Hz	Dielectric loss factor tanδ 10^6 Hz
PE-LD	>10^{17}	10^{14}	2.29	2.28	$1.5 \cdot 10^{-4}$	$0.8 \cdot 10^{-4}$
PE-HD	>10^{17}	10^{14}	2.35	2.34	$2.4 \cdot 10^{-4}$	$2.0 \cdot 10^{-4}$
EVAC	<10^{15}	10^{13}	2.5-3.2	2.6-3.2	0.003-0.02	0.03-0.05
IM	>10^{16}	10^{13}				
PVK	>10^{16}	10^{14}	-	3	$6\text{-}10 \cdot 10^{-4}$	$6/10 \cdot 10^{-4}$
PP	>10^{17}	10^{13}	2.27	2.25	$<4 \cdot 10^{-4}$	$<5 \cdot 10^{-4}$
PB	>10^{17}	10^{13}	2.5	2.2	$7 \cdot 10^{-4}$	$6 \cdot 10^{-4}$
PIB	>10^{15}	10^{13}	2.3	-	$4 \cdot 10^{-4}$	-
PMP	>10^{16}	10^{13}	2.12	2.12	$7 \cdot 10^{-5}$	$3 \cdot 10^{-5}$
PS	>10^{16}	>10^{13}	2.5	2.5	$1\text{-}4 \cdot 10^{-4}$	$0.54 \cdot 10^{-4}$
SAN	>10^{16}	>10^{13}	2.6-3.4	2.6-3.1	$6\text{-}8 \cdot 10^{-3}$	$7\text{-}10 \cdot 10^{-3}$
SB	>10^{16}	>10^{13}	2.4-4.7	2.4-3.8	$4\text{-}20 \cdot 10^{-4}$	$4\text{-}20 \cdot 10^{-4}$
ABS	>10^{15}	>10^{13}	2.4-5	2.4-3.8	$3\text{-}8 \cdot 10^{-3}$	$2\text{-}15 \cdot 10^{-3}$
ASA	>10^{15}	>10^{13}	3-4	3-3.5	0.02-0.05	0.02-0.03
PVC-U	>10^{15}	10^{13}	3.5	3	0.011	0.015
PVC-P	>10^{11}	10^{11}	4.8	4-4.5	0.08	0.12
PTFE	>10^{18}	10^{17}	<2.1	<2.1	$<2 \cdot 10^{-4}$	$<2 \cdot 10^{-4}$
PFEP	>10^{18}	10^{17}	2.1	2.1	$<2 \cdot 10^{-4}$	$<7 \cdot 10^{-4}$
PCTFE	>10^{18}	10^{16}	2.3-2.8	2.3-2.5	$1 \cdot 10^{-3}$	$2 \cdot 10^{-2}$
PETFE	>10^{16}	10^{13}	2.6	2.6	$8 \cdot 10^{-4}$	$5 \cdot 10^{-3}$
PMMA	>10^{15}	>10^{15}	3.3-3.9	2.2-3.2	0.04-0.04	0.004-0.04
POM	>10^{15}	>10^{13}	3.7	3.7	0.005	0.005
PA6	10^{12}	10^{10}	3.8	3.4	0.01	0.03
PA66	10^{12}	10^{10}	8	4	0.14	0.08
PA11	10^{13}	10^{11}	3.7	3.5	0.06	0.04
PA12	10^{13}	10^{11}	4.2	3.1	0.04	0.03
PA6-3-T	10^{11}	10^{10}	4	3	0.03	0.04
PC	>10^{17}	>10^{15}	3	2.9	$7 \cdot 10^{-4}$	$1 \cdot 10^{-2}$
PET	10^{16}	10^{16}	4	4	$2 \cdot 10^{-3}$	$2 \cdot 10^{-2}$
PBT	10^{16}	10^{13}	3	3	$2 \cdot 10^{-3}$	$2 \cdot 10^{-2}$
PPE+PS	10^{16}	10^{14}	2.6	2.6	$4 \cdot 10^{-4}$	$9 \cdot 10^{-4}$
PSU	>10^{16}	-	3.1	3	$8 \cdot 10^{-4}$	$3 \cdot 10^{-3}$
PPS	>10^{16}	-	3.1	3.2	$4 \cdot 10^{-4}$	$7 \cdot 10^{-4}$
PES	10^{17}	-	3.5	3.5	$1 \cdot 10^{-3}$	$6 \cdot 10^{-3}$
PAI	10^{17}	-	-	-	-	-
PEI	10^{18}	-	-	-	-	-
PI	>10^{16}	>10^{15}	-	-	-	-
PEEK	$5 \cdot 10^{16}$	-	-	-	$3 \cdot 10^{-3}$	-
CA	10^{13}	10^{12}	5.8	4.6	0.02	0.03
CP	10^{16}	10^{14}	4.2	3.7	0.01	0.03
CAB	10^{16}	10^{14}	3.7	3.5	0.006	0.021

Continued on next page

Plastic	Properties						
	Specific volume resistance Ωcm	Electric surface resistance Ω	Dielectric constant		Dielectric loss factor $\tan\delta$		
			50 Hz	10^6 Hz	50 Hz	10^6 Hz	
VF	10^{10}	10^8	-	-	0.08	-	
PUR	10^{16}	10^{14}	3.6	3.4	0.05	0.05	
TPU	10^{12}	10^{11}	6.5	5.6	0.03	0.06	
PF	10^{11}	$>10^8$	6	4.5	0.1	0.03	
UF	10^{11}	$>10^{10}$	8	7	0.04	0.3	
MF	10^{11}	$>10^8$	9	8	0.06	0.03	
UP	$>10^{12}$	$>10^{10}$	6	5	0.04	0.02	
DAP	10^{13}-10^{16}	10^{13}	5.2	4	0.04	0.03	
SI	10^{14}	10^{12}	4	3.5	0.03	0.02	
PI	$>10^{16}$	$>10^{15}$	3.5	3.4	$2 \cdot 10^{-3}$	$5 \cdot 10^{-3}$	
EP	$>10^{14}$	$>10^{12}$	3.5-5	3.5-5	0.001	0.01	

	Dielectric strength	Creep resistance		
			Step	
	kV/cm	KA	KB	KC
PE-LD	-	3b	>600	>600
PE-HD	-	3c	>600	>600
EVAC	620-780	-	-	-
IM				
PVK	-	3b	>600	>600
PP	500-650	3c	>600	>600
PB	-	3c	>600	>600
PIB	-	3c	>600	>600
PMP	700	3c	>600	>600
PS	300-700	1-2	140	150-250
SAN	400-500	1-2	160	150-260
SB	300-600	2	>600	>600
ABS	350-500	3a	>600	>600
ASA	360-400	3a	>600	>600
PVC-U	350-500	2-3b	600	600
PVC-P	300-400	-	-	-
PTFE	480	3c	>600	>600
PFEP	550	3c	>600	>600
PCTFE	550	3c	>600	>600
PETFE	400	3c	>600	>600
PMMA	400-500	3c	>600	>600
POM	380-500	3b	>600	>600
PA6	400	3b	>600	>600
PA66	600	3b	>600	>600
PA11	425	3b	>600	>600
PA12	450	3b	>600	>600

Continued on next page

Plastic	Properties			
	Dielectric strength		Creep resistance	
			Step	
	kV/cm	KA	KB	KC
PA6-3-T	350	3b	>600	>600
PC	380	1	120-160	260-300
PET	420	2	-	-
PBT	420	3b	420	380
PPE+PS	450	1	300	300
PSU	425	1	175	175
PPS	595	-	-	-
PES	400	-	-	-
PAI	-	-	-	-
PEI	-	-	-	-
PI	-	1	>300	>380
PEEK	-	-	-	-
CA	400	3a	>600	>600
CP	400	3a	>600	>600
CAB	400	3a	>600	>600
VF	-	-	-	-
PUR	240	3c	-	-
TPU	300-600	3a	>600	>600
PF	300-400	1	140-18	125-175
UF	300-400	3a	>400	>600
MF	290-300	3b	>500	>600
UP	250-530	3c	>600	>600
DAP	400	3c	>600	>600
SI	200-400	3c	>600	>600
PI	560	1	>300	>380
EP	300-400	3c	>300	200-600

Table A.13: Properties of Rigid Foamed Plastics

Foaming process			Raw material		
			Tough-rigid		
			Polystyrene		Polyvinyl chloride
		Particle foam	Extruded foam without foam skin	with foam skin	High pressure foam
Density range	kg/m³	15-30	30-35	25-60	50-130
σ_B (comp.)	MPa	0.06-0.25	>0.15	>0.2	0.3-1.1
σ_B	MPa	0.15-0.5	0.5	>0.2	0.7-1.6
τ_B	MPa	0.09-0.22	0.9	1.2	0.5-1.2
σ_B (flex.)	MPa	0.16-0.5	0.4	0.6	0.6-1.4
E_f	MPa			>15	16-35
k	W/mK	0.032-0.037	0.025-0.035		0.036-0.04
Service T					
Short-term	°C	100	100		80
Long-term	°C	70-80	<75		60
W_w 7 days	Vol.-%	2-3	2	<1	

		Tough rigid			Brittle-rigid
		Polyether sulfone	Poly- urethane	Phenolic resin	Urea resin
		Block foamed	Block foamed without foam skin	with foam skin	Injection foam
Density range	kg/m³	45-55	20-100	40-100	5-15
σ_B (comp.)	MPa	0.6	0.1-0.9	0.2-0.9	0.01-0.05
σ_B	MPa	0.7	0.2-1.1	0.1-0.4	
τ_B	MPa	-	0.1->1	0.1-0.5	
σ_B (flex.)	MPa	0.2	0.2-1.5	0.2-1.0	0.03-0.09
E_f	MPa	3	2-20	6-27	
k	W/mK	0.05	0.018-0.02	0.02-0.03	0.03
Service T					
Short-term	°C	210	>150	>250	>100
Long-term	°C	180	80	130	90
W_w 7 days	Vol.-%	15	1-4	7-10	>20

Table A.14: Properties of Semi-Rigid to Soft-Elastic Foamed Plastics

Foaming process		Raw material		
		Predominantly closed celled		
			Polyethylene	
		Particle foam	Extrusion crosslinked	
Density range	kg/m³	25-40	30-70	100-200
σ_B	kg/m³	0.1-0.2	0.3-0.6	0.8-2.0
ϵ_B	%	30-50	90-110	130-200
Compression hardness (40%)	MPa	0.03-0.06	0.07-0.16	0.25-0.8
Compression set (70°C. 50%)	%	-	10-4	3
Impact resistance	%	40-50	45	-
Service T	°C	up to 100	-70 to 85	-60 to 110
k	W/mK	0.036	0.04-0.05	0.05
W_w 7days	Vol.-%	1-2	0.5	0.4
ϵ_r 50 Hz		1.05	1.1	1.1
$\tan \delta$ 50 Hz		0.0004	0.01	0.01

		Polyvinylchloride		Melamin resin
		High pressure foamed		Band-foamed
Density range	kg/m³	50-70	100	10.5-11.5
σ_B	kg/m³	0.3	0.5	0.01-0.15
ϵ_B	%	80	170	10-20
Compression hardness (40%)	MPa	0.02-0.04	0.05	0.007-0.013
Compression set (70°C. 50%)	%	33-35	32	≈ 10
Impact resistance	%		≈ 50	
Service T	°C	-60 to 50		up to 150
k	W/m/K	0.036	0.041	0.033
W_w 7 days	Vol.%	1-4	3	≈ 1
ϵ_r 50 Hz		1.31	1.45	-
$\tan \delta$ 50 Hz		0.06	0.05	-

		Open celled	
		Polyurethane	
		Block foamed	
		Polyester	Polyether
Density range	kg/m³	20-45	20-45
σ_B	MPa	≈ 0.2	≈ 0.1
ϵ_B	%	200-300	200-270
Compression hardness (40%)	MPa	0.003-0.00	0.002-0.004
Compression set (70°C. 50%)	%	4-20	≈ 4
Impact resistance	%	20-30	40-50
Service T	°C	-40 to 100	
k	W/m/K	0.04-0.05	
W_w 7 days	Vol.%	-	-
ϵ_r 50 Hz		1.45	1.38
$\tan \delta$ 50 Hz		0.008	0.003

Table A.15: Plastic Foam Properties

Plastic	ρ kg/m^3	k W/mK	σ_B comp. 10%ϵ MPa	σ_B comp. MPa	σ_B MPa	Service T °C
PS	15	0.037	0.07-0.12	-	0.15-0.23	70
	20	0.035	0.12-0.16	-	0.25-0.32	80
	30	0.032	0.18-0.26	-	0.37-0.52	70
PVC	50-130	0.036-0.04	-	0.3-1.1	-	60
PES	45-55	0.05	-	0.6	-	180
PF	40-100	0.02-0.03	-	0.2-0.9	-	130
MF	-	0.054	0.0284	-	-	150
	-	-	at 40%	-	-	-
UF spray foam	5-15	0.03	-	0.01-0.05	-	-

Table A.16: Conductive Components for Plastics

Component	In substrate:
Organic antistatic agents	Soluble under certain conditions
Low melting metal alloys	Insoluble
Pigments, fillers	Insoluble, chips, flakes, beads, platelets, etc.
Fibers, fabrics, whisker	Insoluble, length / diameter $\gg$ 1
Intrinsic conductive polymers (ICP)	Insoluble

Table A.17: Electrical Resistance of Various Conductive Additives for Selected Plastics

Product	Specific Resistance Ωcm
Metals: Platinum, iron, tin	0.000 01
copper, gold, aluminum	0.000 002
Compact carbon, graphite, vapor coatings	0.0003-0.006
Condensed carbon blacks: carbon black	0.04
Furnace carbon	0.05-0.1
Gas carbon	0.15-0.6
Metal oxide: SnO$_2$-powder, 1.1 μm	400
SnO$_2$ mit Sb2O5 doped	100
SnO$_2$/ Sb$_2$O$_5$ doped on TiO$_2$, 0.2 μm	10-300
Sb/Sn-oxide, transparent, 0.1 μm	1
Sb/Sn-oxide, transparent, 0.5 μm	10

Table A.18: Characteristic Temperature Values for Plastics

Plastic	Glass fiber content	HDT/A	Service Temperature °C			T_c/T_m	T_g
	%	°C	max. short term	max. long term	min. long term	°C	°C
PE-LD	0	≈ 35	80-90	60-75	-50	110	-30
PE-HD	0	≈ 50	90-120	70-80	-50	-	-
PE-UHMW	0	≈ 50	150	100	-260	-	-
PE-X	0	40-60	200	120	-	-	-
EVAC	0	-	65	55	-60	-	66
COC	0	-	-	-	-	-	60-180
EIM	0	-	120	100	-50	-	-
PP	0	55-70	140	100	0 to -30	160-170	0 to -10
PP	30	120	155	100	0 to -30	160-170	0 to -10
PB	0	55-60	130	90	0	-	-
PIB	0	-	80	65	-40	-	-70
PMP	0	40	180	120	0	245	-
PDCPD	0	90-115	-	-	-	-	-
PS	0	65-85	75-90	60-80	-10	-	95-100
PS-(M)	0	95	-	-	-	270	-
SAN	0	95-100	95	85	-20	-	110
SB	0	72-87	60-80	50-70	-20	-	-
ABS	0	95-105	85-100	75-85	-40	-	80-110
ASA	0	95-105	85-90	70-75	-40	-	100
PVC-U	0	65-75	75-90	65-70	-5	-	85
PVC-C	0	-	100	85	-	-	85
PVC-P	0	-	55-65	50-55	0 to -20	-	≈ 80
PVK	0	150-170	170	150	-100	-	173
PTFE	0	50-60	300	260	-270	327	127
PCTFE	0	65-75	180	150	-40	-	-
PVDF	0	95-110	-	150	-60	140	40
PVF	0	-	-	120	-60	198	-20
ECTFE	0	75	160	140	-75	190	45
ETFE	0	75	200	155	-190	270	-
ETFE	25	210	220	200	-	270	-
FEP	0	-	250	205	-200	290	-
PFA	0	45-50	250	200	-200	-	-
(Teflon AF)	0	-	300/570	260/500	1.16	-	160/240
THV	0	-	-	130	-50	160-180	-
PPE mod.	0	135	120-130	100-110	-	-	-
PPE mod.	30	160	-	-	-	-	-
PPE+PA 66	0	-	210	-	-	-	-
PMMA	0	75-105	85-100	65-90	-40	-	105-115
AMMA	0	73	80	70	-	-	80
POM-H	0	100-115	150	110	-40	175	25
POM-H	30	160	150	110	-60	175	25

Continued on next page

Plastic	Glass fiber content	HDT/A	Service Temperature °C			T_s/T_m	T_g
	%	°C	max. short term	max. long term	min. long term	°C	°C
POM-Copol.	0	110-125	110-140	90-110	-	165	-
POM-Copol.	30	160	110-150	90-110	-	165	-
PA 6	0	55-85	-	-	-	220	55
PA 6	30	190-215	140-180	80-110	-30	220	55
PA 11	0	55	140-150	70-80	-70	185	50
PA 610	0	90	140-180	80-110	-	215	55-60
PA 612	0	-	130-150	80-100	-	-	55-60
PA MXD6	30	228	190-230	110-140	-	240	85-100
PA 6-3-T	0	120	130-140	80-100	-70	240	150
PA 6T	0		120-130	70-90	-	-500	
PA 6/6T	30	250	-	155	-	295	115
PA PACM12	0	105	-	≈ 100	-	250	140
PA 6T/6I	0	-	-	-	-	330	130
PA PDA-T	0	-	-	-	-	500	-
PPA	0	120	-	-	-	-	-
PPA	33	270	-	>160	-	-	-
PPA	45	290	-	-	-	-	-
PMPI	0	-	260	-	-	-	-
PPTA	0	-	>250	>200	-	-	-
PC	0	125-135	115-150	115-130	-150	220-260	150
PC	30	135-150	115-150	115-130	-150		
PC-TMC	0	-	-	150	-	-	150-235
PC+ABS	0	105	-	-	-	-	-
PC+ASA	0	109	-	-	-	-	-
PET	0	80	200	100-120	-20	255	98
PET	30	200-230	220	150		255	98
PBT	0	65	165	100		255	60
PBT	30	200-210	220	150		225	60
PET+PS	0	-	200	100	-20	-	-
PBT+PS	0	-	165	100	-30	-	-
PAR	0	155-175	170	150	-	420	-
PAR15	0	237	200	-	-	-	250
PAR25	0	307	300	-	-	-	325
PPS	0	135	300	200-240	-	285	85
PPS	30	255	300	200-240	-	285	85
PES	0	200-215	180-260	160-200	-	-	225
PES	30	210-225	180-260	160-200	-	-	225
PSU	0	170-175	170	160	-100	-	190
PSU	30	185	180	160	-100	-	190
PPSU	0	-	-	-	-	-	221
PPE mod.	0	135	120-130	100-110	-	-	-

Continued on next page

Plastic	Glass fiber content	HDT/A	Service Temperature °C			T_s/T_m	T_g
	%	°C	max. short term	max. long term	min. long term	°C	°C
PPE mod.	30	160	-	-	-	-	-
PPE+PA 66	0	210	-	-	-	-	-
PPE+PS	0	115-130	120	100	-30	-	140
PPE+PS	30	137-144	130	110	-30	-	140
PAEK	0	200	-	-	-	380	170
PAEK	30	320	-	240-250	-	380	140-170
PEK, PEKEKK	0	170	300	260	-	365-380	175
PEEK	0	140	300	250	-	335-345	145
PEEK	30	315	300	250	-	335-345	145
PEEEK	0	-	-	-	-	324	110
PEEKEK	0	-	-	-	-	345	148
PEEKK	0	103	260	220	-	365	167
PEEKK	30	165	300	250	-	365	167
PEKEK	0	-	-	-	-	384	160
PEKK	0	-	350	260	-	391	165
PI	0	280-360	400	260	-	-	250-270
PI	30	360	400	260	-	-	250-270
PI-molding compound	Various	-	400	260	-240	-	-
PBMI	40	>300	250	190			-
PBO	0		-	<500	-	525	-
PAI	0	280	300	260	-260	-	240-275
PAI	30		300	260	-260	-	240-275
PEI	0	190-200	-	170	-170	-	215
PEI	30	195-215	180	170	-170	-	215
PISO	0	-	>250	210	-	250-350	273
PMI (foam)	0	-	-	180	-	-	-
PMMI	0	130-160	-	120-150	-	-	-
PESI	0		-	200	-	-	-
LCP, Vectra	0	170	-	220	-	285	-
LCP, Vectra	30	230	-	220	-	285	-
LCP	0	180-240	-	185-250	-	275-330	160-190
LCP-A	50	235	-	-	-	-	-
LCP-C	50	250	-	-	-	-	-
PUR-cast resin	0	-	70-100	50-80	-	-	15-90
PF Type 31, 51, 74, 84	Various	160-170	140	110-130	-	-	-
PF Type 13	Various	170	150	120	-	-	-
PF Type 4111	Various	240		170	-	-	-
UF	Various	-	100	70	-	-	-
MF Type 150/52	Various	155	120	80	-	-	-

Continued on next page

Plastic		Glass fiber content	HDT/A	Service Temperature °C			T_s/T_m	T_g
		%	°C	max. short term	max. long term	min. long term	°C	°C
MF Type 156		Various	180	-			-	-
MPF 1206		Various	190	-	160	-	-	-
MPF Type 4165		20-30	165	-	-	-	-	-
UP Type 802/4		20-Oct	250	200	150	-	-	-
UP Type 3620		Various	110	-	-	-	-	-
UP Type 3410		Various	190-200	270	200	-	-	-
EP Type 891		≈ 20		180	130	-	-	-
EP Type 8414		25-35	150	180	130	-	-	-
PDAP		Various	160-280	190-250	150-180	-50	-	-
SI		30	480	250	170-180	-50	-	-

Table A.19: Service Temperatures of Rubbers

Rubber	Service temperature °C		
	maximum short term	maximum long term	minimum long term
NR	90	70	-50
CR	130	110	-40
SBR	120	100	-40
NBR	130	100	-30
IIR	-	130	-40
NBR hydrated	150	-	-
EPDM	150	130	-40
AECM	200	170	-25
EAM	-	150	-10
CSM	-	120	-20
ACM, ABR, ANM		150	-25
FKM	250	200	-20
CO, ECO, ETER	-	120	-40
Q-rubber	≈ 300	≈ 180	-60 to -100
MVFQ	-	175	-60
PNF	-	150	-50

Table A.20: Energy Content of Polymers and Fuels

Energy content kWh/kg	Polymers (without additives)	Fuel (example)
>10	Polyolefins, dienes- and olefin rubbers, PS, SB, SAN, ABS, ASA, EVAC, EVOH, (PPE+SB)	Heating oil, gasoline
>7-10	AMMA, MBS, SMMA, SMAB, polyamides, PVAL, PC, PEC, PBT, PPS, PSU, PPSU, PEI, EP, PF, UP, LCP (copolyester), P	Anthracite
>4-7	PMMA, PUR, POM, PET, PVC, VCVAC, PVF, CSF, TPU, TPA, PBI, PES, cellulose, cellulose-esters and -ethers, starch	Wood, paper, brown coal briquette
>1.5-4	PVC-C, PVDC, PVDF, VDFHFP, ETFE, ECTFE, UF, MF, VF	Raw brown coal
up to 1.5	PTFE, FEP, PCTFE, PFA	No fuels

Table A.21: Electrical Characteristic Values of Conductive Plastic Compounds

Plastic	Filler	ϕ %	ρ_e Ω cm	$\sigma_e c$ Ω	Shield-damping 30-1000 MHz dB
PE	Carbon black	7	7	-	18-20
PP	Carbon black	20	10-20	-	-
PS	Carbon black	25	10	-	-
ABS	Al-chips	30	≈ 10	-	-
PVC	Carbon black	7	2	-	35-40
PVC	Polyaniline	30	1	-	>40
POM	Carbon black	-	20	$3 \cdot 10^3$	
PA 66	C-fibers nickel-plated	40	10^{-2}-10^{-1}	-	64-78
PA 66	Steel fibers	5	1-10	-	30-40
PA 66	Steel fibers	15	10^{-2}-10^{-1}	-	43-68
PA 66	Polyaniline	30	4	-	-
PC	C-fibers	10	10^1-10^4	10^5-10^6	-
PC	C-fibers	20	10^1-10	10^1-10^4	-
PC	C-fibers	30	10^1-10	10^1-10^4	-
PC	C-fibers	40	10^0-10^1	10-10^1	-
PC	Steel fibers	5	1-10	-	28-36
PC	Steel fibers	15	10^{-2}-10^{-1}	-	39-58
PET	C-fibers	30	10^0-10^1	10-10^1	-
PBT	C-fibers	30	10^0-10^1	10-10^1	-
PPS	C-fibers	40	10^0-10^1	10-10^1	-
PPS	Carbon minerals	-	400	$5 \cdot 10^2$	-
PPS	C-fibers nickel-plated	-	10^{-2}-10^{-1}	-	45-70
PEI	C-fibers	30	10-10^1	10^1-10^4	-

Table A.22: Conductivity of Intrinsically Conductive Polymers (ICP)

Plastic	Acronym	Doping	ρ_e Ω cm
Polypyrrol	PPY	BF_4	0.01
Polypyrrol	PPY	SO_3	0.006
Polyacetylene	PAC	AsF_5	0.001
Polyacetylene, stretched	PAC	AsF_5	0.003
Polyaniline	PANI	BF_4	0.1
Poly(p-phenylene)	PPP	AsF_5	0.002
Poly(p-phenylene)	PPP	Na	0.00033
Polyfuran	PFU	CF_3SO_3	0.02
Poly(p-phenylvinylidene), stretched	PPV	AsF_5	0.002
Polythiophene	PT	CF_3SO_3	0.05

Table A.23: Electrostatic Chargeability for Selected Plastics and Their Friction Partners

Plastic	Friction partner	Limiting charge (V/cm) 40% r.h.	65% r.h.	Half life (s) 40% r.h.	65% r.h.
ABS	PA66-fabric	-1300 to -2200	-950 to -1900	28 to 42	9 to 24
	PAN-fabric	+290 to +820	+120 to +600	13 to 45	6 to 30
ABS (antistatic)	PA66-fabric	+1000 to +2000	+1000 to 3000	1000 to 3000	300 to 600
	PAN-fabric	+2000 to +4300	+1000 to +2300	1000 to 3000	500 to 600
SB	PA66-fabric	-7200	-6200	>3600	>3600
	PAN-fabric	-5600	600	>3600	>3600
PC	PA66-fabric		5100		>3600
	PAN-fabric	7200	5600	>3600	>3600
POM	PA66-fabric	5400	3000	3000	1600
	PAN-fabric	5600	5500	3200	1200
CA	PA66-fabric	-3000	-3900	35	3
	PAN-fabric	1200	1100	30	3
CP	PA66-fabric	-3400	-500	>3600	360
	PAN-fabric	6900	5100	>3600	500
CAB	PA66-fabric	3800	-	1100	
	PAN-fabric	5900	5700	850	180
PP (normal)	Wool felt	-3900	-3900	1080	1080
PP (antistatic)	Wool felt	-800	-800	300	300
PMMA	Wool felt	7800	7800	-	-
PF Type 31	Wool felt	1200	1200	60	60
SAN	Leather	4800	4800	1200	1200

Table A.24: Properties of Transparent Plastics

Material	Light transmission	Index of refraction n_o at 20 °C	E_t MPa	σ_Y MPa	ρ g/cm³
Crown glass	Crystal clear	1.4-1.6	-	-	-
Flint glass	Crystal clear	1.53-1.59	-	-	-
Water	Crystal clear	1.33	-	-	-
PE	Transparent to opaque	1.51	200-1400	8-30	0.915-0.96
EIM	Transparent	1.51	150-200	7-8	0.94-0.95
COC	Crystal clear	1.53	3100		1.02
PP	Transparent to opaque	1.5	800-1100	-	0.9
PMP	Crystal clear to opaque	1.46	1200-2000	10-15	0.83-0.84
PS	Crystal clear	1.58-1.59	3100-3300	42-65	1.05
SB	Opaque		2000-2800	25-45	1.03-1.05
ABS	Transparent to opaque	1.52	220-3000	45-65	1.03-1.07
SAN	Crystal clear to opaque	1.57	3600-3900	70-85	1.08
SMMA	Transparent	-	3400	70-83	1.08-1.13
SMSA	Transparent	-	3500	60	1.07-1.17
SBS	Crystal clear	-	1100-1900	-	1.0-1.2
PVC-U	Crystal clear to opaque	1.52-1.54	2900-3000	-	1.37
PVC-HI	Crystal clear to opaque	-	2300-3000	40-55	1.36
PTFE	Opaque	1.35	400-700	-	2.13-2.33
PVDF	Transparent to opaque	1.42	2000-2900	50-60	1.76-1.78
PCTFE	Opaque	1.43	1300-1500	-	2.07-2.12
PBA	Transparent	1.467		-	
PMMA	Crystal clear	1.49	3100-3300	62-75	1.18
MBS	Transparent	-	2000-2600		1.11
MABS	Transparent	-	2000-2100		1.08
PMMA-HI	Crystal clear	-	600-2400	20-60	1.12-1.17
POM	Opaque	1.49	2800-3200	60-75	1.39-1.42
PA 6/11/12/66	Transparent to opaque	1.52-1.53	-	-	-
PA 6-3-T cond.	Transparent	1.57	2800-300	80-90	1.12
PC	Crystal clear	1.58-1.59	2400	55-65	1.2
PBT	Opaque	1.55	2500-2800		1.30-1.32
PET-A	Crystal clear to transparent	1.57	2100-2400	55	1.34
PET	Transparent to opaque	-	2800-3000	60-80	-
PET-G	Opaque	-	1900-2100	-	1.23-1.26
APE (PEC)	Transparent	1.57-1.58	2300	65	1.15-1.18

Continued on next page

Material	Light transmission	Index of refraction n_o at 20 °C	E_t MPa	σ_Y MPa	ρ g/cm^3
PSU	Transparent to opaque	1.63	2500-2700	70-80	1.24-1.25
PES	Transparent to opaque	1.65	2600-2800	80-90	1.36-1.37
PPE+PS	Opaque	-	1900-2700	45-65	1.04-1.06
PF	Transparent	1.63	-	-	-
UP	Crystal clear	1.54-1.58	-	-	-
EP	Crystal clear to opaque	1.47	-	-	-
CA	Crystal clear	1.47-1.50	1000-3000	25-55	1.26-1.32
CAB	Crystal clear	1.48	800-2300	20-55	1.16-1.22
CAP	Crystal clear	1.47	1000-2400	20-50	1.17-1.24
CAP	Crystal clear to transparent	-	1000-2100	-	1.19-1.22

Table A.25: Water Absorption (to Saturation) and Coefficients of Diffusion for Water

Plastic	Water absorption Standard conditions 23°C/50% r.h.	Water 23°C	Coefficient of diffusion D 10^{-6} mm^2/s
PE-LD	-	0.002 to 0.2	≈ 0.14
PE-HD	-	0.002 to 0.2	≈ 0.74
PP	-	>0.02	≈ 0.24
PMP	-	≈ 0.05	-
PS	-	0.2 to 0.3	-
SAN	-	≈ 0.2	-
ABS	-	≈ 0.7	-
PVC-(E)	≈ 0.18	0.5 to 3.5 (60°C)	-
PVC-(S)	-	0.3 (60°C)	-
PMMA	-	1.6 to 2	-
POM	≈ 0.3	≈ 0.6	-
PPE + PS	-	≈ 0.15	-
PPE + PS-GF	≈ 0.03	≈ 0.15	-
PC	≈ 0.2	≈ 0.4	-
PET	≈ 0.35	0.5 to 0.7	-
PET-GF 33	≈ 0.2	0.25	-
PBT	≈ 0.45	≈ 0.45	-
PBT-GF 33	0.1 to 0.2	0.1 to 0.2	-
PSU	≈ 0.25	≈ 0.6	-
PA 6	2.8 to 3.6	9 to 10	≈ 0.4
PA 9		≈ 2.5	-
PA 11	0.8 to 1.2	≈ 1.8	-
PA 12	0.7 to 1.1	1.3 to 1.9	-
PA 66	2.5 to 3.5	7.5 to 9	≈ 0.2
PA 68	≈ 3	4 to 4.5	-
PA 610	1.5 to 2	3 to 4	-
PA 612	1.3 to 2	2.5 to 2.8	-
PA 6-3-T	2.6 to 3	6.2 to 7	-
PA cast	2.3 to 2.7	7 to 8	≈ 0.32
PA 6-GF 30	1.5 to 2	≈ 6	≈ 0.4
PA 66-GF 30	1 to 1.5	≈ 5.5	≈ 0.2
PA 11-GF 30	≈ 0.54	≈ 1.2	-
PA 12-GF 30	≈ 0.45	≈ 1.1	-
CA	-	3.8 to 5	-
CAB	-	2 to 2.5	-
CP	-	2.3 to 2.7	-
UP	-	≈ 0.4	-
UP-GF	-	0.5 to 2.5	-
EP	0.5 to 0.8	0.7 to 1.5	0.2 to 0.3
EP-GF 55	0.3 to 0.5	≈ 0.8	-
PAI	-	0.22 to 0.28	-
PVDF	-	≈ 0.25	-
PES	≈ 0.15	≈ 2.1	-
PI	≈ 1.2	≈ 3	-

Table A.26: Chemical Resistance

Plastic (acronym)	Water	Weak (acid)	Strong (acid)	Hydrofluoric acid	Weak (base)	Strong (base)	Inorganic salts	Halogens	Oxyd. compounds	Paraff. hydrocarbons	Halogens alkanes	Alcohols	Ethers	Esters	Kretones	Aldehydes	Amines	Org. acids	Aromat. compounds	Fuels	Mineral oils	Fats, oils
PE-LLD	+	+	+	+	+	+	+	−	−	*	−	_	−	_	_	_	+	−	*	_	_	o
PE-LD	+	+	+	+	+	+	+	−	−	*	−	+	_	o	o	o	+	+	o	o	o	o
PE-HD	+	+	+	+	+	+	+	−	−	*	+	_	+	+	_	_	+	o	o	_	o	+
PE-C	+	+	+	_	+	+	+	_	_	o	*	+	_	_	*	+	o	*	_	_	*	_
EVA	_	+	*	_	+	+	+	_	*	*	_	o	*	*	_	+	o	o	_	_	_	o
PIB	+	+	+	+	+	+	+	_	_	_	_	+	_	_	*	o	+	+	_	_	_	_
PP	+	+	o	o	+	+	+	*	_	o	*	+	_	o	o	+	+	o	*	o	+	+
PMP	+	+	+	_	+	+	+	_	*	_	*	o	_	*	_	+	_	o	*	_	o	+
PS	+	+	o	o	o	+	+	_	_	*	_	+	*	_	_	*	+	o	*	*	_	+
SB	+	+	_	_	+	+	_	_	_	_	_	o	_	_	_	_	_	_	_	*	_	+
SAN	+	+	o	+	+	+	+	_	_	o	_	o	_	_	_	*	+	o	_	o	+	+
ABS	+	+	o	+	+	+	+	_	*	_	_	o	_	_	_	*	+	+	_	+	+	+
PVC-U	+	+	+	+	o	+	+	_	o	o	*	+	*	_	_	*	o	o	*	*	+	+
PVCVAC	_	+	_	_	+	+	_	_	_	_	_	+	_	_	+	+	o	_	+	_	+	+
PVC-P	+	+	o	+	_	+	_	o	_	*	*	_	_	+	+	o	_	_	_	_	_	_
PTFE	+	+	+	+	+	+	+	+	+	+	+	+	+	+	+	+	+	+	+	+	+	+
PCTFE	+	+	+	+	+	+	+	o	o	*	+	_	*	+	+	_	+	*	*	+	+	+
PMMA	+	o	*	_	+	+	+	_	_	o	_	_	o	_	_	+	+	*	*	o	+	+
AMM A	+	+	+	_	o	o	_	_	*	+	+	+	_	_	_	_	_	_	o	+	+	+
POM	+	o	−	_	+	+	+	−	*	*	+	+	o	+	o	_	o	o	+	o	+	+
PPE + PS	+	+	+	_	+	+	+	_	_	_	_	_	_	_	_	_	_	+	_	_	_	_
CA	+	+	−	−	_	−	+	−	−	+	*	_	+	_	_	_	*	_	o	o	+	+
CTA	+	o	−	−	_	−	+	−	−	+	*	_	+	−	_	+	o	*	o	o	+	+
CAB	+	+	−	−	o	_	+	−	*	+	_	*	*	_	_	+	o	o	_	+	+	+
CP	+	_	−	−	−	−	+	−	−	+	_	−	−	−	_	+	o	o	_	+	+	+
PC	+	+	*	o	−	−	o	+	_	o	_	o	−	−	−	−	_	_	_	_	*	+	+
PET	+	+	o	−	o	−	+	+	_	+	*	o	+	+	+	_	_	+	+	o	+	+
PBT	_	*	_	+	+	+	_	_	+	*	+	+	_	*	_	_	_	o	o	+	+	+
PA 6	+	−	−	−	o	+	+	−	−	o	o	+	o	+	+	o	+	o	o	+	+	+
PA 12	+	−	−	−	o	o	+	−	−	o	+	+	o	+	+	_	+	o	o	+	+	+
PA 66	+	−	−	−	o	+	+	−	−	o	o	+	o	+	+	o	+	o	o	+	+	+
PA 610	+	−	−	−	o	+	+	−	−	o	*	+	o	+	+	o	+	o	o	+	+	+
PA arom.	+	*	*	_	o	+	+	_	+	o	o	*	+	+	o	_	_	*	o	+	+	+
PSU	o	+	+	_	+	+	+	_	+	o	_	+	_	_	_	_	_	_	_	_	_	_
PF	+	+	_	_	_	_	_	_	o	o	o	o	o	_	_	_	_	o	o	_	+	+
UF	+	*	_	_	_	_	_	_	o	+	o	o	o	_	_	_	_	o	o	_	+	+
MF	+	_	_	_	o	_	_	_	o	+	o	o	o	_	_	_	_	o	o	*	+	+
UP	+	o	_	_	−	−	+	*	*	*	+	o	*	*	*	o	*	_	+	+	+	+
EP	+	o	_	+	o	o	+	+	_	_	*	+	+	*	o	*	o	+	*	+	+	+
TPU	+	o	_	−	o	o	+	−	*	+	*	o	o	+	+	+	_	o	+	+	+	+

+ Resistant; o Resistant to conditionally resistant; _ Conditionally resistant; * Conditionally resistant to non-resistant;
− Non-resistant

Table A.27: Chemical Resistance of Sheet and Sealing Strips

C = Commercially available, SS = Saturated solution, TP = technically pure, S = Suspension or slurry;
+ = Resistant, O = Conditionally resistant, – = Non-resistant

Corroding medium	Chemical notation	Concentration %	PVC-U 20°C	PVC-U 40°C	PVC-U 60°C	PVC-P 20°C	PVC-P 40°C	PE-HD 20°C	PE-HD 60°C	PP 20°C	PP 60°C	PP 100°C	PE-HD Sealing strips[1] 23°C±2°C	PS 20°C	PS 50°C
Acetic acid	CH_3COOH	≤ 60%	+	+	–	O		+	+	+	+	+	+	+	+
Acetic acid	CH_3COOH	100%	O	–	–			+	O	+	O	–		O	–
Alumium chloride	$AlCl_3$	≤ SS	+	+	+	+	O	+	+	+	+	+	+	+	+
Alumium sulfate	$Al_2(SO_4)_3$	≤ SS	+	+	+	+	O	+	+	+	+	+	+	+	+
Ammonium chloride	NH_4Cl	≤ SS	+	+	+	+	+	+	+	+	+	+	+		
Ammonium nitrate	NH_4NO_3	≤ SS	+	+	+	+	+	+	+	+	+	+	+	+	+
Ammonium sulfate	$(NH_4)_2SO_4$	≤ SS	+	+	+	+	+	+	+	+	+	+	+	+	+
Ammonium sulfide	$(NH_4)_2S$	≤ SS	+	+	+			+	+	+	+		+		
Arsenic acid	H_3AsO_4	≤ 30%	+	+	+	+		+	+	+	+				
Battery acid	H_2SO_4	≤ C	+	+	+	+		+	+	+	+				
Bleaching lye (natrium hypochlorite)	NaOCl	12% active chlorine	+	+	O	+		O	–	O	+	–	+	+	+
Boracic acid	H_3BO_3	≤ TP	+	+				+	+	+	+	+		+	+
Borax (natrium tetraborate)	$Na_2B_4O_7$	≤ SS	+	+	+	+	+	+	+	+	+	+		+	+
Calcium chloride	$CaCl_2$	≤ SS	+	+	+	+	+	+	+	+	+	+	+		
Calcium nitrate	$CA(NO_3)_2$	≤ SS	+	+	+	+	+	+	+	+	+	+	+		
Carbon dioxide gaseous	CO_2	Any	+	+	+	+	+	+	+	+	+			+	+
Caustic soda	NaOH	> 40–60%	+	+	O	–	–	+	+	+	+	+	+	+	+
Caustic soda	NaOH	15%	+	+	O			+	+	+	+			+	+

Continued on next page

Table A.27: Chemical Resistance of Sheet and Sealing Strips

C = Commercially available, SS = Saturated solution, TP = technically pure, S = Suspension or slurry;
+ = Resistant, O = Conditionally resistant, − = Non-resistant

Corroding medium	Chemical notation	Concentration %	PVC-U 20°C	PVC-U 40°C	PVC-U 60°C	PVC-P 20°C	PVC-P 40°C	PE-HD 20°C	PE-HD 60°C	PP 20°C	PP 60°C	PP 100°C	PE-HD Sealing strips[1] 23°C±2°C	PS 20°C	PS 50°C
Chromate	H_2CrO_4	≤ 10%	+	+	+	+	+								
Chromium sulfuric acid	$CrO_3 + H_2SO_4$	150 g/l +50 g/l	+	+				−		−				+	O
	85.5 vol% H_2SO_4	96%												O	
Chromium sulfuric acid	4.5 vol% H_2CrO_4	50%	+	O	−			−		−				O	
	10 vol% H_2O														
Citric acid (industrial)	$C_6H_8O_7$	≤ SS	+	+	+	+	O	+	+	+	+	+	+	+	+
Cloroacetic acid (monochloroacetic acid)	$CH_2ClCOOH$	≤ 20%	+	+	O			+	+	+	+			+	O
Copper(II)-chloride	$CuCl_2$	≤ SS	+	+	+			+	+	+	+		+		
Copper(II)-sulfate	$CuSO_4$	≤ SS	+	+	+			+	+	+	+	+	+	+	+
Diesel fuel			+	O	−			+	O	+	O		+		
Fatty acids	RCOOH	TP	+	+				+	O	+	+				
Ferric(II) chloride	$FeCl_2$	≤ SS	+	+	+	+	+	+	+	+	+		+	+	+
Ferric(III) chloride	$FeCl_3$	≤ SS	+	+	+	+	+	+	+	+	+	+	+	+	+
Formaldehyde	HCHO	≤ 15%	+	+	O			+	+	+	+			+	+
Formic acid	HCOOH	≤ 85	+	+	O			+	+	+	O		+	+	+
Foto developer		C	+	+		O		+	+	+	+		+	+	O

Continued on next page

Table A.27: Chemical Resistance of Sheet and Sealing Strips

C = Commercially available, SS = Saturated solution, TP = technically pure, S = Suspension or slurry;
+ = Resistant, O = Conditionally resistant, – = Non-resistant

Corroding medium	Chemical notation	Concentration %	PVC-U 20°C	PVC-U 40°C	PVC-U 60°C	PVC-P 20°C	PVC-P 40°C	PE-HD 20°C	PE-HD 60°C	PP 20°C	PP 60°C	PP 100°C	PE-HD Sealing strips[1] 23°C±2°C	PS 20°C	PS 50°C
Foto fixer		C	+	+	+	+	+	+	+	+	+		+		–
Gasoline (C$_5$–C$_{12}$-mixture)		≤ C	+			–	–	+	O	O	–			O	–
Glycerine	C$_3$H$_8$O$_3$	≤ 100%	+	+	+	O	+	+	+	+	+		+	+	+
Glycole (ethylene glycole)	C$_2$H$_6$O$_2$	TP	+	+	+	+	+	+	+	+	+	+		+	+
Glycolic acid	C$_2$H$_4$O$_3$	≤ SS	+	+	+			+	+	+			+		
Heating oil EL		100%	+	+	+	O	–	+	O	+	O		+		
Hydrochloric acid	HCl	≤ 35%	+	+	+	+	O	+	+	+	+	O	+	+	O
Hydrofluoric acid	HF	≤ 40%	+	+	O	+	O	+	O	+	+			+	+
Hydrogen peroxide	H$_2$O$_2$	≤ 70%	+	+	O	+	O	O	–	+	O	+	+		
Hydroxylamine sulfate	(NH$_3$OH)$_2$SO$_4$	≤ 12%	+	+	+			+	+	+	+		+		
Lead acetate	Pb(CH$_3$COO)$_2$	≤ SS	+	+	+			+	+	+	+		+	+	+
Liquid ammonia	NH$_4$OH	≤ SS	+	+	O	+	O	+	+	+	+		+	+	+
Magnesium chloride	MgCl$_2$	≤ SS	+	+	+			+	+	+	+	+	+	+	+
Magnesium sulfate	MgSO$_4$	≤ SS	+	+	+	+	+	+	+	+	+		+	+	+
Mercury nitrate	Hg(NO$_3$)$_2$	S	+	+	+	+	+	+	+	+	+		+		
Methylalcohol	CH$_3$OH	All	+	+	O	–	–	+	+	+	+		+	+	+
Natrium chlorate	NaClO$_3$	≤ SS	+	+	+	+	+	+	+	+	+	+	+	+	+
Natrium chloride	NaCl	≤ SS	+	+	+	+	+	+	+	+	+	+	+	+	+

Continued on next page

Table A.27: Chemical Resistance of Sheet and Sealing Strips

C = Commercially available, SS = Saturated solution, TP = technically pure, S = Suspension or slurry;
+ = Resistant, O = Conditionally resistant, − = Non-resistant

Corroding medium	Chemical notation	Concentration %	PVC-U 20°C	PVC-U 40°C	PVC-U 60°C	PVC-P 20°C	PVC-P 40°C	PE-HD 20°C	PE-HD 60°C	PP 20°C	PP 60°C	PP 100°C	PE-HD Sealing strips[1] 23°C±2°C	PS 20°C	PS 50°C
Natrium hypochlorite	NaOCl	12% active	+	+	O	+		O	−	O	O	−	+	+	+
Nickel(II)-sulfate	NiSO$_4$	≤ SS	+	+	O	+	+	+	+	+	+		+	+	+
Nitric acid	HNO$_3$	65%	+	O	−								+	O	−
Nitric acid	HNO$_3$	50%	+	O	O	−		O	−	O	−		+	O	−
Nitric acid	HNO$_3$	30%	+	+	O	O	O	+	+	+	−		+	+	O
Nitric acid	HNO$_3$	10%	+	+	+	O	O			+	−		+	+	O
Oleic acid		TP	+	+	+	−	−	+	O	+	O	−		+	+
Oxalic acid	(COOH)$_2$	≤ SS	+	+	+	+	O	+	+	+	+	+		+	+
Phenolic	C$_6$H$_5$OH	50%	+	O	−	O	−	+	+	+	+			O	−
Phosphoric acid	H$_3$PO$_4$	≤ 75%	+	+	+			+	+	+	+	+	+	+	+
Potash lye	KOH	≤ 50%	+	+	O	O	−	+	+	+	+	+	+	+	+
Potassium borate	K$_3$BO$_3$	≤ SS	+	+	O			+	+	+	+		+		
Potassium bromate	KbrO$_3$	≤ SS	+	+	O			+	+	+	+	+	+	+	+
Potassium bromide	KBr	≤ SS	+	+	+			+	+	+	+	+	+	+	+
Potassium carbonate	K$_2$CO$_3$	≤ SS	+	+	+	+		+	+	+	+	+	+	+	+
Potassium chloride	KCl	≤ SS	+	+	+	+	+	+	+	+	+	+	+	+	+
Potassium nitrate	KNO$_3$	≤ SS	+	+	+			+	+	+	+			+	+
Potassium permanganate	KMnO$_4$	≤ 10%	+	+				+	+	+				+	O

Continued on next page

Table A.27: Chemical Resistance of Sheet and Sealing Strips

C = Commercially available, SS = Saturated solution, TP = technically pure, S = Suspension or slurry;
+ = Resistant, O = Conditionally resistant, − = Non-resistant

Corroding medium	Chemical notation	Concentration %	PVC-U 20°C	PVC-U 40°C	PVC-U 60°C	PVC-P 20°C	PVC-P 40°C	PE-HD 20°C	PE-HD 60°C	PP 20°C	PP 60°C	PP 100°C	PE-HD Sealing strips[1] 23°C±2°C	PS 20°C	PS 50°C
Potassium persulfate	$K_2S_2O_8$	≤ SS	+	+	O	+	O	+	+	+	+				+
Silver nitrate	$AgNO_3$	≤ SS	+	+	O			+	+	+	+	+	+	+	+
Starch (industrial)	$(C_6H_{10}O_5)_x$	C	+	+	+			+	+	+	+		+		
Stearic acid	$C_{18}H_{36}O_2$	TP	+	+	+			+	O	+	O			+	+
Sulfuric acid	H_2SO_4	≤ 90%	+	+	+	+		+	+	+	O		+	+	+
Sulfuric acid	H_2SO_4	≤ 96%	+	+	O			O	−	O	−		+		−
Tartaric acid	$C_4H_6O_6$	≤ SS	+	+	+	+		+	+	+	+		+	+	+
Tin(IV)-chloride	$SnCl_4$	≤ SS	+	O		+	+	+	+	+	+		+	+	+
Urea	$CO(NH_2)_2$	≤ SS	+	+	+	+	O	+	+	+	+	+	+	+	+
Zinc sulfate	$ZnSO_4$	≤ SS	+	+	+	+	+	+	+	+	+	O	+	+	+

[1] Water endangering liquids that may be stored in a storage room when using Saxolen-PE-HD-sealing strips as sealants. The complete list of corroding media is contained in PA-VI 222.270

Table A.28: Media that May Cause Environmental Stress Cracking

Media causing stress-cracking	ABS	AMMA	PA	PC	PE	PMMA	PP	PS	PVC	SAN	SB
Acetone	x		x	x				x		x	x
Ethyl alcohol	x	x				x		x		x	x
Ether	x				x			x		x	x
Alcohol					x						
Aniline					x		x				
Gasoline	x		x		x			x		x	x
Petroleum					x						
Acetic acid					x		x				
Ester					x						
Glycerine		x				x					
Heating oil					x						
Heptane	x							x		x	x
Hexane	x							x		x	x
Isopropanol	x							x		x	x
Potassium hydroxide					x						
Ketone				x	x						
Hydrocarbon, aromatic				x							
Metal halogenide			x								
Methanol	x							x	x	x	x
Sodium hydroxide		x			x	x					
Sodium hypochloride					x		x				
Paraffine oil		x				x					
Vegetable oil	x							x		x	x
Swelling agents, chlorine-containing				x							
Nitric acid					x		x				
Silicone acid					x						
Sulfuric acid							x				
Tenside					x						
Terpentine			x				x				
Carbon tetrachloride			x	x	x						
Water		x			x	x					

Table A.29: Recommended Media for Stress-Cracking Tests for Selected Plastics

Plastic acronym	Medium causing stress-cracking	Immersion time
PE	Tenside solution (2%), 50°C	>50 h
	Tenside solution (2%), 70°C	48 h
	Tenside solution (5%), 80°C	4 h
PP	Chromic acid, 50°C	-
PS	n-Heptane	-
	Petroleum/gasoline, boiling range 50-70°C	-
	n-Heptane : n-propanol (1:1)	-
SB	n-Heptane	-
	Petroleum/gasoline, boiling range 50-70°C	-
	n-Heptane : n-Propanol (1:1)	-
	Oleic acid	-
SAN	Toluene : n-propanol (1:5)	15 min
	n-Heptane	-
	Carbon tetrachloride	-
ABS	Dioctylphthalate	-
	Toluene : n-propanol (1:5)	-
	Methanol	15 min
	Acetic acid (80%)	20 min
	Toluene	1 h
PMMA	Toluene : n-heptane (2:3)	15 min
	Ethanol	-
	n-methylformamide	-
PVC	Methanol	-
	Methylene chloride	30 min
	Acetone	3 h
POM	Sulfuric acid (50%), local wetting	up to 20 min
PC	Toluene : n-propanol (1:3 to 1:1	3-15 min
	Carbon tetrachloride	1 min
	Caustic soda (5%)	1 h
PC + ABS	Methanol : ethylacetate (1:3)	-
	Methanol : acetic acid (1:3)	-
	Toluene : n-propanol (1:3)	-
PPE + PS	Tributylphosphate	10 min
PBT	ln-caustic soda	-
PA 6	Zinc chloride solution (35%)	20 min
PA 66	Zinc chloride solution (50%)	1 h
PA 6-3-T	Methanol	-
	Acetone	1 min
PSU	Ethyleneglycolmonoethylether	1 min
	Acetic acid-ethylester	-
	1,1,1-Trichlorethane : n-heptane	-
	Methylglycolacetate	-
	Carbon tetrachloride	-
	1,1,2-Trichloroethane	1 min
	Acetone	1 min

Continued on next page

Plastic acronym	Medium causing stress-cracking	Immersion time
PES	Toluene	1 min
	Ethylacetate	1 min
PEEK	Acetone	-
PAR	Caustic soda (5%)	1 h
	Toluene	1 h
PEI	Propylene carbonate	-

Table A.30: Resistance of Plastics

Plastic	Weathering		Resistance	
	Non stabilized	Stabilized	Micro-organisms	Macro-organisms
PE-LD	3	1 to 2	1 to 2	2 to 3
PE-HD	3	1 to 2	1 to 2	2 to 3
PP	3	2	1 to 2	2 to 3
PB	3	2	-	-
PMP	3	-	-	-
PS	3	2	1	2 to 3
SB	3	-	-	-
SAN	3	-	-	-
ABS	3	2	1	2 to 3
ASA	3	2	-	-
PVC-U	2	1 to 2	1 to 2	1
PVC-P	3	2	2 to 3	3
PVDC	2 to 3	-	-	
PTFE	1	-	1	1 to 2
PCTFE	1	-	1	1 to 2
PVF	-	-	1	1 to 2
PVDF	1	-	-	-
PMMA	1	-	1 to 2	1 to 2
POM	3	-	-	-
PPO	3	-	-	-
CA	2 to 3	-	2 to 3	3
CAB	2 to 3	-	2 to 3	3
PC	2	-	-	-
PET	2	-	-	-
PA	3	-	1 to 2	1 to 2
PSU	3	-	-	-
PI	3	-	-	-
PUR	3	-	3	1
GF-UP	1 to 2	-	1 to 2	1
EP	2	-	1	1 to 2
PF	1 to 2	-	1 to 3	1
MF	1 to 2	-	2 to 3	1
UF	3	-	1 to 3	3

Resistance: 1: Excellent, 2: Average, 3: Little resistant

Table A.31: Radiation Resistance of Thermoplastics; Half life (Elongation at Break) at 20 °C

Plastic	Radiation under O_2-exclusion			Radiation in air	
	Decomposition or cross- D/C	Gas generation mm^3/kg·Gy	Half life kGy	Half life at 500 Gy/h kGy	Half life at 50 Gy/h kGy
PE-HD	C	-	60 to 300	25 to 95	10 to 40
PE-LD	C	7	400 to 1300	180	130
PE-LD-X	C	-	600 to 1000	780	640
PE-C	-	-	250 to 600	650 to 1000	450 to 1000
PE-LD-X	-	-	-	-	-
Flame-retardant	C	-	250	450	450
EPDM	C	-	100 to 1000	200 to 1000	200 to 600
PP	C	6	30	10 to 25	6 to 15
PS	C	0.05	10000	590	560
SAN	C	-	2000	-	-
SB	C	0.2	2000	-	550
PVC-U	C	10 to 30	9000	-	-
PVC-P	C	-	2000	-	-
PVDC	D	-	-	750	370
PA	C	2	140 to 430	85	47
PC	D	-	500		
PET	C	0.4	2100	750	400
POM	D	3	200		-
PPE + PS	D	9	26	26	-
PTFE	D	-	-	1.4 to 4.0	1.4 to 4.0
PCTFE	D	-	400		-
PVAL	C	-	330	330	-
PVB	C	-	2000	-	-
CA	D	2	160	160	-
CAB	D	3	200	-	-
CN	D	13	130	-	-
CP	D	3	120	-	-
EC	D	3	10	-	-

Table A.32: Water Vapor (DIN 53 122) and Gas Permeability (DIN 53 380)

Plastic	Temperature °C	Film thickness μm	Water vapor g/m²·d	N_2	Air	O_2
PE-LD	23	100	1	700	1100	2000
PE-LD	20	25	5	-	-	5400 (23°C)
PE-HD	25	40	0.9	525	754	1890
PE-HD	23	100	0.06-1	700	1100	1600-2000
EVAC, 20% VAC	23	100	455	1400		4000
EVAL	20	20	-	-	-	0.2-1.8
EIM	25	25	25	-	-	9300
PP	23	100	0.7-0.8	-	-	600
PP	20	25	≈ 3.5	2300 (23°C)	-	-
PP-O	20	25	≈ 1.3	-	-	
PP	25	40	2.1	430	700	1900
PS	23	100	12	2500	-	1000
PS-O	25	50	14	27	80	235
ABS	23	100	27-33	100-200	-	400-900
ASA	20	100	30-35	60-70	-	150-180
PS-HI	23	100	13	4000	-	1600
SBS	23	250	6	230	-	830
PVC-U	23	100	2.5	2.7-3.8	-	33-45
PVC-U	20	40	7.6	12	28	87
PVC-P	20	40	20	350	550	1500
PVDC	25	25	0.1-0.2	1.8-2.3	5-10	1.7-11
PVDF	23	60	2.4 (90μm)	28	-	94
PVF	23	25	50	3.8	-	4.7
PCTFE	40	25	0.4-0.9	39	-	110-230
ECTFE	23	25	9	150	-	39
ETFE	23	25	0.6	470	-	1560
PTFE	23	300	0.03	60-80	80-100	160-250
PAN	20	20	≈ 80	-	-	15
POM	23	80	12	- 5	8	24
PA 6 and 66	23	100	10-20	1-2	-	2-8
PA 6	20	20	30	-	-	30 (23°C)
PA 11 and 12	23	100	2.4-4	0.5-0.7	-	2-3.5
PC	23	25	4	680	-	4000
PET	23	40	4.5-5.5 (5	6.6	12	30
PET	23	100	5	4	-	25
PET-0	23	25	0.6	9-15	-	80-110
PSU	23	25	6	630	-	3600
PI	23	25	25	94	-	390
PBI	25	100	6	500		1800
PF	20	40	45			

Continued on next page

Plastic	Temperature °C	Film thickness μm	Water vapor g/m₂·d	N₂	Air	O₂
PF Type 31	-	-	300-560	-	-	-
MF Type 152	20	40	400	-	-	-
TPU	23	25	13-25	550-1600	-	1000-4500
CH	20	25	500	-	-	250 (23°C)
CH coated	20	25	5.5	-	-	150 (23°C)
CA	25	25	150-600	470-630	1800-2300	13000-15000
CAB	25	25	460-600	3800	-	15000
SBS+PS 50/50	23	250	4.4	110	-	530
PET-G/ SBS+PS/ PET-G 10/230/10	23	250	4	-	30190	700
	-	-	-	-	-	-
	-	-	-	-	-	-

Plastic	CO₂ cm³/m²·d	H₂	Ar	He	CH₄	
PE-LD	10000	8000	-	-	-	-
PE-LD	-	-	-	-	-	-
PE-HD	7150	6000	-	-	-	-
PE-HD	10000	8000	-	-	-	-
EVAC, 20% VAC	17000		-	-	-	
EVAL	-	-	-	-	-	-
EIM	-	-	-	-	-	-
PP	-	-	-	-	-	-
PP	-	-	-	-	-	-
PP-O	-	-	-	-	-	- -
PP	6100	17700	1480	1920	-	
PS	5200	-	-	-	-	-
PS-O	800	1260		-	-	
ABS	-	-	-	-	-	-
ASA	6000-8000	50	-	-	100-110	-
PS-HI	10000	-	-	-	-	-
SBS	4300	-	-	-	-	-
PVC-U	120-160	-	-	-	-	-
PVC-U	200	-	-	-	-	-
PVC-P	8500	-	-	-	-	-
PVDC	60-700	630-1400	-	-	-	-

Continued on next page

Plastic	CO_2 $cm^3/m^2 \cdot d$	H_2	Ar	He	CH_4	
PVDF	-	345	-	975	-	-
PVF	170	900	-	-	-	-
PCTFE	250-620	3400-5200	-	-	-	-
ECTFE	1700	-	-	-	-	-
ETFE	3800	-	-	-	-	-
PTFE	450-700	-	-	1700-2100	-	-
PAN	-	-	-	-	-	-
POM	470	210	-	-	12	-
PA 6 and 66	80-120	-	-	-	-	-
PA 6			-	-	-	-
PA 11 and 12	6-13	-	-	-	-	-
PC	14500	22000	-		-	-
PET	140	850	16	1170	8	-
PET	90	-	-	-	-	-
PET-0	200-340	1500	-	-	-	-
PSU	15000	28000	-	-	-	-
PI	700	3800	-	-	-	-
PBI	6000	7100	-	-	-	-
PF	-	-	-	-	-	-
PF Type 31	-	-	-	-	-	-
MF Type 152	-	-	-	-	-	-
TPU	6000-22000		-	-	-	-
CH	-	-	-	-	-	-
CH coated	-	-	-	-	-	-
CA	14000	-	-	-	-	-
CAB	94000	-	-	-	-	-
SBS+PS 50/50	2300	-	-	-	-	-
PET-G/	-	-	-	-	-	-
SBS+PS/	-	-	-	-	-	-

Table A.33: Oxygen Permeability of Barrier Films Depending on Temperature and Moisture

Film	Temperature °C	\multicolumn{6}{c}{O_2 permeability [$cm^3/m^2 \cdot d \cdot bar$] at rel. humidity [%]}
		0	50	75	85	95	100
EVOH	23	0.1	0.24	-	1.5	-	15
PVDC	23	2	2.2	2.2	2.2	2.2	2.2
PAN	23	9	12	12	12	12	12
Composite	23	0.3	0.45	1.5	-	15	-
PE-LD/EVAL	30	0.8	1.4	5	-	38	-
20/10/30μm	40	1.7	4	16	-	80	-

Table A.34: Permeation Values for Blends of Amorphous PA (Selar PA) and EVOH and PA 6. Respectively; Film Thickness 25 μm

Selar-PA-content: 0	\multicolumn{3}{c}{Mol.-% ethylene in EVO}
O_2 ($cm^3/m^2 \cdot d \cdot bar$)	32	38	44
Dry: 23°C/35% r.h.	< 0.1 -0.2	0.2 -0.4	0.4 -1.8
Wet: 23°C/80% r.h.	1.1 -2.3	2.9 -4.0	5.3 -5.7

	\multicolumn{6}{c}{% Selar PA blended with}
O_2 ($cm^3/m^2 \cdot d \cdot bar$)	0	20	30	50	80	100
0°C/0-5% r.h.	15	15	15	15	15	15
0°C/95-100% r.h.	55	30	20	8	6	4
30°C/0-5% r.h.	60	59	59	59	59	57
30°C/95-100% r.h.	225	210	180	136	84	-
H_2O ($g/m^2 \cdot d$)	0	20	30	50	80	100
23°C/95% r.h.	186	92	77	44	30	28

Table A.35: Diffusion Resistance Factors for Selected Construction and Insulation Materials

Material	Density kg/m^3	Diffusion resistance factor μ
Clay brick	1360 to 1860	6.8 to 10.0
Roofing tile	1880	37 to 43
Vitrified brick	2050	384 to 469
Lime sand brick, concrete	1500 to 2300	8 to 30
Gas- and foam concrete	600 to 800	3.5 to 7.5
Sheets from phenolic resin foam	23 to 95	30 to 50
polystyrene foam	14 to 40	32 to 125
polyurethane foam	40	51
polyvinyl chloride foam	43 to 66	170 to 328
urea formaldehyde resin	12	1.7
Sheets from polyester resin	2000	6180
polystyrene	1050	21300
polyvinyl chloride	1400	52000
Polyvinyl chloride varnish	1400	25000 to 50000

Table A.36: Flammability Classification According to UL 94

Plastic acronym	With flame retardant	Classification	Specimen thickness (mm)
PE-LD	-	HB	-
PE-LD	X	V2	-
PE-HD	-	HB	-
PE-HD	X	V2	-
PP	-	HB	-
PP	X	V2	-
PS	-	HB	-
SAN	-	HB	-
SB	-	HB	-
SB	X	V2	-
SB	X	V0	-
ABS	-	HB	0.8 to 3.2
ABS	X	V0	0.8 to 3.2
ASA	-	HB	-
PVC	-	-	-
POM	-	HB	-
PA 6	-	V2	-
PA 6	X	V0	-
PA 6-GF	-	HB	0.8 to 3.2
PA 66	-	V2	≥ 0.8
PA 66	X	V0	-
PA 61	-	V2	-
PC	-	V2	1.6 to 3.2
PC	X	V2/V0	1.6 to ≥ 1.6
PC-TMC	-	HB	1.6 to 3.2
PC-TMC	X	V2/V0	1.6 to 3.2
PC + ABS	-	HB	1.6
PC + ABS	X	V0	1.6
PBT	-	HB	0.8 to 3.2
PBT	X	V2/V0	$\geq 0.8/0.8$
PBT-GF	-	HB	0.8 to 3.2
PBT-GF	X	V0	-
PES	-	V0	-
PES-GF	-	V0	-
PAI	-	V0	≥ 0.2
PAR	-	V2	1.6 to 3.2
PAR	-	V0	≥ 0.4
PPS	X	V0	≥ 0.4
PPS-GF40	-	V0/5V	$\geq 0.4/\geq 1.7$
PPS-M65 (metal)	-	V0/5V	$\geq 0.8/\geq 6.1$
PEEK	-	V0	≥ 2
PEK	-	V0	≥ 0.8
PEI	-	V0	≥ 0.4
PEI-GF	-	V0	≥ 0.25
PSU	-	HB/V0	$\geq 1.5/\geq 4.47$
PSU	X	V0	≥ 1.5
PPE + PS	-	HB	≥ 1.6
PPE + PS	X	V1/V0	≥ 1.6

Table A.37: Characteristic Values of Sand Blasting Wear for a Jet Angle of $\alpha = 45°$

Material	Hardness Shore D	Hardness Vickers	Steel sand Nr.*	Abrasion rate** V/VSt 37
Steel TH 8	-	590	2	0.109
PUR elastomer	18	-	2	0.143
PVC-P	5	-	2	0.143
PUR elastomer	34	-	2	0.403
PVC-P	10	-	2	0.42
Rubber	17	-	2	0.57
PVC-P	14	-	2	0.96
Steel St37	-	126	1	1
PE-HD	60	-	1	1.06
Steel St34	-	124	1	1.07
PVC-P	17	-	2	1.12
PA 6	62	-	1	1.33
PA 6	64	-	1	1.33
Copper	-	99	1	1.36
PE-LD	42	-	2	1.4
PE-HD	58	-	2	1.4
PA 11	71	-	1	1.81
PE-HD	58	-	1	2
PE-HD	60	-	2	2
PA 6	70	-	1	2.21
Aluminum	-	39	2	2.68
Messing	-	150	1	2.67
Aluminum	-	29	2	3.23
PA 11	69	-	1	3.31
PE-HD	52	-	2	4.2
PE-HD	78	-	2	6.3
PF-laminate	89	-	2	8.2
PVC-U	76	-	2	8.5
Glass	(6...7 Mohs)	-	2	9.7
Lead	(4 Brinell)	-	2	10.5
PMMA	85	-	2	10.75
PF-laminated fabric	92	-	2	18.5
EP-glass fiber	86	-	2	19.5
EP-silica flour	84	-	2	31

* Nr. 1: Vickers hardness 500, grain size 0.9 mm;
Nr. 2: Vickers hardness 720 to 810, grain size 0.3 to 0.5 mm
** Abrasion rate referring to steel St. 37

Table A.38: Relative Volumetric Abrasion Values for Selected Materials in a Sand Slurry

Material	ρ (g/cm^3)	Rel. volumetric abrasion value
PE-HD	0.95	330
PE-MD	0.92	600
PE-HD-UHMW	0.94	100
PE-HD-UHMW (filled with glass beads)	1.14	165
PE-HD-UHMW (flame retarded)	0.99	150
PE-HD-UHMW (antistatic)	1.02	150
POM-copolymer	1.42	700
Pertinax	1.4	2500
PMMA	1.31	1800
PVC-U	1.33	920
PP	0.9	440
PTFE	1.26	530
PTFE-GF 25	2.55	570
PTFE + 25% carbon black	2.04	960
PTFE-GF 25 + metal compounds	2.33	750
PET	1.4	610
Beech wood	0.83	2700
EP + 50 silica flour	1.53	3400
PA 66	1.13	160
PA 12	1.02	260
Cast polyamide	1.14	150
Steel St 37	7.45	160

Table A.39: Erosion Velocity for Rain Erosion

Plastic	Erosion velocity E_{m2} μm/s
PUR elastomer, 79 Shore	0.2
PUR elastomer, 93 Shore	0.58
PUR elastomer, 80 Shore	0.77
PUR elastomer, 92 Shore	0.96
PUR elastomer, 91 Shore	3.64
PA 616	1.49
PA 6	1.92 to 2.37
PA 610	4.17
PA 11	8.63
PE-LD	3.56 to 18.3
PE-HD	2.65 to 19.5
PE-X	10.5
PP	8.48 to 8.93
PMMA	26.8 to 113.7
AMMA	100
PVC	26.2
PVC (2% plasticizer)	38.45
PVC (4% plasticizer)	111
POM-copolymer	2.4
PC	20.3
ABS	20.35 to 21.4
CAB	61.5
UP-GF	244
PS	384.5

Table A.40: Bearing Strength and Wear of Journal Bearings Sliding on Steel

Plastic	p · v-Value [N/mm² · m/s]			Relative wear factor (PA 6 = 1)
	Sliding velocity			
	0.05 m/s	0.5 m/s	5 m/s	
PE-HD	-	0.2	-	-
PE + PTFE (80+20%)	-	-	-	0.22
PS	0.02	0.05	0.02	15
ABS	-	-	-	17.5
PTFE	0.09 -0.03	0.09 -0.03	0.09 -0.03	5
POM	0.13	0.12	0.008 -0.05	0.32
PPE + PS	0.02	0.02	0.02	-
PC	0.02	0.02	<0.02	12.5
PBT	0.14		0.1- 0.08	1.05
PA 6	0.04	0.08	0.07	1
PA + PE 90-10	-	0.25		-
PA + PTFE 80-20	0.4	0.7	0.5	0.075
PA 11; PA 12	0.06	-	0.03	-
PA 66	0.1	0.08	0.05	1
PA+PTFE 80-20	0.5	0.6	0.3	
PA 610; PA 612	0.08	0.07	<0.07	0.9
PSU	0.2	0.2	0.1	7.5
PES	0.6	1	0.57	0.3
PPS	0.08	0.1	0.13	2.7
PI	-	1	-	-
PUR elastomer	0.07	0.05	<0.05	1.7

Table A.41: Influence of Additives on Friction Behavior of PA 6 and PTFE

Additive (wt.-%) to PA 6				Relative values				
Lubricants		Reinforcements						
PTFE	Silicone	Glass fibers	Carbon fibers	p · v-value Slip velocity v			Friction	Wear
				0.05 m/s	0.5 m/s	5 m/s		
-	-	-	-	1	1	1	1	1
20	-	-	-	5	10	4	0.7	0.075
-	2	-	-	1.5	2	5	0.5	0.25
18	2	-	-	5	12	7	0.4	0.055
-	-	30	-	4	4	4	1.2	0.45
15	-	30	-	7	10	9	1	0.085
15	2	30	-	8	8	10	0.8	0.05
-	-	-	30	7	11	5	0.8	0.15
Additives (wt.-%) to PTFE				p · v-value Slip velocity v				
Glass fibers	Bronze	Graphite	MoS$_2$	0.05 m/s	0.5 m/s	5 m/s	Friction	Wear
-	-	-	-	1	1	1	1	-
15	-	-	-	8	7	6	1.2	-
25	-	-	-	8	7	6	1.3	-
-	60	-	-	12	9	11	1.2	-
20	-	5	-	9	8	9	1.3	-
15	-	-	5	9	8	7	1.2	-

Table A.42: Coefficient of Dynamic Friction for Plastics

Plastic	Coefficient of dynamic friction
PE-HD-HMW	0.29
PE-HD	0.25
PE-LD	0.58
PP	0.3
PS	0.46
SAN	0.52
PTFE	0.22
PMMA	0.54
PET	0.54
POM homopolymer	0.34
POM copolymer	0.32
POM + graphite	0.29
POM + 22% PTFE fiber	0.26
PPE + PS	0.35
PA 6	0.38 - 0.45
PA 6-cast	0.36 - 0.43
PA 6-GF 35	0.30 - 0.35
PA 11	0.32 - 0.38
PA 66	0.25 - 0.45
PA 66 + 11% PE	0.19
PA 66 + 3% MoS_2	0.32 - 0.35
PA 66-GF 35	0.32 - 0.36
PA 610	0.36 - 0.44
PES	0.3
PES + 15% PTFE + 30% GF	0.2
PES + 15% PTFE	0.14
PSU (0.5 to 5 m/s)	0.50 - 0.23
PBMI + 25% graphite	0.25
PBMI + 40% graphite	0.2
PBMI + PTFE	0.17

Table A.43: Permissible Lubrication Film Temperatures (Source: D.Bopp)

Plastic	Permissible temperature, °C
PBI	350
PI	300
PAI	260
PEEK	250
PTFE, sintered	250
PPS	220
PEI	200
PK	180

Table A.44: Lubricants for Plastics (Source: Fuchs Lubritech)

Group	Base oil	Type of lubricant	Thickened/Grease	Service $T\,°C$
1	Silicone	Oil	-	-40 to +200
1	Silicone	Paste	MoS_2	-40 to +175
1	Silicone	Paste	Highly dispersed silicic acid	-40 to +200
1	Silicone	Oil	Li-complex soap	-40 to +200
2	Ester	Oil	Silica gel-PTFE	-40 to +180
3	Perfluorated polyether	Oil	PTFE	-25 to +250 (280)
4	Polyglycol	Oil	-	to 200
5	Polyglycol	Paste	MoS_2	-30 to +400
6	Poly-α-olefin (PAO)/ester	Oil	Li-soap + grease	-45 to +130
6	Poly-α-olefin (PAO)/ester	Paste	Li-soap + grease	-45 to +110
7	Poly-α-olefin (PAO)/ester	Paste	Silica gel + grease	-45 to +110 (180)

Table A.45: Compatibility of Lubricants (see Table A.44) with Plastics and Rubbers (Source: Fuchs Lubritech)

Plastic	1	2	3	4	5	6	7
PE-LD	+		+	+			−
PE-HD	+		+	+	+	+	+
PP	+		+	+			+
ABS	o		+			o	+
PVC			+	+			+
PTFE	+	+	o	+	+	+	+
PET	+		+			+	+
PoM	o	+	+	+	+	+	+
PA	+	+	+	+	+	+	+
PPE	+		+				
PUR	+		+	+		+	+
Elasto							
NR	−		+	+	+	−	+
CR	o	−	+	+	+	+	+
SBR			+	+	+	−	+
NBR	o	−	+	+	+	+	+
EPDM	+	+	+	+	+	−	o
FKM	+	+		+	o	+	o
MVQ	−	+	+	+	+	+	+

+: Resistant, o: Somewhat resistant, -: not resistant

APPENDIX B

LITERATURE

B.1 BOOKS IN THE PLASTICS TECHNOLOGY FIELD

Hundreds of monographs and edited books have been published in the long history of plastics technology. This appendix presents English language books that are currently available from various publishers. With the exception of a couple of classic books, in this list we incorporated books that are at most ten years old.

A Beginner's Guide To Rubber Technology, R. J. Del Vecchio, Technical Consulting Services, 2001.

Adhesion & Adhesive Technology 2nd Ed., A.V. Pocius, Hanser Publisher, 2002.

Advancing Sustainability Through Green Chemistry and Engineering, Eds. R. L. Lankey and P. T. Anastas, Oxford University Press, 2005.

Aging and Chemical Resistance, C. Bonten, Hanser Publisher, 2001.

Basic Injection Molding, W. J. Tobin, SPE, 2000.

Basic Requirements To Achieve Quality and Productivity In The Molding Process, P. N. Canovi, Processing New Technologies Consulting, 2004.

Billmeyer and Saltzman's Principles of Color Technology, 3rd Edition, R. S. Berns, John Wiley & Sons, 2000.

Biocatalysis In Polymer Science, R. A. Gross and H. N. Cheng, Oxford Press, 2003.

Blow Molding Handbook 2nd Ed., D.V. Rosato, Hanser Publisher, 2003.

Coatings of Polymers and Plastics, Eds. R. Ryntz and P. Yaneff, Marcel Dekker, 2003.

Color: A Multidisciplinary Approach, H. Zollinger, John Wiley & Sons, 2000.

Coloring of Plastics Fundamentals, 2nd Edition, R. A. Charvat, John Wiley & Sons, 2004.

Coloring of Plastics, A. Muller, Hanser Publisher, 2003.

Coloring Technology for Plastics, Edited By R. M. Harris, SPE/Plastics Design Library, 1999.

Compression Molding, B.A. Davis, P.J. Gramann, C.A. Rios and T.A. Osswald, Hanser Publisher, 2003.

Computer-Aided Injection Mold Design and Manufacture, J.Y.F. Fuh, Y.F. Zhang, A.Y.C. Nee, M.W. Fu and F.Y.H. Fuh, Marcel Dekker, 2004.

Concise Encyclopedia of Plastics, D.V. Rosato, D.V. Rosato, M. G. Rosato, Kluwer Academic Publishers, 2000.

Design Data for Plastics Engineers, Hanser Publisher, 1998.

Design formulas Plastics Engineers, 2nd Ed., N. Rao, Hanser Publisher, 2004.

Designing Plastic Parts for Assembly, 6th Ed., P.A. Tres, Hanser Publisher, 2006.

Designing With Plastics, G. Erhard, Hanser Publisher, 2006.

Die Makers Handbook, J. Arnold, Industrial Press, 2000.

Discovering Polyurethanes, K. Uhlig, Hanser Publisher, 1999.

Dynamics of Polymeric Liquids, Volume 1, Fluid Mechanics, 2nd Edition, R. B. Bird, R.C. Armstrong and O. Hassager, John Wiley & Sons, 1987.

Dynamics of Polymeric Liquids, Volume 2, 2nd Ed., R. B. Bird, C. F. Curtiss, R. C. Armstrong, and O. Hassager, John Wiley & Sons, 1989.

Elastomer Molding Technology, J. G. Sommer, Elastech, 2003.

Electrical Properties of Polymers, E. Riande and R. Diaz-Calleja, Marcel Dekker, 2004.

Engineering Design With Polymers and Composites, James Gerdeen, Harold Lord, Ronald Rorrer, CRC Press, 2005.

Engineering Thermoplastics, L. Bottenbruch, Hanser Publisher, 1996.

Engineering With Polymers 2nd Edition, P.C. Powell and A.J.I. Housz, Stanley Thornes Publishers,1998.

Engineering With Rubber 2nd Ed., A.N. Gent, Hanser Publisher, 2001.

Experimental Design for Injection Molding , J. P. Lahey and R. G. Launsby, Launsby Consulting , 1998.

Extrusion Blow Molding, M. Thielen, Hanser Publisher, 2001.

Extrusion Control, H.E. Harris, Hanser Publisher, 2004.

Extrusion Dies for Plastics and Rubber 3rd Ed., W. Michaeli, Hanser Publisher, 2003.

Extrusion of Polymers, C. Chung, Hanser Publisher, 2000.

Extrusion Processing Data, A. Naranjo, Hanser Publisher, 2001.

Extrusion: The Definitive Processing Guide and Handbook, Harold F. Giles, Jr., John R. Wagner, Jr., Eldridge M. Mount, Iii, William andrew Publishing, 2005.

Feeding Technology for Plastics Processing, D.H. Wilson, Hanser Publisher, 1998.

Film Processing, T. Kanai, Hanser Publisher, 1999.

Fire and Polymers: Materials and Solutions for Hazard Prevention, G. L. Nelson, C. A. Wilie, Oxford University Press , 2001.

Fluoropolymers Applications In Chemical Processing Industries - The Definitive User's Guide and Databook , Sina Ebnesajjad, Pradip R. Khaladkar, William andrew Publishing, 2004.

Foam Extrusion: Principles and Practice, Ed. S.-T. Lee, CRC Press , 2000.

Functional Fillers for Plastics , Editor: Marino Xanthos, John Wiley & Sons, 2005.

Fundamentals of Injection Molding, 2nd Edition, William J. Tobin, Wjt Associates , 2005.

B.1 Books in the Plastics Technology Field

Giant Molecules: Essential Materials for Everyday Living and Problem Solving, 2nd Edition, Charles E. Carraher, Jr., John Wiley & Sons , 2003.

Glossary of Plastics Terminology, W. Glenz, Hanser Publisher, 2004.

Handbook of Antiblocking, Release and Slip Additives, G. Wypych, Chemtec Publishing , 2005.

Handbook of Elastomers, 2nd Edition, Eds. A. K. Bhowmick and H. L. Stephens, Marcel Dekker , 2001.

Handbook of Fillers–The Definitive User's Guide and Databook of Properties, Effects and Uses, 2nd Edition , G. Wypych, Plastics Design Library , 1999.

Handbook of Molded Part Shrinkage and Warpage, J. M. Fischer, Plastics Design Library , 2003.

Handbook of Plastics Testing Technology, 2nd Ed., V. Shah, John Wiley & Sons , 1998.

Handbook of Polyethylene: Structures, Properties, and Applications, A. J. Peacock, Marcel Dekker, 2000.

Handbook of Polymer Degradation, 2nd Edition, Ed. S. Halim Hamid, Marcel Dekker, 2000.

Handbook of Polymer Testing, Ed. R. Brown, Marcel Dekker , 1999.

Handbook of Polypropylene and Polypropylene Composites, 2nd Edition, Edited By Harutun G. Karian, Marcel Dekker , 2003.

Handbook On Basics of Coating Technology, Artur Goldschmidt, Hans-Joachim Streitbeger. Vincentz Network , 2003.

Hollow Plastic Parts, G.L. Beall and J.L. Throne, Hanser Publisher, 2004.

How To Improve Rubber Compounds, J.S. Dick, Hanser Publisher, 2004.

How To Make Injection Molds 3rd Ed, G. Menges, Hanser Publisher, 2001.

Industrial Design of Plastics Products, M. J. Gordon, Jr., John Wiley & Sons, 2002.

Industrial Inorganic Pigments, 2nd Edition, Ed. G. Buxbaum, John Wiley & Sons, 1998.

Industrial Plastics: Theory and Applications, 3rd Edition, T. L. Richardson and E. Lokensgard, Delmar, 1997.

Injection Mold Tooling Standards: A Guide for SPEcifying, Purchasing and Qualifying Injection Molds, WJT Associates, 1993.

Injection Molding Handbook, T.A. Osswald, P.J. Gramann and L.S. Turng, Hanser Publisher, 2001.

Injection Molding Handbook, D.V. Rosato, D.V. Rosato and M.G. Rosato, 3rd Edition, Kluver Academic Publishers, 2000.

Injection Molding, G. Pštsch, Hanser Publisher, 1995.

Injection Molding Processing Data, A. Naranjo C., M. d. P. Noriega E., J. R. Sanz and J. D. Sierra, Hanser Publisher, 2001.

Injection Molding Reference Guide, 4th Edition, J. Carender, Advanced Process Engineering, 1997.

Injection Molding Troubleshooting Guide, 2nd Edition, J. Carender, Advanced Process Engineering, 1996.

Injection Molds - 130 Proven Designs, H. Gastrow, E. Linder and P. Unger, Hanser Publisher, 2002.

Inorganic and Organometallic Polymers, R. D. Archer, John Wiley & Sons, 2001.

Introduction To Polymers, 2nd Edition, R.J. Young and P.A. Lovell, Chapman & Hall, 1996.

Introduction To Structural Foam, S. Semerdjiev, SPE, 1982.

B.1 Books in the Plastics Technology Field

Introduction To The Dimensional Stability of Composite Materials, E. G. Wolff, Destech Publications, Inc., 2004.

Joining of Plastics 2nd Ed., J. Rotheiser, Hanser Publisher, 2004.

Managing Variation for Injection Molding, J. Carender, Advanced Process Engineering, 2003.

Materials Science of Polymers for Engineers, 2nd Ed, T.A. Osswald and G. Menges, Hanser Publisher, 2003.

Math Skills for Injection, J. Carender, Advanced Process Engineering, 1998.

Metallocene Technology In Commercial Applications, Ed. Dr. G. M. Benedikt, SPE/Plastics Design Library, 1999.

Metallocene-Based Polyolefins, Preparation, Properties and Technology, 2-Volume Set, Ed. J. Scheirs and W. Kaminsky, John Wiley & Sons, 2000.

Microcellular Processing, K. Okamoto, Hanser Publisher, 2003.

Modeling in Materials Processing, J.A. Dantzig and C.L. Tucker Iii, Cambridge University Press, 2001.

Modern Plastics Handbook, Eds. Modern Plastics and Charles A. Harper, Modern Plastics, 2000.

Mold Design for Plastics Injection Molding, 3rd Edition, E. P. Allyn, Allyn Air Publication, 1998.

Mold Engineering 2nd Ed., H. Rees, Hanser Publisher, 2002.

Mold Finishing & Polishing Manual, SPE, 1999.

Mold Making Handbook, K. Stoeckhert, Hanser Publisher, 1998.

Molecular Simulation Methods for Predicting Polymer Properties, V. Galiatsatos, John Wiley & Sons, 2005.

Multi-Scale Modeling of Composite Material Systems, Eds. C. Soutis and P. W. R. Beaumont, CRC Press, 2005.

Nylon Plastics Handbook, M.I. Kohan, Hanser Publisher, 1995.

Permeability Properties of Plastics and Elastomers, 2nd Ed., L. K. Massey, William andrew Publishing, 2003.

Plastic Injection Molding–Manufacturing Process Fundamentals, D. M. Bryce, SME, 1996.

Plastic Injection Molding–Manufacturing Startup and Management, D. M. Bryce, SME, 1999.

Plastic Injection Molding–Material Selection & Product Design Fundamentals, D. M. Bryce, SME, 1997.

Plastic Injection Molding–Mold Design and Construction Fundamentals, D. M. Bryce, SME, 1998.

Plastic Part Design for Injection Mold, R. Malloy, Hanser Publisher, 1994.

Plastics Ð Materials and Processing, A.B. Strong, Prentice Hall, 2000.

Plastics Additives Handbook, 5th Ed., H. Zweifel, Hanser Publisher, 2000.

Plastics and Composites Welding Handbook, D. Grewell and A. Benatar, Hanser Publisher, 2003.

Plastics and The Environment, Ed. A. L. Landry, John Wiley & Sons, 2003.

Plastics Design Handbook, D.V. Rosato, D.V. Rosato, M. G. Rosato, Kluwer Academic Publishers, 2001.

Plastics Engineering, 3rd Ed., R.J. Crawford, 1998.

Plastics Failure Analysis and Prevention, J. Moalli, Plastics Design Library, 2001.

Plastics Failure Guide, M. Ezrin, Hanser Publisher, 1996.

Plastics Flammability Handbook, 3rd Ed., J. Troitzsch, Hanser Publisher, 2004.

Plastics Institute of America Plastics Engineering, Manufacturing and Data Handbook- Two Volumes, D. V. Rosato, N. R. Schott, D. V. Rosato, M. G. Rosato, Kluwer Academic Publishers, 2001.

Plastics Packaging, 2nd Ed., S. Selke, J. Culter and R. Hernandez, Hanser Publisher, 2004.

Plastics Processing, W. Michaeli, Hanser Publisher, 1995.

Plastics: How Structure Determines Properties, G. Gruenwald, Hanser Publisher, 1992.

Pocket Injection Mold Engineering Standards, J. Carender, Advanced Process Engineering, 1999.

Pocket Performance SPEcs for Thermoplastics, 4th Ed., IDES, 2005.

Pocket SPEcs for Injection Molding, 6th Ed., IDES, 2004.

Polymer Crystallization: The Development of Crystalline Order In Thermoplastic Polymers, J. Schultz, Oxford University Press, 2001.

Polymer Extrusion, 4th Ed, C. Rauwendaal, Hanser Publisher, 2001.

Polymer Mixing - A Self Study Guide, C. Rauwendaal, Hanser Publisher, 1998.

Polymer Mixing, Technology & Engineering, J. White, Hanser Publisher, 2001.

Polymer Modification: Principles, Techniques, and Applications, J. Meister, Marcel Dekker, 2000.

Polymer Modifiers and Additives, J. T. Lutz, Jr. and R. F. Grossman, Marcel Dekker, 2000.

Polymer Processing : Principles and Design, D. G. Baird, and D. I. Collias, John Wiley & Sons, 1998.

Polymer Processing Fundamentals, T.A. Osswald, Hanser Publisher, 1998.

Polymer Processing Instabilities: Control and Understanding, Ed. S. Hatzikiriakos and K. Migler, Marcel Dekker, 2005.

Polymer Science and Technology, 2nd Ed., J.R. Fried, Prentice Hall, 2003.

Polymer Surfaces and Interfaces III, Ed. R. W. Richards, S. K. Peace, John Wiley & Sons, 1999.

Polymeric Compatibilzers, S. Datta, Hanser Publisher, 1996.

Polymeric Foams and Foam Technology, D. Klempner, Hanser Publisher, 2004.

Polymeric Materials, G.W. Ehrenstein, Hanser Publisher, 2001.

Polymers From The Inside Out: An Introduction To Macromolecules, A. E. Tonelli and M. Srinivasarao, John Wiley & Sons, 2001.

Polypropylene Handbook, 2nd Edition, N. Pasquini, Hanser Publisher, 2005.

Powder Coatings Chemistry and Technology, 2nd Edition, P.G. De Lange, Vincentz Network, 2004.

Principles of Polymer Processing 2nd Ed., Z. Tadmor and C.G. Gogos, John Wiley & Sons, 2006.

Properties of Polymer 3rd Ed., D.W. Van Krevelen, Elsevier, 2003.

PVC Handbook, C.E. Wilkes, Hanser Publisher, 2005.

Qualifications, Startups and Tryouts of Injection Molds, W. J. Tobin, WJT Associates LLC, 2003.

Quality Control Manual for Injection Molding, W. J. Tobin, T/C Press, 2000.

Radiation Technology for Polymers, J.G. Drobny, CRC Press, 2003.

Rapid Prototyping, A. Gebhardt, Hanser Publisher, 2003.

Reactive Extrusion, M. Xanthos, Hanser Publisher, 1992.

Reactive Polymer Blending, S. Baker, Hanser Publisher, 2001.

Recycling and Recovery of Plastics, J. Brandrup, Hanser Publisher, 1996.

Rheology In Plastics Quality Control, J.M. Dealy, Hanser Publisher, 2000.

Rotational Molding, G. Beall, Hanser Publisher, 1998.

Rubber Processing - An Introduction, P.S. Johnson, Hanser Publisher, 2001.

Rubber Processing, J.L. White, Hanser Publisher, 1995.

Rubber Technology, J.S. Dick, Hanser Publisher, 2001.

Runner and Gating Design Handbook, J.P. Beaumont, Hanser Publisher, 2004.

Selecting Injection Molds, B. Catoen and H. Rees, Hanser Publisher, 2006.

Simulation Methods for Polymers, Ed. D. N. Theodorou and M. Kotelyanskii, Marcel Dekker, 2004.

Solid Phase Processing of Polymers, I.M. Ward, Hanser Publisher, 2000.

SPC In Injection Molding and Extrusion, C. Rauwendaal, Hanser Publisher, 2000.

SPE Guide on Extrusion Technology & Troubleshooting, SPE, 2001.

SPE Handbook on Single Screw Extrusion, SPE, 2003.

SPE Handbook on Twin Screw Extrusion, SPE, 2003.

Strategic Management for the Plastics Industry, R. F. Jones, CRC Press, 2002.

Structure and Rheology of Molten Polymers, J.M Dealy and R.G. Larson, Hanser Publisher, 2006.

Successful Injection Molding, J. Beaumont, Hanser Publisher, 2002.

Surfactants, K.R. Lange, Hanser Publisher, 1999.

Synthesis and Properties of Silicones and Silicone-Modified Materials, Ed. S. Clarson, J. Fitzgerald, M.Owen, S. Smith and M. Van Dyke, Oxford Press, 2003.

Synthetic Fibers, F. Fourne, Hanser Publisher, 1999.

Technology of Thermoforming, J. Throne, Hanser Publisher, 1996.

The Effects of Sterilization Methods On Plastics and Elastomers, 2nd Edition, L. Masses, William Andrew Publishing, 2005.

The First Snap-Fit Handbook, 2nd Ed., P. R. Bonenberger, Hanser Publisher, 2005.

The Handbook of Advanced Materials - Enabling New Designs, J. K. Wessel, Wiley Interscience, 2004.

Thermal Analysis of Plastics, G.W. Ehrenstein, Hanser Publisher, 2004.

Thermoforming - A Practical Guide, A. Illig, Hanser Publisher, 2000.

Thermoforming, A Plastics Processing Guide, 2nd Ed., G. Gruenwald, CRC Press, 1998.

Thermoforming: Improving Process Performance, S. R. Rosen, SME, 2002.

Thermoplastic Elastomers 3rd Ed., G. Holden, Hanser Publisher, 2004.

Thermoplastic Foam Extrusion, J.L. Throne, Hanser Publisher, 2004.

Thermoplastic Foam Processing: Principles and Application, Ed. R. Gendron, CRC Press, 2005.

Toughening of Plastics: Advances In Modeling and Experiments, Ed. R. A. Pearson, H.J. Sue, A. F. Yee, Oxford University Press, 2000.

Training In Injection Molding 2nd Ed., W. Michaeli, Hanser Publisher, 2001.

Training In Plastics Technology 2nd Ed., W. Michaeli, Hanser Publisher, 2000.

Transport Phenomena, 2nd Ed., R.B. Bird, W. E. Stewart, and E. N. Lightfoot, John Wiley & Sons, 2002.

Troubleshooting Injection Molded Parts, W. J. Tobin, WJT Associates, 1996.

Troubleshooting The Extrusion Process, M.d.P.Noriega, and C. Rauwendaal, Hanser Publisher, 2001.

Understanding Compounding, R.H. Wildi, Hanser Publisher, 1998.

Understanding Design of Experiments, R.J. DelVecchio, Hanser Publisher, 1997.

Understanding Extrusion, C. Rauwendaal, Hanser Publisher, 1998.

Understanding Injection Mold Design, H. Rees, Hanser Publisher, 2001.

Understanding Injection Molding Technology, H. Rees, Hanser Publisher, 1994.

Understanding Plastics Packaging Technology, S. Selke, Hanser Publisher, 1997.

Understanding Plastics Testing, D. Hylton, Hanser Publisher, 2004.

Understanding Product Design for Injection Molding, H. Rees, Hanser Publisher, 1996.

Understanding Rheology, F.A. Morrison, Oxford University Press, 2001.

Understanding Thermoforming, J. Throne, Hanser Publisher, 1999.

Volume Polymers In North America and Western Europe, W. C. Kuhlke, Rapra Technology Ltd., 2001.

Wear In Plastics Processing, G. Mennig, Hanser Publisher, 1995.

What Is A Mold? An Introduction To Plastic Injection Molding and Injection Mold Construction, Tech Mold Inc., 1998.

B.2 JOURNALS AND TRADE MAGAZINES

B.2.1 Trade Magazines

English

- British Plastics & Rubber Online, www.polymer-age.co.uk
- European Plastics News, www.emap.com
- Modern Plastics, www.modplas.com
- Modern Plastics Worldwide, www.modplas.com
- Plast Europe, www.kunststoffe.de/pe
- Plastics Engineering (SPE trade journal), www.4spe.org
- Plastics Machining & Fabricating, www.plasticsmachining.com
- Plastics News, www.plasticsnews.com
- Plastics News International, www.plasticsnews.com
- Plastics and Rubber Weekly, www.prw.com
- Plastics Technology, www.plasticstechnology.com
- Rubber & Plastics News, www.rubbernews.com

French

- Caoutchoucs et Plastiques
- Plastiques & Elastomères, www.plastiquesmagazine.com

German

- Kunststoffe, www.kunststoffe.de
- Plastverarbeiter, www.plastverarbeiter.de
- Kunststoffberater, www.giesel.de

Italian

- Macplas, www.macplas.it

Spanish

- Plásticos Modernos (Spain), www.plastunivers.es
- Revista de Plásticos Modernos (Spain), www.revistaplasticosmodernos.com
- Tecnología del Plástico (Latin America), www.tecnologiadelplastico.com

B.2.2 Archival Journals

Society of Plastics Engineers

- Journal of Polymer Engineering & Science
- Journal of Polymer Composites
- Journal of Vinyl & Additive Technology

http://www.4spe.org/pub/journals/index.php

Polymer Processing Society

- International Polymer Processing

www.poly-eng.uakron.edu/pps

Society of Rheology

- Journal of Rheology

www.rheology.org

Scientific Alliance of Polymer Technology (WAK)

- Journal of Plastics Technology (Zeitschrift Kunststofftechnik)

http://files.hanser.de/files/fachzeitschriften/wak/informationen.htm

Other

- Journal of Polymer Engineering

www.freundpublishing.com/Journal_Polymer_Engineering/polprev.htm

APPENDIX C

POLYMER RESEARCH INSTITUTES

POLYMER RESEARCH CENTERS IN ARGENTINA

Centro de Investigación Técnica de la Industria del Plástico (CITIP)
Argentina
www.inti.gov.ar/citip

POLYMER RESEARCH CENTERS IN AUSTRALIA

Centre for Advanced Macromolecular Deisgn (CAMD)
Professor Tom P. Davis, Director
Scool of Chemical Engineering and Industrial Chemistry
The University of New South Wales
UNSW SYDNEY NSW 2052
Tel: 61 (2) 9385 4371 Fax: 61 (2) 9385 6250
www.ceic.unsw.edu.au/centers/CAMD/

Centre for Applied Polymer Science
Professor Robert P. Burford, Director
School of Chemical Engineering and Industrial Chemistry
The University of New South Wales
Sydney 2052 Australia
Tel: 61 (2) 9385 4308 Fax: 61 (2) 9385 5966
www.ceic.unsw.edu.au/centers/polyceic.htm

Intelligent Polymer Research Institute
Professor Gordon Wallace, Director
University of Wollongong
Wollongong NSW 2522, Australia
Tel: 02-42213127 Fax: 02-42213114
http://www.uow.edu.au/science/research/ipri/

Key Centre for Polymer Colloids
Professor Robert G. Gilbert, Director
School of Chemistry, F11
University of Sydney, NSW 2006
Australia
Tel: 61-2-9351 3366 Fax: 61-2-9351 8651
www.kcpc.usyd.edu.au

POLYMER RESEARCH CENTERS IN BELGIUM

Mechanische Materiaalkunde
Herr Prof. Dr. Ir. I. Verpoest, Leuven, Belgien
K.U.Leuven, Kasteelpark Arenberg 44, 3001 Leuven, Belgium
Tel: 32 16/32-1306, Fax: -1990
www.mtm.kuleuven.ac.be

POLYMER RESEARCH CENTERS IN BRAZIL
Instituto do PVC
Rua James Watt, 142 - 12 andar - conj. 122 - CEP 04576-050 Brooklin
São Paulo - SP - Brazil
Tel/Fax: (550 - 11 - 5506-5211
www.institutodopvc.org

POLYMER RESEARCH CENTERS IN CANADA

Institute for Polymer Research
Alex Penlidis, Director
University of Waterloo
Waterloo, Ontario, Canada
N2L 3G1
Tel: 519-888-4567 Ext. 6634 Fax: 519-746-4979
http://www.ipruw.com

POLYMER RESEARCH CENTERS IN CHILE

Centro de Investigación de Polímeros Avanzados (CIPA)
Chile
www.cipachile.cl
e-mail: csilva@udec.cl

POLYMER RESEARCH CENTERS IN COLOMBIA

Instituto de Capacitación e Investigación del Plástico y del Caucho (ICIPC)
The Rubber and Plastic Institute for Training and Research
Carrera 49 # 5 Sur-190
Medellín, Colombia
Tel: (57) 4 - 311-64-78, Fax:(57) 4- 311-63-81
www.icipc.com

Centro de Investigación en Procesamiento de Polímeros (CCIPP)
Carrera 65B #17A-11, Bogotá D.C., Colombia
Tel: (57) 1 - 4055831, Fax: (57) 1 - 4055862
cipp.uniandes.edu.co

POLYMER RESEARCH CENTERS IN FRANCE

Geopolymer Institute
16, rue Galilée - 02100 Saint-Quentin - France
Fax: 33/ 323 678 949
www.geopolymer.org

POLYMER RESEARCH CENTERS IN GERMANY

FB Ingenieurwissenschaft Kunststoftechnik
Herr Prof. Dr.-Ing. H.-J. Radusch
Institut für Werkstofftwissenschaft
Martin-Luther-Universität Halle-Wittenberg, 06099 Halle
Tel: 03461/46 2792, Fax: -3891
www.iw.uni-halle.de/index.htul

Fraunhofer Institut für Angewandte Polymerforschung IAP
Dr. Ulrich Buller, Institute director
Wissenschaftspark Golm
Geiselbergstraße 69, 14476 Potsdam
Tel: 49 (0) 331/ 568 - 10 Fax: 49 (0) 331/ 568 - 3000
www.iap.fraunhofer.de

Institut für Flugzeugbau
Herr Prof. Dr.-Ing. K. Drechsler,
Universität Stuttgart, Pfaffenwaldring 31, 70569 Stuttgart
Tel: 0711/685 2411 Fax: -2449
www.ifb.uni-stuttgart.de

Institut für Allgemeinen Maschinenbau und Kunststofftechnik (IMK)
Herr Prof. Dr.-Ing. G. Mennig
Technische Universität Chemnitz
Reichenhainer Strasse 70, 09126 Chemnitz
Tel: 0371/531-2383, Fax: -3776
www.tu-chemnitz.de/mbv/kunstStTechn/

Institut für Kunststofftechnik
Herr Prof. Dr.-Ing. H. Potente
Universität-GH Paderborn, Warburgerstr. 100, 33100 Paderborn
Tel: 05251/60-2451, Fax: -3821
www.ktp.cc

Institut für Kunststofftechnologie (IKT),
Herr Prof. Dr.-Ing. H.-G. Fritz,
Universität Stuttgart, Böblinger Strasse 70, 70199 Stuttgart
Tel: 0711/641-2317, Fax: -2335
www.ikt.uni-stuttgart.de

Institut für Leichtbau und Kunststofftechnik ILK, TU Dresden
Herr Prof. Dr.-Ing. habil. W. Hufenbach
Dürerstrasse 26, 01062 Dresden
Tel: 0351/463-8142, Fax: -8143
www.tu-dresden.de/mw/ilk

Institut für Polymertechnik/Kunststofftechnikum TU Berlin
Herr Prof. Dr.-Ing. M.H. Wagner,
Fasanenstrasse 90, 10623 Berlin
Tel: 030/314-24271, Fax: -21108
www.tu-berlin.de/fb6/polymer

Institut für Polymerwerkstoffe und Kunststofftechnik
Herr Prof. Dr.-Ing. G. Ziegmann
Technische Universität Clausthal-Zellerfeld, 38678 Clausthal-Zellerfeld
Tel: 05323/72-2090, Fax: -2324
www.puk.tu-clausthal.de

Institut für Produkt Engineering
Herr Prof. Dr.-Ing. J. Wortberg
Mercator-Universität-GH Duisburg, Lotharstrasse 1, 47048 Duisburg
Tel: 0203/379-3252, Fax: -4379
www.ipe.uni-duisburg.de

Institut für Verbundwerkstoffe GmbH
Herr Prof. Dr.-Ing. A. K. Schlarb
Erwin-Schrödinger-Strasse 58, 67663 Kaiserslautern
Tel: 0631/2017-101, Fax: -199, www.ivw.uni-kl.de

Institut für Verbundwerkstoffe GmbH
Herr Prof. Dr.-Ing. Dr. h.c. K. Friedrich,
Erwin-Schrödinger-Strasse 58, 67663 Kaiserslautern
Tel: 0631/2017-102, Fax: -199
www.ivw.uni-kl.de

Lehrgebiet Kautschuktechnik
Herr Prof. Dr.-Ing. E. Haberstroh
Institut für Kunststoffverarbeitung in Industrie und Handwerk (IKV)
RWTH Aachen, Pontstrasse 49, 52056 Aachen
Tel: 0241/803806, Fax: /8888262
www.rwth-aachen.de/ikv

Lehrgebiet Zerstörungsfreie Prüfung
Herr Prof. Dr. rer. nat. habil. G. Busse,
Institut für Kunststoffprüfung und Kunststoffkunde (IKP)
Universität Stuttgart, Pfaffenwaldring 32, 70569 Stuttgart
Tel: 0711/685-2657, Fax: -2066
www.ikp.uni-stuttgart.de

Lehrstuhl für Kunststofftechnik (LKT)
Herr Prof. Dr.-Ing. E. Schmachtenberg,
Demonstratrionszentrum für Faserverbundwerkstoffe,
Universität Erlangen-Nürnberg,
Am Weichselgarten 9, 91058 Erlangen-Tennenlohe
Tel: 09131/85-29700, Fax: -29709
www.lkt.uni-erlangen.de

Lehrstuhl für Kunststoff- und Recyclingtechnik,
Herr Prof. Dr.-Ing. A. K. Bledzki,
Institut für Werkstofftechnik
Uni-GH Kassel, Möncheberstrasse 3, 34109 Kassel
Tel: 0561/80436-90, Fax: -92
www.kutech-kassel.de

Institut für Kunststoffverarbeitung in Industrie und Handwerk (IKV)
Herr Prof. Dr.-Ing. Dr.-Ing. E.h. W. Michaeli
RWTH Aachen, Pontstrasse 49, 52056 Aachen
Tel: 0241/803806, Fax: /8888262
www.rwth-aachen.de/ikv

Lehrstuhl für Polymere Werkstoffe
Herr Prof. Dr.-Ing. V. Altstädt
Universität Bayreuth, Universitätss trasse 30, 95440 Bayreuth
Tel: 0921/55 7470 Fax: 0921/55 7473
www.polymer-engineering.de

Lehrstuhl für Polymerwerkstoffe
Herr Prof. Dr. rer. nat. H. Münstedt
Martenstrasse 7, 91058 Erlangen
Tel: 09131/85-28593, Fax: -28321
www.lsp.uni-erlangen.de

Lehrstuhl für Recyclinggerechte Produktgestaltung/Entfertigung
Herr Prof. Dr.-Ing. R. Renz,
Universität Kaiserslautern, Postfach 3049, 67653 Kaiserslautern
Tel.: 0631/205-3960 Fax: -3963
http://recycling5.mv.uni-ki.de

Lehrstuhl für Werkstoffkunde der Metalle und Kunststoffe
Herr Prof. Dr.-Ing. P. Eyerer,
Universität Stuttgart, Pfaffenwaldring 32, 70569 Stuttgart
Tel: 0711/685-2667, Fax: -2066
www.ikp.uni-stuttgart.de

POLYMER RESEARCH CENTERS IN IRAN
Iran Polymer and Petrochemical Institute
Professor Hamid Mirzadeh, Director
Pazhoohesh Blvd., Km 17, Tehran-Karaj Hwy
Tehran, I.R. Iran
P.O. Box: 14965/115 , 14185/458
Tel: 98-21-44580000 Fax: 98-21-44580021-23
www.iranpolymerinstitute.org/

POLYMER RESEARCH CENTERS IN ISRAEL

Department of Chemical Engineering
Professor Zehev Tadmor
Technion - Israel Institute of Technology
Haifa 32000, Israel
http://chemeng.technion.ac.il

POLYMER RESEARCH CENTERS IN ITALY

Institute of Chemistry and Technology of Polymers
Viale Andrea Doria, 6 -95125 Catania - Italy
Tel: 095-339926 Fax: 095-221541
http://www.ictmp.ct.cnr.it/ictmp/index.jsp

POLYMER RESEARCH CENTERS IN MEXICO

Centro de Investigación en Química Aplicada (CIQA)
México
www.ciqa.mx

POLYMER RESEARCH CENTERS IN THE NETHERLANDS

Eindhoven Polymer Laboratories
Eindhoven University of Technology
Prof.dr.ir. H.E.H. Meijer
PO Box 513, WH 4.140, 5600 MB Eindhoven - The Netherlands
http://www.epl.nu/

Polymer Materials and Engineering
http://www.polymers.tudelft.nl/
Delft University of Technology
P.O. Box 5045, 2600 GA Delft - The Netherlands
http://www.polymers.tudelft.nl

Polymer Service Centre Groningen
Mr. dr. Theo A.C. Flipsen, Managing Director
Kadijk 7D, P.O. Box 70033, 9704 AA GRONINGEN, The Netherlands
Tel: 31-50-368-0777 Fax: 31-50-36-0779
http://www.pscg.nl/

POLYMER RESEARCH CENTERS IN SLOVENIA

Center for Experimental Mechanics
Herr Prof. Dr.-Ing. Igor Emri
Cesta na Brdo 49, SI-1000 Ljubljana, Slovenien
Tel: 386 1/4771-660, Fax: -670
www.uni-lj.si/ cem/

POLYMER RESEARCH CENTERS IN SPAIN

Grupo de Ciencia y Tecnología de Polímeros
Enrique Giménez y José-María Lagarón
Campus Riu Sec, Universitat Jaume I, Castellón, 12071
Tel: (34) 964 728138, Fax: (34)964 728106
http://www.iata.csic.es/čonlag/grupopolimerosuji.html

Institute of Polymer Science and Technology
Maria Soledad Alvarez Gonzalez, Administrator
Instituto de Ciencia y Tecnología de Polímeros
CSIC, Juan de la Cierva 3, 28006-Madrid, Spain
Tel: (34) 915 622 900 Fax: (34) 915 644 853
www.ictp.csic.es

POLYMER RESEARCH CENTERS IN TURKEY

Polymer Research Center
Professor Turkan Haliloglu, Director
Bogazici University, Bebek 80815, Istanbul, Turkey
Tel: (90) 212 359 70 02 Fax: (90) 212 257 50 32
http://klee.bme.boun.edu.tr/

POLYMER RESEARCH CENTERS IN THE UNITED KINGDOM

Interdisciplinary Research Centre in Polymer Science and Technology
Professor T.C.B. McLeish
University of Leeds, Leeds, LS2 9JT, United Kingdom
Tel: (0113) 3433810 Fax: (0113) 3433846
www.dur.ac.uk/irc.web/index.htm

Polymer and Colloid Group
Professor Brian Vincent, Director
Physical/Theoretical Section
University of Bristol, Cantock's Close, Bristol, BS8 1TS, Great Britain
Tel: 44-117-3317157 Fax: 44-117-925 0612
www.tlchm.bris.ac.uk/vincent/bvhome.htm

Polymers and Colloids Group
University of Cambridge
19, J J Thomson Avenue

Cambridge CB3 0HE, United Kingdom
Tel: 44 (0)1223 337007 or 337423 Fax: 44 (0)1223 337000
http://www.poco.phy.cam.ac.uk/

Polymers and Complex Fluids
IRC in Polymer Science and Technology
Department of Physics and Astronomy
University of Leeds
Leeds LS2 9JT, United Kingdom
Tel: 44 (0)113 34 33810 Fax: 44 (0)113 34 33846
www.physics.leeds.ac.uk/pages/PolymersAndComplexFluids

Polymers at Interfaces Group
Professor Terence Cosgrove
School of Chemistry
University of Bristol
Cantock's Close
Bristol BS8 1TS
United Kingdom
Tel: (44) (0)117 928 7663 Fax: (44) (0)117 925 0612
www.chm.bris.ac.uk/pt/polymer/index.shtml

POLYMER RESEARCH CENTERS IN THE UNITED STATES

Center for Advanced Polymer and Composite Engineering (CAPCE)
L. James Lee, Director
Ohio State University
437 Koffolt Labs
140 W. 19th Ave.
Columbus, OH 43210-1180
Tel: 614-292-9271
www.capce.ohio-state.edu/

Center for Composite Materials
Professor John W. Gillespie, Jr.
University of Delaware
201 Composites Manufacturing Science Laboratory
Newark, DE 19716-3144
Tel: 302- 831-8149 Fax: 302- 831-8525
http://www.ccm.udel.edu

Center for Polymer Studies
Boston University
590 Commonwealth Avenue
Boston, MA 02215
Tel: 617-353-8000
http://polymer.bu.edu/

Center for Research on Polymers
James D. Capistran, Director
University of Massachusetts Armherst
Silvio O. Conte Center for Polymer Research
Armherst, MA 01003-4530
Tel: 413-577-1518
www.pse.umass.edu/cumirp/

Center for Responsive-Driven Polymeric Films
Sara Bayley, Facilities Coordinator
University of Southern Mississippi
School of Polymers and High Performance Materials
118 College Drive #10076
Hattiesburg, MS 39406
Tel: 601-266-5581
www.usm.edu/mrsec/about_us/index.htm

Department of Macromolecular Science and Engineering
Case Western Reserve University
2100 Adelbert Road, Kent Hale Smith Bldg
Cleveland, OH 44106
Tel: 216-368-4172
http://polymers.case.edu/

Department of Plastics Engineering
Professor Robert Malloy, Chair
University of Massachusetts Lowell
Ball Hall 207
One University Avenue
Lowell, MA 01854
Tel: 978-934-3420 Fax: 978-934-4141
http://www.uml.edu/college/Engineering/plastics

Florida Advanced Center for Composite Technologies (FAC^{2}T)
Dr. Ben Wang, Director

2525 Pottsdamer Street
Tallahassee, FL 32310-6046
Tel: 850-410-6345 Fax: 850-410-6342
http://www.fac2t.eng.fsu.edu/

Institute of Polymer Engineering (IPE)
Dr. Lloyd Goettler, Institute Director
University of Akron
Department of Polymer Engineering
250 South Forge Street
Akron, OH 44325-0301
Tel: 330-972-7467
www2.uakron.edu/cpspe/dpe/web/IPE.htm

Materials Research Science and Engineering Center (MRSEC)
Professor Thomas P. Russell, Director
Materials Research Science and Engineering Center
at University of Massachusetts Amherst
Silvio O. Conte National Center for Polymer Research
120 Governors Drive
Amherst, MA 01003
Tel: 413-545-0433 Fax: 413-545-0082
www.pse.umass.edu

Polymer Engineering Center
Professor Tim Osswald, Co-Director
Professor L.S. (Tom) Turng, Co-Director
University of Wisconsin-Madison
1513 University Avenue
University of Wisconsin-Madison
Madison, WI 53706-1572
Tel: 608-265-2316
http://pec.engr.wisc.edu/

Polymer Processing Institute
Kun Sup Hyun, President
New Jersey Institute of Technology
Suite 3901, Guttenberg Information Technologies Center
New Jersey Institute of Technology
University Heights
Newark, NJ 07102-1982

Tel: 973-596-3267 Fax: 973-642-4594
www.polymers-ppi.org/

Polymer Processing Lab
Professor Charles L. Tucker III
Department of Mechanical and Industrial Engineering
University of Illinois, Urbana-Champaign
1206 W. Green Street
Urbana IL, 61801-2906
www.mie.uiuc.edu/content/about/research/research_lab_profile.php?lab_id=24

Polymer Research Institute
Polytechnic University
Six MetroTech Center
Brooklyn, NY 11201
Tel: 718-260-3600 Fax: 718-260-3136
www.poly.edu/researchcenters/pri/index.php

Polymer Synergies, LLC
Polymer Synergies, LLC
P.O. Box 456
Mullica Hill, New Jersey 08062
Tel: 856-981-4381
http://www.polymersynergies.com/index.htm

Rheology Research Center
Professor A. Jeffrey Giacomin, Chair
University of Wisconsin-Madison
1513 University Avenue
Madison, WI 53706-1572
Tel: 608-262-7473
http://rrc.engr.wisc.edu/

POLYMER RESEARCH CENTERS IN VENEZUELA

INDESCA (Investigaci—n y Desarrollo C.A.)
Departamento de Productos Pl‡sticos
Complejo Petroqu'mico El Tablazo, Maracaibo, Venezuela
Tel: (58) 261-7909483, Fax:(58) 261-7909481
www.indesca.com

APPENDIX D

TRADENAMES

D.1 INTRODUCTION

This appendix presents the tradenames used around the world for polymeric raw materials, as well as semi-finished products. The list contains the tradename, polymer name, form of delivery, company and webpage for each material. The raw material and semi-finished product form of delivery nomenclature used is presented in Table D.1.

Table D.2, which presents the tradenames is current at the time this handbook is published. However, a current list is kept with the electronic version of the book at http://www.hanser.de/plasticshandbook

Table D.1: Nomenclature for Delivery Form of Raw Materials

Nomenclature	Delivery Form
1	Raw material
1.1	Pre-product, liquid
1.2	Dispersion, aqueous
1.3	Solution
1.4	Paste
1.5	Resin
2	Molding material
2.1	Powder
2.2	Pellets
2.3	Prepregs, mats
2.4	Masterbatch
3	Semi-finished product
3.1	Profile, rod
3.2	Pipe, tube
3.3	Films
3.4	Sheet, blocks
3.5	Composite
3.6	Fiber, non-woven
3.7	Foam

D.2 TRADENAMES TABLE

Table D.2: Tradenames

Tradename	Polymer	Delivery Form	Manufacturer Supplier	Web Address
Ablebond	EP	1.4, 2.2	Ablestik Laboratories	www.ablestik.com
Ablefilm	EP	2.2, 3.3	Ablestik Laboratories	www.ablestik.com
ABS Proquigel	ABS	2.2	Proquigel	www.proquigel.com.br/
Absolac	ABS	2.2	Lanxess	www.lanxess.com
Absolan	SAN	2.2	Lanxess	www.lanxess.com
Absrom	ABS	2.1	Daicel	www.daicel.co.jp
Abstat	ABS	2.2	Mitech Corporation	216-425-1634
Absylux	ABS	3.1, 3.3, 3.4	Westlake Plastics Company	www.westlakeplastics.com/
Acarb	PC	2.2	Aquafil	www.aquafil.com
Acclear	PP	2.2	BP	www.bppetrochemicals.com
Accpro	PP	2.2, 2.2	BP	www.bppetrochemicals.com
Acctuf	PP	2.2, 2.2	BP	www.bppetrochemicals.com
ACCUCOMP[a]	ABS	2.2	ACLO Compounders Inc.	www.aclocompounders.com/
ACCUGUARD[a]	ABS	2.2	ACLO Compounders Inc.	www.aclocompounders.com/
ACCULOY[a]	ABS+PC	2.2	ACLO Compounders Inc.	www.aclocompounders.com/
ACCUTECH[a]	ABS	2.2	ACLO Compounders Inc.	www.aclocompounders.com/
Acester	PBT	2.2	Aquafil	www.aquafil.com
Acetron	POM	3.1	Quadrant Engineering Plastic Product	www.quadrantepp.com/
Achieve	PP-MC	2.2	Exxon	www.exxonchemical.com
Aclar	PCTFE	3.3	Honeywell	www.honeywell.com
Acnor	PPE	2.2	Aquafil	www.aquafil.com
ACP	PVC	2.2	Alpha Gary Corporation	www.alphagary.com/
Acrigel	PMMA	2.2	Proquigel	www.proquigel.com.br/
Acriglas	PMMA	3.4	Acrilex	www.acrilex.com
Acrolex	PC+PMMA+PS	2.2	Ferro	www.ferro.com
Acrolex	PC+Acrylic Alloy	2.2	Ferro Corporation	www.ferro.com/
Acronal	PAA	1.1, 1.2	BASF	www.basf.com
Acrosol	PAA	1.2	BASF	www.basf.com

Continued on next page

Tradename	Polymer	Delivery Form	Manufacturer Supplier	Web Address
Acryalloy	PMMA+PVC	2.2	Mitsubishi Rayon	www.mrc.co.jp
AcrycalMP	PMMA	2.2	Plaskolite-Continental Acrylics	www.plaskolitecontinental.com/
Acrycon	PMMA	2.2	Mitsubishi Rayon	www.mrc.co.jp
Acryester	PMMA	1.1	Mitsubishi Rayon	www.mrc.co.jp
Acryft	EMA	2.2	Sumitomo	www.sumitomo-chem.co.jp
Acrylamac	A-Cop	1.1	Eastman	www.eastman.com
Acrylite	PMMA	2.2	Cyro	www.cyro.com
Acrypanel	PMMA	3.4	Mitsubishi Rayon	www.mrc.co.jp
Acrypet	PMMA	2.2	Mitsubishi Rayon	www.mrc.co.jp
Acrypoly	PMMA	3.4	Chi Mei	www.chimei.com.tw
Acryrex	PMMA	2.2	Chi Mei	www.chimei.com.tw
Acrysol	PMMA	1.2	Rohm & Haas	www.rohmhaas.com
Acrysteel	PMMA	3.4	Aristech	www.aristechchem.com
Acrystex	MBS	2.2	Chi Mei	www.chimei.com.tw
Acrythene	EMA	-	Equistar Chemicals, LP	www.equistarchem.com/
ACS	PE-C	2.2	Showa	www.sdk.co.jp
Acsium	CSM	2.2	DuPont Dow Elastomers	www.dupont-dow.com
Acstyr	ABS	2.2	Aquafil	www.aquafil.com
Acudel	PPSU	2.2	Solvay	www.solvay.com
Adder	PF	3.5	Tufnol	www.tufnol.co.uk
Addilene	PP	2.2	Addiplast	www.addiplast.com
Addinyl	PA	2.2	Addiplast	www.addiplast.com
Adell	PBT	2.2	Adell Plastics, Inc.	www.adellplas.com/
Adflex	PP	2.2	Basell	www.basell.de
Adiprene	PUR	1.1	DuPont	www.dupont.com
Admer	TPO	2.2	Mitsui	www.mitsui.co.jp
Adstif	PP	2.2	Basell	www.basell.de
Adsyl	PP	2.2	Basell	www.basell.de
Ad-Tech	EP	1.4, 2.2, liquid	Ad-Tech Plastic Systems Corp.	www.adtechplastics.com/
Advex	PVC	2.2	PolyOne Th. Bergmann	www.polyone.com
Aecithene	PE-LLD	2.2	AECI	www.aeci.co.za
Aegis	PA-Cop	2.2	Honeywell	www.honeywell.com
Affinity	E-Cop	2.2	Dow	www.dow.com
Aflas	FKM	2.2	Dyneon	www.dyneon.com
Ahlstrom	PP	2.3	Ahlstrom	www.ahlstrom.com
Aicarfen	PF	2.2	Aicar	www.aicar.es
Aim	PS	2.2	Dow	www.dow.com
Airex	PEI, PVC	3.5, 3.7	Airex	www.alcanairex.com
Airflex	EVAC, VCE	1.1	Air Products	www.airproducts.com
Airpreg	PF	2.3	Isovolta	www.isovolta.com

Continued on next page

D.2 Tradenames Table

Tradename	Polymer	Delivery Form	Manufacturer Supplier	Web Address
Akrolen	PP	2.2	Akro-Plastic	www.feddersen.de
Akromid	PA 6, PA 66	2.2	Akro-Plastic	www.feddersen.de
Akulon	PA 6, PA 66	2.2	DSM Eng. Plastics	www.dsmep.com
Akylux	PP	3.4	Kaysersberg	www.kaysersberg-packaging.fr
Akyplen	PP	3.4	Kaysersberg	www.kaysersberg-packaging.fr
Akyver	PC	3.4	Kaysersberg	www.kaysersberg-packaging.fr
Alamid	PA 6, PA 66	2.2	Leis	www.leispt.de
Alastian	PE	2.2	Basell	www.basell.de
Alathon	PE-HD	2.2	Equistar	www.equistar.com
Albester	UP	1.1	Eastman	www.eastman.com
Albis ABS	ABS	2.2	Albis Plastics Corporation	www.albisna.com/
Alcom	Thermopl. allg.	2.2	Albis	www.albis.com
Alcryn	EVAC+PVDC	2.2	DuPont	www.dupont.com
Alcudia	PE	2.2	Repsol	www.repsol.com
ALFATER	TPV	2.2	Lavergne Group	www.lavergneusa.com/
Alflon	PTFE	-	Quadrant Engineering Plastic Product	www.quadrantepp.com/
Alflow	PP	2.3	Ahlstrom	www.ahlstrom.com
Algoflon	PTFE	2.1, 2.2	Solvay	www.solvay.com
Alkathene	PE-LD, PE-LLD	2.2	Qenos	www.qenos.com
Alkorcell	PO	3.3	Solvay	www.solvay.com
Alkorflex	PVC	3.3	Solvay	www.solvay.com
Alkorfol	PVC-P, PVDF	3.3	Solvay	www.solvay.com
Alkorpack	PVC-P	3.3	Solvay	www.solvay.com
Alkorplan	PVC	3.3	Solvay	www.solvay.com
Alkorprop	PP	3.3	Solvay	www.solvay.com
Alkortop	PO, PVC	3.3	Solvay	www.solvay.com
ALM PC	ABS+PC	2.2	Custom Resins Group	www.customresins.com/
Almatex	UP	1.1	Mitsui	www.mitsui.co.jp
Alpha PVC	PVC	2.2, cube	Alpha Gary Corporation	www.alphagary.com/
Alphalac	S-Cop	2.2	LG	www.lgchem.com
Alphamid	PA	2.2	Putsch	www.putsch.de
Alphatec	TPE	2.2	Alpha Gary Corporation	www.alphagary.com/
Alruna	SEBS, SEPS	2.2	Allod	www.allod.com
Alstamp	PP	2.3	Ahlstrom	www.ahlstrom.com
Altair	PMMA	3.4	Aristech	www.aristechchem.com
Altech	Thermopl. allg.	2.2	Albis	www.albis.com

Continued on next page

Tradename	Polymer	Delivery Form	Manufacturer Supplier	Web Address
Altuglas	PMMA	2.2, 3.4	Atoglas	www.atoglas.fr
Amalloy	Proprietary	2.2	Amco Plastic Materials Inc.	www.amco.ws/
AmberGuard	PET	2.2	Eastman	www.eastman.com
AMC	Phenolic	-	Quantum Composites Inc.	www.quantumcomposites.com/
Ameripol	CR	2.2	Goodrich	www.goodrich.com
Amilan	PA	2.2	Toray	www.toray.co.jp
Amilus	POM	2.2	Toray Resin Company	www.toray.com/
Amoco	PAI	2.2	BP Amoco Polymers, Inc.	www.bp.com/
Amodel	PPA	2.2	Solvay	www.solvay.com
Amotech	PAI	2.2	Solvay	www.solvay.com
Ampal	MF+UP	2	Raschig	www.raschig.de
Amtuf	PP	2.2	BP	www.bppetrochemicals.com
Andrez	SB	2.1, 2.2	Anderson	www.andersondevelopment.com
Andur	PUR	liquid	Anderson Development Company	www.andersondevelopment.com/
Anjablend	PC+ABS, PC+PBT	2.2	J & A Plastics	www.j-a.de
Anjacryl	PMMA	2.2	J & A Plastics	www.j-a.de
Anjadur	PBT, PET	2.2	J & A Plastics	www.j-a.de
Anjaflor	PVDF	2.2	J & A Plastics	www.j-a.de
Anjaform	POM	2.2	J & A Plastics	www.j-a.de
Anjalin	ABS, SAN	2.2	J & A Plastics	www.j-a.de
Anjalon	PC	2.2	J & A Plastics	www.j-a.de
Anjamid	PA 6, PA 66	2.2	J & A Plastics	www.j-a.de
Anjapur	TPU	2.2	J & A Plastics	www.j-a.de
Anso	PA	3.6	Honeywell	www.honeywell.com
Antiflex	PMMA	3.4	Tupaj	www.go-ttv.com
Antron	PA	3.6	DuPont	www.dupont.com
Apec	PC-TMC	2.2	Bayer	www.bayer.com
Apel	COC	2.2	Mitsui	www.mitsui.co.jp
Apex	PVC	2.2	Teknor Apex	www.teknorapex.com
API PS	PS	2.2	American Polymers, Inc.	www.americanpolymers.com/
Apical	PI	3.3	Kaneka	www.kaneka.co.jp
Apifive	EVAC	2.2	API	www.apiplastic.com
Apiflex	PVC	2.2	API	www.apiplastic.com
Apigo	TPO	2.2	API	www.apiplastic.com
Apilon	PVC+TPU, TPU	2.2	API	www.apiplastic.com
Apizero	EVAC	2.2	API	www.apiplastic.com
Appeel	E-Cop, EVAC	2.2	DuPont	www.dupont.com
Appeel	EVA	2.2	DuPont Packaging & Industrial Polym	www.dupont.com/packaging/

Continued on next page

D.2 Tradenames Table

Tradename	Polymer	Delivery Form	Manufacturer Supplier	Web Address
Appryl	PP	2.2	Atofina	www.atofina.de
Aqua-Link	EVS (Ethylene Vinyl Silane Copolymer)	-	AT Plastics Inc.	www.atplastics.com/
Aqualoy	Thermopl. allg.	2.2	Schulman	www.aschulman.com
Aquamid	PA 6, PA 66	2.2	Aquafil	www.aquafil.com
Araldite	EP	1.5	Huntsman	www.huntsman.com
Araloy	PA+PPO	-	MRC Polymers, Inc.	www.mrcpolymers.com/
Arbelac	ABS	2.2	Connell Bros Company LTD	webprod.wecon.com/ WECOCBC/ WECO/index.h
Arcel	PS-E+PE-E	2.2	Nova	www.novachem.com
ARCO PP	PP	-	ARCO Polypropylene	
Ardel	PAR	-	BP Amoco Polymers, Inc.	www.bp.com/
Ardylan	PE	2.2	Nova	www.novachem.com
Ardylux	SAN	2.2	Nova	www.novachem.com
Arlon	PEEK	2.2	Greene, Tweed Engineered Plastics	www.gtweed.com/
Arnite	PBT, PET	2.2	DSM Eng. Plastics	www.dsmep.com
Arnitel	TPC	2.2	DSM Eng. Plastics	www.dsmep.com
Aropol	UP	1.5	Ashland	www.ashspec.com
Arotech	VE	1.5	Ashland	www.ashspec.com
Arpro	PP-E	3.7	JSP	www.jsp.com
Artlex	PPE	2.2	Sumitomo	www.sumitomo-chem.co.jp
Artley	PPE	2.2	Sumitomo	www.sumitomo-chem.co.jp
Arylon	PAR	2.2	DuPont	www.dupont.com
Asahi Thermof	ABS	2.2	Asahi-Thermofil	www.asahithermofil.com/
Asaprene	SBS	2.2	Asahi	www.asahithermofil.com
Ascend	PA66	2.2	Solutia Inc.	www.solutia.com/
Ashlene	PA 6, PA 612, PA 66	2.2	Ashley	www.ashleypolymers.com
Asp	PF	3.5	Tufnol	www.tufnol.co.uk
Astem	PPE	2.2	Sumitomo	www.sumitomo-chem.co.jp
Astrel	PAR	2.2	3 M	www.3m.com
Astro Turf	PA	3.3	Dow	www.dow.com
Astryn	PP	2.2	Basell	www.basell.de
AT PE	PE	-	AT Plastics Inc.	www.atplastics.com/

Continued on next page

Tradename	Polymer	Delivery Form	Manufacturer Supplier	Web Address
Ateva	EVA	2.2	AT Plastics Inc.	www.atplastics.com/
Atlac	UP, VE	1.5	DSM Composite Resins	www.dsm.com
ATRATE	ABS	2.2	Nippon A&L Inc.	www.n-al.co.jp/ing/index.html
Attane	PE-ULD	1.5	Dow	www.dow.com
Aurum	Thermopl. allg.	2.2	Mitsui	www.mitsui.co.jp
Austrapol	BR	2.2	Qenos	www.qenos.com
Austrex	PS	2.2	Polystyrene Australia Pty Ltd	www.psa.com.au/
Avalon	TPU	2.2	Huntsman	www.huntsman.com
Avantra	PS-HI	2.2	BASF Corporation	corporate.basf.com/
Avimid	PI	2.2	DuPont	www.dupont.com
Avotone	PAEK	2.2	DuPont	www.dupont.com
AVP	ABS	2.2	GE Polymerland	www.gepolymerland.com/
Axpet	PET	3.4	Bayer Sheet	www.bayersheeteurope.com
Axprint	PP	3.3	British Vita	www.britishvita.com
Azdel	PP	2.3, 3.5	Azdel	www.azdel.com
Azfab	PTP	2.3	Azdel	www.azdel.com
Azloy	PC+PBT, PTP	2.3	Azdel	www.azdel.com
Azmet	PBT, PET	2.3	Azdel	www.azdel.com
B&M ABS	ABS	2.2	B&M Plastics, Inc.	www.bmplastics.com/
Bach	ABS+PC	2.2	Bach Plastic Works Inc.	
Bacon Epoxy R	EP	2.2, liquid	Bacon Industries Inc.	www.baconindustries.com/
Badadur	PBT	2.2	Bada	www.bada.de
Badaflex	TPS	2.2	Bada	www.bada.de
Badamid	PA 6, PA 66	2.2	Bada	www.bada.de
Bakelite	MPF, PF, UP	2.2	Bakelite	www.bakelite.
Bapolan	ABS, PS, SAN	2.2	Bamberger	www.bambergerpolymers.com
Bapolene	PE-LD, PE-LLD, PP	2.2	Bamberger	www.bambergerpolymers.com
Bapolon	PA	3.3	Bamberger	www.bambergerpolymers.com
Barex	PAN	2.2	BP	www.bppetrochemicals.com
Basicpet	PET	2.2	La Seda	www.laseda.es
Basonat	PUR-R	1.1	BASF	www.basf.com
Basotect	MF	3.7	BASF	www.basf.com
Bayblend	PC+ABS, PC+SAN	2.2	Bayer	www.bayer.com
Baydur	PUR-R	1.1	Bayer	www.bayer.com
Bayfill	PUR-R	1.1	Bayer	www.bayer.com
Bayfit	PUR-R	1.1	Bayer	www.bayer.com
Bayflex	PUR-R	1.1	Bayer	www.bayer.com

Continued on next page

D.2 Tradenames Table

Tradename	Polymer	Delivery Form	Manufacturer Supplier	Web Address
Bayfol	PC+PBT	3.3	Bayer	www.bayer.com
Baygal	PUR-R	1.1	Bayer	www.bayer.com
Bayloy	PC-Blend	3.4	Bayer Sheet	www.bayersheeteurope.com
Baymer	PUR-R	1.1	Bayer	www.bayer.com
Baymidur	PUR-R	1.1	Bayer	www.bayer.com
Baynat	PUR-R	1.1	Bayer	www.bayer.com
Baypreg	PUR	2.3	Bayer	www.bayer.com
Baypren	CR	1.2, 2.2	Lanxess	www.lanxess.com
Baysilone	Q	1.1	GE Silicones	www.gesilicones.com
Baytec	PUR	2.3	Bayer	www.bayer.com
Baytherm	PUR-R	1.1	Bayer	www.bayer.com
BCC Resins	Polyester, Thermoset	1.4, liquid, pre-formed parts	BCC Products Inc.	www.bccproducts.com/
Bear	PF	3.5	Tufnol	www.tufnol.co.uk
Beaulon	PB-1	2.2	Mitsui	www.mitsui.co.jp
Beetle	PA, PF, PI, PTP, UF	2.2	BIP	www.bip.co.uk
Begra	PVC	2.2	Begra	www.begra.de
Bellpearl	PF	2.2	Kanebo	www.kanebo.co.jp
Benecor	PVC-P	3.3	Benecke	www.benecke-kaliko.de
Beneflex	PVC, TPO, TPU	3.3	Benecke	www.benecke-kaliko.de
Benefol	PVC-P	3.3	Benecke	www.benecke-kaliko.de
Benelit	PS, PVC, PVC-P	3.3	Benecke	www.benecke-kaliko.de
Benepur	ABS, PUR	3.3, 3.5	Benecke	www.benecke-kaliko.de
Beneron	ABS, PVC	3.3, 3.5	Benecke	www.benecke-kaliko.de
Benova	PUR	3.3	Benecke	www.benecke-kaliko.de
Benvic	PVC	2.2	Solvin	www.solvinpvc.com
Bergacell	CA	2.2	PolyOne Th. Bergmann	www.polyone.com
Bergadur	PBT	2.2	PolyOne Th. Bergmann	www.polyone.com
Bergaflex	SEBS	2.2	PolyOne Th. Bergmann	www.polyone.com
Bergaform	POM	2.2	PolyOne Th. Bergmann	www.polyone.com
Bergamid	PA	2.2	PolyOne Th. Bergmann	www.polyone.com
Bergaprop	PP	2.2	PolyOne Th. Bergmann	www.polyone.com
Bester	PUR-R, TPU	1.1, 2.2	Rohm & Haas	www.rohmhaas.com
Betamid	PA	2.2	Putsch	www.putsch.de
Bexloy	E-Cop, PA, PTP	2.2	DuPont	www.dupont.com

Continued on next page

Tradename	Polymer	Delivery Form	Manufacturer Supplier	Web Address
BFI	PE-HD	3.3	Blueridge Films Inc.	www.blueridgefilms.com/
Bicor	PP	3.3	Exxon	www.exxonchemical.com
Bifan	PP	3.3	Showa	www.sdk.co.jp
Biomax	PET	2.2	DuPont	www.dupont.com
Biopar	PLA	2.2	Biop	www.biopag.de
Bioparen	PLA	2.2	Biop	www.biopag.de
Biophan	PLA	3.3	Treofan	www.treofan.com
Bisco	SI	1.1	Rogers	www.rogerscorporation.com
Blaze Master	VC-Cop	2.2	PolyOne Th. Bergmann	www.polyone.com
Blendex	ABS	2.2	GEP	www.geplastics.com
BMC	Polyester, Thermoset	2.2	Bulk Molding Compounds, Inc.	www.bulkmolding.com/
Boltaron	PVC	3.4	GenCorp Polymer Products	740-498-5900
Bondfast	PP	1.5	Sumitomo	www.sumitomo-chem.co.jp
Borclear	PP	2.2	Borealis	www.borealis.com
Borcom	PP	2.2	Borealis	www.borealis.com
Borecene	PE	2.1	Borealis	www.borealis.com
Borflow	PP	2.2	Borealis	www.borealis.com
Borpact	PP	2.2	Borealis	www.borealis.com
Borsoft	PE	2.2	Borealis	www.borealis.com
Borstar	PE	2.2	Borealis	www.borealis.com
Bralen	PE-LD	2.2	Slovnaft	www.slovnaft.sk
Breon	CR	2.2	Zeon	www.zeon.co.jp
Breox	E-Cop	2.2	Zeon	www.zeon.co.jp
Bricling	PE-LD, PE-LLD	3.3	Zeon	www.zeon.co.jp
Brilen	PTP	3.6	Brilén	www.brilen.com
Brithene	PE-LD, PE-LLD	3.3	Brilén	www.brilen.com
Budene	BR	2.2	Goodyear	www.goodyearchemical.com
Buflon	PVC-P	3.3	Solvay	www.solvay.com
Bulana	PMMA	3.6	Lukoil	www.lukoil.bg
Bulen	PE	2.2	Lukoil	www.lukoil.bg
Bultex	PUR	3.7	Recticel	www.recticel.com
Buna	BR, EPDM, SBR	2.2	Lanxess	www.lanxess.com
Buplen	PP	2.2	Lukoil	www.lukoil.bg
Bustren	SB	1.1	Lukoil	www.lukoil.bg
Butacite	PVB	3.3	DuPont	www.dupont.com
Butaclor	CR	1.2, 2.2	Polimeri	www.polimerieuropa.com
Butaprene	CR	2.2	Firestone	www.firesyn.com
Butofan	SB	1.2	BASF	www.basf.com
Butvar	PVB	2.2	Dow	www.dow.com

Continued on next page

D.2 Tradenames Table

Tradename	Polymer	Delivery Form	Manufacturer Supplier	Web Address
BVC	PVC	2.2	Bayshore Vinyl Compounds (BVC) Inc.	www.bayshoregroup.com/bvchome.html
Bynel	E-Cop, EVAC	2.2	DuPont	www.dupont.com
Bynel	EVA	2.2	DuPont Packaging & Industrial Polym	www.dupont.com/packaging/
C&C Tech Com	PO	-	Color & Composite Technologies, Inc	
Cabelec	Thermopl. allg.	2.2	Cabot	www.cabot-corp.com
Cadon	SMAH	2.2	Bayer	www.bayer.com
Calibre	PC	2.2	Dow	www.dow.com
Calthane	TSU (Polyurethane Thermoset Elastomer)	1.4, liquid	Cal Polymers, Inc.	calpolymersinc.com/
Capilene	PP	2.2	Carmel	www.carmel-olefins.co.il
Caprez	PP	2.2	Alloy	www.alloypolymers.com
Capron	PA 6, PA 66	2.2	BASF	www.basf.com
Carbaicar	UF	2.2	Aicar	www.aicar.es
Carboblend	PC+PBT	2.2	Aquafil	www.aquafil.com
Carboglass	PC	3.4	Aquafil	www.aquafil.com
Carb-o-life	PC	3.4	Aquafil	www.aquafil.com
Carboloy	PC+ABS	2.2	Aquafil	www.aquafil.com
Carbolux	PC	3.4	Aquafil	www.aquafil.com
Carbotex	PC+PBT	2.2	Kotec Corporation	www.hmplastic.com/kotec_en.htm
Carbothane	TPU	2.2	Thermedics	www.thermedicsinc.com
Carilon	PK	2.2	Shell Chemical Company	www.shellchemicals.com/
Carom	CR	2.2	Oltchim	www.oltchim.ro
Carp	PF	3.5	Tufnol	www.tufnol.co.uk
Cashmilon	PAN	3.6	Asahi	www.asahithermofil.com
Castall	EP	1.4, 2.2, liquid	Lord Corporation	www.lordcorp.com/
Castomax	PUR	liquid	Fluid Polymers, Inc.	None
Castomer	PUR-R	1.1	Baxenden	www.baxchem.co.uk
Catalloy	PP	2.2	Basell	www.basell.de
Cefor	PP-Cop.	2.2	Dow Plastics	plastics.dow.com/
Celanese	PA6	-	Ticona	www.topas.com/index.htm
Celanex	PBT	2.2	Ticona	www.ticona.com
Celcon	POM	2.2	Ticona	www.ticona.com
Cellasto	TPU	2.2, 3.7	Elastogran	www.elastogran.de
Cellidor	CAB, CP	2.1, 2.2	Albis	www.albis.com
Cello	CA	3.3	Albis	www.albis.com

Continued on next page

Tradename	Polymer	Delivery Form	Manufacturer Supplier	Web Address
Cellobond	Phenolic	liquid	Blagden Chemicals Ltd.	www.blagdenspecchem.co.uk/
Cellolux	CA	2.2	LA/ES	www.la-es.com
Celmar	PP	3.3, 3.4	Plastech	www.vtsplastech.co.uk
CEL-SPAN	PE-HD	2.2	Phoenix Plastics Co., Inc.	www.phoenixplastics.com/
Celstran	Thermopl. allg.	2.2	Ticona	www.ticona.com
Centrex	AES, ASA, ASA+PC	2.2	Lanxess	www.lanxess.com
Ceramer	PPSU	2.1	Ticona	www.topas.com/index.htm
CERTENE[a]	PA6	2.2	Channel Polymers	www.channelpa.com
Cevian	ABS, SAN	2.2	Daicel	www.daicel.co.jp
C-Flex	SEBS	2.2	Consolidated Polymer Technologies, I	www.c-flex-cpt.com/
Chemfluor	FEP (Perfluoroethylene Propylene Copolymer)	3.2	Saint Gobain - Norton	www.plastics.saint-gobain.com/
Chemigum	ACM	2.2	Goodyear	www.goodyearchemical.com
Chemiton	SEBS, SEPS	2.2	Franplast	www.franplast.it
Chemlon	PA66	2.2	Chem Polymer Corporation	www.teknorapex.com/
Chempex	PE-X	3.2	Golan	www.golan-plastics.com
Chemraz	PTFE	2.2	Greene, Tweed Engineered Plastics	www.gtweed.com/
Chissonyl	PVAC	2.2	Chisso	www.chisso.co.jp
Cisamer	BR	2.2	Indian Petrochemicals	www.ipcl.co.in
Civic	UP	1.1	Nordkemi	www.nordkemi.dk
Claradex	MBS	2.2	Shin-A Corporation	www.shinasys.co.kr/
Clariant ABS	ABS	2.2	Clariant Performance Plastics	www.clariant-northamerica.com/
Clarix	EIM	2.2	Schulman	www.aschulman.com
Clearen	SB	2.2	Denka	www.denka.co.jp
Clearflex	PE-LD, PE-ULD	2.2	Polimeri	www.polimerieuropa.com
Clearlac	S-Cop	2.2	Mitsubishi Rayon	www.mrc.co.jp
Clearstrength	MBS	2.2	Atofina	www.atofina.de
Cleartuf	PET	2.2	M&G	www.mgpolymers.com
Cleartuf	PET	2.2	Goodyear Tire & Rubber Co.	www.goodyear.com/
Clocel	PUR	1.1	Baxenden	www.baxchem.co.uk
Clyrell	PP	2.2	Basell	www.basell.de

Continued on next page

D.2 Tradenames Table

Tradename	Polymer	Delivery Form	Manufacturer Supplier	Web Address
Clysar	PO	3.3	DuPont	www.dupont.com
Cobifoam	PET	2.2	M&G	www.mgpolymers.com
Cobitech	PET	2.2	M&G	www.mgpolymers.com
Cobiter	PET	2.2	M&G	www.mgpolymers.com
Colorcomp	Thermopl. allg.	2.2	LNP	www.lnp.com
Colorite 66 S	PVC	2.2	Colorite Plastics Company	www.coloriteplastics.com/
ComAlloy	ABS+PVC	2.2	Schulman	www.aschulman.com
Combidur	VC-Cop	3.1	Profine	www.koemmerling.de
Combithen	PE	3.3	Wolff Walsrode	www.wolff-walsrode.de
Combitherm	PE	3.3	Wolff Walsrode	www.wolff-walsrode.de
Comoglas	UP	2.3	Kuraray	www.kurarayamerica.com
Comp	PP	2.2	Putsch	www.putsch.de
Compax	PC	3.4	Palram	www.palram.com
Compel	Thermopl. allg.	2.2	Ticona	www.ticona.com
Compodic	PA	2.2	Dainippon	www.dic.co.jp
Comshield	Thermopl. allg.	2.2	Schulman	www.aschulman.com
Comtuf	Thermopl. allg.	2.2	Schulman	www.aschulman.com
Conap	PMMA	liquid, 1.4	Conap, Inc	www.conap.com/
Conapoxy	EP	1.4	Conap, Inc	www.conap.com/
Conastic	PUR	liquid, 1.4	Conap, Inc	www.conap.com/
Conathane	PUR	liquid	Conap, Inc	www.conap.com/
Confor	PUR	2.2	E-A-R Specialty Composites	www.earsc.com/
Conoptic	PUR	liquid	Conap, Inc	www.conap.com/
Conpol	EMA	2.2	DuPont	www.dupont.com
Conpol	EMAAA (Ethylene Methyl Acrylate Acrylic Acid	2.2	DuPont Packaging & Industrial Polym	www.dupont.com/packaging/
CoolPoly	LCP	2.2	Cool Polymers, Inc.	www.coolpolymers.com/
CoorsTek	PEI	preformed parts	CoorsTek	www.coorstek.com/
Copolene	E-Cop	2.2	Asahi	www.asahithermofil.com
COPRO	PP-Cop.	2.2	Ohio Valley Plastics Group	www.ohiovalleyplastics.com/
Cordopreg	UP	2.3	Ferro	www.ferro.com
Cordura	PA	3.6	DuPont	www.dupont.com
Corian	PMMA	3.4	DuPont	www.dupont.com
Coroplast	PO, PP, Q, S-Cop	3.3	Coroplast	www.coroplast.de

Continued on next page

Tradename	Polymer	Delivery Form	Manufacturer Supplier	Web Address
Corterra	PTT	2.2	Shell	www.shellchemicals.com
Corton	PP	2.2	PolyPacific Pty. Ltd.	www.polypacific.com.au/
Corvic	PVC	2.2	EVC	www.evc-int.com
Corzan	PVC-C	3.4	Poly-Hi Div. Menasha Corp.	www.polyhisolidur.com/
Cosmax	PMMA	3.3	Asahi	www.asahithermofil.com
Cosmic DAP	DAP (Diallyl Phthalate)	flakes	Cosmic Plastics, Inc.	www.cosmicplastics.com/
COSMOLEX	PE-LLD	2.2	TPC, The Polyolefin Company (Singapo	www.tosoh.com/
Cosmonate	PUR-R	1.1	Mitsui	www.mitsui.co.jp
COSMOPLENE	PP-Cop.	2.2	TPC, The Polyolefin Company (Singapo	www.tosoh.com/
COSMOTHENE	EVA	2.2	TPC, The Polyolefin Company (Singapo	www.tosoh.com/
Courtelle	PAN	3.6	Mitsui	www.mitsui.co.jp
CP PRYME ABS	ABS	2.2	Chase Plastics Services Inc.	www.chaseplastics.com/
Crastin	PBT, PET	2.2	DuPont	www.dupont.com
Crayamid	PA	1.1	DuPont	www.dupont.com
Creablend	ABS+PC, PC+PBT	2.2	PTS	www.pts-marketing.de
Crealen	PP	2.2	PTS	www.pts-marketing.de
Creamid	PA 6, PA 66	2.2	PTS	www.pts-marketing.de
Createc	PBT	2.2	PTS	www.pts-marketing.de
Crestomer	UP	1.1	Scott Bader	www.scottbader.com
Cri-Line	Fluoroelastomer	2.2	Cri-Tech, Inc.	www.critechinc.com/
Cristalite	PMMA	1.1	Schock	www.schock.de
Cristamid	PA 11, PA 12	2.2	Atofina	www.atofina.de
Cronar	PET	3.3	DuPont	www.dupont.com
Crow	PF	3.5	Tufnol	www.tufnol.co.uk
Crylac	PVC	-	Silac	None
Crystal PS	PS	2.2	NOVA Chemicals	www.novachemicals.com/
Crystallite	PMMA	2.2	Thai Petrochemical Industry Co., Ltd	www.tpigroup.co.th/
Crystalor	PMP	2.2	Chevron Phillips Chemical Co.	www.cpchem.com/
Crystar	PET	2.2	DuPont	www.dupont.com
Crystic	UP	1.1	Scott Bader	www.scottbader.com
Cuticulan	PE-LD	3.3	Odenwald	www.odenwald-chemie.de
Cutilan	PE-HD	3.3	Odenwald	www.odenwald-chemie.de
Cutipylen	PP	3.3	Odenwald	www.odenwald-chemie.de
CX-Serie	PA	2.2	Unitika	www.unitika.co.jp

Continued on next page

D.2 Tradenames Table

Tradename	Polymer	Delivery Form	Manufacturer Supplier	Web Address
Cybercell	PVC	2.2	Cybertech Polymers	www.cybertechpolymers.com/
Cybertuff	PVC	2.2	Cybertech Polymers	www.cybertechpolymers.com/
Cycolac	ABS, ABS+PBT	2.2	GEP	www.geplastics.com
Cycoloy	PC+ABS	2.2	GEP	www.geplastics.com
Cycovin	ABS+PVC	2.2	GEP	www.geplastics.com
Cyglas	Polyester, Thermoset	2.2	Cytec Industries Inc.	www.cytec.com/
Cylon	PVC+PA	2.2	Cybertech Polymers	www.cybertechpolymers.com/
Cymel	MF	2.2	Cytec Industries Inc.	www.cytec.com/
Cyrex	PC+PMMA+PS	2.2	Cyro	www.cyro.com
Cyrolite	MBS, PMMA	2.2, 3.4	Röhm	www.cyro.com
Cyrolon	PC	3.4	Cyro	www.cyro.com
CYROVU	PMMA	2.2	CYRO Industries	www.cyro.com/NewCYRO/flash.html
Dacron	PTP	3.6	DuPont	www.dupont.com
Daelim	PE-HD	2.2	DAELIM INDUSTRIAL CO., LTD.	www.daelim.co.kr/
Daiamid	PA	2.2	Daicel	www.daicel.co.jp
Daicel	PS-HI	2.2	PlastxWorld Inc.	www.plastxworld.com/
Daiel	FKM	2.2	Daikin	www.daikin.com
Daiflon	ECTFE	2.2	Daikin	www.daikin.com
Daltorez	TPU	2.2	Huntsman	www.huntsman.com
Daplen	PP	2.2	Borealis	www.borealis.com
Daploy	PP, PP-E	2.2	Borealis	www.borealis.com
Daron	UP, VE	1.5	DSM Composite Resins	www.dsm.com
Dart	PS	2.2	Dart Polymers, Inc.	800-899-9113
Dartek	PA 66	3.3	DuPont	www.dupont.com
Decelith	PVC, VC-Cop	2.1, 2.2	ECW-Eilenburger	www.ecw-compound.de
Degadur	PMMA	1.5	Degussa	www.degussa.de
Degalan	PAA, PMMA	1.1	Degussa	www.degussa.de
Degaroute	PMMA	1.1	Degussa	www.degussa.de
Deglas	PMMA	3.3	Röhm	www.cyro.com
Delmer	PMMA	1.1	Asahi	www.asahithermofil.com
Delpet	PMMA	2.2	Asahi	www.asahithermofil.com
Delrin	POM	2.2	DuPont	www.dupont.com
Delta ABS	ABS	2.2	Delta FRP MFG Inc	www.deltapoly.com/
Deltech PS	PS	2.2	Deltech Polymers Corporation	www.deltechcorp.com/

Continued on next page

Tradename	Polymer	Delivery Form	Manufacturer Supplier	Web Address
Denka	EVAC, VC-Cop	1.1	Denka	www.denka.co.jp
Denka ABS	ABS	2.2	DENKA	www.denka.co.jp/eng/top.htm
Denka Arena	PTP	2.2	Denka	www.denka.co.jp
Denkastyrol	S-Cop	2.2	Denka	www.denka.co.jp
Denkavinyl	PVC, VC-Cop	2.2	Denka	www.denka.co.jp
Depron	PS	3.3	Mitsubishi Polyester Film	www.m-petfilm.com
Derakane	EP	1.1	Dow	www.dow.com
Desmocap	PUR-R	1.1	Bayer	www.bayer.com
Desmocoll	PUR-R	1.1, 1.5	Bayer	www.bayer.com
Desmodur	PUR-R	1.1	Bayer	www.bayer.com
Desmoflex	TPU	2.2	PTS	www.pts-marketing.de
Desmopan	TPU	2.2	Bayer	www.bayer.com
Desmophen	PUR-R	1.1	Bayer	www.bayer.com
Desmotherm	PUR-R	1.1	Bayer	www.bayer.com
Dexflex	TPO	2.2	Solvay	www.solvay.com
DiaAlloy	PC	2.2	Mitsubishi Rayon America Inc.	www.mrany.com/
Diafoil	PET	3.3	Mitsubishi Polyester Film	www.m-petfilm.com
Diakon	PMMA	2.2	Lucite	www.lucitecp.com
Diamid	PA	2.2	Daicel	www.daicel.co.jp
Diamond	ABS	2.2	Diamond Polymers, Inc.	www.diamondpolymers.com/
Dianite	PBT	2.2	Mitsubishi Rayon	www.mrc.co.jp
Diapet	ABS	2.2	Mitsubishi Rayon	www.mrc.co.jp
Diaprene	PP+EPDM	2.2	AES	www.santoprene.com
DIC.PPS	PPS	2.2	Dainippon Ink and Chemicals, Incorpo	www.dic.co.jp/eng/index.html
Dielectrite	Polyester, Thermoset	2.2, BMC	Industrial Dielectrics Inc.	www.idiplastic.com/
Diene	CR	2.2	Firestone	www.firesyn.com
Diexter	PUR-R	1.1	Coim	www.coim.it
Dilamid	PA 6, PA 66	2.2	Dilaplast	www.dilaplast.com
Dinalen	PE-LD	2.2	Dioki	www.dioki.hr
Diofan	PVDC	2.2	Solvay	www.solvay.com
Dion	Polyester, Thermoset	liquid	Reichhold Chemicals, Inc.	www.reichhold.com/
Diprane	TPU	2.2	Hyperlast	www.hyperlast.com
Disperbond	PUR-R	1.1	Merquinsa	www.merquinsa.com
Disperdur	PUR-R	1.1	Merquinsa	www.merquinsa.com
Doki	PS	2.2	Dioki	www.dioki.hr
Domamid	PA 6, PA 66	2.2	Domo	www.domo.be
Domolen	PP	2.2	Domo	www.domo.be

Continued on next page

D.2 Tradenames Table

Tradename	Polymer	Delivery Form	Manufacturer Supplier	Web Address
Dorlastan	PUR	3.6	Lanxess	www.lanxess.com
Dorlyl	PVC	2.2	Dorlyl	www.dorlyl.com
Dowlex	PE-LLD	2.2	Dow	www.dow.com
DPE	PE-LD	2.2	DuPont Packaging & Industrial Polym	www.dupont.com/packaging/
Dryflex	SBS, SEBS	2.2	British Vita	www.britishvita.com
Dryton	CR, EPDM	2.2	Dow	www.dow.com
Dupanel	PUR	3.7	Recticel	www.recticel.com
Duracap	PVC	2.2	PolyOne Th. Bergmann	www.polyone.com
Duracarb	PUR-R	2.2	PPG	www.ppg.com
Duraclear	PUR	3.3	IMS	www.ims-plastic.com
Duracon	POM	2.2	Polyplastics	www.polyplastics.com
Duradene	SBR	2.2	Firestone	www.firesyn.com
Duraflec	PVC	2.2	PolyOne Corporation	www.polyone.com/
Dural	PVC	2.2, cube	Alpha Gary Corporation	www.alphagary.com/
Duralast	PPS	-	Quadrant Engineering Plastic Product	www.quadrantepp.com/
DURALENE	PE-LD	2.2	Ohio Valley Plastics Group	www.ohiovalleyplastics.com/
DURALON	PA66	2.2	Ohio Valley Plastics Group	www.ohiovalleyplastics.com/
DURALOX	PBT	2.2	Ohio Valley Plastics Group	www.ohiovalleyplastics.com/
Duramac	UP	1.1	Eastman	www.eastman.com
Duramax	PBT	2.2	PolyOne Th. Bergmann	www.polyone.com
Duramid	EP	2.3	Isola	www.isola.de
Duranex	PBT	2.2	Polyplastics	www.polyplastics.com
Durastar	PTP	2.2	Eastman	www.eastman.com
DuraStar	PCTA	2.2	Eastman Chemical Company	www.eastman.com/
Durastrength	PMMA	2.2	Atofina	www.atofina.de
Durasyn	PAO	2.2	BP	www.bppetrochemicals.com
Duratron	PI	3.1, 3.2, 3.4	Quadrant Engineering Plastic Product	www.quadrantepp.com/
Duraver	EP	2.3	Isola	www.isola.de
Durel	PAR	-	Ticona	www.topas.com/index.htm
Durethan	PA 6, PA 66	2.2	Lanxess	www.lanxess.com
Durez	Phenolic	granules	Durez Corporation	www.durez.com/
Durlex	PBT	2.2	Chem Polymer Corporation	www.teknorapex.com/

Continued on next page

Tradename	Polymer	Delivery Form	Manufacturer Supplier	Web Address
Durmax	POM	-	PolyOne Corporation	www.polyone.com/
Durocron	PMMA	2.2	Mitsubishi Rayon	www.mrc.co.jp
Durodet	PP, UP	2.3, 3.5	Mitras	www.mitras-materials.com
Durolite	UP	2.3, 3.5	Mitras	www.mitras-materials.com
Durolon	PC	2.2	Policarbonatos do Brasil	www.policarbonatos.com.br
Durolux	PC+ABS	2.2	LA/ES	www.la-es.com
Duropal	MF	3.4	Wodego	www.duropal.com
Durostone	EP, PF, UP, VE	2.3, 3.1, 3.2, 3.4	Röchling Haren	www.roechling-haren.de
Dutral	EPDM, EPM	2.2	Polimeri	www.polimerieuropa.com
Dyflor	PVDF	2.2	Degussa	www.degussa.de
Dylark	S-Cop, SMAH	2.2	Nova	www.novachem.com
Dylene	PS	2.2	Nova	www.novachem.com
Dylite	PS-E	2.2	Nova	www.novachem.com
Dylon	TPU	2.2	Dahin Group	
Dynaflex	TPS	2.2	GLS	www.glscorp.com
Dynamar	FKM	2.2	Dyneon	www.dyneon.com
Dynaset	Phenolic	granules	Durez Corporation	www.durez.com/
Dyneema	PE	3.6	DSM Dyneema	www.dsm.com
Dyneon	PTFE	2.1, 2.2	Dyneon	www.dyneon.com
Dyneon ETFE	ETFE	2.2	Dyneon	www.dyneon.com
Dyneon FEP	FEP	2.2	Dyneon	www.dyneon.com
Dyneon PFA	PFA	2.2	Dyneon	www.dyneon.com
Dyneon THV	TFEHFPVDF	2.2	Dyneon	www.dyneon.com
DYNEX[a]	PE-LLD	2.2	Chevron Phillips Chemical Co.	www.cpchem.com/
Dytherm	PPE+PS	2.2	Nova	www.novachem.com
Dytron	TPE	2.2	AES	www.santoprene.com
Eastalloy	PC+PBT, PC+PET	2.2	Eastman	www.eastman.com
Eastapak	PET	2.2	Eastman	www.eastman.com
Eastar	PTP	2.2	Eastman	www.eastman.com
Eastobond	PTP	2.2	Eastman	www.eastman.com
Eastoflex	EPM	2.2	Eastman	www.eastman.com
Easypoxy	EP	1.4	Conap, Inc	www.conap.com/
Ecdel	TPC	2.2	Eastman	www.eastman.com
Ecocarb	PC	2.2	Aquafil	www.aquafil.com
Ecoflex	PLA	2.2	BASF	www.basf.com
Ecoform	POM	2.2	Aquafil	www.aquafil.com
Ecomass	PA12	2.2	PolyOne Corporation	www.polyone.com/

Continued on next page

D.2 Tradenames Table

Tradename	Polymer	Delivery Form	Manufacturer Supplier	Web Address
Ecomid	PA 6, PA 66	2.2	Nilit	www.nilit.com
Econol	PTP	2.2	Sumitomo	www.sumitomo-chem.co.jp
Econyl	PA 6, PA 66	2.2	Aquafil	www.aquafil.com
Ecoprene	TPE	2.2	Rubber & Plastics Solutions, Inc.	423-436-3167
Ecoruf	PP	3.4	Palram	www.palram.com
Edgetek	Thermopl. allg.	2.2	PolyOne Th. Bergmann	www.polyone.com
Edistir	PS, PS-HI	2.2	Polimeri	www.polimerieuropa.com
EEI	PUR	liquid	Elastomer Engineering Inc.	www.eeicopoly.com/
Efweko	PUR	2.2	Degussa	www.degussa.de
Ekadur	PBT	2.2	Sattler	www.sattlerkunststoffwerk.de
Ekadure	UP	1.5	Ashland	www.ashspec.com
Ekalon	PC	2.2	Sattler	www.sattlerkunststoffwerk.de
Ekaloy	POM	2.2	Sattler	www.sattlerkunststoffwerk.de
Ekamid	PA 66	2.2	Sattler	www.sattlerkunststoffwerk.de
Ekanyl	ABS	2.2	Sattler	www.sattlerkunststoffwerk.de
Ekatal	POM	2.2	Sattler	www.sattlerkunststoffwerk.de
Elamed	PTP	2.2	Elana	www.elana.pl
Elan	PP	2.2	Putsch	www.putsch.de
Elana	PTP	3.6	Elana	www.elana.pl
Elapor	PE-X, PUR	3.7	EMW	www.emw.de
Elaslen	PE-C	2.2	Showa	www.sdk.co.jp
Elastamax	TPE	2.2	PolyOne Th. Bergmann	www.polyone.com
Elastamide	PA6 Elastomer	2.2	Elastomer Engineering Inc.	www.eeicopoly.com/
Elastan	PUR	1.1	Elastogran	www.elastogran.de
Elaster	VC-Cop	2.2	Zeon	www.zeon.co.jp
Elastocell	TPU	2.2	Elastogran	www.elastogran.de
Elastocoat	PUR-R	1.5	Elastogran	www.elastogran.de
Elastoflex	PUR-R	1.1	Elastogran	www.elastogran.de
Elastofoam	PUR-R	1.1	Elastogran	www.elastogran.de
Elastolit	PUR-R	1.1	Elastogran	www.elastogran.de
Elastollan	TPU	2.2	Elastogran	www.elastogran.de
Elaston	PUR-R	3.6	Chemitex	www.chemitex.com
Elastonat	PUR	1.1	Elastogran	www.elastogran.de
Elastopal	TPU	3.7	Elastogran	www.elastogran.de
Elastopan	PUR-R	1.1	Elastogran	www.elastogran.de
Elastopor	PUR-R	1.1	Elastogran	www.elastogran.de
Elastopreg	PP	2.3	BASF	www.basf.com
Elastosil	Q	1.1	Wacker	www.wacker.com
Elastotec	TPU	2.2	Elastogran	www.elastogran.de
Elastotherm	PUR-R	1.1	Elastogran	www.elastogran.de
Elasturan	PUR-R	1.1	Elastogran	www.elastogran.de
Electra	PE-UHMW	3.3	IMS	www.ims-plastic.com

Continued on next page

Tradename	Polymer	Delivery Form	Manufacturer Supplier	Web Address
Elegante	PET	2.2	Eastman	www.eastman.com
Elektrafil	Thermopl. allg.	2.2	DSM Eng. Plastics	www.dsmep.com
Elexar	SEBS	2.2	Teknor Apex	www.teknorapex.com
Elit	PTP	2.2	Elana	www.elana.pl
Elite	PPMS	2.2	Dow	www.dow.com
Elitel	PTP	2.2	Elana	www.elana.pl
Elix	S-Cop	2.1	Dow	www.dow.com
Eltex	PE-HD, PP	2.2	Solvay	www.solvay.com
Elvacite	A-Cop	2.2	Lucite	www.lucitecp.com
Elvaloy	EBA, EEA, EMA	2.2	DuPont	www.dupont.com
Elvamide	PA	1.1, 2.2	DuPont	www.dupont.com
Elvanol	PVAL	1.1	DuPont	www.dupont.com
Elvax	EVAC	2.2	DuPont	www.dupont.com
Emac	EMA	2.2	Chevron	www.chevrontexaco.com
Emarex	PA 6, PA 66	2.2	MRC Polymers	www.mrcpolymers.com
Emblem	PVDC+PVAC	3.3	Unitika	www.unitika.co.jp
Embrace	PTP	3.3	Eastman	www.eastman.com
Emdicell	TPU	2.2, 3.7	Elastogran	www.elastogran.de
Emerge	PC+ABS	2.2	Dow	www.dow.com
Emiclear	PS	2.2	Toshiba Chemical Corporation	www.toshiba.co.jp/
Emi-X	ABS, PA 66, PBT, PC, PESU	2.2	LNP	www.lnp.com
Empera	PS	2.2	BP	www.bppetrochemicals.com
Emulprene	SB	2.2	INSA	52-5 726-1800
Enable	EBA	1.5	Exxon	www.exxonchemical.com
EnCom	PC	2.2	EnCom, Inc.	www.encompolymers.com/
Encore	POM	-	Ticona	www.topas.com/index.htm
Endura	ABS	2.2	PMC Engineered Plastics, Inc.	www.pmc-group.com/
Enduran	PBT, PBT+PET	2.2	GEP	www.geplastics.com
Enerlon	PTFE	2.2	Greene, Tweed Engineered Plastics	www.gtweed.com/
Engage	TPO	2.2	DuPont Dow Elastomers	www.dupont-dow.com
Enplex	ABS+PVC	2.2	Kaneka	www.kaneka.co.jp
Ensicar	PC	3.1, 3.2, 3.4	Ensinger Inc.	www.shopforplastics.com/
Ensidur	ABS	3.1, 3.2, 3.4	Ensinger Inc.	www.shopforplastics.com/
Ensifide	PPS	3.1, 3.2, 3.4	Ensinger Inc.	www.shopforplastics.com/
Ensifone	PSU	3.1, 3.2, 3.4	Ensinger Inc.	www.shopforplastics.com/

Continued on next page

D.2 Tradenames Table

Tradename	Polymer	Delivery Form	Manufacturer Supplier	Web Address
Ensikem	PVDF	3.1, 3.2, 3.4	Ensinger Inc.	www.shopforplastics.com/
Ensilon	PA66	3.1, 3.2, 3.4	Ensinger Inc.	www.shopforplastics.com/
Ensipro	PP	3.1, 3.2, 3.4	Ensinger Inc.	www.shopforplastics.com/
Ensital	POM	3.1, 3.2, 3.4	Ensinger Inc.	www.shopforplastics.com/
Ensitep	PBT	3.1, 3.2, 3.4	Ensinger Inc.	www.shopforplastics.com/
Envelon	EMA	2.2	Rohm & Haas	www.rohmhaas.com
Envex	PA	3.1, 3.2, 3.4	Rogers Corporation	www.rogers-corp.com/
Enviraloy	PC+PBT	2.2	ACI Plastics	www.aciplastics.com/
Envirez	UP	1.1	PPG	www.ppg.com
EPABS	ABS	2.2	Engineered Plastics Corporation	www.engineered-plastics.com/
Epalex	PP	2.2	PolyPacific Pty. Ltd.	www.polypacific.com.au/
EPalloy	ABS+PC	2.2	Engineered Plastics Corporation	www.engineered-plastics.com/
Epcar	EPDM	2.2	Goodrich	www.goodrich.com
Eperan	PE, PP	3.7	Kaneka	www.kaneka.co.jp
EPI	ABS	2.2	Engineered Polymers Industries	www.osterman-co.com/epi.html
Epic	EP	1.4	Epic Resins	www.epicresins.com/
Epic-Cast	EP	liquid	Epic Resins	www.epicresins.com/
Epic-Lam	EP	liquid	Epic Resins	www.epicresins.com/
Epocast	EP	2.2	Ciba Specialty Chemicals	www.cibasc.com/
Epofriend	TPE	2.2	Polyplastics	www.polyplastics.com
Epolene	PE, PP	1.5	Eastman	www.eastman.com
Epolite	EP	2.2	Fiber Resins	None
Epomic	EP	1.1	Mitsui	www.mitsui.co.jp
Eponac	EP	2.2	Sprea	www.sirindustriale.com
Epotal	PE	1.2	BASF	www.basf.com
Epo-Tek	EP	1.4, 2.2	Epoxy Technology Inc.	www.epotek.com/
Epotuf	EP	1.1	Reichhold	www.reichhold.com
EPPC	PC	2.2	Engineered Plastics Corporation	www.engineered-plastics.com/
EPPCPET	PC+PET	2.2	Engineered Plastics Corporation	www.engineered-plastics.com/
Epsilon	PP-Cop.	2.2	Epsilon Products Company	www.sunocochem.com/
Equistar	PP	2.2	Equistar	www.equistar.com
Eraclene	E-Cop, PE-HD	2.2	Polimeri	www.polimerieuropa.com

Continued on next page

Tradename	Polymer	Delivery Form	Manufacturer Supplier	Web Address
Eref	PA+PP	2.2	Solvay	www.solvay.com
Ertalyte	PET	3.1, 3.4	Quadrant Engineering Plastic Product	www.quadrantepp.com/
Esall	PP+TPS	2.2	Sumitomo	www.sumitomo-chem.co.jp
Esbrid	PA 6	2.2	Asahi	www.asahithermofil.com
Esbrite	PS	2.2	Sumitomo	www.sumitomo-chem.co.jp
Escalloy	PP	-	ExxonMobil Chemical Company	www.exxonmobilchemical.com/
Escor	EAA	2.2	Exxon	www.exxonchemical.com
Escorene	EVAC, PE-LD, PE-LLD, PP	2.2	Exxon	www.exxonchemical.com
Eska	PMMA	3.6	Mitsubishi Rayon	www.mrc.co.jp
Esmedica	PVC	2.2	Sekisui Chemical Company, Ltd.	www.sekisuichemical.com/
Espanex	PI	1.1	Nippon Steel	www.nscc.co.jp
Espladur	PF	2.2	Raschig	www.raschig.de
ESPREE	ABS	2.2	GE Polymerland	www.gepolymerland.com/
Esprene	EPDM	2.2	Sumitomo	www.sumitomo-chem.co.jp
Establend	PC+ABS	2.2	Cossa	www.cossapolimeri.it
Estacarb	PC	2.2	Cossa	www.cossapolimeri.it
Estadiene	ABS	2.2	Cossa	www.cossapolimeri.it
EstaGrip	TPE	2.2	Noveon, Inc.	www.noveon.com/
Estal	PBT	2.2	PolyOne Th. Bergmann	www.polyone.com
Estaloc	TPU	2.2	Goodrich	www.goodrich.com
Estane	TPU	2.2	Goodrich	www.goodrich.com
Estaprop	PP	2.2	Cossa	www.cossapolimeri.it
Estar	UP	1.1	Mitsui	www.mitsui.co.jp
Estasan	SAN	2.2	Cossa	www.cossapolimeri.it
Estastir	PS	2.2	Cossa	www.cossapolimeri.it
Esteform	UP	2.2	Chromos	www.chromos.hr
Estemab	UP	2.3	Mitsui	www.mitsui.co.jp
Estemix	UP	2.2	Chromos	www.chromos.hr
Estyrene	ABS, SMAH	2.2	Nippon Steel	www.nscc.co.jp
Ethavin	TPO	2.2	Vi-Chem Corporation	www.vichem.com/
Ethocel	CA	2.2	DuPont	www.dupont.com
Etinox	PVC	2.2	Aiscondel	www.grupoaragonesas.es
ET-Polymer	PE	2.2	Borealis	www.borealis.com
Etronax	CF, EP, MF, PPE, SI, UP	3.4	Elektro-Isola	www.elektro-isola.com
Etronit	CF, EP, MF	3.5	Elektro-Isola	www.elektro-isola.com

Continued on next page

D.2 Tradenames Table

Tradename	Polymer	Delivery Form	Manufacturer Supplier	Web Address
Eurodrain	PVC	3.2	Hegler	www.hegler-plastik.de
Europhan	PVC	3.3	Huhtamaki	www.4pfolie.de
Europlex	PSU	3.4	Röhm	www.cyro.com
Europrene	BR, NBR, SBR, SBS, SEBS, SIS	2.2	Polimeri	www.polimerieuropa.com
Eutan	TPU	3.1, 3.4	Acla	www.acla-werke.de
Evacon	PMMA	2.2	Lucite	www.lucitecp.com
Eval	EVAL	2.2	Eval Company of America	www.eval.be/
Evalastic	PP+EPDM	3.3	Alwitra	www.alwitra.de
Evalon	EVAC+PVC	3.3	Alwitra	www.alwitra.de
Evaloy	A-Cop, EBA, EEA, EMA	2.2	DuPont	www.dupont.com
Evamelt	EVAC	2.2	Lanxess	www.lanxess.com
Evatane	EVAC	2.2	Atofina	www.atofina.de
Evatate	E-Cop	2.2	Sumitomo	www.sumitomo-chem.co.jp
Evazote	E-Cop	3.7	Zotefoams	www.zotefoams.com
Evicom	PVC	2.2	EVC Compounds Ltd.	www.ineos.com/index.php
Evipol	PVC	2.2	EVC	www.evc-int.com
Evolue	PO	2.2	Mitsui	www.mitsui.co.jp
Evoprene	SEBS, TPS	2.2	Alphagary	www.alphagary.com
Exac	PCTFE	3.1, 3.2	Saint Gobain - Norton	www.plastics.saint-gobain.com/
Exact	TPO	2.2	Dex-Plastomers	www.dexplastomers.com
Exceed	PE-MC	2.2	Exxon	www.exxonchemical.com
Excellen	PE-ULD	2.2	Sumitomo	www.sumitomo-chem.co.jp
Exceval	CP	1.1	Kuraray	www.kurarayamerica.com
Exter	UP	1.1	Coim	www.coim.it
Extir	PS-E	2.2	Polimeri	www.polimerieuropa.com
Extron	PP	2.2	PolyPacific Pty. Ltd.	www.polypacific.com.au/
Extrupet	PET	2.2	La Seda	www.laseda.es
EXXELOR	PP	2.2	ExxonMobil Chemical Company	www.exxonmobilchemical.com/
Exxonmobil	PE, PP	2.2	Exxon	www.exxonchemical.com
Exxtral	PP	2.2	Exxon	www.exxonchemical.com
Fabelnyl	PA	2.2	Exxon	www.exxonchemical.com
Fabeltan	TPU	2.2	Exxon	www.exxonchemical.com
Factor	PA	3	Fact	
Faradex	Thermopl. allg.	2.2	LNP	www.lnp.com
Favorite	PE	2.2	Favorite Plastics	www.favoriteplastics.com/
Fawocel	PP	3.7	Gefinex	www.gefinex.de

Continued on next page

Tradename	Polymer	Delivery Form	Manufacturer Supplier	Web Address
Fawolen	PE	3.7	Gefinex	www.gefinex.de
Fawolit	PP	3.7	Gefinex	www.gefinex.de
Fawotop	PP	3.7	Gefinex	www.gefinex.de
Fenoform	CF	2.2	Chromos	www.chromos.hr
Ferrene	E-Cop	2.2	Ferro	www.ferro.com
Ferrex	PP	2.2	Ferro	www.ferro.com
Ferrocon	PP	2.2	Ferro	www.ferro.com
Ferro-Flex	PP+EPDM	2.2	Ferro	www.ferro.com
Ferroflo	PS	2.2	Ferro	www.ferro.com
Ferropak	PP	2.2	Ferro	www.ferro.com
Fiberfil	Thermopl. allg.	2.2	DSM Eng. Plastics	www.dsmep.com
Fiberite	EP	2.2, granules	ICI Fiberite	507-454-3611
Fiberloc	PVC	2.2, 2.3	PolyOne Th. Bergmann	www.polyone.com
Fibrelam	UP	2.3	Hexcel	www.hexcel.com
Fibrodux	EP, PF	2.3	Hexcel	www.hexcel.com
Finacene	PE-MC	2.2	Atofina	www.atofina.de
Finaclear	SBS	2.2	Atofina	www.atofina.de
Finaprene	SBS	2.2	Atofina	www.atofina.de
Finapro	PP	2.2	Atofina	www.atofina.de
Finathene	PE-HD, PE-LD	2.2	Atofina	www.atofina.de
Fire	PF	2.2	Indspec Chemical Corporation	www.indspec-chem.com/
Fireguard	PVC	2.2	Teknor Apex Company	www.teknorapex.com/
Firestone	PA6	2.2	Firestone Textiles Company	www.firestone-textiles.com/
Fixmaster	EP	2.2	Fel-Pro Chemical Products, Inc.	None
Flexalen	PE-HD	3.2	Pipelife	www.pipelife.at
Flexalloy	PVC	2.2	Teknor Apex	www.teknorapex.com
Flexathene	TPO	2.2	Equistar	www.equistar.com
Flexbond	PVAC	1.1	Air Products	www.airproducts.com
Flexchem	PVC	2.2	Colorite Plastics Company	www.coloriteplastics.com/
Flexirene	PE-LLD	2.2	Polimeri	www.polimerieuropa.com
Flexivolt	PVC	3.2	Solvay	www.solvay.com
Flexloy	PA+PO	2.2	Sumitomo	www.sumitomo-chem.co.jp
Flexocel	PUR	1.1	Baxenden	www.baxchem.co.uk
Flex-O-Film	CAB	2.2	Flex-O-Glass, Inc.	www.flexoglass.com/
FLEXOMER	PE-LLD	2.2	Dow Plastics	plastics.dow.com/
Flexprene	TPE	2.2	Teknor Apex Company	www.teknorapex.com/

Continued on next page

D.2 Tradenames Table

Tradename	Polymer	Delivery Form	Manufacturer Supplier	Web Address
Flo-Blen	PO	2.2	Baxenden	www.baxchem.co.uk
Fluoraz	PTFE	2.2	Greene, Tweed Engineered Plastics	www.gtweed.com/
Fluorel	FKM	2.2	Dyneon	www.dyneon.com
Fluran	Fluoroelastomer	3.2	Saint Gobain - Norton	www.plastics.saint-gobain.com/
Folitherm	PP	3.3	Huhtamaki	www.4pfolie.de
Foraflon	PTFE, PVDF	2.2	Atofina	www.atofina.de
Foralkyl	PFA	2.2	Atofina	www.atofina.de
Forco	PP	3.3	Huhtamaki	www.4pfolie.de
Forex	PVC	3.4, 3.7	Airex	www.alcanairex.com
Forflex	TPO	2.2	PolyOne Th. Bergmann	www.polyone.com
Formax	POM	2.2	Chem Polymer Corporation	www.teknorapex.com/
Formion	EIM	2.2	Schulman	www.aschulman.com
Formolene	PO	2.2	Formosa Plastics	www.fpcusa.com
Formolon	PVC	2.2	Formosa Plastics Corporation USA	www.fpcusa.com/
Formosacon	POM	2.2	Formosa Plastics	www.fpcusa.com
Formula	PP	2.2	Putsch	www.putsch.de
Formvar	PVFM	1.1	Dow	www.dow.com
Forprene	TPE, TPV	2.2	So.F.Ter	www.softerspa.com
Forsan	ABS	2.2	Kaucuk	www.kaucuk.cz
Fortiflex	PE-HD	2.2	BP	www.bppetrochemicals.com
Fortilene	PP	2.2	Solvay	www.solvay.com
Fortron	PPS	2.2	Ticona	www.ticona.com
Fostalink	PA12 Elastomer	2.2	Foster Corporation	www.fostercomp.com/
Fostalon	PA12	2.2	Foster Corporation	www.fostercomp.com/
Fostamid	PA12	2.2	Foster Corporation	www.fostercomp.com/
Franprene	TPO, TPS	2.2	Franplast	www.franplast.it
Frialen	PE-HD	3.2	Friatec	www.friatec.de
Frianyl	PA 6, PA 66	2.2	Frisetta	www.frisetta-polymer.de
Friedola	PP, PVC-P	3.3	Friedola	www.friedola.de
Frila	PVC	2.2	Norsk Hydro	www.hydro.com
Fulton	F-Pol.+POM	2.2	LNP	www.lnp.com
Fundopal	MF	3.4	Funder	www.funder.at
Fürkadur	PA+PC	2.2	Solvadis	www.solvadis.de
Fürkaform	POM	2.2	Solvadis	www.solvadis.de
Fürkalon	TPE	2.2	Solvadis	www.solvadis.de
Furnidur	VC-Cop	3.3	Mitsubishi Polyester Film	www.m-petfilm.com
Furnit	PVC	3.3	Hornschuch	www.hornschuch.de
Fusabond	E-Cop	2.2	DuPont	www.dupont.com

Continued on next page

Tradename	Polymer	Delivery Form	Manufacturer Supplier	Web Address
Gabotherm	PB-1	3.2	Hornschuch	www.hornschuch.de
Gaflon	PTFE	3	Hornschuch	www.hornschuch.de
GAFONE	PES		Gharda Chemicals Limited	www.ghardapolymers.com/
Galirene	PS	2.2	Carmel	www.carmel-olefins.co.il
Gammaflex	PE	2.2	PTS	www.pts-marketing.de
Gapex	PP, PP+PA	2.2	Ferro	www.ferro.com
Garaflex	TPE	2.2	Alphagary	www.alphagary.com
GATONE	PEEK	2.2	Gharda Chemicals Limited	www.ghardapolymers.com/
Gealan	PVC-P	3.1, 3.2	Gealan	www.gealan.de
Geberit	PE-HD	3.2	Geberit	www.geberit.com
Gechron	CO	2.2	Zeon	www.zeon.co.jp
Gedexcel	PS-E	2.2	Nova	www.novachem.com
Gekaplan	PVC-P	3.3	Gehr	www.gehr.de
Gelkyd	UP	1.1	Gehr	www.gehr.de
Gelon	PA 6, PA 66	2.2	GEP	www.geplastics.com
Geloy	ASA, ASA+PC, PC+SAN	2.2	GEP	www.geplastics.com
Genesis	PO, PS	2.2	Nova	www.novachem.com
Genotherm	PVC	3.3	Klöckner-Pentaplast	www.kpfilms.com
Gensil	Q	2.2	GE Silicones	www.gesilicones.com
Geolast	PP+BR	2.2	AES	www.santoprene.com
Geon	PVC	1.2, 2.1, 2.2	PolyOne Th. Bergmann	www.polyone.com
Georgia	PVC	2.2	Georgia Gulf	www.ggc.com/
Gepax	PC	3.4	GEP	www.geplastics.com
Germadur	PBT	2.2	Kukoha	www.kunststoff-kontor-hamburg.de
Germamid	PA 6, PA 66	2.2	Kukoha	www.kunststoff-kontor-hamburg.de
Gerodur	PE, PP, PVC	3.2	Haka Gerodur	www.haka.ch
Gerofit	PE-HD	3.2	Haka Gerodur	www.haka.ch
GESAN	SAN	2.2	GE Plastics	www.geplastics.com
Getec	EP+PPE	2.3	GEP	www.geplastics.com
Gitto/Global	PP	-	The Gitto/Global Corporation	978-537-8261
Glaskyd	Alkyd	2.2	Cytec Industries Inc.	www.cytec.com/
Glastic	Polyester, Thermoset	2.2, BMC	Glastic Corporation	www.glastic.com/
Globalpet	PET	2.2	Selenis	www.selenis.com
Godifin	TPO	2.2	Godiplast	www.godiplast.com
Godiflex	TPS	2.2	Godiplast	www.godiplast.com
Godigum	PVC+NBR	2.2	Godiplast	www.godiplast.com
Godiplast	PVC-P, PVC-U	2.2	Godiplast	www.godiplast.com

Continued on next page

D.2 Tradenames Table

Tradename	Polymer	Delivery Form	Manufacturer Supplier	Web Address
Godiprene	TPU	2.2	Godiplast	www.godiplast.com
Gohsenol	PVAL	2.2	Nippon Gohsei	www.nippongohsei.com
Gohsenyl	PVAC	2.2	Nippon Gohsei	www.nippongohsei.com
Goldadur	PBT	2.2	Goldmann	www.gold-mann.de
Goldaform	POM	2.2	Goldmann	www.gold-mann.de
Goldalon	PC	2.2	Goldmann	www.gold-mann.de
Goldamid	PA 6, PA 66	2.2	Goldmann	www.gold-mann.de
Goldaprop	PP	2.2	Goldmann	www.gold-mann.de
Goldrex	PMMA	2.2	Hanyang Chemical Corp.	hcc.hanwha.co.kr/English/
Gravitech	Thermopl. allg.	2.2	PolyOne Th. Bergmann	www.polyone.com
Greenflex	EVAC	2.2	Polimeri	www.polimerieuropa.com
Gremodur	FF, PF	1.1	Gremolith	www.gremolith.ch
Gremopal	UP	1.1	Gremolith	www.gremolith.ch
Gremothan	PUR-R	1.1	Gremolith	www.gremolith.ch
Grilamid	PA 12, PA-Cop	2.1, 2.2	Ems	www.ems-chemie.com
Grilon	PA 6, PA 66, PA-Cop, TPA	2.2	Ems	www.ems-chemie.com
Grilpet	PBT, PET	2.2	Ems	www.ems-chemie.com
Griltex	PA	1.5	Ems	www.ems-chemie.com
Grivory	PA 6/6-T, PPA	2.2	Ems	www.ems-chemie.com
Gumiplast	PVC-P	2.2	Ems	www.ems-chemie.com
GUR	PE-UHMW	2.1, 2.2	Ticona	www.ticona.com
Gurit	ABS, CA, CP, PO, PS, PVC	3.3	Gurit	www.gurit-worbla.com
Guron	PE-LD	3.7	Koepp	www.koepp-ag.de
Hagulen	PE-HD	3.2	Hagusta	www.gwe-gruppe.de
Haiplen	PP	2.2	Taro	www.taroplast.com
Hakathen	PO	3.2	Haka Gerodur	www.haka.ch
Halar	ECTFE	2.2	Solvay	www.solvay.com
Halon	ECTFE	2.2	Solvay	www.solvay.com
Hanacelan	PS-+PE	2.2	Kumho Chemicals, Inc.	www.kkpc.co.kr/
Hartex	IR	2.2	Firestone	www.firesyn.com
Hawiflex	PUR	3, 3.1, 3.2, 3.3, 3.4	Habermann	www.habermann-gmbh.de
Haysite	Polyester, Thermoset	3.4, BMC	Haysite Reinforced Plastics	www.haysite.com/
Heglerflex	PE, PP, PVC	3.2	Hegler	www.hegler-plastik.de
Heglerplast	PVC	3.2	Hegler	www.hegler-plastik.de
Hekaplast	PE-HD	3.2	Hegler	www.hegler-plastik.de
Heliflex	PVC	3.2	Heliflex	www.heliflex.pt
Helithen	PE-HD	3.2	Heliflex	www.heliflex.pt
Helitherm	PP	3.2	Heliflex	www.heliflex.pt
Heraflex	SEBS	2.2	Radici	www.radiciplastics.com

Continued on next page

Tradename	Polymer	Delivery Form	Manufacturer Supplier	Web Address
Heraform	POM	2.2	Radici	www.radiciplastics.com
Heramid	PA 6, PA 66	2.2	Radici	www.radiciplastics.com
HERAMID	PA66	2.2	Radici Plastics	www.radicigroup.com/plastics/
Herex	PVC	3.5, 3.7	Airex	www.alcanairex.com
Hersalem	PVC	3.2	Solvay	www.solvay.com
Hetron	VE	1.5	Ashland	www.ashspec.com
Hexene	PE-HD, PE-LD	2.2	Ashland	www.ashspec.com
Hexlite	UP	2.3	Hexcel	www.hexcel.com
Hicor	PP	3.3	Exxon	www.exxonchemical.com
Hifax	PP	2.2	Basell	www.basell.de
Hifor	PE-LLD	2.2	Eastman	www.eastman.com
Higlas	PP	2.2	Basell	www.basell.de
Hilex	PE-HD	3.3	Plastech	www.vtsplastech.co.uk
Hiloy	Thermopl. allg.	2.2	Schulman	www.aschulman.com
Himol	PIB	1.5	Nippon Petro-chemicals	www.npcc.co.jp
Hipec	TSE (Thermoset Elastomer)	2.2	Dow Corning Corporation	www.dowcorning.com/
HIPOL	PP	2.2	Mitsui Chemicals America, Inc.	www.mitsuichemicals.com/
Hiprene	TPU	2.2	Mitsui	www.mitsui.co.jp
Hival	POM	2.2	General Polymers	www.generalpolymers.com/
Hivalloy	PP+A-Cop, PP+S-Cop	2.2	Basell	www.basell.de
Hi-Zex	PE-HD, PE-LD	2.2	Mitsui	www.mitsui.co.jp
Hofalon	PF	3.4	Hornitex	www.hornitex.de
Hoffman	PVC	2.2	Hoffman Plastic Compounds Inc.	323-636-3346
Homadur	PF	3.4	Homanit	www.homanit.de
Homepet	PET	2.2	Selenis	www.selenis.com
Horda	PE Cop.	2.2	Horda Cable Compounds	www.wireworld.com/trelleborg/
Hornit	MF	3.4	Hornitex	www.hornitex.de
Hornitex	MF	3.5	Hornitex	www.hornitex.de
Hostacom	PP	2.2	Basell	www.basell.de
Hostaflon	ETFE	-	Ticona	www.topas.com/index.htm
Hostaform	POM	2.2	Ticona	www.ticona.com
Hostaglas	PET	3.4	Hagedorn	www.hagedorn.de
Hostalen	PE, PP	2.2	Basell	www.basell.de
Hostalit	PVC, PVC-U, VC-Cop	2.2	Vinnolit	www.vinnolit.de
Hostalloy	PE-UHMW	2.1, 2.2	Ticona	www.ticona.com
Hostapet	PET	3.3	Klöckner-Pentaplast	www.kpfilms.com

Continued on next page

D.2 Tradenames Table

Tradename	Polymer	Delivery Form	Manufacturer Supplier	Web Address
Hostaphan	PET	3.3	Mitsubishi Polyester Film	www.m-petfilm.com
Howelon	PVC-P	3.3	Hornschuch	www.hornschuch.de
HSPP	PP	2.2	Chisso America Inc.	www.chisso.co.jp/
HTP	ABS	2.2	HiTech Polymers, Inc.	www.hitechpolymers.com/
Huber	PPA		Texas Polymer Services	(330)666-3751
Huntsman	PE, PP	2.2	Huntsman	www.huntsman.com
Huntsman	PS	2.2	Huntsman Corporation	www.huntsman.com/
Hy-Bar	E-Cop, PE, PE-LLD, PP	3.3	Huntsman	www.huntsman.com
Hybrar	TPE	2.2	Kuraray	www.kurarayamerica.com
Hybri-Chem	PUR	2.2	Hybri-Chem Inc.	800-235-4201
Hybrid	ABS+PC, PC+PBT	2.2	Entec	www.entecresins.com
Hycar	CR, IR	2.2	Goodrich	www.goodrich.com
Hyd-Cast	PA6	2.2, 3.1, 3.4	A.L. Hyde Company	www.alhyde.com/
HYDCOR	PVC	3.1	A.L. Hyde Company	www.alhyde.com/
Hyde	PMMA	3.1, 3.2	A.L. Hyde Company	www.alhyde.com/
HYDEL	POM	2.2, 3.1	A.L. Hyde Company	www.alhyde.com/
Hydex	TPU	2.2, 3.1, 3.4	A.L. Hyde Company	www.alhyde.com/
Hydlar	PA66	3.1, 3.4	A.L. Hyde Company	www.alhyde.com/
Hydrex	UP	2.2	Reichhold	www.reichhold.com
Hydrin	CO	2.2	Zeon	www.zeon.co.jp
Hydrocell	PUR	3.7	Koepp	www.koepp-ag.de
Hydrolar	PUR-R	1.2	Coim	www.coim.it
Hydropor	PUR	3.7	Koepp	www.koepp-ag.de
Hyflon	MFA, PFA	2.1, 2.2	Solvay	www.solvay.com
Hylac	ABS	2.2	Entec	www.entecresins.com
Hylar	ECTFE, PVDF	2.1, 2.2	Solvay	www.solvay.com
Hylene	TPU	2.2	DuPont	www.dupont.com
Hylex	PC	2.2	Entec	www.entecresins.com
Hylon	PA 6, PA 66	2.2	Entec	www.entecresins.com
Hylox	PBT	2.2	Entec	www.entecresins.com
Hypalon	CSM	2.2	DuPont Dow Elastomers	www.dupont-dow.com
Hypel	PE-HD, PE-LD, PE-LLD	2.2	Entec	www.entecresins.com

Continued on next page

Tradename	Polymer	Delivery Form	Manufacturer Supplier	Web Address
Hyperlite	PUR-R	1.1	Bayer	www.bayer.com
Hypro	PP	2.2	Entec	www.entecresins.com
Hyrene	PS-HI	2.2	Entec	www.entecresins.com
Hysol	EP	2.3	Entec	www.entecresins.com
Hysun	ASA, ASA+PC	2.2	Entec	www.entecresins.com
Hytemp	CR	2.2	Zeon	www.zeon.co.jp
Hytrel	TPC	2.2	DuPont	www.dupont.com
Hyvex	PPS	2.2	Ferro	www.ferro.com
Hyvin	PVC	2.2	Norsk Hydro	www.hydro.com
Idemitsu	PC	2.2	Idemitsu	www.idemitsu.co.jp
IDI	Polyester, Thermoset	SMC, BMC	Industrial Di-electrics Inc.	www.idiplastic.com/
IE PUR	TSU (Polyurethane Thermoset Elastomer)	liquid	Innovative Polymers, Inc.	www.innovative-polymers.com/
Illmid	PI	3.7	Illbruck	www.illbruck.de
Illtec	MF	3.7	Illbruck	www.illbruck.de
Imipex	PESI	2.2	GEP	www.geplastics.com
Impact	TPO	2.2	ACI Plastics	www.aciplastics.com/
Impet	PET	2.2	Ticona	www.ticona.com
Implex	PMMA	3.4	ATOFINA, Atoglas Division	www.arkema-inc.com/
Impranil	PUR-R	1.1	Bayer	www.bayer.com
Indopol	PB-1	2.2	BP	www.bppetrochemicals.com
Indothene	PE-LD, PE-LLD	2.2	Indian Petrochemicals	www.ipcl.co.in
Indovin	PVC	2.2	Indian Petrochemicals	www.ipcl.co.in
Induvil	PVC	2.2	Solvay	www.solvay.com
Infinity	PA 6	2.2	Honeywell	www.honeywell.com
Innovex	PE-LLD	2.2	BP	www.bppetrochemicals.com
Inspire	PP	2.2	Dow	www.dow.com
Insular	PVC	2.2	Occidental	www.oxychem.com
Insulcast	EP	2.2	American Safety Technologies (Permag	www.astantislip.com/
Insulgel	EP	gel	American Safety Technologies (Permag	www.astantislip.com/
Insulstruc	Polyester, Thermoset	2.2, BMC	Industrial Di-electrics Inc.	www.idiplastic.com/
Intene	BR	2.2	Polimeri	www.polimerieuropa.com
Interpol	PUR	2.2	Cook Composites and Polymers	www.c-flex-cpt.com/
Intol	SBR	2.2	Polimeri	www.polimerieuropa.com
Intrile	CR	3.3	Polimeri	www.polimerieuropa.com

Continued on next page

D.2 Tradenames Table

Tradename	Polymer	Delivery Form	Manufacturer Supplier	Web Address
Invision	TPO	2.2	Schulman	www.aschulman.com
Iotek	EIM	2.2	Exxon	www.exxonchemical.com
IPC	ABS	2.2	International Polymers Corporation	800-526-0953
Ipethene	PE	2.2	Carmel	www.carmel-olefins.co.il
Iradur	PBT	2.2	Godiplast	www.godiplast.com
Irodur	TPU	2.2	Huntsman	www.huntsman.com
Irogran	TPU	2.2	Huntsman	www.huntsman.com
Irostic	PUR-R	1.1	Huntsman	www.huntsman.com
Isoexter	PUR-R	1.1	Coim	www.coim.it
Isolac	ABS	2.2	GEP	www.geplastics.com
Isoloss	TSU (Polyurethane Thermoset Elastomer)	2.2	E-A-R Specialty Composites	www.earsc.com/
Isonate	PUR-R	1.1	Dow	www.dow.com
Isopak	ABS	2.2	Great Eastern	www.greco.com.tw
Isoplast	TPU	2.2	Dow	www.dow.com
Isorene	SBS, SEBS, TPE	2.2	GEP	www.geplastics.com
Isosan	SAN	2.2	Great Eastern	www.greco.com.tw
Isoschaum	UF	3.7	Schaum-Chemie	www.schaum-chemie.de
Isotal	POM	2.2	GEP	www.geplastics.com
Isplen	PP	2.2	Repsol	www.repsol.com
Iupiace	PPE	2.2	Mitsubishi Engineering-Plastics Corp	www.m-ep.co.jp/
Iupilon	PC	2.2	Mitsubishi Engineering-Plastics Corp	www.m-ep.co.jp/
Iupital	POM	2.2	Mitsubishi Engineering-Plastics Corp	www.m-ep.co.jp/
Ivelit	PO	3.3	Solvay	www.solvay.com
Ixan	PVDC	1.2, 2.2	Solvay	www.solvay.com
Ixef	PARA	2.2	Solvay	www.solvay.com
Jambolen	PTP	3.6	Lukoil	www.lukoil.bg
Jazz	PP	2.2	PolyPacific Pty. Ltd.	www.polypacific.com.au/
J-Bond	TPE	2.2	J-Von Incorporated	www.jvon.com/
Jeffol	PUR-R	1.1	Huntsman	www.huntsman.com
J-Flex	TPE	2.2	J-Von Incorporated	www.jvon.com/
J-Last	SBS	2.2	J-Von Incorporated	www.jvon.com/
Jonylon	PA	2.2	BIP	www.bip.co.uk

Continued on next page

Tradename	Polymer	Delivery Form	Manufacturer Supplier	Web Address
J-Prene	TPV	2.2	J-Von Incorporated	www.jvon.com/
J-Soft	SEBS	2.2	J-Von Incorporated	www.jvon.com/
Jupiace	PPE	2.2	Mitsubishi	www.m-ep.co.jp
Jupilon	PC, PC+ABS	2.2, 3.3, 3.4	Mitsubishi	www.m-ep.co.jp
Jupital	POM	2.2	Mitsubishi	www.m-ep.co.jp
Kadel	PAEK	2.2	Solvay	www.solvay.com
Kaifa	PTP	2.2	Solvay	www.solvay.com
Kaladex	PEN	3.3	DuPont	www.dupont.com
Kalrez	FKM	3.1, 3.3	DuPont Dow Elastomers	www.dupont-dow.com
Kamax	PMMA	-	ATOFINA, Atoglas Division	www.arkema-inc.com/
KaneAce	MBS	2.1	Kaneka	www.kaneka.co.jp
Kanebian	PVAL	3.6	Kaneka	www.kaneka.co.jp
Kaneka	ABS	3.4	Kaneka Corporation	www.kaneka.co.jp/kaneka-e/index.html
Kanelite	PS	3.3, 3.7	Kaneka	www.kaneka.co.jp
Kanevinyl	VC-Cop	2.2	Kaneka	www.kaneka.co.jp
Kapex	TPU	3.5	Airex	www.alcanairex.com
Kapton	PI	3.3	DuPont	www.dupont.com
Karlex	PTP	2.2	Ferro	www.ferro.com
Kauramin	MF	1.5	BASF	www.basf.com
Kauresin	PF, RF	1.3, 1.5	BASF	www.basf.com
Kaurit	MUF	1.5	BASF	www.basf.com
K-Bin	PVC	2.2	K-Bin, Inc.	www.shinetsu.co.jp/e/profile/group_kai
Kelburon	PP	2.2	Sabic	www.sabic.com
Keldax	E-Cop	2.2	DuPont	www.dupont.com
Kelon	PA 6, PA 66	2.2	Lati	www.lati.com
Keltan	EPDM	2.2	DSM Elastomers	www.dsm.com
Kematal	POM	2.2	Ticona	www.ticona.com
Kemcor	PE-HD	2.2	Qenos Pty Ltd	www.qenos.com/
Kemid	PEI	3.3	Saint Gobain - Norton	www.plastics.saint-gobain.com/
Kemlex	POM	2.2	Ferro Corporation	www.ferro.com/
Kemplex	POM	2.2	Ferro	www.ferro.com
Kepamid	PA 6, PA 66	2.2	Korea Eng. Plastics	www.kepital.com
Kepex	PBT	2.2	Korea Eng. Plastics	www.kepital.com
Kepital	POM	2.2	Korea Eng. Plastics	www.kepital.com
Keripol	UP	2.2	Bakelite	www.bakelite.
Ketron	PEEK	3	Polytron	www.polytron-gmbh.de

Continued on next page

D.2 Tradenames Table

Tradename	Polymer	Delivery Form	Manufacturer Supplier	Web Address
Kevlar	PPTA	3.6	DuPont	www.dupont.com
Keyflex	TPC	2.2	LG	www.lgchem.com
Keysor	PVC	2.2	Keysor-Century Corporation	805-259-2360
KF...	PVDF, PVF	2.2, 3.3	Kureha	www.kureha.co.jp
K-Flex	PVDC	3.3	Kureha	www.kureha.co.jp
Kibipol	BR	3.4	Chi Mei	www.chimei.com.tw
Kibisan	SAN	2.2	Chi Mei	www.chimei.com.tw
Kibiton	SBS, SEBS	2.2	Chi Mei	www.chimei.com.tw
Kite	PF	3.5	Tufnol	www.tufnol.co.uk
Kobaloy	PC, PC-Blend	2.2	Mitsubishi	www.m-ep.co.jp
Kobatron	PA	2.2	Mitsubishi	www.m-ep.co.jp
Koblend	PE+PS	2.2	Polimeri	www.polimerieuropa.com
Kocetal	POM	2.2	Kolon	www.kolon.co.kr
Kohinor	PVC	2.2	Pantasote Polymers, Inc.	201-393-0888
Kollidon	PVP	2.2	BASF	www.basf.com
Kömabord	PVC	3.4	Profine	www.koemmerling.de
Kömacel	PVC	3.4	Profine	www.koemmerling.de
Kömadur	VC-Cop	3.4	Profine	www.koemmerling.de
Kömapan	PVC	3.1	Profine	www.koemmerling.de
Kömatex	PVC	3.4	Profine	www.koemmerling.de
Konduit	Thermopl. allg.	2.2	LNP	www.lnp.com
Konlux	UP	3.4	LNP	www.lnp.com
Kopa	PA	2.2	Kolon	www.kolon.co.kr
Kopel	TPE	2.2	Kolon	www.kolon.co.kr
Kopet	PET	2.2	Kolon	www.kolon.co.kr
Koplen	PS	1.1	Kaucuk	www.kaucuk.cz
Korad	PMMA	3.3	Polymer Extruded Products, Inc.	973-344-2700
Koresin	PF	1.1	BASF	www.basf.com
Korex	F-Pol., PF, PF	2.2, 3.5, 3.5	DuPont	www.dupont.com
Korton	ECTFE	3.3	Saint Gobain - Norton	www.plastics.saint-gobain.com/
KoSa	PET	2.2	KoSa	www.invista.com/kosa.shtml
Kostil	SAN	2.2	Polimeri	www.polimerieuropa.com
Kostrate	Polyester	pellets, spheres	Plastic Selection Group, Inc.	www.go2psg.com/
Koylene	PP	2.2	Indian Petrochemicals	www.ipcl.co.in
Koylene	PP	granules	Indian Petrochemicals Corporation Lt	www.ipcl.co.in/
KRALASTIC	ABS	2,2	Nippon A&L Inc.	www.n-al.co.jp/ing/index.html

Continued on next page

Tradename	Polymer	Delivery Form	Manufacturer Supplier	Web Address
Krasten	PS	2.2	Kaucuk	www.kaucuk.cz
Kraton D	SBS, SIS	2.2	Kraton	www.kraton.com
Kraton G	SEBS, SEPS	2.2	Kraton	www.kraton.com
Kraton IR	IR	2.2	Kraton	www.kraton.com
K-Resin	SB, S-Cop	2.2	Chevron Phillips	www.cpchem.com
Krylene	SBR	2.2	Lanxess	www.lanxess.com
Krynac	NBR	2.2	Lanxess	www.lanxess.com
Krynol	SBR	2.2	Lanxess	www.lanxess.com
Krystalflex	PUR	3.3	Morton International	www.huntsman.com/
Kumho	ABS	2.2	Kumho Chemicals, Inc.	www.kkpc.co.kr/
KW Plastics	PP	2.2	KW Plastics	www.kwplastics.com/
Kydex	Acrylic+PVC	3.4	Kleerdex Company	www.kydex.com/
Kynar	PVDF	2.1, 2.2, 3.3	Atofina	www.atofina.de
Labellyte	PP	3.3	Exxon	www.exxonchemical.com
Lacbloc	SEBS	2.2	Silac	None
LACEA	PLA	2.2	Mitsui Chemicals America, Inc.	www.mitsuichemicals.com/
Lacovyl	PVC	1.4, 2.2	Atofina	www.atofina.de
Lacqrene	PS, S-Cop	2.2	Atofina	www.atofina.de
Lacqtene	PE-HD, PE-LD, PE-LLD	2.2	Atofina	www.atofina.de
Lactron	PLA	3.6	Kanebo	www.kanebo.co.jp
Ladene	PE, PP	2.2	Sabic	www.sabic.com
Laestra	PS	2.2	Lati	www.lati.com
Lamide	PA	2.2	Lamplast	www.lamplast.it
Laminac	Polyester, Thermoset	2.2	Ashland Chemical Company	www.ashchem.com/
Lanoform	POM	2.2	Godiplast	www.godiplast.com
Lanomid	PA	2.2	Godiplast	www.godiplast.com
Lanoprop	PP	2.2	Godiplast	www.godiplast.com
Lapex	PESI	2.2	Lati	www.lati.com
Laprene	SBS, SEBS, TPE	2.2	So.F.Ter	www.softerspa.com
Laramid	PPA	2.2	Lati	www.lati.com
Larflex	PP+EPDM	2.2	Lati	www.lati.com
Laricol	PUR	2.2	Coim	www.coim.it
Laril	PPE	2.2	Lati	www.lati.com
Laripur	TPU	2.2	Coim	www.coim.it
Larithane	PUR-R	1.1	Coim	www.coim.it
Lariver	UP	3.4	Coim	www.coim.it
Larpeek	PEEK	2.2	Lati	www.lati.com
Larton	PPS	2.2	Lati	www.lati.com
Lastane	PUR	2.2	Lati	www.lati.com
Lastil	SAN	2.2	Lati	www.lati.com

Continued on next page

D.2 Tradenames Table

Tradename	Polymer	Delivery Form	Manufacturer Supplier	Web Address
Lastilac	ABS, ABS+PC	2.2	Lati	www.lati.com
Lastirol	PS	2.2	Lati	www.lati.com
Lasulf	PSU	2.2	Lati	www.lati.com
Latamid	PA 6, PA 66	2.2	Lati	www.lati.com
Latan	POM	2.2	Lati	www.lati.com
Latene	PE-HD, PP	2.2	Lati	www.lati.com
Later	PBT	2.2	Lati	www.lati.com
Latiblend	ABS+PA, PC+PBT	2.2	Lati	www.lati.com
Latilon	PC	2.2	Lati	www.lati.com
Latilub	PA, POM	1.1, 2.2	Lati	www.lati.com
Latishield	PA, PC, PTP	1.1, 2.2	Lati	www.lati.com
Latistat	PA, PC, PTP	1.1, 2.2	Lati	www.lati.com
Laxtar	LCP	2.2	Lati	www.lati.com
Leben	PVC	2.2	Dainippon	www.dic.co.jp
Lemalloy	PPE+PA, PPE+PBT, PPE+PP	2.2	Mitsubishi	www.m-ep.co.jp
Lemapet	PET	2.2	Mitsubishi	www.m-ep.co.jp
Lennite	PE-UHMW	3.2, 3.4	Westlake Plastics Company	www.westlakeplastics.com/
Leona	PA 66	2.2	Asahi	www.asahithermofil.com
Lerille	PTP	3.6	Asahi	www.asahithermofil.com
Levapren	EVAC	2.2	Lanxess	www.lanxess.com
Lexan	PC, PC+PPC, PPC	3.3	GEP	www.geplastics.com
Lexgard	PC	3.4	GEP	www.geplastics.com
Lighter	PET	2.2	Equipolymers	www.equipolymers.com
LinTech	PE-LLD	2.2	Politeno	www.politeno.com.br/
Listac	ASA	2.2	Godiplast	www.godiplast.com
LITAC-A	AS (Acrylonitrile Styrene Copolymer)	2.2	Nippon A&L Inc.	www.n-al.co.jp/ing/index.html
Liten	PE-HD	2.2	Chemopetrol	www.chemopetrol.cz
Lomod	TPC	2.2	GEP	www.geplastics.com
LongLite	Phenolic	2.2	Dowell Trading Company Ltd.	www.dowell.com.hk/
Lotader	E-Cop	2.2	Atofina	www.atofina.de
Lotrene	PE-LD	2.2	QAPCO	www.qapco.com/
Lotryl	EBA, EMA	2.2	Atofina	www.atofina.de
Lubmer	PE-UHMW	2.2	Atofina	www.atofina.de
Lubricomp	Thermopl. allg.	2.2	LNP	www.lnp.com
Lubrilon	Thermopl. allg.	2.2	Schulman	www.aschulman.com
Lubriloy	Thermopl. allg.	2.2	LNP	www.lnp.com

Continued on next page

Tradename	Polymer	Delivery Form	Manufacturer Supplier	Web Address
Lubri-Tech	Thermopl. allg.	2.2	PolyOne Th. Bergmann	www.polyone.com
Lucalen	E-Cop	1.1	Basell	www.basell.de
Lucalor	PVC-U	2.2	Atofina	www.atofina.de
Lucarex	PVC	2.2	Atofina	www.atofina.de
Lucel	POM	2.2	LG	www.lgchem.com
Lucent	ABS	2.2	Lucent Polymers, Inc.	www.lucentpolymers.com/
Lucet	POM	2.2	LG	www.lgchem.com
Lucite	PMMA	2.2	DuPont	www.dupont.com
Lucky	ABS	2.2	LG	www.lgchem.com
Lucobay	PVC	3.1	Atofina	www.atofina.de
Lucobit	E-Cop	1.1, 3.3	Lucobit	www.lucobit.de
Lucofin	TPO	3.3	Lucobit	www.lucobit.de
Lucolit	E-Cop	3.3	Lucobit	www.lucobit.de
Lucon	Thermopl. allg.	2.2	LG	www.lgchem.com
Lucopren	TPO	3.3	Lucobit	www.lucobit.de
Luflexen	PE-LLD	2.2	Basell	www.basell.de
Lumax	PBT+PS	2.2	LG	www.lgchem.com
Lumiflon	Fluoropolymer	2.2	BELLEX INTERNATIONAL CORP.	www.bellexinternational.com/
Lumitac	PE-ULD	2.2	TOSOH Corporation	www.tosoh.com/
Lupan	SAN	2.2	LG	www.lgchem.com
Lupital	POM	2.2	Ticona	www.ticona.com
Lupol	PP	2.2	LG	www.lgchem.com
Lupolen	E-Cop, PE-HD, PE-LD, PE-LLD	2.2	Basell	www.basell.de
Lupolex	PE-LLD	2.2	Basell	www.basell.de
Lupon	PA 66	2.2	LG	www.lgchem.com
Lupos	ABS, SAN	2.2	LG	www.lgchem.com
Lupox	PBT, PC+PBT	2.2	LG	www.lgchem.com
Lupoy	PC, PC+ABS	2.2	LG	www.lgchem.com
Lupranat	PUR-R	1.1	Elastogran	www.elastogran.de
Lupranol	PUR-R	1.1	Elastogran	www.elastogran.de
Lupraphen	PUR-R	1.1	Elastogran	www.elastogran.de
Luprene	SBS	2.2	LG	www.lgchem.com
Luran	ASA, SAN	2.2	BASF	www.basf.com
Luranyl	PPE+PS	2.2	Romira	www.romira.de
Luraskin	ASA	3.3	BASF	www.basf.com
Lusep	PPS	2.2	LG	www.lgchem.com
Lustran	ABS, SAN	2.2	Lanxess	www.lanxess.com
Lustropak	ABS	2.2	Lanxess	www.lanxess.com
Lutene	PE-HD, PE-LD	2.2	LG	www.lgchem.com

Continued on next page

D.2 Tradenames Table

Tradename	Polymer	Delivery Form	Manufacturer Supplier	Web Address
Lutex	PS	2.2	LG	www.lgchem.com
Lutrel	PBT	2.2	LG	www.lgchem.com
Luvicross	PVP	2.2	BASF	www.basf.com
Luviskol	PVP	2.2	BASF	www.basf.com
Luvocom	Thermopl. allg.	2.2	Lehmann & Voss	www.lehvoss-purelast.de
Luvoflex	TPU	2.2	Lehmann & Voss	www.lehvoss-purelast.de
Luwax	PE	2.1, 2.2	BASF	www.basf.com
Luxacryl	PMMA	3.4	Tupaj	www.go-ttv.com
Luxamid	PA 6, PA 66	2.2	Dilaplast	www.dilaplast.com
Lycra	PUR	3.6	DuPont	www.dupont.com
Lynex-T	POM	2.2	Asahi Kasei Corporation	www.asahi-kasei.co.jp/asahi/en/index.h
Lynx	PF	3.5	Tufnol	www.tufnol.co.uk
Lytex	EP	3.4	Quantum Composites Inc.	www.quantumcomposites.com/
Lytron	PS, S-Cop	1.1	Dow	www.dow.com
Mablex	PC+ABS	2.2	MP	www.mpcom.it
Madipur	PUR-R	1.1	Hexcel	www.hexcel.com
Mafill	PP	2.2	MP	www.mpcom.it
Magnacomp	PA	2.2	LNP	www.lnp.com
Magnobond	EP	1.4, liquid	Magnolia Plastics, Inc.	www.magnapoxy.com/
Magnum	ABS	2.2	Dow	www.dow.com
Makroblend	PC+PBT, PC+PET	2.2	Bayer	www.bayer.com
Makrofol	PC	3.3	Bayer	www.bayer.com
Makrolon	PC	2.2	Bayer	www.bayer.com
Makrolon mult	PC	3.4	Bayer Sheet	www.bayersheeteurope.com
Malon	PTP	2.2	M.A. Industries	www.maind.com
Mamax	PET	2.2	M.A. Industries	www.maind.com
Manner	PVC	2.2	Manner Plastics, L.P.	www.mannerplastics.com/
Mantopex	PE-X	3.2	Golan	www.golan-plastics.com
Maranyl	PA66	2.2	ICI Fibres	800-849-4424
MarFlex	PE-HD	2.2	Chevron Phillips Chemical Co.	www.cpchem.com/
Margard	PC	3.4	GEP	www.geplastics.com
Marlex	PE-HD, PE-LD	2.2	Chevron Phillips	www.cpchem.com
Marnot	PC	3.3	Bayer	www.bayer.com
Marvyflo	PVC	2.1	Tessenderlo	www.tessenderlo.com
Marvylan	PVC	2.2	Tessenderlo	www.tessenderlo.com
Marvylex	VC-Cop	2.2	Tessenderlo	www.tessenderlo.com
Marvyloy	VC-Cop	2.2	Tessenderlo	www.tessenderlo.com
MASPLAS	PE-HD	2.2	Lavergne Group	www.lavergneusa.com/
MASPOLENE	PP	2.2	Lavergne Group	www.lavergneusa.com/

Continued on next page

Tradename	Polymer	Delivery Form	Manufacturer Supplier	Web Address
Mastalloy	TPO	2.2	Mitsui	www.mitsui.co.jp
Mater-Bi	PO	2.2	Novamont	www.novamont.com
Matrixx	PC	2.2	The Matrixx Group, Inc.	www.matrixxgroup.com/
Maxlen	PE	2.2	MRC Polymers	www.mrcpolymers.com
Maxnite	PBT, PET	2.2	MRC Polymers	www.mrcpolymers.com
Maxprene	IR	2.2	MRC Polymers	www.mrcpolymers.com
Maxpro	PP	2.2	MRC Polymers	www.mrcpolymers.com
Maxtel	TPO	2.2	MRC Polymers	www.mrcpolymers.com
Maxxam	PE-HD, PE-LD, PP	2.2	PolyOne Th. Bergmann	www.polyone.com
MC	PA6	3.1, 3.2, 3.4	Quadrant Engineering Plastic Product	www.quadrantepp.com/
MCX-A	PA	2.2	Mitsui Chemicals America, Inc.	www.mitsuichemicals.com/
MDE	ABS	2.2	Michael Day Enterprises	www.mdayinc.com/
MDI	PC	2.2	Modern Dispersions, Inc.	www.moderndispersions.com/
Megablend	PC+ABS	2.2	Megapolymers	www.megapolymers.com
Megaflex	TPE	2.2	Megapolymers	www.megapolymers.com
Megalac	ABS	2.2	Megapolymers	www.megapolymers.com
Megalon	PC	2.2	Megapolymers	www.megapolymers.com
Megamid	PA	2.2	Megapolymers	www.megapolymers.com
Megaprop	PP	2.2	Megapolymers	www.megapolymers.com
Megater	PBT	2.2	Megapolymers	www.megapolymers.com
Megol	SEBS	2.2	API	www.apiplastic.com
Melaform	MF	2.2	Chromos	www.chromos.hr
Melaicar	MF	2.2	Aicar	www.aicar.es
Meldin	PI	-	Saint Gobain - Furon	www.plastics.saint-gobain.com/
Melfeform	MF	2.2	Chromos	www.chromos.hr
Melinar	PET	2.2	DuPont	www.dupont.com
Melinex	PET	3.3	DuPont	www.dupont.com
Melmex	MF	2.2	BIP	www.bip.co.uk
Melochem	MF	2.2	Chemiplastica	www.chemiplastica.it
Melopas	MF, MPF	2.2	Raschig	www.raschig.de
Melos	EPDM	2.2	PolyOne Th. Bergmann	www.polyone.com
Melsprea	MF	2.2	Sprea	www.sirindustriale.com
Melthene	EVA	2.2	TOSOH Corporation	www.tosoh.com/
Metablen	A-Cop, MBS	2.2	Mitsubishi Rayon	www.mrc.co.jp
Metalcap	PP	2.2	Polyram	www.polyram.co.il
Metallyte	PP	3.3	Exxon	www.exxonchemical.com
Metamarble	ABS+PC, PC, PMMA+PC	2.2	Teijin	www.teijin.co.jp

Continued on next page

D.2 Tradenames Table

Tradename	Polymer	Delivery Form	Manufacturer Supplier	Web Address
Metocene	PP	2.2	Basell	www.basell.de
Metton	DCPD (Dicyclopentadiene)	liquid	Metton America	
Metzoplast	Thermopl. allg.	3.3, 3.4	Metzeler	www.metzelerplastics.de
Microline	ABS	3.4	LA/ES	www.la-es.com
Microthene	PE	2.1	Equistar	www.equistar.com
Milastomer	TPO	2.2	Mitsui	www.mitsui.co.jp
Mindel	PSU	2.2	Solvay	www.solvay.com
Minlon	PA	2.2	DuPont	www.dupont.com
Miraepol	PP-Cop.	2.2	Basell Polyolefins	www.basell.com/
Miramid	PA	2.2	Leuna	www.leuna-polymer.de
Mirason	PE-LD, PE-LLD	2.2	Mitsui	www.mitsui.co.jp
Miravithen	EVAC	2.2	Leuna	www.leuna-polymer.de
MIRREX	PVC	3.3	VPI, LLC	www.vpicorp.com/
Modar	MMA	1.1	Ashland	www.ashspec.com
Modified	PA66	2.2	Modified Plastics	714-546-4667
Moltopren	PUR-R	3.7	Bayer	www.bayer.com
Monosol	EC, MC, PVAL	3.3	Dow	www.dow.com
Monprene	TPE	2.2	Teknor Apex	www.teknorapex.com
Montac	PA, PESI	1.5, 2.2	Dow	www.dow.com
MonTor	PA6	2.2	Toray Resin Company	www.toray.com/
Moplefan	PP	3.3	Treofan	www.treofan.com
Moplen	PP	2.2	Basell	www.basell.de
Morthane	TPU	1.1	Basell	www.basell.de
Mosten	PP	2.2	Chemopetrol	www.chemopetrol.cz
mPact	PE-LLD	2.2, 3.3	Chevron Phillips Chemical Co.	www.cpchem.com/
MRC	PP	-	MRC Polymers, Inc.	www.mrcpolymers.com/
Muehlstein	PE-MD	2.2	Muehlstein Compounded Products	www.muehlstein.com/
Multi-ABS	ABS	2.2	Multibase, Inc.	www.multibase.com/
Multi-Flam	PP	2.2	Multibase, Inc.	www.multibase.com/
Multiflex	PP+EPDM, TPE	2.2	Multibase	www.multibase.com
Multiprene	TPE	2.2	Multibase	www.multibase.com
Multipro	PP	2.2	Multibase	www.multibase.com
Multitec	PUR-R	1.1	Bayer	www.bayer.com
Muralen	PE-UHMW	3.1, 3.4	Murtfeldt	www.murtfeld.com
Murdopol	PA 12	3.1, 3.4	Murtfeldt	www.murtfeld.com
Murdotec	PPS	3.1, 3.4	Murtfeldt	www.murtfeld.com
Murflor	PTFE	3.1, 3.4	Murtfeldt	www.murtfeld.com

Continued on next page

Tradename	Polymer	Delivery Form	Manufacturer Supplier	Web Address
Murinyl	PVDF	3.1, 3.4	Murtfeldt	www.murtfeld.com
Murlubric	PA 6	3.1, 3.4	Murtfeldt	www.murtfeld.com
Murpec	PEEK	3.1, 3.4	Murtfeldt	www.murtfeld.com
Murtex	PF	3.1, 3.4	Murtfeldt	www.murtfeld.com
Murylat	PET	3.1, 3.4	Murtfeldt	www.murtfeld.com
Murylon	PA 6, PA 66	3.1, 3.4	Murtfeldt	www.murtfeld.com
Murytal	POM	3.1, 3.4	Murtfeldt	www.murtfeld.com
Mutilon	PC+ABS	2.2	Teijin	www.teijin.co.jp
Mxsite	PE	2.2	Voridian Company, a division of East	www.eastman.com/About_Eastman/Division
Mxsten	PE-LD	2.2	Eastman	www.eastman.com
Mylar	PET	3.3	DuPont	www.dupont.com
Mytex	PP	2.2	ExxonMobil Chemical Company	www.exxonmobilchemical.com/
Nafion	F-Pol., PF, PF	2.2, 3.5, 3.5	DuPont	www.dupont.com
Nakan	PVC	2.2	Atofina	www.atofina.de
Nan	PBT	-	Nan Ya Plastics Corporation	www.npcusa.com/
NAS	SMMA	2.2	Nova	www.novachem.com
Natsyn	IR	2.2	Goodyear	www.goodyearchemical.com
NatureWorks	PLA	2.2	Cargill	www.cdpoly.com
NatureWorks[a]	PLA	2.2	Cargill Dow LLC	www.natureworksllc.com/
Naxaloy	PC-Blend	2.2	MRC Polymers	www.mrcpolymers.com
Naxell	PC	2.2	MRC Polymers	www.mrcpolymers.com
Neoflon	ECTFE, PTFE	1.2	Daikin	www.daikin.com
Neopolen	PE-E, PP-E	3.7	BASF	www.basf.com
Neopor	PS-E	2.2, 3.7	BASF	www.basf.com
Neoprene	CR	2.2	DuPont Dow Elastomers	www.dupont-dow.com
Neostar	TPC	2.2	Eastman	www.eastman.com
Nepol	PP	2.2	Borealis	www.borealis.com
Neste	PE	-	Borealis Compounds Inc.	www.borealisgroup.com/
Network	ABS	2.2	Network Polymers, Inc.	www.diamondpolymers.com/
Nexprene	TPV	2.2	Solvay	www.solvay.com
Niblan	PBT	2.2	Soredi	www.soredi.it
Niblend	PC+ABS	2.2	Soredi	www.soredi.it
Niform	POM	2.2	Soredi	www.soredi.it
Nilac	ABS	2.2	Soredi	www.soredi.it
Nilamid	PA-Cop	2.2	Nilit	www.nilit.com
Nilamon	PA 6, PA 66	2.2	Nilit	www.nilit.com
NILENE	PP-Cop.	2.2	SORI S.P.A.	www.sorispa.it/
Nilitop	PBT	2.2	Nilit	www.nilit.com

Continued on next page

D.2 Tradenames Table

Tradename	Polymer	Delivery Form	Manufacturer Supplier	Web Address
Niloy	PC+PBT	2.2	Soredi	www.soredi.it
Nimpact	PC	3.4	GEP	www.geplastics.com
Nipeon	PVC	2.2	Zeon	www.zeon.co.jp
Niplene	PP	2.2	Soredi	www.soredi.it
Nipoflex	EVA	2.2	TOSOH Corporation	www.tosoh.com/
Nipol	ACM, BR, IR, NBR, SBR	2.2	Zeon	www.zeon.co.jp
Nipolit	PVC, VC-Cop	2.2	Chisso	www.chisso.co.jp
Nipolon	PE-HD	2.2	TOSOH Corporation	www.tosoh.com/
Niretan	PA 6, PA 66	2.2	Soredi	www.soredi.it
Nirion	PC	2.2	Soredi	www.soredi.it
Nisseki	BR, PIB	1.5	Nippon Petrochemicals	www.npcc.co.jp
Nistil	SAN	2.2	Soredi	www.soredi.it
Nitrovin	TPV	2.2	Vi-Chem Corporation	www.vichem.com/
Nivionplast	PA66	2.2	EniChem	www.eni.it/home/home_en.html
Noblen	PP	2.2	Sumitomo	www.sumitomo-chem.co.jp
Nomex	PARA	3.3, 3.6	DuPont	www.dupont.com
Nopla	PEN, PET	2.2	Kolon	www.kolon.co.kr
Nordbak	EP	2.2	Fel-Pro Chemical Products, Inc.	None
Nordel	EPDM	2.2	DuPont Dow Elastomers	www.dupont-dow.com
Norpex	PPE	2.2	Custom Resins Group	www.customresins.com/
Norprene	TPE	3.2	Saint Gobain - Norton	www.plastics.saint-gobain.com/
Norslide	PTFE	3.2	Pampus	www.saint-gobain.de
Norsophen	Phenolic	BMC, liquid, SMC	Norold Composites Inc.	800-563-2089
Norsorex	EPDM	2.2	Atofina	www.atofina.de
Norvic	PVC	2.2	Braskem	www.braskem.com.br
Norvinyl	PVC	2.2	Norsk Hydro	www.hydro.com
Noryl	PPE+PA, PPE+PBT, PPE+PE, PPE+PP, PPE+PS	2.2	GEP	www.geplastics.com
Noryl EF	PPE + PS-E	2.2	GEP	www.geplastics.com
Noryl GTX	PPE+PA	2.2	GEP	www.geplastics.com
Nova	PS	2.2	Nova	www.novachem.com

Continued on next page

Tradename	Polymer	Delivery Form	Manufacturer Supplier	Web Address
Novablend	PVC	2.2	Novatec Plastics & Chemicals Co. In	www.novatecplastics.com/
Novaccurate	LCP	2.2	Mitsubishi	www.m-ep.co.jp
NOVACCURATE	LCP	2.2	Mitsubishi Engineering-Plastics Corp	www.m-ep.co.jp/
Novacor	PS	2.2	Nova	www.novachem.com
Novacycle	PVC	2.2	Novatec Plastics & Chemicals Co. In	www.novatecplastics.com/
Novaduran	PBT	2.2	Mitsubishi	www.m-ep.co.jp
Novaflex	PVC	-	Novatec Plastics & Chemicals Co. In	www.novatecplastics.com/
Novalite	PVC	2.2	Novatec Plastics & Chemicals Co. In	www.novatecplastics.com/
Novalloy	ABS+PA, ABS+PBT, ABS+PC	2.2	Daicel	www.daicel.co.jp
Novaloy	PVC	2.2	Novatec Plastics & Chemicals Co. In	www.novatecplastics.com/
Novamid	PA 6, PA 66	2.2	Mitsubishi	www.m-ep.co.jp
Novapet	PET	2.2	Mitsubishi	www.m-ep.co.jp
Novapol	PE-HD, PE-LD, PE-LLD	2.2	Nova	www.novachem.com
Novapps	PPS	2.2	Mitsubishi	www.m-ep.co.jp
Novarex	PC	2.2	Mitsubishi	www.m-ep.co.jp
Novatemp	PVC	-	Novatec Plastics & Chemicals Co. In	www.novatecplastics.com/
Novex	EVAC, PE-LD	2.2	BP	www.bppetrochemicals.com
Novoaccurate	LCP	2.2	Mitsubishi	www.m-ep.co.jp
Novocarb	PC	2.2	Godiplast	www.godiplast.com
Novodur	ABS	2.2	Lanxess	www.lanxess.com
Novolen	PP	2.2	Basell	www.basell.de
Novolux	PC+ABS	2.2	LA/ES	www.la-es.com
NOVOPLAS	PP-Cop.	2.2	Lavergne Group	www.lavergneusa.com/
Nucrel	EAA, EMA	2.2	DuPont	www.dupont.com
NuSil	SI	liquid	NuSil Technology	www.nusil.com/
NWP	PP	2.2	North Wood Plastics, Inc.	www.northwoodplastics.com/
Nybex	PA	2.2	Ferro	www.ferro.com
Nycast	PA6/12	2.2	Cast Nylons Ltd.	www.castnylon.com/

Continued on next page

D.2 Tradenames Table

Tradename	Polymer	Delivery Form	Manufacturer Supplier	Web Address
Nycoa	PA6	2.2	Nycoa (Nylon Corporation of America)	www.nycoa.net/
Ny-Kon	PA	2.2	LNP	www.lnp.com
Nylaforce	PA 6, PA 66	2.2	Leis	www.leispt.de
Nylamid	PA66	2.2	ALM Corporation	www.alphagary.com/
Nylatron	PA	2.2	DSM Eng. Plastics	www.dsmep.com
Nylene	PA66	2.2	Custom Resins Group	www.customresins.com/
Nylex	PA+PP	2.2	Multibase	www.multibase.com
Nylind	PA	2.2	DuPont	www.dupont.com
Nyloy	ABS	2.2	Nytex Composites Co., Ltd. (USA)	www.nytex.com.tw/en-nytex/index.phtml
Nymax	PA 6, PA 66	2.2	PolyOne Th. Bergmann	www.polyone.com
Nypel	PA 6	2.2	BASF	www.basf.com
Nyrim	PA	1.1	DSM RIM	www.rimnylon.com
Nytron	ABS	2.2	Nytex Composites Co., Ltd. (USA)	www.nytex.com.tw/en-nytex/index.phtml
Oasis	PI	3.5	DuPont	www.dupont.com
Okiten	PE-LD	2.2	Dioki	www.dioki.hr
Olaprene	PUR	3.4	Philippine	www.philippine.de
Oldoflex	PUR-R	1.1	Büsing & Fasch	www.buefa.de
Oldopal	UP	1.1	Büsing & Fasch	www.buefa.de
Oldopren	TPU	2.1	Büsing & Fasch	www.buefa.de
Oldopur	PUR-R	1.1	Büsing & Fasch	www.buefa.de
Oleform	PP	2.2	Chisso America Inc.	www.chisso.co.jp/
Olehard	PP	2.2	Chisso America Inc.	www.chisso.co.jp/
Olesafe	PP	2.2	Chisso America Inc.	www.chisso.co.jp/
Oltvil	PVC	2.2	Oltchim	www.oltchim.ro
Ondex	PVC	3.3	Solvay	www.solvay.com
Onflex	TPO	2.2	PolyOne Th. Bergmann	www.polyone.com
Ongrodur	PVC	3.4	Borsod	www.borsodchem.hu
Ongrofol	PVC	3.3	Borsod	www.borsodchem.hu
Ongrolit	PVC	2.2	Borsod	www.borsodchem.hu
Ongromix	PVC	2.2	Borsod	www.borsodchem.hu
Ongronat	PUR-R	1.1	Borsod	www.borsodchem.hu
Ongropur	PUR-R	1.1	Borsod	www.borsodchem.hu
Ongrovil	PVC	2.2	Borsod	www.borsodchem.hu
Ontex	PP	2.2	Solvay	www.solvay.com

Continued on next page

Tradename	Polymer	Delivery Form	Manufacturer Supplier	Web Address
OP	POM	2.2	Oxford Polymers	www.oxfordpolymers.com/
Oppalyte	PP	3.3	Exxon	www.exxonchemical.com
Oppera	PP	-	ExxonMobil Chemical Company	www.exxonmobilchemical.com/
Optema	EMA	2.2	Exxon	www.exxonchemical.com
Optix	PMMA	2.2	Plaskolite-Continental Acrylics	www.plaskolitecontinental.com/
Optum	PO	2.2	Ferro	www.ferro.com
Orel	PET	3.6	DuPont	www.dupont.com
Orevac	E-Cop	1.1, 2.2	Atofina	www.atofina.de
Orgalloy	PP+PA	2.2	Atofina	www.atofina.de
Orgasol	PA 612	2.1	Atofina	www.atofina.de
Orit	PVC-P	3.3	Atofina	www.atofina.de
Ormecon	PANI	1.1	Ormecon	www.ormecon.de
Oromid	PA	2.2	Rhodia	www.rhodia.com
Or-on	PVC-P	3.3	Rhodia	www.rhodia.com
Osstyrol	ABS, PE-HD, PP, PS, SAN	3.3, 3.4	Hagedorn	www.hagedorn.de
Oxnilon	PA6	2.2	Oxford Polymers	www.oxfordpolymers.com/
Oxy	PVC	2.2	Occidental Chemical Corp. (OxyChem)	www.oxychem.com/
Oxyvinyls	PVC	2.2	Occidental	www.oxychem.com
Palapreg	UP, VE	1.5	DSM Composite Resins	www.dsm.com
Palatal	UP, VE	1.5	DSM Composite Resins	www.dsm.com
Palblend	PC+ABS	2.2	Palplast	www.palplast.de
Palclear	PVC	3.4	Palram	www.palram.com
Paldoor	PVC	3.4	Palram	www.palram.com
Paldur	PBT	2.2	Palplast	www.palplast.de
Palflex	TPU	2.2	Palplast	www.palplast.de
Palform	POM	2.2	Palplast	www.palplast.de
Palgard	PC	3.4	Palram	www.palram.com
Palglas	PMMA	3.4	Palplast	www.palplast.de
Palglaze	PET-G	3.4	Palram	www.palram.com
Palight	PVC	3.4	Palram	www.palram.com
Pallaflon	PTFE	3, 3.4	Schieffer	www.schieffer.de
Palmid	PA	2.2	Palplast	www.palplast.de
Palopaque	PVC	3.4	Palram	www.palram.com
Palpet	PET	2.2	Palplast	www.palplast.de
Palprop	PP	2.2	Palplast	www.palplast.de
Palran	ABS	2.2	Palplast	www.palplast.de
Palruf	PVC	3.4	Palram	www.palram.com
Palsafe	PC	2.2	Palplast	www.palplast.de
Palsan	SAN	2.2	Palplast	www.palplast.de

Continued on next page

D.2 Tradenames Table

Tradename	Polymer	Delivery Form	Manufacturer Supplier	Web Address
Palstyrol	PS	2.2	Palplast	www.palplast.de
Palsun	PC	3.4	Palram	www.palram.com
Palthene	PE	2.2	Palplast	www.palplast.de
Paltile	PC	3.4	Palram	www.palram.com
Palvinyl	PVC	2.2	Palplast	www.palplast.de
Pamflon	PTFE	2.2	Pampus	www.saint-gobain.de
Panaflex	PVC-P	3.3	3 M	www.3m.com
Pandex	TPU	2.2	Dainippon	www.dic.co.jp
Panlite	PC	2.2	Teijin	www.teijin.co.jp
Pantarin	PUR	3.7	Koepp	www.koepp-ag.de
Papi	PUR-R	1.1	Dow	www.dow.com
Paraglas	PMMA	3.3, 3.4	Degussa	www.degussa.de
Paraloid	AMMA, MBS, PMMA	1.5, 2.1, 2.2	Rohm & Haas	www.rohmhaas.com
Parel	EPDM	2.2	Zeon	www.zeon.co.jp
Paricarb	PC	2.2	Tecnopolimers	www.tecnopolimers.com
Paridur	PBT	2.2	Tecnopolimers	www.tecnopolimers.com
Pariform	POM	2.2	Tecnopolimers	www.tecnopolimers.com
Parilene	PP	2.2	Tecnopolimers	www.tecnopolimers.com
Parinil	PA 6, PA 66	2.2	Tecnopolimers	www.tecnopolimers.com
Parisab	ABS	2.2	Tecnopolimers	www.tecnopolimers.com
Paristirol	PS	2.2	Tecnopolimers	www.tecnopolimers.com
Paxon	PE-HD	2.2	Exxon	www.exxonchemical.com
Pax-Plus	PE-HD	2.2	ExxonMobil Chemical Company	www.exxonmobilchemical.com/
P-Blend	PC+ABS	2.2	Putsch	www.putsch.de
PCLight	PC	3.4	Policarbonatos do Brasil	www.policarbonatos.com.br
P-Comp	PP	2.2	Putsch	www.putsch.de
Pearlbond	TPU	2.2	Merquinsa	www.merquinsa.com
Pearlcoat	TPU	2.2	Merquinsa	www.merquinsa.com
Pearlstick	TPU	2.2	Merquinsa	www.merquinsa.com
Pearlthane	TPU	2.2	Merquinsa	www.merquinsa.com
Pebax	TPA	2.1, 2.2	Atofina	www.atofina.de
Pedigree	Polyester, Thermoset	2.2	P.D. George Company	www.pdgeorge.com/
Peek	PAEK	2.2	Victrex	www.victrex.com
Pektran	PAEK	2.2	Quadrant Engineering Plastic Product	www.quadrantepp.com/
P-Elan	PP+EPDM	2.2	Putsch	www.putsch.de
Pellethane	TPU	2.2	Dow	www.dow.com
Pentaclear	PVC	3.3	Klöckner-Pentaplast	www.kpfilms.com
Pentadur	PVC	3.3	Klöckner-Pentaplast	www.kpfilms.com
Pentaform	PVC	3.3	Klöckner-Pentaplast	www.kpfilms.com

Continued on next page

Tradename	Polymer	Delivery Form	Manufacturer Supplier	Web Address
Pentapharm	COC, PET, PP, PVC	3.3	Klöckner-Pentaplast	www.kpfilms.com
Pentaprint	PVC	3.3	Klöckner-Pentaplast	www.kpfilms.com
Pentaprop	PP	3.3	Klöckner-Pentaplast	www.kpfilms.com
Pentastat	PET, PET-G, PP, PVC	3.3	Klöckner-Pentaplast	www.kpfilms.com
Pentatherm	PVC	3.3	Klöckner-Pentaplast	www.kpfilms.com
Pentex	PUR-R	1.1	Hüttenes-Albertus	www.huettenes-albertus.com
Peranyl	PPE+PS	2.2	Godiplast	www.godiplast.com
Perbunan	NBR	2.2	Lanxess	www.lanxess.com
Perfluorogum	FKM	2.2	Daikin	www.daikin.com
Peripor	PS	3.7	BASF	www.basf.com
PermaStat	PP	2.2	RTP Company	www.rtpcompany.com/
Permasted	Thermopl. allg.	2.2	RTP	www.rtpcompany.com
Perspex	PMMA	2.2, 3.4	Lucite	www.lucitecp.com
Perstorp	MF	2.2	Perstorp Compounds, Inc.	www.thermosets.com/
Petal	PET	2.2	PolyOne Th. Bergmann	www.polyone.com
PETLON	PET	granules, 2.2	Albis Plastics Corporation	www.albisna.com/
Petra	PET	2.2	BASF	www.basf.com
Petromont	PE-MD	2.2	Petromont	www.petromont.qc.ca/
Petrothene	PE, PP	2.2	Equistar	www.equistar.com
Pevikon	PVC	2.2	Norsk Hydro	www.hydro.com
Pexgol	PE-X	3.2	Golan	www.golan-plastics.com
P-Fib	PP	2.2	Putsch	www.putsch.de
P-Flex	ABS	2.2	Putsch	www.putsch.de
PharMed	TPE	3.2	Saint Gobain - Norton	www.plastics.saint-gobain.com/
Philan	PUR	3.4	Philippine	www.philippine.de
Phtalopal	PDAP	1.1	BASF	www.basf.com
Pibiflex	TPE	2.2	P-Group	www.p-group.de
Pibiter	PBT, PC, PET	2.2	P-Group	www.p-group.de
Pinnacle	PP	2.2	Pinnacle Polymers	www.pinnaclepolymers.com/
PLANAC	PBT+PS	2.2	Dainippon Ink and Chemicals, Incorpo	www.dic.co.jp/eng/index.html
Plas-Glas	PTP	-	Plaslok Corporation	None
Plaskolite	PMMA	2.2	Plaskolite, Inc.	www.plaskolite.com/

Continued on next page

D.2 Tradenames Table

Tradename	Polymer	Delivery Form	Manufacturer Supplier	Web Address
Plaskon	EP	liquid	Cookson Electronics - Semiconductor	www.cooksonelectronics.com/
Plaslok	Phenolic	-	Plaslok Corporation	None
Plaslube	Thermopl. allg.	2.2	DSM Eng. Plastics	www.dsmep.com
Plastazote	PE-X	3.7	Zotefoams	www.zotefoams.com
Plastelene	PE-HD	3.2	Solvay	www.solvay.com
Plastidro	PVC	3.2	Solvay	www.solvay.com
Plastilit	PVC	3.2	Solvay	www.solvay.com
Plastin	PE-HD	3.3	Huhtamaki	www.4pfolie.de
Plastistrengt	PMMA	2.2	Atofina	www.atofina.de
Plastolen	PO	2.2	Pongs & Zahn	www.pongsundzahn.de
Plastomid	PA 6	2.2	Pongs & Zahn	www.pongsundzahn.de
Plastopil	PE, PVC-P	3.3	Pongs & Zahn	www.pongsundzahn.de
Plastor	PVC-P	3.3	Pongs & Zahn	www.pongsundzahn.de
Plastotrans	PE-LD	3.3	Huhtamaki	www.4pfolie.de
Platabond	EVAC	2.1	Atofina	www.atofina.de
Platamid	PA	2.2	Atofina	www.atofina.de
Platherm	PA, PTP, PUR	1.5	Atofina	www.atofina.de
Platilon	PA, PP, TPU	3.3	Wolff Walsrode	www.wolff-walsrode.de
Plenco	MF, PF, UP	1.1, 2.2	Plastics Engineering	www.petsinc.net
PlenStar	PVC	2.2	PolyOne Corporation	www.polyone.com/
Plexalloy	PMMA+ABS	2.2	Röhm	www.cyro.com
Plexifix	PMMA	1.5	Röhm	www.cyro.com
Plexiglas	PMMA	2.2, 3.1, 3.2, 3.4	Röhm	www.cyro.com
Plexigum	PMMA	1.1	Röhm	www.cyro.com
Plexileim	PAA	1.1, 1.3	Röhm	www.cyro.com
Pleximid	PMMI	2.2	Röhm	www.cyro.com
Plioflex	CR	2.2	Goodyear	www.goodyearchemical.com
Pliolite	ACM, CR, SBR	1.2, 2.2	Goodyear	www.goodyearchemical.com
Pluracol	PUR-R	1.1	BASF	www.basf.com
Plytron	PO	2.2	Borealis	www.borealis.com
PMC	PA6	2.2	PMC Engineered Plastics, Inc.	www.pmc-group.com/
Pocan	PBT, PET	2.2	Lanxess	www.lanxess.com
Pokalon	PC	3.3	Lofo	www.lofo.de
Polaris	MMA	1.5	Ashland	www.ashspec.com
Polene	PO	2.2	Ashland	www.ashspec.com
Polidux	PS	2.2	Aiscondel	www.grupoaragonesas.es
Polifil	PP	2.2	Ipiranga	www.ipiringa.com.br
Polifin	PE-LD	2.2	SASOL Polymers	www.sasol.com/
Poliform	POM	2.2	Aquafil	www.aquafil.com

Continued on next page

Tradename	Polymer	Delivery Form	Manufacturer Supplier	Web Address
Polisul	PE-HD, PE-LD	2.2	Ipiranga	www.ipiringa.com.br
Politeno	PE	2.2	Politeno	www.politeno.com
Polva	PE-HD, PE-LD, PVC	3.2	Solvay	www.solvay.com
Polyabs	ABS	2.2	Polykemi	www.polykemi.se
Polyasa	ASA	2.2	Polykemi	www.polykemi.se
Polyaxis	EVAC, PE-HD, PE-LLD, PP	2.1	Schulman	www.aschulman.com
Polyblend	PC+ABS	2.2	Polykemi	www.polykemi.se
Polybutene-1	PB	2.2	Basell Polyolefins	www.basell.com/
Polyclad	PC	3.4	GEP	www.geplastics.com
Polycoat	PUR	1.5	EMW	www.emw.de
Polycomp	PP-Cop.	2.2	PolyPacific Pty. Ltd.	www.polypacific.com.au/
Polycor	Vinyl Ester	SMC, BMC	Industrial Dielectrics Inc.	www.idiplastic.com/
Polycure	PE	2.2	Nova-Borealis Compounds LLC	www.borealisgroup.com/public/
Polydet	UP	3.4, 3.5	Mitras	www.mitras-materials.com
Polydux	ABS, SAN	2.2	Repsol	www.repsol.com
Polyelast	TPE	2.2	Polykemi	www.polykemi.se
Polyfabs	ABS	2.2	Polykemi	www.polykemi.se
Polyfast	ASA	2.2	MP	www.mpcom.it
Polyfill	PP	2.2	Polykemi	www.polykemi.se
Polyflam	Thermopl. allg.	2.2	Schulman	www.aschulman.com
Polyflon	PTFE	2.2	Daikin	www.daikin.com
Polyform	POM	2.2	Polykemi	www.polykemi.se
Polyfort	EVAC, PE, PE-HD, PP	2.2	Schulman	www.aschulman.com
Polyglad	PC	3.4	GEP	www.geplastics.com
Polykarbonat	PC	2.2	Polykemi	www.polykemi.se
Polylac	ABS	2.2	Chi Mei	www.chimei.com.tw
Polylite	UP	2.2	Reichhold	www.reichhold.com
Polyloy	PA	2.2	Ems	www.ems-chemie.com
Polylux	MABS	2.2	Polykemi	www.polykemi.se
Polyman	Thermopl. allg.	2.2	Schulman	www.aschulman.com
Polymer Resou	ABS	2.2	Polymer Resources Ltd.	www.prlresins.com/
Polymist	PTFE	2.1, 2.2	Solvay	www.solvay.com
Polynil	PA-Cop	2.2	Nilit	www.nilit.com
Polypenco	PEEK	3.1, 3.4	Quadrant Engineering Plastic Product	www.quadrantepp.com/

Continued on next page

D.2 Tradenames Table

Tradename	Polymer	Delivery Form	Manufacturer Supplier	Web Address
Polyplex	PMMA	2.2	Polykemi	www.polykemi.se
Polypro	PP	3.4	Doeflex	www.vtsdoeflex.co.uk
Polypur	TPU	2.2	Schulman	www.aschulman.com
Polyram	PBT	-	Polyram Ram-On Industries	www.polyram.co.il/
Polyrex	PS	2.2	Chi Mei	www.chimei.com.tw
Polysan	SAN	2.2	Polykemi	www.polykemi.se
Polyshine	PBT	2.2	Polykemi	www.polykemi.se
Polyspeed	EP	2.3	Hexcel	www.hexcel.com
Polystat	Thermopl. allg.	2.2	Schulman	www.aschulman.com
Polystruc	Polyester, Thermoset	SMC, BMC	Industrial Di-electrics Inc.	www.idiplastic.com/
Polytherm	PVC	3.3	Huhtamaki	www.4pfolie.de
Polytron	Thermopl. allg.	2.2	Polyram	www.polyram.co.il
Polytrope	TPO	2.2	Schulman	www.aschulman.com
POLYVIC	PVC	2.2	Connell Bros Company LTD	webprod.wecon.com/ WECOCBC/WECO/ index.h
Polyvin	PVC, VC-Cop	2.2	Schulman	www.aschulman.com
Polyviol	PVAL	2.2	Wacker	www.wacker.com
Polywood	PS	2.2	Polykemi	www.polykemi.se
Pomalux	POM	3.3	Westlake Plastics Company	www.westlakeplastics.com/
Ponacom	Thermopl. allg.	2.2	Pongs & Zahn	www.pongsundzahn.de
Ponaflex	TPE	2.2	Pongs & Zahn	www.pongsundzahn.de
Ponalen	PO	2.2	Pongs & Zahn	www.pongsundzahn.de
Ponamid	PA	2.2	Pongs & Zahn	www.pongsundzahn.de
Porene	ABS, PS, SAN	2.2	Pongs & Zahn	www.pongsundzahn.de
Porene	ABS	2.2	Thai Petrochemi-cal Industry Co., Ltd	www.tpigroup.co.th/
Poret	PUR	3.7	EMW	www.emw.de
Poron	PUR, PVC-P	3.7	Rogers	www.rogerscorporation.com
Poval	PVAL	2.2	Denka	www.denka.co.jp
Powersil	Q	1.1	Wacker	www.wacker.com
PPO	PPE	2.2	GEP	www.geplastics.com
Pre-Elec	Thermopl. allg.	2.2	Premix	www.premix.com
Pregnit	PI	2.3, 3.5	Krempel	www.krempel.com
Premi-Glas	Polyester, Thermoset	2.2	Premix, Inc.	www.premix.com/
Premi-Ject	Polyester, Thermoset	2.2	Premix, Inc.	www.premix.com/

Continued on next page

Tradename	Polymer	Delivery Form	Manufacturer Supplier	Web Address
Prevail	ABS+TPU, TPU	2.2	Dow	www.dow.com
Prima	VC-Cop	2.2	EVC	www.evc-int.com
PRIMABLEND	PMMA+PC	-	Prima Plastics, LLC	Discontinued
Primacor	EAA, EMA	2.2	Dow	www.dow.com
Primal	PMMA	1.2	Rohm & Haas	www.rohmhaas.com
PRIMAMID	PA66	-	Prima Plastics, LLC	Discontinued
PRIMANATE	PC	-	Prima Plastics, LLC	Discontinued
PRIMANEX	POM	-	Prima Plastics, LLC	Discontinued
PRIMANITE	PET	-	Prima Plastics, LLC	Discontinued
PRIMAPRO	PP	-	Prima Plastics, LLC	Discontinued
PRIMAPRON	PA6	-	Prima Plastics, LLC	Discontinued
PRIMATEL	PO	-	Prima Plastics, LLC	Discontinued
PRIMATHON	PE-HD	-	Prima Plastics, LLC	Discontinued
PRIMATRAN	ABS	-	Prima Plastics, LLC	Discontinued
Prime	PVC	2.2	Prime PVC	www.primepvc.com/
Primeace	POM	2.2	Prime Source Polymers, Inc.	586-757-5777
Primecarb	PC	2.2	Prime Source Polymers, Inc.	586-757-5777
Primef	PPS	2.2	Solvay	www.solvay.com
Primefin	PP	2.2	Prime Source Polymers, Inc.	586-757-5777
Primeflex	TPO	2.2	Prime Source Polymers, Inc.	586-757-5777
Primelene	PE-HD	2.2	Prime Source Polymers, Inc.	586-757-5777
Primerene	PS	2.2	Prime Source Polymers, Inc.	586-757-5777
Primesan	SAN	-	Prime Source Polymers, Inc.	586-757-5777
Primethane	PA66	-	Prime Source Polymers, Inc.	586-757-5777
Primolac	ABS	-	Prime Source Polymers, Inc.	586-757-5777
Prism	PUR-MDI	liquid	Bayer Corporation, Polyurethanes Div	www.rimmolding.com/

Continued on next page

D.2 Tradenames Table

Tradename	Polymer	Delivery Form	Manufacturer Supplier	Web Address
Prismex	PMMA	3.4	Lucite	www.lucitecp.com
Profax	PP	2.2	Basell	www.basell.de
Profilen	PTFE	2.1, 3.6	Lenzing	www.lenzing.at
Proflex	TPU	2.2	PolyOne Th. Bergmann	www.polyone.com
Prolen	PP	granules	Polibrasil Resinas S.A.	www.polibrasil.com.br/
Propak	PP-Cop.	2.2	PolyPacific Pty. Ltd.	www.polypacific.com.au/
Propazote	PP	3.7	Zotefoams	www.zotefoams.com
Propilco	PP	2.2	PROPILCO S.A.	www.propilco.com/
Propilven	PP	2.2	Propilven	www.trade-venezuela.com/propliven
Proppet	PET	2.2	Braskem	www.braskem.com.br
Propyform	PP	3.3, 3.4	Plastech	www.vtsplastech.co.uk
Propylex	PP, PP+EPDM	3.3, 3.4	Plastech	www.vtsplastech.co.uk
Propylux	PP	3.1, 3.4	Westlake Plastics Company	www.westlakeplastics.com/
Proteus	PP-Cop.	3.4	Poly-Hi Div. Menasha Corp.	www.polyhisolidur.com/
ProTherm	PVC-C	2.2	Georgia Gulf	www.ggc.com/
Provista	PET-G	2.2	Eastman	www.eastman.com
PSC	PUR	2.2	Polyurethane Specialties Company	201-438-2325
PSG	ABS	2.2	Plastic Selection Group, Inc.	www.go2psg.com/
P-Tex	PE-UHMW	3.3	IMS	www.ims-plastic.com
PTS	ABS	2.2	Polymer Technology and Services, LLC	www.ptsllc.com/
Pulse	PC+ABS	2.2	Dow	www.dow.com
Purell	PP	2.2	Basell	www.basell.de
Purenit	PUR	3.7	Puren	www.puren.com
Purex	PET	3.3	Teijin	www.teijin.co.jp
PUR-fect	PUR	2.2	Ciba Specialty Chemicals	www.cibasc.com/
PVC Film	PVC	3.3	Teknor Apex Company	www.teknorapex.com/
Pyro-Chek	PS	2.2	Ferro	www.ferro.com
Pyrofil	PAN	3.6	Mitsubishi Rayon	www.mrc.co.jp
Quarite	PMMA	2.2, 3.4	Aristech	www.aristechchem.com
Quelflam	PUR-R	1.1	Baxenden	www.baxchem.co.uk
Questra	PS	2.2	Dow	www.dow.com
Quinn	ABS, PC, PMMA, PP, PS, PS-HI, SAN	3.4	Quinn	www.quinn-plastics.com

Continued on next page

Tradename	Polymer	Delivery Form	Manufacturer Supplier	Web Address
Radel A	PESU	2.2	Solvay	www.solvay.com
Radel R	PPSU	2.2	Solvay	www.solvay.com
Radicron	PET-G	2.2	Radici	www.radiciplastics.com
Radiflam	PA, PBT	2.2	Radici	www.radiciplastics.com
Radilon	PA 6, PA 66, PA-Cop	2.2	Radici	www.radiciplastics.com
Raditer	PBT	2.2	Radici	www.radiciplastics.com
Radlite	Thermopl. allg.	3.6	Azdel	www.azdel.com
Rail-Lite	PEI	3	Azdel	www.azdel.com
Ralupol	UP	2	Raschig	www.raschig.de
RapidCast	TSU (Polyurethane Thermoset Elastomer)	liquid	Innovative Polymers, Inc.	www.innovative-polymers.com/
RapidVAC	TSU (Polyurethane Thermoset Elastomer)	liquid	Innovative Polymers, Inc.	www.innovative-polymers.com/
Raplan	SBS	2.2	API	www.apiplastic.com
Ravamid	PA 6, PA 66	2.2	Ravago	www.ravago.de
Ravatal	POM	2.2	Ravago	www.ravago.de
RC Plastics	PA6	2.2	RC Plastics, Inc.	Defunct www.rcplastics.com/
Reconyl	PA 6, PA 66	2.2	Frisetta	www.frisetta-polymer.de
Recticel	PUR	3.7	Recticel	www.recticel.com
Regalite	PVC	3.3	PolyOne Th. Bergmann	www.polyone.com
Relon	PA 6, PA 66	2.2	Tecnopolimeri	www.fbtecnopolimeri.com
Remex	PBT, PC+PBT	2.2	GEP	www.geplastics.com
Remid	PA 6, PA 66	2.2	Tecnopolimeri	www.fbtecnopolimeri.com
Remix	PC+ABS	2.2	Tecnopolimeri	www.fbtecnopolimeri.com
Ren	EP	liquid, 1.4, 2.2	Ciba Specialty Chemicals	www.cibasc.com/
Ren:c:o-thane	TSU (Polyurethane Thermoset Elastomer)	liquid	Ciba Specialty Chemicals	www.cibasc.com/
Renodur	PVC	3.2	Solvay	www.solvay.com
Renofort	PVC	3.2	Solvay	www.solvay.com
Reny	PA 6/6-T	2.2	Mitsubishi	www.m-ep.co.jp
Repak	PO	3.3	Mitsubishi	www.m-ep.co.jp
Repete	PET	2.2	M&G	www.mgpolymers.com
Replay	PS	2.2	Nova	www.novachem.com
Repolem	PVAC	2.2	Atofina	www.atofina.de
Resilt	PC	2.2	Tecnopolimeri	www.fbtecnopolimeri.com
ResinDirect	PP-Cop.	2.2	ResinDirect, LLC	www.resindirect.com/

Continued on next page

D.2 Tradenames Table

Tradename	Polymer	Delivery Form	Manufacturer Supplier	Web Address
Resinoid	Phenolic	2.2	Resinoid Engineering Corporation	www.resinoid.com/
Resinol	PF	2.2	Raschig	www.raschig.de
Resirene	PS-HI	2.2	Resirene, S.A. de C.V.	www.resirene.com.mx/
Restil	SAN	2.2	Restil	www.restil.com
Restiran	ABS	2.2	Restil	www.restil.com
Restirolo	PS, S-Cop	2.2	Restil	www.restil.com
Retain	PE	2.2	Dow	www.dow.com
Retipor	PUR	3.7	EMW	www.emw.de
Retpol	PP-Cop.	2.2	PolyPacific Pty. Ltd.	www.polypacific.com.au/
Revil	EPDM	2.2	Lamplast	www.lamplast.it
REXell	PE-MD	2.2	Huntsman Corporation	www.huntsman.com/
REXflex	PP	-	Huntsman Corporation	www.huntsman.com/
REXOMER	TPE	2.2	GVR Complast, Ltd. Co.	www.gvrcomp.com/
Rextac	PAO	2.2	Huntsman	www.huntsman.com
Rezibond	PF	2.2	Chromos	www.chromos.hr
Rhetech	PP	2.2	RheTech, Inc.	www.rhetech.com/
Rhodorsil	SI	2.2	Rhodia	www.rhodia.com
Ribetak	PF	1.5	Ceca	www.ceca.fr
Riblene	PE-LD	2.2	Polimeri	www.polimerieuropa.com
Rigidex	PE-HD	2.2	BP	www.bppetrochemicals.com
Rigipore	PS-E	2.2	BP	www.bppetrochemicals.com
Rilsan	PA 11, PA 12	2.1, 2.2, 3.6	Atofina	www.atofina.de
Rimplast	PO, Q, TPU	1.1, 2.2	Degussa	www.degussa.de
Riteflex	TPC	2.2	Ticona	www.ticona.com
Rodrun	PE-C	2.2	Unitika	www.unitika.co.jp
Rogers	Alkyd	2.2	Rogers Corporation	www.rogers-corp.com/
Rohacell	PMMI	3.7	Röhm	www.cyro.com
Romiloy	ABS+PA, ABS+PC, ASA+PC	2.2	Romira	www.romira.de
Ropol	PE-LD	2.2	Oltchim	www.oltchim.ro
Ropoten	PE	2.2	Oltchim	www.oltchim.ro
Roscom	PVC	2.2	Roscom, Inc.	www.vinylcompounds.com/
Rosevil	PVC	2.2	Oltchim	www.oltchim.ro
Rosite	Polyester, Thermoset	BMC	Rostone Corporation	www.rostone.com/
Rotec	ABS, ASA, SAN	2.2	Romira	www.romira.de

Continued on next page

Tradename	Polymer	Delivery Form	Manufacturer Supplier	Web Address
Rotolon	PTFE	2.2	Greene, Tweed Engineered Plastics	www.gtweed.com/
Rotuba	CA	2.2	Rotuba Plastics	www.rotuba.com/
Rovel	TPE	2.2	Dow	www.dow.com
Royalcast	PUR	2.2	Uniroyal Chemical Group	www.chemtura.com/chem/default.htm
Royalex	ABS, PVC	3.4	Royalite	www.vtsroyalite.co.uk
Royalite	ABS, ABS+PVC	3.1, 3.3, 3.4	Royalite	www.vtsroyalite.co.uk
Royalstat	ABS+PVC	3.4	Royalite Thermoplastics Division	www.spartech.com/
RTP	PP	2.2	RTP Company	www.rtpcompany.com/
RTV-2	SI	liquid	Silicones, Inc.	www.silicones-inc.com/
Rubiflex	PUR-R	1.1	Huntsman	www.huntsman.com
Rubinate	PUR-R	1.1	Huntsman	www.huntsman.com
Rulon	PTFE	-	Saint Gobain - Furon	www.plastics.saint-gobain.com/
Rütamid	PA	2.2	Bakelite	www.bakelite.
Rütaphen	FF, PF, RF	1.5	Bakelite	www.bakelite.
Rütapox	EP	1.5	Bakelite	www.bakelite.
Rütapur	PUR-R	1.1	Bakelite	www.bakelite.
RxLOY	PP	2.2	Ferro Corporation	www.ferro.com/
RYLENE	PE	2.2	GVR Complast, Ltd. Co.	www.gvrcomp.com/
Rynite	PET	2.2	DuPont	www.dupont.com
Ryton	PPS	2.2	Chevron Phillips	www.cpchem.com
Ryulex	ABS+PC	2.2	Dainippon	www.dic.co.jp
Sabre	PC+PET	2.2	Dow	www.dow.com
Saduren	MF	1.1	BASF	www.basf.com
Safe-FR	PO	2.2	UVtec, Inc.	(972) 991-0600
Saflex	PVB	3.3	Dow	www.dow.com
Sangel	SAN	2.2	Proquigel	www.proquigel.com.br/
SANTAC	ABS	2.2	Nippon A&L Inc.	www.n-al.co.jp/ing/index.html
Santoprene	PP+EPDM	2.2	AES	www.santoprene.com
Sapelec	PVC	2.2	Exxon	www.exxonchemical.com
Saran	PVDC	2.2	Dow	www.dow.com
Sarlink	TPO	2.2	DSM Elastomers	www.dsm.com
Sarnafil	PO, PVC	3.3	Sarna	www.sarna.com
Sarnatex	PVC-P	3.3	Sarna	www.sarna.com
Sasolen	PP	2.2	Sasol	www.sasol.com
Saterflex	TPS	2.2	Sasol	www.sasol.com
Satinflex	PVC	2.2	Alpha Gary Corporation	www.alphagary.com/
Satinflo	PVC	2.2	Vi-Chem Corporation	www.vichem.com/

Continued on next page

D.2 Tradenames Table

Tradename	Polymer	Delivery Form	Manufacturer Supplier	Web Address
Satran	ABS, PS, SAN	2.2	MRC Polymers	www.mrcpolymers.com
Saxene	EPDM, PP, PP	2.2	ECW-Eilenburger	www.ecw-compound.de
Saxomer	TPE	2.2	ECW-Eilenburger	www.ecw-compound.de
Scanamid	PA	2.2	Polykemi	www.polykemi.se
Scancomp	PA, PP	2.2	Polykemi	www.polykemi.se
Scanrex	PPS	2.2	Polykemi	www.polykemi.se
Scantec	PC	2.2	Polykemi	www.polykemi.se
Scarab	UF	2.2	BIP	www.bip.co.uk
Schuladur	PBT, PET	2.2	Schulman	www.aschulman.com
Schulaform	POM	2.2	Schulman	www.aschulman.com
Schulamid	PA 6, PA 66	2.2	Schulman	www.aschulman.com
Schulink	PE-HD	2.1	Schulman	www.aschulman.com
Sclair	PE-HD, PE-LLD, PE-ULD	2.2	Nova	www.novachem.com
Sclaircoat	PE-HD	2.2	Nova	www.novachem.com
Sclairfilm	PE	3.3	Nova	www.novachem.com
Sclairlink	PE	2.1	Nova	www.novachem.com
Sclairpipe	PE	3.2	Nova	www.novachem.com
Scolefin	PP	2.2	RP	www.rpcompounds.de
Sconablend	TPE	2.2	Ravago	www.ravago.de
Scotchcast	EP	2.1	3 M	www.3m.com
Scotchkote	EP	1.1	3 M	www.3m.com
Scotchpak	PTP	3.3	3 M	www.3m.com
Scotchpar	PET	3.3	3 M	www.3m.com
Scotchply	EP, PF	2.2, 2.3	3 M	www.3m.com
Scotchshield	PET	3.3	3 M	www.3m.com
Sebiform	PA6	2.2	Sebi Innovative Compounds	966-1-225-8000
Sebimid	PA6	2.2	Sebi Innovative Compounds	966-1-225-8000
Sedapet	PET	2.2	La Seda	www.laseda.es
Selar	PA-Cop, PET	2.2	DuPont	www.dupont.com
Selectrofoam	PUR-R	1.1	PPG	www.ppg.com
Selectron	UP	1.1, 2.2	PPG	www.ppg.com
Semicon	PE+IR	2.2	Borealis	www.borealis.com
Senosan	ABS, ASA, PET-G, PMMA, PS	3.3	Senoplast	www.senoplast.com
Sentrixx	PP	2.2	The Matrixx Group, Inc.	www.matrixxgroup.com/
SEP	PUR	2.2	Foster Corporation	www.fostercomp.com/
Septon	TPE	2.2	Kuraray	www.kurarayamerica.com
Sequel	TPO	2.2	Solvay	www.solvay.com

Continued on next page

Tradename	Polymer	Delivery Form	Manufacturer Supplier	Web Address
Serfene	PVDF	2.1	Solvay	www.solvay.com
Setilithe	CA	2.2	Solvay	www.solvay.com
Sevrene	TPE	2.2	Vi-Chem Corporation	www.vichem.com/
Sevrite	TPE	2.2	Vi-Chem Corporation	www.vichem.com/
Shafting	EP	2.2	Club-Kit, Inc.	
Shinblend	PC+PBT	2.2	Shinkong	www.shinkong.com.tw
Shincor	SI	2.2	Shincor Silicones, Inc.	www.shincor.com/
Shinite	PBT	2.2	Shinkong	www.shinkong.com.tw
Shinkolac	ABS	2.2	Mitsubishi Rayon	www.mrc.co.jp
Shinkolite	PMMA	1.1, 2.2, 3.4	Mitsubishi Rayon	www.mrc.co.jp
Shinpak	PET	3.4	Shinkong	www.shinkong.com.tw
Shinpet	PET	2.2	Shinkong	www.shinkong.com.tw
Shinpex	PET	3.3	Shinkong	www.shinkong.com.tw
Sho-Allomer	PP	2.2	Showa	www.sdk.co.jp
Sholex	PE-HD, PE-LD	2.2	Showa	www.sdk.co.jp
Shorko	PP	3.3	Treofan	www.treofan.com
Shuman	ABS	2.2	Shuman Plastics, Inc.	www.shuman-plastics.com/
Siamvic	PVC	2.2	Solvay	www.solvay.com
Sibrflex	SIR	2.2	Goodyear	www.goodyearchemical.com
Sicalit	CA	2.2	EVC	www.evc-int.com
Sicobox	PVC	3.3	EVC	www.evc-int.com
Sicodex	PVC	3.4	EVC	www.evc-int.com
Sicofarm	PVC	3.3	EVC	www.evc-int.com
Sicoffset	PVC	3.3	EVC	www.evc-int.com
Sicoflex	ABS	2.2	MP	www.mpcom.it
Sicoklar	PC	2.2	MP	www.mpcom.it
Sicolene	PE, PVC	3.3	EVC	www.evc-int.com
Sicoplast	PVC	3.3	EVC	www.evc-int.com
Sicoprint	PVC	3.3	EVC	www.evc-int.com
Sicoran	SAN	2.2	MP	www.mpcom.it
Sicoreg	PVC	3.3	EVC	www.evc-int.com
Sicostirolo	PS	2.2	MP	www.mpcom.it
Sicoter	PBT, PPE	2.2	MP	www.mpcom.it
Sicovimp	VC-Cop	3.3	EVC	www.evc-int.com
Sicovinil	PVC, VC-Cop	3.3, 3.4, 3.6	EVC	www.evc-int.com
Silacron	PMMA	1.1	Schock	www.schock.de
Silastic	SI	liquid	Dow Corning Corporation	www.dowcorning.com/
Silmar	Polyester, Thermoset	liquid	Interplastic Corporation	www.interplastic.com/

Continued on next page

D.2 Tradenames Table

Tradename	Polymer	Delivery Form	Manufacturer Supplier	Web Address
Silon	PTFE+PDMS (Polydimethylsiloxane)	3.3	Bio Med Sciences, Inc	www.silon.com/
Silres	Q	1.1	Wacker	www.wacker.com
Silsoft	SI	1.1	GE Silicones	www.gesilicones.com
Siltem	PEI	2.2	GEP	www.geplastics.com
Simona	Thermopl. allg.	3.1, 3.2, 3.3, 3.4	Simona	www.simona.de
Sinkral	ABS	2.2	Polimeri	www.polimerieuropa.com
Sinotherm	PF	1.1	Hüttenes-Albertus	www.huettenes-albertus.com
Sinvet	PC	2.2	EniChem	www.eni.it/home/home_en.html
Siroplan	PVC	3.2	Hegler	www.hegler-plastik.de
Siroplast	PE-HD	3.2	Hegler	www.hegler-plastik.de
Sirowell	PVC	3.2	Hegler	www.hegler-plastik.de
Siveras	LCP	2.2	Toray	www.toray.co.jp
Skai	PVC-P	3.3	Hornschuch	www.hornschuch.de
Skailan	PUR	3.3, 3.7	Hornschuch	www.hornschuch.de
Skybond	PI	1.1	Dow	www.dow.com
Skygreen	PET-G	2.2	SK Chemicals	www.skchemicals.com
Skypel	TPE	2.2	SK Chemicals	www.skchemicals.com
Skypet	PET	2.2	SK Chemicals	www.skchemicals.com
Skythane	TPU	2.2	SK Chemicals	www.skchemicals.com
Skyton	PBT	2.2	SK Chemicals	www.skchemicals.com
SLCC	PC	2.2	GE Polymerland	www.gepolymerland.com/
Smartan	SMAH	1.1	Atofina	www.atofina.de
Sniamid	PA	2.2	Rhodia	www.rhodia.com
Sniatal	POM	2.2	Rhodia	www.rhodia.com
Soarblen	E-Cop	2.2	Nippon Gohsei	www.nippongohsei.com
Soarlex	E-Cop	2.2	Nippon Gohsei	www.nippongohsei.com
Soarnol	EVAL	2.2	Atofina	www.atofina.de
Socarex	PE-LD	3.2	Solvay	www.solvay.com
Sofprene	SBS, SEBS	2.2	So.F.Ter	www.softerspa.com
Softell	PP	2.2	Basell	www.basell.de
Softlex	E-Cop	2.2	Nippon Petrochemicals	www.npcc.co.jp
Solef	PVDF	2.2	Solvay	www.solvay.com
Solflex	SBR	2.2	Goodyear	www.goodyearchemical.com
Solidur	PE-HD, PE-UHMW, PP	3.1, 3.4	Poly Hi Solidur	www.polyhisolidur.com
Solmed	PO, PVC	3.3	Solvay	www.solvay.com
Solprene	SB	bale, granules	INSA	52-5 726-1800
Soltub	PVC	3.2	Solvay	www.solvay.com
Soluforce	PE	3.2	Pipelife	www.pipelife.at

Continued on next page

Tradename	Polymer	Delivery Form	Manufacturer Supplier	Web Address
Soluphene	PFA	1.5	Atofina	www.atofina.de
Solvic	PVC	2.2	Solvay	www.solvay.com
Solvin	PVC, VC-Cop	1.4, 2.1, 2.2	Solvin	www.solvinpvc.com
Sorona	PTT	2.2, 3.3, 3.6	DuPont	www.dupont.com
Spandal	PUR	3.5	Baxenden	www.baxchem.co.uk
Spandofoam	PUR	3.7	Baxenden	www.baxchem.co.uk
Spartech	PP	2.2	Spartech Polycom	www.spartech.com/
Specialmid	PA6	-	Aquafil Technopolymers S.p.A.	www.aquafil.com/
Spectar	PTP	2.2	Eastman	www.eastman.com
Spectrum	PE-HD, PE-LD, PP	3.6	Eastman	www.eastman.com
Spilac	VE	1.3	Showa	www.sdk.co.jp
Stabilux	PBT	3.1, 3.2, 3.3, 3.4	Westlake Plastics Company	www.westlakeplastics.com/
Stamax	PP	2.3	Sabic	www.sabic.com
Stamylan	PE, PP	2.2	Sabic	www.sabic.com
Stamylan UH	PE-UHMW	2.2	DSM Eng. Plastics	www.dsmep.com
Stamylex	PE-HD, PE-LLD	2.2	Sabic	www.sabic.com
Stamyroid	PP	2.2	Sabic	www.sabic.com
Stamytec	PP	2.2	Sabic	www.sabic.com
Stanuloy	PET	-	MRC Polymers, Inc.	www.mrcpolymers.com/
Stanyl	PA 46	2.2	DSM Eng. Plastics	www.dsmep.com
Stapron	ABS+PA, PC+ABS, PC+PET	2.2	DSM Eng. Plastics	www.dsmep.com
Staramide	PA 6, PA 66	2.2	LNP	www.lnp.com
Staren	PE	2.2	Samsung, a division of Cheil Industr	www.samsungstarex.com/
Starex	ABS	2.2	Cheil	www.cii.samsung.co.kr
Starflam	PA 6, PA 66	2.2	GEP	www.geplastics.com
Staroy	PC+ABS	2.2	Cheil	www.cii.samsung.co.kr
Staroy	ABS+PC	2.2	Samsung, a division of Cheil Industr	www.samsungstarex.com/
Starpylen	PP	2.2	LNP	www.lnp.com
Stat-Kon	Thermopl. allg.	2.2	LNP	www.lnp.com
Stat-Loy	Thermopl. allg.	2.2	LNP	www.lnp.com

Continued on next page

D.2 Tradenames Table

Tradename	Polymer	Delivery Form	Manufacturer Supplier	Web Address
Stat-Rite	TPU	2.2	Noveon, Inc.	www.noveon.com/
Stereon	SBS	2.2	Firestone	www.firesyn.com
Sterocoll	PAA, PMMA	1.2, 1.3	BASF	www.basf.com
Stevens	TPU	2.2	JPS Elastomerics Corp.	www.jpselastomerics.com/
Stirolan	PS	2.2	Soredi	www.soredi.it
Strabusil	PE	3.2	Fränkische Rohrwerke	www.fraenkische.de
Strafil	EP, PF	2.3	Hexcel	www.hexcel.com
Strandfoam	PP	3.7	Dow	www.dow.com
Strasil	PVC	3.2	Fränkische Rohrwerke	www.fraenkische.de
Structon	PUR	liquid	Hercules Incorporated	www.herc.com/
Stylac	ABS	2.2	Asahi	www.asahithermofil.com
Styraclear	PS	-	Westlake Plastics Company	www.westlakeplastics.com/
Styroblend	PS+PE	2.2	BASF	www.basf.com
Styrocell	PS-E	2.2	Nova	www.novachem.com
Styrodur	PS	3.4, 3.7	BASF	www.basf.com
Styrofan	SB	1.2	BASF	www.basf.com
Styroflex	SBS	2.2	BASF	www.basf.com
Styrofoam	PS-E	2.2	Dow	www.dow.com
Styrolux	SBS, S-Cop	2.2, 3, 3.4	BASF	www.basf.com
Styron	PS	2.2	Dow	www.dow.com
Styronal	SB	1.2	BASF	www.basf.com
Styropor	PS-E	2.2	BASF	www.basf.com
Styrosun	PS-HI	2.2	Nova	www.novachem.com
STYRYLIC	SMMA	2.2	Deltech Polymers Corporation	www.deltechcorp.com/
Stystat	PS-HI	2.2	United Composites, Inc.	817-468-2929
Styvex	PS	2.2	Ferro Corporation	www.ferro.com/
Sumidur	PUR-R	1.1	Sumitomo	www.sumitomo-chem.co.jp
Sumiflex	VC-Cop	2.2	Sumitomo	www.sumitomo-chem.co.jp
Sumigraft	EVAC+PVC	2.2	Sumitomo	www.sumitomo-chem.co.jp
Sumikadel	PMMA+PVAL	2.2	Sumitomo	www.sumitomo-chem.co.jp
SUMIKAEXCEIPES		2.2	Sumitomo Chemical America, Inc.	www.sumitomo-chem.co.jp/english/index.
Sumikaflex	E-Cop, TPO	2.2	Sumitomo	www.sumitomo-chem.co.jp
Sumikagel	E-Cop	3.3	Sumitomo	www.sumitomo-chem.co.jp

Continued on next page

Tradename	Polymer	Delivery Form	Manufacturer Supplier	Web Address
Sumikasuper	LCP	2.2	Sumitomo	www.sumitomo-chem.co.jp
SUMIKASUPER	LCP	2.2	Sumitomo Chemical America, Inc.	www.sumitomo-chem.co.jp/english/index.
Sumikathene	PAO	2.2	Sumitomo	www.sumitomo-chem.co.jp
Sumikon	EP	liquid	Sumitomo Bakelite Co., Ltd.	www.sumibe.co.jp/english/index.html
Sumilit	PVC	2.2	Sumitomo	www.sumitomo-chem.co.jp
Sumilite	PESU	3.3	Sumitomo	www.sumitomo-chem.co.jp
Sumipex	PMMA	2.2, 3.3	Sumitomo	www.sumitomo-chem.co.jp
SUMIPLOY	PES	2.2	Sumitomo Chemical America, Inc.	www.sumitomo-chem.co.jp/english/index.
Sunfrost	VC-Cop	2.2	Schulman	www.aschulman.com
Sunglas	PMMA	3.4	Palram	www.palram.com
Sunigum	TPE	2.2	Goodyear	www.goodyearchemical.com
Sunlite	PC	3.4	Palram	www.palram.com
Sunoco	PP	2.2	Sunoco Chemicals, Polymers Division	www.sunocochemicals.com/
Sunprene	VC-Cop	2.2	Schulman	www.aschulman.com
Suntec	PE-HD	2.2	Asahi	www.asahithermofil.com
Suntra	PPS	2.2	SK Chemicals	www.skchemicals.com
Suntuf	PC	3.4	Palram	www.palram.com
Supazote	E-Cop	3.7	Zotefoams	www.zotefoams.com
Supec	PPS	2.2	GEP	www.geplastics.com
Superkleen	PVC	2.2	Alpha Gary Corporation	www.alphagary.com/
Superlinear[a]	PO	2.2	A. Schulman Inc.	www.aschulman.com/
Superlite	PP	2.3	Azdel	www.azdel.com
SUPLEX	PE	2.2	Politeno	www.politeno.com.br/
Supra-Carta	EP, PF	3.5	Isola	www.isola.de
Supradel	PPSU	2.2	Solvay	www.solvay.com
Supraplast	UP	2	Raschig	www.raschig.de
Suprasec	PUR-R	1.1	Huntsman	www.huntsman.com
Suprel	PET	3.6	DuPont	www.dupont.com
Suprel	SVA (Styrenic + Vinyl + Acrylonitrile Alloy)	2.2	Georgia Gulf (CONDEA Vista Company)	www.ggc.com/
Supreme	PS	2.2	Supreme Petrochem Ltd.	www.supremepetrochem.com/
Surell	MF	3.4	Formica	www.formica-europe.com
Surlyn	E-Cop	1.5, 3.3	DuPont	www.dupont.com
Surpass	PE-HD, PE-LLD	2.2	Nova	www.novachem.com

Continued on next page

D.2 Tradenames Table

Tradename	Polymer	Delivery Form	Manufacturer Supplier	Web Address
Sustadur	PET	3.1, 3.3	Sustaplast	www.sustaplast.com
Sustamid	PA	3.1, 3.3	Sustaplast	www.sustaplast.com
Sustanat	PC	3.1, 3.3	Sustaplast	www.sustaplast.com
Sustarin	POM	3.1, 3.3	Sustaplast	www.sustaplast.com
Sustatec	PET, PVDF	3.1, 3.3	Sustaplast	www.sustaplast.com
Swan	PF	3.5	Tufnol	www.tufnol.co.uk
Sylgard	SI	liquid, 2.2	Dow Corning Corporation	www.dowcorning.com/
Sylvin	PVC+PUR	2.2	Sylvin Technologies Incorporated	www.sylvin.com/
Symalen	PE	3.2	Symalit	www.symalit.com
Symalit	PO	2.3, 3.1, 3.2, 3.4	Symalit	www.symalit.com
Syncure	PE	-	PolyOne Corporation	www.polyone.com/
Synolac	UP	1.1	Symalit	www.symalit.com
Synolite	UP, VE	1.5	DSM Composite Resins	www.dsm.com
Synprene	SBS, SEBS	2.2	PolyOne Th. Bergmann	www.polyone.com
Syntegum	PP	2.2	Lamplast	www.lamplast.it
Systanat	PUR-R	1.1	Elastogran	www.elastogran.de
Systol	PUR-R	1.1	Elastogran	www.elastogran.de
Ta-adin	PVC-P	3.3	Elastogran	www.elastogran.de
Tacphan	CTA	3.3	Lofo	www.lofo.de
Tactene	BR	2.2	Lanxess	www.lanxess.com
Tafmer	TPO	2.2	Mitsui	www.mitsui.co.jp
Tairilac	ABS	2.2	Formosa Chemicals	www.fcfc.com.tw
Tairipro	PP	2.2	Formosa Chemicals	www.fcfc.com.tw
Tairisan	SAN	2.2	Formosa Chemicals	www.fcfc.com.tw
Taitalac	ABS	2.2	Taita Chemical Company, Ltd.	www.ttc.com.tw/
Taktene	BR	2.2	Bayer	www.bayer.com
Talcoprene	PP	2.2	P-Group	www.p-group.de
Talnex	POM	2.2	MRC Polymers	www.mrcpolymers.com
Ta-or	PVC-P	3.3	MRC Polymers	www.mrcpolymers.com
Taradal	PVC-P	3.3	MRC Polymers	www.mrcpolymers.com
Taraflex	PVC-P	3.3	MRC Polymers	www.mrcpolymers.com
Taralay	PVC-P	3.3	MRC Polymers	www.mrcpolymers.com
Tarflen	PTFE	2.2	Tarnow	www.azozy.tarnow.pl
Tarflon	PC	2.2	Formosa Chemicals	www.fcfc.com.tw
Tarisan	SAN	2.2	Formosa Chemicals	www.fcfc.com.tw
Tarnaform	POM	2.2	Tarnow	www.azozy.tarnow.pl

Continued on next page

Tradename	Polymer	Delivery Form	Manufacturer Supplier	Web Address
Tarnamid	PA 6	2.2	Tarnow	www.azozy.tarnow.pl
Taroblend	PC+ABS	2.2	Taro	www.taroplast.com
Tarodur	ABS	2.2	Taro	www.taroplast.com
Tarolon	PC	2.2	Taro	www.taroplast.com
Tarolox	PBT, PET	2.2	Taro	www.taroplast.com
Taroloy	PC+PBT	2.2	Taro	www.taroplast.com
Taromid	PA 6, PA 66	2.2	Taro	www.taroplast.com
Tatren	PP	2.2	Slovnaft	www.slovnaft.sk
Tecast	PA	1.1, 3	Ensinger	www.shopforplastics.com
TECHNIACE	ABS+PC	2.2	Nippon A&L Inc.	www.n-al.co.jp/ing/index.html
Technoflex	PA	3.2	Hegler	www.hegler-plastik.de
Technogel	PUR	1.1	Technogel	www.technogel.it
Technora	PPTA	3.6	Teijin	www.teijin.co.jp
Technyl	PA 6, PA 66, PA-Cop	2.2	Rhodia	www.rhodia.com
Technylstar	PA 6	2.2	Rhodia	www.rhodia.com
Techpet	PET	2.2	Selenis	www.selenis.com
Techtron	PPS	3	Polytron	www.polytron-gmbh.de
Tecnoflon	FKM	2.2	Solvay	www.solvay.com
Tecnoline	PA 6, PA 66, PP	2.2	Domo	www.domo.be
Tecnoprene	PP	2.2	P-Group	www.p-group.de
Tecoflex	TPU	2.2	Thermedics	www.thermedicsinc.com
Tecophilic	TPU	2.2	Thermedics	www.thermedicsinc.com
Tecoplast	TPU	2.2	Thermedics	www.thermedicsinc.com
Tecothane	TPU	2.2	Thermedics	www.thermedicsinc.com
Tectur	PBT	2.2	Leis	www.leispt.de
Tediflex	PUR-R	1.1	Dow	www.dow.com
Tedilast	PUR-R	1.1	Dow	www.dow.com
Tedimon	PUR-R	1.1	Dow	www.dow.com
Tedirim	PUR-R	1.1	Dow	www.dow.com
Teditherm	PUR-R	1.1	Dow	www.dow.com
Tedlar	PVF	3.3	DuPont	www.dupont.com
Tedur	PPS	2.2	Albis	www.albis.com
Téfabloc	SBS, SEBS	2.2	Tessenderlo	www.tessenderlo.com
Téfanyl	PVC	2.2	Tessenderlo	www.tessenderlo.com
Téfaprene	TPS	2.2	Tessenderlo	www.tessenderlo.com
Teflon	FEP, PFA, PTFE	2.2	DuPont	www.dupont.com
Tefzel	ETFE	2.2	DuPont	www.dupont.com
Tegocoll	EP	1.5	Goldschmidt	www.goldschmidt.com
Tegophan	PMMA, UP	3.3	Goldschmidt	www.goldschmidt.com
Tego-Tex	MF, PF	2.2	Goldschmidt	www.goldschmidt.com
Tejinconex	PPTA	3.6	Teijin	www.teijin.co.jp
Tekbond	TPE	2.2	Teknor Apex	www.teknorapex.com
Teklamid	PA	2.2	PolyOne Th. Bergmann	www.polyone.com

Continued on next page

D.2 Tradenames Table

Tradename	Polymer	Delivery Form	Manufacturer Supplier	Web Address
Teknoplen	PP	2.2	PolyOne Th. Bergmann	www.polyone.com
Tekol	PIB	2.2	Kraiburg	www.kraiburgtpe.com
Tekron	TPE	2.2	Teknor Apex	www.teknorapex.com
Tekudur	PBT	2.2	Tekuma	www.tekuma.de
Tekuform	POM, PTFE	2.2	Tekuma	www.tekuma.de
Tekulon	PC	2.2	Tekuma	www.tekuma.de
Tekumid	PA 6, PA 66	2.2	Tekuma	www.tekuma.de
Tekusan	SAN	2.2	Tekuma	www.tekuma.de
Telalloy	PMMA+S-Cop	2.2	Kaneka	www.kaneka.co.jp
Telcar	TPO	2.2	Teknor Apex	www.teknorapex.com
Telene	PVC, PVC	2.2	PolyOne Th. Bergmann	www.polyone.com
Telprene	TPE		Teknor Apex Company	www.teknorapex.com/
Tempalloy	PP	2.2	Schulman	www.aschulman.com
Tempalux	PEI	3.1, 3.3, 3.4	Westlake Plastics Company	www.westlakeplastics.com/
TempRite	PVC-U, VC-Cop	2.2	PolyOne Th. Bergmann	www.polyone.com
Tempur	PUR	3.7	Gefinex	www.gefinex.de
Tenac	POM	2.2	Asahi	www.asahithermofil.com
Tenex	CAB	2.2	Teijin	www.teijin.co.jp
Tenite	CA, CAB, PE, PE-LLD, PP, PTP	1.1, 2.1, 2.2, 3.7	Eastman	www.eastman.com
Teonex	PEN	3.3	Teijin	www.teijin.co.jp
Tepcon	POM	2.2	Polyplastics	www.polyplastics.com
Tepeo	TPO	3.3	Benecke	www.benecke-kaliko.de
Teplagum	NBR	2.2	Lamplast	www.lamplast.it
Terate	PUR-R	1.1	Invista	www.invista.com
Terathane	PUR-R	2.2	DuPont	www.dupont.com
Teratron	PET	3	Polytron	www.polytron-gmbh.de
Terblend	ABS+PA	2.2	BASF	www.basf.com
Tercarol	PUR-R	1.1	Dow	www.dow.com
Terlac	SBS	2.2	Silac	None
Terluran	ABS, MABS	2.2	BASF	www.basf.com
Terlux	MABS	2.2	BASF	www.basf.com
Termaloy	ABS+PC	2.2	Proquigel	www.proquigel.com.br/
Terpalex	PA-Cop	2.2	Ube	www.ube.com
Teslin	PE	3.3	PPG Industries, Inc.	www.ppg.com/
Tetoron	PET	3.3	Teijin	www.teijin.co.jp
Tetralene	PE-UHMW	2.2	Coors	www.coorstek.com
Tetralon	PTFE	2.2	Coors	www.coorstek.com
Tetrax	PIB	1.5	Nippon Petrochemicals	www.npcc.co.jp

Continued on next page

Tradename	Polymer	Delivery Form	Manufacturer Supplier	Web Address
Texicryl	PMMA, S-Cop	1.2	Scott Bader	www.scottbader.com
Texigel	PMMA	1.2	Scott Bader	www.scottbader.com
Texin	TPU	2.2	Bayer	www.bayer.com
Texipol	PAA	1.2	Scott Bader	www.scottbader.com
Textolite	EP, MF, PF, SI, UP	3.5	GEP	www.geplastics.com
Therban	HNBR	2.2	Lanxess	www.lanxess.com
Thermalate	Polyester, Thermoset	3.4	Haysite Reinforced Plastics	www.haysite.com/
Thermalux	PSU	3.3	Westlake Plastics Company	www.westlakeplastics.com/
Thermassiv	PMMA	3.1	Schock	www.schock.de
Thermex-1	Thermopl. allg.	2.2	Schulman	www.aschulman.com
Thermoclear	PC	3.4	GEP	www.geplastics.com
Thermocomp	Thermopl. allg.	2.2	LNP	www.lnp.com
Thermodet	ABS, SAN, S-Cop	3.3	Mitras	www.mitras-materials.com
Thermofile	Thermopl. allg.	2.2	Asahi	www.asahithermofil.com
Thermoflex	SEBS, SEPS	2.2	PTS	www.pts-marketing.de
Thermolast	TPS	2.2	Kraiburg	www.kraiburgtpe.com
Thermoset	EP	2.2	Thermoset, Lord Chemical Products	www.lord.com/
Thermotuf	Thermopl. allg.	2.2	LNP	www.lnp.com
Thermovin	TPV	2.2	Vi-Chem Corporation	www.vichem.com/
Thermx	PTP	2.2	DuPont	www.dupont.com
Thermylene	PP	2.2	Asahi	www.asahithermofil.com
Thermylon	PA 6, PA 66	2.2	Asahi	www.asahithermofil.com
Thoprene	PPS	2.2	Asahi	www.asahithermofil.com
Tipcofil	PA6	2.2	Tipco Industries Ltd.	www.tipco-india.com/
Tipcolene	PE-LD	2.2	Tipco Industries Ltd.	www.tipco-india.com/
Tipcolite	Phenolic	2.2	Tipco Industries Ltd.	www.tipco-india.com/
Tipelin	PE-HD	2.2	TVK	www.tvk.hu
Tipolen	PE-LD	2.2	TVK	www.tvk.hu
Tipplen	PP	2.2	TVK	www.tvk.hu
Titan	LCP	2.2	DuPont	www.dupont.com
Tivilon	TPV	2.2	API	www.apiplastic.com
Tone	PCL (Polycaprolactone)	2.2	Dow Plastics	plastics.dow.com/

Continued on next page

D.2 Tradenames Table

Tradename	Polymer	Delivery Form	Manufacturer Supplier	Web Address
Tool-A-Thane	TSU (Polyurethane Thermoset Elastomer)	liquid	Urethane Tooling & Engineering Corp	www.urethanetooling.com/
Topas	COC	2.2	Ticona	www.ticona.com
Topet	PET	2.2	Tong Yang	www.tongyang.co.kr
Topex	PBT	2.2	Tong Yang	www.tongyang.co.kr
Topilene	PP	2.2	Tong Yang	www.tongyang.co.kr
Toplamid	PA	2.2	Tong Yang	www.tongyang.co.kr
Toplex	PC+ABS	2.2	Multibase	www.multibase.com
Toray	PBT	2.2	Toray Resin Company	www.toray.com/
Toraycon	PBT	2.2	Toray	www.toray.co.jp
Torlon	PAI	2.2	Solvay	www.solvay.com
Tospearl	SI	2.1	GE Silicones	www.gesilicones.com
Toughlon	PC	2.2	Idemitsu	www.idemitsu.co.jp
Toyolac	ABS, SAN	2.2	Toray	www.toray.co.jp
Toyolacparel	ABS	2.2	Toray Resin Company	www.toray.com/
TPX	PMP	2.2	Mitsui	www.mitsui.co.jp
TPX	PMP Cop.	granules	Mitsui Chemicals America, Inc.	www.mitsuichemicals.com/
Trancend	E-Cop	2.2	DuPont	www.dupont.com
Transilwrap	PS-HI	3.4	Transilwrap Company, Inc.	www.tosoh.com/
Traytuf	PET	2.2	M&G	www.mgpolymers.com
Trefsin	PP+IR	1.1	AES	www.santoprene.com
Trespaphan	PP	3.3	Treofan	www.treofan.com
Trevira	PET	3.6	Invista	www.invista.com
Triax	ABS+PA	2.2	Lanxess	www.lanxess.com
Tribit	PBT	2.2	Sam Yang	www.samyang.com
Triloy	PC+ABS	2.2	Sam Yang	www.samyang.com
Tripet	PET	2.2	Sam Yang	www.samyang.com
Trirex	PC	2.2	Sam Yang	www.samyang.com
Trivin	PVC	2.2	Vi-Chem Corporation	www.vichem.com/
Trivoltherm	EP	2.3	Krempel	www.krempel.com
Trixene	PMMA+PUR, PUR-R	1.1	Baxenden	www.baxchem.co.uk
Trocal	PVC	3.1, 3.3	HT Troplast	www.ht-troplast.com
Trocellen	PE, PE-X	3.7	HT Troplast	www.ht-troplast.com
Trogamid	PA 6/6-T	2.2	Degussa	www.degussa.de
Trolac	ABS, ABS+PA, ABS+PC, ABS+SAN	2.2	Godiplast	www.godiplast.com
Trosifol	PVB	3.3	HT Troplast	www.ht-troplast.com
Trovidur	PE, PP	3.3, 3.4	Metzeler	www.metzelerplastics.de

Continued on next page

Tradename	Polymer	Delivery Form	Manufacturer Supplier	Web Address
Trycite	PS	3.3	Dow Plastics	plastics.dow.com/
Tufel	Q	1.1	GE Silicones	www.gesilicones.com
Tuffak	PC, PMMA	2.2	Atoglas	www.atoglas.fr
TUFLIN	PE-LLD	2.2	Dow Plastics	plastics.dow.com/
Tufnol	EP, MF, PF, PI, SI	3.5	Tufnol	www.tufnol.co.uk
Tufpet	PBT	2.2	Mitsubishi Rayon	www.mrc.co.jp
Tufprene	SBS	2.2	Asahi	www.asahithermofil.com
Tufset	PUR	3.1, 3.4	Tufnol	www.tufnol.co.uk
Tuf-Stif	PVC	2.2	Georgia Gulf	www.ggc.com/
Tuftec	SEBS	2.2	Asahi	www.asahithermofil.com
Twaron	PPTA	3.6	Teijin	www.teijin.co.jp
Tygon	PVC	3.2	Saint Gobain - Norton	www.plastics.saint-gobain.com/
Tygothane	PUR	3.2	Saint Gobain - Norton	www.plastics.saint-gobain.com/
TYLON	PA66	2.2	Tyne Plastics LLC.	www.tyne.com/
TYNAB	ABS	2.2	Tyne Plastics LLC.	www.tyne.com/
TYNE	ABS	2.2	Tyne Plastics LLC.	www.tyne.com/
TYNEA	POM	2.2	Tyne Plastics LLC.	www.tyne.com/
TYNEC	PC	2.2	Tyne Plastics LLC.	www.tyne.com/
TYNEL	PSU	2.2	Tyne Plastics LLC.	www.tyne.com/
TYNELOY	ABS+PC	2.2	Tyne Plastics LLC.	www.tyne.com/
TYNEP	PET	2.2	Tyne Plastics LLC.	www.tyne.com/
Tynex	PA	3.6	DuPont	www.dupont.com
Typar	PP	3.6	DuPont	www.dupont.com
Tyril	SAN	2.2	Dow	www.dow.com
Tyrin	PE-C	2.2	DuPont Dow Elastomers	www.dupont-dow.com
Tyvek	PE	3.6	DuPont	www.dupont.com
U Polymer	PAR	2.2	Unitika America Corporation	www.unitika.co.jp/e/
Ubatol	PMMA	1.2	DuPont	www.dupont.com
UBE	PA6	2,2	UBE Industries, Ltd.	www.ube-ind.co.jp/english/index.htm
Ube Nylon	PA 12, PA 6, PA 66, PA-Cop	2.2	Ube	www.ube.com
Ubepol	BR	2.2	Ube	www.ube.com
Ubesta	PA	2.2	Ube	www.ube.com

Continued on next page

D.2 Tradenames Table

Tradename	Polymer	Delivery Form	Manufacturer Supplier	Web Address
Udel	PSU	2.2	Solvay	www.solvay.com
Ultem	PEI, PEI+PCE	2.2	GEP	www.geplastics.com
Ultra Wear	PE, PP	3	Polytron	www.polytron-gmbh.de
Ultracast	PUR-R	1.1	Baxenden	www.baxchem.co.uk
Ultracel	PUR-R	1.1	Bayer	www.bayer.com
Ultradur	PBT	2.2	BASF	www.basf.com
Ultraform	POM	2.2	BASF	www.basf.com
Ultralen	PP	3.3	Lofo	www.lofo.de
Ultralight	PVC	3.3	PolyOne Th. Bergmann	www.polyone.com
Ultramid	PA 6, PA 66	2.2	BASF	www.basf.com
Ultraprene	PVC	2.2	Teknor Apex	www.teknorapex.com
Ultrashield	PVC	3.3	PolyOne Th. Bergmann	www.polyone.com
Ultrason	PESU, PSU	2.2	BASF	www.basf.com
Ultrastyr	ABS	2.2	EniChem	www.eni.it/home/home_en.html
Ultrathene	EVAC	2.2	Equistar	www.equistar.com
ULTRATRAC	Polyester, Thermoset	3.4	Haysite Reinforced Plastics	www.haysite.com/
Ultzex	PE-LLD	2.2	Mitsui	www.mitsui.co.jp
UNIBRITE	AES	2.2	Nippon A&L Inc.	www.n-al.co.jp/ing/index.html
Unichem	PVC	2.2	Colorite Plastics Company	www.coloriteplastics.com/
Uniclene	PE	2.2	Mitsui	www.mitsui.co.jp
Unidene	CR	1.3	Mitsui	www.mitsui.co.jp
Unifill-60	PE-HD	2.2	North Wood Plastics, Inc.	www.northwoodplastics.com/
Uniflex	TPC	2.2	PTS	www.pts-marketing.de
Union	PE-LD	2.2	Dow Plastics	plastics.dow.com/
UNIPOL	PE-LD	2.2	Dow Plastics	plastics.dow.com/
Uniprene	TPV	2.2	Teknor Apex	www.teknorapex.com
Uniset	PE	2.2	Teknor Apex	www.teknorapex.com
Unithane	PUR	1.1	Teknor Apex	www.teknorapex.com
UNIVAL	PE-HD	2.2	Dow Plastics	plastics.dow.com/
Uralane	PUR	2.2	Ciba Specialty Chemicals	www.cibasc.com/
Uralite	PUR	2.2	Fiber Resins	None
Uravin	PVC	2.2	Vi-Chem Corporation	www.vichem.com/
Urecoll	MF, UF	1.1, 1.3	BASF	www.basf.com
Urecom	PUR-R	1.1	Coim	www.coim.it
Urepan	PUR-R	2.2	Bayer	www.bayer.com
Urexter	PUR-R	1.1	Coim	www.coim.it
Urochem	UF	2.2	Chemiplastica	www.chemiplastica.it
Uroform	UF	2.2	Chromos	www.chromos.hr
Uroplas	UF	2.2	Sprea	www.sirindustriale.com

Continued on next page

Tradename	Polymer	Delivery Form	Manufacturer Supplier	Web Address
Utec	PE-UHMW	2.1, 2.2	Braskem	www.braskem.com.br
Valite	Phenolic	2.2	Plastics Engineering Co.	www.plenco.com/
Valmax	PP	2.2	PolyOne Th. Bergmann	www.polyone.com
Valox	PBT, PBT+PET, PC+PBT, PET	2.2	GEP	www.geplastics.com
Valtec	PP	2.2	Basell	www.basell.de
Valtra	PS	2.2	Chevron	www.chevrontexaco.com
Vamac	ACM	2.2	DuPont	www.dupont.com
Vamptech	PA 6, PA 66	2.2	Albis	www.albis.com
Vandar	PBT	2.2	Ticona	www.ticona.com
Vatar	ECTFE	2.2	Solvay	www.solvay.com
Vaycron	TPE	2.2	Norsk Hydro	www.hydro.com
Vector	SBS, SIS	2.2	Exxon	www.exxonchemical.com
Vectra	LCP	2.2	Ticona	www.ticona.com
Vekton	PA6	preformed parts	Ensinger Inc.	www.shopforplastics.com/
Velkor	PS, PVC, PVC-P	3.3, 3.4	Alkor	www.alkor.de
Venipak	PVC	3.3	Alkor	www.alkor.de
Verdur	EP	2.3	Krempel	www.krempel.com
Verex	PS	2.2	Nova	www.novachem.com
Veroplas	PA 6, PA 66	2.2	Nova	www.novachem.com
Versaflex	TPS, TPU, TPV	2.2	GLS	www.glscorp.com
Versalloy	TPV	2.2	GLS	www.glscorp.com
VersaTray	PET	2.2	Voridian Company, a division of East	www.eastman.com/About_Eastman/Division
Versify	EPDM	2.2	Dow	www.dow.com
Versollan	TPU	2.2	GLS	www.glscorp.com
Verton	PA 66, PP, PPA	2.2	LNP	www.lnp.com
Vespel	PI	3, 3.4	DuPont	www.dupont.com
Vestamid	PA 12	1.5, 2.1, 2.2	Degussa	www.degussa.de
Vestenamer	TPO	2.2	Degussa	www.degussa.de
Vestodur	PBT	2.2	Degussa	www.degussa.de
Vestolen	PE-HD, PP	2.2	Sabic	www.sabic.com
Vestolit	PVC	1.1, 2.2	Vestolit	www.vestolit.de
Vestoplast	PAO	2.2	Degussa	www.degussa.de
Vestoran	PPE	2.2	Degussa	www.degussa.de
Vestosint	PA 12	2.1	Degussa	www.degussa.de
Vestowax	PE	1.5	Degussa	www.degussa.de
Vestyron	PS	2.2	BP	www.bppetrochemicals.com

Continued on next page

D.2 Tradenames Table

Tradename	Polymer	Delivery Form	Manufacturer Supplier	Web Address
Vexel	PBT	2.2	Custom Resins Group	www.customresins.com/
Vialkyd	UP	1.1	Lukoil	www.lukoil.bg
Vibrathane	PUR	liquid	Uniroyal Chemical Group	www.chemtura.com/chem/default.htm
Vi-Chem	PVC	2.2	Vi-Chem Corporation	www.vichem.com/
Viclon	PVC	3.6	Kureha	www.kureha.co.jp
Vicotex	EP, PF	2.3	Hexcel	www.hexcel.com
Victrex	PEEK	2.2	Victrex	www.victrex.com
Vidlon	PA	3.6	Lukoil	www.lukoil.bg
Vidux	PVC	2.2	Teknor Apex	www.teknorapex.com
Vilit	PVDC, VC-Cop	1.2	Degussa	www.degussa.de
Villpet	PET	3.3	Klöckner-Pentaplast	www.kpfilms.com
Vinac	PVAC	1.1	Air Products	www.airproducts.com
Vinalkyd	UP	2.2	Lukoil	www.lukoil.bg
Vinidur	VC-Cop	2.2	Solvin	www.solvinpvc.com
Vinika	VC-Cop	2.2	Schulman	www.aschulman.com
Vinnapas	E-Cop, PVAC	1.2, 1.3, 2.2	Wacker	www.wacker.com
Vinnol	PVC	2.2	Wacker	www.wacker.com
Vinnolit	PVC, VC-Cop	2.2	Vinnolit	www.vinnolit.de
Vinofan	PVAC	1.2	BASF	www.basf.com
Vinoflex	PVC	2.2	Solvin	www.solvinpvc.com
Vinophane	PVC-P	3.3	Solvin	www.solvinpvc.com
Vinuran	VC-Cop	2.2	Solvin	www.solvinpvc.com
Vinychlore	PVC-P	2.2	Solvin	www.solvinpvc.com
Vinyl	PVC	2.2	Vinyl Solutions, L.L.C.	www.kpafilms.com/KCC/info_pages/vinyls
Vinylbond	PVC	2.2	Colorite Plastics Company	www.coloriteplastics.com/
Vinylec-F	PVFM	2.2	Chisso	www.chisso.co.jp
Vinyloy	Acrylic+PVC	3.4	Kleerdex Company	www.kydex.com/
Vipolit	PVAC	1.2	Wacker	www.wacker.com
Vipophan	PVC	3.3	Lofo	www.lofo.de
Virgaloy	PC+Acrylic Alloy	-	MRC Polymers, Inc.	www.mrcpolymers.com/
Viscacelle	CA	3.3	Lofo	www.lofo.de
Viscolas	TSU (Polyurethane Thermoset Elastomer)	3.4	E-A-R Specialty Composites	www.earsc.com/
Vista	PVC	-	CONDEA Vista Company	www.ggc.com/

Continued on next page

Tradename	Polymer	Delivery Form	Manufacturer Supplier	Web Address
Vistaflex	TPO	2.2	AES	www.santoprene.com
Vistalon	EPDM	2.2	Exxon	www.exxonchemical.com
Vistamax	EPDM, EPM	2.2	Exxon	www.exxonchemical.com
Vistanex	PIB	1.5	Exxon	www.exxonchemical.com
Vistel	PVC	-	Georgia Gulf (CONDEA Vista Company)	www.ggc.com/
Vitacom	TPO, TPS	2.2	British Vita	www.britishvita.com
Vitafoam	PUR	3.7	British Vita	www.britishvita.com
Vitapol	PVC	2.2	British Vita	www.britishvita.com
Vitaprene	TPE	2.2	British Vita	www.britishvita.com
Vitawrap	PUR-R	3.7	British Vita	www.britishvita.com
Vitax	ASA	2.2	Hitachi Chemical Co., Ltd.	www.hitachi-chem.co.jp/english/index.h
Vitel	PTP	2.2	Bostik	www.bostikfindley-us.com
Vitiva	PET	2.2	Eastman	www.eastman.com
Viton	FKM	2.2	DuPont Dow Elastomers	www.dupont-dow.com
Vivac	PET	3.4	Bayer Sheet	www.bayersheeteurope.com
Vole	PF	3.5	Tufnol	www.tufnol.co.uk
Voloy	PP	2.2	ComAlloy International Corporation	
Voltalef	ECTFE	2.1	Atofina	www.atofina.de
Voranate	PUR-R	1.1	Dow	www.dow.com
Voranol	PUR-R	1.1	Dow	www.dow.com
Voridian	PET	2.2	Eastman	www.eastman.com
Vorvel	EMA, PA 11, PE, PP, PTP	2.1	Rohm & Haas	www.rohmhaas.com
Vova Tec	PE-HD	2.2	Rohm & Haas	www.rohmhaas.com
VPI	PS-HI	3.4	VPI, LLC	www.vpicorp.com/
Vulkollan	PUR	1.1	Bayer	www.bayer.com
Vybex	PBT	2.2	Ferro Corporation	www.ferro.com/
Vydyne	PA 66	2.2	Dow	www.dow.com
VYFLEX	TPO	2.2	Lavergne Group	www.lavergneusa.com/
VYLENE	PP-Cop.	2.2	Lavergne Group	www.lavergneusa.com/
Vynaloy	PVC	3.3	PolyOne Th. Bergmann	www.polyone.com
Vynaprene	PVC	2.2	Georgia Gulf	www.ggc.com/
Vynite	PVC+NBR	2.2	Alpha Gary Corporation	www.alphagary.com/
VYPET	PC+PET	2.2	Lavergne Group	www.lavergneusa.com/
VYPRENE	TPE	2.2	Lavergne Group	www.lavergneusa.com/
VYPRO	PP-Cop.	2.2	Lavergne Group	www.lavergneusa.com/
Vyram	CR, EPDM	2.2	AES	www.santoprene.com
VYTEEN	ABS	2.2	Lavergne Group	www.lavergneusa.com/

Continued on next page

D.2 Tradenames Table

Tradename	Polymer	Delivery Form	Manufacturer Supplier	Web Address
VYTEK	PE-LLD	2.2	Chevron Phillips Chemical Co.	www.cpchem.com/
Vythene	PVC+TPU	2.2	Alphagary	www.alphagary.com
Wacker Sil Ge	SI	1.1	Wacker	www.wacker.com
Wacosit	EP	2.3, 3.1	Krempel	www.krempel.com
Walocel	CMC, EC, MC	1.1	Wolff Walsrode	www.wolff-walsrode.de
Walomelt	PUR	3.3	Wolff Walsrode	www.wolff-walsrode.de
Walomer	PAA	1.3	Wolff Walsrode	www.wolff-walsrode.de
Walomid-Combi	PE, PP, PVAL	3.3	Wolff Walsrode	www.wolff-walsrode.de
Walopur	PUR	3.3	Wolff Walsrode	www.wolff-walsrode.de
Waloran	CN	1.1	Wolff Walsrode	www.wolff-walsrode.de
Walotex	PUR	3.3	Wolff Walsrode	www.wolff-walsrode.de
Walothen	PP	3.3	Wolff Walsrode	www.wolff-walsrode.de
Waterpet	PET	2.2	Selenis	www.selenis.com
WearRing	PEEK	3.2	Greene, Tweed Engineered Plastics	www.gtweed.com/
Wellamid	PA	2.2	CP-Polymer	www.cp-polymer-technik.de
WEP	Polyester, Thermoset	2.2	Ashland Chemical Company	www.ashchem.com/
Westlake	PK	3.1, 3.3, 3.4	Westlake Plastics Company	www.westlakeplastics.com/
Whale	PF	3.5	Tufnol	www.tufnol.co.uk
Witt	PS-HI	3.4	Witt Plastics Inc.	www.kpfilms.com/index_en.asp
Wonderlite	PC	2.2	Chi Mei	www.chimei.com.tw
Wonderloy	PC+ABS	2.2	Chi Mei	www.chimei.com.tw
Wood-Stock	PP	3.5	Solvay	www.solvay.com
Wopadur	PVC	3.3, 3.4	Gurit	www.gurit-worbla.com
Wopavin	PVC	3.3, 3.4	Gurit	www.gurit-worbla.com
Worblex	ABS, CA, CP, PO, PS	3.3	Gurit	www.gurit-worbla.com
WPP	ABS	2.2	Washington Penn Plastic Co. Inc.	www.washpenn.com/default2.htm
Xantar	PC, PC+ABS	2.2	DSM Eng. Plastics	www.dsmep.com
Xarec	PS	2.2	Idemitsu	www.idemitsu.co.jp
Xavan	PP	3.6	DuPont	www.dupont.com
Xenalak	PMMA	2.2	Baxenden	www.baxchem.co.uk
Xenoy	PC+PBT, PC+PET	2.2	GEP	www.geplastics.com
Xmod	PP	2.2	Borealis	www.borealis.com
XT	PMMA	2.2	CYRO Industries	www.cyro.com/NewCYRO/flash.html

Continued on next page

Tradename	Polymer	Delivery Form	Manufacturer Supplier	Web Address
Xtel	PPS	2.2	Chevron Phillips	www.cpchem.com
Xtel	PPS	2.2	Chevron Phillips Chemical Co.	www.cpchem.com/
Xydar	LCP	2.2	Solvay	www.solvay.com
Xylex	PC-Blend	2.2	GEP	www.geplastics.com
Xyron	PPE+PA, PPE+PP, PPE+PS	2.2	Asahi	www.asahithermofil.com
Yery-or	PVC-P	3.3	Asahi	www.asahithermofil.com
Yparex	E-Cop	2.2	DSM Eng. Plastics	www.dsmep.com
Zelux	PC	3.1	Westlake Plastics Company	www.westlakeplastics.com/
Zemid	PE	2.2	DuPont Engineering Polymers / Canada	www.dupont.ca/
Zenite	LCP	2.2	DuPont	www.dupont.com
Zeonex	COC	2.2	Zeon	www.zeon.co.jp
Zeonor	COC	2.2	Zeon	www.zeon.co.jp
Zetpol	HNBR	2.2	Zeon	www.zeon.co.jp
Zonyl	PTFE	2.1	DuPont	www.dupont.com
Zylar	SBMMA	2.2	Nova	www.novachem.com
Zyntar	PS-HI	2.2	Nova	www.novachem.com
Zytel	PA 6, PA 612, PA 66, PPA	2.2	DuPont	www.dupont.com

INDEX

A

abrasion, 710
abrasion
   table, 769
abrasive wear, 307
ABS, 550
ABS+PA (PA6), 550
ABS+PC, 550
ACM, 694
acoustic properties, 266
   pressure effect, 266
   sound absorption, 268
   sound reflection, 266
   sound transmission, 268
   speed of sound, 266
   temperature effect, 266
acronyms
   plasticizers, 14
   plastics, 8
   polymer characteristics, 13
acrylate rubber, 694

acryloyl modified phenols, 704
additives, 701
adhesive bonding, 727
AECM, 693
AEM, 694
AFMU, 698
agglomerates, 274
alcohol ethoxylates, 708
aliphatic polyester, 620
aliphatic polyketones, 662
   applications, 664
   chemistry, 662
   processing, 663
   properties, 663
alkyl phosphites stabilizers, 706
alkyl radical scavengers
   acryloyl modified phenols, 704
   benzofuranone derivatives, 704
   hindered amine stabilizers, 704
   hydroxyl amines, 704
alkylphosphates, 707

alkylsulfonates, 707
alkyltim stabilizers, 706
aminoplastic molding compounds, 679
aminoplastics, 684
  applications, 684
amorphous polymers, 40
anisotorpy predictions, 479
anisotropy, 494
ANM, 694
annular snap fit, 500
antiblocking agents, 701
  limestone, 701
  natural silica, 701
  talc, 701
  zeolityes, 701
antifogging agents, 708
  alcohol ethoxylates, 708
  glycerol esters, 708
  nonyl phenol ethoxylates, 708
  polyglycerol esters, 708
  sorbitan esters, 708
antimicrobial agents, 707
antioxidants, 703
  alkyl radical scavengers, 704
  hydrogen donors, 703
  hydroperoxide decomposers, 703
  metal deactivators, 704
antistatic agents, 706
  alkylphosphates, 707
  alkylsulfonates, 707
  diethanolamides, 707
  ethoxylated alcohol, 707
  ethoxylated alkylamines, 707
  fatty acid esters, 707
applicatiions
  polysilicooxoaluminate, 664
applications
  aliphatic polyketones, 664
  aminoplastics, 684
  butadiene rubber, 690
  cellulose, 651
  chloroprene rubber, 691
  EPDM, 693
  epichlorohydrine, 696
  ethylene-chlorotrifluoroethylene colopymer, 574
  ethylene-tetrafluoroethylene copolymer, 575
  fluorine rubber, 695
  impact resistant PMMA, 582
  liquid crystalline polymers, 632
  methacrylate butadiene styrene copolymers, 581
  methylmethacrylate acrylonitrile butadiene styrene copolymers, 581
  methylmethacrylate copolymers, 581
  methylmethacrylate exo-methylene lactone copolymer, 584
  nitrile-butadiene rubber, 691
  perfluoroalkoxy, 575
  perfluoropropylvinylether copolymer, 575
  PMMA+PC blends, 584
  poly-m-phenylene isophthalamide, 599
  polyamides, 592
  polyamidimide, 627
  polyarylate, 612
  polyarylsulfone, 617
  polybismaleinimide, 626
  polybutene, 544
  polybutylene naphthalate, 612
  polybutylene teraphthalates, 609
  polychlorotrifluoroethylene, 574
  polyetherimide, 628
  polyethylene naphthalate, 613
  polyethylene teraphthalates, 606
  polyfluoroethylenepropylene, 575
  polyglycols, 622
  polyisobutene, 544
  polymethacrylate methylimide, 629
  polymethacrylimide, 629
  polymethacrylmethylimide, 582
  polymethylmethacrylate, 578
  polyoxymethylene, 585
  polyphenylene ether blends, 618
  polyphenylene sulfide, 615
  polypropylene, 538
  polystyrene copolymers/blends, 550
  polystyrene foams, 554
  polytetrafluoroethylene, 573
  polyvinyl carbazole, 568
  polyvinyl fluoride, 574
  Polyvinyl Methyl Ether, 568
  polyvinylchloride copolymers/blends, 565
  polyvinylchloride-plasticized (PVC-P), 562
  polyvinylidene fluoride, 573
  propylene oxide rubber, 696
  PTFE-AF copolymers, 576
  tetrafluoroethylene-hexa-fluoropropylene-copolymers, 575
  tetrafluoroethylene-hexafluoropropylene-vinylidenefluoride copolymer, 576
thermoplastic, 4
thermoset, 4

INDEX

TPE, 666
U-rubbers, 698
unsaturated polyester, 684
aramides, 599
aromatic amines, 703
aromatic diols, 610
aromatic polyesters, 599
aromatic polyether, 617
aromatic polyimides, 622
aromatic polysulfides, 613
aromatic polysulfones, 613
aromatic saturated polyesters
   chemistry, 599
ASA, 550
ASA+PC, 550
ASTM test
   150, 227
   D 1238, 64, 116
   D 149, 243
   D 150, 225
   D 1525, 64, 194
   D 1822, 152
   D 1895, 66
   D 256, 64, 147
   D 2863, 220
   D 2990, 65, 160
   D 3028, 65
   D 3417, 66
   D 3418, 66, 95–96
   D 3638, 243
   D 3763, 64
   D 3801, 220
   D 3835, 66
   D 4473, 67
   D 5045, 65
   D 5279-150, 65
   D 570, 64, 209
   D 5930, 66
   D 635, 221
   D 638, 64–65, 128
   D 648, 64, 194
   D 695, 65
   D 790, 139
   D 792, 64, 66, 81
   D 955, 64, 67, 90
   E 831, 64, 67, 88
AU, 698
auger mixer, 276

B

ball mills, 274
barrier flights, 303
benzofuranone derivatives, 704
benzylcellulose, 31
BIIR, 692
biodegradable polymers, 658
biopolymers, 650
Bird-Carreau model, 111
Bird-Carreau-Yasuda model, 110
birefringence, 251, 459
bismuth-tin alloy core, 330
bladder press, 376
bleeding silicone rubber, 354
blends, 53, 273
   polypropylene, 541
blister ring mixing section, 281
block head mixing section, 278
blooming, 355
blow molding
   blow-outs, 395
   bubbles, 393, 395
   cloudy parison, 393
   curls, 393
   curves, 393
   cycle time, 394
   die lines, 393
   excessive shrinkage, 395
   flash, 394
   injection blow molding, 395
   intermittent feed, 394
   melt fracture, 393
   neck underfills, 395
   non-uniformity, 395
   orange peel, 395
   partial handle, 394
   polyvinylchloride-plasticized (PVC-P), 564
   poor pinch-off, 395
   sag, 393
   scratches, 393
   smoking, 393
   streaks, 393
   thick bottom weld, 394
   thin bottom weld, 394
   troubleshooting, 392
   warpage, 395
   wrinkles, 393
blowing agents, 709
blush, 356
BMC, 369, 371, 676
bonding, 727
BR, 690
branching, 19
bridging, 304
bulk molding compound, 369

bulk molding compounds, 676
butadiene rubber, 690
  applications, 690
butt welding, 413
  heated wedge lamination, 417
  hot tool socket, 416
  pipes, 417
  process description, 414
butyl rubber, 692

C

CA, 650
CAB, 650
CAD, 454
calendering, 400
CAMPUS, 8, 456, 64
cantilever beam, 489
CAP, 650
capillary viscometer, 116
caprolactam, 590
carbon black, 705
carbonates, 710
carboxylic acids, 610
carboxymethyl cellulose, 654
carboxymethyl celluloses, 657
carboxymethylcellulose, 31
cascade extruders, 296
casein, 656
cast film production, 388
cavity transfer mixing section, 280
Cedluka foamed extrusion die, 313
cellulose derivates, 650
cellulose hydrate, 655
cellulose nitrate, 654
cellulose propionate, 654
cellulose, 31
cellulose(tri)acetate, 651
cellulose
  applications, 651
  chemistry, 650
  processing, 651
  properties, 651
celluloseacetate, 31
celluloseacetobutyrat, 31
celluloseacetobutyrate, 654
celluloseacetopropionate, 31
celluloseester, 31
celluloseether, 31
cellulosenitrate, 31
cellulosepropionate, 31
cellulosetriacetate, 31
central gates, 324

ceramics, 664
CF, 670, 683
CH, 650
Chalking, 711
char, 102
Charpy impact test, 152
check valve, 319
chemical blowing agents, 709
  azodicarbonamide, 709
  carbonates, 710
  hydrazine derivatives, 709
  nitroso compounds, 710
  semicarbazides, 709
  tetrazoles, 710
chemical degradation, 214
chemical resistance
  table, 760
chemistry, 17
chlorinated PE rubber, 694
chlorinated PE, 524
chloroprene rubber, 691
  applications, 691
chlorosulfonated PE rubber, 694
choker ring with holes mixing section, 278
chromaticity diagram, 264
CIIR, 692
clamping force prediction, 480
clamping unit, 321
cleaners, 729–730
clogging, 304
CM, 694
CMC, 650, 657
CN, 650
CO, 695
co-injection, 326
co-kneader, 277
co-rotating twin screw extruders, 296
coat-hanger die, 309
coating, 402
  curtain coating, 405
  forward coating, 404
  knife coating, 403
  laminating, 405
  melt roll coating, 405
  metallizing, 406
  plasma coating, 407
  reverse coating, 404
  slide coating, 405
  wire coating, 403
coefficient of friction, 205
  table, 773
cold press forming, 376

INDEX

cold slug, 362
Color Index (CI), 710
color, 262
colorants, 710
colorimeter, 263
comparability, 63
complex modulus
   elastic modulus, 175
   loss modulus, 175
   storage modulus, 175
complex shear modulus, 174,
composites processing, 381
   filament winding, 384
   helical winding, 385
   LCM, 382
   pultrusion, 385
   resin infusion process (RIP), 384
   robot filament winding, 386
   RTM, 382
   SRIM, 382
composites, 54
   Halpin-Tsai model, 496
compounders, 275
compression molding, 369
   abrasion, 381
   applications, 369
   blisters, 378
   BMC, 369, 371
   charge pattern, 480
   chipped edges, 378
   contamination, 378
   cracking, 378
   die forging, 377
   dieseling, 378
   drag marks, 378
   drop forging, 377
   dull part surface, 378
   fiber jamming, 380
   fiber orientation, 464
   fiber-matrix separation, 380
   filler pull, 379
   flash, 379
   flexform press, 376
   flow lines, 379
   fluid cell press, 376
   GMT process, 374
   GMT processing cycle, 375
   GMT, 369, 373
   heat smear, 379
   laking, 379
   LFT process, 375
   LFT, 369, 373
   molds, 370
   mottling, 379
   non-fills, 381
   paint popping, 380
   porosity, 380
   pre-gel, 380
   press, 372
   process simulation, 480
   process, 370, 372
   read-out, 380
   separation/phasing, 380
   short shots, 381
   sink marks, 380
   SMC production, 371
   SMC, 369, 371, 463
   SMCprocessing cycle, 373
   sticking part, 380
   surface waviness, 381
   troubleshooting, 377
   under-cure, 381
   voids, 381
   warpage predictions, 484
   warpage, 381
concurrent engineering, 452
condensation polymerization, 21
conductive additives
   resistance, 740
conductive components, 740
conductive polymers, 659, 747
   chemistry, 659
   electrical properties, 746
   properties, 661
conductivity, 67
cone-plate, 118
configuration, 24
conformation, 24
conical fracture criterion, 492
conical twin screw extruders, 297
consistency index, 110
contact angle, 122
contamination, 355
continuous discharge screw, 287
continuous kneader, 287
conventional extruder, 299
cooled pulverizing system, 291
cooling jacket, 305
cooling lines, 322
copolyamides, 667
copolyester, 667
copolymerization
   bipolymer, 18
   block, 18

graft, 18
random, 18
copolymers, 20
　PP, 539
core pin bending/shift, 358
cost break down, 455
Couette, 118
Coulomb's law of friction, 205
counter helix slotted flight mixing section, 278
counter-rotating twin screw extruders, 296
CR, 691
creep data
　fracture strain, 172
　isochronous, 165
　isometric, 165
　PBT, 163
　PMMA, 169
　PP, 163
　secant modulus, 165
　thermoplastics, 165
creep rupture, 165
　temperature effect, 169
　thermoplastics, 169
creep test, 160
creep
　isochronous curves, 488
　isometric curves, 487
cross head tubing die, 311
cross-linked polymers, 51
crosslinked PE, 523
crystal, 37
crystallinity
　degree, 75
crystallization, 36
　degree, 39
　speed, 39
CSM, 694
CTA, 650
cure, 99
　degree of, 99
　diffusion controlled, 102
　heat activated, 99
　mixing activated, 99
　phase, 100
curing, 99
　rheology, 112
　SMC, 469
　UPE, 468
curtain coating, 405
cycloolefine, 31

D

damping properties, 268
data banks
　materials, 456
　product, 452
Deborah number, 107
deformable toys, 529
deformation, 464
degassing, 295
degradable PE, 524
degree of crystallinity, 94
dehumidifier drying system, 293
density, 76
　measurement, 98
design philosophy, 451
design
　dimensional requirements, 453
　environmental conditions, 453
　FDA, 454
　loading, 453
　material selection, 455
　process selection, 458
　regulations, 454
　requirements, 453
　standards, 454
diallyl phthalate resins, 675
diallyl phthalate, 681
diallylphthalate, 686
diaphragm gate, 325
dicyclopentadiene, 31
die characteristic curves, 299, 301
die drool, 308
die forging, 377
die lines, 306, 308
dielectric behavior, 223
dielectric dissipation, 227
dielectric strength data, 236
　film thickness effect, 242
　PE, 239
　PE-LD, 237
　PF paper, 241
　plasticizer effect, 242
　POM, 239
　PP, 242
　PVC-P, 241
dielectric strength, 238
　thickness dependance, 236
　load time dependance, 237
dieseling, 378
diethanolamides, 707
differential scanning cal-orimeter, 93
differential thermal analysis, 92

diffusion activation energy, 199
diffusion resistance
    table, 766
diffusion, 82, 195
    table, 751
    temperature effect, 199
diffusivity, 82
dimensioning, 453
diode laser welding, 726
dispersive mixing, 280
DMC, 676
double auger mixer, 276
double block head mixing section, 278
double refractance, 251, 459
dough molding compound, 676
draw resonance, 308
driving cylinder, 316
drop forging, 377
drop impact test, 157
drying parameters, 292
drying
    dehumidifier, 293
    in-hopper probe, 293
    recommended drying parameters, 292
DSC, 93
DTA, 92
dynamic friction
    table, 773
dynamic mechanical analysis (DMA), 172

E

EAA, 531
EAM, 694
EAMA, 531
EBA, 531
EC, 650
ECB, 533
ECB/TPO blend, 533
ECO, 695
EEAK, 531
EIM, 532
Einstein's model, 115
ejection pin, 322
ejection plate, 322
ejection ring, 322
ejector bolt, 316
ejector pin, 316
ejector plate, 316
ekonol, 633
elastic effects, 103
elastic modulus
    thermoplastics, 175

electric breakdown, 235
electric charge, 748
electric conductivity, 229
electrical properties, 223
    conductive fillers, 235
    conductive polymers, 746
    dielectric behavior, 223
    dielectric coefficient, 223
    dielectric dissipation factor, 227
    dielectric polarization, 225
    electric conductivity, 229
    electric resistance, 229
    electrical and thermal loss, 227
    filler effect, 233
    frequency dependency, 226
    table, 735
    temperature effect, 226
electrofusion, 442
electromagnetic interference shielding (EMI), 247
electron beam cross-linking PE, 523
electronegativity, 28
electrostatic charge, 245
elongational deformation, 103
elongational viscosity, 111
EMA, 531
EMI shielding, 247
energy content, 745
Engel process, 523
engineering design, 452
enthalpy of fusion, 94
environmental design conditions, 453
environmental stress cracking, 216, 759
    table, 757
EP, 670, 673, 681, 686
EPDM, 540, 693
    applications, 693
epichlorohydrine, 695
    applications, 696
EPM, 693
epoxy resin, 681
epoxy resins, 673, 686
EPP, 536
erucamide, 702
ESC, 757, 759
ET, 697–698
ETER, 695
ethoxylated alcohol, 707
ethoxylated alkylamines, 707
ethyl cellulose, 654
ethylcellulose, 31
ethylene copolymers, 525

884 INDEX

cycloolefin copolymers, 531
ethylene copolymer/bitumen blends, 533
ethylene-acrylic copolymers, 531
EVAC, 528
EVAL, 531
ionomer copolymers, 532
PE-α-olefin copolymers, 531
ultra-light PE, 525
ethylene vinyl acetate copolymer (EVAC), 528
ethylene vinyl acetate copolymer
  applications, 529
  deformable toys, 529
  films, 529
  gaskets, 529
  wire coating, 529
ethylene-acrylic copolymers, 531
ethylene-acrylic ester rubber, 693
ethylene-chlorotrifluoroethylene colopymer, 574
  applications, 574
ethylene-propylene (diene) copolymers (EPDM), 540
ethylene-propylene-(diene) rubber, 693
ethylene-tetrafluoroethylene copolymer, 575
  applications, 575
ethylene-vinylacetate rubber, 694
ethylene/vinyl alcohol copolymers (EVAL), 531
ethylidennorbomene, 31
EU, 698
EVAL, 531
exothermic reaction, 95
extrudate swell, 108
extruder, 278
extrusion blow molder, 392
extrusion blow molding, 389
extrusion die, 107
extrusion dies, 308
extrusion welding, 437
  process description, 439
extrusion, 294
  clearance, 298
  plasticization, 297
  abrasive wear, 307
  barrier flights, 303
  bridging, 304
  bubbles in part, 306
  bubbles in product, 308
  calibration system, 312
  cascade system, 296
  cast film production, 388
  Celuka-type die, 313
  channel depth, 298

  clogging, 304
  co-extrusion system, 392
  coat-hanger die, 309
  compression ratio, 298
  contamination, 305, 307
  conventional extruder, 299
  cooling jacket, 305
  corrosive wear, 307
  cross-heat tubing die, 311
  degassing, 295
  degradation, 306
  die characteristic curves, 299, 301
  die drool, 308
  die lines, 306, 308
  dimensions, 297
  draw resonance, 308
  extrusion blow molding, 389
  extrusion area, 295
  feed hopper, 305
  film blowing, 388
  film production, 388
  fish eyes, 306
  foamed extrusion, 313
  gel formation, 307
  geometry, 298
  grooved feed extruder, 301
  grooved feed section, 301
  helix angle, 298
  Maillefer barrier flights, 303
  manifold, 310
  melt fracture, 308
  melting zone, 297, 301
  metering zone, 297
  plugging, 305
  polyvinylchloride-plasticized (PVC-P), 564
  power consumption, 275
  profiles, 309
  pumping zone, 297
  screen pack, 305
  screw characteristic curves, 299, 301
  sharkskin, 306
  sheeting dies, 309
  smooth barrel extruder, 299
  solids conveying zone, 297
  spider leg tubing die, 311
  spiral die, 312
  surging, 305
  transition zone, 297, 301
  troubleshooting, 302
  tubular dies, 310
  twin screw, 296
  vacuum sleeve, 312

INDEX 885

voids, 308
warpage, 306
wear, 307
weld lines, 308

F

failure critera
  conical fracture criterion, 492
failure criteria, 491
  Huber-van Mises-Henky, 492
  parabolic fracture criterion, 492
  uniaxial stress case, 491
fast open vat mixer, 276
fasteners, 504
fatigue data
  ABS, 182
  PA6-GF, 180
  PA66, 182, 185
  PC, 182
  POM, 180
  PVC-U, 182
  SMC, 185
fatigue test, 178
  creep, 184
  S-N curves, 178
  stress concentration effect, 180
  temperature rise, 179
  thermal failure, 180
fatty acid esters, 707
FDA, 454
FDM, 448
feed hopper, 305
FF, 670, 683
FFKM, 695
fiber attrition, 465
fiber damage, 465
  critical stress, 466
fiber length, 466
fiber orientation, 459
  SMC, 464
fiber reinforced composite theory, 495
fiber spinning, 387
fibers, 49
filament winding, 384
fillers, 712
filling pattern, 478
film blowing, 388
film gate, 325
film production, 388
films
  PE, 522
fish eyes, 306

FKM, 695
flame retardants, 704
flammability
  5VA, 221
  5VB, 221
  HB, 221
  table, 767
  UL, 221
  V-0, 221
  V-1, 221
  V-2, 221
flash point, 218
flash, 318, 360
flat roofs, 533
flexform press, 376
flexible pipes, 529
flexural test, 139
flow density meter, 99
fluid cell press, 376
fluid-assisted injection molding
  barrel cavity, 332
  filled/nearly filled mold, 332
  moving core, 332
  partially filled mold, 332
  side mold cavity, 332
fluorescent whitening agents, 712
fluorine rubber, 695
  apllications, 695
fluoro copolymers, 574
  ECTFE, 574
  ETFE, 574
  FEP, 574
  FFKM, 574
  FKM, 574
  FPM, 574
  PFA, 574
  PTFEAF, 574
  TFEHFPVDF, 574
  TFEP, 574
  THV, 574
fluoropolymers, 569
  chemistry, 569
foamed extrusion calibration, 313
foaming, 407, 709
  hard PUR foam, 409
  soft PUR foam, 409
foams
  properties, 739
Food and Drug Administration, 454
formaldehyde resins, 670
  chemistry, 670
forward coating, 404

FPM, 695
friction data, 205
  surface finish, 205
  temperature effect, 205
friction, 205
  additives, 772
  electric charge, 748
  PA6, 772
  PTFE, 772
  table, 773
fuels, 745
functional elements, 498
  annular snap fit, 500
  fasteners, 504
  living hinges, 499
  press fit assemblies, 498
  snap fit assemblies, 499
  snap fit magnification factor, 502
fungizides, 707
fused deposition modeling (FDM), 448
fusible core injection molding, 329
FZ, 699

G

GAIM cycle, 331
GAIM, 330
gas permeability
  table, 762
gas-assisted injection molding, 330
gate bushing, 324
gates, 324–325
gel formation, 307
gel point, 102
geometric isomers, 26
geometric modeling, 454
glass mat reinforced thermoplastics, 369
glass transition temperature
  table, 740
gloss, 259
glossmeter, 261
glycerol esters, 708
GMC, 676
GMT process, 374
GMT processing cycle, 375
GMT, 369, 373
grafted copolymer, 18
granulated molding compounds, 676
grooved feed section, 301
guiding pin, 322

H

Halpin-Tsai model, 496

HALS, 705
HDT
  apparatus, 189
  data, 191
heat deflection temperature, 185
heat of fusion, 75
Heat resistance, 710
heat
  released during cure, 99
heated wedge lamination, 417
helical filament winding, 385
Henry's law, 195
high temperature plastics, 175
hindered amine stabilizers, 704
HNBR, 693
Hookean solid, 108
hopper mixer, 276
hot face pelletizer, 288
hot gas welding, 435
  joint types, 437
  process types, 436
hot runner systems, 324
hot tool socket welding, 416
HPC, 657
Huber-van Mises-Henky failure criteria, 492
hydrated NBR rubber, 693
hydraulic clamping unit, 322
hydrochloric acid
  polyvinylchloride-plasticized (PVC-P), 564
hydrogen donors, 703
  aromatic amines, 703
  phenols, 703
hydroperoxide decomposers, 703
  phosphites, 703
  phosphonites, 703
  thiosynergists, 703
2-hydroxybenzophenones, 705
hydroxyl amines, 704
2-hydroxyphenylbenzotriazoles, 705
hydroxypropyl celluloses, 657,

I

identification of plastics, 508,
identification of plastics, 508
  appearance, 508
  burn test, 508
  density, 508
  elastic properties, 508
  flame color, 508
  igniteability, 508
  odor, 508
  pyrolisis, 508

INDEX

solvents, 508
identification
    polyvinylchloride-unplasticized (PVC-U), 556
    SPI, 59
IEC test
    60112, 243
    60243-1, 243
    60250, 227
    60259, 225
IIR, 692
impact energy
    filler effect, 159
impact resistant PMMA, 582
    applications, 582
impact strength, 141
    blends, 143
    filler effect, 159
    notch tip radius effect, 155
    processing conditions effect, 157
    rate of deformation effect, 142
    temperature effect, 155
    weathering effect, 211
impact test
    tensile specimen, 154
    Charpy, 152
    Izod, 152
    notched, 155, 159
    test specimen, 152
impingment mixing head, 347
implant induction welding, 439
in-hopper probe drying, 293
in-mold decoration, 336
in-mold lamination, 336
index of refraction, 251
infinite-shear-rate viscosity, 111
infrared spectrometer, 264
infrared spectroscopy, 263
injection blow molding, 395
injection molding, 314
    bismuth-tin alloy core, 330
    black specks, 354
    blisters, 355
    blooming, 355
    blush, 356
    bubbles, 356
    burns, 356
    central gates, 324
    check valve, 319
    clamping unit, 321
    classification, 314
    cloudiness, 357
    co-injection, 326
    cold slug, 362
    color variation, 357
    contamination, 355
    cooling lines, 322
    core pin bending/shift, 358
    cracking, 358
    crazing, 358
    deformation, 464
    delamination, 359
    diaphragm gate, 325
    dimensional variations, 359
    driving cylinder, 316
    ejection pin, 322
    ejection plate, 322
    ejection ring, 322
    ejector bolt, 316
    ejector pin, 316
    ejector plate, 316
    film gate, 325
    flash, 318, 360
    flow, 464
    fusible core, 329
    GAIM cycle, 331
    GAIM, 330
    gas compression cylinder, 332
    gas-assisted, 330
    gate blush, 360
    gate bushing, 324
    gate, 324
    gates, 325
    gloss, 360
    guiding pin, 322
    haze, 357
    hot runner systems, 324
    hot tip stringing, 360
    hydraulic clamping unit, 322
    in-mold decoration, 336
    in-mold lamination, 336
    injection unit drive cylinder, 315
    injection unit guide, 315
    injection-compression, 335
    liquid silicone rubber, 352
    lost core, 329
    LSR bleeding formulation, 354
    LSR mold, 353
    LSR, 352
    microcellular molding, 340
    microinjection molding, 338
    mold build-up/deposits, 361
    mold cavity, 321
    molding cycle, 314

molding diagram, 319
multi-gated co-injection, 329
multicomponent injection molding, 341
non-return valve, 361
nonfill, 364
nozzle drool, 361
nozzle, 319, 322
odor, 362
orange peel, 362
pinking, 362
pinpoint gate, 324
pitting, 363
plasticating screw, 320
plasticating unit, 319
polyvinylchloride-plasticized (PVC-P), 564
poor mixing, 363
powder injection molding, 344
pvT behavior, 472
pvT diagram, 317
racetracking, 363
reaction injection molding, 346
record grooves, 363
runner system, 323
sandwich molding, 326
screw design, 320, 363
screw recovery, 364
sequential co-injection, 327
short shot, 364
short shots, 318
shrinkage, 90, 318, 365, 472
side gates, 324
sink marks, 471
sinks, 365
soluble core, 329
special processes, 326
sprue bushing, 322
sprue sticking, 366
sprue, 323–324
sticking, 367
structural foam, 349
submarine gate, 325
tapered sprue gate, 324
three channel co-injection, 328
tie bar, 315
tie bars, 321
toggle mechanism, 315, 321
trapped air, 368
troubleshooting, 368
two channel co-injection, 328
two stage degassing screw, 320
umbrella gate, 325
warpage prediction, 483

warpage, 368
weldlines, 368
injection unit drive cylinder, 315
injection unit guide, 315
injection-compression molding, 335
inorganic pigments, 711
intermeshing twin screw extruders, 296
internal batch mixer, 285
intramolecular forces, 28
intrinsically conductive polymers (ICP), 747
ionomer copolymers, 532
IR welding, 430
IR, 689
ISO test
 1133, 64, 116
 11357, 95
 11357-2, 66, 96
 11357-3, 66
 11357-4, 66
 11357-5, 67
 11357-7, 67
 11359, 88
 11359-2, 64, 67
 11443, 66
 1183, 64, 81
 1210, 220–221
 178, 64, 139
 179-1, 147
 180, 64
 2577, 67
 294-4, 64, 67, 90
 306, 64, 194
 4589-2, 220
 527-1, 64–65, 128
 527-2, 65
 604, 65
 61, 66
 62, 64, 209
 6603-2, 64
 6721-10, 66
 6721-2, 65, 67
 6721-5, 65
 6721-7, 67
 75-1, 64, 194
 8256, 152
 8295, 65
 899-1, 65, 160
 899-2, 65
isochronous creep curves, 165
isochronous curves, 488
isometric creep curves, 165
isometric curves, 487

INDEX

isoprene rubber, 689
isoprene-styrene rubber, 692
Izod impact test, 152

K
Kenics static mixer, 283
knife coating, 403

L
ladder polymers, 632
ladder polymers
   chemistry, 632
lag time, 203
laminate theory, 495
laminated object manufacturing (LOM), 449
laser welding, 430
   heating, 431
   IR heating, 432
   TPE, 727
   TTIr, 432
laurilactam, 590
LCM, 382
LCP, 45, 629
LCP-PAR, 629
LCP-PBT, 629
LCP-PEC, 629
LCP-PET, 629
LCP-PMPI, 629
LCP-PPTA, 629
LFT process, 375
LFT, 369, 373
Light fastness, 710
limestone, 701
liquid composites molding (LCM), 382
liquid crystalline polymers, 45, 629
   applications, 632
   chemistry, 629
   poly(m-phenylene isophthalamides), 629
   poly(p-phenylene phthalamides), 629
   polyarylates, 629
   polyester carbonates, 629
   polyterephthalates, 629
   processing, 630
   properties, 631
liquid silicone rubber injection molding, 352
living hinges, 499
LOM, 449
long fiber reinforced thermoplastics, 369
lost core injection molding, 329
LSR bleeding formulation, 354
LSR mold, 353
LSR, 352

Lubonyl process, 524
lubricants
   compatibility, 774
   table, 773
lubrication, 773

M
Maddock-type mixing section, 281
magnification factor, 502
Mailefer screw, 303
Maillefer mixing section, 281
manifold, 310
material identification, 508
material selection, 455
materials data bank, 456
MC, 650, 657
measuring
   Charpy impact strength, 145
   creep modulus, 162
   CTI, 244
   density, 81
   dissipation factor, 229
   electric strength, 243
   flammability, 220
   flexural properties, 139
   glass transition temperature, 97
   HDT, 192
   ignitability, 221
   melt flow index, 117
   melting temperature, 96
   notched Charpy impact strength, 147
   permittivity, 225
   rheological properties, 115
   shrinkage, 90
   tensile impact strength, 150
   tensile properties, 129
   thermal devices, 92
   thermal expansion (CLTE), 89
   Vicat softening temperature, 188
   viscosity, 116
   water absorption, 208
mechanical behavior, 485
mechanical fasteners, 504
mechanical properties
   orientation effect, 465
   table, 731
mechanics, 485
Meissner's extensional rheometer, 120
melamine, 680
melt flow index, 24, 116
melt flow indexer, 116
melt fracture, 108, 308

melt roll coating, 405
melting point, 74
melting temperatures
 table, 740
melting zone, 297, 301
memory effects, 107
metallizing, 406
metallocene catalylists polymerization, 19
metallocene PE, 519
metering zone, 297
methacryl copolymers, 576
methacrylate butadiene styrene copolymers, 581
 applications, 581
 processing, 581
methyl cellulose, 654
methyl celluloses, 657
methylcellulose, 31
methylmethacrylate acrylonitrile butadiene
  styrene copolymers, 581
 applications, 581
 processing, 581
methylmethacrylate copolymers, 581
 applications, 581
 processing, 581
 properties, 581
methylmethacrylate exo-methylene lactone
  copolymer, 582
 applications, 584
MF, 670, 679, 684
MFI, 24, 116
MFQ, 696
microcellular molding, 340
microinjection molding, 338
microwave welding, 444
Migration, 710
mixed metal stabilizers, 706
mixing, 272
 sand mills, 274
 twin screw extruders, 295
 auger mixer, 276
 ball mills, 274
 blister ring mixing section, 281
 block head mixing section, 278
 cavity transfer mixing section, 280
 choker ring with holes mixing section, 278
 co-kneater, 277
 continuous kneader, 287
 counter helix slotted flight mixing section,
  278
 dispersion, 274
 dispersive, 280
 double auger mixer, 276

 double block head mixing section, 278
 extruder, 278–280
 fast open vat mixer, 276
 hopper mixer, 276
 impingment mixing, 347
 internal batch mixer, 285
 Maddock-type mixing section, 281
 Maillefer mixing section, 281
 offset pins mixing section, 278
 paddle mixer, 276
 pins on barrel mixing section, 278
 pins on screw mixing section, 278
 planetary mixer, 288
 ploughshare continuous mixer, 276
 plougshare mixer, 276
 polymer blends, 273
 QSM mixer, 278–279
 rhomboidal mixing section, 279
 shear roll mixer, 282
 slotted blister ring mixing section, 280
 slow open vat mixer, 276
 static mixer, 283
 straight channel barrier mixing section, 281
 strata-blend mixing section, 280
 through mixer, 276
 torpedo barrier mixing section, 281
 Troester mixing section, 281
 vertical screw mixer, 276
 wing mixer, 276
mold build-up/deposits, 361
mold cavity, 321
mold filling pattern, 478
mold nonfill, 364
molding diagram, 319
molecular orientation, 459
molecular weight, 22
 measuring, 24
morphology, 35
mottling, 379
MPF, 684
MPQ, 696
MQ, 696
MUF, 670
multi-gated co-injection molding, 329
multicomponent injection molding, 341
MUPF, 670
MVFQ, 696
MVQ, 696

N

natural rubber, 31
natural rubber, 689

natural silica, 701
NBR, 691
NCR, 692
Newtonian fluid, 108
Newtonian plateau, 110
nitrile-butadiene rubber, 691
  applications, 691
nitrile-chloroprene rubber, 692
nitrose rubber, 698
nitroso compounds, 710
non-intermeshing twin screw extruders, 296
non-Newtonian viscosity, 103
nonyl phenol ethoxylates, 708
norbomen, 31
normal stresses, 103
notched impact test, 155
nozzle drool, 361
nozzle, 319, 322
NR, 689
nuclear spin tomograph, 249

O

O-rubbers, 695
offset pins mixing section, 278
oleamide, 702
olyethylene dioxythiopene-polystyrene
    sulfonate, 662
optical properties, 249
  color, 262
  fluorine content effect, 259
  fringe pattern, 255
  gloss, 259
  index of refraction, 250
  polymethylmethacrylate, 580
  reflection, 258
  reflectivity, 261
  table, 749
  temperature effect, 251
  transmittance, 258
  wave length effect, 251
orange peel, 362
organic nickel compounds, 705
orientation
  molecular, 47
oxethyl cellulose, 655
oxygen permeability
  table, 765

P

packaging
  PE, 522
paddle mixer, 276

PAEK, 621
PAI, 627
papplications
  olysulfide rubber, 698
PAR, 610
parabolic fracture criterion, 492
paralynes, 633
PARI, 627
part design, 476
PB, 542
PBN, 612
PC, 600
PCTFE, 569
PDAP, 670, 675, 681, 686
PDCPD, 545
PE-α-olefin copolymers, 531
PE-X, 523
PE-XA, 523
PE-XB, 523
PEDT, 662
PEDT/PSS, 662
PEEEK, 622
PEEK, 621
PEEKK, 622
PEI, 627
PEK, 622
PEKEKK, 622
PEKK, 622
pelletizing
  cold cutting, 288
  hot face pelletizer, 288
  hot plate pelletizer, 288
  rotary knife pelletizer, 289
  shear roll mixer, 283
  underwater pelletizer, 289
  water ring pelletizer, 289
PEN, 612
PEOX, 620, 657
perfluoro rubber, 695
perfluoroalkoxy, 575
  applications, 575
perfluoropropylvinylether copolymer, 575
  applications, 575
permeability, 194, 201
  temperature effect, 199
permeation, 195, 201
  table, 765
peroxide cross-linked PE, 523
PESI, 627
PF, 670, 683
PFU, 662
PHA, 670

phenacrylate resins, 673
phenol-formaldehyde, 21
phenols, 703
phenoplastics, 683
phosphites, 703
phosphonites, 703
photodegradable polymers, 657
photoelasticity, 251
PI, 622
PIB, 542, 544
pinpoint gate, 324
pins on barrel mixing section, 278
pins on screw mixing section, 278
Pipkin diagram, 108
PISO, 627
PK, 662
PLA, 656
plane strain, 487
plane stress, 486
planetary mixer, 288
plasma coating, 407
plastic foam
  properties, 740
plasticating single screw extruder, 297
plasticating unit, 319
plasticization, 297
plasticizers, 14, 561, 702
  polyvinylchloride-plasticized (PVC-P), 561
Plate-out, 711
ploughshare continuous mixer, 276
ploughshare mixer, 276
plug-assist, 397
PMI, 627
PMMA+PC blends, 584
  applications, 584
PMMI, 627
PMP, 545
PMPI, 599
PMS, 546
PNF, 699
PNR, 692
PO, 695–696
Poisson's ratio, 136
  filler effect, 137
  rate of deformation effect, 136
  temperature effect, 137
polariscope, 255
polarity, 28
poly(m-phenylene isophthalamides), 629
poly(oxadiazo)benzimidazole, 626
poly(p-phenylene phthalamides), 629
poly-α-methylstyrene (PMS), 546

poly-α-olefins (PMP
  PDCPD), 545
poly-2
  6-dimethylphenylenether, 31
poly-3
  3-bis-chlormethyl-propylenoxide, 31
poly-4-methylpentene-1 (PMP), 545
poly-4-methylpentene-1
  applications, 545
  chemistry, 545
  properties, 545
poly-m-phenylene isophthalamide, 599
  applications, 599
poly-p-hydroxybenzoate, 633
Poly-p-Methylstyrene (PPMS), 546
poly-p-phenyleneterephthalate, 599
poly-p-xylylenes, 633
polyacetal, 30–31
polyacetaldehyde, 31
polyacetals, 584
polyacryl- copolymers, 576
polyacrylate, 577
polyacrylates, 576
  ANBA, 576
  ANMA, 576
  PAA, 576
  PAN, 576
  PMA, 576
polyacrylonitrile, 576
polyalkenamers, 31
polyamide, 32
polyamides copolymers, 594
  66/6/610, 594
  PA+ABS, 594
  PA+EPDM, 594
  PA+EVA, 594
  PA+PPE, 594
  PA+PPS, 594
  PA6/12, 594
  PA66/6, 594
  rubber, 594
polyamides, 586
  additives, 593
  applications, 592
  aromatic, 599
  caprolactam, 590
  cast, 598
  chemistry, 586
  fillers, 593
  laurinlactam, 590
  modifications, 593
  PA-RIM, 598

PA11, 586
PA12, 586
PA12-C, 598
PA1313, 586
PA46, 586
PA6, 586
PA6-C, 598
PA610, 586
PA612, 586
PA613, 586
PA66, 586
PA69, 586
PA7, 586
PA8, 586
PA9, 586
  processing, 587
  properties, 590
  reaction injection molding, 598
  reinforcements, 593
  special polymers, 597
polyamidimide, 627
  applications, 627
polyarylate, 610
  applications, 612
  chemistry, 610
  modifications, 611
  processing, 611
polyarylates, 629
polyarylenes, 633
polyarylether, 31
polyaryletherketone, 31,
polyarylsulfone, 616
  applications, 617
  chemistry, 616
  processing, 616
  properties, 616
polybismaleinimide, 625
  applications, 626
polybutadiene, 31
polybutene (PB
  PIB), 542
polybutene
  applications, 544
  chemistry, 543
  processing, 543
  properties, 543
polybutylene naphthalate, 612
  applications, 612
polybutylene teraphthalates, 608
  applications, 609
  chemistry, 608
  processing, 608

properties, 608
polycarbonate blends, 603
  applications, 604
  PC+ABS, 603
  PC+AES, 603
  PC+ASA, 603
  PC+HIPS, 603
  PC+PBT, 603
  PC+PET, 603
  PC+PMMA+PS, 603
  PC+PP-CO, 603
  PC+PPE, 603
  PC+PPE+HISB, 603
  PC+SMA, 603
  PC+TPU, 603
polycarbonate copolymers, 602
  aliphatic dicarbonylic acids, 603
  DPC, 603
  PC-TMC, 602
  PPC, 603
polycarbonate, 600
  birefringence, 461
  bisphenol A, 600
  chemistry, 600
  processing, 600
  properties, 600
polychloroprene, 31
polychlorotrifluoroethylene, 574
  applications, 574
  processing, 574
  properties, 574
polycyclone, 634
polydicyclopentadiene (PDCPD), 545
polydicyclopentadiene
  applications, 546
  chemistry, 545
  properties, 546
1
  4-Polydiene, 31
polydispersity index, 22
polyester carbonates, 629
polyester resin, 681
polyester, 33
polyesterimide, 629
polyether, 31
polyetherimide, 627
  applications, 628
polyethers, 30
polyethylene dioxythiophene, 662
polyethylene naphthalate, 612
  applications, 613
  processing, 613

properties, 613
polyethylene oxides, 657
polyethylene teraphthalates, 605
  applications, 606
  chemistry, 605
  processing, 605
  properties, 606
polyethylene, 19, 30, 513
  applications, 521
  Azo cross-linked PE, 524
  chlorinated, 524
  cross-linked PE, 523
  degradable, 524
  derivatives, 523
  electron beam cross-linking, 523
  ethylene copolymers, 525
  Lubonyl process, 524
  metallocene grades, 519
  packaging, 522
  PE-X, 523
  peroxide cross-linked, 523
  polymerization, 514
  post-processing, 517
  processing, 516
  properties, 517
  silane cross-linked, 523
polyfluoroethylenepropylene, 575
  applications, 575
  proceswsing, 575
  properties, 575
polyformaldehyde, 31
polyformaldehydes, 584
polyfurans, 662
polyglycerol esters, 708
polyglycols, 620
  applications, 622
  chemistry, 620
  processing, 621
  properties, 621
polyhetero aromates, 662
polyhydroxybutyric acid, 655
polyimidazopyrrolone, 633
polyimide, 34
polyimides, 622
  properties, 623
  thermoplastic, 627
  thermoset, 623
polyimidesulfone, 628
polyisobutene (PIB), 544
polyisobutene
  applications, 544
  chemistry, 544

  processing, 544
  properties, 544
polyisoprene, 31
polylactic acid, 656
polymer
  filled, 68
  reactive, 99
  reinforced, 115
polymerization, 18–19
  cationic, 19
  chain, 19
  condensation, 21
  free radical, 19
  ionic, 19
  medium and low pressure, 515
  metallocene catalysts, 19
  Phillips, 515
  ring opening, 19
  step, 19
  Ziegler, 515
  Ziegler-Natta, 19
polymethacrylate methylimide, 629
polymethacrylate methylimideapplications, 629
polymethacrylate/modifications/blends, 581
  MMA-EML copolymers, 581
  PMMA+ABS blends, 581
  PMMA-HI, 581
  PMMI, 581
polymethacrylates, 577
  AMMA, 577
  MABS, 577
  MBS, 577
  PMMA, 577
polymethacrylimide, 628
  applications, 629
  delivery form, 628
  properties, 628
  rigid foam, 628
polymethacrylmethylimide, 581
  applications, 582
  chemistry, 582
polymethylene, 30
polymethylmethacrylate, 577
  applications, 578
  chemistry, 577
  optical properties, 580
  processing, 577
  properties, 578
  transmissivity, 580
  UV transmissivity, 580
polynorbornene rubber, 692
Polyoctenamer, 31

polyoctenamer, 31
polyolefin blends, 669
polyolefin elastomers, 668
polyolefins, 508
　copolymers, 508
　derivatives, 508
　polyethylene, 513
polyoxymethylene, 31, 584
　applications, 585
　chemistry, 584
　properties, 584
polyoxymethylene/modifications/blends, 585
　chemistry, 586
polyparaphenylene vinylene, 662
polyparaphenylene, 662
polyphenylene amines, 662
polyphenylene ether blends
　applications, 618
polyphenylene ether, 617
　blends, 618
　chemistry, 617
polyphenylene sulfide, 613
　applications, 615
　chemistry, 613
　processing, 613
　properties, 614
polyphenylenes, 662
polyphenylether, 31
polyphosphazenes, 699
polypropylene homopolymer (H-PP), 533
polypropylene oxides, 657
polypropylene, 30, 533
　applications, 538
　blends, 541
　chemistry, 535
　chlorinated, 539
　copolymers, 539
　expanded(EPP), 536
　foams, 536
　homopolymer, 533
　post processing, 537
　processing, 536
　properties, 537
　special grades, 541
　tacticity, 25
polypropylenoxide, 31
polypyrroles, 662
polysaccharides, 657, 655
polysilicooxoaluminate, 664
　applications, 664
polystyrene (PS), 546
polystyrene copolymers/blends, 547

ABS, 550
ABS+PA, 550
ABS+PC, 550
acronyms, 547
applications, 550
ASA, 550
ASA+PC, 550
chemistry, 547
comonomers-property interaction, 548
processing, 548
properties, 548
PS-HI, 549
SAN, 550
polystyrene foam, 553
polystyrene foams
　applications, 554
polystyrene thermoplastic elestomers, 668
polystyrene, 30
polysulfide rubber, 698
　applications, 698
polysulfides, 613
polysulfone, 35
polysulfones, 613
polyterephthalate blends, 609
　PET+elastomer, 609
　PET+MBS, 609
　PET+PBT, 609
　PET+PMMA, 609
　PET+PSU, 609
polyterephthalates, 629
polytetrafluoroethylene, 569
　applications, 573
　delivery form, 572
　films, 572
Polytetrafluoroethylene
　processing, 569
　properties, 572
polytetrafluoroethylene
　ram extrusion, 572
　sintering, 571
polythiophenes, 662
polytriazine, 626
polytrimethylene terephthalate, 609
polyubenzimidazole, 626
polyurethane elastomers, 669
polyurethane
　hard foam, 409
　processing, 347
　soft foam, 409
polyurethanes, 635
　additives, 641
　chain extenders, 641

chemistry, 635
cross-linking agents, 641
environment, 638
flammability, 638
flexible foam, 642
handling, 638
integral foams, 646
one-shot process, 637
polyamines, 639
polyisocyanates, 639
polymer manufacture, 636
polyols, 639
pre-polymer process, 637
quality assurance, 638
raw materials, 639
recycling, 638
rigid foam, 645
safety, 638
solid, 648
polyviny butyral, 568
polyvinyl acetate, 567
polyvinyl alcohol, 567
polyvinyl alcohols, 657
polyvinyl carbazole, 568
  applications, 568
polyvinyl chloride, 30
polyvinyl fluoride, 573
  applications, 574
  properties, 573
polyvinyl formal, 568
polyvinyl methyl ether, 568
Polyvinyl Methyl Ether
  applications, 568
polyvinylchloride copolymers/blends, 565
  applications, 565
  constitution, 565
polyvinylchloride
  foams, 566
  organosols, 566
  pastes, 566
  plastisols, 566
  stabilizers, 706
polyvinylchloride-plasticized (PVC-P), 560,
polyvinylchloride-plasticized (PVC-P), 560
  applications, 562
  blow molding, 564
  constitution, 560
  delivery form, 560
  extrusion, 564
  hydrochloric acid, 564
  injection molding, 564
Polyvinylchloride-plasticized (PVC-P)

  plasticisers, 561
  processing, 560
polyvinylchloride-plasticized (PVC-P)
  properties, 562
polyvinylchloride-unplasticized (PVC-U), 554
  applications, 557
  delivery form, 554
  identification, 556
  processing, 554
  properties, 557
polyvinylidene chloride, 567
polyvinylidene fluoride, 573
  applications, 573
  processing, 573
  properties, 573
POM, 584
post-processing, 724
powder injection molding, 344
power consumption, 275
power law index, 110
power law model, 110
PP foams, 536
PP-EPDM blend elastomers, 540
PPE, 617, 622
PPMS, 546
PPOX, 620, 657
PPP, 662
PPS, 613
PPTA, 599
PPV, 662
PPY, 662
press fit assemblies, 498
press-fit, 490
pressure predictions, 479
3D printing, 447
process influences, 458
process selection, 458
processing, 271
  raw material, 272
product data bank, 452
product performance, 458
product requirements, 453
production
  thermoplastic, 3
  thermoset, 3
  US, 5
  worldwide, 2
programable parison, 391
properties
  acoustic, 266
  coefficient of linear expansion, 508
  density, 76

dielectric constant, 508
dielectric loss factor, 508
dielectric strength, 508
electrical, 223
elongation at break, 508
elongation at yield, 508
failure strain, 508
failure stress, 508
flammability, 508
friction, 205
HDT, 508
magnetic, 247
mechanical, 126
melt temperature, 508
moisture absorption, 508
optical, 249
orientation, 459
permeability, 194
plastics, 64
rheological, 103
short term tensile, 126
specific surface resistivity, 508
specific volume resistivity, 508
stress at 0.5% strain, 508
surface tension, 122
tensile modulus, 508
tensile stress, 508
thermal, 67
water absorption, 508
wear, 205
propylene oxide rubber, 696
  applications, 696
  properties, 696
propylene-tetrafluoroethylene rubber, 695
PS-E, 553
PS-HI, 549
PSAC, 650, 657
PSIOA, 664
PSU, 616
PT, 662
PTFE, 569
PTFE-AF copolymers, 576
  applications, 576
PTHF, 620
PTT, 609
pultrusion, 385
pulverizing
  cooled pulverizing system, 291
  shear-type pulverizer, 290
pumping zone, 297
PUR-F, 642
PUR-I, 646
PUR-R, 645
PUR-S, 648
PVAC, 567
PVAL, 567, 657
PVB, 567
PVC stabilizers, 706
  alkyl phosphites stabilizers, 706
  alkyltim stabilizers, 706
  mixed metal stabilizers, 706
PVC-U, 554
PVDC, 567
PVDF, 569
PVF, 569
PVK, 567
PVME, 567
PVMQ, 696
PVP, 567
pvT diagram
  shrinkage prediction, 471
pvT-diagram, 79
pyrone, 633
PZ, 699

Q
Q-silicone rubber, 696
QSM mixer, 278–279

R
racetracking, 363
radiation resistance
  table, 760
rain erosion
  table, 770
random copolymer, 18
rapid prototyping, 444
  3D printing, 447
  fused deposition modeling, 448
  laminated object manufacturing, 449
  selective binding, 447
  slective laser sintering, 446
  solid ground curing, 445
  stereo-lithography, 445
raw material preparation, 272
raw material
  delivery, 804
reaction injection molding, 346
  polyamides, 598
recycling, 58
  bottles, 61
  polyurethanes, 638
regulations, 454
relaxation time, 107

residual stresses, 470, 472
  predictions, 474
resin infusion process(RIP), 384
resin transfer molding, 382
reverse coating, 404
reverse draw thermoforming, 398
RF, 670, 683
RF/Dielectric welding, 433
  electrode configuration, 434
rheology, 103
rheometer, 117
  cone-plate, 118
  Couette, 118
  extensional, 120
rheometry, 115
rhomboidal mixing section, 279
ribs
  sink marks, 472
rigid foams
  properties, 738
rigid PVC (PVC-U), 554
RIM, 346
RIP, 384
ripping strength
  rubber, 143
robot filament winding, 386
rotary knife pelletizer, 289
rotational molding, 410
rubber elasticity, 126
rubber membrane press, 376
rubber pad press, 376
rubber
  service temperatures, 744
  filled, 128
  ripping strength, 143
rubbers, 686
  properties, 687
runner system, 323

S

SAN, 550
sand blasting, 768
sand mills, 274
sandwich molding, 326
saturated polyesters, 599
SBR, 691
screen pack, 305
screw characteristic curves, 299, 301
screw design, 320
screw geometry, 298
sealing films, 533
secant creep modulus, 165

secondary shaping, 386
selective binding, 447
selective laser sintering, 446
self-diffusion, 205
semi-crystalline polymers, 43
semi-rigid foams
  properties, 739
semicarbazides, 709
sequential co-injection molding, 327
service temperatures
  rubber, 744
SGC, 445
shark skin, 108
sharkskin, 306
shear modulus
  amorphous polymers, 53
  cross-linked polymers, 53
shear thinning, 103
shear-type pulverizer, 290
sheet molding compound, 369, 676
shish-kebab, 37
short shots, 318
shortshots, 364
shrinkage, 460, 470, 481, 718
  pvT behavior, 471
Si, 670
SI, 683, 686
side gates, 324
silane cross-linked PE, 523
silicone molding compounds, 686
silicone resins, 675, 683
simulation, 476
  air entrapment, 478
  clamping force, 480
  compression molding, 480
  filling pattern, 478
  geometric modeling, 477
  mid-plane representation, 477
  orientation, 479
  pressure, 479
  shrinkage, 481
  warpage, 481
  weldlines, 478
single screw extruder, 294
sink marks, 380, 471
sintering
  polytetrafluoroethylene, 571
SIR, 692
slide coating, 405
slip additives, 702
  erucamide, 702
  oleamide, 702

INDEX

stearamide, 702
slotted blister ring mixing section, 280
slow open vat mixer, 276
SLS, 446
SMC processing cycle, 373
SMC production, 371
SMC, 369, 371, 463, 676
  curing, 469
  fiber orientation, 464, 481
  mechanical properties, 465
  warpage predictions, 484
smooth barrel extruder, 299
snap fit assemblies, 499
soft-elastic foams
  properties, 739
software, 505
  Animator, 505
  Ansys, 505
  AutoCAD, 505
  B-SIM, 506
  CADMOULD, 506
  CADPRESS, 506
  CATIA, 505
  EXPRESS, 506
  Fidap, 505
  Fluent, 505
  HyperWorks, 505
  I-Deas, 505
  LS-Dyna, 505
  Marc, 505
  MCBase, 506
  MOLDEX, 506
  MOLDFLOW, 506
  PAMCASH, 505
  POLYFLOW, 506
  Pro/Engineer, 505
  SIGMASOFT, 506
  solid works, 505
  T-SIM, 506
solid extrusion calibration, 313
solid ground curing (SGC), 445
solidification, 463
solids conveying zone, 297
solubility, 195
soluble core injection molding, 329
solvents, 729
sorbitan esters, 708
sorption constant, 201
sorption, 194
  temperature effect, 199
sound absorption, 268
special injection molding processes, 326

specific enthalpy, 76
specific heat, 74, 76, 94
  filled plastics, 76
specific volume, 79
speed of sound, 266
spherulite, 37,
spider leg tubing die, 311
spin welding, 427
  excess particulate, 430
  excessive flash, 430
  joint design, 429
  part marking, 430
  vibration while spinning, 430
spiral die, 312
spring forward effect, 475
sprue bushing, 322
sprue, 323–324
spurt flow, 108
stabilizers, 702
standards, 454
starch derivates, 650
static mixer, 283
Statistical data, 1
stearamide, 702
stearically hindered amines, 705
stereo-lithography(STL), 445
stick-slip effect, 108
STL, 445
straight channel barrier mixing section, 281
strata-blend mixing section, 280
strength of materials, 485
strength stability under heat, 185
strength
  rate of deformation effect, 131
stress-strain relations, 485
stress-strain
  failure criteria, 491
  rate of deformation effect, 131
  sample calculations, 488, 490
  semi-crystalline, 135
  viscoelastic effect, 134
structural foam injection molding, 349
structural RIM, 382
Structure, 17
structure
  amorphous, 40
  cross-linked, 51
  liquid crystalline, 45
  semi-crystalline, 43
styrene polymers, 546
  chemistry, 546
styrene-butadiene rubber, 691

styrofoam, 553
submarine gate, 325
surface tension, 122
surging, 305
suspension
  rheology, 115
swell, 108
symbols
  plasticizers, 14
  polymer characteristics, 13
  polymers, 8

T

T-rubber, 697
tacticity, 25
talc, 701
tapered sprue gate, 324
TCF, 697–698
TCP, 609, 667
temperature
  glass transition, 94
tensile strength
  filler effect, 144
tensiometer, 123
tetrafluoroethylene-hexa-
    fluoropropylene-copolymers, 575
  applications, 575
  processing, 575
  properties, 575
tetrafluoroethylene-hexafluoropropylene-
    vinylidenefluoride copolymer,
    576
  applications, 576
tetrazoles, 710
TGA, 96
thermal conductivity, 67
  draw effect, 68
  filler effect, 72
  pressure effect, 68
  temperature effect, 68
  thermoplastics, 68
thermal degradation, 216
thermal diffusivity, 82
  temperature effect, 82
thermal evaporation metallizing, 406
thermal expansion, 82, 95
  draw effect, 88
  filled plastics, 85
  polyethylene, 88
thermal properties
  table, 733
thermoforming, 397

blisters, 399
blushing, 400
bubbles, 399
cracking, 400
excessive shrinkage, 400
incomplete forming, 399
mottled surface, 400
packaging, 399
pinhole, 400
plug-assist, 397
reverse draw, 398
rotational vacuum, 399
sheet scorched, 400
tearing, 400
thin corners, 400
troubleshooting, 399
warping, 400
webbing/bridging, 399
thermogravimetry, 96
thermomechanical, 95
thermoplastic applications, 4
thermoplastic elastomes, 664
thermoplastic polyester-elastomers, 609
thermoplastic production, 3
thermoplastic starch, 657
thermoset applications, 4
thermoset production, 3
thermosets, 670
Thiocarbonyldifluoride copolymer rubber, 698
thiosynergists, 703
three channel co-injection molding, 328
through mixer, 276
through transmission infrared welding (TTIr),
    432
tie bar, 315
tie bars, 321
time-temperature-transformation, 101
TM, 697–698
TMA, 95
toggle mechanism, 315, 321
tolerances, 718, 720–721
TOR, 692
torpedo barrier mixing section, 281
torsion test, 173
TPA, 667
TPE, 664
  applications, 666
  chemistry, 666
  copolyamides, 667
  copolyester, 667
  physical constitution, 666
  polyolefin elastomers, 668

polystyrene, 668
properties, 666
TPO, 668
TPS, 657, 668
TPU, 669
TPV, 669
tradenames, 804
trans-polyoctenamer rubber, 692
transition temperatures
　table, 740
transition zone, 297, 301
transparent polymers, 749
Troester mixing section, 281
troubleshooting
　blow molding, 392
　compression molding, 377
　extrusion, 302
　injection molding, 368
　thermoforming, 399
　ultrasound welding, 421
　vibration welding, 426
TTIr welding, 432
TTT, 101
tubular dies, 310
twin screw extruders, 281, 296
two channel co-injection molding, 328
two stage degassing screw, 320

U

U-rubbers, 698
U-rubbers
　applications, 698
UF, 670, 679, 684
UL Subject 94 evaluation, 221
ultra-light PE, 525
ultrasoun welding
　energy director, 421
ultrasound welding, 416
　cracks, 423
　equipment problems, 423
　flash, 422
　frequencies, 419
　inconsistencies, 422
　inconsistent insertion, 423
　joints, 419
　marking, 422
　overwelding, 422
　staking problems, 423
　troubleshooting, 421
　underwelding, 422
　vibration amplitudes, 421
　weak inserts, 422

umbrella gate, 325
underwater pelletizer, 289
uniaxial failure criteria, 491
unsaturated polyester, 671, 684
　applications, 684
　composites, 497
　curing, 468
UP, 670–671, 681, 684
UPE
　composites, 497
　curing, 468
urethane rubber, 698
US production, 5
UV stabilizers, 705
　2-hydroxybenzophenones, 705
　2-hydroxyphenylbenzotriazoles, 705
　carbon black, 705
　HALS, 705
　organic nickel compounds, 705
　stearically hindered amines, 705

V

vacuum sleeve calibration, 312
VE, 670, 673, 681
venting, 478
vertical screw mixer, 276
VF, 650
vibration welding, 424
　flash, 427
　fracture outside of joint, 427
　internal component damage, 427
　internal parts welding, 427
　marking, 427
　melting outside of joint, 427
　misalignment, 427
　nonuniformities, 426
　overwelding, 426
　phases, 425
　process, 425
　troubleshooting, 426
　underwelding, 426
　weld examples, 426
vinyl ester, 681
vinyl polymers, 554, 567
　PVC-U, 554
vinylester resins, 673
vinylester urethanes, 673
viscoelastic, 103
viscoelasticity, 107
viscometer, 109
　capillary, 116
viscosity curves, 103

viscosity
  elongational, 111
  flow model, 109
vitrification line, 102
VU, 673
vulcanization, 99

W

warpaga
  mold thermal imbalance, 474
warpage, 470, 481
  anisotropy, 475
  curvature change, 475
water absorption, 208
  relative humidity dependance, 209
  table, 751
  temperature dependance, 208
water ring pelletizer, 289
wear, 205
  bearings, 771
  rain erosion, 770
  sand blasting, 768
Weather fastness, 710
weathering, 211
  impact test, 211
  laboratory test, 211
  pigment effect, 214
  table, 760
welding, 412
  diode laser, 726
  thermoplastic elastomers, 726
  TPE, 727
weldline prediction, 478
wetting angle, 122
  projector, 123
wing mixer, 276
wire coating, 403
WLF relation, 111
world production, 2

X

XF, 670
XF, 683
XPS, 553

Y

yield stress, 115

Z

zeolityes, 701
zero-shear-rate viscosity, 110
Ziegler-Natta polymerization, 19

ARBURG
Spritzgußmaschinen

50 years of ARBURG
injection moulding machines

50 years of permanent innovations are a great achievement. From the first, manually operated piston injection moulding machine which ARBURG produced in series, right up to the all-electric modular ALLROUNDER A.

Thank you for travelling down this long road with us.

EINE **ARBURG** SPRITZGUSS MASCHINE *keine* BOHRMASCHINE

ARBURG GmbH + Co KG
Postfach 11 09 · 72286 Lossburg
Tel.: +49 (0) 74 46 33-0
Fax: +49 (0) 74 46 33 33 65
e-mail: contact@arburg.com

ARBURG

www.arburg.com

Measuring and drive systems with state-of-the-art technology

The Lab-Station

The new generation of the torque rheometers for all kind of processing technical investigations or processing tasks in laboratories and simulation.

The all-digital motor guarantees full torque over the entire speed range. The inverter drive provides for precise and constant speed.

Profit from our outstanding features of this universal drive unit:

- **Permanent digital communication between all system components**
- **State-of-the-art field bus technology with standard bus system – all components will be available within the long term**
- **Self-recognition and self-validation**
- **Modular configuration- scale up your system whenever needed**
- **Easy handling - just plug & play**
- **Possibility of remote diagnostics**

The comfortable 32-bit-software for all common Windows™ versions offers the complete spectrum of modern test methods and enables a comprehensive process recording.

Brabender® Measurement & Control Systems

Brabender® **GmbH & Co. KG**
Kulturstr. 51 - 55   47055 Duisburg   Germany
Tel.: +49 (0) 203/7788-0
Fax: +49 (0) 203/7788-100
E-Mail: plastics-sales@brabender.com
Internet: www.brabender.com

Beaumont Technologies, Inc.
World Leader in Mold and Process Optimization Technologies

Are you plagued by customer returns for flash & non-fills?

Learn how BTI can add a new level of control to eliminate these problems... and save you money!

Solving Your QC Problems
- Dimensional Variation
- Increased Inspection
- Flash & Short Shots
- Production Sorting
- Part Warping

Melt Management Technologies
- Design for Six Sigma
- Commission your molds faster
- Increase productivity and efficiencies
- Eliminate detrimental filling imbalances
- Reduce your mold/part lead-times

MELTFLIPPER®

Guaranteed 100%; No New Capital Equipment Required

As Albert Einstein Said:
"Insanity is doing the same thing over and over again and expecting a different result."

No Control → **Rheological Control**

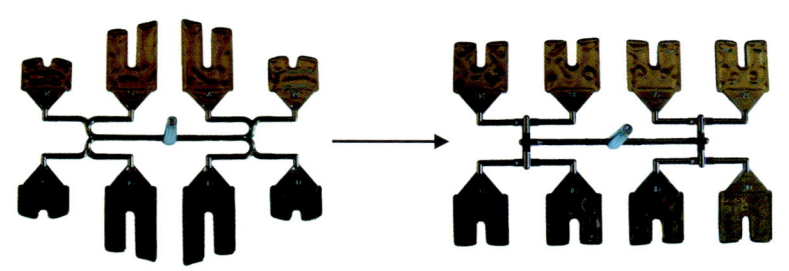

"We implemented MeltFlipper technology in a new 8-cavity mold... this is the first mold in a very long time that started up without any filling problems or part quality issues."
- Damon DeVore, Engineering Manager, Accudyn Products, Erie, PA

5 STEP™ — Mold Balance Analysis Software

CAE by BTI — Mold Filling Analysis

BTI TRAINING™ — "Back to Basics" Seminars

2103 East 33rd Street • Erie, PA 16510-2529 • Tel: 814.899.6390 • Fax: 814.899.7117 • www.beaumontinc.com • meltflipper@beaumontinc.com

INNOVATIVE PROCESS ENGINEERING FROM WEBER

Fibre material extrusion

DS 8.22 FE

Fibre material extrusion

Direct extrusion and compound processing have established themselves within fibre material extrusion.

In direct extrusion the individual mixing components are directly added to the extruder feed opening by means of volumetric dosing. The mixing ratio is variable and can easily be adjusted for the respective application. For this method, the material has to be pelletized. So the pourability and the bulk weight is optimized to the point, that a continuous extrusion process is guaranteed.

Due to the requirement in the market for a low-cost material we have developed for the fibre extrusion a force feed unit, which enables the feeding of loose fibres. Fibres which are extremely fine to powdery ground can also be processed using this technology. The advantages gained are avoiding the expensive preparation of materials prior to extrusion and the implementation of a tight retooling gap. Refitting the extruder with a fibre material force feed unit and a mixing unit for optional processing of loose fibres is both feasible and straightforward.

Contrary to that compound processing is used. The finished processed mixture of fibre material and thermoplastics is referred to as the compound. Mixing ratios are always fixed. Manufacturing can be achieved in several ways. The dosing unit on the extruder is then obsolete. The used thermoplastics are PVC, PP, PE. Fibre materials used can be, other than wood, on principle all natural fibres.

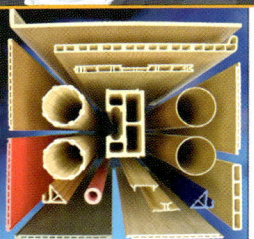

DS 15.22FE with sheet tooling

6-strand tool

Hans Weber
Maschinenfabrik GmbH
Bamberger Straße 19 – 21
D-96317 Kronach
Postfach 18 62
D-96308 Kronach
Phone +49(0) 92 61 4 09-0
Fax +49(0) 92 61 4 09-1 99
email: info@hansweber.de
Internet: www.hansweber.de

WEBER

Testing Calculation Development

- Testing of plastics, semi-finished products and components
- Testing, inspection and certification of plastic building materials: pipes, sandwiches, storage container, district heating pipelines (Free State of Saxony, DIBt)
- Accredited laboratory for plastic material tests and recognition by DVGW, DIN CERTCO, THÜGA, AGFW
- Testing of automotive components
- Design, calculation and development of fibre reinforced plastic components, prototyping and testing
- Development of material databases

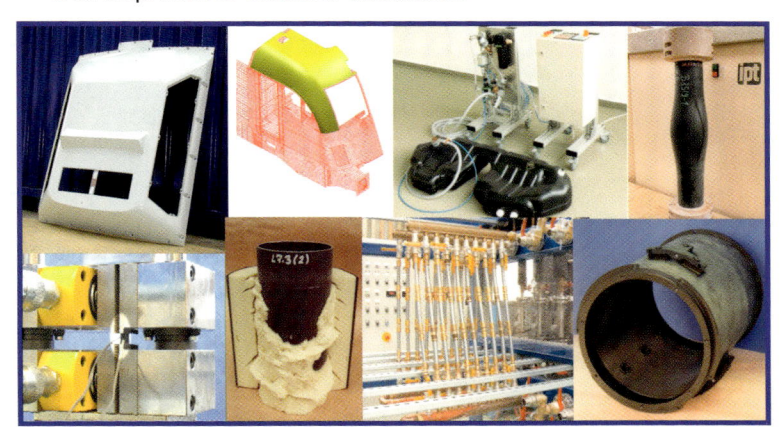

IMA Materialforschung und Anwendungstechnik GmbH

Department of Plastics

Wilhelmine-Reichard-Ring 4
01109 Dresden

Certified according
DIN EN ISO 9001:2000
and DIN EN 9100:2003

DAP-PL-01.062-00

Phone: +49 (0) 351 8837-404
Fax: +49 (0) 351 8837-530
E-Mail: a300@ima-dresden.de
Internet: www.ima-dresden.de

JENOPTIK JENA

Automatisierungstechnik.

JENOPTIK
Automatisierungs
technik GmbH
07739 Jena
Tel. 03641 652570
www.automation-
jenoptik.de

NON METAL LASER TECHNOLOGY

CUTTING
TRENNEN
WELDING
VERBINDEN
PERFORATING
PERFORIEREN
ABLATING
ABTRAGEN

JENOPTIK Group.

TECHNICAL MAGAZINES | TECHNICAL BOOKS | ONLINE SERVICES | SEMINARS

HANSER

First-hand Information

Europe's leading plastics magazine

Access to full-text online archives in English and German including free pdf downloads

E-mail newsletter "trend-setting technology" free of charge

1.200 pages with up-to-date information

Exclusive information on materials, processing and application of plastics

Specials on current subjects

Further information and complete range on plastics at **www.kunststoffe-international.com**

Order your free sample copy now:
www.kunststoffe-international.com

www.kunststoffe-international.com

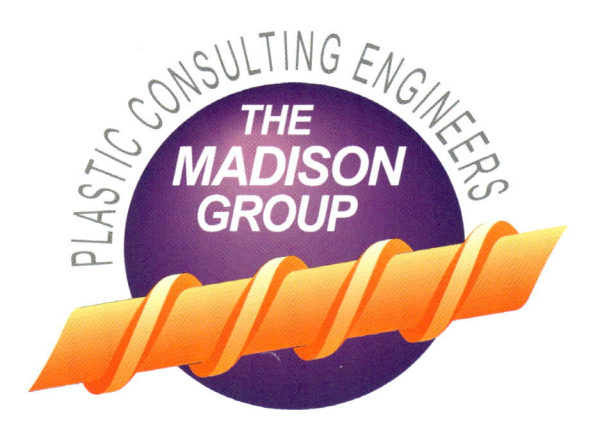

The Madison Group
Consultants for the Plastics Industry

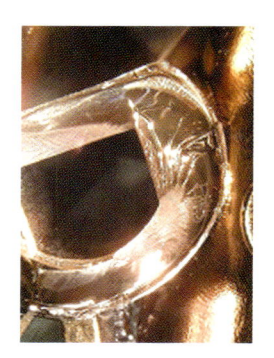

Failure • Design • Processing Analysis

Failures cost money. Engineers at The Madison Group specialize in the analysis, determination of root cause, and prevention of plastic part failure. Using optical, analytical, and physical testing we ascertain contributing factors to failure, including design, processing, quality/type of plastic, and environment. Leverage our expertise to solve and prevent your plastic part failures.

5500 Nobel Dr., Suite 210
Madison, WI 53711 USA
Ph: 1-608-231-1907
Fx: 1-608-231-2694

email: info@madisongroup.com
www.madisongroup.com

M-Base Engineering + Software GmbH
www.m-base.de

Material Data Center

The source for high quality material data and application information

✓ CAMPUS® data
✓ ASTM properties
✓ Search
✓ Table function
✓ Compare
✓ PDF datasheets
✓ Calculation of
  · Snapfits
  · Flow length
  · Cooling time
  · Material model parameters
  · Multi Point Data
✓ Application database
✓ Search for substitutes
✓ Extensive text based search

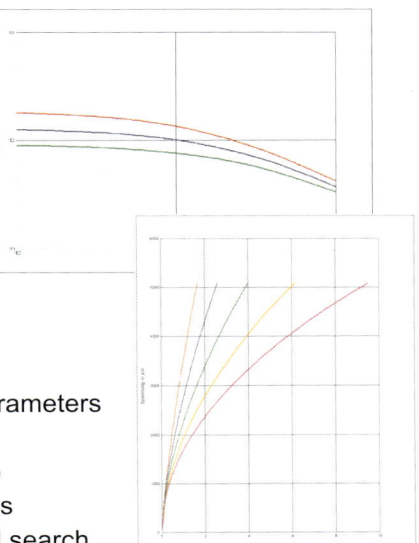

Free temporary access; special discount price for owners of this book
www.materialdatacenter.com

RICO
efficient elastomere projects

✓ **Silicone Moulds**
✓ **Multicomponent Moulds**
✓ **Product development**

Rico Elastomere Projecting GmbH
A-4600 Thalheim/Wels, Am Thalbach 8
Tel.: +43 (0) 7242 76 460
E-Mail: office@rico.at
www.rico.at

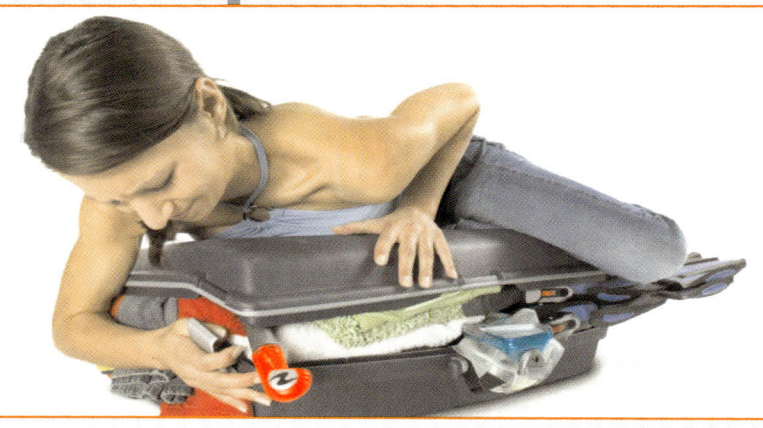

More space
for **your ideas**

Do not unnecessarily waste space in your automation lines. With **components** of RINCO ULTRASONICS AG you have more space for your own ideas. Our versatile range and the individual solutions provide you the support which you need as a **manufacturer of special purpose machines.**

rinco ultrasonics

RINCO ULTRASONICS AG
CH-8590 Romanshorn 1
Tel. +41 71 466 41 00

info@rincoultrasonics.com
www.rincoultrasonics.com

A CREST GROUP COMPANY

PTOnline
The Definitive Web Resource for Plastics Processors

www.ptonline.com
Where processors go to get the information they need

ONLINE RESOURCES FOR PLASTICS PROCESSORS

Materials Database
Machinery Database
Supplier Directory
Industry News
Pricing Pages
Events Calendar
Article Library
Emphasis Zones

Injection Molding
Blow Molding
Extrusion
Compounding
Materials
Auxiliary Eqmt
Tooling

USE PTOnline TO FIND MORE PRODUCT AND TECHNOLOGY INFORMATION FASTER

Review Supplier Product Showrooms

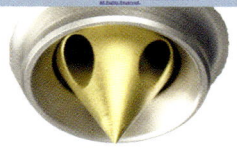

Look for the LEARN MORE online logo at the end of magazine articles and find out where you can get more information online

Search Online Databases

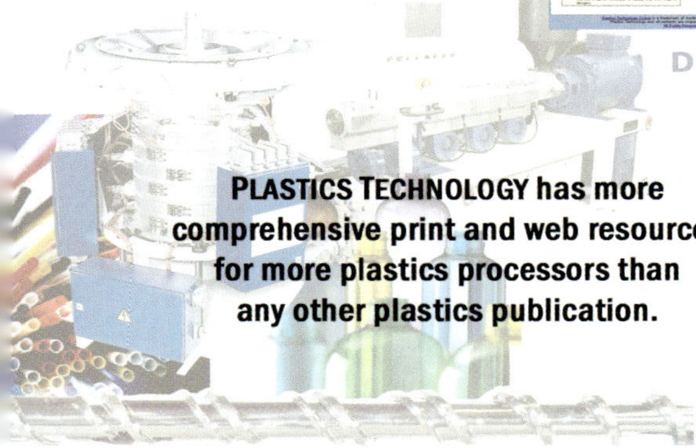

PLASTICS TECHNOLOGY has more comprehensive print and web resources for more plastics processors than any other plastics publication.

Conduct fast, accurate searches with

PT Direct
at www.ptonline.com

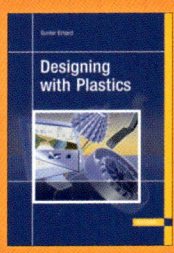

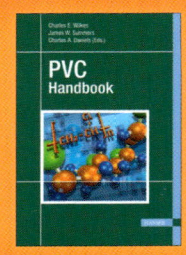

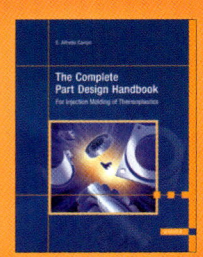

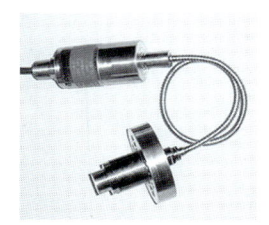